Methods in Enzymology

Volume 346
GENE THERAPY METHODS

METHODS IN ENZYMOLOGY

EDITORS-IN-CHIEF

John N. Abelson   Melvin I. Simon

DIVISION OF BIOLOGY
CALIFORNIA INSTITUTE OF TECHNOLOGY
PASADENA, CALIFORNIA

FOUNDING EDITORS

Sidney P. Colowick and Nathan O. Kaplan

Methods in Enzymology

Volume 346

Gene Therapy Methods

EDITED BY

M. Ian Phillips

COLLEGE OF MEDICINE
UNIVERSITY OF FLORIDA
GAINESVILLE, FLORIDA

ACADEMIC PRESS

An Elsevier Science Imprint

San Diego   San Francisco   New York   Boston   London   Sydney   Tokyo

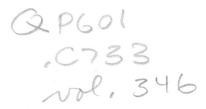

This book is printed on acid-free paper. ∞

Academic Press
An Elsevier Science Imprint.
525 B Street, Suite 1900, San Diego, California 92101-4495, USA
http://www.academicpress.com

Academic Press
32 Jamestown Road, London NW1 7BY, UK
http://www.academicpress.com

International Standard Book Number: 0-12-182247-8

PRINTED IN THE UNITED STATES OF AMERICA
02 03 04 05 06 07 SB 9 8 7 6 5 4 3 2 1

Table of Contents

Section I. Nonviral

v

Section II. Adenovirus

Section III. Adeno-Associated Virus

Section IV. Lentivirus

Section V. Retrovirus

Section VI. Other Strategies

Contributors to Volume 346

Article numbers are in parentheses following the names of contributors.
Affiliations listed are current.

ANDREW H. BAKER (10), *Department of Medicine and Therapeutics, University of Glasgow, Glasgow G11 6NT, United Kingdom*

PARAMITA BANDYOPADHYAY (2), *ValiGen, Inc., Newtown, Pennsylvania 18940*

ANDREA BANFI (9), *Department of Molecular Pharmacology, Stanford University School of Medicine, CCSR 4215, Stanford, California 94305*

CATHERINE BARJOT (13), *Department of Human Genetics, University of Michigan, Ann Arbor, Michigan 48109*[*]

ARTHUR L. BEAUDET (11), *Department of Human and Molecular Genetics, Baylor College of Medicine, Houston, Texas 77030*

HELEN M. BLAU (9), *Department of Molecular Pharmacology, Stanford University School of Medicine, CCSR 4215, Stanford, California 94305*

P. BOROS (12), *Institute for Gene Therapy and Molecular Medicine, Mount Sinai School of Medicine, New York, New York 10029*

O. BOYER (17), *Laboratoire de Biologie et Thérapeutique des Pathologies Immunitaires, Université Pierre et Marie Curie, Hôpital de la Pitié-Salpétriére, 75651 Paris Cedex 13, France*

LARS J. BRANDÉN (6), *Center for BioTechnology, Department of Biosciences, Karolinska Institute, SE-141 57 Huddinge, Sweden*[†]

XANDRA BREAKEFIELD (34), *Molecular Neurogenetics Unit, Massachusetts General Hospital, Charlestown, Massachusetts 02129*

J. S. BROMBERG (12), *Institute for Gene Therapy and Molecular Medicine, Mount Sinai School of Medicine, New York, New York 10029*

VLADIMIR BUDKER (7), *Departments of Pediatrics and Medical Genetics, University of Wisconsin, Madison, Wisconsin 53705*

MARK M. BURCIN (31), *Cardiogene, 40699 Erkrath, Germany*

MASSIMO BUVOLI (8), *Department of Molecular, Cellular, and Developmental Biology, University of Colorado, Boulder, Colorado 80309*

CHERYL A. CARLSON (16), *Department of Medicine, Division of Medical Genetics, University of Washington, Seattle, Washington 98195*

M. G. CASTRO (17), *Molecular Medicine and Gene Therapy Unit School of Medicine, University of Manchester, Manchester M13 9PT, United Kingdom*[‡]

[*] Current affiliation: UMR INRA 703, ENVN—Atlanpole La Chanterie, F-44307 Nantes Cedex 3, France

[†] Current affiliation: Clinical Research Center, Karolinska Institute, S-141 86 Stockholm, Sweden

[‡] Current affiliation: Gene Therapeutics Research Institute, Cedars-Sinai Medical Center, Los Angeles, California 90048

JEFFREY S. CHAMBERLAIN (13), *Department of Neurology, University of Washington School of Medicine, Seattle, Washington 98195*

JULIE CHAO (14), *Department of Biochemistry and Molecular Biology, Medical University of South Carolina, Charleston, South Carolina 29425*

LEE CHAO (14), *Department of Biochemistry and Molecular Biology, Medical University of South Carolina, Charleston, South Carolina 29425*

KYE CHESNUT (24), *Powell Gene Therapy Center, University of Florida, Gainesville, Florida 32610*

JAYANTA ROY CHOWDHURY (2), *Department of Medicine and Molecular Genetics, and Marion Bessin Liver Research Center, Albert Einstein College of Medicine, Bronx, New York 10461*

NAMITA ROY CHOWDHURY (2), *Department of Medicine and Molecular Genetics, and Marion Bessin Liver Research Center, Albert Einstein College of Medicine, Bronx, New York 10461*

YI CHU (15), *Cardiovascular Division, University of Iowa College of Medicine, Iowa City, Iowa 52242*

MARINEE K. L. CHUAH (33), *Flanders Interuniversity Institute of Biotechnology, Center for Transgene Technology and Gene Therapy, University of Leuven, B-3000 Leuven, Belgium*

DESIRE COLLEN (33), *Flanders Interuniversity Institute of Biotechnology, Center for Transgene Technology and Gene Therapy, University of Leuven, B-3000 Leuven, Belgium*

PIETER R. CULLIS (3), *Department of Biochemistry and Molecular Biology, University of British Columbia, Vancouver, British Columbia, Canada V6T 1Z3, and Inex Pharmaceuticals Corporation, Burnaby, Canada V5J 5L8*

BEVERLY L. DAVIDSON (25), *Departments of Internal Medicine, Neurology, Physiology, and Biophysics, College of Medicine, University of Iowa, Iowa City, Iowa 52242*

MICHELE DE PALMA (29), *Institute for Cancer Research and Treatment, Laboratory for Gene Transfer and Therapy, University of Torino Medical School, 10060 Candiolo, Torino, Italy*

Y. DING (12), *Institute for Gene Therapy and Molecular Medicine, Mount Sinai School of Medicine, New York, New York 10029*

J. KEVIN DONAHUE (19), *Institute of Molecular Cardiobiology, Johns Hopkins University School of Medicine, Baltimore, Maryland 21205*

JIAN-YUN DONG (30), *Department of Microbiology, Medical University of South Carolina, Charleston, South Carolina 29425*

DONGSHENG DUAN (20), *Department of Anatomy and Cell Biology, University of Iowa College of Medicine, Iowa City, Iowa 52242*

FIONA M. ELLARD (27), *Department of Biochemistry, Oxford BioMedica (UK) Limited, Oxford OX4 4GA, United Kingdom*

JOHN F. ENGELHARDT (20), *Department of Anatomy and Cell Biology, University of Iowa College of Medicine, Iowa City, Iowa 52242*

DAVID B. FENSKE (3), *Department of Biochemistry and Molecular Biology, University of British Columbia, Vancouver, British Columbia, Canada V6T 1Z3*

TERRY FLOTTE (24), *Powell Gene Therapy Center, University of Florida, Gainesville, Florida 32610*

ANTONIA FOLLENZI (26), *IRCC, Institute for Cancer Research and Treatment, Laboratory for Gene Transfer and Therapy, University of Torino Medical School, 10060 Candiolo, Torino, Italy*

CORNEL FRAEFEL (34), *Institute of Virology, University of Zurich, CH-8057 Zurich, Switzerland*

JASON J. FRITZ (21), *Department of Molecular Genetics and Microbiology, College of Medicine, University of Florida, Gainesville, Florida 32610*

S. FU (12), *Institute for Gene Therapy and Molecular Medicine, Mount Sinai School of Medicine, New York, New York 10029*

MARK R. GALLAGHER (23), *Harvard/ Généthon Joint Laboratory, Harvard Institutes of Medicine, Boston, Massachusetts 02115*

CRAIG H. GELBAND (32), *Department of Physiology, University of Florida College of Medicine and Functional Genomics, Gainesville, Florida 32610*

C. A. GERDES (17), *Molecular Medicine and Gene Therapy Unit School of Medicine, University of Manchester, Manchester M13 9PT, United Kingdom*

JILL GLASSPOOL-MALONE (4), *Gene Delivery Alliance, Inc., Rockville, Maryland 20850*

JOHN T. GRAY (23), *Harvard/Généthon Joint Laboratory, Harvard Institutes of Medicine, Boston, Massachusetts 02115*

WALTER H. GÜNZBURG (35), *Institute of Virology, University of Veterinary Sciences, A-1210 Vienna, Austria*

YUTAKA HANAZONO (22), *Division of Genetic Therapeutics, Center for Molecular Medicine, Jichi Medical School, Kawachi, Tochigi 329-0498, Japan*

KRISTINE HANSON (7), *Departments of Pediatrics and Medical Genetics, University of Wisconsin, Madison, Wisconsin 53705*

DENNIS HARTIGAN-O'CONNOR (13), *Department of Neurology, University of Washington School of Medicine, Seattle, Washington 98195*

WILLIAM W. HAUSWIRTH (21), *Department of Ophthalmology and Powell Gene Therapy Center, College of Medicine, University of Florida, Gainesville, Florida 32610*

DONALD D. HEISTAD (15), *Cardiovascular Division, University of Iowa College of Medicine, Iowa City, Iowa 52242*

MIKKO O. HILTUNEN (18), *University of Kuopio, A. I. Virtanen Institute, FIN-70210 Kuopio, Finland*

MATTHEW J. HUENTELMAN (32), *Department of Physiology and Functional Genomics, College of Medicine, University of Florida, Gainesville, Florida 32610*

NEIL JOSEPHSON (37), *Division of Hematology, University of Washington, Seattle, Washington 98195*

YASUFUMI KANEDA (36), *Division of Gene Therapy Science, Graduate School of Medicine, Osaka University, Suita City, Osaka 565-0871, Japan*

MICHAEL J. KATOVICH (32), *Department of Physiology and Functional Genomics, College of Medicine, University of Florida, Gainesville, Florida 32610*

NOBUFUMI KAWAI (22), *Department of Physiology, Jichi Medical School, Kawachi, Tochigi 329-0498, Japan*

SUSAN M. KINGSMAN (27), *Department of Biochemistry, Oxford BioMedica (UK) Limited, Oxford OX4 4GA, United Kingdom*

D. KLATZMANN (17), *Laboratoire de Biologie et Thérapeutique des Pathologies Immunitaires, Université Pierre et Marie Curie, CNRS, Hôpital de la Pitié-Salpétriére, 75651 Paris Cedex 13, France*

BETSY T. KREN (2), *Department of Medicine, University of Minnesota Medical School, Minneapolis, Minnesota 55455*

*Current affiliation: GlycArt Biotechnology AG, 8093 Zurich, Switzerland

T. KU (12), *Institute for Gene Therapy and Molecular Medicine, Mount Sinai School of Medicine, New York, New York 10029*

AKIHIRO KUME (22), *Division of Genetic Therapeutics, Center for Molecular Medicine, Jichi Medical School, Kawachi, Tochigi 329-0498, Japan*

LESLIE A. LEINWAND (8), *Department of Molecular, Cellular, and Developmental Biology, University of Colorado, Boulder, Colorado 80309*

MICHELLE K. LEPPO (19), *Institute of Molecular Cardiobiology, The Johns Hopkins University School of Medicine, Baltimore, Maryland 21205*

ALFRED S. LEWIN (21), *Department of Molecular Genetics and Microbiology, Powell Gene Therapy Center, College of Medicine, University of Florida, Gainesville, Florida 32610*

ANDRÉ LIEBER (16), *Department of Medicine, Division of Medical Genetics, University of Washington, Seattle, Washington 98195*

DEXI LIU (5), *Department of Pharmaceutical Sciences, University of Pittsburgh School of Pharmacy, Pittsburgh, Pennsylvania 15261*

FENG LIU (5), *Department of Pharmaceutical Sciences, University of Pittsburgh School of Pharmacy, Pittsburgh, Pennsylvania 15261*

J.-MATTHIAS LÖHR (35), *Department of Molecular Gastroenterology, Medical Clinic II, University of Heidelberg, D-68167 Mannheim, Germany*

P. R. LOWENSTEIN (17), *Molecular Medicine and Gene Therapy Unit School of Medicine, University of Manchester, Manchester M13 9PT, United Kingdom**

IAN MACLACHLAN (3), *Protiva Biotherapeutics, Burnaby, British Columbia, Canada V5J 5L8*

ROBERT W. MALONE (4), *Gene Delivery Alliance, Inc., Rockville, Maryland 20850*

EDUARDO MARBÁN (19), *Institute of Molecular Cardiobiology, The Johns Hopkins University School of Medicine, Baltimore, Maryland 21205*

ENCA MARTIN-RENDON (27), *Department of Biochemistry, Oxford BioMedica (UK) Limited, Oxford OX4 4GA, United Kingdom*

LYDIA C. MATHEWS (23), *Harvard/Généthon Joint Laboratory, Harvard Institutes of Medicine, Boston, Massachusetts 02115*

TAKASHI MATSUSHITA (22), *Division of Genetic Therapeutics, Center for Molecular Medicine, Jichi Medical School, Kawachi, Tochigi 329-0498, Japan*

NICHOLAS D. MAZARAKIS (27), *Department of Biochemistry, Oxford BioMedica (UK) Limited, Oxford OX4 4GA, United Kingdom*

PAUL B. MCCRAY, JR. (28), *Departments of Pediatrics and Internal Medicine, University of Iowa, Iowa City, Iowa 52242*

KYRIACOS A. MITROPHANOUS (27), *Department of Biochemistry, Oxford BioMedica (UK) Limited, Oxford OX4 4GA, United Kingdom*

HIROAKI MIZUKAMI (22), *Division of Genetic Therapeutics, Center for Molecular Medicine, Jichi Medical School, Kawachi, Tochigi 329-0498, Japan*

RYUICHI MORISHITA (36), *Division of Gene Therapy Science, Graduate School of Medicine, Osaka University, Suita City, Osaka 565-0871, Japan*

NICHOLAS MUZYCZKA (24), *Powell Gene Therapy Center, University of Florida, Gainesville, Florida 32610*

*Current affiliation: Gene Therapeutics Research Institute, Cedars-Sinai Medical Center, Los Angeles, California 90048

LUIGI NALDINI (26, 29, 33), *Institute for Cancer Research and Treatment, Laboratory for Gene Transfer and Therapy, University of Torino Medical School, 10060 Candiolo, Torino, Italy*

NATHALIE NEYROUD (19), *Institute of Molecular Cardiobiology, The Johns Hopkins University School of Medicine, Baltimore, Maryland 21205*

STUART A. NICKLIN (10), *Department of Medicine and Therapeutics, University of Glasgow, Glasgow G11 6NT, United Kingdom*

TATSUYA NOMOTO (22), *Division of Genetic Therapeutics, Center for Molecular Medicine, Jichi Medical School, Kawachi, Tochigi 329-0498, Japan*

JAMES S. NORRIS (30), *Department of Microbiology and Immunology, Medical University of South Carolina, Charleston, South Carolina 29425*

H. BRADLEY NUSS (19), *Institute of Molecular Cardiobiology, The Johns Hopkins University School of Medicine, Baltimore, Maryland 21205*

TAKASHI OKADA (22), *Division of Genetic Therapeutics, Center for Molecular Medicine, Jichi Medical School, Kawachi, Tochigi 329-0498, Japan*

BERT W. O'MALLEY (31), *Department of Molecular and Cellular Biology, Baylor College of Medicine, Houston, Texas 77030*

KEIYA OZAWA (22), *Division of Genetic Therapeutics, Center for Molecular Medicine, Jichi Medical School, Kawachi, Tochigi 329-0498, Japan*

LUCIO PASTORE (11), *Department of Human and Molecular Genetics, Baylor College of Medicine, Houston, Texas 77030**

M. IAN PHILLIPS (1), *Department of Physiology and Functional Genomics, College of Medicine, University of Florida, Gainesville, Florida 32610*

MARK POTTER (24), *Powell Gene Therapy Center, University of Florida, Gainesville, Florida 32610*

L. QIN (12), *Institute for Gene Therapy and Molecular Medicine, Mount Sinai School of Medicine, New York, New York 10029*

PIPPA A. RADCLIFFE (27), *Department of Biochemistry, Oxford BioMedica (UK) Limited, Oxford OX4 4GA, United Kingdom*

MOHAN K. RAIZADA (32), *Department of Physiology and Functional Genomics, College of Medicine, University of Florida, Gainesville, Florida 32610*

PHYLLIS Y. REAVES (32), *Department of Physiology, College of Medicine, University of Florida, Gainesville, Maryland 32610*

TERESA C. RITCHIE (20), *Department of Anatomy and Cell Biology, University of Iowa College of Medicine, Iowa City, Iowa 52242*

JONATHAN B. ROHLL (27), *Department of Biochemistry, Oxford BioMedica (UK) Limited, Oxford OX4 4GA, United Kingdom*

SEMYON RUBINCHIK (30), *Department of Microbiology, Medical University of South Carolina, Charleston, South Carolina 29425*

DAVID W. RUSSELL (37), *Division of Hematology, University of Washington, Seattle, Washington 98195*

ROBERT SALLER (35), *Bavarian Nordic, D-82152 Martinsried, Austria*

BRIAN SALMONS (35), *Austrian Nordic, A-1210 Vienna, Austria*

*Current affiliation: CEINGE-Biotecnologie Avanzate and Dipartimento di Biochima e Biotecnologie Mediche, Università degli Studi di Napoli "Federico II," 80131 Napoli, Italy

GIOVANNI SALVATORI (13), *Department of Human Genetics, University of Michigan, Ann Arbor, Michigan 48109*[*]

KURT SCHILLINGER (31), *Department of Molecular and Cellular Biology, Baylor College of Medicine, Houston, Texas 77030*

DMITRY M. SHAYAKHMETOV (16), *Department of Medicine, Division of Medical Genetics, University of Washington, Seattle, Washington 98195*

KUNIKO SHIMAZAKI (22), *Department of Physiology, Center for Molecular Medicine, Jichi Medical School, Kawachi, Tochigi 329-0498, Japan*

PATRICK L. SINN (28), *Departments of Pediatrics and Internal Medicine, Program in Gene Therapy, College of Medicine, University of Iowa, Iowa City, Iowa 52242*

C. I. EDVARD SMITH (6), *Center for BioTechnology, Department of Biosciences, Karolinska Institute, SE-141 57 Huddinge, Sweden*[†]

RICHARD O. SNYDER (23), *Harvard/Généthon Joint Laboratory, Division of Molecular Medicine, The Children's Hospital, Department of Pediatrics, Harvard Institutes of Medicine, Boston, Massachusetts 02115*[‡]

YOUNG K. SONG (5), *Department of Pharmaceutical Sciences, University of Pittsburgh School of Pharmacy, Pittsburgh, Pennsylvania 15261*[§]

MATTHEW L. SPRINGER (9), *Department of Molecular Pharmacology, Stanford University School of Medicine, CCSR 4215, Stanford, California 94305*

HARMUT STECHER (16), *Department of Medicine, Division of Medical Genetics, University of Washington, Seattle, Washington 98195*

CLIFFORD J. STEER (2), *Department of Medicine, University of Minnesota Medical School, Minneapolis, Minnesota 55455*

COLLEEN S. STEIN (25), *College of Medicine, University of Iowa, Iowa City, Iowa 52242*

DIRK S. STEINWAERDER (16), *Department of Medicine, Division of Medical Genetics, University of Washington, Seattle, Washington 98195*

R. SUNG (12), *Institute for Gene Therapy and Molecular Medicine, Mount Sinai School of Medicine, New York, New York 10029*

C. E. THOMAS (17), *Molecular Medicine and Gene Therapy Unit School of Medicine, University of Manchester, Manchester M13 9PT, United Kingdom*[∥]

KIYOTAKE TOBITA (22), *Department of Virology, Jichi Medical School, Kawachi, Tochigi 329-0498, Japan*

S. TONDEUR (17), *Laboratoire de Biologie et Thérapeutique des Pathologies Immunitaires, Université Pierre et Marie Curie, Hôpital de la Pitié Salpétriére 75651 Paris Cedex 13, France*

GRANT TROBRIDGE (37), *Division of Hematology, University of Washington, Seattle, Washington 98195*

SOPHIA Y. TSAI (31), *Department of Molecular and Cellular Biology, Baylor College of Medicine, Houston, Texas 77030*

[*]Current affiliation: Department of Immunology, Sigma-Tau S.P.A., 00040 Pomezia, Italy
[†]Current affiliation: Clinical Research Center, Karolinska Institute, S-141 86 Stockholm, Sweden
[‡]Current affiliation: Powell Gene Therapy Center, Department of Molecular Genetics and Microbiology, University of Florida, Gainesville, Florida 32610
[§]Current affiliation: Department of Pharmacology, University of Pittsburgh School of Medicine, Pittsburgh, Pennsylvania 15261
[∥]Current affiliation: Department of Pediatrics and Genetics, Stanford University, Stanford, California 94305

MIKKO P. TURUNEN (18), *A. I. Virtanen Institute, University of Kuopio, FIN-70210 Kuopio, Finland*

P. UMANA (17), *Molecular Medicine and Gene Therapy Unit School of Medicine, University of Manchester, Manchester M13 9PT, United Kingdom**

MASASHI URABE (22), *Division of Genetic Therapeutics, Center for Molecular Medicine, Jichi Medical School, Kawachi, Tochigi 329-0498, Japan*

THIERRY VANDENDRIESSCHE (33), *Flanders Interuniversity Institute of Biotechnology, Center for Transgene Technology and Gene Therapy, University of Leuven, B-3000 Leuven, Belgium*

GEORGE VASSILOPOULOS (37), *Division of Hematology, University of Washington, Seattle, Washington 98195*

T. VERAKIS (17), *Molecular Medicine and Gene Therapy Unit School of Medicine, University of Manchester, Manchester M13 9PT, United Kingdom*

JEAN-MICHEL H. VOS (deceased) (38), *Lineberger Comprehensive Cancer Center, Department of Biochemistry and Biophysics, University of North Carolina at Chapel Hill, Chapel Hill, North Carolina 27599*

CINDY WANG (14), *Departments of Biochemistry and Molecular Biology, Medical University of South Carolina, Charleston, South Carolina 29425*

GUOSHUN WANG (28), *Departments of Pediatrics and Internal Medicine, Program in Gene Therapy, College of Medicine, University of Iowa, Iowa City, Iowa 52242*†

JIANLONG WANG (38), *Lineberger Comprehensive Cancer Center, University of North Carolina at Chapel Hill, Chapel Hill, North Carolina 27599*‡

SAM WANG (34), *Molecular Neurogenetics Unit, Massachusetts General Hospital, Charlestown, Massachusetts 02129*

D. ALAN WHITE (21), *Department of Molecular Genetics and Microbiology, College of Medicine, University of Florida, Gainesville, Florida 32610*

STEVE J. WHITE (10), *Department of Medicine and Therapeutics, University of Glasgow, Glasgow G11 6NT, United Kingdom*

PHILLIP WILLIAMS (7), *Departments of Pediatrics and Medical Genetics, University of Wisconsin, Madison, Wisconsin 53705*

JON A. WOLFF (7), *Departments of Pediatrics and Medical Genetics, University of Wisconsin, Madison, Wisconsin 53705*

LORRAINE M. WORK (10), *Department of Medicine and Therapeutics, University of Glasgow, Glasgow G11 6NT, United Kingdom*

ZIYING YAN (20), *Department of Anatomy and Cell Biology, University of Iowa College of Medicine, Iowa City, Iowa 52242*

XIANGCANG YE (31), *Department of Molecular and Cellular Biology, Baylor College of Medicine, Houston, Texas 77030*

SEPPO YLÄ-HERTTUALA (18), *Department of Molecular Medicine, A. I. Virtanen Institute, University of Kuopio, FIN-70210 Kuopio, Finland*

JOSEPH ZABNER (28), *Departments of Pediatrics and Internal Medicine, Program in Gene Therapy, College of Medicine, University of Iowa, Iowa City, Iowa 52242*

*Current affiliation: GlycArt Biotechnology AG, 8093 Zurich, Switzerland

†Current affiliation: Departments of Medicine and Genetics, Gene Therapy Program, Louisiana State University Health Sciences Center, New Orleans, Louisiana 70112

‡Current affiliation: Division of Hematology/Oncology, Children's Hospital Boston, Boston, Massachusetts 02115

GUISHENG ZHANG (5), *Department of Pharmaceutical Sciences, University of Pittsburgh School of Pharmacy, Pittsburgh, Pennsylvania 15261*

GUOFENG ZHANG (7), *Departments of Pediatrics and Medical Genetics, University of Wisconsin, Madison, Wisconsin 53705*

HESHAN ZHOU (11), *Cell and Gene Therapy Center, Baylor College of Medicine, Houston, Texas 77030*

SERGEI ZOLOTUKHIN (24), *Department of Molecular Genetics and Microbiology, Powell Gene Therapy Center, University of Florida, Gainesville, Florida 32610*

Preface

Gene therapy is less than ten years old and still very much in its infancy. The first clinical gene therapy study was carried out in 1995 by Blaese and colleagues. Although early results on clinical efficacy were disappointing, the logic of gene therapy is irresistibly attractive. As science continues to evaluate the prospects for gene therapy, so the clinical benefits have begun to be demonstrated. Early results were hampered because of inadequate vectors for gene transfer. Most of the clinical studies involved gene addition. However, gene therapy allows both correction and replacement of defective genes. Ultimately, the goal is to have an *in vivo* somatic gene therapy that can deal with not only immediate life-threatening diseases, such as cancer and AIDS, but also chronic diseases that reduce the quality of life, such as hypertension and inflammatory diseases. The basis for gene therapy is understanding which genes are involved in diseased phenotypes and which vectors are appropriate for providing therapeutic genes. The rapid progress in gene discovery has been accelerated by the completion of the human genome project.

This book brings together, for the first time, methods in gene therapy that reflect the development of scientifically grounded systems for delivering genes. DNA can be engineered to carry a therapeutic gene in sufficient quantities for full-scale clinical trials. The methods can be classified as either viral or nonviral. Viral vectors are replication defective viruses with part of their coding sequences replaced by the therapeutic gene. These viral vectors include retroviruses, adenovirus, adeno-associated virus, herpes simplex virus, papillomavirus, and lentivirus. Nonviral vectors are simpler and easier to produce on the large scale. However, each has its advantage. Viral vectors can be engineered to be expressed in specific tissue and only under specific conditions. Nonviral vectors are less easy to control so precisely. Some diseases need gene therapy for a rapid effect, such as killing off tumor cells. Others need the presence of a stable, safe gene delivery system for chronic lifetime diseases. The use of gene therapy could eliminate the need for repeated administrations, improved therapeutic efficacy, and fewer side effects. In hypertension, for example, one of the major problems is the lack of patient compliance in taking current prescribed drugs that have to be administered once a day. The prospect of prolonged effective control of blood pressure and the subsequent reduction in heart attacks, stroke, and end-stage renal disease are an exciting possibility of the true benefits of gene therapy.

In this book we have brought together some of the leading researchers and research methods in gene therapy. There are many ways to classify these chapters: by disease, by the type of method, or the type of delivery system. We have chosen to classify them under the main type of delivery system being investigated. However, each chapter stands on its own, offering scientific insight and experience with a particular approach. In some cases they cross the boundaries of these classifications. Although this is the first volume entitled "Gene Therapy Methods" for the *Methods in Enzymology* series, the increasing number of new methods and the progress of gene therapy will undoubtedly require more volumes in the future.

I wish to thank the authors for their contributions. I also wish to thank Ms. Gayle Butters of the University of Florida, Department of Physiology and Functional Genomics, for her excellent editorial assistance. My thanks also go to Shirley Light of Academic Press for her encouragement to do this volume.

M. IAN PHILLIPS

METHODS IN ENZYMOLOGY

VOLUME 73. Immunochemical Techniques (Part B)
Edited by JOHN J. LANGONE AND HELEN VAN VUNAKIS

VOLUME 74. Immunochemical Techniques (Part C)
Edited by JOHN J. LANGONE AND HELEN VAN VUNAKIS

VOLUME 75. Cumulative Subject Index Volumes XXXI, XXXII, XXXIV–LX
Edited by EDWARD A. DENNIS AND MARTHA G. DENNIS

VOLUME 76. Hemoglobins
Edited by ERALDO ANTONINI, LUIGI ROSSI-BERNARDI, AND EMILIA CHIANCONE

VOLUME 77. Detoxication and Drug Metabolism
Edited by WILLIAM B. JAKOBY

VOLUME 78. Interferons (Part A)
Edited by SIDNEY PESTKA

VOLUME 79. Interferons (Part B)
Edited by SIDNEY PESTKA

VOLUME 80. Proteolytic Enzymes (Part C)
Edited by LASZLO LORAND

VOLUME 81. Biomembranes (Part H: Visual Pigments and Purple Membranes, I)
Edited by LESTER PACKER

VOLUME 82. Structural and Contractile Proteins (Part A: Extracellular Matrix)
Edited by LEON W. CUNNINGHAM AND DIXIE W. FREDERIKSEN

VOLUME 83. Complex Carbohydrates (Part D)
Edited by VICTOR GINSBURG

VOLUME 84. Immunochemical Techniques (Part D: Selected Immunoassays)
Edited by JOHN J. LANGONE AND HELEN VAN VUNAKIS

VOLUME 85. Structural and Contractile Proteins (Part B: The Contractile Apparatus and the Cytoskeleton)
Edited by DIXIE W. FREDERIKSEN AND LEON W. CUNNINGHAM

VOLUME 86. Prostaglandins and Arachidonate Metabolites
Edited by WILLIAM E. M. LANDS AND WILLIAM L. SMITH

VOLUME 87. Enzyme Kinetics and Mechanism (Part C: Intermediates, Stereochemistry, and Rate Studies)
Edited by DANIEL L. PURICH

VOLUME 88. Biomembranes (Part I: Visual Pigments and Purple Membranes, II)
Edited by LESTER PACKER

VOLUME 89. Carbohydrate Metabolism (Part D)
Edited by WILLIS A. WOOD

VOLUME 90. Carbohydrate Metabolism (Part E)
Edited by WILLIS A. WOOD

VOLUME 213. Carotenoids (Part A: Chemistry, Separation, Quantitation, and Antioxidation)
Edited by LESTER PACKER

VOLUME 214. Carotenoids (Part B: Metabolism, Genetics, and Biosynthesis)
Edited by LESTER PACKER

VOLUME 215. Platelets: Receptors, Adhesion, Secretion (Part B)
Edited by JACEK J. HAWIGER

VOLUME 216. Recombinant DNA (Part G)
Edited by RAY WU

VOLUME 217. Recombinant DNA (Part H)
Edited by RAY WU

VOLUME 218. Recombinant DNA (Part I)
Edited by RAY WU

VOLUME 219. Reconstitution of Intracellular Transport
Edited by JAMES E. ROTHMAN

VOLUME 220. Membrane Fusion Techniques (Part A)
Edited by NEJAT DÜZGUÜNES

VOLUME 221. Membrane Fusion Techniques (Part B)
Edited by NEJAT DÜZGÜNES

VOLUME 222. Proteolytic Enzymes in Coagulation, Fibrinolysis, and Complement Activation (Part A: Mammalian Blood Coagulation Factors and Inhibitors)
Edited by LASZLO LORAND AND KENNETH G. MANN

VOLUME 223. Proteolytic Enzymes in Coagulation, Fibrinolysis, and Complement Activation (Part B: Complement Activation, Fibrinolysis, and Nonmammalian Blood Coagulation Factors)
Edited by LASZLO LORAND AND KENNETH G. MANN

VOLUME 224. Molecular Evolution: Producing the Biochemical Data
Edited by ELIZABETH ANNE ZIMMER, THOMAS J. WHITE, REBECCA L. CANN, AND ALLAN C. WILSON

VOLUME 225. Guide to Techniques in Mouse Development
Edited by PAUL M. WASSARMAN AND MELVIN L. DEPAMPHILIS

VOLUME 226. Metallobiochemistry (Part C: Spectroscopic and Physical Methods for Probing Metal Ion Environments in Metalloenzymes and Metalloproteins)
Edited by JAMES F. RIORDAN AND BERT L. VALLEE

VOLUME 227. Metallobiochemistry (Part D: Physical and Spectroscopic Methods for Probing Metal Ion Environments in Metalloproteins)
Edited by JAMES F. RIORDAN AND BERT L. VALLEE

VOLUME 228. Aqueous Two-Phase Systems
Edited by HARRY WALTER AND GÖTE JOHANSSON

Section I

Nonviral

[1] Gene Therapy for Hypertension: The Preclinical Data

By M. IAN PHILLIPS

Introduction

Although many excellent pharmacological agents are available commercially for the treatment of hypertension, the problems of cardiovascular disease related to hypertension continue to affect millions of people throughout the world. Hypertension is a multifactorial, multigenic disease, but the drugs aimed at controlling hypertension are aimed at relatively few targets. They target the renin–angiotensin system, β-adrenergic receptors or α-adrenergic receptors, and calcium channels. They should be very effective, but why is hypertension so widespread and morbid in our society and in the world?

Many of the drugs are expensive, and, therefore, unavailable to poor segments of all societies. Another problem is detection. Hypertension is undetected in about 40% of the population of the United States, according to the NHANES III Report.[1] Of that 40% in whom hypertension has been detected, about half receive treatment. The problem is further confounded because it is estimated that only 27% of those treated hypertensive patients fully comply with their treatment and have their hypertension controlled.[1] Clearly, there is a need for rethinking our approach to the treatment of hypertension. Detection could be increased by education. Nonpharmacological treatment, such as exercise, weight loss, and low salt diets, could provide inexpensive treatment, but it has proved very difficult to achieve compliance for these approaches. For treating hypertension on a world scale, we need something akin to an immunization against hypertension. Since hypertension is polygenic and not a single gene disease, except in very few cases[2] it cannot be immunized against. We need to develop ways that would improve hypertension control by providing longer lasting effects with a single dose and reducing side effects that lead to poor compliance. To do this, we began developing a somatic gene therapy approach in 1993[3,4] with the goal of producing prolonged control of hypertension. There have been two strategies taken: one by Chao and colleagues[5] to increase genes for vasodilation, and the other by Phillips and colleagues to decrease genes for vasoconstriction.[4] They represent the two sides to transferring

---

[1] N. M. Kaplan, "Clinical Hypertension." Williams and Wilkins, Baltimore, 1998.

[2] R. P. Lifton, *Science* **272,** 676 (1996).

[3] R. Gyurko, D. Wielbo, and M. I. Phillips, *Reg. Pept.* **49,** 167 (1993).

[4] M. I. Phillips, D. Wielbo, and R. Gyurko, *Kidney Intl.* **46,** 1554 (1994).

[5] J. Chao and L. Chao, *Immunopharmacology* **36,** 229 (1997).

TABLE I
PRECLINICAL DATA ON GENE THERAPY FOR HYPERTENSION VASODILATOR GENES[a]

Target gene	Delivery	Model	Max Δ BP (mmHg)	Duration of effect	Reference
Human tissue	Adenovirus	Dahl-salt sensitive		4 weeks	Chao et al., 1997[5]
Kallikren		5/6 renal mass	−37	5 weeks	Wolf et al., 2000[42]
		SHR	−30	36 days	
	Adenovirus	SHR		5 weeks	Zhang et al., 1999[43]
	Intramuscular	2K-1C	−26	24 days	
	Adenovirus iv	Doca-salt		23 days	Dobrzynski et al., 1999[9]
Adrenomedullin	Adenovirus	Doca-salt	−41	9 days	Dobrzynski et al., 2000[44]
	iv	SHR			Chao et al., 1997[7]
	iv	Dahl-salt sensitive		4 weeks	Zhang et al., 2000[45]
Atrial natriuretic peptide (ANP)	Adenovirus	Dahl-salt sensitive	−32	5 weeks	Lin et al., 1997[6]
Nitric oxide	Plasmid	SHR	−21	5–6 weeks (1st injection) 10–12 weeks (2nd injection)	Lin et al., 1997[8]

[a] Abbreviations: SHR, spontaneously hypertensive rat; AGT, angiotensinogen; AAV, adeno-associated virus; AT1-R, angiotensin type 1 receptor; ACE, angiotensin-converting enzyme; LNSV, retrovirus; AS-ODN, antisense oligodeoxynucleotide; CIH, cold-induced hypertension.

DNA into cells. One is the sense approach, i.e., the normal DNA sequence direction, and the other is the antisense approach, i.e., the opposite DNA sequence direction.

Sense to Vasodilation Genes

Chao et al. have an extensive series of studies on gene transfer to genes that act to increase vasodilator proteins (Table I). They have used genes such as kallikrein,[5] atrial natriuretic peptide,[6] adrenomedullin,[7] and endothelial nitric oxide synthase.[8] In different rat models of hypertension (SHR, Dahl salt-sensitive, Doca-salt) they showed that they could achieve blood pressure lowering effects for 3–12 weeks with the overexpression of these genes. The decrease in pressure resulting from

[6] K. F. Lin, J. Chao, and L. Chao, *Hum. Gene Ther.* **9**, 1429 (1998).
[7] J. Chao, L. Jim, K. F. Lin, and L. Chao, *Hypertens. Res.* **20**, 2692 (1997).
[8] K. F. Lin, L. Chao, and J. Chao, *Hypertension* **30**, 307 (1997).

these vasodilator proteins ranged from -21 to -41 mmHg. The results of this group are consistent and impressive. Even though the effects were not very prolonged, there were reductions in end organ damage with these therapies.[9] However, the use of adenovirus limits the possibility of translating these strategies to humans. The use of plasmids, however, had very prolonged effects in their hands.

Antisense to Vasoconstrictor Genes

To counter overexpression of a gene as a critical factor contributing to hypertension, we introduced antisense somatic gene therapy. Antisense provides a highly specific, biological approach to produce attenuation of the sense DNA expression which produces too much protein: for example, angiotensin II (Ang II), which is responsible for increased vasoconstriction. Antisense gene therapy involves recombinant antisense DNA to express an antisense mRNA or antisense oligonucleotides to inhibit mRNA designed to specifically reduce an overexpressed protein that is critical to the disease. Since hypertension is a multigene disease, how can we decide on the candidate genes for gene therapy? We have ignored the difficulties of defining all the candidate genes by concentrating on the genes that have already been shown to be successful targets by experience with current drugs. These include beta receptors, angiotensin-converting enzyme (ACE), and angiotensin type 1 receptor (AT_1-R). Other targets follow logically, including angiotensinogen (AGT). Transfer of the antisense genes to somatic cells is achieved by an *in vivo* approach. It would be possible to try an *ex vivo* approach in which target cells are removed from the host, transduced *in vivo,* and then reimplanted as genetically modified cells. However, this strategy has no obvious applicability to hypertension, where the cause of the disease lies in the reaction of blood vessels but not in one specific tissue; even the heart, kidney, and brain are obviously very important in hypertension. The *in vivo* approach is challenging. One challenge is to provide sufficient antisense DNA, either alone or in a vector, to produce a sufficient concentration for uptake in a large number of cells. To do this we have developed two different strategies for hypertension gene therapy based on antisense with (a) antisense oligonucleotides (Table II) and (b) viral vectors to deliver antisense DNA (Table III).

Nonviral Delivery

Antisense Oligonucleotides

Antisense oligonucleotides are short lengths of synthetically made nucleotides (DNA) designed to hybridize with a specific sequence of mRNA. The hybridization has one or two effects: it stimulates RNase H or sterically inhibits the mRNA from translating its message in the read-through process at the ribosome, or it does both.

[9] E. Dobrzynski, H. Yoshida, J. Chao, and L. Chao, *Immunopharmacology* **44,** 57 (1999).

TABLE II
PRECLINICAL DATA ON GENE THERAPY FOR HYPERTENSION, VASOCONSTRICTOR GENES: ANTISENSE OLIGODEOXYNUCLEOTIDES[a]

Target gene	Delivery	Model	Max Δ BP (mmHg)	Duration of effect	Reference
AT1-R	AS-ODN icv	SHR	−30	Unknown	Gyurko et al.[3]
AGT	AS-ODN icv	SHR	−35	Unknown	Phillips et al.[4]
AT1-R	AS-ODN pvn microinjection	MRen2	−24	4 days	Li et al.[32]
TRH (Thyrotropin-releasing hormone)	AS-ODN intrathecal	SHR	−38	Unknown	Suzuki et al.[33]
AGE-2	AS-ODN portal vein	SHR	−20	6 Days	Morishita et al.[34]
Angiotensin-gene activating elements		SHR	−28	7 days	Nishii et al.[35]
Carboxypeptidase Y	AS-ODN HJV liposome	Doca-salt	−15	4 days	Hyashi et al.[36]
c-fos	AS-ODN microinection in RVLM	WKY	−16	4–6 hours	Suzuki et al.[37]
		SD	−17	Unknown	
CYP4A1	AS-ODN continuous infusion	SHR	16	Unknown	Wang et al.[39]
AGT	AS-ODN iv	SHR	−25	Unknown	Wielbo et al.[3]
AGT	AS-ODN hepatic vein HJV-liposome	SHR	−20	4 days	Tomita et al.[40]
AT1-R	AS-ODN icv	SHR	−30	7 days	Gyurko et al.[3]
AGT	AS-ODN with asialoglycoprotein iv	SHR	−30	7 days	Makino et al.[14]
AT1-R	AS-ODN iv	2K-1C acute	−30	>7 days	Galli et al.[41]
AT1-R	AS-ODN icv	2K-1C 6 months	−30	>5 days	Kagiyama et al.[17]
AT1-R	AS-ODN iv in liposomes	CIH	−38	Unknown	Peng et al.[17]
β1-AR	AS-ODN iv in liposomes	SHR	−35	30–40 days	Zhang et al.[10,11]

[a] Abbreviations: icv, intracerebroventricular; pvn, paraventricular nucleus; unknown, recovery of pressure not recorded.

Delivery of antisense oligodeoxynucleotides (AS-ODNs) can be carried out with direct injection of "naked DNA." We have found that direct injection is effective, but the efficiency of uptake is greatly increased by delivering the ODN in cationic liposomes provided the correct ratio has been calculated.[10]

[10] Y. C. Zhang, J. D. Bui, L. P. Shen, and M. I. Phillips, Circulation 101, 682 (2000).

TABLE III
PRECLINICAL DATA ON GENE THERAPY FOR HYPERTENSION, VASOCONSTRICTOR GENES:
VIRAL VECTOR DELIVERY OF ANTISENSE

Target gene	Delivery	Model	Max Δ BP mmHg	Duration of effect	Reference
AGT	AAV-based plasmid	SHR	−22.5	8 days	Tang et al.[21]
AT1-R	AAV ic	SHR adult	−40	9 weeks plus	Phillips[25]
AGT	AAV ic	SHR adult	−40	10 weeks plus	Phillips[25]
AT1-R	LNSV ic	SHR (neonates)	−40	90 days	Iyer et al.[27]
AT1-R	LNSV iv	SHR	−30	36 days	Katovich[38]
ACE	LNSV ic	SHR (neonates)	−15		Reaves et al.[31]
AGT	AAV ic	SHR (neonates)	−30	6 months	Kimura et al.[26]
AT1-R	AAV iv	Double transgenic mice (adult)	−40	6 months	Phillips et al.[29]

β_1-Adrenoceptor Antisense

Nonviral gene delivery, using cationic liposomes such as DOTAP and DOPE, have been successfully used by our group to deliver β_1-adrenoceptor antisense oligonucleotides (β_1-AR-AS-ODN) to act as novel beta blockers with prolonged effects.[10,11] By optimizing the liposome/ODN ratio and the incubation procedure, we are able to produce antihypertensive effects with β_1-AS-ODN for up to 33–40 days with a single dose.[11] The beauty of the β_1-AS-ODN is its specificity. The β_1-AS-ODN reduces β_1-adrenoceptors but does not affect β_2-adrenoceptors. Secondly, the β_1-AS-ODN does not cross the blood–brain barrier, and therefore, the novel β_1 blocker, based on antisense, will have no central nervous system side effects. The strongest uptake sites are in the heart and kidney where the β_1-adrenoceptors play a significant role. In the heart they control the force of contraction and this is reduced by the β_1-adrenoceptor. However, the heart rate is not affected by the β_1-AS-ODN.[11] This is in contrast to the effects of currently available beta blockers that have both β_1 and β_2 actions, and second, reduce heart rate as well as heart contractility. Therefore, the specificity offered by the ODN provides a more precise and accurate way of controlling the mechanisms contributing to high blood pressure without the side effects of bradycardia.[10] Furthermore, since the effect lasts for 30–40 days with a single injection, the antisense ODN is greatly superior to any of the currently available drugs, all of which have to be taken on a daily basis. Repeated injections intravenously (iv) at intervals of 3–4 weeks of β_1-AS-ODN produce prolonged control of high blood pressure without any toxic effects in the liver, blood, or organs.

[11] Y. C. Zhang, B. Kimura, L. Shen, and M. I. Phillips, *Hypertension* **35,** 219 (2000).

Angiotensinogen Antisense ODN

We have also established that angiotensinogen iAS-ODN is effective for antisense ODN for hypertension therapy. In human hypertension, the angiotensinogen gene has been shown to be linked and to play a role in the disease.[12] However, there is no currently available drug to inhibit angiotensinogen. We have designed antisense targeted to AGT mRNA and tested it *in vivo* and *in vitro*.[13] When given iv the angiotensinogen AS-ODN reduces blood pressure significantly when delivered with a liposome. These studies have been confirmed by others independently, showing that AGT-AS-ODN reduces blood pressure for up to 7 days with a single systemic dose.[14]

AT₁-R Antisense ODN

A similar story is true for the effects of AT_1-AS-ODN. This has been tested centrally with intracerebroventricular injections and with intravenous injections. It has been tested in spontaneously hypertensive rats (SHR)[15] and also in 2 kidney-1 clip animals[16] and environmentally induced hypertension.[17] In these three different models of hypertension, genetic, surgical, and environmental, the antisense produces a decrease in blood pressure within 24 hr of administration. The effect lasts for up to 7 days and there is no effect on heart rate.[15] The distribution of antisense is in blood vessels, kidney, liver, and heart.[17] The majority of uptake is in the kidney and liver.[17] A reduction in AT_1 receptors after treatment with the AT_1-AS-ODN reveals reductions in the protein in kidney, aorta, and liver.[17]

In summary, AS-ODNs have proved to be useful in demonstrating in the preclinical setting the power of AS-ODN to target specific genes and to reduce blood pressure for several days (or weeks) with a single administration. Laboratory data indicate that these effects are the result of rapid uptake of the antisense ODN into cells[18] where they migrate to the nucleus and inhibit the production of protein, mostly likely through translational inhibition of messenger RNA.[18,19] This could occur by the hybridization of ODN with specific mRNA, preventing the passage of the mRNA through the ribosome. Alternatively, DNA hybridization to RNA will in some tissues stimulate the production of RNase H for the specific sequence of mRNA bound to the ODN. RNase H destroys the RNA hybridized to DNA

[12] X. Jeunemaitre, F. Soubrier, Y. V. Kotelevetsev, R. P. Lifton, C. S. Williams, A. Charru, S. C. Hunt, P. N. Hopkins, R. R. Williams, and J. M. Lalouel, *Cell* **71**, 169 (1992).

[13] D. Wielbo, A. Simon, M. I. Phillips, and S. Toffolo, *Hypertension* **28**, 147.

[14] K. Makino, M. Sugano, S. Ohtsuka, and S. Sawada, *Hypertension* **31**, 1166 (1998).

[15] R. Gyurko, D. Tran, and M. I. Phillips, *Am. J. Hypertension* **10**, 56S (1997).

[16] S. Kaguyama, A. Varela, M. I. Phillips, and S. M. Galli, *Hypertension* (2001), in press.

[17] J.-F. Peng, B. Kimura, M. J. Fregly, and M. I. Phillips, *Hypertension* **31**, 1317 (1998).

[18] B. Li, J. A. Hughes, and M. I. Phillips, *Neurochem. Intl.* **31**, 393 (1996).

[19] S. T. Crooke, *Methods Enzymol.* **313**, 3 (2000).

and thereby releases the oligonucleotide for further hybridization. This recycling action induced by RNase H may account for the long action of AS-ODNs.

Other useful features that make oligonucleotides attractive for hypertension therapy is that they can be produced relatively cheaply, rapidly, and in large quantities. The demand for oligonucleotides and primers has reduced the cost per base to a few cents. Second, they do not cross the blood–brain barrier and therefore, when given peripherally, will not have central effects.[10] Third, they are most effective when delivered in the right combination of ODN to cationic liposome.[10,11] Treatment of rats with liposome ODN complexes has not shown any toxicity in our experience.

Viral Vector Delivery

To produce very prolonged effects (i.e., several months) with a single injection, we use antisense DNA delivery by viral vector. Several viral vectors are available, but the adeno-associated virus (AAV) is both safe for use in humans and large enough to carry antisense genes with tissue-specific promoters.[20] The AAV is not to be confused with the adenovirus. Adenoviruses, although easy to use in lab animals, have caused a death in a human during trials and are not, in their present form, acceptable vectors. AAV is a parvovirus that does not replicate and does not induce inflammatory reactions. The AAV can be stripped of its rep and gag genes to carry up to 4.5 kb and deliver it to the nuclei of cells where it integrates in the genome.[21] When antisense DNA is used, the AAV allows the continuous production of an RNA that is in the antisense direction. This antisense RNA hybridizes to specific mRNA and inhibits translation. Therefore, we are developing antisense therapy using the AAV as a vector. To construct a viral vector requires the design and production of plasmids and gene packaging into the vector.

Delivery by Plasmids

Plasmids are effective vectors, but last for a shorter time than the viral vector because they do not allow integration into the genome. This is illustrated with the adeno-associated-based vector for angiotensinogen antisense cDNA.[22] A plasmid containing AAV terminal repeats was prepared with a cassette consisting of a CMV promoter, the rat AGT cDNA based on the sequence by Lynch et al.[23] The cDNA is oriented in the antisense direction. In addition, the cassette contains an internal ribosome entry site (IRES) and, as a marker, the green fluorescent protein gene

---

[20] M. I. Phillips, *Hypertension* **29,** 177 (1997).

[21] P. Wu, M. I. Phillips, J. Bui, and E. F. Terwillinger, *J. Virol.* **72,** 5919 (1998).

[22] X. Tang, D. Mohuczy, C. Y. Zhang, B. Kimura, S. M. Galli, and M. I. Phillips, *Am. J. Physiol.* **277,** H2392 (1999).

[23] K. R. Lynch, V. I. Simnad, E. T. Ben Ari, and J. C. Garrison, *Hypertension* **8,** 540 (1986).

(GFP).[24] At 48 hr after transfection into pAAV-AGT-AS, there was clear dominant expression of GFP in the H-4 cells. There was a significant reduction of AGT (120 ± 14 vs 230 ± 20 ng/mg protein, $p < 0.01$). Transgene expression detected by RT-PCR in the H-4 cells started at 2 hr and continued for at least 72 hr.

The plasmid was then tested *in vivo* by injecting the S and AS plasmids iv into SHR rats.[22] AGT-AS expression was positive in heart and lung at 3 days and 7 days. Expression in the kidney was absent or weak. When injected with 3 mg/kg plasmid, pAAV-AGT-AS produced a significant drop in blood pressure ($p < 0.01$) for 6–8 days in SHR. The drop in blood pressure correlated to a drop in plasma angiotensinogen levels that was significant at days 3 and 5 after injection. The decrease in blood pressure with injection of plasmid could be prolonged by injecting the plasmid with cationic liposome (DOTAP/DOPE).

Plasmids are useful for delivery of AS to produce an antihypertensive effect lasting about 1 week. They do not require the more complex packaging needed for recombinant AAV (rAAV).

Delivery by Recombinant AAV Vector

To produce long-term decreases in hypertension, we developed rAAV to deliver antisense to AT_1R in SHR.[20,25] The results showed that single intracardiac injection of rAAV-AT_1R-AS effectively reduced blood pressure by 30 mmHg for at least 5 weeks compared to controls.

To test whether an AAV delivery of an AT_1R antisense would inhibit development of hypertension, we injected 5-day-old SHRs. Hypertension in SHR develops between the eighth and tenth week after birth. Therefore, injecting in 5-day-old SHR allowed us to observe if the development of hypertension would be reduced. A single injection of AAV-AGT-AS in 5-day-old SHR significantly attenuated the full development[26] and level of hypertension for up to 6 months. In 3-week-old SHR rAAV-AT_1R-AS significantly reduced hypertension by about 30 mmHg for at least 5 weeks (the length of the study). However, unlike the reports of the effect of retrovirus delivery of an AT_1R-AS in 5-day-old SHR "curing" hypertension,[27] we did not find a complete inhibition of the rise in blood pressure.

In rAAV-AGT-AS treated SH rats, measures of plasma AGT levels showed a corresponding lack of increase in AGT in the AS treated groups, compared to the significant increase of AGT in the control animals.[26] Correlation of AGT versus blood pressure was significant ($p < 0.05$) in the control treated animals and not significant in the AS treated animals. This shows that angiotensinogen in the SHR

[24] S. Zolotukhin, M. Potter, W. W. Hauswirth, J. Guy, and N. A. Muzyczka, *J. Virol.* **70**, 4646 (1996).

[25] M. I. Phillips, D. Mohuczy-Dominiak, M. Coffey, S. M. Galli, W. Ping, and T. Zelles, *Hypertension* **29**, 374 (1997).

[26] B. Kimura, D. Mohuczy, X. Tang, and M. I. Phillips, *Hypertension* **37**, 376 (2001).

[27] S. N. Iyer, D. Lu, M. Katovich, and M. K. Raizada, *Proc. Natl. Acad. Sci. U.S.A.* **93**, 9960 (1996).

is correlated with an increase in blood pressure. The AAV was expressed in kidney, heart, and liver throughout the time of the reduction in blood pressure. Thus, we concluded that early treatment with a single dose of rAAV-AGT-AS, given systemically, prevents the full development of hypertension in the adult SHR by a prolonged reduction in AGT levels. Similarly, the results with the rAAV-AT$_1$-AS showed a reduction in hypertension development correlated with a consistent reduction in AT$_1$ receptors in VSMC.[25] No toxicity was noted.[26] To prove the potential therapeutic value of rAAV, we have have used a mouse model of hypertension that clearly depends on an overactive renin–angiotenisn system. In this model, which has human renin and human AGT trangenes, rAAV-AS-AT$_1$R reduced high blood pressure for up to 6 months with a single systemic injection.[28] This latest data with rAAV-AT$_1$R-AS confirms the results in adult SHR rats[20] and gives an even clearer picture that the AAV as vector has many advantages for hypertension therapy.

Other Vectors

Other vectors are being tested for hypertension gene therapy. As noted above, adenovirus vectors have been used with kallikrein gene insertion[5] and recently to deliver calcitonin gene-related peptide for hypoxia induced pulmonary hypertension in mice.[29] However, the adenovirus synthesizes proteins that trigger the immune system and cause inflammation, which limits its use in human therapy so far. Raizada and colleagues have worked with LNSV, a retrovirus, with antisense AT$_1$ receptor injected into newborn SHR to prevent the development of hypertension in the adults.[27,30,31] In a series of papers they report evidence that AT$_1$R-AS normalizes blood pressure and prevents organ damage. Retroviruses are appropriate only for dividing cells and therefore are not suitable for hypertension therapy in adults. The idea of injecting infants with AT$_1$R antisense on the chance

[28] M. I. Phillips, B. Kimura, Y.-C. Zhang, and C. H. Gelband, *Hypertension* **36,** P204 (2000).

[29] H. C. Champion, T. J. Bivalacqua, K. Toyoda, D. D. Hystad, A. L. Hyman, and P. J. Kadowitz, *Circulation* **101,** 923 (2000).

[30] M. I. Phillips, *Hypertension* **33,** 8 (1999).

[31] P. Y. Reaves, C. H. Gelband, H. Wang, H. Yang, D. Lu, K. H. Berecek, M. J. Katovich, and M. K. Raizada, *Circ. Res.* **86,** E44 (1999).

[32] P. Li, M. Morris, D. I. Diz, C. M. Ferrario, D. Ganten, and M. F. Callahan, *Am. J. Physiol.* **270,** R1178 (1996).

[33] S. Suzuki, P. Pilowsky, J. Minso, L. Arnolda, I. Llewellyn-Smith, and J. Chalmers, *Circ. Res.* **77,** 679 (1995).

[34] R. Morishita, J. Higaki, N. Tomita, M. Aoki, A. Moriguichi, K. Tamura, K. Murakami, Y. Kaneda, and T. Ogihara, *Hypertension* **27,** 502 (1996).

[35] T. Nishii, A. Moriguchi, R. Morishita, K. Yamada, S. Nakamura, N. Tomita, Y. Kaneda, A. Fukamizu, H. Mikami, J. Higaki, and T. Ogihara, *Circ. Res.* **85,** 257 (1999).

[36] I. Hayashi, M. Majima, T. Fujita, T. Okumura, Y. Kumagai, N. Tomita, R. Morishita, J. Higaki, and T. Ogiwara, *Br. J. Pharmacol.* **131,** 820 (2000).

they might have become hypertensive is questionable, but their studies offer a demonstration of antisense effectiveness. Retroviruses may be useful in treating cardiomyopathy, restenosis, and vascular remodeling, where cells are actively dividing, but retroviruses integrate randomly into the genome and the possibility of tumorigenesis is a high risk. Lentivirus vectors, which can infect dividing cells, are just beginning to be explored for therapeutic value. They offer large gene carrying capacity, are stable, and are easily produced. The disadvantage is the risk of uncontrolled infection and the potential for neoplastic changes. Other vectors, such as herpes simplex virus and Japan Sendai virus, are being tested as vectors, but all vectors are as yet only in limited use by certain laboratories.

Engineering Viruses

In addition to the choice of vectors, the control of transgene needs to be engineered and new promoters need to be explored before viral vectors can be used in humans.[30] The ideal promoter will be active for prolonged periods to maintain transgene expression and specific for a tissue cell type. The vector will need mechanisms to switch it on or off as required. This is being tested with the tetracycline transactivator system (tTA), by which a transgene can be activated in the presence (or absence) of tetracycline. Ultimately the promoters and transactivating factors will have to be so specific that the antisense can be turned on in a specific tissue when the need arises.

Conclusion

Both antisense oligonucleotides and antisense DNA delivered in a vector have advantages for gene therapy. Antisense ODNs can be used as drugs. They have an action that lasts for days or weeks, depending on how they are delivered. They are specific for a target protein and reduce overactive proteins, but because the inhibition is never total, they permit normal physiology. They are not toxic at therapeutic doses. ODNs can be produced in large quantities, relatively cheaply for humans. The challenge for antisense ODNs is to deliver them.

[37] S. Suzuki, P. Pilowsky, J. Minson, L. Arnolda, I. J. Llewellyn-Smith, and J. Chalmers, *Am. J. Physiol.* **266,** R1418 (1994).

[38] M. J. Katovich, C. H. Gelband, P. Reaves, H. W. Wang, and M. K. Raizada, *Am. J. Physiol.* **277,** H1260 (1999).

[39] M. H. Wang, F. Zhang, J. Marji, B. A. Zand, A. Nasjletti, and M. Laniado-Schwartzman, *Am. J. Physiol. Regul. Integr. Comp. Physiol.* **280,** R255 (2001).

[40] Y. Kaneda and T. Ogihara, *Hypertension* **26,** 131 (1995).

[41] S. M. Galli and M. I. Phillips, *Hypertension* **38,** 543 (2001).

[42] W. C. Wolf, H. Yoshida, J. Agata, L. Chao, and J. Chao, *Kidney Int.* **58,** 7130 (2000).

[43] J. J. Zhang, C. Wang, K. F. Lin, L. Chao, and J. Chao, *Clin. Exp. Hypertens.* **21,** 1145 (1999).

[44] E. Dobrzynski, C. Wang, J. Chao, and L. Chao, *Hypertension* **36,** 995 (2000).

[45] J. J. Zhang, H. Yoshida, L. Chao, and J. Chao, *Hum. Gene Ther.* **11,** 1817 (2000).

The AAV vector with antisense DNA has very prolonged action (weeks/months) with a single dose, and is safe, nonpathogenic, and noninflammatory. The AAV is extremely stable. The challenge for clinical use is to increase production of large amounts at reasonable cost and to further engineer the control of the vector, as described above.

This brief review of some of the preclinical data shows that gene therapy for hypertension is possible.[30] The question is, Will these strategies be tested at the clinical level? The rAAV antisense strategy appears to be effective for reducing high blood pressure in different models of hypertension. Its development could provide a new generation of antihypertensive agents that would be administered in a single dose for prolonged effects lasting several months. Alternatively, antisense oligonucleotides are effective and highly specific. They could be used like long-acting drugs to provide sustained control of hypertension with infrequent administration. It seems that of the two strategies, the antisense oligonucleotides will be clinically acceptable first because of our familiarity with drug treatments. The viral vector approach will come much later, when all the basic science has been done to ensure that the patient is safe.

Summary

In spite of several drugs for the treatment of hypertension, there are many patients with poorly controlled high blood pressure. This is partly due to the fact that all available drugs are short-lasting (24 hr or less), have side effects, and are not highly specific. Gene therapy offers the possibility of producing longer-lasting effects with precise specificity from the genetic design. Preclinical studies on gene therapy for hypertension have taken two approaches. Chao et al.[5] have carried out extensive studies on gene transfer to increase vasodilator proteins. They have transferred kallikrein, atrial natriuretic peptide, adrenomedullin, and endothelin nitric oxide synthase into different rat models. Their results show that blood pressure can be lowered for 3–12 weeks with the expression of these genes. The antisense approach, which we began by targeting angiotensinogen and the angiotensin type 1 receptor, has now been tested independently by several different groups in multiple models of hypertension. Other genes targeted include the β_1-adrenoceptor, TRH, angiotensin gene activating elements, carboxypeptidase Y, c-*fos*, and CYP4A1. There have been two methods of delivery antisense; one is short oligodeoxynucleotides, and the other is full-length DNA in viral vectors. All the studies show a decrease in blood pressure lasting several days to weeks or months. Oligonucleotides are safe and nontoxic. The adeno-associated virus delivery antisense to AT_1 receptors is systemic and in adult rodents decreases hypertension for up to 6 months. We conclude that there is sufficient preclinical data to give serious consideration to Phase I trials for testing the antisense ODNs, first and later the AAV.

[2] Oligonucleotide-Mediated Site-Directed Gene Repair

By Betsy T. Kren, Paramita Bandyopadhyay,
Namita Roy Chowdhury, Jayanta Roy Chowdhury,
and Clifford J. Steer

In contrast to traditional gene replacement, oligonucleotides (ONs) can now be designed to correct point mutations in genomic DNA, thereby repairing the endogenous faulty copy of the gene. Correction of the genomic misspelling permits the repaired gene to remain at its native site under its own endogenous regulation. This gene alteration approach was based on studies to elucidate the molecular aspects of DNA repair. It was reported that a significant increase in efficiency of pairing occurred between an ~50 base ON and a genomic DNA target if RNA replaced DNA in a portion of the targeting molecule.[1,2] Other modifications of the chimeric ON were made to increase stability and improve localization to genomic target sites in mammalian cells. Two single-stranded ends consisting of unpaired nucleotide T hairpin caps flank the double-stranded region of the molecule. The 5′ and 3′ ends of the molecule are juxtaposed and sequestered. This design together with 2′-*O*-methylation of the RNA residues, contributes to the enhanced resistance of the hybrid ON to nucleases (Fig. 1).

The ON is designed to be homologous to the target genomic DNA with the exception of a single mismatched nucleotide. Alignment of the hybrid ON with its genomic target generates a mismatch that is thought to initiate certain endogenous DNA repair functions.[3] Yoon *et al.*[4] and Cole-Strauss *et al.*[5] reported that these RNA/DNA ONs were capable of introducing targeted single nucleotide conversions in episomal and genomic DNA in cultured cells. We then demonstrated that the chimeric ONs could introduce a missense mutation in genomic DNA in cultured human hepatoma cells[6] and nonreplicating isolated rat hepatocytes.[7] The high rates of nucleotide conversion in the primary hepatocytes resulted, in part, from a highly efficient delivery of the ONs to the cells using a nonviral, asialoglycoprotein receptor-targeted delivery system described below.

[1] H. Kotani and E. B. Kmiec, *Mol. Cell. Biol.* **14,** 6097 (1994).

[2] H. Kotani and E. B. Kmiec, *Mol. Cell. Biol.* **14,** 1949 (1994).

[3] E. B. Kmeic, B. T. Kren, and C. J. Steer, *in* "The Development of Human Gene Therapy" (T. Friedmann, ed.), p. 643. Cold Spring Harbor Laboratory Press, Cold Spring Harbor, NY, 1999.

[4] K. Yoon, A. Cole-Strauss, and E. B. Kmiec, *Proc. Natl. Acad. Sci. U.S.A.* **93,** 2071 (1996).

[5] A. Cole-Strauss, K. Yoon, Y. Xiang, B. C. Byrne, M. C. Rice, J. Gryn, W. K. Holloman, and E. B. Kmiec, *Science* **273,** 1386 (1996).

[6] B. T. Kren, A. Cole-Strauss, E. B. Kmiec, and C. J. Steer, *Hepatology* **25,** 1462 (1997).

[7] B. T. Kren, P. Bandyopadhyay, and C. J. Steer, *Nature Med.* **4,** 285 (1998).

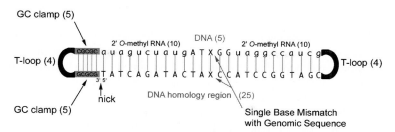

FIG. 1. Basic structure of a chimeric RNA/DNA ON. The hybrid molecules are typically 68 nt in length and contain 20 2'-*O*-methyl modified ribonucleotides in the 25-nt segment homologous to the target gene. The RNA and DNA residues are shown in lowercase and uppercase, respectively. The chimeric ONs contain two hairpins of four T residues each (indicated in black), and a 5 base-pair GC clamp (shown in gray). The 25-nt region of homology is mismatched with the target sequence at a single position (indicated by the X) designated for base substitution, deletion, or addition. The number of nucleotides in the particular region of the chimeric ON is indicated in parentheses. ON, Oligonucleotide.

Cultured CD34+-enriched cells,[8] as well as other nonhematopoietic cells,[6,9] have also been shown to be amenable to targeted genomic alteration at a specific nucleotide position in the β-globin gene. The chimeric ONs have also been used to correct a nonsense mutation of the carbonic anhydrase II gene in nude mouse primary kidney tubular cells.[10] Similarly, correction of a missense mutation in the tyrosinase gene of albino mouse melanocytes reestablished melanin production.[11] In both cases, the cells exhibiting ON-mediated gene correction were expanded and the genotypic and phenotypic changes were shown to be permanent and inherited. Chimeric ONs were also employed *in vivo* to correct the point mutations in *mdx* that are responsible for muscular dystrophy in mice and dogs.[12,13] A recent study reported *in vivo* correction of the carbonic anhydrase II nonsense mutation in renal tubular epithelial cells of nude mice.[14] Chimeric ONs are also effective in inducing targeted nucleotide changes in genomic DNA of plants for improved herbicide resistance.[15–17] Together, these studies indicate that site-specific ON-mediated nucleotide alteration of genomic DNA occurs in numerous cell types.

[8] Y. Xiang, A. Cole-Strauss, K. Yoon, J. Gryn, and E. B. Kmiec, *J. Mol. Med.* **75,** 829 (1997).

[9] E. Santana, A. E. Peritz, S. Iyer, J. Uitto, and K. Yoon, *J. Invest. Derm.* **111,** 1172 (1998).

[10] L.-W. Lai, H. M. O'Connor, and Y.-H. Lien, Conference Proceedings: 1st Annual Meeting of the American Society of Gene Therapy, Seattle, WA, 1998, p. 183a.

[11] V. Alexeev and K. Yoon, *Nature Biotech.* **16,** 1343 (1998).

[12] T. A. Rando, M.-H. Disatnik, and L. Z.-H. Zhou, *Proc. Natl. Acad. Sci. U.S.A.* **97,** 5363 (2000).

[13] R. J. Bartlett, S. Stockinger, M. M. Denis, W. T. Bartlett, L. Inverardi, T. T. Le, N. t. Man, G. E. Morris, D. J. Bogan, J. Metcalf-Bogan, and J. N. Kornegay, *Nature Biotech.* **18,** 615 (2000).

[14] L.-W. Lai, B. Chau, and Y.-H. H. Lien, Conference Proceedings: 2nd Annual Meeting of the American Society of Gene Therapy, Washington, D.C., 1999, p. 236a.

Liposome Encapsulation of a Synthetic 68-Mer RNA/DNA ON

Producing a homogeneous, stable, and reproducible liposome preparation for efficient delivery of nucleic acids into cells involves a number of factors. The choice of lipids, the method and medium of hydration, the target nucleic acid, and the final sizing strategy all play potentially critical roles. The liposome formulation should be optimized for its unique intended purpose. Efficiency of delivery, lack of toxicity, the need for extended shelf-life, the requirement for size selection, intracellular fate, and cell type-specificity must all be considered in formulating the liposome as a delivery vehicle for nucleic acids.

Initially, we were interested in encapsulating a 68-mer chimeric RNA/DNA ON designed for increased nuclease resistance and genomic targeting. In particular, our goal was to deliver the ONs primarily to the liver, and specifically to hepatocytes. Traditionally, this polyanionic molecule would be complexed with a polycation or cationic liposomes.[18,19] However, such complexes were primarily trapped in the lung capillaries[20] and, therefore, did not reach the liver, raising concerns regarding their potential safety and efficiency for human use.[21] Although ONs could be successfully encapsulated in anionic liposomes, incorporating phosphatidylserine as one of the lipid constituents appeared to increase targeting to the liver.[22] The targeting could be further enhanced by also including galactocerebroside,[23] which serves as a ligand to the unique asialoglycoprotein receptor (ASGPr) on hepatocytes. The choice of specific lipid molecules was somewhat influenced by the consideration that the use of phospholipids with unsaturated fatty acid chains yielded more flexible bilayers capable of efficiently capturing large ONs or plasmids.[24] The lipids used in our preparations were dioleoylphosphatidylcholine (DOPC), a neutral lipid; dioleoylphosphatidylserine (DOPS), an anionic phospholipid; and the targeting lipid, galactocerebroside (Gc), in a precise molar ratio.

[15] T. Zhu, K. Mettenburg, D. J. Peterson, L. Tagliani, C. L. Baszczynski, and B. Bowen, *Nature Biotech.* **18,** 555 (2000).

[16] P. R. Beetham, R. B. Kipp, X. L. Sawycky, C. J. Arntzen, and G. D. May, *Proc. Natl. Acad. Sci. U.S.A.* **96,** 8874 (1999).

[17] T. Zhu, D. J. Peterson, L. Tagliani, G. St. Claire, C. L. Baszczynski, and B. Bowen, *Proc. Natl. Acad. Sci. U.S.A.* **96,** 8768 (1999).

[18] H. E. J. Hofland, L. Shephard, and S. M. Sullivan, *Proc. Natl. Acad. Sci. U.S.A.* **93,** 7305 (1996).

[19] X. Gao and L. Huang, *Biochemistry* **35,** 1027 (1996).

[20] N. S. Templeton, D. D. Lasic, P. M. Frederik, H. H. Strey, D. D. Roberts, and G. N. Pavlakis, *Nature Biotech.* **15,** 647 (1997).

[21] C. Plank, K. Mechtler, F. C. Szoka, Jr., and E. Wagner, *Human Gene Ther.* **7,** 1437 (1996).

[22] R. Fraley, R. M. Straubinger, G. Rule, E. L. Sringer, and D. Papahadjopoulos, *Biochemistry* **20,** 6978 (1981).

[23] H. H. Spanjer and G. L. Scherpof, *Biochim. Biophys. Acta* **734,** 40 (1983).

[24] J. H. Felgner, R. Kumar, C. N. Sridhar, C. J. Wheeler, Y. J. Tsai, R. Border, P. Ramsey, M. Martin, and P. Felgner, *J. Biol. Chem.* **269,** 2550 (1994).

Preparation of Liposome-Encapsulated Oligonucleotides

Liposomes were prepared by the film hydration method using DOPC : DOPS : Gc combined at a 1 : 1 : 0.16 molar ratio. Briefly, individual phospholipids (DOPS and DOPC) were obtained as lyophilized powders from Avanti Polar Lipids (Alabaster, AL) and dissolved in chloroform at a concentration of 5 to 10 mg/ml and stored at $-20°$ until further use. Galactocerebroside (Avanti Polar Lipids) was dissolved in warm methanol at a concentration of 1 to 2 mg/ml and also stored at $-20°$. Lipid stocks and Gc solutions were slowly warmed to room temperature prior to use. Specified aliquots of the individual components were pipetted into the bottom of 16×100 mm borosilicate glass tubes (cleaned and sterilized). The total amount of lipid for the formulation was 2 mg per tube. The lipids were dried under a stream of nitrogen, carefully rotating the tube to ensure the formation of a thin uniform film, which was essential for efficient hydration. The dried film was either used immediately or stored in desiccated bottles at $-20°$ under nitrogen and sealed with Parafilm and Teflon tape. When larger amounts of lipids were used ($\sim$5 mg), they were dried under reduced pressure in round bottom flasks using a rotary evaporator.

Lipids were resuspended in 0.15 M NaCl or 5% dextrose containing 0.6 mg/ml ON. The final concentration of lipids was 2 mg/ml. The lipid suspension was hydrated by mild intermittent vortexing and swirling until the lipids dispersed resulting in a smooth milky white suspension free of clumps. The dispersion could be facilitated by warming the mixture in a 37–40° water bath. After lipid dispersion, the hydration mixture was allowed to stand for 30 to 60 min at room temperature before extruding for size reduction and selection.[25]

Extrusion is the process by which the lipid suspension is forced through polycarbonate membranes of defined pore size to obtain a uniform population of liposomes. For small-scale preparation, liposomes were extruded at the bench using a syringe-type extruder consisting of two interlocking syringes with Teflon barrels that are connected by a nut-and-bolt type of assembly which houses a thin polycarbonate membrane. The liposome suspension was passed a minimum of five times back and forth through the membrane sandwiched between the syringes. Sequential extrusion was performed through membranes of 0.8, 0.4, and 0.2 μm pore size. In some cases, the liposomes were extruded through pore sizes down to 0.1 or 0.05 μm. Two pieces of extrusion equipment, the Avestin Liposofast and the Avanti miniextruder, were compared. Both instruments required strong manual pressure to push the suspension back and forth through the membranes. Although the two instruments operate on the same principle, in our hands, the Liposofast model performed significantly better in terms of reliability and reproducibility. Furthermore,

---

[25] F. Olson, C. A. Hunt, F. C. Szoka, W. J. Vail, and D. Papahadjopoulos, *Biochim. Biophys. Acta* **557**, 9 (1979).

only the Avanti extruder required the sample to be warmed to ensure reasonably easy passage across the membranes, especially with the smaller membrane sizes (0.1 and 0.05 μm).

Evaluation of Liposome Size, Phospholipid Recovery, Liposome Stability, and Encapsulation Efficiency

Size Determination

The size distribution of the final liposome formulation was determined by photon-correlation spectroscopy or PCS on a Malvern Zetasizer 3000 instrument using the manufacturer's software. Ten to 20 μl of the liposome suspension was diluted into 500 μl of sterile 0.15 M NaCl in a polystyrene cuvette (Starstedt) and mixed. The analysis was performed in the automatic mode; preparations that yielded a monodisperse population of particles with a Zave mean diameter between 80 and 110 nm following extrusion through 0.1 μm membranes were considered satisfactory.

Phospholipid Determination

Lipid recovery in the final liposome preparation after extrusion was assessed by a colorimetric assay as described by Stewart.[26] Briefly, 20–50 μl of liposome suspension was diluted to 500 μl with water and mixed in a 13 × 100 mm tube. An equal volume of $CHCl_3 : CH_3OH$ (1 : 1 v/v) was added to the diluted suspension and the mixture vortexed vigorously for at least 1 min. The aqueous and organic phases were separated by centrifugation at $\sim$100g for 2 min. The lower aqueous phase was transferred to a clean tube using a Pasteur pipette. The aqueous phase was reextracted as described above followed by two extractions with $CHCl_3$. The organic phases were pooled and dried down under a stream of nitrogen. The lipids were then dissolved in 1 ml of $CHCl_3$ and added to 1 ml of ammonium ferro-thiocyanate (AFT) reagent. The mixture was vortex mixed vigorously for at least 1 min. The aqueous and organic layers were separated by centrifugation and the absorbance of the organic layer was measured at 468 nm. Phospholipid content of liposome aliquots was determined using a standard curve generated with a lipid mixture having the same molar composition as that of the liposomes. Typical lipid recovery following syringe extrusion was 80% or higher.

Determination of Oligonucleotide Encapsulation

Several methods were evaluated for determining the efficiency of ON encapsulation by the liposomes. Encapsulation was initially determined by subjecting

---

[26] J. C. M. Stewart, *Anal. Biochem.* **104**, 10 (1980).

TABLE I
DIALYSIS FOR REMOVING UNENCAPSULATED OLIGONUCLEOTIDES
FROM LIPOSOME PREPARATIONS

Sample	Membrane molecular weight cutoff	Absorbance 260 nm[a]		
		Initial	5 hr	24 hr
150 μl blank	60,000	0.259	0.196	0.145
Liposomes (0.2 μm)	100,000	0.247	0.159	0.162
+30 μg ON	300,000	0.261	0.103	N.D.[b]
30 μg ON	60,000	0.245	0.192	0.160
	100,000	0.272	0.161	0.064
	300,000	0.259	0.173	0.025

[a] 1/20 dilution.
[b] N.D., Not determined.

the liposome preparation to nuclease digestion, assuming that only the nonencapsulated ONs would be degraded. This analysis yielded an estimated 50–60% encapsulation.[27] However, a drawback of this method was that partially degraded products were seen on the gel near the intact material, which could result in overestimation of the encapsulation efficiency.

Two other methods commonly used in estimating encapsulation efficiency, namely, dialysis and size exclusion chromatography, were also done. However, both of these methods have certain limitations. Dialysis of a mixture containing blank liposomes and free radiolabeled ONs demonstrated that the free ON did not dialyze over 24 hr even using 300,000 molecular weight cutoff membranes (Table I). Thus, this method would overestimate the encapsulation efficiency. In addition, the results for the size exclusion chromatography varied significantly depending on how the columns were packed and loaded as well as the centrifuge model. In a large number of experiments, there was significant overlap between the eluted fractions containing free and liposome-encapsulated ONs.

The method that was the most operator independent, and easily reproducible was ultrafiltration through Microcon tubes with 100,000 molecular weight cut-off membrane. A 20 μl sample of liposomes (containing no more than 20–40 mg lipids) was diluted to 400 μl with 0.15 M NaCl or 5% dextrose and carefully pipetted onto the membrane. The Microcon unit was assembled as described by the manufacturer and the suspension was filtered through the membrane by centrifuging at 1500 rpm in a table-top microfuge at 4° for 6–8 min (or until the retentate volume was reduced to ~150 μl). After 10 washes with the appropriate diluent (400 μl per wash), the retentate was collected as described in the manufacturer's protocol and transferred to a new filtering unit in which 10 additional 400 μl washes were

[27] P. Bandyopadhyay, B. T. Kren, X. Ma, and C. J. Steer, *BioTechniques* **25**, 282 (1998).

TABLE II
ENCAPSULATION EFFICIENCY OF LIPOSOMES DETERMINED BY ULTRAFILTRATION

Sample (μl)	Lipid (μg)	ON (μg)	Washes (#)	% Recovered		Encapsulation (%)
				Retentate[a]	Eluate	
Encapsulated ON						
20	40	12	12	10.9	86.4	11.2
50	100	30	12	33.0	52.2	38.7
20	40	12	20	6.0	78.0	7.6
Blank liposomes + free ON	40	12	12	6.6	92.6	6.5
	100	30	12	13.8	70.0	16.5
	40	12	20	1.5	90	1.6

[a] Phospholipid recovery was 80–100% in the retentates and none was detected in the eluates.

carried out. It was important to mix the suspension thoroughly before each spin. Side-by-side controls using free ON (12–20 μg) or free ON (12–20 μg) + blank liposomes (20–40 mg lipid) were also run. Aliquots of the pooled eluates and the retentate were extracted with phenol : chloroform (1 : 1 v/v), chloroform : isoamyl alcohol (1 : 1 v/v), and ether. The aqueous phase was warmed to remove the ether and then dried down in a refrigerated centrifugal evaporator (SpeedVac, Savant Instruments, Farmingdale, NY) under reduced pressure. The ONs were resuspended in water or TE buffer (~50–100 μl) and quantitated by reversed phase HPLC or agarose gel analysis. We found that simple OD readings at 260 nm were often unreliable since some of the blank extractions (sample-free) gave significant absorbance. The accuracy of this method was established using radiolabeled ONs for encapsulation (Table II). Results showed that it was important to keep within our specified limits for lipid and ON loading and number of washes in order to obtain meaningful results. Simultaneous controls as described above should be run whenever possible.

Liposome Scale-Up and Stability

Three methods were examined to determine the feasibility of preparing liposomes at increasing scale: (1) a mini-scale with syringe-type extruders producing 0.5- to 2-ml preparations; (2) a mid-scale production of 1–10 ml using nitrogen-pressure driven extrusion through a cylindrical barrel-type assembly; and (3) large-scale production with high-pressure homogenization followed by extrusion. A comparison of the three methods is shown in Table III.

Liposome Stability

Liposomes were tested for stability upon storage. Those prepared in 0.15 M NaCl or 5% dextrose were stable at 4° for 1 month or longer for both the lipid

TABLE III
Comparison of Mini-, Mid-, and Large-Scale Preparation of Liposomes

Features	Mini-scale	Mid-scale	Large-scale
Equipment	Avanti miniextruder; easy to use.	The Extruder, Lipex Biomembranes Inc.; requires high pressure nitrogen tank and gauge (up to 500 psi).	Avestin C-5 homogenizer/extruder. Requires training and practice and strict maintenance schedule.
Capacity	0.5 to 2 ml preparations using a maximum lipid concentration of 2 mg/ml.	1 to 10 ml of liposome preparations per run. Up to 20 mg/ml lipid has been used with no problems.	15 ml to 300 ml per run with supplied vessel. Scaled up to 3 liters with a customized vessel. Lipid concentrations up to 100 mg/ml can also be processed.
Hydration	Lipids hydrated as a thin film on the bottom of 16×100 mm tubes.	Lipids hydrated as a thin film on the bottom of round-bottom flasks.	As in mid-scale or premixed amounts of lyophilized powdered lipids (Avanti Polar Lipids) hydrated directly in vials.
Extrusion	Manual thumb pressure required for extrusion.	Nitrogen driven extrusion. Sample heating (to $\sim40°$) and high pressures (up to 300 psi) required for extrusion through membranes with pore diameters $<0.2~\mu m$	High-pressure homogenization (10,000 to 15,000 psi) reduces initial size of vesicles, which are then easily extruded at moderate pressures (60 psi).
Liposome quality	Lipid recovery always >80%; reproducible preparations with narrow size distribution. Encapsulation of ON between 10 and 20%[a]	Similar to mini-scale. Encapsulation was typically around 10%.	Lipid recovery and size distribution were comparable to the other methods, but encapsulation efficiency was estimated to be only 5%.

[a] Encapsulation ranged from 40 to 60% with the Avestin Lipofast extruder.

TABLE IV
Liposome Phospholipid Stability in Saline or Dextrose

Diluent and incubation time (weeks)	Phospholipid (mg/ml)	
	4°	55°
0.15 M NaCl		
0	1.52	1.52
2	1.53	1.21
3	1.42	1.31
4	1.54	1.23
5% Dextrose		
0	1.76	1.76
2	1.41	0.40
3	1.61	0.36
4	1.72	0.26

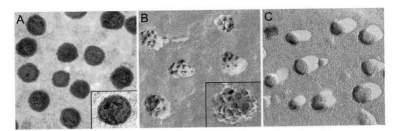

FIG. 2. Transmission electron microscopy and freeze-fracture analysis of anionic liposomes. (A) Anionic liposomes with encapsulated ONs were extruded through 0.05 μm membranes and applied to glow-discharged formvar carbon grids. Following negative staining with 2% ammonium molybdate, they were examined using a JEOL100-CX electron microscope. The 68-mer ONs appear as punctate electron dense irregularities. Original magnification 112,500×; inset 200,000×. Freeze fracture analysis of anionic liposomes extruded to 0.05 μm with (B) or without (C) encapsulated ONs. Magnification, 125,000×; inset 225,000×. ON, Oligonucleotide.

profile and vesicle size distribution. However, storage at 55° significantly affected phospholipid and ON stability in 5% dextrose (Table IV). At 4°, ON degradation was apparent after a few weeks when 5% dextrose was used (data not shown). This was of concern for long-term storage only.

In conclusion, anionic liposomes can be formulated to encapsulate anionic ONs (Fig. 2). These liposomes are stable and can be reproducibly generated with the encapsulation of at least 10% of the ONs (greater at the mini scale). The 10% encapsulation efficiency was >10 times that predicted from theoretical calculations based on lipid concentration and liposome size (100 nm). This would indicate specific, albeit unexpected, interactions between the ON and lipids allowing a higher loading efficiency. It is possible that this interaction is facilitated by the lipid hydration technique and the significantly thinner, more uniform films employed in our small-scale method. DSC and NMR and FT-IR should provide additional information to better characterize the ON interaction with the lipid bilayers.

We have successfully used the RNA–DNA chimeric ONs encapsulated in the anionic liposome formulation prepared by the mini-scale Avestin lipofast extruder. Specifically, we have effected base pair conversion[28] as well nucleotide insertion[29] in the genomic DNA in isolated primary rat hepatocytes and in hepatocytes *in vivo*. Uptake of the encapsulated ONs by hepatocytes both in culture and *in vivo* is competitively inhibited by coadministration of galactose or asialofetuin, which prevent binding of galactocerebroside to the hepatocyte ASGPR. Blocking the

[28] B. T. Kren, B. Parashar, P. Bandyopadhyay, N. R. Chowdhury, J. R. Chowdhury, and C. J. Steer, *Proc. Natl. Acad. Sci. U.S.A.* **96,** 10349 (1999).

[29] P. Bandyopadhyay, X. Ma, C. Linehan-Stieers, B. T. Kren, and C. J. Steer, *J. Biol. Chem.* **274,** 10163 (1999).

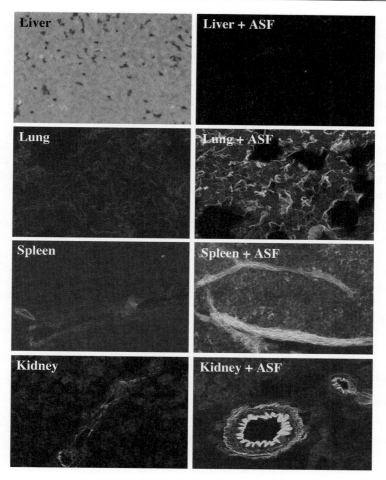

FIG. 3. *In vivo* tissue distribution of Gc targeted anionic liposome with encapsulated fluorescein-labeled ONs. Rats received 200 μg of liposome encapsulated ONs as a single bolus injection with or without coinjection of the competitive inhibitor asialofetuin (ASF). The addition of ASF inhibits ASGPᵣ uptake of the labeled ONs by hepatocytes, and significantly increases the abundance of the fluorescein-labeled ON in lung, spleen, and kidney. Liver sections shown are 2 hr postinjection, and the remaining tissues are 8 hr postinjection. ASGPᴿ, Asialoglycoprotein receptor; Gc, galactocerebroside; ON, oligonucleotide.

receptor-mediated endocytosis of the Gc-containing liposomes by the liver with asialofetuin resulted in significant uptake by kidney, spleen, and, particularly, lung (Fig. 3). Importantly, the fluorescein-labeled ONs were not detectable in the testes with or without asialofetuin administration, suggesting an intact gonadal barrier to the liposome-ON complex (data not shown).

Lactosylated Polyethyleneimine for Delivery of Chimeric RNA/DNA ONs

Delivery of intact nucleic acids from the extracellular environment to the nucleus has remained a major barrier to successful long-term expression of genes transferred to cells both *in vitro* and *in vivo*. Although the entire repertoire of endogenous DNA repair factors involved in the process remains to be described,[30,31] it is clear that effective delivery of ONs to the nucleus is key to successful gene repair. Polycations such as PEI form water-soluble noncovalent electrostatic complexes with anionic nucleic acids and, thus, provide simple and effective delivery systems. Free amino groups present on these compounds also provide the means for easily attaching ligands to provide targeted delivery to specific tissues. Furthermore, the electrostatic binding of the ONs by the free amino groups makes the formation of ON complexes extremely efficient.

There are certain constraints shared by both the polycation complexes and liposome-encapsulated ONs for efficient delivery. In particular, a small particle size is critical for effective access to hepatocytes through the fenestrated endothelial cells of the liver sinusoid.[32] Larger particles are trapped in Kupffer cells and, nonspecifically, at extrahepatic sites, such as the lungs and spleen.[33] Additionally, the complexes must protect the ONs from nuclease degradation and deliver them to the nucleus efficiently. We chose polyethyleneimine (PEI) as a carrier for the ONs because this polycation is an effective nucleic acid delivery agent in cells both *in vitro* and *in vivo*.[34–36] PEI was also shown to promote the transfer of nucleic acids from the cytoplasm to the nucleus in nonreplicating cells,[37] which is important considering that most hepatocytes are quiescent *in vivo*.

Preparation of Targeted Polyethyleneimine

To target the ASGPR on hepatocytes, PEI was lactosylated by modification of previously described methods for oligosaccharide conjugation. The first method

[30] A. Cole-Strauss, H. Gamper, W. K. Holloman, M. Muñoz, N. Cheng, and E. B. Kmiec, *Nucleic Acids Res.* **27**, 1323 (1999).

[31] H. B. Gamper, Jr., A. Cole-Strauss, R. Metz, H. Parekh, R. Kumar, and E. B. Kmiec, *Biochemistry* **39**, 5808 (2000).

[32] T. Hara, Y. Tan, and L. Huang, *Proc. Natl. Acad. Sci. U.S.A.* **94**, 14547 (1997).

[33] J. C. Perales, G. A. Grossmann, M. Molas, G. Liu, T. Ferkol, J. Harpst, H. Odaand, and R. W. Hanson, *J. Biol. Chem.* **272**, 7398 (1997).

[34] O. Boussif, F. Lezoualc'h, M. A. Zanta, M. D. Mergny, D. Scherman, B. Demeneix, and J.-P. Behr, *Proc. Natl. Acad. Sci. U.S.A.* **92**, 7297 (1995).

[35] B. Abdallah, A. Hassan, C. Benoist, D. Goula, J. P. Behr, and B. A. Demeneix, *Human Gene Ther.* **7**, 1947 (1996).

[36] A. Boletta, A. Benigni, J. Lutz, G. Remuzzi, M. R. Soria, and L. Monaco, *Human Gene Ther.* **8**, 1243 (1997).

[37] H. Pollard, J.-S. Remy, G. Loussouarn, S. Demolombe, J.-P. Behr, and D. Escande, *J. Biol. Chem.* **273**, 7507 (1998).

we used for conjugating oligosaccharides to 25 kDa PEI (Aldrich Chemical Co., Milwaukee, WI) relies on the ability of the cyanoborohydride anion to selectively reduce the imminium salt formed between an amine and an aldehyde of a reducing sugar.[38] Briefly, a 0.2 M stock of the monomeric 43 kDa PEI (CH_2CH_2NH) in 0.2 M ammonium acetate, pH 7.6, is prepared as follows. The PEI is transferred to a tare weighed beaker using a glass pipette to spool the sticky material. Sufficient 0.2 M ammonium acetate/hydroxide buffer, pH 7.6, is added to the beaker to yield a final concentration of 0.2 M monomeric PEI and the material stirred at room temperature until it is fully in solution. For conjugation of the lactose to the PEI amines, 3 ml of the 0.2 M monomeric PEI in 0.2 M ammonium acetate/hydroxide buffer, pH 7.6, is incubated with 30 mg of lactose and 8 mg of sodium cyanoborohydride (Sigma Chemical Co., St. Louis, MO) at 37° for 10 days. The stock PEI used for the conjugation as well as 3 ml of the 0.2 M PEI and 30 mg of lactose without sodium cyanoborohydride anion are also incubated at 37° for 10 days. The reaction mixture and controls are dialyzed using 10,000 molecular weight cutoff membranes against Milli Q water at 4° for 48 hr with 2 changes of water per day.

The second method for covalently linking oligosaccharides to the 25 kDa PEI utilized conversion of the carbohydrate hapten to aldonic acid,[39] and subsequent coupling of the derivatized reducing sugar to the primary amines by 1-ethyl-3-(dimethylaminopropyl)carbodiimide (EDAC) (Sigma Chemical Co.).[40] In brief, 0.6 g of lactonic acid is added to 4 ml of a 0.8 M solution of 25 kDa PEI in Milli Q water adjusted to pH 4.75 with HCl, while rapidly stirring at room temperature. One-half gram of EDAC is dissolved in 0.75 ml of Milli Q water and added dropwise over a 30-min period alternating with the dropwise addition of 0.5 M HCl to maintain pH at 4.75. The pH of the reaction mixture is monitored for another 15 min, adding HCl as needed to maintain a pH of 4.75. Once the pH is stabilized, it is left stirring at room temperature for 6 hr, during which the pH of the solution decreases to ∼3.2. The reaction is then quenched by addition of 5 ml of 1 M sodium acetate, pH 5.5, and the modified PEI is dialyzed using 3500 molecular weight cutoff membranes against Milli Q water for 48 hr with 2 changes of water per day at 4°.

The amount of sugar (as galactose) conjugated with PEI is determined by the phenol/sulfuric acid[41] or resorcinol method.[42] We have found that the resorcinol method is easier and more reproducible than the phenol/sulfuric acid determination. In short, resorcinol (Sigma Chemical Co.) 6 mg/ml in Milli Q water is made up every 30 days and stored at 4° in the dark. Analytical grade sulfuric acid

[38] G. Gray, *Arch. Biochem. Biophys.* **163**, 426 (1974).

[39] S. Moore and K. P. Link, *J. Biol. Chem.* **132**, 293 (1940).

[40] J. Lönngren, I. J. Goldstein, and J. E. Niederhuber, *Arch. Biochem. Biophys.* **175**, 661 (1976).

[41] M. Dubois, K. A. Gilles, J. K. Hamilton, P. A. Rebers, and F. Smith, *Anal. Chem.* **28**, 350 (1956).

[42] M. Monsigny, C. Petit, and A.-C. Roche, *Anal. Biochem.* **175**, 525 (1988).

(100 ml) is added to 24 ml of Milli Q water to make a 75% solution, cooled to room temperature, and stored in the dark at room temperature for up to 3 weeks. Galactose (0.2 mg/ml) is dissolved in Milli Q water to generate a standard curve, which was linear from 4 μg (22.2 nmol) to 20 μg (111 nmol). Aliquots of the standard or lactosylated PEI (L-PEI) are diluted to 200 μl in Milli Q water in glass tubes, and then 200 μl of resorcinol (6 mg/ml) and 1 ml of the 75% sulfuric acid are added sequentially to the samples, which are mixed by vortex and heated to 90° for 30 min. After cooling them in a cold-water bath in the dark for 30 min, the optical density of the samples and standards were determined at 430 nm.

The number of moles of free primary amines in the L-PEI was determined using ninhydrin reagent with leucine as the standard. PEI is composed of primary, secondary, and tertiary amines at a ratio of 1 : 2 : 1[43]; thus, each μl of a 0.2 M stock of the monomeric PEI contains 200 nmol of amines, with 25% or 50 nmol primary amines which are detected in the following assay. Leucine (5 mM) dissolved in Milli Q water was used to generate the standard curve, which was linear between 15 and 100 nmol. Aliquots of the standard, 0.2 M stock of the monomeric PEI in Milli Q water or L-PEI were diluted to 90 μl in Milli Q water in 1.5 ml microcentrifuge tubes. To each tube, 10 μl of 1 M HEPES, pH 7.3, is added and mixed by vortex prior to adding 100 μl of ninhydrin reagent (Sigma Chemical Co.). Following vortexing, the samples are heated for 15 min at 100° and then placed on ice. Ice-cold Milli Q water, 300 μl is added quickly to each tube followed by 500 μl of 100% ethanol. The solutions are mixed by vortex and the optical density determined at 570 nm. The 0.2 M stock of the monomeric PEI in Milli Q water is used to validate the concentration of this sample, which is diluted to generate the standard curves for assaying the secondary and total amine concentration of the L-PEI.

To determine the number of moles of free secondary amines in the L-PEI, a standard curve is formed using a 0.02 M solution of PEI in Milli Q water, which is linear between 50 and 3000 nmol of secondary amines. Several aliquots of the stock and L-PEI are diluted to 1 ml using Milli Q water in glass tubes and 50 μl of ninhydrin reagent (Sigma Chemical Co.) is added to each tube. After vortex mixing vigorously for 10 sec, color development is allowed to proceed in the dark at room temperature for 12 min and the optical density determined at 485 nm.

The number of total amines is determined using 2,4,6-trinitrobenzenesulfonic acid (TNBS).[44] A standard curve is generated using a 4 mM solution of PEI in Milli Q water, which is linear between 40 and 400 nmol of amines. Briefly, aliquots of the standard and L-PEI are diluted to 1 ml using sodium borate buffer, pH 9.3, in glass tubes and vortex mixed. To each sample, 25 μl of a 0.03 M TNBS solution in

[43] J. Suh, H.-j. Paik, and B. K. Hwang, *Bioorg. Chem.* **22,** 318 (1994).
[44] S. L. Snyder and P. Z. Sobocinski, *Anal. Biochem.* **64,** 284 (1975).

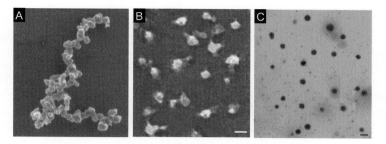

FIG. 4. Electron microscopy of L-PEI complexes. L-PEI-ON complexes were prepared in either 0.15 M NaCl or 5% dextrose and then examined by electron microscopy. NaCl induces aggregation of the individual particles into large complexes by SEM (A). In contrast, complexes prepared in 5% dextrose do not exhibit aggregation of the individual particles either by SEM (B) or TEM, following negative staining with 2% ammonium molybdate (C). L-PEI, Lactosylated polyethylenimine; ON, oligonucleotide; SEM/TEM, scanning/transmission electron microscopy. Bars, 25 nm.

Milli Q water is added and the mixture is agitated. Following a 30 min incubation at room temperature in the dark, the optical density is determined at 420 nm.

Using the above assays, we have established that reductive amination using sodium cyanoborohydride anion covalently attaches the lactose to the secondary amines while the EDAC conjugation of the aldonic acid derivative of lactose couples this oligosaccharide only to the primary amines. Both protocols result in derivatization of ~13% of the total amines of the PEI by the disaccharide.

Evaluation of PEI/L-PEI ON Complexes

Size Determination

The decision to choose PEI as a complexing vehicle for ONs was, in part, based on a number of prerequisites necessary for efficient and safe delivery of the chimeric molecules to hepatocytes *in vivo*. Thus, we determined the characteristics of these complexes in either 0.15 M NaCl or 5% dextrose, either of which solution could be used for intravenous delivery of these complexes. Particle sizes were determined by gas phase electrophoretic mobility molecular analysis (GEMMA),[45] light scattering, and scanning and transmission electron microscopy.[28] Discrete particles of ~20 nm were formed in both solutions. In fact, there was a notable difference in particle size with or without the ONs. Complexes formed in dextrose remained monodispersed, whereas those formed in saline had a size distribution ranging from 20 to 200 nm. The aggregates formed by the ~20 nm diameter particles were visible by electron microscopy only with saline (Fig. 4). Electron microscopy of complexes formed in dextrose indicated that they remained monodispersed for up to two weeks after formation when stored at room temperature or 4°.

[45] S. L. Kaufman, J. W. Skogen, F. D. Dorman, and F. Zarrin, *Anal. Chem.* **68,** 1895 (1996).

Charge and Stability of the L-PEI/ON Complexes

In addition to small size, the characteristic that is associated with efficient receptor-mediated delivery of polycation-complexed nucleic acids is neutral or negative overall charge.[46] Therefore, the amount of L-PEI required in complex with the ONs to reach neutrality was determined by increasing the ratio of L-PEI to ON until migration of the ONs on agarose gel electrophoresis was completely inhibited.[47] The ability of the L-PEI complex to protect the ONs was determined by the amount of nuclease degradation at 37°. The ONs were analyzed by agarose gel electrophoresis following nuclease digestion and phenol : chloroform extraction. Only the ONs complexed with PEI were protected during the 40 min incubation period, whereas uncomplexed ONs were completely degraded.[11]

Receptor Specificity and Transfection Efficiency

To assay transfection efficiency and receptor specificity of the L-PEI delivery system, we examined both fluorescein-labeled ONs and a reporter plasmid PGL3 (Promega Corp., Madison, WI) encoding firefly luciferase. The expression plasmid provides rapid assessment of the receptor-mediated transfection efficicacy while confocal microscopic analysis of the fluorescently labeled ONs establishes cellular localization of the transfected material. Using these two techniques, it was empirically established that mixing unmodified PEI with L-PEI significantly improved the transfection efficiency into both human hepatoma cell lines and primary rat hepatocytes. The optimal ratio of unmodified PEI to L-PEI differed with the various cell types and >80% inhibition of luciferase expression was routinely observed when 100 mM D-galactose was used to inhibit ASGPR-mediated uptake. The transfection efficiency of the PGL3 plasmid was increased significantly, especially for the HuH-7 human hepatoma cell line, when the lactose residues were attached to the primary rather than the secondary amines of the PEI.

To evaluate receptor-mediated uptake and cellular localization of the ONs, cultured cells were transfected with L-PEI/PEI/fluorescein-labeled ONs at ratios found to be optimal in the PGL3 experiments. Confocal microscopy demonstrated that delivery of the fluorescently labeled ONs in complex with PEI/L-PEI significantly improved uptake and nuclear localization of the ONs (Fig. 5). This uptake was competitively inhibited by 100 mM D-galactose and little uptake or nuclear localization was detected in the absence of the delivery system. Similar results of hepatic uptake and tissue distribution were observed *in vivo* using fluroescein-labeled ONs complexed with L-PEI, and asialofetuin as a competitive inhibitor as with the liposomal delivery method (data not shown).

[46] R. J. Lee and L. Huang, *J. Biol. Chem.* **271**, 8481 (1996).
[47] M. A. Findeis, C. H. Wu, and G. Y. Wu, *Methods Enzymol.* **247**, 341 (1994).

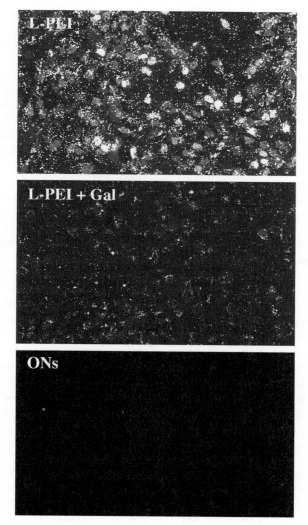

FIG. 5. HuH-7 cell uptake of fluorescein-labeled ONs. Cells were incubated with 1 μg of labeled ONs either complexed with L-PEI (top) or naked (bottom). Confocal microscopy shows significant uptake and nuclear localization of the fluorescently-labeled ONs only with L-PEI (top). Uptake by the ASGPR is significantly inhibited by 100 mM D-galactose (middle). ON, Oligonucleotide; Gal, galactose.

Detection and Characterization of the Site-Directed Genomic DNA Modification

DNA Isolation and Subcloning

The cultured cells transfected with the ONs and control cultures are harvested 48 hr following transfection. The cells are washed 3 times in PBS, then scraped off the dish in 1 ml PBS, and pelleted by centrifugation; the PBS is removed from the cell pellet prior to flash-freezing in liquid nitrogen. The genomic DNA >100 to 150 base pairs is isolated from the cell pellet ($\sim$4.0 $\times$ 10^5 cells) using the High Pure PCR Template Preparation Kit (Roche Molecular Biochemicals, Indianapolis, IN) according to the manufacturer's specifications. For liver tissue, the High Pure PCR Template Preparation Kit is also used with the following modifications. Fifty mg of the liver tissue flash-frozen in liquid nitrogen is minced in a clean weigh boat with a single-edge razor blade on dry ice. The minced tissue is combined with 100 μl of tissue lysis buffer in a 1.5 ml microfuge tube on ice. To the tissue suspension is added 50 μl of grinding resin (Geno Technology, Inc., St. Louis, MO) that has been resuspended in 0.5 M EDTA according to manufacturer's instructions. A cutoff 200 μl pipette tip is used to aliquot the resin and the tissue is homogenized with a disposable pestle. Next, 200 μl of tissue lysis buffer is added to the tissue homogenate and the homogenization is repeated. Binding buffer (300 μl) is added and mixed well, followed by 60 μl of proteinase K (10 mg/ml) and the tube is inverted at least 5 to 10 times. The mixture is incubated at 72° for 10 min, with the tubes mixed by inversion 2–3 times. The samples are then centrifuged at room temperature for 3 min at 8000 $\times$ g to pellet the grinding resin and tissue debris. The supernatant is carefully transferred to a fresh 1.5 ml microfuge tube avoiding the debris. To 0.45 ml of the supernatant, 100 μl of isopropanol is added and mixed by inversion. Following a quick spin to collect material from the cap, the tube is opened and the contents are added to the filter setup supplied in the kit. The manufacturer's protocol is then followed from step #4 as outlined in the kit instructions.

The purified DNA (0.25 μg) is used for PCR amplification of the genomic DNA spanning the targeted nucleotide conversion or insertion site. Following amplification using the Expand High Fidelity PCR System (Roche Molecular Biochemicals) under the most stringent conditions possible, the PCR products are purified to remove the primers and unincorporated nucleotides prior to A-tailing with Taq polymerase. In brief, 1 μl of 10$\times$ Taq polymerase buffer, 1 μl of 25 mM MgCl$_2$, and 5 units of Taq polymerase are added to 2 μl of the purified PCR products, in a final volume of 10 μl. The reaction mixture is incubated at 70° for 15 min, and then 1 to 2 μl is used for the ligation reactions. The A-tailed inserts are mixed with the pGEM-T Easy Vector (Promega Corp.) according to the manufacturer's recommended protocol for optimizing insert : vector molar ratios. After incubation overnight at 4°, the entire ligation mixture is used to transform competent *Escherichia coli*. Following a 50 min recovery at 37°, the transformed bacteria are plated on 150 mm hard luria agar plates containing a minimum of 75 μg ampicillin/ml of agar. After 18 to 20 hr incubation at 37°, the plates are removed (colony size ~1 mm) and chilled at 4°.

Filter Lift Hybridization Analysis

Duplicate nylon filter lifts are used for hybridization to determine the frequency of nucleotide alteration in the genomic DNA. The bottoms of the chilled plates have registration patterns for the filters indicated in permanent marker. The 137 mm nylon 0.22 μm MagnaGraph membranes (Micron Separations Inc., Westboro, MA) are labeled in duplicate with one of each pair being labeled as the replicate. Using sterile forceps, the nonreplicate nylon filter is carefully placed on the surface of the agar plate. The registration marks on the plate bottom are transferred to the membrane by punching through it with a 20 gauge sterile hypodermic needle prior to removing the filter with sterile forceps. This membrane is then placed colony side up on a sheet of filter paper; the replicate membrane is carefully placed on top of the primary membrane that has the bacterial colonies from the plate and then covered with a glass plate. Firm and even pressure is applied to the glass plate, pressing the two filters together so that colonies are transferred from the primary to replicate membranes. Using forceps, the filters are carefully turned over and the registration holes from the primary membrane are transferred to the replicate membrane by punching through both with the needle. The membranes are carefully peeled apart and placed colony side up on filter paper prior to processing for hybridization. It is expected that the majority of the bacterial colonies will be transferred to the replicate filter during this process. The agar plates are then left at room temperature for 24 hr in the dark to permit regrowth of the colonies, prior to storage at 4°. The plates are not returned to 37° for regrowth, because the ampicillin concentration is insufficient to prevent the growth of feeder colonies.

The membrane pairs are then transferred with the colony side up on to filter paper that has been saturated with freshly prepared 0.5 M NaOH. In 10 min, the NaOH solution should dampen the entire membrane, yet not wet it to the extent that the transferred colonies will smear together. Following the 10 min incubation, the membranes are transferred to dry filter paper, colony side up, to remove sufficient moisture that the surface changes from shiny to matte. The membranes are then neutralized by transferring to filter paper saturated with 0.5 M Tris-HCl pH 8.0 containing 0.5 M NaCl for 8 min. Following moisture removal from the membranes after transfer to dry filter paper, the membranes are held by the edge with forceps and washed by agitating vigorously in 200 ml of 2× sodium chloride/sodium citrate (SSC). The membranes are placed colony side up on dry filter paper, transferring as needed until the surface of the membrane is matte. After transfer to a fresh sheet of dry filter paper, the membranes are dried for 15 min at 70–80°. After this step, the membranes may be used immediately for hybridization or stored dried at room temperature between sheets of filter paper.

Detection of the targeted single nucleotide alteration in the subcloned PCR amplicons must be precise and reproducible. In fact, 3 M tetramethylammonium chloride (TMAC) (Sigma Chemical Co.) allows sensitive hybridization to effectively differentiate a single nucleotide mismatch between the 17-mer radiolabeled ON probe and the cloned DNA segments.[48] For the most specific hybridization, the membrane with the majority of each bacterial colony transferred to its surface (usually the replicate membrane) is hybridized with the probe that detects the original genomic DNA sequence. One of each pair of membranes is placed in two separate stacks, colony side up, and each stack of membranes is placed in a heat sealable pouch. For each pouch, containing a maximum of 15 membranes, 5 ml per membrane or a minimum of 40 ml of hybridization fluid ($2 \times$ SSC containing 1% SDS, $5 \times$ Denhardt's solution, and 200 μg/ml denatured sonicated fish sperm DNA) is added; the bags are sealed and incubated at $37°$ for a minimum of 2 hr. A pair of 17-mer ON probes with the ninth nucleotide being either the starting or altered nucleotide are ^{32}P-end-labeled using (γ-^{32}P)ATP ($>$7000 Ci/mmol) (ICN Biochemicals, Inc., Costa Mesa, CA) and T4 polynucleotide kinase (New England Biolabs, Inc., Beverly, MA) according to the manufacturer's recommendations. Following purification of the probes to remove unincorporated nucleotides, the prehybridization fluid is removed and replaced with fresh hybridization fluid, and sufficient radiolabeled-labeled 17-mer probe is added for a final concentration of 0.125 ng probe/ml of hybridization fluid.

The bags, following resealing, are incubated at $37°$ for 18 to 24 hr. After hybridization, the filters are washed twice in $1 \times$ sodium chloride sodium phosphate EDTA (SSPE), 0.5% SDS (33 ml per filter minimum) for 15 min. The membranes are then transferred individually to 50 mM Tris-HCl, pH 8.0, containing 3 M TMAC, 2 mM EDTA, pH 8.0, 0.1% SDS that has been preheated to $52°$ (40 ml per filter) and then washed at $52°$ for 1 hr. Following this wash, the filters are transferred individually to $1 \times$ SSPE, 0.5% SDS (33 ml per filter minimum) and washed for an additional 15 min at room temperature. The filters are then placed individually on Whatman filter papers, and the excess wash solution is blotted from the membranes prior to wrapping them (they should still be damp) in plastic wrap for autoradiography at $-70°$ using X-ray film and an intensifying screen. The membranes in plastic wrap are taped to a paper backing and a glow-in-the-dark crayon (Crayola) is used to make registration marks on the backing. After suitable exposures of the membranes are obtained, the films are aligned using the crayon marks over the taped-down membranes on a light box and the plate registration marks from the filters are transferred to the films using permanent marker. The colonies from the regrown plates that hybridized with the 17-mer radiolabeled probes specific for either the starting or converted nucleotide can then be identified. The overall frequency of conversion of the targeted nucleotide is calculated

---

[48] W. B. Melchoir, Jr. and P. H. Von Hippel, *Proc. Natl. Acad. Sci. U.S.A.* **70**, 298 (1973).

by dividing the number of clones hybridizing with the ON probe specific for the altered nucleotide by the total number of clones hybridizing with both ON probes.

To confirm the specificity of the nucleotide change as well as to establish that only the target base alteration occurred, individual clones that hybridized with the radiolabeled ON probes from the regrown plates are cultured. Plasmid DNA isolated from the individual clones is subjected to automated sequencing spanning the entire cloned PCR amplicon. The only difference in the nucleotide sequence expected between the clones, which hybridized to the 17-mer starting, or altered DNA probes is the single base conversion or insertion.[28,29] If the alteration of targeted nucleotide occurs at a frequency greater than 14%, the PCR amplicons used for the subcloning can be sequenced directly to permit the confirmation of the targeted base change. However, the peak heights of initial and site-directed altered nucleotides cannot be used to establish the frequency since they are sequence and nucleotide dependent.

Further confirmation of the genomic modification can be obtained by isolating RNA from parallel transfected cultures and liver tissue using the High Pure RNA Isolation or Tissues Kit (Roche Molecular Biochemicals) for RT-PCR. RT-PCR amplification is done utilizing the Titan one-tube RT-PCR system (Roche Molecular Biochemicals) according to the manufacture's protocol using primers that amplify genomic DNA that spans both an intron/exon junction as well as the targeted nucleotide alteration. To rule out DNA contamination, the RNA samples are treated with RQ1 DNase free RNase (Promega Corp.) and RT-PCR is performed. For negative control for RT-PCR, RNase-treated RNA samples are used in parallel with the RT-PCR reaction. Following purification and A-tailing, the RT-PCR amplicons are ligated into the same TA cloning vector and transformed into competent *Escherichia coli*. Filter lift hybridization analysis is performed as outlined above.

Restriction Fragment Length Polymorphism Detection

Depending on the targeted nucleotide, the modification of the genomic DNA sequence may affect a restriction endonuclease digestion site making it possible to detect the desired change by restriction fragment length polymorphism (RFLP) analysis.[12] This technique although rapid, can only detect changes that have occurred at a frequency of ∼10% or greater. Its robustness is also dependent on the restriction site altered, i.e., the activity of the particular restriction endonuclease, as well as the ability to resolve the digestion fragments by gel electrophoresis. Thus, although it permits rapid screening of the PCR amplified genomic DNA, it cannot be used to quantify the conversion frequency. However, if sufficient site-directed nucleotide alteration does occur in the genomic DNA, Southern blot hybridization can be used to confirm the genomic DNA alteration.

In summary, L-PEI mixed with unmodified PEI has been shown to be an effective delivery system for ONs both *in vitro* and *in vivo*. When the complexes are

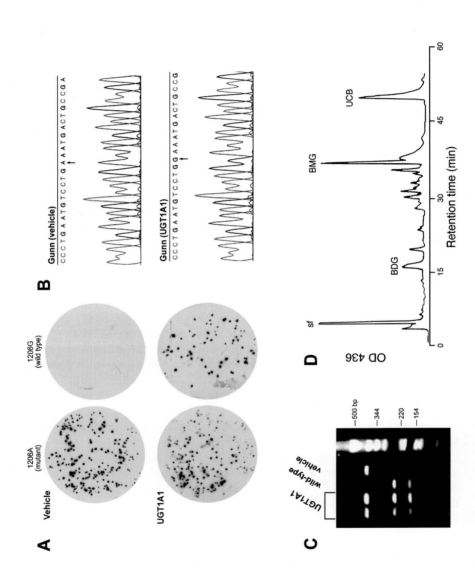

formed in 5% dextrose the particles are small, monodispersed, and remain as such for up to 2 weeks without discernible change. The ONs in these complexes are effectively delivered to the ASGPR on hepatoma cell lines as well as hepatocytes and efficiently translocated from the cytosol to the nucleus. The efficient nuclear localization promoted by these delivery systems is key for the successful implementation of ON-mediated site-directed alteration of hepatic genomic DNA. In fact, both nucleotide conversion and insertion into liver genomic DNA in rodents have been accomplished using ONs delivered with L-PEI.[28,29] The genomic change is stable and has resulted in the long-term production of wild-type UGT1A1 enzyme activity in the Gunn rat model of Crigler-Najjar syndrome type 1. This restoration of enzyme activity by insertion of a deleted nucleotide provides the mechanism for the glucuronidation and excretion of bilirubin, resulting in the excretion of mono- and diglucuronidated bilirubin in the bile and significant reduction of serum bilirubin levels (Fig. 6).

Acknowledgments

We thank the members of the laboratories for their support and encouragement during the course of this work. These studies were supported, in part, by grants from ValiGen, Inc., Newtown, PA, and by Grants RO1-DK39137 (N.R.C.), RO1-DK46057 (J.R.C.), and PO1-HD32652 (B.T.K.) from the National Institutes of Health.

FIG. 6. Long-term genomic DNA and enzymatic correction of the *UGT1A1* gene in the Gunn rat model of Crigler-Najjar syndrome type I. Gunn rats were treated with a synthetic 76-mer RNA/DNA ON (UGT1A1) for site-specific insertion of guanosine and correction of the *UGT1A1* frameshift mutation. (A) Filter lift hybridizations of PCR amplified DNA isolated from livers of Gunn rats treated with chimeric ON or vehicle, and probed with radiolabeled 17-mer ONs for mutant (1206A) or wild-type (1206G) sequence. (B) Automated sequence analysis of individual clones from vehicle (top) or UGT1A1 (bottom) filter lifts. The arrows indicate targeted reinsertion of guanosine with partial restoration of the *Bst*NI restriction site by RFLP of amplicons following PCR amplification of the genomic DNA spanning the target site (C). (D) HPLC analysis of bile pigments from Gunn rat livers. Bile ducts were cannulated and the bile collected for analysis of bilirubin conjugation. Restoration of the enzymatic activity is shown by the appearance of both mono- (BMG) and diglucuronidated (BDG) bilirubin detected in the bile of UGT1A1 treated Gunn rats 2 years after treatment with ONs. ON, Oligonucleotide; UCB, unconjugated bilirubin.

[3] Stabilized Plasmid-Lipid Particles: A Systemic Gene Therapy Vector

By David B. Fenske, Ian MacLachlan, and Pieter R. Cullis

Introduction

Genetic drugs are a class of therapeutic agents with considerable potential for the treatment of human diseases such as cancer and genetic disorders. Although numerous methods exist for effective *in vitro* gene delivery, current systems have limited utility for systemic applications. Viral systems, for example, are rapidly cleared from the circulation, limiting transfection to "first-pass" organs such as the lungs, liver, and spleen. In addition, these systems induce immune responses that compromise transfection resulting from subsequent injections. In the case of nonviral systems such as plasmid DNA–cationic lipid complexes (lipoplexes), the large size and positively charged character of these aggregates also result in rapid clearance, and the highest expression levels are again observed in first-pass organs, particularly the lungs.[1,2–4] Plasmid DNA–cationic lipid complexes can also result in toxic side effects both *in vitro*[5] and *in vivo*.[6,7]

The need for a gene delivery system for treatment of systemic disease is obvious. For example, for cancer gene therapy there is a vital need to access metastatic disease sites as well as primary tumors. Similar considerations apply to other systemic disorders, such as inflammatory diseases. The design features for lipid-based delivery systems that preferentially access such disease sites are increasingly clear. It is now generally recognized that preferential delivery of anticancer drugs to tumor sites following intravenous injection can be achieved by encapsulation of these drugs in large unilamellar vesicles (LUVs) exhibiting a small size (<100 nm diameter) and extended circulation lifetimes (circulation half-life in mice >5 h).[8–10]

[1] L. Huang and S. Li, *Nat. Biotechnol.* **15,** 620 (1997).

[2] N. S. Templeton, D. D. Lasic, P. M. Frederik, H. H. Strey, D. D. Roberts, and G. N. Pavlakis, *Nat. Biotechnol.* **15,** 647 (1997).

[3] A. R. Thierry, Y. Lunardi-Iskandar, J. L. Bryant, P. Rabinovich, R. C. Gallo, and L. C. Mahan, *Proc. Natl. Acad. Sci. U.S.A.* **92,** 9742 (1995).

[4] H. E. Hofland, D. Nagy, J. J. Liu, K. Spratt, Y. L. Lee, O. Danos, and S. M. Sullivan, *Pharm. Res.* **14,** 742 (1997).

[5] G. S. Harrison, Y. Wang, J. Tomczak, C. Hogan, E. J. Shpall, T. J. Curiel, and P. L. Felgner, *Biotechniques* **19,** 816 (1995).

[6] S. Li and L. Huang, *Gene Ther.* **4,** 891 (1997).

[7] P. Tam, M. Monck, D. Lee, O. Ludkovski, E. C. Leng, K. Clow, H. Stark, P. Scherrer, R. W. Graham, and P. R. Cullis, *Gene Ther.* **7,** 1867 (2000).

[8] R. T. Proffitt, L. E. Williams, C. A. Presant, G. W. Tin, J. A. Uliana, R. C. Gamble, and J. D. Baldeschwieler, *Science* **220,** 502 (1983).

The accumulation of these drug delivery systems at disease sites, including sites of infection and inflammation as well as tumors, has been attributed to enhanced permeability of the local vasculature in diseased tissue.[11]

A gene delivery system containing an encapsulated plasmid for systemic applications should therefore be small (<100 nm diameter) and must exhibit extended circulation lifetimes to achieve enhanced delivery to disease sites. This requires a highly stable, serum-resistant plasmid-containing particle that does not interact with cells and other components of the vascular compartment. In order to maximize transfection after arrival at a disease site, however, the particle should readily interact with cells at the site and should have the ability to destabilize cell membranes to promote intracellular delivery of the plasmid. In this work we describe a straightforward detergent dialysis procedure used for the production of stabilized plasmid-lipid particles (SPLP) that satisfy the demands of plasmid encapsulation, small size, and serum stability. We will also indicate how the transfection properties of these systems can be modulated by changing the composition of the encapsulating lipid bilayer. The behavior of SPLP can be modified by employing poly(ethyleneglycol) (PEG) coatings which can dissociate from the SPLP,[12] by increasing the cationic lipid content,[13,14] by including other lipids such as cholesterol, or by incorporating cationic PEG-lipids (CPL) that enhance intracellular delivery.[15] Although most of this review will focus on technical aspects of SPLP production and characterization, we will also provide examples of their use in gene delivery and expression. A detailed protocol for producing SPLP is provided in the Appendix.

Construction and Characterization of Stabilized Plasmid-Lipid Particles

Encapsulation of Plasmid DNA within Liposomal System Using Detergent Dialysis Procedure

In recent years, a large number of liposomal systems have been designed for the systemic delivery of conventional drugs (including chemotherapeutic agents and antibiotics[16]), several of which have shown promise in the treatment of specific

[9] A. Gabizon and D. Papahadjopoulos, *Proc. Natl. Acad. Sci. U.S.A.* **85**, 6949 (1988).

[10] A. Chonn and P. R. Cullis, *Curr. Opin. Biotechnol.* **6**, 698 (1995).

[11] S. Kohn, J. A. Nagy, H. F. Dvorak, and A. M. Dvorak, *Lab Invest.* **67**, 596 (1992).

[12] J. J. Wheeler, L. Palmer, M. Ossanlou, I. MacLachlan, R. W. Graham, Y. P. Zhang, M. J. Hope, P. Scherrer, and P. R. Cullis, *Gene Ther.* **6**, 271 (1999).

[13] Y. P. Zhang, L. Sekirov, E. G. Saravolac, J. J. Wheeler, P. Tardi, K. Clow, E. Leng, R. Sun, P. R. Cullis, and P. Scherrer, *Gene Ther.* **6**, 1438 (1999).

[14] E. G. Saravolac, O. Ludkovski, R. Skirrow, M. Ossanlou, Y. P. Zhang, C. Giesbrecht, J. Thompson, S. Thomas, H. Stark, P. R. Cullis, and P. Scherrer, *J. Drug Target.* **7**, 423 (2000).

[15] T. Chen, K. F. Wong, D. B. Fenske, L. R. Palmer, and P. R. Cullis, *Bioconjug. Chem.* **11**, 433 (2000).

cancers or other diseases and are now either in clinical trials or approved for use in humans.[10] Plasmid delivery systems such as cationic lipid-plasmid complexes have proved to be effective *in vitro* transfection agents,[17] but have limited ability for systemic applications. It is obviously desirable to develop liposomal systems for the *systemic* delivery of genetic drugs. This requires encapsulation of plasmid in small lipid vesicle systems. Unfortunately, techniques that had proved so useful for loading small, weakly basic drugs into LUVs[16] could not be adopted to the encapsulation of large molecules of plasmid DNA.

The solution came with the development of a detergent dialysis method for encapsulating plasmid in unilamellar lipid vesicles.[12] Previous work had shown that incubation of plasmid DNA with cationic lipids could result in a hydrophobic particle which was soluble in organic solvent.[18] This suggested the possibility that such a hydrophobic particle could be surrounded by an outer coating of lipid, which would then result in small, plasmid-containing particles stabilized in an aqueous medium. Detergent dialysis was recognized as a logical technique for achieving this, as the detergent was expected to solubilize the hydrophobic plasmid DNA–cationic lipid particles. The addition of phospholipid and subsequent removal of detergent by dialysis could then result in the exchange of the solubilizing detergent with phospholipid, leaving particles that were stable in aqueous suspension.

Initial experiments employed the cationic lipid DODAC, the plasmid pCMVCAT, the non-ionic detergent OGP, and the bilayer-forming lipid palmitoyloleoylphosphatidylcholine (POPC). When DODAC was added to plasmid in distilled water, the formation of large (> 1000 nm diameter) precipitates was observed. However, the subsequent addition of OGP (200 mM) resulted in solubilization of the precipitate, forming an optically clear suspension consistent with entrapment of hydrophobic plasmid DNA–cationic lipid particles within detergent micelles. This optically clear quality was maintained when POPC solubilized in OGP was added. However, when dialysis was attempted to facilitate removal and substitution of the detergent associated with the particles for POPC, extensive precipitation of the suspension was observed. A method of stabilizing the plasmid-containing particles preventing aggregation and precipitation during the dialysis process was therefore required.

Previous studies had shown that a PEG-lipid coating could prevent aggregation and fusion of LUVs induced by covalent coupling of protein to the vesicle

[16] P. R. Cullis, M. J. Hope, M. B. Bally, T. D. Madden, L. D. Mayer, and D. B. Fenske, *Biochim. Biophys. Acta* **1331,** 187 (1997).

[17] P. L. Felgner, T. R. Gadek, M. Holm, R. Roman, H. W. Chan, M. Wenz, J. P. Northrop, G. M. Ringold, and M. Danielsen, *Proc. Natl. Acad. Sci. U.S.A.* **84,** 7413 (1987).

[18] D. L. Reimer, Y. Zhang, S. Kong, J. J. Wheeler, R. W. Graham, and M. B. Bally, *Biochemistry* **34,** 12877 (1995).

surface.[19,20] This suggested that the stabilizing properties of a PEG coating could prevent aggregation during dialysis. However, the standard PEG-phosphatidyl-ethanolamine (PEG-PE) could not be used because the PEG-PE molecule bears a net negative charge and would be expected to displace the cationic lipid from the plasmid as had been noted for other negatively charged lipids.[21] To address this issue, ceramide was used as a neutral hydrophobic anchor that was in turn linked to PEG_{2000} to produce a neutral molecule. Three ceramide anchors were synthesized, differing in the length of the ceramide acyl chain ($CerC_8$, $CerC_{14}$, and $CerC_{20}$). When 10 mol% PEG-$CerC_{20}$ was incorporated in the detergent mixture with POPC, DODAC, and plasmid DNA, precipitation was no longer observed during detergent dialysis (against 5 mM HEPES, 150 mM NaCl, pH 7.4). Further, a proportion of the plasmid was encapsulated, as measured by recovery of DNA after elution on a DEAE-Sepharose CL-6B anion exchange column, or by Picogreen associated fluorescence in the absence and presence of Triton X-100. The extent of plasmid encapsulation was found to be a sensitive function of the DODAC content, with encapsulation levels of 30% or higher at about 9 to 12% DODAC. Addition of plasmid to preformed vesicles with the same lipid composition, followed by DEAE chromatography, yielded no plasmid encapsulation as measured by complete plasmid retention on the column.

These results suggest that "stabilized plasmid-lipid particles" (SPLP) could be produced by detergent dialysis employing a POPC/DODAC/PEG-$CerC_{20}$ (79:11:10; mol:mol:mol) lipid mixture, as diagrammatically represented in Fig. 1. However, it has been shown that when POPC is employed as a "helper" lipid in plasmid DNA–cationic lipid complexes, very low transfection rates are observed, whereas when dioleoylphosphatidylethanolamine (DOPE) is present, much higher transfection rates are achieved.[22] The encapsulation properties of DOPE/DODAC/PEG-$CerC_{20}$ lipid mixtures were therefore investigated. As shown in Fig. 2, high levels of encapsulation were obtained only for a narrow range of DODAC content (6–7 mol%), as observed for the POPC-containing systems. Significant differences are that maximum encapsulation was greater (~70%) for the DOPE-containing system and that optimum encapsulation was observed at about 6 mol% DODAC, compared with approximately 9% DODAC for the POPC-containing particles. If PEG-$CerC_{14}$ was substituted for PEG-$CerC_{20}$ very similar plasmid encapsulation behavior was observed.

The lower encapsulation observed for POPC-containing SPLP may be partially attributed to the ionic strength of the dialysis medium. As will be discussed

[19] T. O. Harasym, P. Tardi, S. A. Longman, S. M. Ansell, M. B. Bally, P. R. Cullis, and L. S. Choi, *Bioconjug. Chem.* **6**, 187 (1995).

[20] J. W. Holland, C. Hui, P. R. Cullis, and T. D. Madden, *Biochemistry* **35**, 2618 (1996).

[21] Y. Xu and F. C. J. Szoka, *Biochemistry* **35**, 5616 (1996).

[22] H. Farhood, N. Serbina, and L. Huang, *Biochim. Biophys. Acta* **1235**, 289 (1995).

High Cationic Lipid Content

Critical Cationic Lipid Content

Low Cationic Lipid Content

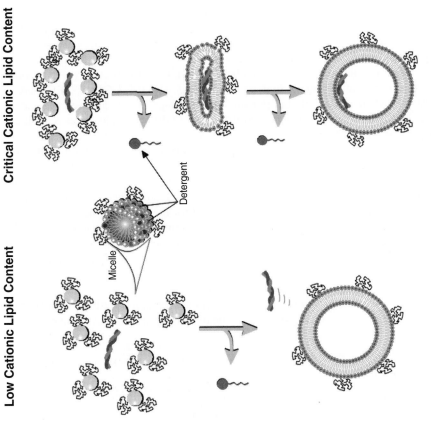

Detergent

Micelle

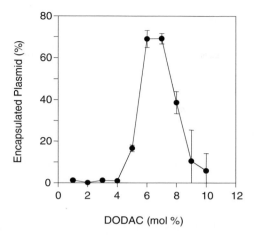

FIG. 2. Effect of DODAC concentration on the encapsulation efficiency of plasmid DNA (pCMV-CAT) in SPLP composed of DOPE, DODAC, and 10 mol% PEG-CerC$_{20}$. Lipid (10 mg/ml total), dissolved in octylglucoside (0.2 M), was mixed with plasmid DNA (50 μg/ml) in a total volume of 1 ml to form an optically clear solution. This was then placed in a dialysis tube (12–14,000 molecular weight cutoff) and dialyzed against HBS for 36 h at 20°. Encapsulation efficiency was determined following removal of unencapsulated plasmid by anion exchange chromatography (see text). Reprinted from J. J. Wheeler, L. Palmer, M. Ossanlou, I. MacLachlan, R. W. Graham, Y. P. Zhang, M. J. Hope, P. Scherrer, and P. R. Cullis, *Gene Ther.* **6,** 271 (1999).

in greater detail below, variation in lipid composition can alter the solute concentrations at which optimal encapsulation efficiency is achieved. Furthermore, the total lipid and plasmid concentrations have a significant effect on encapsulation efficiency. This was first shown for formulations of DOPE/DODAC/PEG-CerC$_8$

FIG. 1. Model of the formation and possible structure of SPLP. The first stage of dialysis is proposed to result in formation of macromolecular lipid intermediates, which may be in the form of lamellar sheets, cylindrical micelles, or leaky vesicles [M. Ollivon, O. Eidelman, R. Blumenthal, and A. Walter, *Biochemistry* **27,** 1695 (1988); P. K. Vinson, Y. Talmon, and A. Walter, *Biophys. J.* **56,** 669 (1989)]. If the cationic lipid content is too low (left panel), plasmid does not associate with these intermediates as dialysis proceeds, leading to formation of empty vesicles and free plasmid. At higher cationic lipid contents, plasmid associates with the lipid intermediates, drawn here as a bilayer sheet wrapped around the plasmid. If the cationic lipid content is at a critical level the presence of the plasmid reduces the net positive surface charge of the lipid intermediate to the extent that further association of plasmid is inhibited. As dialysis proceeds further, additional lipid would be expected to condense on this structure, leading to formation of a vesicle containing encapsulated plasmid, as indicated. In addition, empty vesicles and free plasmid would be expected. At high cationic lipid contents (right panel), the surface charge on the lipid intermediate structures is so high that two or more plasmids can associate with a given membrane sheet, leading to the formation of large aggregates. Reprinted from J. J. Wheeler, L. Palmer, M. Ossanlou, I. MacLachlan, R. W. Graham, Y. P. Zhang, M. J. Hope, P. Scherrer, and P. R. Cullis, *Gene Ther.* **6,** 271 (1999).

$(42.5 : 42.5 : 15)$.[14] The percent plasmid encapsulation increased from roughly 20 to 70% as the total lipid concentration was increased from 1 to 10 mg/ml. Increasing the plasmid concentration from 100 to 1000 μg/ml resulted in a drop in encapsulation efficiency from 60 to 20%. Similar trends have been observed for other lipid compositions.

Isolation of Stabilized Plasmid-Lipid Particles by Density Centrifugation

The detergent dialysis process clearly results in plasmid-containing particles in which the plasmid is protected from the external environment. In addition, however, a population of empty vesicles is produced. To remove empty vesicles from plasmid containing SPLP a density gradient purification process was established taking advantage of the difference in density between SPLP and empty vesicles. The density gradient profile of a DOPE/DODAC/PEG-CerC$_{20}$ (84 : 6 : 10; mol : mol : mol) SPLP preparation (initial plasmid-to-lipid ratio of 200 μg DNA to 10 mg lipid) was therefore examined employing sucrose density step gradient centrifugation. As shown in Fig. 3, after centrifugation at 160,000g for 2 h, the encapsulated DNA is present as a band localized at the 2.5% sucrose–10% sucrose interface in the step gradient. It is interesting to note that less than 10% of the total lipid (as assayed by the ^{3}H-CHE lipid marker) is associated with the encapsulated plasmid DNA, which in turn corresponds to 55% of the total input DNA. The plasmid-to-lipid ratio of these purified SPLP was determined to be 62.5 μg plasmid per μmol lipid. SPLP generated by detergent dialysis and purified by density gradient centrifugation may be concentrated by either ultrafiltration or dialysis against carboxymethyl cellulose to achieve plasmid concentrations of 1 mg/ml or higher.

Narrow Size Distribution of Stabilized Plasmid-Lipid Particles

The sizes of the empty lipid vesicles in the upper band and the isolated SPLP in the lower band of the sucrose density gradient were examined by quasi-elastic light scattering (QELS) and freeze-fracture electron microscopy techniques. As shown in Fig. 4, the QELS analysis indicated that the mean diameter of the empty vesicles was approximately 44 nm ($\chi^2 = 0.48$), whereas the isolated SPLP were larger, with a mean diameter of 75 nm ($\chi^2 = 0.14$). Freeze-fracture electron microscopy studies gave similar results. A size analysis of the particles in these micrographs indicated a size of 36 $\pm$ 15 nm for the empty vesicles and 64 $\pm$ 9 nm for the isolated SPLP.

Stabilized Plasmid-Lipid Particles: Plasmid Trapped inside Bilayer Lipid Vesicle

As mentioned above, purified SPLP composed of DOPE/DODAC/PEG-CerC$_{20}$ (84 : 6 : 10) were determined to have a plasmid-to-lipid ratio of 62.5 μg plasmid per μmol lipid. When considering a 4.49 kb plasmid, such as pCMVCAT, this

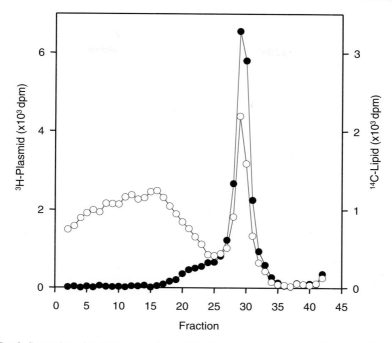

FIG. 3. Separation of SPLP from empty vesicles by discontinuous sucrose density gradient centrifugation. The solid circles indicate the behavior of the ^{3}H-labeled plasmid (pCMVLuc), whereas the open circles indicate the distribution of lipid as reported by the ^{14}C-labeled CHE lipid marker. SPLP (DOPE/DODAC/PEG-CerC$_{20}$; 84 : 6 : 10; mol : mol : mol) were prepared as indicated in the legend to Fig. 2, and an aliquot (1.5 ml containing approximately 50 μg of ^{3}H-plasmid DNA) was applied to a discontinuous sucrose density gradient (3 ml 10% sucrose, 3 ml 2.5% sucrose, 3 ml 1% sucrose; all in HBS). The gradient was then centrifuged at 160,000g for 2 h. [Reprinted from J. J. Wheeler, L. Palmer, M. Ossanlou, I. MacLachlan, R. W. Graham, Y. P. Zhang, M. J. Hope, P. Scherrer, and P. R. Cullis, *Gene Ther.* **6**, 271 (1999).]

corresponds to a plasmid to particle ratio of 0.97 (for an SPLP with a diameter of 70 nm).[12] Thus each SPLP contains one plasmid molecule.

SPLP structure has been further characterized employing cryoelectron microscopy. Purified SPLP were prepared using the lipids DOPE : DODAC : PEG-CerC$_{20}$ (83 : 7 : 10) and pCMVluc plasmid DNA. Large unilamellar vesicles (LUV) with the same lipid composition were prepared by extrusion of the hydrated lipid mixture through 100 nm pore size filters. As shown in Fig. 5a, cryoelectron micrographs clearly reveal SPLP to consist of a lipid bilayer surrounding an internal structure consistent with entrapped plasmid DNA molecules. The minority of small (diameter ~30 nm), empty vesicles formed during the detergent dialysis

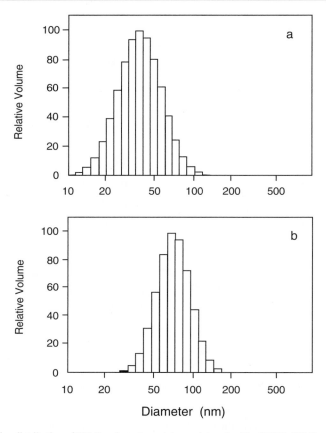

FIG. 4. Size distribution of SPLP and empty vesicles as determined by QELS. SPLP were prepared containing pCMVLuc as indicated in the legend to Fig. 2, and separated from empty vesicles by discontinuous sucrose density gradient centrifugation. (a) Size distribution for empty vesicles (upper band). (b) Size distribution for SPLP (lower band). The sizes were determined by quasi-elastic light scattering using a Nicomp Model 370 Sub-Micron particle sizer operating in the "solid particle" mode. [Reprinted from J. J. Wheeler, L. Palmer, M. Ossanlou, I. MacLachlan, R. W. Graham, Y. P. Zhang, M. J. Hope, P. Scherrer, and P. R. Cullis, *Gene Ther.* **6,** 271 (1999).]

process[23] that were not removed by density centrifugation do not exhibit the electron dense internal structure (see arrows in Fig. 5a). Nor is this internal structure observed in the LUV produced by extrusion (Fig. 5b). It may also be noted that SPLP visualized by cryoelectron microscopy have a remarkably homogeneous size (diameter 72 ± 5 nm), in close agreement with measurements of SPLP employing

[23] L. T. Mimms, G. Zampighi, Y. Nozaki, C. Tanford, and J. A. Reynolds, *Biochemistry* **20,** 833 (1981).

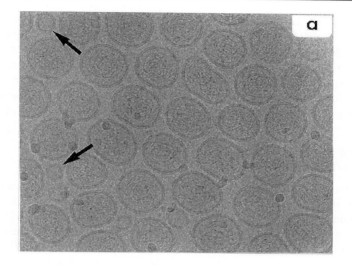

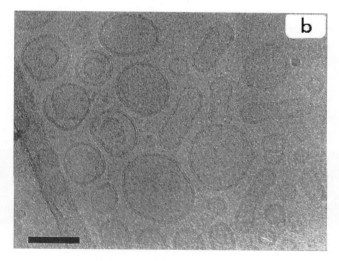

FIG. 5. Cryoelectron micrographs of (a) purified SPLP and (b) LUV prepared by extrusion. SPLP were prepared from DOPE : DODAC : PEG-CerC$_{20}$ (83 : 7 : 10; mol : mol : mol) and pCMVluc and purified employing DEAE column chromatography and density gradient centrifugation. LUV were prepared from DOPE : DODAC : PEG-CerC$_{20}$ (83 : 7 : 10; mol : mol : mol) by hydration and extrusion through filters with 100 nm diameter pore size. The arrows in panel (a) indicate the presence of residual "empty" vesicles formed during the detergent dialysis process that were not removed by the density centrifugation purification step. The bar in panel (b) indicates 100 nm. [Reprinted from P. Tam, M. Monck, D. Lee, O. Ludkovski, E. C. Leng, K. Clow, H. Stark, P. Scherrer, R. W. Graham, and P. R. Cullis, *Gene Ther.* **7,** 1867 (2000).]

freeze-fracture electron microscopy (diameter 64 ± 9 nm).[12] The homogeneous size and morphology of SPLP contrasts with the irregular morphology and large size distribution of the extruded vesicles. The narrow size distribution of SPLP was also reflected by QELS measurements (data not shown) indicating a mean diameter of 83 ± 4 nm. Plasmid DNA–cationic liposome complexes made from DOPE : DODAC (1 : 1; mol : mol) LUV exhibited a large, heterogeneous size distribution as determined by QELS (diameter 220 ± 85 nm, data not shown).

Stabilized Plasmid-Lipid Particle Protection of DNA from DNase and Serum Nucleases

It is important to demonstrate that the encapsulated plasmid in the particles obtained by the detergent dialysis process is, in fact, fully protected from the external environment. As a first measure of protection, the ability of recombinant DNase I to digest plasmid DNA in SPLP can be determined. Figure 6 illustrates the results of an experiment in which SPLP were prepared using DOPE/DODAC/PEG-CerC$_{20}$ (84 : 6 : 10; mol : mol : mol) and pCMVLuc (200 μg/ml). Protection of SPLP plasmid is compared to protection of free plasmid and plasmid in plasmid cationic lipid complexes prepared with DODAC-DOPE (1 : 1) LUVs at a charge ratio of ± 3.0. Samples containing 1 μg plasmid were exposed to 0, 100, and 1000 units of DNase I in a total volume of 1.0 ml HBS for 30 min at $37°$. After incubation the plasmid was isolated and characterized by agarose gel electrophoresis. Figure 6 shows that

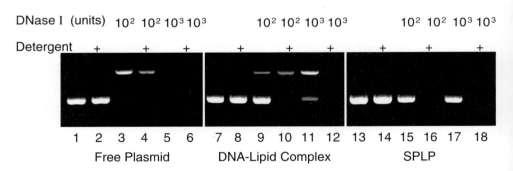

FIG. 6. Stability of free plasmid (lanes 1–6), plasmid encapsulated in SPLP (lanes 13–18), and plasmid in plasmid DNA–cationic lipid complexes (lanes 7–12) in the presence of DNase I. One μg of plasmid DNA was subjected to no treatment (lanes 1, 7, 13), exposure to detergent alone (1% Triton X-100) (lanes 2, 8, 14), exposure to 100 and 1000 units of DNase I alone (lanes 3, 9, 5 with 100 units and lanes 5, 11, 17 with 1000 units), and exposure to both detergent and DNase I (lanes 4, 10, 16 with 100 units and lanes 6, 12, 18 with 1000 units). The plasmid DNA–cationic lipid complexes consisted of DODAC : DOPE (50 : 50; mol : mol) LUVs (100 nm diameter) complexed to plasmid at a 3 : 1 charge ratio (positive-to-negative). [Reprinted from J. J. Wheeler, L. Palmer, M. Ossanlou, I. MacLachlan, R. W. Graham, Y. P. Zhang , M. J. Hope, P. Scherrer, and P. R. Cullis, *Gene Ther.* **6**, 271 (1999).]

free plasmid is completely digested by incubation with either 100 or 1000 units of DNase I. The plasmid in complexes (lipoplex) formed with cationic LUVs is marginally protected when compared to free DNA exposed to 100 units of DNase I, but is almost entirely digested by incubation with 1000 units. In contrast, plasmid DNA encapsulated in SPLP is completely protected from nuclease digestion unless detergent is added to disrupt the SPLP lipid bilayer prior to incubation with DNase.

A rigorous test of SPLP stability and protection of encapsulated plasmid involves incubation in serum. Serum contains a variety of nucleases, and serum proteins can rapidly associate with lipid systems,[24] resulting in enhanced leakage and rapid clearance of liposomal systems. The ability of serum nucleases to degrade plasmid is illustrated in Fig. 7. Intact pCMVCAT elutes in the void volume of the Sepharose CL-4B column, whereas after incubation with mouse serum (90%) at 37° for 30 min the plasmid is degraded into fragments that elute in the included volume (Fig. 7a). The behavior of the DOPE/DODAC/PEG-CerC$_{20}$ (84 : 6 : 10) SPLP system where nonencapsulated plasmid has not been removed is shown in Fig. 7b. In this particular preparation, 53% of the plasmid DNA elutes with the lipid in the void volume, and 47% of the DNA, which represents degraded plasmid, elutes in the included volume. This indicates that 53% of the plasmid is encapsulated and protected from the external environment, in good agreement with a 55% trapping efficiency of this sample as determined by DEAE ion exchange chromatography.

A final test of the stability of the SPLP formulation is given in Fig. 7c, which illustrates the elution profile of the DOPE/DODAC/PEG-CerC$_{20}$ (84 : 6 : 10; mol : mol : mol) SPLP system following removal of the external plasmid by DEAE chromatography and incubation in 90% mouse serum (30 min at 37°). In this case more than 95% of plasmid applied to the column eluted in the void volume, demonstrating the stability and the plasmid protection properties of the SPLP formulation. It should also be noted that SPLP containing PEG-CerC$_{14}$ in place of PEG-CerC$_{20}$ exhibit similar plasmid protection properties.

Variation of Cationic Lipid Content of Stabilized Plasmid-Lipid Particles over Wide Range

The SPLP systems described above fulfill many requirements of a systemic gene delivery system, but when used to transfect mammalian cells *in vitro* were found to have low transfection potency when compared to plasmid DNA–cationic lipid complexes. Initial attempts to increase the transfection potency of SPLP involved developing methods for increasing the DODAC content of the particles. Utilizing a formulation consisting of DOPE/DODAC/PEG-CerC$_8$, Zhang *et al.*[13]

---

[24] A. Chonn, S. C. Semple, and P. R. Cullis, *J. Biol. Chem.* **267**, 18759 (1992).

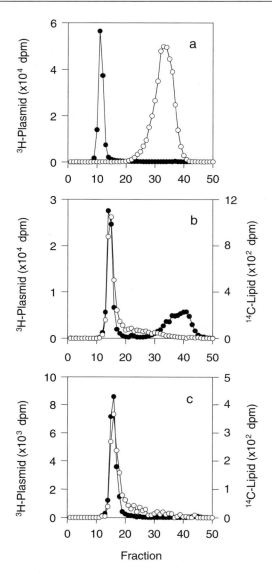

FIG. 7. Plasmid in SPLP is protected from serum nuclease cleavage. The stability of plasmid (pCMVCAT) in the free form or encapsulated in SPLP was determined in the presence of serum. The SPLP (DOPE/DODAC/PEG-CerC$_{20}$; 84 : 6 : 10; mol : mol : mol) were prepared as indicated in the legend to Fig. 2 and contained ^{14}C-labeled CHE as a lipid marker. Samples with 5 μg of ^{3}H-labeled plasmid DNA were incubated in the presence of HBS or 90% mouse serum for 30 min at 37° and eluted on a Sepharose CL-4B column equilibrated in HBS. (a) Elution profile of nucleic acid resulting from incubation of free plasmid in HBS (●) or 90% mouse serum (○). (b) Elution profile of nucleic

found that DODAC concentrations ranging from 7 to 30 mol% could be achieved by detergent dialysis against HEPES-buffered saline if appropriate concentrations of citrate were included in the dialysis medium. The encapsulation of DNA via cationic lipid requires ionic interactions between the positively charged lipid and the negatively charged DNA (Fig. 1). SPLP appear to form only when lipid structures possessing an appropriate surface charge are available for interactions with DNA.[12] If these interactions are too strong, as occurs at low ionic strength, encapsulation efficiency is high but is accompanied by the formation of large aggregates. If the ionic strength is too high, the DNA does not associate with the lipid, leading to the formation of small, empty vesicles and low encapsulation efficiencies. Therefore, it was reasoned that formation of SPLP with higher cationic lipid content might be possible if the ionic strength of the dialysis medium was raised to shield the higher net surface charge on the intermediate lipid structures. It was found that aggregation in samples containing higher DODAC concentrations could not be prevented simply by increasing the NaCl concentration. An effective solution to this problem was to use polyvalent anionic counterions such as citrate to produce stronger shielding effects. A study of plasmid encapsulation as a function of citrate concentration was performed for a lipid mixture composed of DODAC/DOPE/PEG-CerC$_8$ (20 : 65 : 15) and the pCMVLuc plasmid. At concentrations up to 60 mM citrate the dialyzed samples contained large (diameter > 150 nm) and polydisperse (χ^2 > 3.0) particles. However, small, monodisperse particles (82 $\pm$ 40 nm) exhibiting high encapsulation efficiencies of 50–70% were formed when the dialysis medium contained 65–80 mM citrate. Increasing the citrate concentration further also resulted in formation of small particles, but the encapsulation efficiency decreased dramatically.

These results suggest two criteria for determining the optimum ionic strength for plasmid encapsulation: (1) formation of monodisperse (χ^2 < 3.0) particles with diameter smaller than 100 nm and (2) an encapsulation efficiency greater than 50%. Studies to determine citrate concentrations that satisfy these criteria over a range of DODAC concentrations were performed, and the results are summarized in Fig. 8. The range of citrate concentrations giving rise to particles with diameter smaller than 100 nm and with encapsulation efficiencies of 50% or higher is represented by the solid circles. Higher citrate concentrations give rise to low plasmid encapsulation efficiencies of 30% or less, whereas citrate concentrations below the optimum levels resulted in large, polydisperse aggregates (χ^2 > 5).

acid (●) and lipid (○) following incubation of SPLP in 90% mouse serum. (c) Elution profile of nucleic acid (●) and lipid (○) following incubation of SPLP with mouse serum where unencapsulated plasmid was removed by anion exchange chromatography prior to the serum treatment. [Reprinted from J. J. Wheeler, L. Palmer, M. Ossanlou, I. MacLachlan, R. W. Graham, Y. P. Zhang, M. J. Hope, P. Scherrer, and P. R. Cullis, *Gene Ther.* **6,** 271 (1999).]

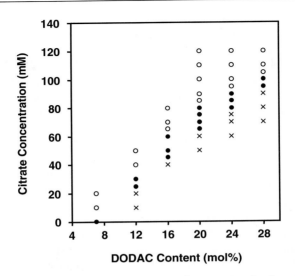

FIG. 8. Determination of the optimal citrate concentration range as a function of SPLP DODAC content to achieve maximum plasmid encapsulation in combination with minimum aggregation. Formulations were composed of DODAC/DOPE/PEG-CerC$_8$ (x : 85-x : 15; mol : mol : mol) and pCMVLuc (10 mg lipid and 100–200 μg plasmid per ml) and were prepared by detergent dialysis where the dialyzate contained the indicated sodium citrate concentrations as well as 150 mM NaCl, 10 mM HEPES (pH 7.2). The solid circles (●) indicate formulations that exhibited plasmid encapsulation efficiencies greater than 50% and a small, monodisperse size distribution as determined by QELS (diameter <100 nm, $\chi^2 < 3$). The open circles (○) indicate formulations that exhibited plasmid encapsulation efficiencies of less than 40% in combination with a small, monodisperse size distribution (diameter <100 nm, $\chi^2 < 3$). The crosses (×) indicate polydisperse formulations with large size distributions (diameter >100 nm, $\chi^2 > 3$). [Reprinted from Y. P. Zhang, L. Sekirov, E. G. Saravolac, J. J. Wheeler, P. Tardi, K. Clow, E. Leng, R. Sun, P. R. Cullis, and P. Scherrer, *Gene Ther.* **6**, 1438 (1999).]

It was not possible to obtain satisfactory formulations for preparations containing 30 mol% DODAC simply by varying the citrate concentration. Aggregation persisted in formulations dialyzed in 70–90 mM citrate buffer, whereas at higher concentrations the plasmid encapsulation was less than 25%. Improved results were achieved, however, by leaving the citrate concentration constant at 100 mM and decreasing the NaCl concentration from 150 mM to 120–140 mM. This resulted in an increase in the plasmid encapsulation efficiency to 55–70% while the particle size remained small (<100 nm diameter).

Other multivalent salts can be used in the formulation of SPLP with high DODAC content. Utilizing a formulation consisting of DOPE/DODAC/PEG-CerC$_8$, Saravolac *et al.*[14] found that DODAC concentrations ranging from 7 to 42.5 mol% could be achieved by detergent dialysis against a sodium phosphate buffer.

Both of these latter examples highlight an important point in formulation of SPLP, especially when higher quantities of cationic lipid are required. Small variations in the cationic lipid to DNA charge ratio result in changes in the buffer concentrations required for optimal size and encapsulation. In practice, it is usually necessary to perform a citrate or salt titration in order to determine optimal conditions for SPLP formation. This is often necessary even for lower DODAC concentrations, where minor variations in lipid stock solution concentrations can lead to changes in final DODAC levels, and thus to buffer or salt concentrations required for optimal results. Recall that at a salt concentration of 150 mM, efficient encapsulation was only observed over a range of 6–7% DODAC. There are two general methods available for determining the appropriate conditions for preparing SPLP by detergent dialysis: one can select a buffer system and dialyze a range of lipid formulations that differ in cationic lipid content, or one can fix a formulation's lipid composition and perform citrate and/or salt titrations as necessary.

A second point to note in this context involves the challenges that can be encountered in the production of larger preparations of SPLP, such as would be required for a series of *in vivo* experiments. For example, when performing *in vitro* transfection experiments, one would typically begin with 10 mM lipid and 400 μg plasmid DNA in a 1 ml volume (solubilized at an OGP concentration of 200 mM). However, *in vivo* experiments require scale-up to starting quantities of 50–250 mg plasmid DNA in a volume of 125–625 ml. In order to confirm the optimal buffer concentrations for detergent dialysis, one would perform a series of test dialyses using 1 ml aliquots of the lipid-detergent mixture, and assay for size and encapsulation efficiency following overnight dialysis. Even so, it is possible that aggregation, or low encapsulation, may occur following dialysis of a large-scale preparation. If this occurs, one can resolubilize the plasmid-lipid mixture with OGP, adjust the buffer solute concentrations, and redialyze. The criteria for determining successful formulation is an average SPLP diameter less than 100 nm with a low χ^2 (<3).

All of the examples discussed above involve SPLP formed with the cationic lipid DODAC. It should be noted that SPLP have been formed from a number of other cationic lipids as well, including DOTMA, DODMA-AN, DSDAC, and DC-Chol,[25] giving rise to particles that exhibit a range of transfection potencies.

Formation of Stabilized Plasmid-Lipid Particles from Various Lipids

An effective systemic gene delivery system must possess traits that at first appear contradictory. Perhaps the clearest example of this is the need for particles that are stable while in the blood compartment yet are able to release their

[25] K. W. Mok, A. M. Lam, and P. R. Cullis, *Biochim. Biophys. Acta* **1419,** 137 (1999).

plasmid payload once they have accumulated at disease sites or are taken up by cells. This can potentially be accomplished through the use of systems capable of programmable fusion. Programmable fusion can be achieved by preparing SPLP with PEG-Cer molecules that exchange out of the SPLP at an optimized rate. The original SPLP particle is protected from serum nucleases and the proteins that lead to vesicle clearance by the PEG-Cer coating. Once sufficient PEG-Cer is lost from the particle, the remaining lipid bilayer becomes unstable and becomes more able to fuse with other lipid bilayers, enhancing the intracellular delivery of the plasmid to cells. To date we have focused on SPLP formed using PEG-CerC$_{20}$, PEG-CerC$_{14}$, or PEG-CerC$_8$. (The concentration of the last lipid must be adjusted to 15 mol% if stable particles are to be formed.) Particles made with PEG-CerC$_{20}$ have long circulation lifetimes due to the slow rate of release of the PEG-lipid from SPLP, whereas those made with the shorter chain PEGs have circulation lifetimes and lipid exchange rates that are greatly reduced. Both *in vitro*[12] and *in vivo*[13,14] studies have demonstrated that transfection rates are increased for SPLP containing shorter-chain PEG-ceramides.

In addition to forming SPLP with different PEG-Cer molecules, we have made particles containing different phosphatidylcholines (POPC and DOPC), and DOPE-SPLP containing various concentrations of cholesterol. It is worth noting that the formation of SPLP from these or other different lipids may occur over a wide range of ionic strength conditions in the dialysis buffer. For example, SPLP composed of DOPC/DODAC/PEGCerC$_{20}$/Rho-PE (82.5 : 7.5 : 10 : 1) would only form at very low salt conditions ([NaCl] = 80–85 mM). In similar manner, SPLP composed of DOPE/DODAC/cholesterol/PEG-CerC$_{20}$/rhodamine-PE (30.5 : 14 : 45 : 10 : 0.5) were formed following dialysis against 20 mM HEPES, 30 mM citrate, 130 mM NaCl, pH 7.2. Both the citrate and NaCl concentrations were lower than obtained for the analogous system without cholesterol (DOPE/DODAC/PEG-CerC$_{20}$/rhodamine-PE (75.5 : 14 : 10 : 0.5), for which the optimal dialysis buffer was 20 mM HEPES, 40 mM citrate, 150 mM NaCl, pH 7.2. Thus, even the addition of a neutral molecule such as cholesterol may alter the optimized buffer solute concentrations for SPLP formation.

Post-Insertion of Cationic PEG Lipids into Stabilized Plasmid-Lipid Particles

A key finding of *in vitro* transfection studies involving cationic lipid-plasmid DNA complexes and SPLP relates to the significantly lower transfection levels achieved with the latter.[25] This stems from the lower levels of lipid and DNA delivered to cells by these particles, and highlights the need for increased transfection potency. Although some of the modifications listed above have significant effects on transfection potency, they may be insufficient.

We have described a new class of cationic PEG lipids (CPL) that were designed to enhance interactions of liposomes with cells by increasing nonspecific ionic interactions.[15] These lipids consist of a hydrophobic distearoylphosphatidyleth-anolamine (DSPE) anchor coupled to a highly fluorescent N_ε-dansyl lysine moiety, which is in turn attached to a hydrophilic poly(ethylene glycol) (PEG) spacer linked to a cationic headgroup made of lysine residues. The most effective CPL, designated CPL_4 (Fig. 9A), has three lysine residues in the headgroup, giving a charge of +4 at the chain terminus. We have shown that these CPL_4 can be incorporated into pre-formed vesicles and into SPLP using a post-insertion technique (Figure 9B) leading to greatly enhanced uptake and gene expression in mammalian cells (L. Palmer, manuscript submitted).

Briefly, an aliquot of CPL_4 (in methanol), corresponding to approximately 5 mol% of the SPLP lipid, is incubated with SPLP (in HEPES-buffered saline) at 60° for 2–3 h. This results in the incorporation of about 80% of the CPL_4 into the SPLP, corresponding to a concentration of 4 mol% (relative to total lipid). Nonincorporated CPL is removed by gel filtration chromatography on a column of Sepharose CL-4B equilibrated in HBS. The $SPLP-CPL_4$ elute from the column near the void volume and can be easily identified by fluorescence of both the dansyl (CPL) and rhodamine (lipid) markers. Unincorporated CPL is retained by the column matrix and elutes later, effectively separated from SPLP.

When eluted in HBS, $SPLP-CPL_4$ form large aggregate structures, readily visualized by fluorescence microscopy using a rhodamine filter. Aggregation of $SPLP-CPL_4$ can be prevented by the addition of Ca^{2+}, which can be present during the entire insertion and isolation procedure, or can be added to the SPLP-CPL following removal of excess CPL. In the presence of 40 mM Ca^{2+}, the size of SPLP was determined by QELS prior to and following insertion of CPL_4, giving values of 80 ± 19 nm and 76 ± 15 nm, respectively. The values determined from freeze-fracture electron microscopy were 68 ± 11 nm and 64 ± 14 nm, respectively.

In Vitro Transfection Properties of Stabilized Plasmid-Lipid Particles

In previous sections we have established that SPLP encapsulate DNA in small uniform particles that confer significant protection from DNase. These properties, essential for a systemic delivery system, may actually hinder cell-surface binding, uptake, and concomitant intracellular delivery of plasmid. Several studies were undertaken to assess the transfection potency of SPLP. Initially, SPLP consisting of $DOPE/DODAC/PEG-CerC_{20}$ (84 : 6 : 10) and the plasmid pCMVLuc coding for the luciferase reporter gene were prepared. As shown in Fig. 10, incubation of these SPLP with COS-7 cells for 24 h resulted in little if any transfection activity. This was attributed to the presence of the PEG coating on the SPLP, which is expected to inhibit the association and fusion of the SPLP with cells in the same manner that PEG coatings inhibited fusion between lipid vesicles.[20] In this regard, previous studies on LUVs with PEG coatings attached to phosphatidylethanolamine (PE) anchors had demonstrated that, for PE anchors containing short acyl chains, the

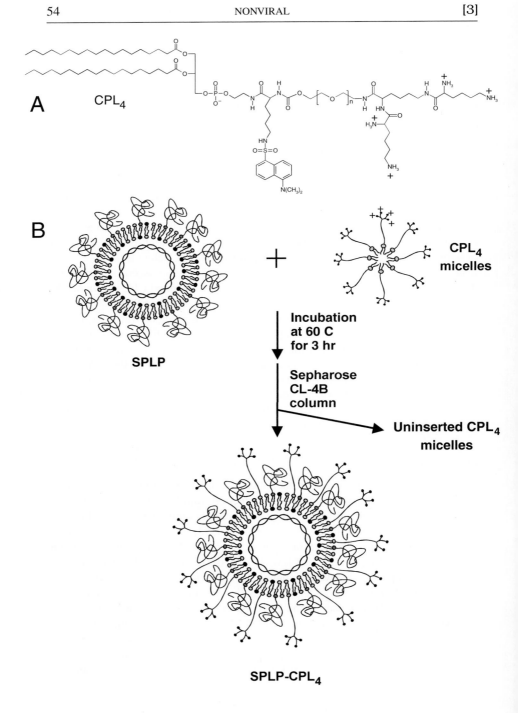

A CPL$_4$

B

SPLP + CPL$_4$ micelles

Incubation at 60 C for 3 hr

Sepharose CL-4B column

Uninserted CPL$_4$ micelles

SPLP-CPL$_4$

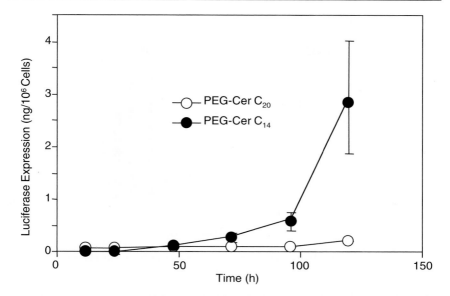

FIG. 10. Effect of PEG-Cer coating of SPLP on transfection activity *in vitro*. Plasmid (pCMVLuc) was encapsulated in SPLP (DOPE/DODAC/PEG-Cer; 84 : 6 : 10; mol/mol/mol) containing PEG-$CerC_{20}$ (○) or PEG-$CerC_{14}$ (●). Nonencapsulated plasmid was removed by anion exchange chromatography. The SPLP preparation (1 μg plasmid) was then added to COS-7 cells at a density of 2×10^4 per 24-well plate. The cells were incubated with the SPLP for the times indicated, following which the luciferase activity was measured. [Reprinted from J. J. Wheeler, L. Palmer, M. Ossanlou, I. MacLachlan, R. W. Graham, Y. P. Zhang , M. J. Hope, P. Scherrer, and P. R. Cullis, *Gene Ther.* **6,** 271 (1999).]

PEG-PE could rapidly exchange out of the LUV, rendering the LUVs increasingly able to interact and fuse with each other. To test this hypothesis we compared the transfection properties of SPLP containing PEG-$CerC_{20}$·or PEG-$CerC_{14}$. As shown in Fig. 10, after incubation with COS-7 cells for 24 h, the SPLP containing PEG-$CerC_{14}$ exhibited substantially higher levels of transfection compared with the system containing PEG-$CerC_{20}$. As indicated earlier, subsequent studies have revealed that SPLP containing PEG-$CerC_8$ possess even higher transfection

FIG. 9. Insertion protocol for the production of SPLP-CPL_4. (A) Structure of dansylated CPL_4. CPL_4 possesses four positive charges at the end of a PEG_{3400} molecule attached to a lipid achor, DSPE. (B) Protocol for insertion of CPL_4 into preformed SPLP. The SPLP are composed of DOPE (light headgroups), DODAC (black headgroups), and PEG-$CerC_{20}$ (lipids with attached polymer). SPLP and CPL_4 are incubated together at 60° for 3 h, during which time CPL_4 monomers transfer from micelles and insert into the external monolayer of SPLP. Following insertion, unincorporated CPL_4 is removed using Sepharose CL-4B column chromatography. See text for further details.

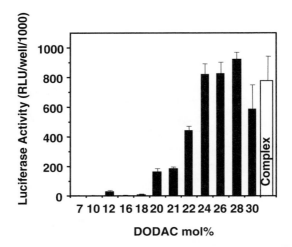

FIG. 11. Effect of DODAC content in SPLP on transfection activity *in vitro*. Plasmid (pCMVLuc) was encapsulated in SPLP containing 7-30 mol% DODAC, as described in the caption for Fig. 8. SPLP isolation by density gradient centrifugation was conducted as described in the caption for Fig. 3. SPLP were added to COS-7 cells (1 μg plasmid per well) and then incubated for 24 h, following which the luciferase activity was determined. The "complex" bar illustrates the transfection activity achieved with complexes of pCMVLuc and DODAC/DOPE (1 : 1; mol : mol) LUV at a charge ratio of 1.5 : 1 (+/−). [Reprinted from Y. P. Zhang, L. Sekirov, E. G. Saravolac, J. J. Wheeler, P. Tardi, K. Clow, E. Leng, R. Sun, P. R. Cullis, and P. Scherrer, *Gene Ther.* **6,** 1438 (1999).]

potency. This was consistent with the ability of the shorter chain PEG-Cers to exchange off the SPLP surface at a greater rate than the PEG-CerC$_{20}$.

The effect of cationic lipid content in the SPLP on *in vitro* transfection was investigated in COS-7 and HepG2 cell lines.[13] The luciferase activities detected in COS-7 cells following transfection are shown in Fig. 11, where it is clear that the transfection activity was strongly dependent on the DODAC content in the SPLP. Luciferase activity was low for SPLP containing 7–8 mol% DODAC but increased substantially to reach a plateau value between 24 and 28 mol% DODAC. Importantly, the luciferase activities detected for SPLP containing high DODAC levels were comparable with those obtained for the plasmid-lipid complexes formed with pCMVLuc and DODAC/DOPE (1 : 1; mol : mol) liposomes.

The most dramatic enhancement in *in vitro* transfection has been observed for SPLP containing 4–5 mol% of CPL$_4$ in the presence of 8 mM Ca^{2+} (L. Palmer, manuscript submitted). Previously, we have shown that the incorporation of CPL$_4$ into LUVs can lead to a 50-fold increase in cell binding and uptake.[15] We have found that the presence of Ca^{2+} alone can increase tranfection potency of SPLP several hundredfold (A. Lam, manuscript submitted[26]). This suggests a possible

[26] A. M. Lam, Ph.D. Thesis (2000).

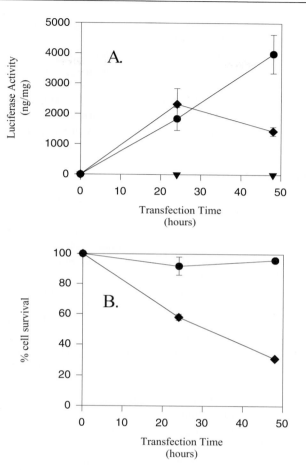

FIG. 12. (A) The transfection potency of SPLP-CPL$_4$ (●) containing 4 mol% CPL$_4$ and and lipofectin lipoplexes (◆) following extended transfection times with BHK cells. BHK cells were transfected in DMEM containing 10% FBS for 24 and 48 h with SPLP-CPL$_4$ and lipofectin lipoplexes (charge ratio of 1.5 : 1) containing 5.0 μg/ml pCMVLuc. Following transfection the luciferase expression levels and cell protein levels were determined in the cell lysate. The luciferase activity was normalized for protein content in the lysate and plotted as a function of transfection time. (B) The toxicity of SPLP-CPL$_4$ (●) containing 4 mol% CPL$_4$ and lipofectin lipoplexes (◆) as a function of transfection time, as assayed by cell survival based on the protein concentration in the cell lysate.

synergistic effect of Ca^{2+} and CPL$_4$, especially in light of the ability of Ca^{2+} to prevent aggregation of CPL$_4$-LUVs. Therefore, 4 mol% of CPL$_4$ was postinserted into SPLP composed of DOPE/DODAC/PEGCerC$_{20}$ (84/6/10), and transfection studies were performed on BHK cells in the presence of Ca^{2+}. As shown in Fig. 12A, a dramatic increase in transfection was observed for CPL$_4$-SPLP as compared with

SPLP alone, with a 10^6 enhancement observed following a 48 h incubation. While the transfection levels of CPL_4-SPLP were roughly equal to lipofectin lipoplexes at 24 h, by 48 h they were double that of lipoplex. Over this time period, the cytotoxic effects of lipoplex became apparent in lipoplex transfected cells. Such effects were not observed for the CPL_4-SPLPs.

It is likely that the increased positive surface charge of SPLP-CPL_4 will result in rapid serum clearance upon intravenous administration. However, this approach demonstrates the potential utility of postinsertion of a targeting ligand into SPLP. The pharmacology of native SPLP is discussed in the next section.

Circulation Lifetime of Stabilized Plasmid-Lipid Particles

The properties of small size, serum stability, and low levels of cationic lipid and the presence of the PEG coating suggest that SPLP should exhibit extended circulation lifetimes and disease site targeting properties following intravenous administration. A direct test of the pharmacokinetic properties of SPLP particles can be made by preparing SPLP containing trace amounts of ^{3}H-cholesteryl hexadecylether (^{3}H-CHE), a nonexchangeable lipid marker routinely used to label liposomes or vesicle preparations. The serum clearance of intravenously administered SPLP can then be determined by collecting blood at various time points and subjecting it to analysis for ^{3}H-CHE lipid by liquid scintillation analysis. This experiment reveals that PEG-$CerC_{20}$ containing SPLP are cleared from serum gradually with a measured serum half-life of 8.0 ± 1.1 h (Fig. 13). In contrast to the behavior of the SPLP system, ^{3}H-CHE labeled plasmid DNA-lipoplexes are rapidly cleared from the circulation ($t_{1/2} \ll 15$ min), appearing predominantly in the lung and liver (data not shown). The serum half-life of unprotected plasmid DNA is known to be less than 5 min.[27]

The levels of intact plasmid DNA in the circulation following intravenous administration of SPLP can also be determined directly by Southern blot hybridization and phosphorimaging analysis.[7] Quantitative analysis confirms that when naked plasmid is administered systemically, less than 0.01% of the injected dose remains intact in the circulation after 15 min. Only a small fraction of plasmid ($<2\%$) is still intact in the circulation at 15 min following the administration of lipoplexes. In contrast, following intravenous injection of SPLP, approximately 85% of the injected plasmid DNA remains intact in the circulation at 15 min. The circulation half-life of intact plasmid DNA following injection of SPLP as measured by this method is calculated to be 7.2 ± 1.6 h, in good agreement with the circulation half-life of ^{3}H-CHE labeled SPLP, confirming the highly stable nature of SPLP in the circulation and validating the use of the ^{3}H-CHE lipid label as a surrogate marker for SPLP plasmid DNA in pharmacokinetic studies.

---

[27] A. R. Thierry, P. Rabinovich, B. Peng, L. C. Mahan, J. L. Bryant, and R. C. Gallo, *Gene Ther.* **4**, 226 (1997).

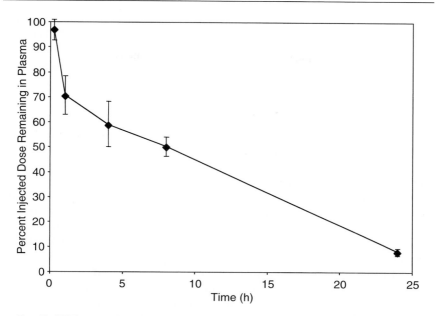

FIG. 13. SPLP serum clearance following a single intravenous administration in neuro-2a tumor-bearing A/J mice. On day 0, 1.5×10^6 cells were injected subcutaneously in the hind flank of each mouse (injection volume: 50 μl). When tumors were an appropriate size (about day 9), [^{3}H]CHE-SPLP (100 ug DNA) was administered i.v. in a total volume of 200 μl. The specific activity of the [^{3}H]CHE was 1 μCi/mg lipid. Each time point reflects the average results from 4 mice. [Reproduced from D. B. Fenske, I. MacLachlan, and P. R. Cullis, *Curr. Opin. Mol. Ther.* **3,** 153 (2001), with permission of PharmaPress Ltd.]

SPLP pharmacokinetics vary dramatically depending on the species of PEG-Cer employed for stabilization. Whereas SPLP containing PEG-CerC$_{20}$ remain in the circulation for hours after injection, SPLP-CerC$_{14}$ and SPLP-CerC$_8$ exhibit circulation half-lives of approximately 30 min and <5 min, respectively.[7] The pharmacokinetic behavior of SPLP containing different PEG-ceramides can be readily understood on the basis of the ability of these molecules to dissociate from the SPLP at different rates. In particular, it would be expected that SPLP-CerC$_8$ and SPLP-CerC$_{14}$ would rapidly shed their PEG coating in the blood compartment, whereas the SPLP-CerC$_{20}$ system should exhibit much longer PEG-Cer retention times. PEG-CerC$_8$ and PEG-CerC$_{14}$ exhibit half-times for dissociation from liposomes of <1.2 min and ~1.1 h, respectively, under *in vitro* conditions, whereas PEG-CerC$_{20}$ exhibits a dissociation half-time greater than 13 days.[12] The absence of a PEG coating would be expected to facilitate serum protein adsorption to the SPLP surface, leading to enhanced uptake by the mononuclear phagocytes of the reticuloendothelial system (RES). Thus the pharmacokinetic behavior of SPLP-CerC$_8$, SPLP-CerC$_{14}$, and SPLP-CerC$_{20}$ is consistent with the ability of the

PEG coating to dissociate at rates dependent on the acyl chain component of the ceramide anchor. It should be noted that the rates of removal of the PEG coating *in vivo* are likely to be faster than under *in vitro* conditions. Thus a quantitative correlation between PEG dissociation rates *in vitro* and SPLP clearance rates would not be expected.

Tumor Accumulation of Stabilized Plasmid-Lipid Particles

^{3}H-CHE labeled SPLP may also be used to determine the biodistribution resulting from intravenous administration of SPLP. The biodistribution of SPLP can be determined by collecting tissue at various time points after administration and subjecting it to analysis for ^{3}H-CHE lipid. This approach reveals that SPLP containing PEG-CerC$_{20}$ behave in a manner analogous to small, long circulating liposomes containing small molecule drugs such as doxorubicin. Whereas approximately 3–12% of the injected SPLP dose accumulates for every gram of tumor at the tumor site over 24 h, liposomes containing doxorubicin accumulate to the extent of 5–10% of the injected dose per gram of tumor in mouse models.[28]

The stability and long circulation lifetimes of the SPLP-CerC$_{20}$ system would be expected to lead to maximum plasmid delivery to a distal tumor site when compared to other SPLP systems. The extent of DNA delivery to tumor tissue does vary greatly with the PEG-Cer in SPLP. Not surprisingly, relatively low levels of intact plasmid are delivered by SPLP-CerC$_8$ and SPLP-CerC$_{14}$. Preferential accumulation at the tumor site of the SPLP-CerC$_{20}$ is consistent with the well-characterized behavior of liposomes that avoid immediate uptake by the RES and preferentially accumulate at disease sites such as tumors because of the permeable nature of the tumor vasculature. The benefits of this disease site targeting are clear in that the SPLP-CerC$_{20}$ formulation delivers 43-fold and 1200-fold more plasmid to the tumor site than the more rapidly cleared SPLP-CerC$_{14}$ and SPLP-CerC$_8$ formulations, respectively. The amounts of intact plasmid delivered to the tumor site by the SPLP-CerC$_{20}$ system is substantial, corresponding to greater than 10% of the total injected dose per gram of tumor at the 24 h time point (Fig. 14).

The differential ability of the PEG-ceramides to dissociate from SPLP has also been shown to modulate the biodistribution of SPLP in other organs. The clearance of SPLP from the circulation and accumulation in the tumor is mirrored by accumulation in the liver. Whereas only 14% of the injected SPLP-CerC$_{20}$ dose accumulates in the liver 3 h after SPLP-CerC$_{20}$ administration, 85% and nearly 100% of SPLP-CerC$_8$ and SPLP-CerC$_{14}$, respectively, accumulate in the liver within 3 h of injection.[7] It is well known that rapid clearance of liposomes from the circulation is mediated by RES uptake, primarily the Kuppfer cells of the liver. Thus the accumulation of SPLP-CerC$_8$ and SPLP-CerC$_{14}$ in the liver can be

[28] L. D. Mayer, M. B. Bally, P. R. Cullis, S. L. Wilson, and J. T. Emerman, *Cancer Lett.* **53**, 183 (1990).

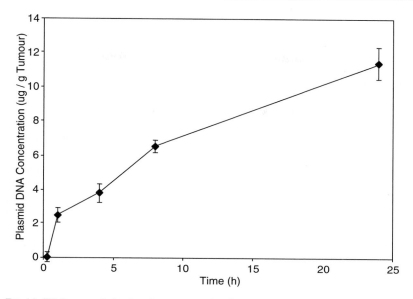

FIG. 14. SPLP accumulation in subcutaneous neuro-2a tumors following a single intravenous administration in A/J mice. For experimental details, see legend to Fig. 13. [Reproduced from D. B. Fenske, I. MacLachlan, and P. R. Cullis, *Curr. Opin. Mol. Ther.* **3,** 153 (2001), with permission of PharmaPress Ltd.]

attributed to dissociation of the PEG coating, opsonization by serum proteins, and uptake by the liver phagocytes.

A noteworthy feature of SPLP-CerC$_{20}$ is the ability to bypass the lung, unique among cationic lipid containing gene delivery systems. This is likely due to the properties required for disease site targeting, namely small uniform size and low surface charge. A number of studies have characterized the transfection properties of plasmid DNA–cationic liposome complexes following intravenous administration.[2,29] High levels of transgene expression are usually observed in the lungs, with lower levels of expression in the spleen, liver, heart, and kidneys. Lipoplex-mediated gene expression in the lung appears to arise from deposition in lung microvasculature and reflects the rapid clearance of plasmid DNA–cationic lipid complexes from the circulation due to their large size (>200 nm diameter) and high cationic lipid content. This is consistent with the observation that murine B16 tumors seeded in the pulmonary vascular compartment can be transfected by intravenous administration of lipoplex.[30] The ability of SPLP to bypass the

[29] N. Zhu, D. Liggitt, Y. Liu, and R. Debs, *Science* **261,** 209 (1993).
[30] X. Zhou and L. Huang, *J. Contr. Release* **19,** 269 (1992).

lung may be considered a predictor for the transfection of distal tumors following systemic administration.

Gene Expression Following Intravenous Injection of Stabilized Plasmid-Lipid Particles

Although PEG-containing SPLP are promising with respect to their ability to deliver intact plasmid DNA to disease sites, SPLP exhibit relatively low transfection efficiencies *in vitro* (Fig. 7) due to the ability of the PEG coating to inhibit cell association and uptake of PEG-containing liposomes.[20–22] However, the use of diffusible PEG-ceramides facilitates the formulation of stable particles containing a high percentage (79 to 84 mol%) of the fusogenic lipid DOPE. As the PEG-ceramide dissociates from the particle it is expected to become increasingly fusogenic. In particular, the SPLP lipid composition of DOPE, DODAC, and PEG-Cer is stabilized in the bilayer organization by the presence of the PEG-Cer component and will tend to assume the hexagonal (H_{II}) organization preferred by DOPE when the PEG-Cer dissociates. Liposomes having lipid compositions favoring H_{II} organization rapidly aggregate and fuse, releasing their entrapped contents. The inclusion of PEG-ceramides in SPLP may help to resolve the previously mentioned conflicting demands imposed upon carriers for systemic gene therapy. First, the carrier must be stable and circulate long enough to facilitate accumulation at disease sites. Second, the carrier must be capable of interacting with target cells in order to facilitate intracellular delivery. PEG coatings that dissociate from the carrier at the disease site are a potential solution to this problem.

A direct assessment of the *in vivo* transfection potential of SPLP reveals that SPLP-CerC$_{20}$ are indeed transfection competent at disease sites following systemic administration (Fig. 15). A single intravenous administration of SPLP yields significant levels of gene expression within 24 h after administration. Marker gene expression increases considerably over the following 48 h, reaching maximal levels more than 72 h after administration. Remarkably, the greatest levels of gene expression observed upon systemic administration of SPLP-CerC$_{20}$ are found in tumor tissue. Although moderate levels of marker gene expression are observed in all tissues assayed, at later time points the tumor gene expression is two orders of magnitude greater than that observed in other organs. (The maximal level of luciferase gene expression measured, 1.6 ± 0.18 ng/g corresponds to $\sim$20 copies of active luciferase protein per tumor cell.) Most notable among the other organs assayed is the lung. While all other nonviral gene delivery systems are known to predominantly transfect lung upon intravenous administration, SPLP are a notable exception to this rule. The lack of marker gene expression in the lung correlates well with the ability of SPLP to bypass the lung on systemic administration. Unlike the lung, the liver does accumulate significant levels of SPLP. It is notable that although liver accumulation of SPLP-CerC$_{20}$ may reach as much as 25% of the total

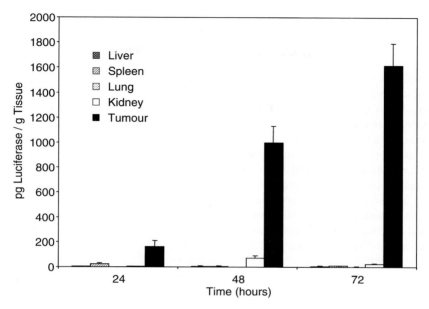

FIG. 15. Luciferase gene expression following a single intravenous administration of SPLP in neuro-2a tumor-bearing A/J mice. For experimental details, see legend to Fig. 13. [Reproduced from D. B. Fenske, I. MacLachlan, and P. R. Cullis, *Curr. Opin. Mol. Ther.* **3,** 153 (2001), with permission of PharmaPress Ltd.]

injected dose at later time points, the liver is relatively refractory to SPLP-mediated transfection.

The reasons for the marked preference for transfection of tumor tissue become apparent when one considers the various barriers that must be overcome in transfection process. The first requirement for transfection is delivery of the plasmid DNA to the disease site, readily facilitated by systems capable of disease site targeting. Once at the disease site, the transfection reagent must be interact with and be internalized by the target cell. On internalization the plasmid DNA needs to escape the endosomal compartment and translocate to the nucleus before gene expression can occur. It is probable that tumor tissue is more readily transfected than other tissues because tumor cells are more rapidly dividing than normal tissues. Tumor cells are highly endocytic, have a high mitotic index, and are highly active at the transcriptional and translational level. It has been shown that SPLP (in common with other nonviral gene delivery systems) require cell division for efficient transfection.[31]

[31] I. Mortimer, P. Tam, I. MacLachlan, R. W. Graham, E. G. Saravolac, and P. B. Joshi, *Gene Ther.* **6,** 403 (1999).

PEG-ceramide has an expected effect on the transfection potency of SPLP. Preliminary studies have determined that SPLP-CerC$_{20}$ produce significantly higher levels of marker gene expression than SPLP-CerC$_{14}$ and SPLP-CerC$_8$ when administered systemically in tumor-bearing mice. Neither SPLP-CerC$_{14}$ nor SPLP-CerC$_8$ are able to mediate significant levels of luciferase gene expression at tumor sites, suggesting that the amount of intact plasmid DNA delivered to the tumor by SPLP-CerC$_{14}$ and SPLP-CerC$_8$ is not sufficient to produce detectable levels of gene expression.[7] The SPLP-CerC$_{20}$ mediated transfection levels are significant and offer a basis on which to optimize SPLP properties to reach higher levels. For example, the observation that the SPLP-CerC$_{20}$ formulation is the only formulation that results in transfection at the tumor site emphasizes the obvious fact that little transfection can be achieved unless adequate levels of plasmid are delivered to the target tissue. SPLP-CerC$_{20}$ are, however, the most stable preparations and the least intrinsically transfection potent of the formulations investigated when evaluated *in vitro*. In turn, this suggests methods for improving the tumor transfection potential of SPLP. For example, the utilization of PEG-ceramides containing acyl groups with intermediate chain lengths such as C$_{16}$ or C$_{18}$ may allow greater SPLP stability than SPLP-CerC$_8$ and SPLP-CerC$_{14}$, thus promoting greater delivery of intact plasmid to tumor sites. These compositions should also render the SPLP more transfection potent than SPLP-CerC$_{20}$ after arrival at the tumor site because of an improved ability of the PEG coating to dissociate. Alternatively, methods for actively targeting SPLP to tumor cells after arrival at the tumor site may also facilitate target cell uptake and transfection. These and other methods for achieving high levels of SPLP-mediated gene expression at distal tumors following systemic administration are currently under investigation.

Summary

The ability of a systemically administered gene therapy vector to exhibit extended circulation lifetimes, accumulate at a distal tumor site, and enable transgene expression is unique to SPLP. The flexibility and low toxicity of SPLP as a platform technology for systemic gene therapy allows for further optimization of tumor transfection properties following systemic administration. For example, the PEG coating of SPLP is necessary to engender the long circulation lifetimes required to achieve tumor delivery. However, PEG coatings have also been shown to inhibit cell association and uptake required for transfection. The dissociation rate of the PEG coating from SPLP can be modulated by varying the acyl chain length of the ceramide anchor, suggesting the possibility of developing PEG-Cer molecules that remain associated with SPLP long enough to promote tumor delivery, but which dissociate quickly enough to allow transfection. Alternatively, improvements may be expected from inclusion of cell-specific targeting ligands in SPLP to promote cell association and uptake. Finally, the nontoxic properties of SPLP allow the possibility of higher doses. A dose of 100 μg plasmid DNA per mouse corresponds to a dose of approximately 5 mg plasmid DNA per kg body weight. This compares well to small molecules used for cancer therapy, which typically are used at dose levels of 10 to 50 mg per kg body weight.

In summary, SPLP consist of plasmid encapsulated in a lipid vesicle that, in contrast to naked plasmid or complexes, exhibit extended circulation lifetimes following intravenous injection, resulting in accumulation and transgene expression at a distal tumor site in a murine model. The pharmacokinetics, biodistribution, and tumor transfection properties of SPLP are highly sensitive to the nature of the ceramide anchor employed to attach the PEG to the SPLP surface. The SPLP-CerC$_{20}$ system in which the PEG-Cer does not readily dissociate exhibits good serum stability, long circulation lifetimes, and high levels of tumor accumulation and mediates marker gene expression at the tumor site. The flexibility of the SPLP system offers the potential of further optimization to achieve therapeutically effective levels of gene transfer and clearly has considerable potential as a nontoxic systemic gene therapy vehicle with general applicability.

These features of SPLP contrast favorably with previous plasmid encapsulation procedures. Plasmid DNA has been encapsulated by a variety of methods, including reverse phase evaporation,[32-34] ether injection, [35,36] detergent dialysis in the absence of PEG stabilization,[34,35] lipid hydration and dehydration–rehydration techniques,[37-39] and sonication,[40-42] among others. The characteristics of these protocols are summarized in Table I. None of these procedures yields small, serum-stable particles at high plasmid concentrations and plasmid-to-lipid ratios in combination with high plasmid-encapsulation efficiencies. Trapping efficiencies comparable with the SPLP procedure can be achieved employing methods relying on sonication. However, sonication is a harsh technique that can shear nucleic acids. Size ranges of 100 nm diameter or less can be achieved by reverse-phase techniques; however, this requires an extrusion step through filters with 100 nm or smaller pore size which can often lead to significant loss of plasmid. Finally,

[32] R. Fraley, S. Subramani, P. Berg, and D. Papahadjopoulos, *J. Biol. Chem.* **255**, 10431 (1980).

[33] P. Soriano, J. Dijkstra, A. Legrand, H. Spanjer, D. Londos-Gagliardi, F. Roerdink, G. Scherphof, and C. Nicolau, *Proc. Natl. Acad. Sci. U.S.A.* **80**, 7128 (1983).

[34] M. Nakanishi, T. Uchida, H. Sugawa, M. Ishiura, and Y. Okada, *Exp. Cell Res.* **159**, 399 (1985).

[35] R. T. Fraley, C. S. Fornari, and S. Kaplan, *Proc. Natl. Acad. Sci. U.S.A.* **76**, 3348 (1979).

[36] C. Nicolau and S. Rottem, *Biochem. Biophys. Res. Commun.* **108**, 982 (1982).

[37] P. F. Lurquin, *Nucleic Acids Res.* **6**, 3773 (1979).

[38] S. F. Alino, M. Bobadilla, M. Garcia-Sanz, M. Lejarreta, F. Unda, and E. Hilario, *Biochem. Biophys. Res. Commun.* **192**, 174 (1993).

[39] M. Baru, J. H. Axelrod, and I. Nur, *Gene* **161**, 143 (1995).

[40] D. G. Jay and W. Gilbert, *Proc. Natl. Acad. Sci. U.S.A.* **84**, 1978 (1987).

[41] C. Puyal, P. Milhaud, A. Bienvenue, and J. R. Philippot, *Eur. J. Biochem.* **228**, 697 (1995).

[42] M. Ibanez, P. Gariglio, P. Chavez, R. Santiago, C. Wong, and I. Baeza, *Biochem. Cell Biol.* **74**, 633 (1996).

TABLE I

PROCEDURES FOR ENCAPSULATING PLASMID IN LIPID-BASED SYSTEMS

Procedure[a]	Lipid composition	Length of DNA	Trapping efficiency	DNA-to-lipid ratio	Diameter
Reverse phase evaporation[1]	PS or PS : Chol (50 : 50)	SV40 DNA	30–50%	<4.2 μg/μmol	400 nm
Reverse phase evaporation[2]	PC : PS : Chol (40 : 10 : 50)	11.9 kbp plasmid	13–16%	0.23 μg/μmol	100 nm to 1 μm
Reverse phase evaporation[3]	PC : PS : Chol (50 : 10 : 40)	8.3 kbp, 14.2 kbp plasmid	10%	0.97 μg/μmol	ND[b]
Reverse phase evaporation[4]	EPC : PS : Chol (40 : 10 : 50)	3.9 kbp plasmid	12%	0.38 μg/μmol	400 nm
Ether injection[5]	EPC : EPG (91 : 9)	3.9 kbp plasmid	2–6%	<1 μg/μmol	0.1–1.5 μm; Aug = 230 nm
Ether injection[6]	PC : PS : Chol (40 : 10 : 50) PC : PG : Chol (40 : 10 : 50)	3.9 kbp plasmid	15%	15 μg/μmol	ND
Detergent dialysis[7]	EPC : Chol : stearylamine (43.5 : 43.5 : 13)	sonicated genomic DNA (~250,000 MW)	11%	0.26 μg/μmol	50 nm
Detergent dialysis, extrusion[8]	DOPC : Chol : oleic acid or DOPE : Chol : oleic acid (40 : 40 : 20)	4.6 kbp plasmid	14–17%	2.25 μg/μmol	180 nm (DOPC) 290 nm (DOPE)
Lipid hydration[9]	EPC : Chol (65 : 35) or EPC	3.9 kbp, 13 kbp plasmid	ND	ND	0.5–7.5 μm
Dehydration–rehydration, extrusion (400 or 200 nm filters)[10]	Chol : EPC : PS (50 : 40 : 10)	ND	ND	0.83 μg/μmol (200 nm) 1.97 μg/μmol (400 nm)	142.5 nm (200 nm filter) 54.6 nm (400 nm filter, ultracentrifugation)
Dehydration–rehydration[11]	EPC	2.96 kbp, 7.25 kbp plasmid	35–40%	2.65–3.0 μg/μmol	1–2 μm
Sonication (in the presence of lysozyme)[12]	asolectin (soybean phospholipids)	1.0 kbp linear DNA	50%	0.08 μg/μmol	100–200 nm
Sonication[13]	EPC : Chol : lysine-DPPE (55 : 30 : 15)	6.3 kb ssDNA 1.0 kb dsRNA	60–95% ssDNA 80–90% dsRNA	13 μg/μmol ssDNA; 14 μg/μmol dsRNA	100–150 nm

Spermidine-condensed DNA, sonication, extrusion[14]	EPC:Chol:PS (40:50:10) EPC:Chol:EPA (40:50:10) or EPC:Chol:CL (50:40:10)	4.4 kbp, 7.2 kbp plasmid	46–52%	2.53–2.87 $\mu g/\mu mol$	400–500 nm
Ca^{2+}-EDTA entrapment of DNA–protein complexes[15]	PS:Chol (50:50)	42.1 kbp bacteriophage	52–59%	22 $\mu g/\mu mol$	ND
Freeze–thaw, extrusion[16]	POPC:DDAB (99:1)	3.4 kbp linear plasmid	17–50%	ND	80–120 nm
SPLP (this work)	DOPE:PEG-Cer:DODAC (84:10:6)	4.4–10 kbp plasmid	60–70%	62.5 $\mu g/\mu mol$	75 nm (QELS); 65 nm (freeze-fracture)

[a] Key to References: [1] R. Fraley, S. Subramani, P. Berg, and D. Papahadjopoulos, *J. Biol. Chem.* **255**, 10431 (1980); [2] P. Soriano, J. Dijkstra, A. Legrand, H. Spanjer, D. Londos-Gagliardi, F. Roerdink, G. Scherphof, and C. Nicolau, *Proc. Natl. Acad. Sci. U.S.A.* **80**, 7128 (1983); [3] M. Nakanishi, T. Uchida, H. Sugawa, M. Ishiura, and Y. Okada, *Exp. Cell. Res.* **159**, 399 (1985); [4] A. Cudd and C. Nicolau, *Biochim. Biophys. Acta* **845**, 477 (1985); [5] R. T. Fraley, C. S. Fornari, and S. Kaplan, *Proc. Natl. Acad. Sci. U.S.A.* **76**, 3348 (1979); [6] C. Nicolau and S. Rottem, *Biochem. Biophys. Res. Commun.* **108**, 982 (1982); [7] J. C. Stavridis, G. Deliconstantinos, M. C. Psallidopoulos, N. A. Armenakas, D. J. Hadjiminas, and J. Hadjiminas, *Exp. Cell Res.* **164**, 568 (1986); [8] C. Y. Wang and L. Huang, *Proc. Natl. Acad. Sci. U.S.A.* **84**, 7851 (1987); [9] P. F. Lurquin, *Nucleic Acids Res.* **6**, 3773 (1979); [10] S. F. Alino, M. Bobadilla, M. Garcia-Sanz, M. Lejarreta, F. Unda, and E. Hilario, *Biochem. Biophys. Res. Commun.* **192**, 174 (1993); [11] M. Baru, J. H. Axelrod, and I. Nur, *Gene* **161**, 143 (1995); [12] D. G. Jay and W. Gilbert, *Proc. Natl. Acad. Sci. U.S.A.* **84**, 1978 (1987); [13] C. Puyal, P. Milhaud, A. Bienvenue, and J. R. Philippot, *Eur. J. Biochem.* **228**, 697 (1995); [14] M. Ibanez, P. Gariglio, P. Chavez, R. Santiago, C. Wong, and I. Baeza, *Biochem. Cell. Biol.* **74**, 633 (1996); [15] J. Szelei and E. Duda, *Biochem. J.* **259**, 549 (1989); [16] P. A. Monnard, T. Oberholzer, and P. Luisi, *Biochim. Biophys. Acta* **1329**, 39 (1997).

it may be noted that the plasmid DNA-to-lipid ratios that can be achieved for SPLP are significantly higher than those achievable by any other encapsulation procedure.

Appendix: SPLP Formulation Protocol

Scope

This sample protocol describes the preparation of SPLP for systemic gene delivery and expression. The batch described initially contains 40 mg of plasmid DNA in a total volume of 100 ml. This protocol provides detailed steps for

> Preparation of lipids
> Salt curve analysis
> Removal of unencapsulated DNA by DEAE-Sepharose chromatography
> Concentration by Amicon ultrafiltration
> Sucrose density isolation procedure
> Sterilization and storage

Materials

1. Sterile bottles or round bottom flasks
2. Sterile graduated cylinders
3. Sterile crimp-top glass vials
4. Spectrum Spectra/Por Molecular porous 25 mm and 45 mm membrane tubing 12,000–14,000 MWCO
5. Large chromatography column (2 cm radius)
6. Amicon ultrafiltration apparatus (350 ml and 180 ml)
7. Diaflo 76 mm and 62 mm ultrafiltration membranes
8. Seton 1 inch × 3.5 inches open top polyclear ultracentrifuge tube
9. Assorted sterile syringes and 18G 1/2 inch needles
10. Millipore 0.22 μm syringe filtration unit
11. Sterile Pasteur pipettes
12. Picogreen DNA quantitation reagent

Preparation of Lipids

1. Prepare 20 l of 10× HBS (1× = 10 mM HEPES and 150 mM NaCl pH 7.4).

2. Weigh out the required lipids and OGP on an analytical balance, or alternatively use lipid stock solutions previously prepared in methanol. For a 100 ml preparation with an initial lipid concentration of 10 mg/ml, 200 mM OGP, and the composition DOPE : DODAC : PEGCerC$_{20}$ (82.5 : 7.5 : 10), the following would be required: 664 mg DOPE, 47 mg DODAC, 289 mg PEGCerC$_{20}$, 5848 mg OGP, and 40 mg plasmid DNA.

3. In a sterile bottle or round bottom flask, add 1/3 total formulation volume of filter sterilized 1× HBS and add the DOPE, PEGC$_{20}$, OGP, and DODAC. Dissolve the lipids and OGP by magnetic stirring.

4. While waiting for the lipids to dissolve, cut six 8 cm strips of dialysis tubing so that there are a few extra centimeters of bag on both sides. Presoak the dialysis bags in filter-sterilized water for at least 30 min in preparation for the salt curve.

5. Once the lipids have dissolved add the DNA at 400 μg DNA/10 mg lipid.

6. Using a sterile graduated cylinder, add 1× HBS to bring the total volume to 100 ml and mix.

Test Formulate: Preparing for Salt Curve

1. Dialyze 1 ml aliquots of the above preparation overnight in 3 liters of the following buffers prepared using the 10× HBS stock:

[NaCl] (mM)
120
130
140
150
160
170

2. The next day, set aside 500 μl of each sample for QELS analysis and a PicoGreen assay.

PicoGreen Assay

(i) Prepare 1/40 dilutions of all samples (i.e., 25 μl stock sample +975 μl 1× HBS).

(ii) Prepare duplicates of 1/400 dilutions of all samples (i.e., 100 μl of above dilution +900 μl of 1× HBS).

(iii) Prepare plasmid DNA standard curve. Add 200 μl of a 10 μg/ml DNA standard to 1.8 ml HBS, and use this solution to prepare a serial dilution from 1 to 0.0625 μg/ml. Each standard and sample will have a volume of 1 ml.

(iv) Carry out PicoGreen analysis according to manufacturer's instructions in the absence and presence of 0.1% Triton X-100. The excitation and emission wavelengths of PicoGreen are 495 and 525 nm, respectively. Slit width = 4 nm. For each standard and sample, the steps are
- Add 2 μl PicoGreen to standard or sample.
- Transfer to cuvette, and measure fluorescence for 10–20 s (−Triton value).

- Add 10 μl 10% Triton X-100 to cuvette and mix thoroughly.
- Measure fluorescence for 10–20 s (+Triton value).

The % encapsulation = (([+Triton] − [−Triton]) × 100)/[+Triton]

3. The optimal NaCl concentration is determined by the percent of plasmid encapsulation and particle size. The optimal NaCl provides for 50–80% encapsulation and a single population with a particle size of ≤100 nm. Particle sizes greater than 120 nm indicate a "crashed" preparation where the vesicles have aggregated. Determine the optimal NaCl concentration and prepare 4 × 4 liters of dialysis buffer for a 48-h dialysis.

4. Dialyze the initial lipid solution for 48 h using the optimal buffer conditions. Transfer the dialysis bags to fresh dialysis buffer every 12 h.

5. Perform a PicoGreen assay and QELS on the resulting material. Expect a percent encapsulation of approximately 50–80%. If this is not achieved, perform another salt curve before proceeding to the next step.

Removal of Unencapsulated DNA by DEAE-Sepharose Chromatography

1. For every 0.64 mg of DNA loaded on the DEAE-Sepharose CL-6B (Sigma) column, 1 mL of stationary volume will be required. Pour the column and wash with 10 column volumes of filter sterilized 1× HBS.

2. Once the column has settled, slowly load the formulation suspension against the column wall in a circular motion and allow it to flow through the resin. Elute with HBS.

3. Use small glass test tubes to determine the point at which the cloudy formulation begins to elute from the column. Once the formulation appears, collect into a sterile flask or bottle. Collect a final volume equal to 1.5 times the sample volume to completely elute all of the formulation from the column.

Concentration by Amicon

1. Assemble the Amicon apparatus.
   (a) Place the membrane shiny side up into the bottom plate and the rubber seal on top of the membrane.
   (b) Screw on the bottom plate tightly.
   (c) Hydrate the filter with a steady stream of filter sterilized water for 10–20 min.
   (d) Add the sample and adjust the pressure so as not to exceed 50 psi, with a flow rate of 1 drop/sec.
   (e) Concentrate the sample down to the desired volume (typically 4× concentration).

Sucrose Density Isolation Procedure

1. Prepare desired volumes of 2.5%, 5.0%, and 10.0% w/v sucrose in 1× HBS and filter sterilize into a sterile container. Store at 4°.

2. Pull a long glass pipette to a small point using forceps and a Bunsen burner. Place the elongated pipette into the ultracentrifuge tube (Seton 1 inch × 3.5 inches open top polyclear) and pour 14 ml 2.5% sucrose solution into the pipette using a second sterile pipette.

3. Pour the sucrose layers in order of increasing density: 14 ml of 2.5%, 10 ml of 5.0%, and 7 ml of 10.0%. Load 5 to 7 ml of SPLP on top of the gradients.

4. Balance all the tubes with 2.5% sucrose or 1× HBS to within 0.01 g and place the tubes in the SW-28 buckets. Run for 18 h at 28,000 rpm at 20°.

5. The following day, slowly aspirate the lower band using an 18-gauge needle with a 10 ml syringe. Pool all of the lower bands and place the formulation into a dialysis bag overnight in 1× HBS to remove the sucrose.

6. Perform a PicoGreen assay to determine the percent encapsulation and DNA concentration.

Final Concentration by Amicon

1. Concentrate the sample until the desired concentration of DNA is achieved (typically >0.5 mg DNA/ml).

Sterilization and Storage

1. Filter sterilize the final volume through a sterile Millipore 0.22 μm filter unit in a biological safety cabinet and adjust the final volume so that the DNA concentration is exactly 0.5 mg/ml. Store SPLP in sterile vials at 4° for up to 2 years. Analyze the particle size, percent encapsulation, and lipid concentration via QELS analysis, PicoGreen assay, and HPLC analysis, respectively. Agarose gel electrophoresis can also be used to verify the integrity of the plasmid.

[4] Enhancing Direct *in Vivo* Transfection with Nuclease Inhibitors and Pulsed Electrical Fields

By JILL GLASSPOOL-MALONE and ROBERT W. MALONE

Introduction

Gene therapy may be defined as providing clinical benefit via the administration of recombinant viruses or nucleic acid formulations that encode transgenic proteins. Implementation of this paradigm requires (1) identification of a disease-therapeutic transgene linkage and (2) a method for delivering and expressing the therapeutic transgene within appropriate cells. Simple and compelling in concept, the promise of gene therapy is generally perceived as unfulfilled and in many cases oversold. It is our belief that the lack of effective, reproducible, and safe gene therapies are the consequence of a lack of polynucleotide delivery technologies that are effective, reproducible, and of no or low toxicity in humans. Therefore, for the past decade our work has focused on the *in vivo* development, testing, and improvement of gene delivery systems rather than the development of "gene therapies."

Both viral and nonviral delivery systems may be safe at low doses, but are often toxic at the higher doses required to achieve therapeutic levels of transgene expression. In general, recombinant viral vector-based treatments have been associated with complications that reflect evolved viral biology and/or host/parasite interactions. Nonviral delivery systems are often compromised by inefficiency, *in vivo* clearance, and formulation/manufacture complexities.

In contrast to the challenges associated with the effective delivery of nucleic acid medicines, identification and characterization of the genetic basis of disease has proceeded rapidly and efficiently. Potentially therapeutic "genes," open reading frames, mRNAs, ribozymes, and oligonucleotides are being identified rapidly and efficiently. For the near term, the rich diversity of potentially therapeutic nucleic acids will continue to expand at an apparently exponential rate due to advances in genomics research. These observations indicate that current gene delivery technology is a principal rate-limiting factor in the translation of molecular knowledge to molecular therapeutics.

In the late 1980s, when testing the *in vivo* efficacy of various cationic lipid : polynucleotide formulations in a rat model, a negative control resulted in the serendipitous discovery that intramuscular injection of reporter plasmid or mRNA resulted in transgene expression.[1] Surprisingly, higher levels of transgene

---

[1] J. A. Wolff, R. W. Malone, P. Williams, W. Chong, G. Acsadi, A. Jani, and P. L. Felgner, *Science* **247,** 1465 (1990).

expression were observed with the "naked" polynucleotide injection than with lipoplexes incorporating the same polynucleotides (unpublished data, R. Malone, J. Wolff, and P. Felgner, 1989). This observation, compared by some to "cold fusion" at the time of initial publication, has turned out to be robust and reproducible. First reported to be restricted to muscle, subsequent studies demonstrated that a wide range of tissues are transfected by direct polynucleotide injection.[2-8] Since that time, many different viral vector systems have been developed, and a new subdiscipline known as nonviral gene therapy has emerged. However, no other method yet approaches the simplicity and safety of direct polynucleotide (naked DNA) injection. Unfortunately, direct delivery systems are typically inefficient, and hence clinical development has been largely restricted to use for vaccination (e.g., DNA vaccines). Factors that reduce the efficiency of direct *in vivo* transfection include nuclease-mediated clearance of the polynucleotide (both extra- and intracellularly), and the transport of high molecular weight charged polynucleotides across cytoplasmic membranes. By addressing both of these issues using methods described herein, levels of transgene expression and transfection efficiency may be markedly enhanced.[9,10]

No single polynucleotide delivery system is likely to meet all clinical needs. Time has shown that each system has strengths and weaknesses including differences in the host response to retreatment, tissue trophism, duration of transgene expression, and toxicity.[11] One approach to the complex opportunity landscape formed by mapping gene delivery technologies to "gene therapy" opportunities is to focus on identifying and demonstrating which opportunities are supported by the efficacy/toxicity profile of a given delivery system as they are developed. This approach recognizes that the field is at the beginning of what is hoped will be a long and productive path of molecular medicine development. Direct *in vivo* transfection in its current form will not support many classic "gene therapy" applications. One example of a currently unsupported application would be therapeutic treatment of

[2] H. Lin, M. S. Parmacek, G. Morle, S. Bolling, and J. M. Leiden, *Circulation* **82**, 2217 (1990).

[3] G. Acsadi, S. S. Jiao, A. Jani, D. Duke, P. Williams, W. Chong, and J. A. Wolff, *New Biol.* **3**, 71 (1991).

[4] R. W. Malone, M. A. Hickman, K. Lehmann-Bruinsma, T. R. Sih, R. Walzem, D. M. Carlson, and J. S. Powell, *J. Biol. Chem.* **269**, 29903 (1994).

[5] M. A. Hickman, R. W. Malone, K. Lehmann-Bruinsma, T. R. Sih, D. Knoell, F. C. Szoka, R. Walzem, D. M. Carlson, and J. S. Powell, *Hum. Gene Ther.* **5**, 1477 (1994).

[6] M. L. Sikes, B. W. J. O'Malley, M. J. Finegold, and F. D. Ledley, *Hum. Gene Ther.* **5**, 837 (1994).

[7] J. P. Yang and L. Huang, *Gene Ther.* **3**, 542 (1996).

[8] H. Li, R. J. Bartlett, R. Pastori, N. S. Kenyon, C. Ricord, and R. Alejandro, *Transplant Proc.* **29**, 2220 (1997).

[9] J. Glasspool-Malone and R. W. Malone, *Hum. Gene Ther.* **10**, 1703 (1999).

[10] J. Glasspool-Malone, S. Somiari, J. J. Drabick, and R. W. Malone, *Mol. Ther.* **2**, 140 (2000).

[11] R. W. Malone, *in* "Advanced Gene Delivery: From Concepts to Pharmaceutical Products" (A. Rolland, ed.). Harwood Academic Publishers, Amsterdam, 1998.

the life-threatening aspects of severe muscular dystrophy. However, the combination of safety and modest efficacy does enable some applications such as DNA vaccination. The methods described below have been found to improve the efficacy of direct *in vivo* transfection, and hence are likely to expand the domain of potential applications for the technology. With these improvements as well as those described separately within this volume, the opportunity landscape for gene therapies based on direct *in vivo* transfection may now include the expression of angiogenic factors for the treatment of claudication, local cutaneous disorder therapies, chronic low-level expression of cytokines, coagulation or growth factor expression, and nucleic acid vaccines. If the skin transfection methods are as benign as initial studies indicate, focal cosmetic applications may even become practical.

In the short term, improved direct *in vivo* transfection may be most useful for functional genomics research. Biology is moving into an era where high-throughput gene expression profiling and sequencing capabilities are routinely available, and the genomes of many species have been defined. Unfortunately, it is often difficult to discriminate between genetic correlation and causation using these methods. *In vivo* gene transfer methods that enable high-throughput screening of transgene function, or the routine production of antibody libraries, will have significant value in a post-genomics era. Although transfected plasmid or RNA may elicit biologic responses that are unrelated to transgene expression (for example, via inflammatory cytokine induction or the titration of scarce intracellular factors), direct transfection avoids the introduction of foreign lipids, polymers, or antigens, viral proteins (both expressed and from the particle), and other undesired "baggage" that may complicate experimental design and interpretation. Based on our own experience as well as that of collaborating laboratories, we suggest that the efficiency, technical simplicity, and reduced interpretive complexity associated with enhanced direct *in vivo* transfection will enable many functional genomics research and antibody-related applications.

Use of Broad Spectrum Nuclease Inhibitor Aurintricarboxylic Acid to Enhance Direct Gene Transfer

In vivo biologic clearance of gene delivery preparations is a critical determinant of transfection or transduction efficiency.[12-14] Assuming that the biological pathway involved in transfection or transduction is not otherwise critically limiting, reduced *in vivo* clearance of a gene delivery formulation may enhance gene transfer efficiency by exposing tissue to a higher concentration of functional

[12] R. P. Rother, S. P. Squinto, J. M. Mason, and S. A. Rollins, *Hum. Gene Ther.* **6**, 429 (1995).

[13] S. Kuriyama, K. Tominaga, M. Kikukawa, T. Nakatani, H. Tsujinoue, M. Yamazaki, S. Nagao, Y. Toyokawa, A. Mitoro, and H. Fukui, *Anticancer Res.* **18**, 2345 (1998).

[14] M. Y. Levy, L. G. Barron, K. B. Meyer, and F. C. Szoka, *Gene Ther.* **3**, 201 (1996).

polynucleotide for a longer period of time. Free polynucleotides are susceptible to endo- or exonucleolytic cleavage, and there are a number of pharmaceutical agents that directly or indirectly interfere with nucleolytic clearance of polynucleotides. The most commonly used are divalent cation chelators including ethylenedi-aminetetraacetic acid (EDTA) and citrate. These agents indirectly inhibit nuclease activity by reducing the effective concentration of enzymatically required diva-lent cation cofactors. Unfortunately, when used *in vivo,* these potent chelators are typically quite toxic at the doses required to reduce extracellular nuclease activ-ity, presumably as a consequence of the reduction of free calcium required for many biologic functions including cardiac contraction. Agents that directly inhibit nucleases are less common, and include various competitive or noncompetitive inhibitors such as nucleotide analogs and aurintricarboxylic acid (ATA).[15]

Aurintricarboxylic acid (ATA) inhibits many endonucleases including DNase I, RNase A, S1 nuclease, exonuclease III, and a variety of restriction nucleases.[15] ATA is a triphenylmethane-derivative dye first synthesized in 1892, and initially prepared as a pure compound in 1949. The molecular mass of the polycarboxylated molecule is 473 Da, and it can polymerize into larger complexes of up to 6000 Da.[16] It is reported that ATA does not permeate intact cell membranes.[17] Thus, in the absence of a permeabilizing agent or process (such as electroporation) it must presumably either bind to extracellular factors such as nucleases or interact with cell membrane-associated molecules in order to influence intracellular events. Our interest in ATA began with a group of studies aimed at improving pulmonary lung transfecton.

Our laboratory has achieved some success in development and testing lipo-plexes for the *in vivo* transfection of lung tissue, providing a 60- to 70-fold enhance-ment of reporter transgene expression in murine lung relative to direct or "naked" DNA injection.[18–23] Unfortunately, when testing the resulting formulations in the lungs of rhesus macaques (via bronchoscopic administration), it was observed that

[15] R. B. Hallick, B. K. Chelm, P. W. Gray, and E. M. Orozco, Jr., *Nucleic Acids Res.* **4,** 3055 (1977).

[16] Z. Guo, M. J. Weinstein, M. D. Phillips, and M. H. Kroll, *Thromb. Res.* **71,** 77 (1993).

[17] D. Apirion and D. Dohner, *in* "Antibiotics" (J. W. Corcoran and F. E. Hahn, eds.), pp. 327–340. Springer-Verlag, New York, 1975.

[18] M. J. Bennett, R. W. Malone, and M. H. Nantz, *Tetrahedron Lett.* **36,** 2207 (1995).

[19] M. J. Bennett, M. H. Nantz, R. P. Balasubramaniam, R. P. Gruenert, and R. W. Malone, *Biosci. Rep.* **15,** 47 (1995).

[20] R. P. Balasubramaniam, M. J. Bennett, A. M. Aberle, J. G. Malone, M. H. Nantz, and R. W. Malone, *Gene Ther.* **3,** 163 (1996).

[21] M. J. Bennett, A. M. Aberle, R. P. Balasubramaniam, M. J. Bennett, M. H. Nantz, and R. W. Malone, *Liposome Res.* **6,** 545 (1996).

[22] M. J. Bennett, R. P. Balasubramaniam, A. M. Aberle, J. G. Malone, M. H. Nantz, and R. W. Malone, *J. Med. Chem.* **40,** 4069 (1997).

[23] A. M. Aberle, M. J. Bennett, R. W. Malone, and M. H. Nantz, *Biochim. Biophys. Acta.* **1299,** 281 (1996).

the murine optimized lipoplexes were less active than directly administered plasmid, and also provoked an exceptionally strong acute and chronic inflammatory infiltrate in the macaques, with considerable associated epithelial necrosis and sloughing. This result prompted a reconsideration of our approach to pulmonary transfection, and a shift in focus to improving direct DNA transfection activity. Although the viscosity of necrotic lung fluids associated with late-stage cystic fibrosis is partially due to high molecular weight DNA, it was determined that normal lung fluids from mice,[9] macaques, and humans contain significant amounts of DNAse activity (unpublished data, P. Steenland and R. Malone, 1999). Reasoning that reduced extracellular nuclease-mediated plasmid clearance would likely result in enhanced delivery, a variety of nuclease inhibitors that were shown to inhibit pulmonary nucleases *in vitro* were tested using direct pulmonary transfection. Using a fixed dose of a luciferase-coding plasmid and the murine intratracheal model, titrated doses of various nuclease inhibitors were analyzed for effect on both protection of extracellular plasmid and enhancement of reporter gene expression. While the divalent cation chelators were very toxic *in vivo,* a seemingly benign dose of ATA resulted in a 50-fold enhancement of transgene expression relative to the "free" or "naked" plasmid ($p = 0.0001$). Furthermore, in contrast to the experience with lipoplexes, no histopathology was observed in the lungs of treated mice, and the LD_{50} of ATA delivered to mouse lung was observed to be 8–10 times that of the optimal dose for increased transfection efficiency. The toxicity at higher doses appears to be caused by systemic damage but not by damage to pulmonary tissue. However, we have also observed a 50-fold enhancement and lack of toxicity in the lungs of macaques treated with a similar dose/body mass as that employed in the murine studies with no apparent toxicity (unpublished data, R. MacDonald, J. Glasspool-Malone, and R. Malone, 1999).

Since the initial serendipitous observation of direct *in vivo* transfection of naked DNA, one of the primary applications of the method has been for nucleic acid vaccination (NAV). Typically, direct transfection-mediated NAV is performed by either intradermal or intramuscular injection of an antigen-coding plasmid. More efficient methods for expressing antigens in skin, muscle, and/or the antigen presenting cells therein might enhance NAV efficacy, and so coadministration of ATA and plasmid to skin and muscle was tested. The enhancing effects of ATA have not been observed in muscle, either in our hands or others.[14] As globular actin is well known to inhibit DNase activity, the lack of ATA enhancement and relatively high levels of transfection activity associated with direct intramuscular plasmid injection may reflect the extracellular and/or intracellular effects of muscle-derived globular actin. In contrast, intradermal injection of rats, mice, pigs, and macaques with plasmid and ATA resulted in an approximately 10-fold enhancement of transgene expression (varying from $p < 0.02$ to $p < 0.04$).[10] This difference in intradermal transgene expression is also associated with an increase and modulation in vaccine response (unpublished data, J. Glasspool-Malone and R. Malone, 2000). We have not observed either systemic or local toxicity attributable to ATA with intradermal injection (unpublished data, J. Glasspool-Malone and R. Malone, 2000).

Use of Pulsed Electrical Fields to Enhance Direct *in Vivo* Transfection and Nucleic Acid Vaccination

Although nuclease inhibitors such as ATA may be used to significantly enhance direct *in vivo* transfection, presumably by reducing polynucleotide clearance, they are unlikely to augment the basic processes by which polynucleotides are taken up into cells after direct injection. The basic mechanism by which this occurs remains obscure, and hence it is difficult to develop and test methods that might enhance the process. Initial efforts to improve the method were biased by the belief that the phenomenon was restricted to muscle, but the observation of direct liver transfection, particularly with large volume injection, resulted in the hypothesis that nonspecific pressure or mechanical distortion effects might be involved.[4,5] This hypothesis was subsequently supported and extended by the more recent findings of Liu *et al.*[24] In contrast to the mechanistic ambiguity of "naked DNA" transfection, the pioneering work of Heller *et al.*[25] involving *in vivo* liver electroporation after direct plasmid injection has provided both a method and access to well-developed experimental theory that may be used to markedly enhance direct *in vivo* transfection (see Ref. 26 for a review of the theory and application of electroporation for *in vivo* gene delivery).

A variety of tissues may be efficiently transfected using direct plasmid injection and electroporation. In practice, the method is usually limited to those tissues that may be readily accessed with a needle and an electrode. Although endoscopic and catheter-guided electrodes have been developed, in most cases the electrodes are of either the flat plate (caliper) or penetrating (pin or pin array) type. Typically, a small needle (28 or 30 gauge) is used to inject the polynucleotide (dissolved in isotonic saline or water for injection) into tissue, which is then subjected to a train of electrical pulses administered by the electrodes. Under the influence of the applied electrical field, an increased charge differential develops across the cytoplasmic membrane. At a transmembrane potential that exceeds the dielectric strength of the cell membrane (typically about 0.5 V), one or more reversible pores may develop. Polynucleotides and other charged molecules may pass through these pores, particularly when driven by the electrophoretic forces associated with subsequent pulses. As the electrodes are in contact with a conducting electrolyte (the extracellular fluid), considerable power is often required to maintain the electrical field

[24] F. Liu, Y. Song, and D. Liu, *Gene Ther.* **6,** 1258 (1999).

[25] R. Heller, M. Jaroszeski, A. Atkin, D. Moradpour, R. Gilbert, J. Wands, and C. Nicolau, *FEBS Lett.* **389,** 225 (1996).

[26] S. Somiari, J. Glasspool-Malone, J. J. Drabick, R. A. Gilbert, R. Heller, M. J. Jaroszeski, and R. W. Malone, *Mol. Ther.* **2,** 178 (2000).

gradient within the tissue. Therefore, pulse amplitude, pulse duration, waveform, and total number of administered electrical pulses must be carefully optimized to avoid tissue damage (burning). The electrical field strength required for poration is inversely proportional to cell size, and therefore skeletal muscle cells are readily transfected by applying modest field strength pulses (100 to 200 V/cm). Many laboratories (including ours) now use this method for transfecting skeletal muscle, achieving in the range of a 1000-fold enhancement of transgene expression relative to direct DNA injection without applied pulses (see Fig. 1 for data from our laboratory). However, muscle is readily damaged by higher field strengths, and with 300 to 400 V/cm pulses many investigators including ourselves have observed substantial muscle tissue necrosis as well as significant decreases in transgene expression.

Administering pulsed electrical fields to directly transfected muscle not only enhances transgene expression, but also substantially improves vaccine responses to plasmid-encoded antigens (Fig. 2). The mechanism by which direct intramuscular transfection elicits immune responses was initially proposed to involve T-cell

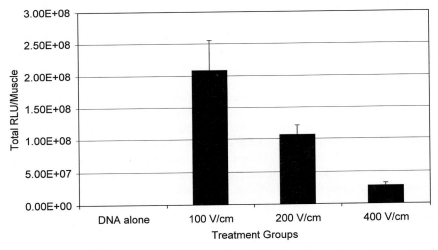

FIG. 1. Luciferase transgene expression obtained via direct *in vivo* muscle transfection with or without electrical field treatment. Groups of 10 anterior tibialis muscles were treated with 15 μg of plasmid expressing the luciferase reporter gene under the control of the CMV promoter dissolved in 30 μl of water for injection using the described open injection methods. In those groups that were treated with the indicated field strength (e.g., 100 V/cm), a caliper electrode was placed over the muscle immediately after DNA injection and the field pulses were applied. In addition to the indicated field strength, electrical pulse parameters consisted of a train of eight 20 ms pulses administered at 1 Hz frequency. To calculate the indicated field strength, the muscle diameter was measured using the caliper electrode at the time of electrode application (typically 3 to 4 mm), and the applied voltage was modulated accordingly. Results demonstrate that, for open treatment, reporter protein levels were optimized at a 100 V/cm field strength. Data are displayed as mean + SEM total relative luciferase light units per muscle.

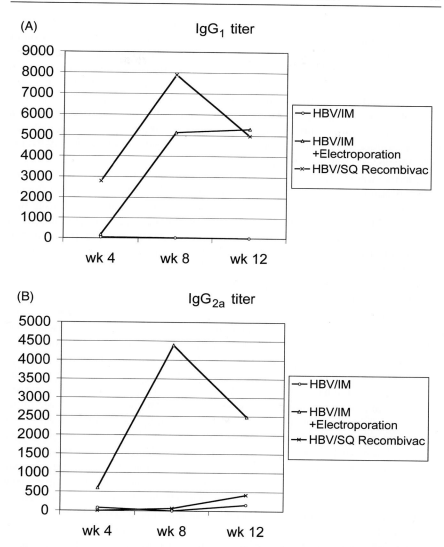

FIG. 2. Humoral immune responses to murine intramuscular DNA vaccination. Two groups of seven mice received 15 μg HBV DNA diluted in WFI via IM injection, and were boosted at 4 weeks in the same fashion. One group received electroporation after inoculation and one did not. Electroporation parameters were eight (20 ms) pulses at a 1 Hz frequency (100 V/cm muscle tissue) using the muscle electroporation methods described. Electroporated mice were primed in one TA, and boosted in the other. The positive control group received 0.05 μg Recombivac in 50 μl WFI SQ (human recombinant hepatitis B protein vaccine) with no electroporation. Prevaccination serum samples were negative for HBV responses. Points represent mean IgG$_1$ (A) and IgG$_{2A}$ (B) titers for different groups. Serum for each animal was run separately by ELISA (not pooled) and mean titers calculated.

priming by transfected muscle cells via class 1 MHC presentation (D. Carson, personal communication, 1989) but more recent data suggests direct transfection of antigen presenting cells residing within or around the muscle tissue, and/or cross priming of antigen presenting cells by transfected muscle (see ref. 27 for review of this issue). Eighteen hours after direct muscle transfection with EGFP involving electrical fields, we routinely observe transfected monocytic cells with elaborate dendritic processes along the fascia of treated muscle. This suggests that efficient transfection of antigen-presenting cells may explain the markedly improved vaccine responses. We do not observe these transfected cells when DNA is delivered without electroporation.

Dermal transfection and nucleic acid vaccination is also enhanced by pulsing injected skin with electrical fields.[10,28] The plasmid is diluted in isotonic saline or water for injection, intradermally injected with a 30-gauge needle resulting in a cutaneous bleb, the injected skin is either pinched between caliper electrodes or a penetrating electrode array is placed so that it penetrates through the epidermis, and a series of electrical pulses are administered. Although some investigators prefer long-duration, lower voltage pulses such as are typically employed for muscle electroporation, we have optimized pulsing conditions around the short, high-voltage pulsing parameters employed in clinical electrochemotherapy trials.[29] These parameters result in a 20- to 40-fold increase in transgene expression relative to injection without pulsing, but what may be more important is the distribution and types of cells transfected by this process. Direct intradermal plasmid injection (CMV promoter) typically results in the productive transfection of basal, suprabasal, and follicular epidermal cells. In contrast, when electrical pulses are administered after identical injection treatments, the majority of transgene expression is observed in dermis and subdermal adipose tissue. Of particular interest is the large number of transfected monocytic cells with dendritic morphology that are observed within the electroporated dermis. Consistent with the hypothesis that enhanced transfection of such skin monocytes will result in enhanced responses to nucleic acid vaccination, an increased immune response is obtained relative to intradermal DNA vaccination.[28] Furthermore, coinjection of ATA and DNA with subsequent skin electroporation increases skin transgene expression by over 115-fold relative to direct plasmid injection.[10] We hypothesize that this synergism between ATA and skin electroporation may reflect the inhibition of both extracellular and intracellular degradation of the administered polynucleotide, as with electroporation the charged ATA molecule is likely to gain access to the cytoplasm via the same pores and electrophoretic effects that enable enhanced polynucleotide uptake.

[27] K. A. Berlyn, S. Ponniah, S. A. Stass, J. G. Malone, G. Hamlin-Green, J. K. Lim, M. Fox, G. Tricot, R. B. Alexander, D. L. Mann, and R. W. Malone, *J. Biotechnol.* **73,** 155 (1999).

[28] J. J. Drabick, J. Glasspool-Malone, S. Somiari, and R. W. Malone, *Mol. Ther.* **3,** 249 (2001).

[29] R. Heller, M. J. Jaroszeski, D. S. Reintgen, C. A. Puleo, R. C. DeConti, R. A. Gilbert, and L. F. Glass, *Cancer* **83,** 148 (1998).

Use of Dielectric Insulating Material to Reduce Tissue Current during Electrical Field Pulsing

Both theory and experimental analysis indicate that the critical parameters governing electroporation include the electrical field strength across a given cell, cell size, and the length of time the field is administered. Field strength may be defined as the negative electric potential gradient. In the case of traditional electrodes, the local field experienced by any one cell during poration is therefore a function of applied voltage, electrode architecture, and the conductivity or resistance of regional tissue elements and extracellular fluids. To generate the electrical field, most electroporation devices employ electrodes that are in electrical contact with the electrolyte surrounding the cells. Because of this design, current flows between the electrodes while the field is being applied. Consequently, significant power is required to maintain the field strength, and the flow of current through the extracellular electrolyte (in this aspect acting as a resistor) produces heat and associated cell damage, which limit the efficacy of electroporation. However, there is no generally accepted evidence that externally applied current provides any beneficial effects. Therefore, it may be most desirable to minimize current flow through the extracellular matrix while applying the electrical field required for cellular membrane dielectric breakdown.

A method for reducing current flux through tissue while enabling the application of an electrical field has been developed and tested. This method involves the coating of one electrode with a dielectric insulating material of high dielectric strength and coefficient. Imposing a dielectric insulating material between the tissue or cell suspension and one electrode to reduce current flow will result in a drop in the field strength experienced by the tissue due to the charge shielding that occurs within the insulator. The extent of this drop, and hence the extent to which the applied voltage must be increased, is a function of the interactions between the electric field and the dielectric insulator. The interactions between applied electric fields and dielectric materials may be characterized by the field susceptibility and dielectric coefficient of the dielectric material. These constants may be used to estimate the field strength within the dielectric material based on the applied external field. To estimate the applied field strengths required for tissue electropermeabilization when a dielectric material is placed in series with tissue, we have modeled the system as two dielectric materials placed in series between flat plate electrodes.

The most significant problem in developing a theoretical base for predicting the effects of dielectric materials placed between an electrode and the tissue to be electropermeabilized is the complex electrical nature of the tissue. Tissues consist of a mixture of solids, water, electrolytes, and cells surrounded by semipermeable membranes. This complex mixture may be modeled as resembling an electrolytic cell (battery), as a mixture of resistors, capacitors, dielectric materials and conductors, or in the case below as a dielectric material similar to water (the primary component). Clearly, tissue is not a simple, homogeneous dielectric material. Unfortunately, precise theoretical models and data are not available for the development of more complex models. Therefore, in order to estimate the voltage range to be tested in developing the novel current-minimizing electrode design discussed above, we have applied equations that predict the electrical fields within dielectric materials subjected to an external electrical field. Based on these assumptions, we calculate that such fields may need to be approximately 10- to 40-fold stronger than those administered in the absence of an additional dielectric insulating material between the cathode and the tissue. Currently available commercial electroporation pulse generators do not provide the required voltage, and therefore the experiments performed to date, while demonstrating the utility of the concept, have only tested field strengths that are known to be suboptimal based on standard electrode designs.

Figure 3a provides a diagram of the field interactions between a single dielectric material placed between a pair of flat plate electrodes. In this diagram, the dielectric is assumed to extend from one plate to another, but to improve clarity the diagram has been drawn with a gap between the dielectric material and the electrodes, and the edge effects of the field have been disregarded. As summarized in the diagram, the surface density on the electrodes (σ) will induce a surface charge on

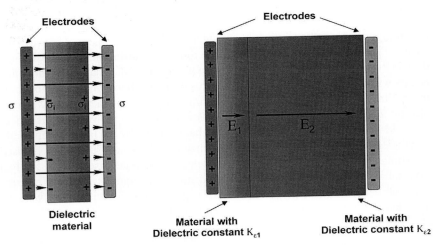

3a. Induced charges on the faces of a dielectric in an electrical field.

3b. Two dielectrics of different coefficients between oppositely charged plates

FIG. 3. Diagram illustrating charge-shielding effects associated with dielectric insulation materials. See text for explanation. [From F. W. Sears, "Electricity and Magnetism." Addison-Wesley, Cambridge, MA, 1946.]

the dielectric (σ_i). The induced surface charge on the dielectric will neutralize a portion of the plate charge, reducing the effective surface charge density by the amount ($\sigma - \sigma_i$). As a result, the electric field intensity (E) within the dielectric may be described by Eq. (1):

$$E = 1/\varepsilon_0(\sigma - \sigma_i) = 1/\varepsilon_0(\sigma) - 1/\varepsilon_0(\sigma_i) \tag{1}$$

In this equation, the term $1/\varepsilon_0(\sigma)$ represents the component of the field internal to the dielectric material set up by the external field, and the term $1/\varepsilon_0(\sigma_i)$ represents the opposing field created by the induced charges. This opposing field is a function of both the applied external electrical field and the intrinsic electrical properties of the dielectric material. This relationship is termed the electric susceptibility of the material (η) and is defined as the ratio of the induced charge density (σ_i) to the resultant electric intensity (E), resulting in Eq. (2):

$$\eta = (\sigma_i)/E, \quad \text{or} \quad \sigma_i = \eta(E) \tag{2}$$

For most materials at moderate fields and temperature ranges, it has been found that the susceptibility of a dielectric is a constant, independent of the applied field, and so may be considered a primary descriptive characteristic of any given dielectric material. Therefore, Eq. (1) may be redefined in terms of the susceptibility (η) to yield Eq. (3):

$$E = 1/\varepsilon_0(\sigma) - 1/\varepsilon_0(\eta)E \quad \text{or} \quad E = \sigma/[(1 + \eta/\varepsilon_0)\varepsilon_0] \tag{3}$$

Based on this equation, a second dielectric material property known as the dielectric coefficient (K_ε) (e.g., dielectric constant or specific inductive capacity) may be defined as indicated in Eq. (4):

$$K_\varepsilon = (1 + \eta/\varepsilon_0) \tag{4}$$

Both (η) and (ε_0) are expressed in similar units, and therefore the dielectric coefficient is a pure number.

To a first approximation, one may model the effects of placing a dielectric material in series with tissue as being analogous to placing two dielectrics of different coefficients in series between flat plate electrodes (Fig. 3b). If the dielectric coefficients of the dielectric materials are known, the electrical field intensities within each of the two dielectric materials may be determined, as the electric intensities are inversely proportional to the corresponding dielectric coefficients. This relationship is expressed in Eq. (5):

$$(E_1/E_2) = (K_{\varepsilon 2}/K_{\varepsilon 1}) \tag{5}$$

Mylar and Kapton films (0.1 mm thickness) are commonly used materials with high dielectric strength, and are readily available for application to the surface of conductive electrodes. Therefore, these materials may be placed over one electrode to impede electron flux through the treated tissue, and modeled as one of the two dielectric materials in Fig. 3b. The dielectric coefficient of these materials ranges from 3.2 to 3.3.

To approximate the range of external electrical fields required to evoke an appropriate poration field within tissue isolated from one electrode by a dielectric material, the tissue may be modeled as the second dielectric material in Fig. 3b. With this assumption, if the dielectric coefficient of the tissue is known then the electrical field within the tissue may be calculated from Eq. (5). Our experience based on oscilloscope monitoring of tissue electropermeabilization under a wide range of conditions teaches that the electrical properties of tissues vary significantly. Consistent with this observation, studies of the penetration of radio-frequency and microwave fields into tissue have documented that tissue does not behave strictly as a classic dielectric with a fixed dielectric coefficient. Instead, it has been determined that the measured effective dielectric coefficient varies considerably with wavelength, generally decreasing with increasing frequency. Tissue dielectric coefficients for the pulsing parameters typically employed for electropermeabilization are completely unknown.

Given this ambiguity, for the purposes of developing an initial estimate to guide experimentation, we have chosen to initially model tissue by substituting the dielectric coefficient of pure water, the major component of tissue. Clearly, this assumption is a very crude approximation, but it is suggested that it will serve for initial analysis and demonstration purposes. The dielectric coefficient of water has been characterized as $K_{\varepsilon water} = 81$, and given the high water content of most tissues this may serve as an initial estimate of $K_{\varepsilon tissue}$. From this information, we may easily calculate an estimate of the difference between applied electrical field and the field intensity within the tissue under such conditions. The ratio of the fields within the two dielectric materials simplifies to Eq. (6):

$$(E_{water}/E_{kapton}) = (K_{\varepsilon kapton}/K_{\varepsilon water}) = 3/81 = 0.037 \tag{6}$$

To solve the remaining two unknowns requires a second equation, which may be derived by recognizing that the sum of the two internal fields equals the external applied field, yielding Eq. (7):

$$E_{kapton} = (E_{total} - E_{water}) \tag{7}$$

In turn, substitution of Eq. (7) into Eq. (6) yields the reduced Eq. (8):

$$E_{water} = [(E_{total})(0.037)]/1.037 \quad \text{or} \quad E_{water} = 1/27 E_{total} \tag{8}$$

From this expression, it is easy to calculate the approximate applied field strength required. For instance, for a field of 400 V per cm (optimal for skeletal muscle) one would require an applied pulse of 10,800 V per cm tissue. Likewise, to administer a pulse of 1750 V per cm to tissue, an applied external field of 47,250 V per cm would be required.

Other, more exotic dielectric insulating materials with much higher dielectric coefficients have been developed and are typically used for high-voltage capacitor design where light weight and small size are required (such as cardiac defibrillators). Examples of such films include Kynar (dielectric coefficient = 6.4, electrical strength 7000 V/mil) and polyvinylidene fluoride (pVdF, dielectric coefficient = 12, electrical strength 7000 V/mil). pVdF is commercially available with a metallized backing (CSI Inc., San Marcos, CA) and we have used this material to test the hypothesis that application of electrical fields while minimizing externally applied current will augment transfection. The significance of such materials may be appreciated by comparing the required applied voltage to that obtained in the calculation listed above. Based on the calculation below, to administer a pulse of 1,750 V per cm to tissue across a pVdF dielectric insulator, an applied external field of 13,650 V per cm would be required, providing an obvious advantage relative to kapton.

$$(E_{water}/E_{PVDF}) = (K_{\varepsilon PVDF}/K_{\varepsilon water}) = 12/81 = 0.148$$

$$E_{water} = [(E_{total})(0.148)]/1.148 \quad \text{or} \quad E_{water} = 1/7.8 E_{total}$$

We have tested the utility of a pVdF-coated caliper electrode for muscle electroporation. In these studies, optimizing pulsing parameters using the maximum voltage supported by the pulse generator provided a 40-fold enhancement of transgene expression relative to DNA injection alone (Fig. 4). Although a far cry from the thousandfold enhancement typically observed with optimized muscle pulsing parameters and standardized electrodes, the results are consistent with the modest calculated applied field strengths. We suggest that these data indicate that the development of specialized pulse generators and further optimization of electrode design, film thickness, and pulsing parameters are likely to provide the benefits of pulsed electrical field-enhanced direct *in vivo* transfection without the problems of substantial current flux through the treated tissue.

Materials and Methods

Animals

Five- to 7-week-old BALB/c mice were purchased from Charles River, Wilmington, MA, and were anesthetized by an intraperitoneal injection of ketamine hydrochloride, xylazine hydrochloride, and acepromazine maleate before treatment. Four-week-old Yorkshire pigs were obtained from Tom Morris Inc., Reistertown, MD. Before treatment, pigs were anesthetized intramuscularly with ketamine and acepromazine. To harvest tissues for gene expression, mice were sacrificed by CO_2 inhalation and pigs anesthetized as detailed above and given a lethal pentobarbital IV injection. All animal work was approved by an institutional animal use review board.

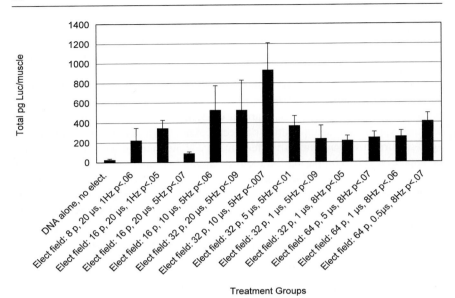

FIG. 4. Enhanced direct *in vivo* transfection using a pVdF-coated electrode and pulsed electrical fields. Groups of 10 anterior tibialis muscles were treated with 15 μg of plasmid expressing the luciferase reporter gene under the control of the CMV promoter dissolved in 30 μl of water for injection using the described open injection methods. In those groups that were treated with field pulses, a caliper electrode with a pVdF-coated cathode was placed over the muscle immediately after DNA injection and the field pulses were applied. The pulsing conditions are listed with each group, with 1100 V applied across the electrodes using the indicated number of pulses, pulse duration, and frequency. Results of *t* test analyses comparing groups to DNA injection without pulsing are listed. Data are displayed as mean + SEM total relative luciferase light units per muscle.

Plasmid DNA

For gene transfer and reporter protein expression analysis, CMV IE driven plasmids encoding an enhanced *Protinus pyralis* luciferase, a red-shifted variant of the *Aequorea victoria* green fluorescent protein (pEGFP-C1, Clontech), and a modified β-galactosidase that incorporates a nuclear localization signal were used (pNGVL1-ntβ-Gal-National Core Vector Library). The plasmids were transformed into JM101, amplified, and purified (MonsterPrep, Merlin Core Services, Bio 101, Vista, CA). Plasmid encoding the hepatitis B surface antigen, pRc/CMV-HBs(S), was purified by Aldevron, Fargo, ND, and was a gift from Dr. R. Whalen.

Luciferase Analysis

Relative luciferase activity was determined via the Enhanced Luciferase Assay Kit (#556-866, Pharmingen, San Diego, CA) and Monolight luminometer (#2010,

Analytical Luminescence Laboratories, San Diego, CA) as per the manufacturer's recommendations. Mice were sacrificed 2 days after plasmid treatment via CO_2 inhalation; the treated anterior tibialis or lung tissue tissue was dissected free and homogenized in 1 ml of 1× lysis buffer. After homogenization, samples were incubated on wet ice for 30 min to enhance the recovery of luciferase activity. Luciferase light emissions from 40 μl of the lysate were integrated over a 10-sec period, and results were expressed as a function of the total lysate volume.

Nuclease Inhibitor Preparation

The triammonium salt of aurintricarboxylic acid (#A-0885, Sigma, St. Louis, MO) was prepared as a 1 mg or 5 mg/ml solution in water for injection and filtered through a 0.2 μm cellulose acetate filter. It is important to order the triammonium salt, as other derivatives are not soluble in water and are more toxic. Solution can be stored at room temperature with no known loss of nuclease inhibition, but for constancy and sterility, a new solution is typically made prior to each experiment.

Intratracheal Plasmid Instillation

To ensure reproducible application of gene transfer reagents to lung paren-chyma, we rely on direct intratracheal instillation of polynucleotide/nuclease in-hibitor mixtures. Typically, 100 μg of the plasmid pND2Lux is diluted into sterile water for injection (#NDC 0517-3020-25, Luitpold Pharmaceuticals, Shirley, NY) with or without ATA.

To instill the reagents into mouse lung, 1-ml tuberculin syringes are fitted with a 24 gauge needle, and approximately 500 μl of room air is drawn (air bolus) followed by 100 μl of the polynucleotide formulations (fluid bolus). Chasing the fluid bolus down the trachea with the air bolus helps to ensure that the anesthetized animal can clear the fluid from the main branches of the conducting airway. Opti-mal mixtures for pulmonary transfection consisted of 100 μg plasmid in 100 μl of water for injection into which is dissolved 0.5 to 4 μg ATA/gram animal weight. For the treatment, the 24-gauge needle is replaced with a 1/2-inch 30-gauge nee-dle. Mice are anesthetized, and the procedure is initiated by making a 1-cm medial incision through the skin at the ventral site of the neck using forceps retraction and Isis scissors. The neck is then held in an extended position, and by dissecting overlying musculature and salivary tissues with blunt technique, the trachea is ex-posed. With the trachea visualized, the needle is inserted between the third to fifth intratracheal ring space (inferior to the larynx) directed toward the lung. The fluid and then air bolus are rapidly instilled so that the air bolus clears fluid remaining within the large conducting airway. We have found that this method is more repro-ducible and is associated with reduced postoperative mortality relative to intranasal or intratracheal instillation sans air bolus. After drug instillation, the salivary gland is placed over the tracheal defect, and the incision is closed with staples

(#7630, 9 mm MikRon Autoclip applier, #427631, 9 mm wound clips, Becton Dickinson and Co., Clay Adams division, Parsippany, NJ).

Muscle Electroporation

Mice are anesthetized prior to the procedure as described above using a combination of ketamine and acepromazine. The posterior limbs and adjacent abdomen are decontaminated using 70% ethanol. A small surface cut (<1 cm) is made midline between the knee joint and the ankle. The tibialis anterior (TA) muscle is visualized and injected with 50 μl of DNA solution resulting in a 15 μg plasmid dose per muscle. To inject plasmid DNA, a 30 gauge, 1/2-inch long needle attached to a 0.1 ml Hamilton syringe is used. Plasmid is injected into the muscle with the needle almost perpendicular to the tibia. Caliper-type electrodes (BTX/Genetronics, San Diego, CA) are applied parallel to the tibia in a fashion so as to flank the injected tibialis anterior. A train of eight 20-ms pulses of the indicated field strength are then applied at a 1 Hz frequency (typically 100 V/cm muscle tissue). The fascia and skin over the TA is then closed with a single wound clip staple.

Intradermal DNA Injection and Electroporation

The required concentration of each plasmid DNA (pND2Lux; pEGFP; lacZ, HBs) are intradermally injected into rodent or swine. The sites on mice are typically limited to the thicker skin just above the base of the tail. The site was shaved prior to injection and electroporation. For intradermal injection, the sides and flanks on the swine are used. Swine are an ideal animal model for intradermal gene transfer experiments as a total of 48 sites per pig may be attained. The skin on young pigs (weanlings) is very similar to that of humans, and it is easy to miocroscopically distinguish cell types and dermal layers within the tissue. However, upon aging, the skin becomes too thick and is difficult to intradermally inject and work with. Total fluid volume per injection is typically 100 μl in injection grade water. A pin electrode consisting of 2 rows of seven 7-mm pins (1 mm × 5.4 mm gaps) is typically used to transfer the electric field to the injection site, although we have used caliper electrodes with EKG electroconductive jelly to transfect pinched skin. The pin electrode typically penetrates about 2.5 mm or less into the animal skin. Electromediated experiments are typically performed using PulseAgile electroporation equipment and software (CytoPulse Sciences, Hanover, MD) or a Genetronics electroporation device (BTX Genetronics, San Diego, CA).

Quantitation of Antibody Response and Immunoglobulin Subclass Analysis

We typically employ a readily available plasmid expression vector that encodes the hepatitis B virus surface antigen as a "test antigen" much as reporter genes are typically used. The vector (pRc/CMV-HBs(S)) was developed by Heather Davis and Robert Whalen, has been used in many published DNA vaccine studies, and is available from Aldevron (Fargo, ND). In this way, immune responses may be compared to those reported with other methods, although different ELISA assays are certainly not as consistent as luciferase assays have become. We typically immunize groups of mice with ID or IM injection of either pRc/CMV-HBs(S) as transgenic antigen, or pND2-lux as a negative control followed by EP with pin electrode, caliper electrode, or sham treatment as appropriate. As a positive control, we typically inoculate mice with human recombinant hepatitis B vaccine (Recombivax-HB) with a dosing regimen of 0.05 μg in 50 μl per mouse. Antibody levels to hepatitis B surface antigen are determined by ELISA as previously described[30] using hepatitis B surface antigen purchased from Biodesign International. IgG subclass analysis is performed using HRP conjugated goat anti-mouse IgG, IgG_1, IgG_{2a}, and IgG_{2b} (Southern Biotechnology Associates, Inc.).

Testing Dielectric Insulating Material-Coated Electrode

The tibialis anterior muscles of 20-g female Balbc mice were surgically exposed and injected with 30-μl samples containing 15 μg of a CMV-driven luciferase expression plasmid using a Hamilton syringe fitted with a 30 gauge needle. Groups of four such mice (eight muscles) were treated immediately after injection with the indicated applied voltage and pulsing parameters (see Fig. 4) using flat plate caliper electrodes. In the case of animals treated with the pVdF-coated electrode, the cathode of the flat plate caliper device was coated with electroconductive paste and a layer of 4.5 μm 100 ohm/square pVdF film coated with a thin layer of stabilizing material of unknown dielectric coefficient, which was placed across the electrode prior to treatment. The pVdF film was obtained from CSI Technologies and consisted of a prototype material obtained from Trephane Inc. (polyvinylidene fluoride, Trephane, Bloomfield, NY) typically used for capacitor development. Twenty-four hours after treatment, tissue was harvested, homogenized, and analyzed for luciferase expression. Muscle was chosen for this test as it is known to have an unusually low poration threshold due to large cell size. The escalation of applied field strength was limited by the pulse generator design.

Perspective and Conclusions

Simple, nontoxic, and pharmaceutically defined methods for genetic modification of tissues are required for the development of a variety of molecular medicines. Similarly, the emerging scientific discipline known as functional genomics requires improved, high-throughput methods for genetically modifying

[30] J. G. Malone, P. J. Bergland, P. Liljestrom, G. H. Rhodes, and R. W. Malone, *Behring Inst. Mitt.* **98,** 63 (1997).

tissue so that the phenotypes associated with both wild-type protein and aberrant protein expression may be determined. Finally, both functional genomics and proteomics research will benefit from methods that enable production of polyclonal antisera and monoclonal antibodies directed against the products of open reading frames.

Direct *in vivo* transfection is the simplest method for delivering and expressing transgenes in tissues, but is relatively inefficient. Plasmid-based expression vectors are generally safe and flexible, and are widely accepted as the standard for both cell culture experimentation and transgenic animal development. Direct *in vivo* transfection methods extend the use of robust and well-developed plasmid expression technology to animal models, and in some cases to clinical treatment (DNA vaccines). Given the obvious advantages associated with simplicity, it is desirable to design and discover methods that might improve the efficiency of direct *in vivo* transfection while retaining the simplicity. Unfortunately, a decade of investigation has yet to define the biological mechanism that supports direct transfection. It is difficult to design a rational strategy that will enhance the activity of an undefined mechanism. However, basic pharmaceutical principles may be applied to augment the process. One such principle is that the dose-response of a drug may be improved by reducing clearance. As the predominant mechanism of *in vivo* polynucleotide clearance involves intra- and extracellular nuclease activity, inhibition of nuclease activity is one simple method by which direct transfection might be enhanced. As discussed above, the promiscuous low molecular weight nuclease inhibitor aurintricarboxylic acid provides a simple and effective method to validate this principle. Based on the transfection enhancement observed with ATA, the development of other, more potent and specific nuclease inhibitors is expected to lead to further improvements in direct transfection efficacy. Furthermore, more specific nuclease inhibitors would allow experimental analysis of the role of different nucleases during the transfection process. Additional mechanistic information is likely to result in further improvements in direct *in vivo* transfection efficacy.

The *in vivo* application of pulsed electrical fields after direct polynucleotide injection improves transfection activity up to 1000-fold and provides a controlled and theoretically defined method for modulating the process. The leading mechanistic hypothesis for direct *in vivo* transfection involves the action of mechanical forces on cell membrane integrity. Unfortunately, in practice this results in significant interoperator and intraexperimental variability and/or the need to manipulate tissues by injecting large fluid volumes into closed tissue compartments. Application of pulsed electrical fields also results in the creation of transient membrane defects through which polynucleotides may pass, but the process relies on defined and reproducible electrical parameters. Electroporation theory posits that the key parameter for cellular poration is the applied electrical field. "State of the art" *in vivo* electroporation relies on electrodes that are in electrical continuity with the tissue, and therefore current must be continually supplied during pulsing to maintain the field strength across the cells. This current flux through tissue resistors results in heat, and it is therefore no surprise that tissue electroporation efficacy is limited by the associated toxicity. Hypothesizing that application of electrical fields without significant current flux through the tissue will enhance direct *in vivo* transfection, we have designed a fundamentally different electrode. This new design places a sheet of dielectric insulating material between one electrode and the tissue, and relies on new materials developed for high-voltage capacitor applications. Initial testing of this new design demonstrates marked enhancement of direct *in vivo* transfection, supporting the hypothesis that the electroporative process involves field- rather than current-mediated effects. Further development will require the assembly of novel pulse generators designed to generate the required high-voltage, low-power pulses, as well as testing a variety of dielectric insulating materials of differing thickness using *in vivo* and *in vitro* models.

Acknowledgments

We thank the many laboratory workers and collaborators who have assisted us with these studies over the past 15 years, many of whom are listed as co-authors on publications cited in this chapter. Bronchoscopic macaque studies were performed in collaboration with Dr. Ruth MacDonald. Development of the theory and design of dielectric insulating material-coated electroporation electrodes greatly benefited from the advice and assistance of E. W. Malone. This work has been supported by the University of Maryland Medical System, NIH Grants NIH K02AI01370 and NIH-R01RR12307, USAMRAA award DAMD17-94-J-4436, UC TRDRP Grant KT-0205, Cystic Fibrosis Foundation award S884, a grant from the Arthur Ashe Foundation for the Defeat of AIDS, a Bank of America Giannini fellowship award, and sponsorship from Promega Corporation, Boehringer Mannheim GmbH, and the Clinical Breast Care Project of the Uniformed Services University. In some cases, our electroporation studies have involved devices provided by CytoPulse, Inc., of Hanover, Maryland. "CytoPulse and University of Maryland, Baltimore have had a collaboration agreement under which CytoPulse's equipment has been used by Robert Malone, M.D., in scientific research and development, principally in the areas of delivering DNA using pulsed electrical fields."

[5] Hydrodynamics-Based Transfection: Simple and Efficient Method for Introducing and Expressing Transgenes in Animals by Intravenous Injection of DNA

By YOUNG K. SONG, FENG LIU, GUISHENG ZHANG, and DEXI LIU

Introduction

The ability to introduce genetic materials (usually DNA) into cells has played a critical role in enhancing our understanding of biological systems and enabling the establishment of the modern biotechnology industry. Initial studies in this field focused on expressing recombinant proteins in prokaryotic and eukaryotic cells and were designed to study protein function with respect to mechanism of action, involvement in metabolism, regulation of gene expression, and cell regeneration and differentiation. This type of work, although mostly *in vitro,* has led to the discovery of many genes of medical significance, and has provided the basis for considering the use of genetic materials for treatment of human diseases—commonly called gene therapy. Although some of the techniques initially developed for *in vitro* gene transfer are applicable to gene therapy, new techniques have been actively sought to satisfy the needs of the new field because gene therapy is usually performed *in vivo* and whole animals, instead of established cell lines, are the target.

The currently used methods for *in vivo* gene transfer can be generally classified as direct and indirect methods. Direct methods involve the introduction of genetic materials into cells using physical principles. These include direct DNA injection,[1-9] gene gun,[10-17] and electroporation[18-26] techniques. Indirect methods

[1] J. A. Wolff, R. W. Malone, P. Williams, W. Chong, G. Acsadi, A. Jani, and P. L. Felgner, *Science* **23,** 1465 (1990).

[2] G. Acsadi, S. S. Jiao, A. Jani, D. Duke, P. Williams, W. Chong, and J. A. Wolff, *New Biol.* **3,** 71 (1991).

[3] A. Turkay, T. Saunders, and K. Kurachi, *Gene Ther.* **6,** 1685 (1999).

[4] M. L. Sikes, B. W. O'Malley, M. J. Finegold, and F. D. Ledley, *Hum. Gene Ther.* **5,** 837 (1994).

[5] M. A. Hickman, R. W. Malone, K. Lehmann-Bruinsma, T. R. Sih, D. Knoell, F. C. Szoka, R. Walzem, D. M. Carlson, and J. S. Powell, *Hum. Gene Ther.* **5,** 1477 (1994).

[6] G. Zhang, D. Vargo, V. Budker, N. Armstrong, S. Knechtle, and J. A. Wolff, *Hum. Gene Ther.* **8,** 1763 (1997).

[7] V. Budker, G. Zhang, S. Knechtle, and J. A. Wolff, *Gene Ther.* **3,** 593 (1996).

[8] K. B. Meyer, M. M. Thompson, M. Y. Levy, L. G. Barron, and F. C. Szoka, *Gene Ther.* **2,** 450 (1995).

[9] K. A. Choate and P. A. Khavari, *Hum. Gene Ther.* **8,** 1659 (1997).

[10] W. H. Sun, J. K. Burkholder, J. Sun, J. Culp, J. Turner, X. G. Lu, T. D. Pugh, W. B. Ershler, and N. S. Yang, *Proc. Natl. Acad. Sci. U.S.A.* **92,** 2889 (1995).

[11] L. Cheng, P. R. Ziegelhoffer, and N. S. Yang, *Proc. Natl. Acad. Sci. U.S.A.* **90,** 4455 (1993).

involve the active or passive uptake of genetic materials by cells. The most commonly used methods for indirect gene transfer include the use of viral or nonviral vectors, through which the genetic materials are taken up by active cellular processes. The use of viral and nonviral vectors has become quite popular in the past 10 years and has been dealt with extensively in many recent articles,[27–34] including those in this volume. The discussion of procedures in this article is restricted to a new method belonging to the direct methods category. This method, called the "hydrodynamics-based procedure" to reflect the involvement of fluid dynamics during transfection, uses hydrostatic pressure induced by a rapid intravenous injection of a large volume of DNA solution into an animal to accomplish gene

[12] B. I. Loehr, P. Willson, L. A. Babiuk, and van Drunen Littel-van den Hurk, *J. Virol.* **74,** 6077 (2000).

[13] T. Sakai, H. Hisaeda, Y. Nakano, H. Ishikawa, Y. Maekawa, K. Ishii, Y. Nitta, J. Miyazaki, and K. Himeno, *Immunology* **99,** 615 (2000).

[14] A. L. Rakhmilevich, J. Turner, M. J. Ford, D. McCabe, W. H. Sun, P. M. Sondel, F. Grota, and N. S. Yang, *Proc. Natl. Acad. Sci. U.S.A.* **93,** 6291 (1996).

[15] T. M. Pertmer, M. D. Eisenbraun, D. McCabe, S. H. Prayaga, D. H. Fuller, and J. R. Haynes, *Vaccine* **13,** 1427 (1995).

[16] R. S. Williams, S. A. Johnston, M. Riedy, M. J. DeVit, S. G. McElligott, and J. C. Sanford, *Proc. Natl. Acad. Sci. U.S.A.* **88,** 2726 (1991).

[17] S. Kuriyama, A. Mitoro, H. Tsujinoue, T. Nakatani, H. Yoshiji, T Tsujimoto, M. Yamazaki, and H. Fukui, *Gene Ther.* **7,** 1132 (2000).

[18] J. M. Wells, L. H. Li, A. Sen, G. P. Jahreis, and S. W. Hui, *Gene Ther.* **7,** 541 (2000).

[19] H. Maruyama, M. Sugawa, Y. Moriguchi, I. Imazek, Y. Ishikawa, K. Ataka, S. Hasegawa, Y. Ito, N. Higuchi, J. J. Kazama, F. Gejyo, and J. I. Miyazaki, *Hum. Gene Ther.* **11,** 429 (2000).

[20] L. M. Mir, M. F. Bureau, J. Gehl, R. Rangara, D. Rouy, J. M. Caillaud, P. Delaere, D. Branellec, B. Schwartz, and D. Scherman, *Proc. Natl. Acad. Sci. U.S.A.* **96,** 4262 (1999).

[21] Y. Oshima, T. Sakamoto, I. Yamanaka, Y. Nishi, T. Ishibashi, and H. Inomata, *Gene Ther.* **5,** 1347 (1998).

[22] R. Heller, M. Jaroszeski, A. Atkin, D. Moradpour, R. Gilbert, J. Wands, and C. Nicolau, *FEBS Lett.* **389,** 225 (1996).

[23] T. Susuki, B. C. Shin, K. Fujikura, T. Matsuzaki, and K. Takata, *FEBS Lett.* **423,** 436 (1998).

[24] M. P. Rols, C. Delteil, M. Golzio, P. Dumond, S. Cros, and J. Teissie, *Nat. Biotechnol.* **16,** 168 (1998).

[25] S. Somiari, J. G. Malone, J. J. Drabick, R. A. Gibert, R. Heller, M. J. Jaroszeski, and R. W. Malone, *Mol. Ther.* **2,** 178 (2000).

[26] J. M. Vicat, S. Boisseau, P. Jourdes, M. Laine, D. Wion, R. Bouali-Benazzouz, A. L. Benabid, and F. M. Berger, *Hum. Gene Ther* **11,** 909 (2000).

[27] L. Huang, M. C. Hung, and E. Wagner, "Nonviral Vectors for Gene Therapy." Academic Press, San Diego, 1999.

[28] P. D. Robbins, H. Tahara, and S. C. Ghivizzani, *Trends Biotechnol.* **16,** 35 (1998).

[29] A. D. Miller, D. G. Miller, J. V. Garcia, and C. M. Lynch, *Methods Enzymol.* **217,** 581 (1993).

[30] K. L. Berkner, *Curr. Top. Microbio. Immunol.* **58,** 39 (1992).

[31] K. F. Kozarsky and J. M. Wilson, *Curr. Opin. Genetic. Dev.* **3,** 499 (1993).

[32] B. J. Carter, *Curr. Opin. Biotechnol.* **3,** 533 (1992).

[33] D. J. Fink, N. A. DeLuca, W. F. Goins, and J. C. Glorioso, *Annu. Rev. Neurosci.* **19,** 245 (1996).

[34] L. Naldini, U. Blomer, P. Gallay, D. Ory, R. Mulligan, F. H. Gage, I. M. Verma, and D. Trono, *Science* **272,** 263 (1996).

transfer into cells.[35,36] The discussion of this new transfection technique in this article is focused on the issues related to gene transfer efficiency, toxicity, and the patterns of transgene expression in mice.

Design of Hydrodynamics-Based Transfection

In this method, a large volume of DNA solution is rapidly injected into the tail vein of a mouse. Although the injected DNA solution will eventually reach all the organs through regular blood circulation, the large volume injected in a short period of time exceeds the cardiac output, resulting in an accumulation of DNA solution in the inferior vena cava and a higher pressure within this specific venous section of the circulation. Such pressure in the inferior vena cava creates a reflux of DNA solution into the liver and kidneys, which are directly connected to the vena cava through the hepatic or renal veins. Since the liver is the largest organ in the body and has an expandable structure, a large portion of DNA solution enters it in a direction opposite to its regular blood flow, thereby making the liver a primary target for gene transfer. Although it is unlikely that the elevated pressure will last for long as the heart continues to pump the solution out of the inferior vena cava, the transient increased hydrostatic pressure in these organs allows DNA molecules enter cells. The transferred genes can then be expressed. Fig. 1 details the anatomy of major vesicular structures involved.

General Procedure

DNA Preparation

Standard molecular biology procedures were used for the construction, analysis, and amplification of the plasmids from *Escherichia coli*.[37] Plasmids used in our studies contain human cytomegalovirus (CMV) immediate early promoter. Plasmid DNA was purified by the CsCl–ethidium bromide gradient centrifugation methods and kept in saline at $-20°$ until use.

Hydrodynamics-Based Transfection

DNA is dissolved in saline (0.9% NaCl) in an appropriate concentration and injected into the tail vein of a mouse using a 3 ml latex-free syringe with a 27 gauge 1/2 needle. A brief (1–2 min) warming of animals with an infrared heating lamp is usually helpful in assisting visualization and needle insertion into the tail vein. Mice

[35] F. Liu, Y. K. Song, and D. Liu, *Gene Ther.* **6**, 1258 (1999).

[36] G. Zhang, V. Budker, and J. A. Wolff, *Hum. Gene Ther.* **10**, 1735 (1999).

[37] J. Sambrook, E. F. Fritch, and T. Maniatis, "Molecular Cloning: A Laboratory Manual." Cold Spring Harbor Laboratory Press, New York, 1989.

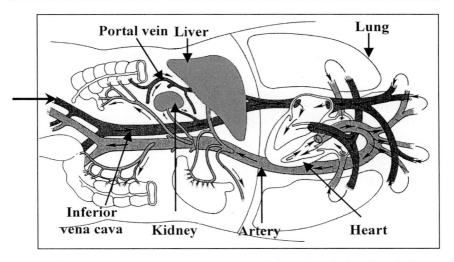

FIG. 1. Schematic presentation of major vessels involved in hydrodynamics-based transfection.

are held steady during the injection in a standard mouse restrainer, a transparent acrylic tube with holes in its sides. The volume of DNA solution used is calculated based on a ratio of 0.08 to 0.12 ml per gram of body weight and is injected into a mouse in 5–6 sec.

Preparation of Tissue Extracts and Luciferase Assay

Organs including the liver, spleen, lung, heart, and kidney were dissected from a sacrificed animal following standard surgical procedures and kept on dry ice or at $-80°$ until use. One ml of ice cold lysis buffer (0.1 M Tris-HCl, 2 mM EDTA, and 0.1% Triton X-100, pH 7.8) was added to each organ with the exception of liver, of which approximately 200 mg was used. Tissues were homogenized for 15–20 sec with a Tissue Tearor (Biospec Products, Bartlesville, OK) operated at maximal speed. Tissue homogenates were then centrifuged in a microcentrifuge for 10 min at its maximal speed (13,000g) at $4°$. The supernatants as the tissue extracts were used for gene product analysis. Protein concentration of these tissue extracts was determined using a Coomassie Plus Protein assay kit (Pierce, Rockford, IL). For luciferase assay, liver extract was further diluted 60-fold using HEPES buffer (70 mM HEPES, 7 mM MgSO$_4$, 3 mM dithiothreitol, 0.1% BSA, pH 7.8). Ten μl of tissue extracts from the lung, heart, spleen, or diluted liver extract were mixed with 100 μl of luciferase assay reagent (Luciferase Assay System, Promega, Madison, WI). Luciferase activity in each sample, obtained as relative light units (RLU), was measured in a luminometer (AutoLumant LB 953, EG&G, Salem, MA) for 10 sec. The obtained luciferase activity was then converted to luciferase protein amount using a standard curve established using reagents and procedure recommended by Analytic Luminescence Laboratory (ALL, San Diego, CA). The amount of luciferase protein was calculated using the equation derived from the standard curve in which luciferase protein (pg) $= 7.88 \times 10^{-5} \times \text{RLU} + 0.092$ ($r^2 = 0.9999$).

Establishment of Optimal Conditions for Hydrodynamics-Based Transfection

Optimization of Injection Volume

The volume of DNA solution required for hydrodynamics-based transfection will vary with animal weight, the ultimate level of gene product desired, and the types of animal models used. Data shown in Fig. 2 represent a set of experiments designed to establish the co-relationship between the injection volume of DNA solution and level of luciferase gene expression in CD-1 mice (male) with different body weights. In these experiments, each mouse was injected within 5 sec with 10 μg of pCMV-Luc plasmid DNA in varying volume of solution. The level of luciferase protein in different organs (heart, lung, liver, kidney, and spleen) was

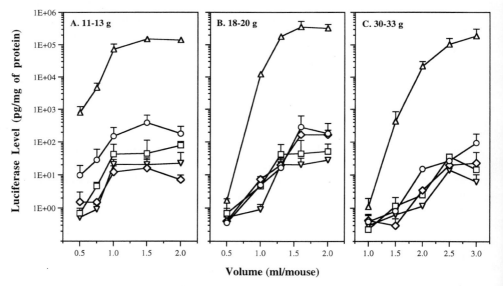

FIG. 2. Effect of injection volume on the level of gene expression in liver (triangle), kidney (circle), spleen (diamond), lung (square), and heart (inverted triangle). Three groups of mice with different weight (11–13 g, 18–20 g, and 30–32 g) were intravenously injected with 10 μg of pCMV-Luc plasmid in various volume of saline. Luciferase protein in different organs was determined 8 hr post DNA injection. Error bars represent the standard error of the mean (SEM) from three mice. [Reproduced from F. Liu, Y. K. Song, and D. Liu, *Gene Ther.* **6,** 1258 (1999), with permission.]

determined by luciferase assay 8 hr after transfection. Luciferase gene expression was detected in all examined organs with the highest seen in the liver. Maximal gene expression was obtained at approximately 1.2, 1.6, and 3.0 ml for CD-1 mice with a body weights of 11–13, 18–20, and 30–32 g, respectively. The volume to body weight ratio determined for optimal transfection is around 0.1 ml/g.

Optimization of Injection Time

Data in Fig. 3 show the importance of injection speed for achieving a high level of transgene expression. Using an optimal injection volume, a 6-fold increase in injection time (from 5 to 30 sec) results in more than a 4500-fold decrease in the level of luciferase gene expression in the liver. Thus, the optimal injection time is approximately 5 sec.

Evaluation of Toxicity

The toxic effects of a transfection method on animals are important criteria in evaluating whether the transfection procedure is acceptable. Animal toxicity can be induced by either the gene product or the method itself. Although there are

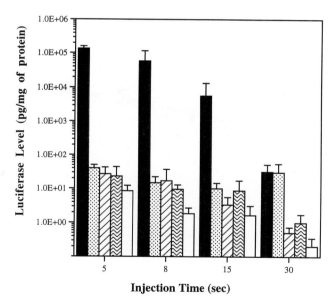

FIG. 3. Effect of injection time on the level of gene expression. Mice (18–20 g) were intravenously injected with 10 μg of plasmid DNA in 1.6 ml saline. The injection time was varied from 5 to 30 sec. Mice were sacrificed 8 hr post-injection and luciferase protein in the extracts of liver (filled), kidney (dotted), spleen (hatched), lung (waved), and heart (open) was measured. Error bars represent SEM from three mice. [Reproduced from F. Liu, Y. K. Song, and D. Liu, *Gene Ther.* **6**, 1258 (1999), with permission.]

many types of assays to evaluate toxic effect on animals, we have taken two general approaches to determine whether the hydrodynamics-based transfection under the optimal conditions for CD-1 mice causes any significant toxicity to animals. The first approach employs the measurement of animal growth rate. The rationale of this approach is that mouse growth rate should be affected if the hydrodynamics-based procedure causes significant toxic effects in animals. Five animals in each of three groups were injected with saline with or without 10 μg of pCMV-Luc plasmid DNA. The body weight of each animal was measured every day for a total of 6 days post-injection. Data shown in Fig. 4 suggest that growth rates between the control group and the groups injected with either saline or saline containing 10 μg of pCMV-Luc plasmid DNA were not statistically different as analyzed by the one-way ANOVA test ($p > 0.05$).

The second approach involves serum biochemistry. These tests are designed to evaluate the effect of hydrodynamics-based transfection on the major biochemical parameters in serum including the concentration of major ions (Na^+, K^+, and Cl^-); major protein concentration (albumin, the total proteins); and the concentration of liver-specific enzymes including alkaline phosphatase (ALP),

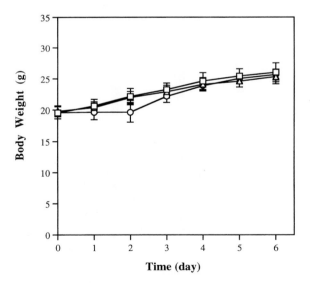

Fig. 4. Effect of hydrodynamics-based transfection on animal growth. Each mouse was injected with 1.6 ml saline or saline containing 10 μg of pCMV-Luc plasmid DNA. Animal weight was measured before and after the injection. Control group (triangle) represents mice not receiving injection. Testing animals include those receiving 1.6 ml saline per mouse with (square) or without (circle) 10 μg of pCMV-Luc plasmid DNA. Error bars represent SEM from five mice. [Reproduced from F. Liu, Y. K. Song, and D. Liu, *Gene Ther.* **6**, 1258 (1999), with permission.]

TABLE I
EFFECT OF HYDRODYNAMICS-BASED TRANSFECTION ON SERUM BIOCHEMISTRY[a]

Time after injection (day)	Serum concentration[b]								
	Na^+ (mM)	K^+ (mM)	Cl^- (mM)	Alb (mg/l)	T. Prot (mg/l)	ALP (U/l)	AST (U/l)	ALT (U/l)	T. Bili (mg/l)
1	145 (1)	6.5 (0.2)	107 (2)	23 (2)	46 (3)	176 (39)	183 (17)	177 (56)	2 (1)
3	ND	ND	ND	ND	ND	ND	ND	58 (14)	ND
7	147 (2)	7.6 (0.4)	107 (1)	26 (2)	52 (3)	201 (32)	164 (29)	46 (9)	4 (1)
Normal mice	144 (1)	7.7 (0.8)	107 (1)	23 (3)	48 (4)	227 (40)	157 (43)	45 (3)	3 (2)

[a] Serum was prepared by separation of the coagulated whole blood of animals at indicated time after injection of 1.6 ml of saline or DNA solution.

[b] Serum concentrations of Na^+, K^+, Cl^-, albumin (Alb), total protein (T. Prot), alkaline phosphatase (ALP), aspartate aminotransferase (AST), alanine aminotransferase (ALT), and total bilirubin (T. Bili) were measured by automated analyzer (Vitros 950, Johnson & Johnson Clinical, Rochester, NY). ND; Not determined. Data represent mean (SEM) from 5 mice. [Reproduced from F. Liu, Y. K. Song, and D. Liu, *Gene Ther.* **6**, 1258 (1999), with permission.]

aspartate transaminase (AST), alanine aminotransferase (ALT), and total bilirubin. Both short (1 day) and long (7 days) term effect of DNA administration on these parameters was evaluated. Data presented in Table I show that all the biochemical parameters evaluated were in the normal range as compared to those of normal animals with the exception of the ALT value. A transient increase in ALT value was seen in the animal groups injected with either saline or saline containing plasmid DNA, suggesting that the increased ALT value is not caused by either plasmid DNA or gene product. It was noted that the ALT value falls into a normal range 3 days after the injection.

It is important to point out here that caution should be taken when injection volume and rate exceed the aforementioned optimal conditions. Histological examination of liver samples revealed significant tissue damage when higher than necessary volumes of DNA solution and injection speeds were used.

Efficiency of Hydrodynamics-Based Transfection

Transfection efficiency is normally defined by the number (or percentage) of cells expressing the transgene with respect to the total cells transfected or the level of transgene expression as a function of the amount of DNA used. The transfection efficiency of hydrodynamics-based procedure in CD-1 mice was evaluated using three separate constructs with luciferase, human α1-antitrypsin (hAAT), or Lac Z gene as a reporter. Different amounts of plasmid DNA were injected into animals following the hydrodynamics-based procedure. Data shown in Fig. 5 suggest that as little as 5 μg of pCMV-Luc DNA per mouse is sufficient to generate approximately

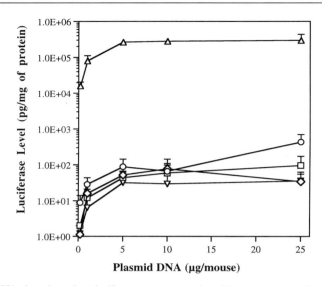

FIG. 5. DNA dose-dependent luciferase gene expression. Various amounts of plasmid DNA (pCMV-Luc) in 1.6 ml saline were injected into each mouse within 5 sec. The level of luciferase gene expression was determined 8 hr post-injection in liver (triangle), kidney (circle), spleen (diamond), lung (square), and heart (inverted triangle). Error bars represent SEM from three mice. [Reproduced from F. Liu, Y. K. Song, and D. Liu, *Gene Ther.* **6,** 1258 (1999), with permission.]

300 ng of luciferase protein per mg of extracted protein for liver, 0.1 ng for kidney, 0.08 ng for spleen, 0.07 ng for lung, and 0.03 ng for heart. For liver, this level represents 45 μg of luciferase protein per gram of liver in a mouse with a body weight of 18–20 g.

A dose response curve was also established using plasmid DNA (pCMV-hAAT) containing the hAAT gene that encodes a serum protein. Data in Table II show that the serum level of hAAT increased with an increase in the amount of pCMV-hAAT injected. The highest level obtained was 200–500 μg/ml of serum in animals injected with 10–50 μg of pCMV-hAAT plasmid DNA.

The percentage of cells transfected in the liver as a function of the amount of plasmid DNA administered was evaluated by using a histochemical technique employing Lac Z containing plasmids (pCMV-Lac Z). It is evident in Fig. 6 that the density of blue cells in liver sections increases with increasing administered doses of plasmid DNA. About 40% of liver cells can be identified as β-galactosidase positive under our experimental conditions using an injection dose of 25 μg per mouse. At higher magnification, it appeared that these cells were mainly hepatocytes identifiable by their polygonal shape and round nuclei. The transfected cells are grouped around the hepatic vein and located at the perivenular region in the cell plate. Figures 6B, 6D, 6F, and 6H show sections stained with hematoxylin and

TABLE II
DNA DOSE-DEPENDENT EXPRESSION OF HUMAN
α1-ANTITRYPSIN GENE IN MICE[a]

Dose (μg/mouse)	Serum hAAT concentration (ng/ml)
0	<1
0.05	1,664 (411)
0.5	14,655 (11,509)
5.0	154,694 (81,303)
10.0	215,562 (84,837)
25.0	524,774 (223,144)
50.0	223,143 (195,436)

[a] Each mouse was transfected with various amounts of pCMV-hAAT plasmid DNA. Serum concentration of human α1-antitrypsin in mice was determined 24 hr post-injection using ELISA. Data represent mean (SEM) from 5 mice.

eosin to identify potential liver damage. Compared to uninjected mice, no obvious liver damage can be seen in animals injected with either saline (Fig. 6A) or saline containing different amounts of pCMV-LacZ (Figs. 6C, 6E, and 6G).

These results confirm that genes encoding either cellular proteins (e.g., luciferase and β-galactosidase) or secretory protein (e.g., hAAT) can be efficiently expressed in mice by hydrodynamics-based transfection. Depending on the type of transgene, the level of expression reaches the maximum at a dose of 5–50 μg of plasmid DNA per mouse. It is hard to make a fair comparison between these results and those of virus-mediated gene transfer because of differences in experimental conditions. However, the level of transgene expression obtained by the hydrodynamics-based procedure appears comparable to that of adenovirus vectors administered at the maximal tolerated dose and 2- to 8-fold higher than that obtained using adeno-associated viruses when a systemic gene transfer was performed in mice.

Characterization of Transgene Expression

Persistency of Transgene Expression in Transfected Animals

The level of gene product expressed in animals transfected by the hydrodynamics-based method exhibits a biphasic pattern as shown in Fig. 7. In these experiments, animals were injected with 10 or 50 μg of either pCMV-Luc or pCMV-hAAT plasmid DNA and the levels of transgene product measured at appropriate time points. For animals transfected with pCMV-Luc, the luciferase level reached maximum approximately 4–12 hr post-injection, declined thereafter

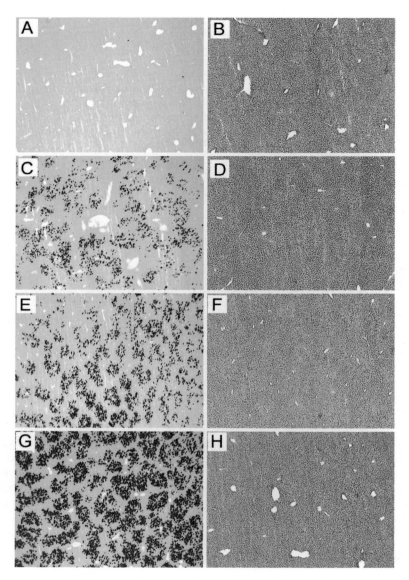

FIG. 6. Histochemical analysis of β-galactosidase gene expression in liver. Mice were injected with 1.6 ml saline containing various amounts of pCMV-LacZ plasmid DNA. Animals were sacrificed 8 hr post-injection and liver sections were made using cryostat. Sections (A, C, E and G) were stained with X-Gal solution followed by an eosin counterstain. Sections (B, D, F, and H) were stained by a standard hematoxylin/eosin staining method. Sections were made from animals each receiving 0 (A, B), 0.5 (C, D), 2.5 (E, F), and 25 μg (G, H) of pCMV-LacZ. (Original magnification: 25×.) [Reproduced from F. Liu, Y. K. Song, and D. Liu, *Gene Ther.* **6,** 1258 (1999), with permission.]

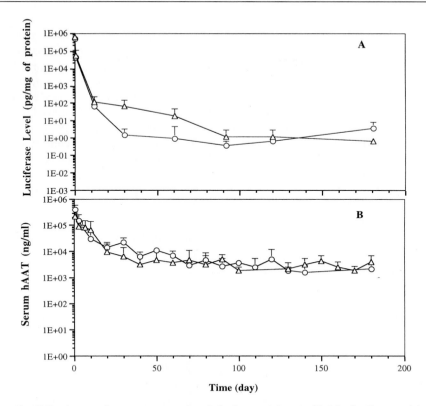

FIG. 7. Persistence of transgene expression. Animals were injected with 1.6 ml saline containing 10 μg (triangle) or 50 μg (circle) of either pCMV-Luc (A) or pCMV-hAAT (B) plasmid DNA. For luciferase gene expression, animals were sacrificed at various time intervals and luciferase activity in the liver was determined by a standard luciferase assay. For hAAT gene expression, blood samples were collected at different time and serum concentration of hAAT was determined by a standard ELISA. Error bars represent SEM from five mice.

for about 30 days, and then persisted at a lower level for more than 6 months (Fig. 7A). A similar pattern of gene product level with time was also seen in animals transfected with plasmids containing the hAAT gene. Serum hAAT level dropped approximately 100-fold from day 1 to day 30 and persisted at lower level for more than 6 months. The serum concentration 6 months post transfection was approximately 5–10 μg/ml in animals each injected with 10–50 μg of pCMV-hAAT plasmid DNA (Fig. 7B).

Southern blot analysis of extracted DNA from mouse liver 6 months after the transfection showed that the pCMV-hAAT plasmid that remained was in the episomal form.[38] RT-PCR analysis of the same set of liver samples confirmed that

[38] G. Zhang, Y. K. Song, and D. Liu, *Gene Ther.* **7,** 1344 (2000).

the hAAT genes are still transcriptionally active 6 months after the transfection.[38] In contrast to the general perception that persistent gene expression cannot be achieved without using viral vectors, results from these, as well as other experiments,[39–41] clearly show that a long-term and persistent gene expression can be achieved *in vivo* with methods not involving viruses. The remaining issue with respect to persistent transgene expression using a nonviral method is at what level the gene product can be maintained. A study by Miao *et al.* showed that a near peak level of transgene expression with human factor IX as a reporter can be maintained for more than 7 months by incorporating the apolipoprotein E locus control region, first intron (intron A) and 3'-untranlated region (3'-UTR) of human factor IX gene into the plasmid construct.[41]

Effect of Repeated Administration

The potential for repeated transfection using a hydrodynamics-based procedure was explored in our previous studies.[35,38] Data in Fig. 8 represent a typical pattern

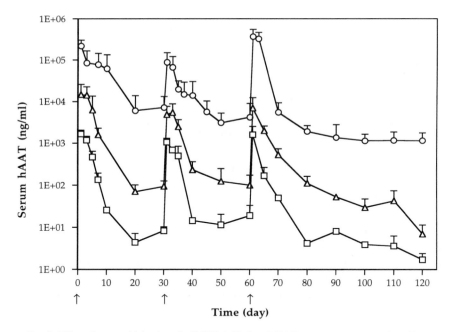

FIG. 8. Effect of repeated injection of pCMV-hAAT plasmid DNA on serum concentration of human α1-antitrypsin. Mice were injected with 0.05 μg (square), 0.5 μg (triangle), or 10 μg (circle) of pCMV-hAAT plasmid DNA, respectively, on days zero, 30, and 60. Blood samples were collected at appropriate time before and after each injection. The serum concentration of hAAT was determined using standard ELISA. Arrows indicate the time of injection. Error bars represent SEM from five mice. [Reproduced from G. Zhang, Y. K. Song, and D. Liu, *Gene Ther.* **7,** 1344 (2000), with permission.]

of the dose response curve. In these experiments, mice were injected on day 0 with 0.05, 0.5, or 10 μg of pCMV-hAAT plasmid DNA, respectively. Repeated administrations were performed on days 30 (2nd injection) and 60 (3rd injection). The serum concentration of hAAT was examined at 1, 3, 5, 10, 20, and 30 days after each injection. Animals responded to each of the three injections in an identical dose-dependent manner. The serum level of hAAT peaked at day 1 after each DNA administration with the highest level in animals injected with 10 μg of pCMV-hAAT plasmid DNA, followed by that of animals receiving 0.5 μg and 0.05 μg of the same type of plasmid DNA.

Conclusions

Introduction of genetic material into cells has proved to be a valuable system for gene function analysis with regard to mechanism of action and involvement in maintaining physiological or inducing pathophysiological conditions. The unique contribution of the hydrodynamics-based procedure is that this new transfection method provides a convenient and efficient tool for a researcher to conduct gene function analysis in whole animals. Continued refinements of the current technique and development of new protocols should encourage widespread adaptation of this strategy for the aforementioned applications. As a basic tool for gene therapy studies, development of the hydrodynamics-based transfection procedure is an important advance since it is now possible to introduce genetic materials into cells and obtain sufficient levels of transgene expression in mice by a simple tail-vein injection of DNA.

Acknowledgment

We thank Dr. Joseph E. Knapp for critical review of the manuscript. This work was supported by the National Institutes of Health Grant CA72925 and a research contract from Target Genetics Corporation.

[39] J. A. Wolff, J. J. Ludtke, G. Acsadi, P. Williams, and A. Jani, *Hum. Mol. Genet.* **1,** 363 (1992).

[40] S. P. Yant, L. Meuse, W. Chiu, Z. Ivics, Z. Izsvak, and M. A. Kay, *Nat. Genet.* **25,** 35 (2000).

[41] C. H. Miao, K. Ohashi, G. A. Patijn, L. Meuse, X. Ye, A. R. Thompson, and M. A. Kay, *Mol. Ther.* **1,** 522 (2000).

[6] Bioplex Technology: Novel Synthetic Gene Delivery System Based on Peptides Anchored to Nucleic Acids

By Lars J. Brandén and C. I. Edvard Smith

Introduction

This article describes a novel synthetic technology allowing the coupling of functional peptide/protein entities to nucleic acids in a sequence-specific fashion. More specifically, we have developed a system based on hybridizing peptide nucleic acid (PNA) anchors. This technology allows the inclusion of single, or multiple, functions attached directly to DNA and represents a versatile system for the transport, biochemical modulation and processing of nucleic acids. One aspect of this technology, integrating multiple components, is the development of a synthetic alternative to viral gene transfer.

Gene transfer can be defined as the cellular uptake of nucleic acids. Nucleic acid transfection[1] complexes taken up by cells are compartmentalized and processed in a cell-specific vesicle transport machinery.[2] The central dogma of genetics is that DNA can be copied into RNA, a process that may be reversed under certain circumstances. Specificity is achieved by base complementarity. This can take place in trans, as when DNA serves as a template for mRNA or when tRNA recognition of the primer-binding site (PBS) initiates retroviral reverse transcription. Alternatively, cognate sequences may interact in cis, allowing, for example, functional 3D structures to be created (such as internal ribosomal entry sites, IRES) or the formation of RNA catalytic enzymes, ribozymes.

Transfection is a complex process involving several steps: adsorption of the transfection complex to the cellular surface, uptake into the cell, escape from the endosome/lysosome, nuclear translocation with or without chromosomal integration, and finally the effect of the nucleic acid, whether it is expression of a reporter/therapeutic gene or induction of mutation(s). The different steps which limit transfection have been solved in nature. Viruses perform these tasks with great efficacy,[3] whereas standard methods of plasmid transfection rely on suboptimal transfer techniques. The efficiency frequently varies by several orders of magnitude among different viruses and methods of transduction. This is a complex issue, however, since many particles in virus stocks are noninfectious. Thus the titers used in comparisons between transfection and transduction should ideally correspond to the total number of virus particles in the stock.

[1] D. Luo and W. M. Saltzman, *Nature Biotechnol.* **18,** 33 (2000).
[2] P. Erbacher, J.-S. Remy, and J.-P. Behr, *Gene Ther.* **6,** 138 (1999).
[3] M. I. Bukrinsky and O. K. Haffar, *Mol. Med.* **4,** 138 (1998).

TABLE I
TRANSFECTION NOMENCLATURE

Transfection reagent	Function	Name	Sequence specificity	Selected examples
Cationic lipid	DNA condensing activity	Lipoplex[5]	No	Lipofectamine
Cationic polymer	DNA condensing activity	Polyplex[5]	No	Polyethyleneimine
Nucleic acid fused to functional moiety	Multiple functions including or excluding DNA condensing activity	Bioplex[18]	Yes	Nucleic acid–nuclear localization signal, (NLS) fusion[17,18,29]

Although the existing techniques for plasmid-based gene transfer and oligonucleotide delivery are insufficient, they often represent viable alternatives to viral vectors. In animals or humans, these approaches avoid the negative aspects of viral gene transfer, such as the immune response to viral proteins and the potential formation of replication competent viruses. Moreover, in contrast to viral vectors, naked DNA shows virtually no size limitations for inserts. Furthermore, the making of viral vectors is frequently a cumbersome process, involving complex cloning and cell cultivation steps and, subsequently, when viral supernatants are used for gene delivery to humans, the extensive testing is very expensive and often the limiting factor. Thus, improved nonviral gene delivery systems would offer great benefits provided that the negative aspects of viral-based transfer could be avoided. It is therefore to be expected that entirely synthetic systems would be preferred, since they are known to display a high degree of consistency and are more amenable to rigorous quality control at multiple levels. Synthetic systems would thus mean a lower cost for individual research groups involved in clinical trials. The formulation of these systems is important for the ease-of-use to clinicians, and advances are being made in this respect.[4]

In this article, we describe a new platform technology which we have established. We refer to it as Bioplex (biological complex) in analogy with the nomenclature for other transfection reagents,[5] since we attach the biological functions to the nucleic acid complex via sequence-specific hybridization (Table I). The technology is based on the use of two particular aspects of PNA[6] (or analog thereof; Fig. 1), namely, the high T_m value,[7] making these molecules most suitable as genetic anchors, and the possibility of making a continuous peptide synthesis, allowing PNA

[4] H. Talsma, J.-Y. Cherng, H. Lehrmann, M. Kursa, M. Ogris, W. E. Hennink, M. Cotten, and E. Wagner, *Int. J. Pharm.* **157**, 233 (1997).

[5] P. L. Felgner, Y. Barenholz, J.-P. Behr, S. H. Cheng, P. Cullis, L. Huang, J. A. Jessee, L. Seymore, P. Szoka, A. R. Thierry, E. Wagner, and G. Wu, *Hum. Gene Ther.* **8**, 511 (1997).

[6] P. E. Nielsen, M. Egholm, J. B. Christensen, and O. Buchardt, *Science* **254**, 1497 (1991).

[7] H. Knudsen and P. E. Nielsen, *Nucleic Acids Res.* **24**, 494 (1996).

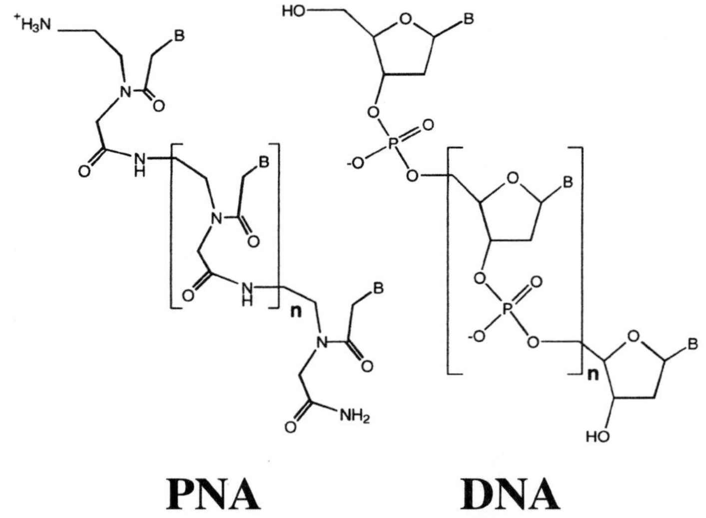

PNA   **DNA**

FIG. 1. Comparison of PNA and DNA structure. The PNA backbone consists of repeated *N*-(2-aminoethyl) glycine units linked by amide bonds. The purine and pyrimidine bases are attached to the backbone by means of a methylene carbonyl linkage. The uncharged nature of the PNA molecule as compared to the DNA with negatively charged phosphate groups is illustrated here.

and a peptide moiety to become adjacent without involving cumbersome linking chemistry and purification of the conjugated peptides. We use the anchor property (trans) as a means of incorporation, in a sequence-specific fashion, of various functional elements in *cis* relative to the PNA. This enables modulation of nucleic acids, including multifunctional approaches that permit improved gene delivery. PNA is a synthetic compound based on the linkage of purines or pyrimidines to a neutral pseudo-peptide backbone instead of the charged pentose phosphate moiety, which is used in DNA or RNA (Fig. 1). The neutral charge contributes to the increased T_m value for PNA–DNA/RNA interactions versus "homotypic" DNA (DNA–DNA) or RNA (RNA–RNA), or RNA–DNA duplex formation. Under certain circumstances, these properties will permit "strand invasion"[8] even at room temperature, i.e., allow a PNA molecule to invade a DNA duplex, a phenomenon dependent on the inherent DNA breathing activity.[9] Although it has been hypothesized that PNA may once have served as an ancestor for nucleic acids, the general notion is that PNA does not exist in any form in nature. This idea is supported by the fact that there are no naturally occurring enzymes which degrade PNA,

[8] P. Wittung, P. E. Nielsen, and B. Nordén, *J. Am. Chem. Soc.* **118,** 7049 (1996).
[9] T. Bentin and P. E. Nielsen, *Biochemistry* **35,** 8863 (1996).

making this compound very stable. To date, PNA has mainly been studied as an antisense reagent[10] with the aim of neutralizing certain activities, such as microbial RNA. There is a vast literature describing the underlying chemistry, thermodynamics, and duplex structure[11] of PNA–nucleic acid interactions, hybridization-based techniques, and other aspects[12,13] of PNA technology.[14]

In the main, recombinant DNA technology is based on the use of plasmids. However, other vector systems, such as bacmids, yeast artificial chromosomes (YACs), and various viruses, are important tools with particular qualities. Thus, although the production of plasmids allows large quantities of genetic material to be generated in a short time and at low cost, viruses permit highly efficient gene transfer, since these organisms have evolved a number of features promoting the delivery of genetic material. Although many tools have been developed for the handling of nucleic acids in the laboratory, these technologies could be improved in multiple ways by taking advantage of the way nature has come up with solutions. Thus, in this article we describe how to link peptide functions directly to DNA via hybridization instead of covalent linkage[15] or via nonspecific charge interactions[16] (Fig. 2). This technique is applicable both *in vitro*[17] and *in vivo*[18] and avoids the hazards of vectors based on viruses and other microorganisms.

PNA

Peptide nucleic acid is a DNA analog that was first synthesized by P. Nielsen *et al.* in 1991.[6] The backbone consists of repeating *N*-(2-aminoethyl) glycine units linked by amide bonds. The purine and pyrimidine bases are attached to the backbone by means of a methylene carbonyl linkage. Because of the achiral physical characteristics of the PNA backbone, it can bind complementary nucleic acids in both parallel and antiparallel orientation.[19] To achieve the best possible interaction between the PNA peptide and the target nucleotides, the PNA sequence should be designed to bind in antiparallel orientation to its cognate binding site. The affinity of PNA for pentose phosphate backbone nucleic acids is higher than that of

[10] H. Knudsen and P. E. Nielsen, *Nucleic Acids Res.* **24**, 494 (1996).

[11] M. Egholm, O. Buchardt, L. Christensen, C. Behrens, S. M. Freier, D. A. Driver, R. H. Berg, S. K. Kim, B. Nordén, and P. E. Nielsen, *Nature* **365**, 566 (1993).

[12] P. E. Nielsen, *Pharmacol. Toxicol.* **36**, 3 (2000).

[13] B. Hyrup and P. E. Nielsen, *Bioorg. Med. Chem.* **4**, 5 (1996).

[14] P. Nielsen and M. Egholm (eds.), "Peptide Nucleic Acids: Protocols and Applications." Horizon Scientific Press (1999).

[15] M. G. Sebestyen, J. J. Ludtke, M. C. Bassik, G. Zhang, V. Budker, E. A. Lukhtanov, J. E. Hagstrom, and J. A. Wolff, *Nat. Biotechnol.* **16**, 80 (1998).

[16] P. Collas, H. Husebye, and P. Alestrom, *Transgenic Res.* **5**, 451 (1996).

[17] L. J. Brandén, A. J. Mohamed, and C. I. E. Smith, *Nat. Biotechnol.* **17**, 784 (1999).

[18] L. J. Brandén, B. Christensson, and C. I. E. Smith, *Gene Ther.* **8**, 84 (2001).

[19] M. Egholm, O. Buchardt, P. E. Nielsen, and R. H. Berg, *J. Am. Chem. Soc.* **114**, 895 (1992).

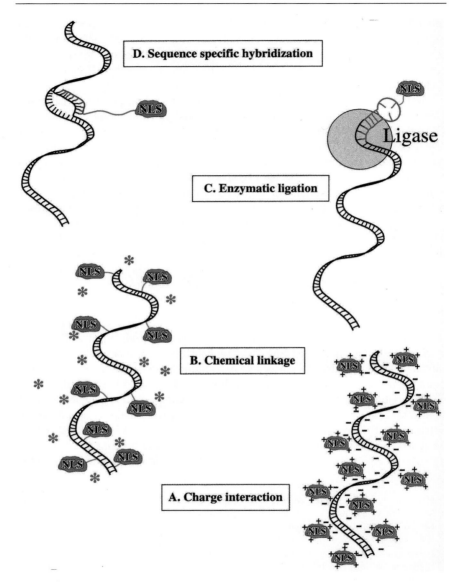

Fig. 2. Technological evolution of linking peptide functions to nucleic acids. Initially, DNA and free NLS peptides were mixed to form NLS/nucleic acid complexes by charge interactions (panel A).[16] Subsequently, methods were developed to enhance the peptide association with the nucleic acid (B–D). Because chemical linkage is used (panel B),[15] the location of the peptide binding is not sequence specific. Chemical reagents are depicted as asterisks. In panel C, sequence specificity is achieved by hybridizing and subsequently ligating a peptide-conjugated DNA oligonucleotide to the plasmid.[29] In panel D, the NLS is anchored to its cognate site by PNA–DNA duplex formation.[17,18]

naturally occurring nucleic acid duplexes. Because PNA lacks the phosphate group of nucleic acids, it is devoid of the inherent repulsion of DNA–DNA, RNA–RNA, or RNA–DNA duplexes. The lack of repulsion between the two strands makes the duplex highly insensitive to salt concentration. This allows binding of a PNA molecule to its cognate nucleic acid sequence even though the target sequence for the PNA molecule has secondary structures that would normally impair the formation of a duplex. The T_m for a PNA–DNA duplex can be calculated from the following formula[20]:

$$T_m(\text{pred}) = c_0 + c_1^* T_m(\text{nnDNA}) + c_2^* f_{\text{pyr}} + c_3^* \text{ length}$$

$T_m(\text{nnDNA})$ is the T_m for DNA calculated with the nearest neighbor model for the DNA–DNA duplex corresponding to the DNA–PNA. To determine the T_m according to this method, the values of $\Delta H°$ and $\Delta S°$ calculated by SantaLucia et al.[21] should be used. f_{pyr} denotes the fraction of pyrimidines; length is the number of bases in the PNA sequence. The constants c_0–c_3 are 20.79, 0.83, −26.13, and 0.44, respectively.

Dual/Multifunctional PNA Hybrids

By anchoring peptide functions to nucleic acids via the PNA molecule, it is possible to add many functional entities directly to the genetic material. The dual/multifunction PNA peptide is synthesized as a normal peptide. This means that it is easy to extend the PNA sequence with amino acid residues, thus integrating a peptide function or property into the PNA molecule. Examples of such sequences are cell receptor ligands[22] such as the tripeptide RGD,[2] which binds to integrins, nuclear translocation signals (NLS),[23–25] and membrane-disruptive peptides such as the influenza hemagglutinin peptide HA2.[26] As is the case for many cellular receptors, cross-linking is often beneficial for endocytosis. There are different ways of achieving this. One way is to have multiple PNA target sequences on the nucleic acid that hybridizes to the PNA molecule, thus enabling binding of multiple PNA entities carrying a functional moiety. By using standard techniques

[20] U. Gieser, W. Kleider, C. Berding, A. Geiger, H. Orum, and P. E. Nielsen, *Nucleic Acids Res.* **26,** 5004 (1998).

[21] J. SantaLucia, Jr., H. T. Allawi, and P. A. Seneviratne, *Biochemistry* **35,** 3555 (1996).

[22] R. Kircheis, A. Kichler, G. Wallner, M. Kursa, M. Ogris, T. Felzmann, M. Buchberger, and E. Wagner, *Gene Ther.* **4,** 409 (1997).

[23] Y. Yoneda, T. Semba, Y. Kaneda, R. L. Noble, Y. Matsuoka, T. Kurihara, Y. Okada, and N. Imamoto, *Exp. Cell Res.* **201,** 13 (1992).

[24] Y. Miyamoto, N. Imamoto, T. Sekimoto, T. Tachibana, T. Seki, S. Tada, T. Enomoto, and Y. Yoneda, *J. Biol. Chem.* **272,** 26375 (1997).

[25] N. Michaud and D. S. Goldfarb, *Exp. Cell Res.* **208,** 128 (1993).

[26] J. Chen, J. J. Skehel, and D. C. Wiley, *Biochemistry* **37,** 13643 (1998).

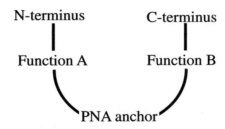

FIG. 3. Dual/multifunctional PNA hybrids. Functions A and B are synthesized at the N-terminal and C-terminal parts of the molecule, respectively, with the PNA sequence located in between.

of peptide branching, it is possible to create PNA molecules with multivalent entities linked to them. By this approach, it is possible to use only one PNA binding site and still have multiple peptide entities linked via the single PNA molecule. By using the N terminus and the C terminus of the PNA molecule, it is possible to make a PNA molecule with function A at the N terminus and function B at the C terminus (Fig. 3). The sequence-specific binding of the PNA is then one function and that of A and B two more functions. The combinatorial possibilities of this type of molecule have only now begun to be manifest as we are investigating different combinations of peptide moieties and their effect(s) on gene transfer. We are presently investigating the combination of the integrin binding RGD peptide, the SV40 NLS,[27] and endosome-breaking peptides. By synthesizing the PNA peptide with a unique sequence, we can hybridize each PNA molecule sequentially to a carrier oligonucleotide labeled with a fluorescence marker.[28] By altering the number of target sequences and their internal order, we can investigate the influence of spatial distribution on functional activity of the hybridized PNA molecules. An oligonucleotide labeled with a different fluorescent dye represents an essential internal control. It can be designed to hybridize to a nonfunctional PNA peptide. By introducing the test and the control complexes into the same target cell, it is possible to compare the effects of the PNA peptide (functional) vs a PNA peptide (mutant).

Functional Moieties

To deliver nucleic acids to cells *in vitro* or *in vivo,* certain criteria need to be fulfilled to varying degrees:

1. Nuclear entry
2. Packaging of the nucleic acid

[27] C. Y. Xiao, S. Hubner, and D. A. Jans, *J. Biol. Chem.* **272,** 22191 (1997).
[28] R. E. Mujumdar, L. A. Ernst, S. R. Mujumdar, C. J. Lewis, and A. S. Waggoner, *Bioconj. Chem.* **4,** 05 (1993).

3. Cell adhesion
4. Cell membrane penetration
5. Avoidance/escape of endocytic vesicles

Nuclear Entry. As described previously,[17,18,29] it is possible to use a preexisting nuclear transport pathway to enhance the nuclear delivery of transfected genetic material. The nuclear pore is capable of transporting large organic molecules such as an IgM molecule.[23] Different tissues contain different subsets of nuclear transport proteins. This is thought to be part of the regulatory functions of a cell.[30] Yoneda *et al.* reported that the α- and β-subunit types of the nuclear import machinery differ in expression levels in various tested tissue types. This allows us to enhance the possibilities of tissue type specific gene expression by tailor-making the reagents used to deliver the genetic material to a defined set of cells through the selection of a combination of cell-specific receptor ligand, the correct NLS, and cell-type restricted promoter elements.

Packaging of the Nucleic Acid. To minimize exposure to deleterious factors in serum, it is desirable to package and thus protect the nucleic acid. This can be accomplished in various ways with different types of reagents. Transfection is divided into different categories, depending on what type of compound is used. Each reagent has different properties as regards its ability to condense the nucleic acid, the cell adhesion capacity, membrane disruption, etc. One of the more widely used reagents today is the cationic polymer polyethyleneimine (PEI[31]). The branched variants of the molecule have one protonable nitrogen on every third amine group. Linear variants have become more widely used and are showing promising results.[32] This type of compound is often used to condense nucleic acids prior to transfection. The proton acceptor function is believed to work as a buffer in the endosomal compartment as the endosome converts into a lysosome. By buffering the vesicle containing the transfection complex, inhibition of the majority of lysosomal nucleases is accomplished, thus protecting the genetic material from degradation. The transfection reagent itself seems to be protective of the condensed genetic material as well. It has been shown that transfected material condensed with PEI is stable for prolonged periods of time when introduced into cells. We have used 70-mer oligonucleotides with ^{32}P-end labeling in combination with 25 kDa PEI to study the intracellular integrity of transfected nucleic acids and found minor nucleic acid degradation 3 hr after transfection in the nuclear cell fraction and the cytoplasmic fraction[33] (Fig. 4). The degradation increased over

[29] M. A. Zanta, P. Belguise-Valladier, and J.-P. Behr, *Proc. Natl. Acad. Sci. U.S.A.* **96,** 91 (1999).
[30] Y. Kamei, S. Yuba, T. Nakayama, and Y. Yoneda, *J. Histochem. Cytochem.* **47,** 363 (1999).
[31] O. Boussif, F. Lezoualc'h, M. A. Zanta, M. D. Mergny, D. Scherman, B. Demeneix, and J.-P. Behr, *Proc. Natl. Acad. Sci. U.S.A.* **92,** 7297 (1995).
[32] J.-P. Coll, P. Chollet, E. Bambilla, D. Desplanques, J.-P. Behr, and Marie Favrot, *Hum. Gene Ther.* **10,** 1659 (1999).
[33] L. J. Brandén, L. Yurchenko, and C. I. E. Smith, in preparation.

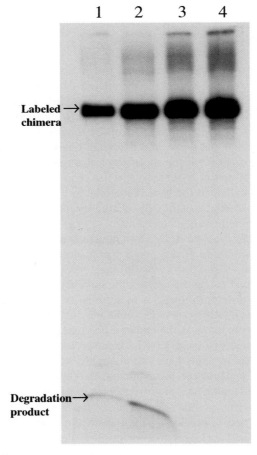

FIG. 4. Stability of transfected material. Stability of RNA–DNA chimeric oligonucleotide in living cells. Cos-7 and 3T3 cells were transfected by ^{32}P-Cy5 RNA–DNA chimera (100 nM). The cells were harvested after transfection. DNA extracted from cytosolic and nuclear fractions was separated by 20% polyacrylamide gel electrophoresis (PAGE) (lanes 1, 2 and lanes 3, 4, respectively). Lanes 1 and 3 are from Cos-7, and 2 and 4 are from 3T3 cells.

time, but there still existed a major fraction of intact 70-mer oligonucleotides in the cytoplasm as well as in the nucleus after 24 hr. This indicates not only protection from degrading enzymes but also a high internal stability of the transfected complexes. By using different charge ratios of condensing reagent to nucleic acid, it is possible to control the net charge of the transfection complex. It is desirable to have a positive net charge for transfection *in vitro* as this aids in cellular adhesion and thereby triggers endocytosis of the transfection complex. For transfection *in vivo*, it is beneficial to accomplish a more neutral net charge so as to avoid inactivation by serum factors. The contradiction is that a less charged transfection complex is more stable when injected intravenously but at the same time less likely to become endocytosed by the target cells.

Cell Adhesion. The ability of the transfection complex to adhere to the target cell is vital in order to achieve functional gene delivery. Lipoplexes and polyplexes use different methods of cell entry depending on the specific reagent used. Lipopexes can be divided into pH-dependent and cationic lipids with more subtypes. The mechanism behind the cell entry of lipoplexes has not been conclusively determined as yet, but hypotheses have been put forward.[34] To enhance the cellular uptake of lipoplexed nucleic acids, mixtures of different types of lipids are used for the transfection reagent. Lipids of the transfection complex are believed to interact with the lipid bilayer in the cell membrane and gradually switch position between the transfection complex and the membrane, as a vesicle forms around the lipoplex. As the invagination proceeds, an equilibrium forms with the zwitterionic components of the lipids in the lipid–nucleic acid complex. This process gradually releases the genetic material inside the cytoplasm of the cell where it can enter the nuclear compartment.[34] To enhance the endocytosis of the transfection complexes or to target the complex to a certain type of tissue or cells, a cellular receptor ligand needs to be added to the transfection reagent, or linked to the outside of the preformed condensate.[35] Transferrin has been linked to different types of polymeric transfection reagent such as poly-L-lysine and PEI. The DNA/transferrin–PEI complex can then dock to the transferrin receptor of cells harboring transferrin receptors, such as hematopoetic cells. Different types of organic molecule can be used to target the complexes to tissue and organs *in vivo*. A good example is to target the asialoglycoprotein receptor on hepatocytes.[36] By adding galactose molecules to PEI molecules, it is possible to create a transfection reagent, which can be used *in vivo* to target the transfection complex to the liver. The uptake mechanism for the complex is governed by the asialoglycoprotein receptor-coupled endocytosis. It has been shown that a large portion of injected galactosylated DNA/PEI complexes is rapidly taken up by the hepatocytes. It has become increasingly apparent that by selecting a specific uptake mechanism for the transfection complex, it is possible to avoid the lysosomal degradation pathway. The very same deleterious functions of the lysosome can still be used to aid in the active intracellular release of transfected nucleic acids if pH-dependent membrane-disruptive peptides are used.

[34] Y. Xu and F. C. Szoka, Jr., *Biochemistry* **35,** 5616 (1996).

[35] S. Simões, V. Slepushkin, P. Pires, R. Gaspar, M. C. Pedroso de Lima, and N. Düzgünes, *Gene Ther.* **6,** 1798 (1999).

[36] M. A. Zanta, O. Boussif, A. Adib, and J.-P. Behr, *Bioconj. Chem.* **8,** 839 (1997).

Cell Membrane Penetration. To penetrate the cellular membrane, it is important either to penetrate by physical force, as in the case of ballistic gene delivery where DNA coated gold particles penetrate the cell membrane by kinetic energy content, or to target a discrete cellular uptake mechanism. In ballistic gene delivery the nucleic acid covered particle is ejected from the muzzle of a thin tube and penetrates the surface cell layer of the tissue. In contrast, transfection reagents have to be designed so that the nucleic acid is encapsulated in such a way as to enable the release of the genetic material into the nucleus, or at a location which maximizes the possibility of nuclear entry.

Avoidance/Escape of Endocytic Vesicles. As has been shown, it is very important either to avoid interception by the endosome/lysosome pathway or to be able to escape it. By targeting non-clathrin-coated pits, as is the case for, e.g., cholera toxin, it is possible to avoid lysosomal degradation. Secondary intracellular adaptor proteins involved in the vesicle transport machinery play distinct roles in the conversion of endosomes into lysosomes. This fact highlights the importance of selecting appropriate cellular receptors as targets for transfection. Another method of avoiding lysosomal degradation is used by certain viruses such as adenovirus and influenza virus. These viruses utilize membrane-disruptive functions inherent in their physical structure. Adenoviruses use the penton base on the capsid. As a response to the lowering pH in the lysosome, it changes conformation and becomes inserted into the lysosomal membrane, thereby disrupting it. This action allows the adenovirus to escape the vesicle and subsequently enables the virus capsid to dock to the nuclear pore via capsid protein moieties. It is less than desirable to utilize the penton base as a membrane-disruptive reagent for gene delivery[37,38] due to the fact that a large fraction of the human population has preexisting immunity against adenovirus. The penton base is also highly immunogenic. Another peptide that is more promising is the HA2 peptide from influenza virus.[26,39] It becomes active at pH 5–6[40] and forms membrane-penetrating complexes. In this way it is possible to target the DNA to a protective vesicle out of the way of cytoplasmic nucleases. As soon as the environment gets hostile, as indicated by the lowering pH, the HA2 peptide disrupts the forming lysosome and frees its contents into the cytoplasm.

Synthesis of PNA-NLS Peptides

The rationale for making a molecule with a sequence-specific hybridizing ability and an active peptide moiety is to convey specific functions to nucleic acids,

[37] P. Zhang, P. Andreassen, P. Fender, E. Geissler, J.-F. Hernandez, and J. Chroboczek, *Gene Ther.* **6**, 171 (1999).

[38] S. S. Diebold, H. Lehrman, M. Kursa, E. Wagner, M. Cotton, and M. Zenke, *Hum. Gene Ther.* **10**, 775 (1999).

[39] R. P. Epand, J. C. Macosko, C. J. Russell, Y.-K. Shin, and R. M. Epand, *J. Mol. Biol.* **286**, 489 (1999).

[40] D.-K. Chang, S.-F. Cheng, V. D. Trivedi, and S.-H. Yang, *J. Biol. Chem.* **275**, 19150 (2000).

their derivatives, or their analogs via a "piggyback" system where the peptide/ protein function is anchored to the nucleic acid of interest via a PNA moiety. The sequence of the PNA is to be chosen with the criterion of being excluded from the plasmids as well as from known eukaryotic DNA sequences to avoid possible nonspecific binding. The PNA peptides were attached with a hydrophilic spacer, Fmoc-AEEA-OH, to a stretch of amino acid residues, PKKKRKV, the SV40 core NLS. The complete sequence is PKKKRKV-linker-GCGCTCGGCCCTTCC. In later plasmid experiments, we have used three linkers instead of one to ensure minimal steric interference. Like peptides, PNA is synthesized on a polyethylene glycol polystyrene (PEG-PS) support with a peptide amide linker, yielding a PNA amide on cleavage of the final product.[41,42]

Fluorescence Microscopy Detection of Fluorophore-Labeled Oligonucleotides and Data Processing

To analyze transfected cells for nuclear fluorescence, fluorescence microscopy was performed with a Leica DMRXA microscope (Leica Microsystems Gmbh, Wetzlar, Germany) equipped with a cooled CCD camera (Model S/N 370 KL 0565, Cooke Corporation, NY). Filter sets for DAPI/Hoechst, FITC, Cy-3, and Cy-5 were obtained from Chroma Technology (Brattleboro, VT). The images were acquired and analyzed using the image processing software SlideBook 2.1.5 (Intelligent Imaging Innovations Inc., Denver, CO). Images were postprocessed and mounted in Adobe Photoshop 5.0 (Adobe Systems Inc., San Jose, CA) and printed on a Kodak DS 8650 printer (Rochester, NY) (Fig. 5). The color-separated images from the subcutaneous/intracutaneous injection were imported into NIH image software and the histograms along the indicated line of measurement were assembled, thus showing the Cy-3, Cy-5, and DAPI fluorescence intensities plotted against one another.

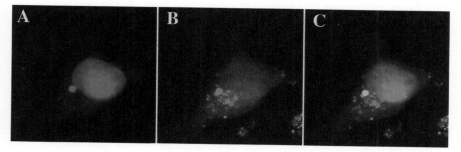

FIG. 5. Differential localization of PNA-NLS and PNA-revNLS hybridized oligonuceotides. (A) The Cy-3 labeled oligonucleotide hybridized to PNA-NLS and subsequently translocated to the nuclei of exposed cells. (B) The cellular spread of Cy-5 labeled oligonucleotide hybridized to PNA-NLS containing an inverted NLS sequence. (C) A merger of A and B channels where the preferential nuclear translocation of the Cy-3 labeled oligonucleotide is visible.

Plasmid Shift Analysis

The plasmids containing PNA target sites were analyzed by agarose gel electrophoresis. Two percent agarose gels were used with 1× TBE buffer running at 100 V for 6–24 hr. The plasmids were visualized with ethidium bromide staining.

Oligonucleotide Shift Analysis

Oligonucleotides were end-labeled with T4-polynucleotide kinase using $[\gamma\text{-}^{32}\text{P}]$ATP, >5000 mCi/mmol. The labeled oligonucleotides were incubated at room temperature with varying amounts of PNA–NLS. For separation of oligonucleotide/PNA–NLS complexes 15% low ionic strength nondenaturing polyacrylamide gels were used[43] (Fig. 6).

Plasmid DNA Preparation

Plasmid DNA was amplified in *Escherichia coli* strain XL-Blue and prepared with the Qiagen Giga, Maxi, or Midi prep kit. It is imperative to have a good plasmid preparation since the transfection includes a prolonged incubation at room temperature. If any nucleases contaminate the preparation, this will invariably result in a degradation of the plasmid to varying degrees depending on the contamination levels.

Hybridization of PNA–NLS to PNA Target-Containing Plasmids

To minimize the hybridization time of the PNA–NLS with the reporter plasmid, a relatively high concentration of DNA solution was used, 0.5 μg/μl. A tenfold molar excess of PNA–NLS was used in the hybridization to plasmids. The DNA sample was deposited in a 1.5 ml Eppendorf tube and the PNA–NLS was added into the lid of the tube. The sample was denatured at 96° for 10 min and put on ice for transport to a 4° precooled centrifuge. The samples were centrifuged for 30 sec, briefly mixed, and thereafter centrifuged for 30 sec again. The samples were then left at room temperature for 8 hr.

---

[41] PerSeptive Biosystems Inc., (PNA) Synthesis: Custom PNA Synthesis, http://www.pbio.com/cat/synth/pna/custmpna.htm.

[42] Oswel Research Products Ltd., Lab 5005, Medical and Biological Sciences Building, University of Southampton, Boldrewood, Bassett Crescent East, Southampton, SO167PX U.K. http://www.oswel.com.

[43] F. M. Ausubel, R. Brent, R. E. Kingston, D. D. Moore, J. G. Seidman, J. A. Smith, and K. Struhl (eds.), *in* "Short Protocols in Molecular Biology," 2nd Ed., Greene Publishing Associates and John Wiley & Sons, New York, 1992.

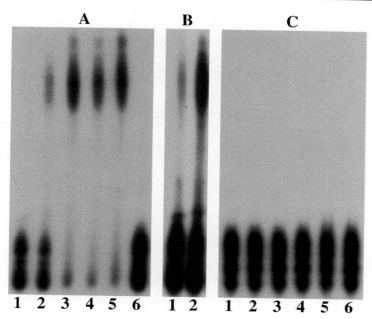

FIG. 6. Oligo shift analysis. Shift assay of antisense and sense oligonucleotides with PNA-NLS dual-function peptide. (A) Oligonucleotide with a target sequence (AS) for the PNA. (B) As for (A), but with prolonged exposure to show the weaker shift of the 1 : 10 hybridization. (C) Oligonucleotide without a target sequence (S) for the PNA. Oligonucleotide concentration was 2.64 pmol in all lanes. PNA–NLS was added to 5′-labeled oligonucleotides at different molar ratios. Lane 1, 1 : 10; lane 2, 1 : 1; lane 3, 10 : 1; lane 4, 100 : 1; lane 5, 1000 : 1; lane 6, 0 : 1. [Reproduced with permission from L. J. Brandén, A. J. Mohamed, and C. I. E. Smith, *Nat. Biotechnol.* **17,** 784 (1999).]

Hybridization of Oligonucleotides to PNA–NLS for in Vitro Transfection

Oligonucleotides were hybridized at 0.05 μg/μl with PNA–NLS at a 10-fold excess. The oligonucleotide solution was deposited in a 1.5 ml Eppendorf tube and the PNA–NLS solution was added. The mixture was allowed to hybridize for at least 8 hr before transfection. The resulting Bioplexes can be either used directly for transfection (Fig. 7) or complexed further with other transfection reagents (Fig. 8).

Plasmid Transfection of Bioplexes in Polyplex Conjugation

An empirical protocol was developed for the Bioplex/Polyplex transfections. 0.1 mM 25 kDa PEI at pH7 was used to complex the PNA–NLS hybridized plasmids. For 5 μg DNA, 2.4 μl of 0.1 mM PEI solution was used. The complex

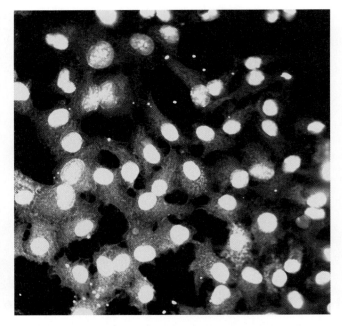

FIG. 7. *In vitro* transfection of HeLa cells. Bioplex transfection of HeLa cells without other transfection reagents. Cy-3 labeled oligonucloetides were hybridized to PNA-NLS and added to the cell culture medium. The cells were assayed after 6 hr.

formation was performed in 0.05 $\mu g/\mu l$ DNA. The transfection mix was added to the cells and incubated for 24–72 hr, depending on the readout system used (Fig. 9).

In Vivo Transfection of Oligonucleotides/PNA-NLS Bioplexes with PEI

The PNA–NLS (33 μM) and Cy-3 labeled oligonucleotide (33 μM) were mixed at a ratio of 1 : 1 in water. An equal amount and concentration of Cy-5 labeled control oligonucleotide (33 μM) was added to the hybridization mixture after 4 hr of incubation at room temperature followed by incubation at room temperature for 12 hr. The final volume was 50 μl. The hybridization mix was complexed with 25 kDa PEI at a charge ratio of 3 : 1, taking the charge of the PNA–NLS into account. The transfection mixture was incubated for 20 min to allow for complex formation, and supplemented with phosphate-buffered NaCl to physiological levels immediately prior to injection. For the intradermal/subcutaneous injection the volume was 10 μl, corresponding to 4 μg of oligonucleotide (Fig. 10).

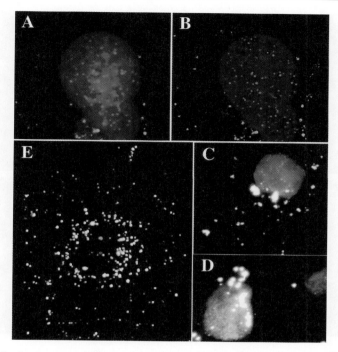

FIG. 8. Confocal images of cells transfected with PNA–NLS or/and PNA-revNLS hybridized Cy-3 and Cy-5 labeled oligonucleotides. Panel A shows a fluorescent image of the nuclear region of a cell transfected with PNA-revNLS (Cy-3, green) and PNA-NLS (Cy-5, red) hybridized oligonucleotides. Panel B shows the same cell after deconvolution. Panels C and D show a confocal image of the nuclear distribution of PNA–NLS hybridized Cy-3 labeled oligonucleotide. Panel E shows the Cy-3 labeled oligonucleotide used in the transfections in panels A–D hybridized with PNA–revNLS.

In Vitro Transfection of Bioplexes

The PNA-NLS (3.3 μM) and Cy-3 labeled oligonucleotide (3.3 μM) were mixed at a ratio of 1 : 1 in water. In the case of competition assays, an equal amount and equal concentration of Cy-5 labeled control oligonucleotide (3.3 μM) was added to the hybridization mixture after 4 hr of incubation at room temperature followed by incubation at room temperature for 12 hr. The final volume was 50 μl. The hybridization mix was complexed with 25 kDa PEI at a charge ratio of 3 : 1, taking the charge of the PNA–NLS into account. The transfection mixture was incubated for 20 min to allow for complex formation. The cells were transfected by adding transfection mixture to the complete cell culture medium and mixing by pipetting up and down gently. The petri dishes were then swirled momentarily for optimal distribution of the transfection mixture.

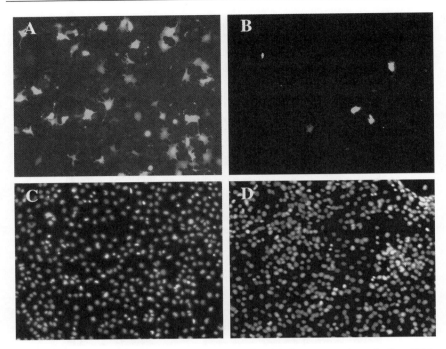

FIG. 9. Plasmid transfections of confluent HeLa cells. Panel A shows a normal transfection of HeLa cells by a GFP plasmid PNA target containing plasmid. Panel B indicates the effect on transfection efficacy after PNA–NLS has been hybridized to the plasmid, thus enhancing the transfection. Panels C and D indicate the positions of the nuclei stained with DAPI from panels A and B, respectively.

Plasmid Constructs

The reporter constructs used have been lacZ (β-galactosidase), luciferase, dsRed-mito (red fluorescent protein fused to a mitochondrial targeting sequence), and EGFP. In the case of luciferase and EGFP, they have been used as a fusion reporter gene. All constructs have been based on the backbone of commercially available plasmids from Clontech and placed into the vector immediately after the poly A site. The PNA target sequences were cloned into the plasmid constructs via oligonucleotides ligated into concatemeric fragments prior to plasmid ligation. Although the number of PNA target sites in the plasmid seems to play an important role for the enhancement of the transfection efficacy, there exists a clonal variation between the constructs. This variation cannot be explained entirely by variation in the number of PNA-target sites.

Cell Lines

The cell lines used were HeLa, 3T3, and Cos-7.

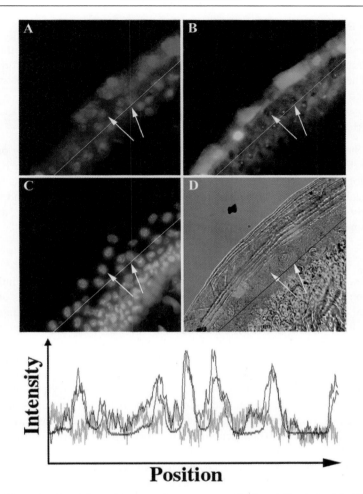

FIG. 10. *In vivo* skin section: subcutaneous/intracutaneous injection of oligonucleotides. The PNA–NLS hybridized oligonucleotide is shown in red (Cy-3), panel A, while the sense oligonucleotide in green (Cy-5) is shown in panel B. The nuclei are shown in panel C in purple. In panel D, a composite image can be seen consisting of gray-scale differential interference contrast (DIC), blue DAPI-stain, red Cy-3 conjugated antisense oligonucleotide, and green Cy-5 conjugated sense oligonucleotide. The fluorescence along the white/black line was measured and the intensities from the different color channels (red, green, and blue) were plotted against one another. The colors from panel A to C correspond to the colors indicated in the histogram. The two fluorochrome-labeled oligonucleotides were synthesized at DNA Technology (Science Park Aarhus, Aarhus, Denmark); antisense (AS) Cy-3–labeled, antisense to the PNA–NLS dual-function peptide; and sense (S) Cy-5 labeled, sense to the PNA–NLS dual-function peptide. [Reproduced with permission from L. J. Brandén, B. Christensson, and C. I. E. Smith, *Gene Ther.* **8,** 84 (2001).]

Sample Preparation

The cell lines were cultivated in 6-well dishes on coverslips prior to transfection. Transfected cells were analyzed with or without fixation in paraformaldehyde. The fixation solution used contained 3% paraformaldehyde and 3% sucrose dissolved in 1× phosphate-buffered saline.

Concluding Remarks

We have developed a novel platform technology allowing anchoring of multiple biological functions directly to nucleic acids. We believe that the Bioplex technology is a versatile tool which could be utilized in basic research as well as in clinical protocols.[44] The technology is still in its infancy, however, and we are presently addressing a number of key aspects relating both to anchoring and the inclusion of additional functional entities.

Future Potential

We see Bioplex technology as a toolbox for any aspect of gene manipulation, including delivery. Our concept is based on utilizing what nature has already provided in terms of functional entities and linking these to the anchoring PNA moiety. Initially, we have concentrated our efforts on small peptides, but clearly polypeptides could also be used. Crucial to future development will be the inclusion of functional moieties in a dynamic fashion, allowing temporal expression of functional entities in a concerted way. Also, we are likely to use microorganisms as a source of inspiration. In viral-mediated gene delivery, a great number of concerted concepts have been deciphered in which hidden functions are freed and subsequently aborted once they have fulfilled their unique task. Given the unprecedented speed by which genome sequences of complete organisms are being revealed, there is certainly no shortage of candidates. The possibility of using components from different organisms provides us with great flexibility, although the complexity of such interspecies interactions may not be apparent until the actual experiment is carried out. Although it is a major challenge, we believe that combinations of various functions using Bioplex technology could eventually provide us with a gene delivery system which is superior to that of viruses. As an example, we envisage that tropism, cellular uptake, and inducible gene expression are all aspects that can be controlled.

Acknowledgments

We thank the Swedish Medical Research Council, the Swedish Cancer Foundation, and the Clas Groschinsky Foundation for their support. We also thank Dr. Jean-Paul Behr for helpful discussions.

[44] R. J. Bartlett, *Nat. Biotechnol.* **16,** 312 (1998).

[7] Surgical Procedures for Intravascular Delivery of Plasmid DNA to Organs

By GUOFENG ZHANG, VLADIMIR BUDKER, PHILLIP WILLIAMS, KRISTINE HANSON, and JON A. WOLFF

Introduction

Gene therapy has been heralded for its potential to revolutionize modern medicine through the use of recombinant DNA for the treatment of inherited and acquired diseases. In the past decade innumerable attempts have been made to effect efficient gene transfer to somatic cells, using both viral and nonviral vectors. Following the discovery by Wolff and his colleagues, which reported that genes could be expressed following direct injection into skeletal muscle,[1] there has been an intense interest in using this method to produce therapeutic protein *in vivo* and to generate antiviral immune response. Our group continues to investigate nonviral delivery methods and has developed a highly efficient method of delivering naked plasmid DNA to many organs via an intravascular route. Using this method, DNA cannot only be transferred to muscle cells, but also to cells of other organs, such as liver, kidney, and intestine.

Intravascular Injection of Plasmid DNA for Liver Gene Transfer

Liver is an attractive organ for gene therapy owing to its important role in metabolism and in many inherited and acquired diseases. Various vehicles have been used for liver gene transfer, including recombinant adenovirus,[2] retrovirus,[3] adeno-associated virus,[4] liposomes,[5] and receptor-mediated methods.[6] We found that hepatic cells can take up naked DNA and express reporter genes when delivered intravascularly. Expression is relatively very high when the intravascular injection of DNA is performed at high pressure and in a large volume.[7] Delivering DNA through the portal vein is an obvious choice, but the hepatic vein, bile duct,[8] and even the tail vein[9,10] are also viable routes.

[1] J. A. Wolff, R. W. Malone, P. Williams, W. Chong, G. Acsadi, A. Jani, and P. L. Felgner, *Science* **247**, 1465 (1990).

[2] M. Suzuki, R. N. Singh, and R. G. Crystal, *Hepatology* **27**, 160 (1998).

[3] O. Kittn, F. L. Cosset, and N. Ferry, *Hum. Gene Ther.* **8**, 1491 (1997).

[4] D. D. Koeberl, I. E. Alexander, C. L. Halbert, D. W. Russell, and A. D. Miller, *Proc. Natl. Acad. Sci. U.S.A.* **94**, 1426 (1997).

[5] S. Kawakami, S. Fumoto, M. Nishikawa, F. Yamashita, and M. Hashida, *Pharm. Res.* **17**, 306 (2000).

[6] T. Bettinger, J. S. Remy, and P. Erbacher, *Bioconjug. Chem.* **10**, 558 (1999).

[7] V. Budker, G. Zhang, S. Knechtle, and J. A. Wolff, *Gene Ther.* **3**, 593 (1996).

Intraportal Injection for Liver Gene Transfer in Mouse

ICR mice of 25–30 g are suitable for use. The animal is anesthetized by intramuscular injection of ketamine (80 mg/kg) and xylazine (2 mg/kg); metofane inhalation may be added if necessary. A mid-sagittal incision from the xiphoid process to the lower third of the abdomen is made to open abdominal cavity. For the purpose of clearly exposing the liver, the xiphoid process is pulled in the cephalic direction with a hemostat and the tail is taped on the table. Both sides of the abdominal wall are pulled laterally using retractors. The front part of the falciform ligament from the front edge to the vena cava is cut carefully to enable clamping of the vena cava and hepatic vein. A curved microvessel clamp is held by a forceps and then is placed on the vena cava and hepatic vein through the upper side of the liver. There are several branches of hepatic vein that feed into the vena cava so the clamp should be placed close to the liver and as deep as possible. If the clamp does not block all of the hepatic veins, the transfection will not be successful.

After blocking of the hepatic vein, the portal vein is injected immediately. The portal vein is easily found by pushing the intestine to left side of the abdominal cavity or by removing the intestine from the abdominal cavity. Fixing the portal vein using a cotton-tipped applicator, 1 ml of a normal saline solution containing 15% mannitol, 2.5 units of heparin, and 10 to 100 μg plasmid DNA is injected into portal vein over 30 sec using a 1 ml syringe and 30-gauge needle. A piece of gelfoam (Pharmacia and Upjohn Co., Kalamazoo, MI) (4 mm × 4 mm) is placed on the injection site before the needle is removed, and pressure is maintained on the gelfoam using the applicator. Bleeding should stop 1 to 2 min after removal of the needle. The clamp is taken off 2 min after DNA injection. The abdominal cavity is closed by suturing. Injection velocity, solution volume, successful blocking of the hepatic vein, and mannitol solution each affect the transfection efficiency greatly.[7]

Intra-Vena Cava Injection for Liver Gene Transfer in Mouse

The anesthesia and liver exposure are the same as that for portal vein injection. The vena cava is separated upstream of the renal vein and the falciform ligament is cut as in the intraportal injection. Two clamps are put on the vena cava; one is located downstream of the hepatic vein and one is upstream of the renal vein. The hepatic artery and portal vein can be left free or blocked. One ml of 15% mannitol–normal saline solution containing 2.5 units of heparin and 10 to 100 μg of plasmid DNA is injected through a 1 ml syringe and 30-gauge needle. The vena

[8] G. Zhang, D. Vargo, V. Budker, N. Armstrong, S. Knechtle, and J. A. Wolff, *Hum. Gene Ther.* **8,** 1763 (1997).

[9] G. Zhang, V. Budker, and J. A. Wolff, *Hum. Gene Ther.* **10,** 1735 (1999).

[10] F. Liu, Y. Song, and D. Liu, *Gene Ther.* **6,** 1258 (1999).

cava is fixed and cannot be moved because of the connective tissue and peritoneum; therefore two key points are very important for good results. First, a 30 degree angle should be bent into the needle so that during the injection the needle can be kept almost parallel to the vessel and thereby avoid penetrating the back wall of the vena cava. Second, when performing the injection, while one hand operates the syringe, the other hand should secure the needle. The clamps are released 2 min after injection. Closing and recovery are the same as that for intraportal injection. Solution volume, injection velocity, leakage prevention, and mannitol solution are all critical as in the intraportal injection.

Intra-Bile Duct Injection for Liver Gene Transfer in Mouse

Anesthesia and exposure are the same as that for intraportal injection. The bile duct can be easily found by pushing the intestine to the left side of the abdominal cavity. A microvascular clamp is put on the distal part of the bile duct to block bile flow. The bile duct should be extended with bile after blocking and comparatively easy to inject. If it is not extended enough, the gall bladder should be pressed using two cotton-tipped applicators. The bile duct is fixed by using a cotton-tipped applicator and a 30-gauge needle is inserted into the bile duct in the retrograde direction, after which the DNA in 15% mannitol–saline solution can be injected. The volume of DNA solution can be 0.3 to 1 ml and injection can be varied. After injection, a small piece of gelfoam is put on the injection site and pressure is maintained using an applicator. Two minutes after injection the clamp is taken off and the injection site is checked for bile leakage. No special care is needed after surgery. A modification has been made for the intra-bile duct injection; 1 ml of DNA 15% mannitol–saline solution is injected into the bile duct with the hepatic vein blocked, as for the intraportal injection. This method can enhance expression and decrease variability.

Tail Vein Injection for Liver Gene Transfer in Mouse

The animal is lightly anesthetized by metofane inhalation. Other types of anesthesia usually are not appropriate, as recovery times are often lengthened and there is increased mortality. For a 20- to 30-g mouse we usually inject 2 to 3 ml of DNA solution, or about 1 ml per 10 g of body weight. DNA should be in Ringer's solution containing 5 units heparin since injection of normal saline increases mortality. The DNA dose is 10 to 100 μg per animal, but we have used as little as 100 ng and as much as 200 μg. After dilating the tail vein using warm water or a light, the thumb and ring finger are used to hold the distal part of the tail and blood flow is blocked at the root of the tail using the index finger and median finger of the same hand. Then 2–3 ml of solution is injected into the tail vein through a 3 ml syringe and 27-gauge needle. Meanwhile pressure at the root of the tail is relaxed slightly to allow the solution to pass through the vessel.

The velocity of injection is very important; if the injection takes 1 min or more, the expression is low. Best results are achieved when the injection is completed within 7 to 10 sec. After tail vein delivery of plasmid DNA the expression can be found not only in the liver, but also in most organs, including kidney, adrenals, gonads, pancreas, spleen, stomach, intestine, lung, and muscle. Kidney and spleen, in particular, have very good expression.

Some modifications for tail vein injection are useful for enhancing organ-specific expression. For example, blocking the vena cava downstream of the hepatic vein increases expression in most abdominal organs, including the liver. Likewise, blocking the vena cava between the kidney and liver markedly increases expression in kidney, adrenal glands, and gonads.

Discussion

The gene expression levels are comparable between injection methods, although the vena cava injection method often yields slightly higher expression. The tail vein injection method is very easy to handle and expression is very efficient as it requires delivery of only a very small amount of plasmid DNA. Because of the case of operation and the high sensitivity, this method allows the mouse liver to be used like an *in vivo* test tube.

The intra-vena cava injection and intra-bile duct injection methods are comparably more difficult. The vena cava is very thick and easily injected, but it is difficult to prevent leakage due to the mobility of the vena cava. However, if a good injection is achieved, the expression is usually somewhat higher than that with other injection methods. The main advantage of this method is its adaptability for future clinical trials using the intravascular balloon technique to deliver DNA into the hepatic vein. The intra-bile duct injection is more difficult because of the relative thinness of the duct, but is also potentially adaptable for clinical trials.

All of the above methods for mouse liver gene transfer have proved to be suitable for larger animals such as rat, rabbit, and dog. The solution volume and DNA concentration should be adjusted, and in some cases the blocking method should be changed. For example, in rats we use 10 to 15 ml of DNA solution and place two clamps on the vena cava; one is located between the renal vein and hepatic vein, the other is downstream of the hepatic vein. This blocking method is suitable for both intra-vena cava and intra-portal injection.

Intravascular Delivery of Naked Plasmid DNA to Skeletal Muscle

Since the discovery that reporter genes could be expressed following direct intramuscular injection of naked DNA, this technique has been widely used. Many attempts have been made to improve the transfection efficiency of the intramuscular injection technique, including incorporating minor tissue damage[11] and electroporation.[12] Although these efforts did improve expression levels, the expression itself was still very localized. Based on our success with intravascular injection of naked DNA to the liver, arterial delivery of DNA to rat muscle was attempted with very positive results. High levels of reporter gene expression were found throughout the hindleg and foot muscles.[13] This intraarterial injection technique has since been applied to large animals such as rabbit, monkey, and dog.

Intraarterial Injection of Naked DNA for Rat Muscle Gene Transfer

A 150-g rat is anesthetized using intramuscular injection of ketamine (80–100 mg/kg) and xylazine (2 mg/kg). A mid-sagittal incision from the upper third of the abdomen to the end of lower abdomen is made. Each side of the abdominal wall is pulled laterally and the internal organs in the lower part of the abdominal cavity are pulled to the left side by retractors. The peritoneum and connective tissue are pushed away from the front of the iliac artery using cotton-tipped applicators until the vessel surface becomes clear. The external iliac artery and vein are separated from the surrounding tissue using curved Micro Tying Forceps and suture silk is placed around the two vessels. The caudal epigastric vessels (artery and vein), internal iliac vessels, vessels of the deferent duct, and caudal gluteal vessels are separated using the same method. Clamps are put on the separated vessels to block blood flow. The external iliac vessels are blocked last, clamping as close as possible to the origin of the artery. The external iliac artery is fixed by pulling the two ends of the silk which was placed there previously and then 3 ml of normal saline containing 0.5 mg of papaverine (Sigma Co.) is injected into the artery through a 3 ml syringe and 30-gauge needle. Five minutes later, 10 ml of normal saline containing 500 μg of plasmid DNA is injected within 7 to 15 sec through the original site of the papaverine injection. A 10 ml syringe and 27-gauge butterfly needle are used for this DNA injection. A 6 mm × 6 mm gelfoam is placed on the injection site and pressure is maintained using a cotton-tipped applicator. Two minutes after injection the clamps are removed and the pressure on the gelfoam is continued until all bleeding has stopped. The animal is put back in its normal environment, and no special care is needed after suturing the abdominal cavity.

Vessel blocking, solution volume, injection velocity, animal size, DNA dose, and vascular permeability are important parameters for gene transfer in this protocol. Our experience indicates that 10 ml of solution is suitable for a 130- to 160-g rat.

[11] I. Danko, J. D. Fritz, S. Jiao, K. Hogan, J. S. Latendresse, and J. A. Wolff, *Gene Ther.* **1,** 114 (1994).

[12] H. Aihara and J. I. Miyazaki, *Nature Biotechol.* **16,** 867 (1998).

[13] V. Budker, G. Zhang, I. Danko, P. Williams, and J. A. Wolff, *Gene Ther.* **5,** 272 (1998).

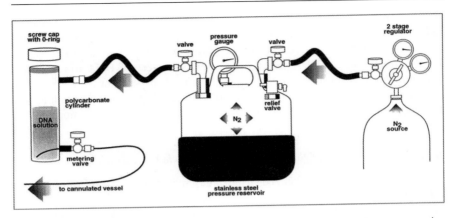

FIG. 1. Pressure injection system for intravascular delivery of plasmid DNA. The system consists of a two-stage regulator (Fisher Scientific), a pressure reservoir (Millipore Corp.) equipped with a relief valve and pressure gauge, and a polycarbonate cylinder (W. A. Hammond Drierite) holding the DNA solution. A metering valve (Cole-Parmer) is manually opened to allow the DNA solution to flow into the catheterized vessel.

If the animal is larger than this, the expression decreases unless the solution volume is increased. Effective blocking of the blood flow between the leg and the body, sufficient solution volume, and high injection velocity work together to create very high pressure inside the vessels, enabling pDNA extravasation. Changes in any one of these parameters greatly affect expression. Increasing vessel permeability, in particular, clearly improves transfection efficiency.

Intraarterial Injection for Monkey Muscle Gene Transfer

In order to accomplish the injection of larger volumes of DNA in the relatively short time span needed for optimum transfection in large animals, such as monkey, we designed a pressure injection system using compressed nitrogen (Fig. 1). A two-stage regulator (Fisher Scientific) is used to decrease tank pressure to that desired for injection. This controls the speed of injection, i.e., lower pressure for longer time of injection. A pressure reservoir (Millipore Corp.) equipped with a relief valve and pressure gauge is used to ensure that constant nitrogen pressure and flow are delivered to a polycarbonate cylinder (W. A. Hammond Drierite) holding the DNA solution. A metering valve (Cole-Parmer) is manually opened to allow the DNA solution to flow into the catheterized vessel. The flow of the solution is visually monitored and the valve closed when the desired volume has been injected.

The injection procedure for the monkey is as follows: the monkey is anesthetized by intramuscular injection of ketamine followed by halothane inhalation.

For a forearm injection, a longitudinal incision, 3 cm in length, is made on the skin along the inside edge of the biceps brachii and 2 cm above the elbow. For a lower leg injection, the incision is located on the upper edge of the popliteal fossa. A retractor is useful for exposing the vessels. A 20-gauge catheter is inserted into the brachial or popliteal artery anterogradely and ligated in place with a silk suture after the surrounding tissue and veins are separated. Blood flow is blocked by a sphygmomanometer belt surrounding the arm or leg proximal to the injection site. After the sphygmomanometer is inflated above 300 mmHg, 30 ml of normal saline containing 5 mg papaverine (Sigma Co.) is injected into the catheterized vessels. After 5 min, 120 ml of DNA solution (for the arm) or 180 ml (for the leg) is injected rapidly through an air pump driven by nitrogen gas. The injection typically takes 30 to 60 sec. A piece of gelfoam is placed on the injection site and pressure is maintained until the catheter is removed and the bleeding has stopped (or, if necessary, until the artery has been repaired). Two minutes after injection, the sphygmomanometer belt is deflated and the incision is sutured.

A modification has been made in large animals for intraarterial injection of plasmid DNA. A long cannula is inserted from the proximal part of the artery, past the position of the sphygmomanometer belt, to the distal portion of the artery. The blocking and injection are the same as that described above. The cannula modification makes it possible to transfect multiple extremities or organs through a single incision, which is very attractive for clinical applications.

Discussion

The two methods described above are applicable to most animals. Small animals or small extremities are best suited to the rat artery injection method, whereas larger animals and larger extremities are transfected well using the monkey intraarterial method.

Limb mobility was normal in all animals following recovery from surgery. Muscle slides show no obvious damage except for very rare necrosis. Slight arterial damage has been observed but seems to be transient. The blood CPK level does increase after injection, but only twofold higher than that which typically occurs following exercise, and returns to normal within several days. The intraarterial delivery method may have clinical utility because of its greater efficiency and wide range of transfection.

Intravascular Injection of Plasmid DNA to Other Organs

As mentioned before, many organs can express reporter genes following intra-tail vein injection. Therefore, the intravascular injection technique should be widely applicable as a basic method of gene transfer to many organs. Because there is large variation from organ to organ with respect to the structure, blood supply, cell biological characteristics, etc., optimum conditions for gene transfer will also vary from organ to organ. Strategies for optimization are discussed below.

Intravascular Injection for Mouse Kidney Gene Transfer

The anesthesia and organ exposure are the same as described in the intraportal injection protocol. The vena cava is separated from surrounding tissue both upstream and downstream of the renal vein. Two clamps are used to block the vena cava blood flow: one located downstream of the renal veins, and another upstream about 0.5 to 1 cm away from the renal veins. Pressure is best maintained if the inferior phrenic and spermatic vessels are blocked. A 30-gauge needle bent at a 30 degree angle is inserted into the vena cava upstream of the renal vein and is pushed forward to position the tip of the needle close to, but not over, the juncture point of the renal vein and the vena cava. Then, 0.4–0.6 ml of DNA solution (10–100 μg of DNA in 15% mannitol–normal saline solution containing 2.5 units of heparin per ml) is injected at about 1 ml/20 sec. A piece of gelfoam is used to stop bleeding and clamps are removed 2 min post-injection. No special care is needed after closing the incision.

The same protocol can be used for bilateral gene transfer to rat kidney. However, because the renal artery and vein are longer and thicker, a single kidney is usually selected as a target. If the renal vein is selected for gene delivery, the vein must be clamped at the juncture point of the renal vein and the vena cava. The renal artery can be kept free or blocked. If the artery is selected, both the artery and the vein should be blocked. It is important to remember that when the left kidney is selected as the target organ the inferior phrenic and spermatic vessels should be clamped as well.

Intravascular Injection for Intestine Gene Transfer in Mouse

Anesthesia and organ exposure are performed as described previously. The superior mesenteric vein is clamped at its juncture with the splenic vein. A 30-gauge needle is inserted into the mesenteric vein in the retrograde direction of blood flow. One to 2 ml of DNA solution is then injected at high velocity. Gelfoam is used to stop bleeding and the clamp is released 2 min later.

Intravascular gene transfer to rat intestine is somewhat different from that in mice. Usually we select a segment of intestine as the target for gene transfer. The two ends of the targeted segment of intestine are clamped, including the intestine itself and the collateral vessels which connect the target segment and the neighboring area. The main trunk of the artery and vein which supply the target or drain blood from the segment should be clamped, as well as any branches which are concerned with the collateral circulation between the segment and neighboring area. DNA solution is injected through the artery anterogradely or through the vein retrogradely. The volume of DNA solution is varied according to the length of the selected intestine segment.

Intravascular Injection for Ovary Gene Transfer in Rat

Anesthesia and exposure are the same as that described for the intraportal injection. The ovary receives its blood supply from two sources, the ovarian artery and the uterine artery, and there is rich collateral circulation between these two systems. Therefore the DNA solution can be delivered through either of them. If DNA is delivered through the ovarian vessel, the ovarian artery and vein should be clamped at a high position and all collateral vessels between the ovary and the uterus should be clamped as well. To inject through the uterine vessel, the ovarian vessels are clamped close to the ovary. Uterine vessels should be clamped far away from the ovary to leave enough space for injection. The ovarian artery or ovarian vein and uterine vein can be selected for delivering 0.3–0.5 ml of DNA solution, but not the uterine artery since it is too thin. Closing and recovery are the same as described previously.

Intravascular Injection for Testis Gene Transfer in Rat

Two positions can be selected for DNA delivery to testis. One is at a high position of the testicular artery and vein; in this case the artery and vein should be separated and clamped at the original position and also should block the artery and vein of the deferent duct. Both the testicular artery and vein can be used for injection and 0.5–0.8 ml of DNA mannitol solution should be suitable. Another choice is the surface vessels of the testis; before injection all inflow and outflow vessels should be blocked close to the testis. Then, 0.3–0.5 ml of DNA solution can be injected into the surface artery.

Discussion

It is clear that for most organs if high pressure is obtained inside the vessels, highly efficient gene transfer occurs. It is possible that the high pressure enables pDNA extravasation, and the pDNA is then picked up by cells.[14] Alternatively, the high pressure may directly affect the membrane of targeted cells so as to enable pDNA uptake. Regardless of the mechanism, the intravascular injection offers an efficient and safe method for gene delivery to many organs. Additionally, our data show that immunosuppression promotes long-term expression following intra-arterial delivery of plasmid DNA to rat muscle.[15] Other groups have demonstrated long-term gene expression in liver using liver-specific promoters.[16] These are exciting advances which hold great promise for use in gene therapy.

[14] V. Budker, T. Budker, G. Zhang, V. Subbotin, A. Loomis, and J. A. Wolff, *J. Gene Med.* **2,** 76 (2000).
[15] G. Zhang, V. Budker, P. Williams, V. Subbotin, and J. A. Wolff, *Hum. Gene Ther.*, in press (2001).
[16] C. H. Miao, K. Ohashi, G. A. Patijn, L. Meuse, X. Ye, A. R. Thompson, and J. A. Wolff, *Mol. Ther.* **1,** 522 (2000).

[8] Direct Gene Transfer into Mouse Heart

By MASSIMO BUVOLI and LESLIE A. LEINWAND

Introduction

Direct gene transfer into skeletal and cardiac muscles is most notable for its technical simplicity. This approach involves injection of purified naked DNA or recombinant viruses into the muscle tissues with a narrow-gauge hypodermic needle.[1] The possible mechanisms underlying the DNA uptake remain poorly understood.[1] However, it may be that the unique anatomy and physiology of muscle are accountable for the efficient uptake of polynucleotides, since other types of tissues do not express injected DNA at nearly the same levels as muscle.[2]

DNA injection into the myocardium has been reported in different animal models.[2–7] In general, following DNA injection, recombinant gene expression can be detected as early as ~12 hr, peaks in 2–3 weeks, and is generally stable for a few months with the DNA remaining as a circular or linear episome.[8] Histological studies have shown that transgene expression is restricted to cardiac myocytes and cannot be detected in endothelial cells or connective tissue.[3] Thus, this natural tissue specificity represents a worthy safety issue when direct intramyocardial injection of naked DNA is contemplated for gene therapy aims.

The major drawback of this methodology resides in the low efficiency of gene transfer. In fact, only a relatively low number of cardiac cells around the needle track can be genetically reprogrammed.[2,3,7] For this reason, this technique has limited, although useful, therapeutic applications. For example it can be used to study the effect of gene products that do not need to be widely expressed. Direct DNA injection into the myocardium has been successfully employed for testing the ability of recombinant angiogenic factors (as VEGF) to stimulate collateral vessel formation in areas of ischemic myocardium.[9] Moreover, we have employed

[1] V. Budker, T. Budker, G. Zhang, V. Subbotin, A. Loomis, and J. A. Wolff, *J. Gene Med.* **2,** 76 (2000).

[2] G. Acsadi, S. Jiao, A. Jani, D. Duke, P. Williams, C. Wang, and J. A. Wolff, *New Biol.* **3,** 71 (1991).

[3] H. Lin, M. S. Parmacek, G. Morle, S. Bolling, and J. M. Leiden, *Circulation* **82,** 2217 (1990).

[4] R. Kitsis, P. Buttrick, E. McNally, M. Kaplan, and L. A. Leinwand, *Proc. Natl. Acad. Sci. U.S.A.* **88,** 4138 (1991).

[5] D. Gal, L. Weir, G. LeClerc, J. G. Pickering, J. Hogan, and J. M. Isner, *Lab. Invest.* **68,** 18 (1993).

[6] R. von Harsdorf, R. J. Schott, Y. T. Shen, S. F. Vatner, V. Mahdavi, and N. Nadal-Ginard, *Circ. Res.* **72,** 688 (1993).

[7] K. Li, R. E. Welikson, K. L. Vikstrom, and L. A. Leinwand, *J. Mol. Cell. Cardiol.* **29,** 1499 (1997).

[8] M. Buttrick, A. Kass, R. N. Kitsis, M. L. Kaplan, and L. A. Leinwand, *Circ. Res.* **6,** 616 (1992).

[9] R. A. Tio, T. Tkebuchava, T. H. Scheuermann, C. Lebherz, M. Magner, M. Kearny, D. D. Esakof, J. M. Isner, and J. F. Symes, *Hum. Gene Ther.* **10,** 2953 (1999).

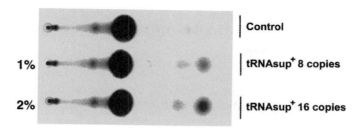

FIG. 1. *In vivo* suppression of a premature termination codon achieved by injection of plasmids carrying multicopy tRNA suppressor genes. The hearts of transgenic mice expressing a reporter gene (chloramphenicol acetyltranferase) containing an ocher mutation were injected with different constructs carrying 8 or 16 copies of a suppressor tRNA gene. CAT activities produced after the rescue of the CAT ochre mRNA are shown beside each construct. The control lane corresponds to the CAT ochre-expressing transgenic mouse injected with normal saline alone. Fifty μg of each tRNA suppressor plasmid was injected in 20 μl of normal saline (modified from Buvoli *et al.*[10]).

this technique to test tRNA suppression as a gene therapy approach for nonsense mutations *in vivo* (Fig. 1).[10]

In addition to its potential medical applications, direct DNA injection also represents a powerful molecular tool for studying cardiac gene transcription.[4,11] This approach, bypassing the limitations of transient transfections of cultured cardiac myocytes, can be used to investigate the influence of complex physio/pathological states (e.g., volume overload, hypertension) on cardiac gene expression *in vivo*.

In this article the procedure we have developed for direct DNA injection into the mouse myocardium is outlined. The availability of a large number of mouse genetic models for human diseases make this approach an excellent choice for the evaluation of new molecular drugs as well as for studying the molecular pathways involved in cardiovascular disorders.

Anesthesia

A stock solution of 100% avertin is prepared by dissolving 10 g of 2,2,2-tribromoethanol (Aldrich) in 10 ml of *tert*-amyl alcohol (Aldrich). The solution is vortexed for several minutes to completely dissolve the crystals produced after mixing the two components. The stock is then diluted to 2.5% working solution in sterile phosphate-buffered saline (PBS) (20×: 160 g/liter NaCl, 4 g/liter KCl, 43 g/liter $Na_2HPO_4 \cdot H_2O$). The appropriate general dose of avertin, corresponding to ~12, 15 μl/g body weight (see below), is determined each time a new stock solution is prepared.

---

[10] M. Buvoli, A. Buvoli, and L. A. Leinwand, *Mol. Cell. Biol.* **20,** 3116 (2000).

[11] M. S. Parmacek, A. J. Vora, T. Shen, E. Barr, F. Jung, and J. M. Leiden, *Mol. Cell. Biol.* **12,** 1967 (1992).

In general, although both solutions can be stored protected from light at 4° for a few months, the 2.5% solution is made fresh just before the surgery.

DNA

We routinely inject naked DNA isolated by a standard double CsCl–ethidium bromide gradient ultracentrifugation.[12] After the ethidium bromide is removed by isoamyl alcohol, the DNA is dialyzed overnight at 4° against a large excess (3000×) of Tris-HCl 10 mM, pH 7.5, to completely eliminate the cesium chloride and other toxic heavy metal ions. The DNA is then ethanol precipitated and directly resuspended in sterile USP saline or water. Buffers containing EDTA should be avoided, since this component can cause an increased inflammatory response.

DNA concentration is measured by absorption at 260 nm and 280 nm and the quality of the DNA analyzed by agarose gel. DNA is aliquoted and stored at −20°.

Adenovirus

Virus stocks expanded according to standard procedures[13] are aliquoted in small volumes and stored at −80°. Before injections, viruses are dialyzed in 10 mM Tris-Cl pH 7.6, 135 mM NaCl, 1 mM MgCl$_2$, 10% glycerol for 3 hr at 4° and then diluted in cold PBS.

Surgical Procedure

The optimal dose of avertin has to be determined for each mouse since different mouse genetic backgrounds react to small overdoses of anesthetic by critically depressing the respiratory functions.

After the mouse is positioned on its back (by holding the mouse's tail between the little finger and the hand and the skin behind the ears between the thumb and the index finger) avertin is injected intraperitoneally to the left of the mouse midline at about one-third of the distance from the umbilicus to the pubis. The needle (26 G 1/2) is inserted at a 30–45° angle for 4–5 mm to avoid the risk of puncturing the abdominal organs. The injection is administered after a brief aspiration if no blood or fluid enters the syringe. In general, the anesthesia takes effect within 5 min, lasts for ∼20–30 min, and is followed by a variable period of postoperative partial anesthesia.

When working with new mouse strains, it is advisable to start with suboptimal doses of avertin and carefully monitor the frequency of respiration. If necessary,

---

[12] J. Sambrook, E. F. Fritsch, and T. Maniatis, "Molecular Cloning: a Laboratory Manual," 2nd Ed., p. 142. Cold Spring Harbor Laboratory Press, Cold Spring Harbor, NY, 1989.

[13] J. Schaack, S. Langer, and X. Guo, *J. Virol.* **69,** 3920 (1995).

the administration(s) of a booster, corresponding to 10% of the initial dose, is then repeated until the absence of response to painful stimulation (tail pinch). This scheme can also be repeated during the surgery, if the level of anesthesia becomes inadequate. Although additional doses may result in additive effects, we found that this procedure can reduce dramatically the mortality of several transgenic strains.

After anesthesia, animals are placed in a supine position with anterior and posterior limbs extended and taped to the surgical table. The hair from the area around the sternum is then closely shaved and the skin treated with iodophor swabs. A left lateral thoracotomy is carried out as follows. A 1- to 1.5-cm vertical incision is made along the side of the sternum. A microdissecting straight forceps (Roboz RS-5130, tip width 0.8 mm, length 4 inches) is then gently pushed under the transverse pectoral muscle until its complete dissection from the other layers of the chest musculature. The same process is repeated to dissect the deep pectoral muscle. It is important to bear in mind that maintaining the integrity of the two muscles can reduce chances of pneumothorax at the time of the chest closure. In preparation for a fast suture of the muscle layers during the last phases of the surgery a 4-0 chromic gut (Ethicon) is then inserted through the transverse pectoral and deep pectoral muscle. When the ribs are exposed, the apical impulse corresponding to the most pronounced cardiac pulsation is individuated and the rib just below it is cut with a small pair of straight scissors (Roboz RS-5882) oriented perpendicularly to the chest wall. If an intercostal artery is visualized along the chosen rib, it is necessary to cauterize it (Roboz Cautery RS-208) before the incision. In addition, caution should be taken to avoid damage to the internal mammary artery running medially along the chest wall. Small curved hemostatic forceps (Roboz Mosquito, RS-7111) are subsequently introduced in the chest cavity perpendicularly to the chest with the tips in line with the incision. During this step it is crucial to prevent injures to the heart or lungs. When the incision is widened by turning (45° clockwise turn) and opening the hemostatic forceps, the lungs deflate due to the equalization of intrapleural and external pressures. After the index and the middle finger of the left hand are gently pushed against the mouse shoulders, the heart is pushed out of the thoracic cavity with the left thumb (Fig. 2A). The hemostatic forceps are then removed from the incision and DNA–saline solution or recombinant adenoviruses are injected through a 30-gauge needle into the left ventricular wall using the right hand and 50-μl Hamilton syringe (Fig. 2B). Since the speed of the injection does not affect transgene expression, solutions are delivered within few seconds. Following injection, the heart is gently pushed back in the thoracic cavity with the right index finger and immediately the muscle layers are closed using the pre-inserted chromic gut. The chest cavity is then squeezed to expel any trapped air, a thin coat of Superglue (commercial cyanoacrylate adhesive) is applied over the muscles, and before the glue dries, the remaining air is expelled by squeezing the chest for a second time. The Superglue treatment, perfectly sealing the thoracic cavity, drastically reduces the formation of pneumothorax. The skin is closed with

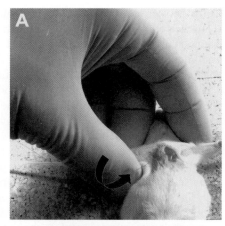

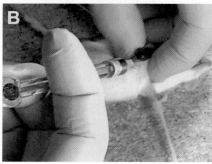

FIG. 2. Exteriorization of the mouse heart, and direct DNA injection. (A) The maneuver for the heart exteriorization is shown with the arrow indicating the direction of the pressure applied to the mouse abdomen. (B) During the injection the heart is immobilized between the thumb and the left index finger.

one staple (Roboz clip applier RS-9260). At the end of the surgery, mice are usually left under a heat lamp until they begin to react to mild painful stimulations, then they are transferred back to the cage. No postoperative analgesia seems to be necessary since the mice return to their normal activities in less than 2 hr without any impairment. Sometimes, 10–20 min after surgery, a small percentage of mice show dramatic reduction and/or irregularities of the respiratory functions. When this condition arises, a few tail pinches (repeated when necessary) can restore the normal respiratory rhythm.

Finally, intraperitoneal injection of 0.1–0.2 mg of dexamethasone just after the surgery can improve the postoperatory recovery of strains particularly sensitive to the anesthesia.

Our laboratory has also performed several heart injections via a subdiaphragmatic approach. Although this method is apparently simple, it does not appear to be reliable, since just a small percentage of mice expressed low amounts of reporter genes.

Critical Parameters

When manual dexterity is achieved, the entire surgery can be done in less than 10 min. Pushing the heart out of the thoracic cavity is the most difficult step of the procedure. In particular, care has to be taken to avoid damage of the great vessels, exteriorization of the left lung, or formation of a long tear along the intercostal muscles flanking the rib incision. Since at this point the respiration is abolished, the exteriorization of the heart has to be achieved in just a few seconds or the number of survivors will decrease dramatically.

If the surgery is carried out in the correct manner, the survival rate (with 3 week- to 15-month-old mice) ranges between 60 and 90%, depending on the mouse strain used. Mortality due to depressed respiration occurs primarily during the first 2 hr, as well as mortality due to the extremely toxic effect of traces of CsCl or EDTA still present in the DNA preparation.

Twenty to 40 μg of DNA in a volume of 15–25 μl is usually injected. Although the expression levels of the transgene do not appear to be linear in this range, we found that variability (see below) can be reduced if amounts of DNA close to the high end of the range are injected.

DNA is usually injected in normal saline solution although we did not observe a decrease in expression when PBS was used. Gene expression generally peaks at 5–12 days post-injection, then slightly declines but remains stable for 6–9 weeks.

Since chemical DNA carriers improve gene transduction *in vivo,* we tested the ability of the cationic polymer polyethyleneimine (PEI) to enhance the uptake and expression of a plasmid carrying a CAT reporter gene. However, when the polymer–plasmid complexes were injected into the heart (10 μg of DNA was incubated in 150 mM NaCl with PEI 25 kDa with a ratio of PEI nitrogen to DNA phosphate corresponding to 10 : 1 and 15 : 1) a deleterious effect on gene expression was observed.

Variability among animals (15–20%, as calculated by standard deviation) represents one important issue that needs to be carefully considered. In fact, the functional comparison between promoters, or their response to physiological or induced pathological conditions, cannot be easily studied if the differences in activity are small.

The primary source of variability lies in the difficulty in consistently injecting the DNA into the thin cardiac wall of the left ventricle avoiding any DNA loss in the ventricular chamber. For this reason we advise those with a minimal surgical background to perform several injections using a solution containing 3% Evan's blue dye that allows a transient visualization of the area injected.

Although to a lesser extent, variable degradation of the injected DNA or loss of reporter gene activity during the preparation of the tissue extract (see below) can also affect the reproducibility of this technique.

The coinjection of a second reporter gene (whose product can be used to normalize for transduction efficiency) can be employed to correct the variability caused by poor injections. However, we found that the results obtained after normalization do not always coincide with those obtained when we just increased the number of animals injected. There are different possible explanations that account for this inconsistency. As previously reported, the activity of an internal control plasmid can be reduced or suppressed by the presence of another plasmid.[14,15] Therefore, if the cardiocytes take up the two plasmids randomly and, as a result, do not fully colocalize, it is likely that a variable percentage of the control construct can escape the potential negative transcriptional effects of the other plasmid. This would, in turn, erratically change the activity of the internal control, affecting the interpretation of the results. Moreover, when different cellular conditions are tested, it is possible that the two promoters can react differently to the new stimuli. Based on these potential problems, we usually do not include an internal control plasmid in our experiments.

Variability can be substantially reduced if recombinant adenoviruses are employed. This probably reflects the remarkable efficiency of gene delivery shown by these vectors. However, we found that the injection of adenovirus can dramatically affect mouse mortality. We have noticed that shortly after the delivery of the virus, signs of viral toxicity are present in a large percentage of the mice: uncontrollable muscle contractions and spasms, deep depression of respiration, and paralysis of the posterior limbs. To avoid additive lethal effects due to the presence of potential toxic compounds in the virus storage buffer (i.e., BSA, high concentration of glycerol, traces of CsCl) viruses are usually dialyzed a few hours before the injection. With the optimized viral dose of $1-2 \times 10^8$ plaque-forming units (pfu), mortality was around 60–65%.

Harvesting and Homogenization of the Heart

Mice are normally sacrificed by cervical dislocation. Ribs are then cut along the left side of the sternum, and the heart exposed and ventricles separated from the atria and great vessels. Before homogenization, the blood is removed by gently squeezing the ventricles in a large volume of ice-cold PBS. For reporter gene activity, the tissue is first minced in a small sterile petri dish, transferred in 5 ml polypropylene Falcon tubes (#2063), and then homogenized in 600–800 μl of buffer H (25 mM glycylglycine pH 7.8, 15 mM MgSO$_4$, 4 mM EGTA pH 7.8, 1 mM dithiothreitol). Homogenization is performed immediately after the harvesting,

[14] A. Farr and A. Roman, *Nucleic Acids Res.* **20**, 920 (1991).
[15] T. Lee, M. E. Bradley, and J. Walowitz, *Nucleic Acids Res.* **26**, 3215 (1998).

with an IKA-Turrax T 25 (IKA Labortechnik, Germany) for 20–25 sec at maximum speed with tubes kept on ice. Samples are then centrifuged at 5000g for 30 min at 4°. Supernatants are removed and their protein concentration determined using the Bio-Rad Protein Assay.

Reporter Gene Choice

To monitor the efficiency of gene transfer (evaluated as the number of transfected cardiac cells) we usually employ the green fluorescent protein (GFP) linked to the CMV promoter. After harvesting, hearts are places in a Tissue-Tek cryomold (Miles Inc.) embedded in OCT, quickly frozen in liquid nitrogen, and stored at −80°. Ten to 15-μm thick sections are then cut at −20° in a cryostat. In general just one out of every 20–30 sections is transferred to a poly-L-lysine-coated slide and analyzed under the microscope. If the slides are stored at −20°, the GFP fluorescence is detectable for several weeks. Although the majority of positive cells are usually found around the central part of the needle track, scattered recombinant gene expression can also be detected far from the injection site.

For quantitative assays we have used the reporter genes firefly luciferase and chloramphenicol acetyltransferase linked to both viral (RSV LTR and CMV) and eukaryotic (α-cardiac myosin heavy chain promoter) transcriptional elements.

In transfected cells, luciferase protein has a half-life of about 3 hr, whereas CAT has a half-life of 50 hr.[16] The relative stability of the two reporter genes might be also comparable *in vivo*. In fact, CAT expression from identical vectors is usually 5–10 times higher than luciferase expression and appears to be more protracted. Therefore, luciferase may be the best candidate if inducible systems or real-time changes in expression need to be measured; on the other hand, CAT becomes more useful for detecting very low amounts of gene expression.

Although CAT assays are time consuming, have a narrow linear range (2 orders of magnitude in contrast to 4 orders of luciferase), and are not the most sensitive (with a limit of molecule detection of $1-2.5 \times 10^5$ in contrast to $5-10 \times 10^7$ molecules of luciferase),[17] we found that injection of CAT reporter plasmids gives more reproducible results.

Reporter Gene Assay

For the CAT assay, samples are first heated for 10 min at 65° to inactivate cellular deacetylases and subsequently centrifuged at 12,000g for 5 min at 4° to pellet the particulate material usually formed after this step. Ten to 60 μl of each sample

---

[16] J. F. Thompson, L. S. Hayes, and D. B. Lloyd, *Gene* **103**, 171 (1991).

[17] S. R. Kain and S. Ganguly, *in* "Current Protocols in Molecular Biology" (F. M. Ausubel, R. Brent, R. E. Kingston, D. D. Moore, J. G. Seidman, I. A. Smith, and K. Struhl, eds.), p. 9.6.8, supplement 29. John Wiley & Sons, New York, 1995.

diluted to 103 μl in homogenization buffer is then assayed by addition of 47 μl of a cocktail containing 4 μl of [^{14}C]chloramphenicol (50–60 mCi/mmol, NEN), 20 μl of 4 mM acetyl-coenzyme A, and 23 μl of 2 M Tris-Cl (pH 7.4). Depending on the promoter strength, reactions are incubated 1–2 hr at 37°. At the end of the incubation 1 ml of ethyl acetate is added, samples are vortexed for 20 sec, and centrifuged at 12,000g for 5 min at room temperature. The organic phase (top layer), containing the acetylated forms of chloramphenicol, is next transferred to a new tube and dried down in a rotating evaporator (Speed Vac Concentrator, Savant) for approximately 45 min. Reactions are resuspended in 15 μl of ethyl acetate and a few microliters at a time are applied to the bottom of a thin-layer chromatography (TLC) plate (Baker-flex 20 × 20 cm). Chromatography is carried out in a TLC chamber containing 95% chloroform, 5% methanol for 1.45 hr (with samples spotted just above the level of chloroform/ methanol solution). Plates are air dried for 10 min, wrapped in Saran Wrap, and exposed on a phosphorimager cassette. Extracts are diluted in 1% bovine serum albumin if the CAT activities are greater than 60–70% and consequently out of linear range. Since a minimal acetylation of chloramphenicol occurs in absence of extract, we routinely include a negative control (homogenization buffer alone) that is subtracted from the CAT activity of each sample.

For luciferase assays, 10–60 μl of each sample is diluted to 100 μl of homogenization buffer. Nexts 360 μl of reaction buffer [25 mM glycylglycine pH 7.8, 15 mM MgSO$_4$, 4 mM EGTA pH 7.8, 1 mM dithiothreitol, 15 mM potassium phosphate pH 7.8, 2 mM ATP, and 0.27% (v/v) Triton X-100] is added to each sample. The enzymatic reaction is initiated by mixing 100 μl of buffer H containing 0.2 mM luciferin (sodium salt) with each sample. Light production is in general measured over a 20 sec period following addition of luciferin solution. As for the CAT assay, different dilutions are assayed to determine whether sample activities are exceeding the linear range of the light detection method.

During the procedure samples and buffers (except the luciferin solution) are kept on ice and the light-sensitive luciferin solution is protected with aluminum foil. Samples are usually assayed in duplicate. In general for a given extract, the variation corresponds to ~10%.

Finally, one important technical point has to be considered when measuring the transcription activity of weak promoters linked to the luciferase gene. More than 25% of luciferase can be lost during the homogenization step if foam or heat is generated. Better recoveries are achieved if the samples are immediately frozen and ground into a fine powder using a porcelain mortar and pestle.

Acknowledgments

We thank A. Buvoli for reading the manuscript, J. Carnes for helpful suggestions, and S. Zeiler for assistance with the figures. M. Buvoli was supported by a grant from the Muscular Dystrophy Association and L. A. Leinwand by a grant from the N.I.H.

Section II

Adenovirus

[9] Myoblast-Mediated Gene Transfer for Therapeutic Angiogenesis

By ANDREA BANFI, MATTHEW L. SPRINGER, and HELEN M. BLAU

Introduction

Therapeutic angiogenesis has emerged in the past few years as one of the most promising applications of gene therapy. The goal is to deliver growth factors that control the formation of new vasculature to ischemic tissues in order to restore the blood supply that has been pathologically disrupted or reduced. This strategy has the potential to affect the treatment of diseases that afflict millions of patients each year and are disabling and often fatal, such as stroke, myocardial infarction, limb ischemia, and diabetic vascular complications. Vascular endothelial growth factor (VEGF) and basic fibroblast growth factor (bFGF/FGF2) have been used in clinical trials and the first results have been encouraging. For example, plasmid DNA[1] or adenovirus[2] carrying the cDNA for VEGF has been directly injected into the myocardium of patients with severe coronary artery disease, resulting in improvement of anginal symptoms and cardiac function.

Early studies focused on growth factors that activate endothelial cells and induce them to sprout new structures from existing vessels. However, it has become clear that the angiogenic process is highly complex and that the subsequent stages of remodeling and stabilization are crucial to attaining stable and functional vessels. Macrophages, endothelial cells, and smooth muscle cells/pericytes all play a fundamental role in the development of mature blood vessels and regulate each other through cell–cell contact and the orchestrated expression of secreted molecules. VEGF recruits macrophages, which in turn produce this factor themselves and secrete metalloproteinases that also release VEGF that is normally bound to the extracellular matrix.[3] PDGF-BB is produced by activated endothelial cells and mediates the recruitment of pericytes to nascent vascular structures, without which new vessels would not be stable and could not mature.[4] The necessity for pericyte recruitment derives from several lines of evidence. In a tumor model in which VEGF production can be regulated, vessels that are associated with pericytes

[1] P. R. Vale, D. W. Losordo, C. E. Milliken, M. Maysky, D. D. Esakof, J. F. Symes, and J. M. Isner, *Circulation* **102,** 965 (2000).
[2] T. K. Rosengart, L. Y. Lee, S. R. Patel, T. A. Sanborn, M. Parikh, G. W. Bergman, R. Hachamovitch, M. Szulc, P. D. Kligfield, P. M. Okin, R. T. Hahn, R. B. Devereux, M. R. Post, N. R. Hackett, T. Foster, T. M. Grasso, M. L. Lesser, O. W. Isom, and R. G. Crystal, *Circulation* **100,** 468 (1999).
[3] G. Bergers, R. Brekken, G. McMahon, T. H. Vu, T. Itoh, K. Tamaki, K. Tanzawa, P. Thorpe, S. Itohara, Z. Werb, and D. Hanahan, *Nat. Cell Biol.* **2,** 737 (2000).
[4] D. C. Darland and P. A. D'Amore, *J. Clin. Invest.* **103,** 157 (1999).

persist in a VEGF-independent manner, whereas those lacking pericytes regress following its withdrawal.[5] Mice in which the PDGF-B gene has been knocked out die during embryogenesis because of the instability and breakage of vessels in the vascular beds where pericytes are absent; furthermore, at such sites endothelial cells do not mature and proliferation persists unabated.[6] In human disease, a similar defect occurs in the pathogenesis of diabetic retinopathy in which hypoxia leads to an excessive production of VEGF and the formation of pericyte-poor vessels that are prone to hemorrhage and ultimately lead to blindness.[4] Moreover, tumors often overexpress VEGF in order to induce their own blood supply, but the vessels produced have few and dysfunctional pericytes, are leaky, and are permanently immature,[7] an observation that has led H. Dvorak to define a tumor as a "wound that never heals."[8]

We have put forward the concept of the "well-tempered vessel,"[9] or controlled angiogenesis, as the goal of gene-therapy approaches. This concept highlights the need for a complex balance of cell types and signaling molecules in order to induce mature, stable vessels that are capable of regulating their permeability and response to inflammation in a physiological manner. Indeed, a number of "vessel maturation strategies" have been proposed (Fig. 1). One approach involves delivery of VEGF alone at the appropriate level and timing. A second approach involves the combined delivery of pro-angiogenic and stabilizing factors. VEGF at high levels often produces vessels that are hyperpermeable and leaky, but this effect can be tempered by the action of angiopoietin-1.[10] An alternative involves co-delivery of VEGF with PDGF-B to ensure appropriate pericyte recruitment. A third strategy exploits the hypoxia-inducible factor-1α (HIF-1α), a pleiotropic transcriptional regulator that mediates the physiologic response to hypoxia in ischemic tissues.

HIF-1α increases expression of several factors, including VEGF, angiopoietin-2, and PDGF-BB, among others.[7] At least three methods have been developed to overcome the fact that endogenous HIF-1α is tightly regulated by ubiquitination and consequently rapidly degraded. First, the delivery of a plasmid encoding an HIF-1α/VP16 fusion protein to rabbit ischemic muscle induces the formation of new collateral circulation as evidenced by angiography and increased blood flow.[11] Second, deletion of the ubiquitination domain may suffice. Indeed, transgenic mice carrying a HIF-1α construct with a deleted degradation domain

[5] L. E. Benjamin, D. Golijanin, A. Itin, D. Pode, and E. Keshet, *J. Clin. Invest.* **103**, 159 (1999).

[6] P. Lindahl, B. R. Johansson, P. Leveen, and C. Betsholtz, *Science* **277**, 242 (1997).

[7] P. Carmeliet and R. K. Jain, *Nature* **407**, 249 (2000).

[8] H. F. Dvorak, L. F. Brown, M. Detmar, and A. M. Dvorak, *Am. J. Pathol.* **146**, 1029 (1995).

[9] H. M. Blau and A. Banfi, *Nat. Med.* **7**, 532 (2001).

[10] G. Thurston, J. S. Rudge, E. Ioffe, H. Zhou, L. Ross, S. D. Croll, N. Glazer, J. Holash, D. M. McDonald, and G. D. Yancopoulos, *Nat. Med.* **6**, 460 (2000).

[11] K. A. Vincent, K. G. Shyu, Y. Luo, M. Magner, R. A. Tio, C. Jiang, M. A. Goldberg, G. Y. Akita, R. J. Gregory, and J. M. Isner, *Circulation* **102**, 2255 (2000).

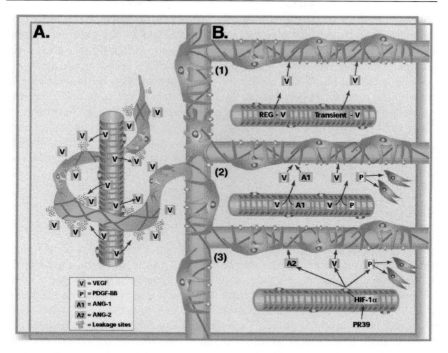

FIG. 1. Gene therapy strategies for "well-tempered" angiogenesis. (A) Overexpression of VEGF induces the growth of numerous new blood vessels which are tortuous, hyperpermeable, and covered with few and dysfunctional pericytes, like those characteristic of the abnormal vasculature induced by tumors. (B) Gene therapy approaches that take into account the complex interplay of molecules and cell types essential to the development of mature vessels. (1) VEGF alone delivered either transiently at low levels or long-term at the appropriate dose and time with regulatable vectors. (2) Co-delivery of VEGF and a gene such as Ang-1, which stabilizes endothelium, or PDGF-B, which recruits pericytes. (3) Delivery of HIF-1α, a pleiotropic transcription factor that activates a number of genes involved in the physiological response to hypoxia. A secreted peptide (PR39) increases HIF-1α half-life by inhibiting its ubiquitination. [Modified from H. M. Blau and A. Banfi, *Nat. Med.* **7,** 532 (2001), with permission from Nature Publishing Group.]

displayed robust angiogenesis, with non-leaky vessels that, however, retained the physiological ability to increase their permeability in response to inflammatory stimuli.[11a] Finally, the PR39 peptide, which is produced by macrophages, has the ability to inhibit HIF-1α degradation and caused improvement of coronary blood flow when delivered by gene therapy.[12]

[11a] D. A. Elson, G. Thurston, L. E. Huang, D. G. Ginzinger, D. M. McDonald, R. S. Johnson, and J. M. Arbeit, *Genes Dev.* **15,** 2520 (2001).

[12] J. Li, M. Post, R. Volk, Y. Gao, M. Li, C. Metais, K. Sato, J. Tsai, W. Aird, R. D. Rosenberg, T. G. Hampton, F. Sellke, P. Carmeliet, and M. Simons, *Nat. Med.* **6,** 49 (2000).

Whereas the unregulated constitutive expression of VEGF, as in tumors or by high-efficiency delivery systems, can be deleterious,[13-15] significant clinical results have been obtained by transient expression using naked DNA or adenoviral vectors.[1,2] Clearly, the ability to deliver the appropriate dosage at the appropriate time is critical in order to exploit the therapeutic window of this remarkably potent angiogenic factor. As depicted in Fig. 1B, this is currently being pursued by the use of intrinsically transient delivery systems, or through the development of regulatable expression systems, which will afford the ability to modulate both dosage and timing of expression of the therapeutic gene product. Further study is required, but it may prove true that, given such fine control, VEGF alone may lead to "well-tempered vessels."

Myoblast-Mediated Gene Transfer: Advantages and Outline of Procedure

In the past decade our laboratory has developed a method for *ex vivo* gene transfer to muscle based on the retroviral transduction of autologous myoblasts. Although this procedure is the focus of this chapter, many of the features of muscle that make it an ideal target for both therapeutic angiogenesis and the study of its basic biology also make it amenable to *in vivo* direct delivery of adenoviral, adeno-associated viral (AAV), lentiviral, and naked DNA vectors. Muscle is a very accessible tissue in which molecular and cellular biology have been very well studied, allowing the design of effective strategies to deliver the gene of interest and to control its expression. Differentiated myofibers are long-lived and stable and therefore permit prolonged constant production of the therapeutic protein, if desirable. This also allows the study of its long-term effects in preclinical studies. Further, although rich in contractile apparatus and not obviously a secretory tissue, genetically altered skeletal muscle has proved to be a surprisingly efficient "factory" for the production of recombinant soluble factors.[16]

Myoblasts are mononucleate progenitor cells that are responsible for muscle tissue formation during development and that persist in the adult as satellite cells. These cells remain alongside myofibers, between their plasma membrane and the basal lamina, and continue to fuse to neighboring fibers during regeneration following injury.[17] They can be purified, cultured, and manipulated *in vitro*, retaining the ability, after extensive passaging, to fuse stably with preexisting myofibers or with

[13] M. L. Springer, A. S. Chen, P. E. Kraft, M. Bednarski, and H. M. Blau, *Mol. Cell* **2,** 549 (1998).

[14] R. J. Lee, M. L. Springer, W. E. Blanco-Bose, R. Shaw, P. C. Ursell, and H. M. Blau, *Circulation* **102,** 898 (2000).

[15] M. L. Springer, G. Hortelano, D. M. Bouley, J. Wong, P. E. Kraft, and H. M. Blau, *J. Gene Med.* **2,** 279 (2000).

[16] H. M. Blau and M. L. Springer, *New Engl. J. Med.* **333,** 1554 (1995).

[17] D. R. Campion, *Int. Rev. Cytol.* **87,** 225 (1984).

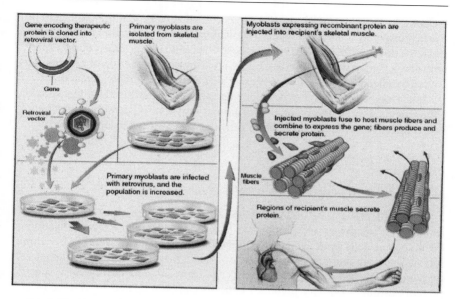

FIG. 2. Myoblast-mediated gene transfer: outline of the procedure. [Modified from H. M. Blau and M. L. Springer, *N. Engl. J. Med.* **333,** 1554 (1995), with permission from the Massachusetts Medical Society.]

each other to form new fibers, making them an ideal vehicle to deliver therapeutic genes to muscle tissue. The myoblast-mediated gene transfer procedure is outlined in Fig. 2. Primary myoblasts can be readily isolated from postnatal skeletal muscle that will proliferate extensively *in vitro*. The gene of interest (e.g., VEGF or another angiogenic factor) is cloned into a retroviral vector, which is then used to infect the pure myoblast population. Since the retrovirus integrates into the cellular chromatin, its genetic material cannot be lost and this ensures stable long-term expression of the transgene if appropriate promoters are used. The myoblasts are then injected into the muscle, where they fuse into mature fibers often containing hundreds of nuclei. Every donor nucleus will influence a domain of sarcoplasm encompassing numerous recipient nuclei, thereby enlisting a large surface area for secretion of the angiogenic factor of interest.

The use of myoblast-mediated gene transfer has some advantages specific to this method. Transduced myoblasts can be thoroughly characterized *in vitro* to ensure appropriate levels of production of the secreted protein before being implanted. Furthermore, the infection step takes place *in vitro* so that only myoblasts will carry the retrovirus and there is no risk of transducing other cell types present in the muscle *in vivo,* such as antigen-presenting cells. For the specific application to the study of angiogenesis, myoblast-mediated gene transfer is a powerful tool to investigate the complex interplay of factors that directs blood vessel formation. In fact, it produces transgenic fibers directly in the adult animal (a form of "adult transgenesis"), thereby allowing the study of genes whose constitutive expression during embryonic development would be lethal. Further, the area of expression of the transgene is very localized and, therefore, systemic side effects are largely avoided.

In the next section a series of detailed protocols are described for the isolation, culture, infection, and implantation of primary myoblasts, as well as methods to assess the angiogenic response to the gene delivered using this approach.

Procedures

Mouse Primary Myoblast Isolation and Culture

The cells that can be cultured from skeletal muscle primary isolates consist primarily of myoblasts and fibroblasts. This protocol results in the purification of the myogenic cells from the mixed population. During the first 2 weeks of culture, selective growth and passaging conditions make the myoblast the dominant cell type and, by 3 weeks, nearly 100% of the cells are positive for the myoblast-specific protein desmin.[18] Primary myoblast cultures can be prepared from mice of any age, but neonatal animals give a greater yield.

One to five neonatal mice are sacrificed by decapitation, and the limbs are rinsed with 70% ethanol and removed with sterile scissors. The muscle is dissected away under a stereo dissecting microscope in a sterile hood and successive tissue is stored in a culture dish under several drops of sterile phosphate-buffered saline (PBS, pH 7.4) on ice. The accumulated tissue is minced to a slurry with a sterile razor blade and then 2 ml per gram (minimum 0.5 ml) is added of a solution containing 1.5 U/ml of collagenase D (Roche Molecular Biochemicals, Indianapolis, IN), 2.4 U/ml of dispase II (Roche Molecular Biochemicals), and 2.5 mM CaCl$_2$. The mincing is continued for several minutes and then the preparation is transferred to a sterile tube and incubated at 37° for approximately 20 min until it is reduced to a fine slurry, with occasional trituration with a 1 ml pipette. If desired, a filtration step can be carried out through a piece of 80 μm nylon mesh (Nitex; Tetko Inc., Monterey Park, CA). The cells are centrifuged at 350g for 5 min, resuspended in growth medium, and plated in a 60 mm collagen-coated culture dish. Growth medium is F-10 (Gibco BRL, Gaithersburg, MD) supplemented with 20% FCS (Hyclone, Logan, UT) and 2.5 ng/ml human bFGF (Promega, Madison, WI). F-10 gives a growth advantage to myoblasts over fibroblasts. Collagen-coated dishes are made by incubating with a 0.01% solution of calf skin collagen (Sigma, St. Louis, MO) in 0.1 N acetic acid overnight and rinsed in sterile water, after which they can be stored dry. Cells are incubated at 37° in 5% CO$_2$ and medium is changed every other day.

[18] T. A. Rando and H. M. Blau, *J. Cell Biol.* **125**, 1275 (1994).

When the cells reach about 80% confluence, or if they have been in culture for 5 days, the population should be passaged and enriched for myoblasts by taking advantage of their differential adherence to the dish compared to fibroblasts. Cells are removed from the dish using only PBS without any trypsin or EDTA; after washing, a small quantity of PBS is left in the plate and, after several minutes, myoblasts are knocked off the surface of the plate by hitting the dish very firmly sideways against a tabletop. Under these conditions, fibroblasts will remain adherent, but myoblasts will freely detach. To further enrich for myoblasts, cells are preplated on a collagen-coated dish for 15 min and then transferred to the new dishes, discarding all the cells that adhere rapidly, which are predominantly fibroblasts. The split must be at no more than a 1 : 5 dilution. This way of passaging cells is repeated for the first week of culture expansion, after which cells are passaged by conventional trypsinization with pre-plating until the contaminating fibroblasts have disappeared from the population. This can be confirmed by desmin staining. When the culture contains only myoblasts, the growth medium can be switched to 50% F-10/50% DMEM (Gibco BRL), always supplemented with 20% FCS and 2.5 ng/ml recombinant human bFGF, which allows myoblasts to grow faster and to higher density, but also removes the growth advantage over fibroblasts. This medium can be changed every 3 days. Primary myoblasts can be frozen using standard cell culture protocols, e.g., in 90% serum and 10% DMSO.

If desired, the differentiation potential of the primary myoblasts can be tested *in vitro* by culturing them in fusion medium: 95% DMEM and 5% horse serum (Hyclone, Logan, UT). Cells must be 30–50% confluent at the start and medium is changed daily. Within several days to 1 week large multinucleated myotubes will form and sometimes will be seen to twitch randomly.

Retroviral Infection of Primary Myoblasts

High-titer retroviral supernatants can be obtained from the helper-free, ecotropic ϕNX packaging cell line,[19,20] which is grown in high-glucose (4.5 mM) DMEM (Gibco BRL) supplemented with 10% FCS, at 37° in 5% CO_2. For best virus production, cells should be transfected while in logarithmic growth phase, at about 50% confluence. Transfection of the retroviral plasmid DNA is routinely carried out using Fugene 6 reagent (Roche Molecular Biochemicals) according to the manufacturer's instructions. One day after transfection, medium is changed and the cells are placed at 32°, since virus half-life is only 4 hr at 37°. After 12 hr at this temperature virus can be harvested every 6–8 hr for about 3 days. The supernatant is passed through a 0.45 μm syringe filter to eliminate the contaminating

[19] W. S. Pear, G. P. Nolan, M. L. Scott, and D. Baltimore, *Proc. Natl. Acad. Sci. U.S.A.* **90**, 8392 (1993).

[20] S. Swift, J. Lorens, P. Achacoso, and G. P. Nolan, *in* "Current Protocols in Immunology," (J. E. Coligan, A. M. Kruisbeek, D. H. Margulies, E. M. Shevach, and W. Strober, eds.), Unit 10.17C. John Wiley and Sons, New York, 2001.

producer cells (do not use 0.2 μm filters, because virus can be damaged during filtration). Virus is best used fresh, but it can also be stored at $-80°$ and no special fast or slow freezing is necessary. The titer may drop after freezing and thawing, but a drop in viability after one freeze–thaw cycle has not been observed.

Unlike established myogenic cell lines, for example C2C12, primary myoblasts are not readily infected with retroviruses under standard conditions. We have developed a high-efficiency transduction protocol[21] with an infection efficiency greater than 99%. Primary myoblasts must be in logarithmic growth phase for the duration of the procedure (2–3 days), but not too sparse: a starting condition of 10–20% confluence reproducibly gives good results. Undiluted viral supernatant is supplemented to a final concentration of 8 μg/ml polybrene (Sigma, St. Louis, MO). Myoblast medium is removed and replaced with supplemented supernatant and the plate is returned to the incubator for 15 min to allow CO_2 equilibration to take place. The dish is then carefully wrapped in Parafilm, to contain the retrovirus in the event of splashing and to retard loss of CO_2, and centrifuged in a microplate carrier at 1100g for 30 min, ideally at $32°$, although no decline in efficiency was observed when the procedure was carried out at room temperature. This step increases infection efficiency 3- to 10-fold. After centrifugation, viral supernatant is removed and cells are returned to $37°$ and 5% CO_2 in fresh myoblast growth medium. This procedure can be repeated every 8 hr with fresh viral supernatant for as many times as desired. After the first infection, the efficiency ranges from 60% to 90% and usually reaches greater than 99% after 3 to 4 rounds of infection, which is considered optimal.

Myoblast Implantation into Skeletal Muscle

The injection of transduced myoblasts into skeletal muscle allows them to fuse with the recipient fibers and results in stable, long-term expression of the exogenous gene product.[22,23] If allogeneic cells are used, or the transgene encodes a non-self protein, immunodeficient mice or immunosuppression must be used.[24,25] If syngeneic cells are implanted without immunosuppression and the gender of the myoblasts is male, mixed, or unknown, only male mice should be used as recipients because the H-Y minor histocompatibility antigen on the Y chromosome can still cause immunological rejection of the implanted cells.

Primary myoblasts should be passaged no more than 48 hr before injection. Myoblasts should be given fresh medium between 12 and 24 hr prior to injection.

[21] M. L. Springer and H. M. Blau, *Somat. Cell Mol. Genet.* **23,** 203 (1997).

[22] J. Dhawan, L. C. Pan, G. K. Pavlath, M. A. Travis, A. M. Lanctot, and H. M. Blau, *Science* **254,** 1509 (1991).

[23] E. Barr and J. M. Leiden, *Science* **254,** 1507 (1991).

[24] D. J. Watt, T. A. Partridge, and J. C. Sloper, *Transplantation* **31,** 266 (1981).

[25] G. K. Pavlath, T. A. Rando, and H. M. Blau, *J. Cell Biol.* **127,** 1923 (1994).

Cells should be resuspended at a concentration of 10^8/ml in sterile PBS +0.5% BSA and can be stored on ice in a 1.5 ml snap-cap tube for up to 1 hr. To make sure that cells did not die before being injected, viability can be checked both at the time of trypsinization and after the last injection by Trypan Blue exclusion: in a healthy population less than 5% of the cells should take up the dye.

Mice can either be anesthetized with intraperitoneal (i.p.) sodium pentobarbital (Nembutal, Abbott Laboratories, North Chicago, IL) or with an inhaled compound such as methoxyfluorane or isofluorane. For Nembutal, depending on the mouse strain and age, a range of dosages can be used from 55 to 75 μg/g of body weight. A 1 ml insulin-type syringe fitted with a 27-gauge needle can be used to inject the anesthetic in the lower right quadrant of the abdomen, taking care to avoid any bubbles. While injecting, it is safer to hold the mouse on its back with the head tilted downward, so that the internal organs will shift toward the diaphragm and the chances of accidental perforation by the needle will be minimized. This anesthetic has a potent depressing effect on respiratory centers in the medulla oblongata and it is easy to overdose, causing the death of the animal. As a result, it is preferable to use the lowest possible amount and, if anesthesia is not deep enough, to adjust its level with a safer drug, such as inhaled methoxyfluorane.

Methoxyfluorane can also be used alone, which is safer for the mice, but requires the researcher to work with one mouse at a time because the animals will awaken as soon as the anesthetic is removed. A simple inhalation chamber can be built by placing a methoxyfluorane-soaked gauze at the bottom of a 50 ml conical tube and having the mouse enter it. Methoxyfluorane must be used in a fume hood. After the mouse is asleep, it should be kept lying flat on its back with its head just inside the tube. Care must be taken not to let the face of the mouse touch the gauze, as methoxyfluorane is irritating to the skin, and the amount inhaled can be regulated by how deep inside the tube the mouse is kept. While the sleeping mouse is maintained under methoxyfluorane, the tube and mouse can be moved out of the hood to a well-ventilated work area as long as the animal's body blocks most of the tube's opening. [Note: at least temporarily, methoxyfluorane is not being produced in the United States, but its use is legal; it is currently available from Medical Developments Australia (Springvale, Victoria)].

The region of the leg to be injected is shaved with an electric trimmer and sterilized with 70% ethanol. If immunodeficient mice are used, all operations must be carried out in a sterile tissue culture hood. Cell clumps in the myoblast suspension should be gently dispersed by pipetting up and down. The amount to be injected is then removed with a Hamilton syringe of appropriate volume (Hamilton, Reno, NV). A typical injection consists of 5 μl of a 10^8 cells/ml suspension, for a total of 5×10^5 total myoblasts, although smaller amounts can be used. If a syringe with a dead space is used, to ensure precise dosing the needle reservoir must be filled with cell suspension before it is attached to the syringe barrel. In the hind leg, muscles that are easily accessible are the lateral gastrocnemius, the tibialis anterior, and the quadriceps. Holding the syringe at a 45° angle, the needle is gently pushed 1–2 mm into the muscle. The myoblasts are gradually injected and, after a pause of a couple of seconds to prevent the backward leak of the cells, the needle is slowly retracted.

Histochemical Assessment of Implantation and Angiogenic Response

In order to be able to evaluate the incorporation of the implanted myoblasts, cells that have also been transduced with a retrovirus carrying the *Escherichia coli β-galactosidase* reporter gene (*lacZ*) are commonly used. The implanted muscle is harvested and serial sections are cut, which are then subjected to X-Gal staining to identify the transgenic fibers, hematoxylin and eosin (H&E) staining to monitor the general tissue architecture, and immunofluorescent staining with appropriate primary antibodies to assess the angiogenic response induced by the gene of interest carried by the myoblasts. All staining procedures are carried out according to standard specifications.

The mice are sacrificed by CO_2 inhalation and the injected muscles are dissected. The harvested tissue is placed in a vinyl cryomold (Miles, Elkhart, IN) covered with an adequate volume of O.C.T. compound (Sakura Finetek USA, Torrance, CA) and quickly frozen in freezing isopentane (2-methylbutane, Aldrich). This is achieved by hanging a metal beaker two-thirds full of isopentane from a clamp stand into liquid nitrogen. When solid isopentane starts appearing on the sides, the plastic cryomold with the tissue is lowered into the beaker with metal tongs and kept there for 1 min. Once frozen, tissue can be stored at −80°. Serial sections are cut with a cryostat: 10 μm thickness is appropriate for H&E and immunostaining, whereas 20 μm is needed for X-Gal. A typical angiogenic response to VEGF produced by implanted myoblasts is shown in Fig. 3: panel A shows transgenic fibers by X-Gal staining, whereas in panel B the vessels are stained with an antibody against the endothelial marker CD31.

Assessment of Vascular Continuity by Fluorescent Microbead Perfusion

To assess the angiogenic response to a given molecule, it is of paramount importance to be able to show whether the endothelial structures that have formed possess a lumen and display continuity with the general circulation. This parameter can be studied by perfusing the animal with fluorescent microspheres of appropriate size before harvesting the tissues. High molecular weight tracers, such as ink or dextran, have been used to study vascular structures in other settings. However, as discussed in the introduction, since VEGF causes vessels to become hyperpermeable,[8] these molecules leak out and therefore cannot be used to study vasculature in regions of high VEGF concentration. Indeed, extravasation of Evans Blue dye has been effectively exploited to assess and quantify VEGF-induced vascular leakage.[10, 26]

[26] A. A. Miles and E. M. Miles, *J. Physiol. (London)* **118**, 228 (1952).

A   B

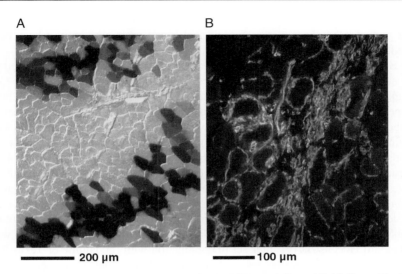

■■■■■ 200 μm   ■■■■100 μm

FIG. 3. Angiogenic response to myoblast-mediated VEGF gene delivery. (A) All fibers with which transduced myoblasts have fused express the marker gene *lacZ* and stain (dark) with the substrate X-Gal. (B) The angiogenic response to the delivered VEGF is assessed by immunofluorescent staining with an antibody against the endothelial-specific marker PECAM-1. All capillaries appear as fluorescently stained threads and can be seen wrapping around individual muscle fibers. The muscle fiber pattern can be recognized by faint autofluorescence.

Fluorescent microspheres that are too large to leak out even of hyperpermeable vessels, however, can be used to prove the vascular continuity of the structures they fill. The microspheres used in conventional blood flow studies are typically about 7–16 μm in diameter and are large enough to lodge in the capillary bed, allowing the subsequent quantitation of microspheres in tissue sections or extracts.[27,28] However, because they are intended to lodge in the capillaries, these microspheres are useful for quantitation only and do not provide a visual picture of the vessels being studied. We have examined the continuity of VEGF-induced vessels by perfusing mice with a suspension of 0.2 μm fluorescent microspheres, much smaller than those used for standard flow studies.[29] The small size of these microspheres allows the suspension to behave like a fluid and completely fill up the lumens of the vessels without leaking, even if those same vessels are permeable to large fluorescent dextrans. If performed properly, the vast majority of capillaries are filled with the microsphere suspension, although it is advisable to counterstain

[27] M. S. Jasper, P. McDermott, D. S. Gann, and W. C. Engeland, *J Auton. Nerv. Syst.* **30**, 159 (1990).
[28] P. Kowallik, R. Schulz, B. D. Guth, A. Schade, W. Paffhausen, R. Gross, and G. Heusch, *Circulation* **83**, 974 (1991).
[29] M. L. Springer, T. K. Ip, and H. M. Blau, *Mol. Therapy* **1**, 82 (2000).

with antibodies to vascular endothelium, such as CD31/PECAM-1, to confirm that perfusion has been successful and that blood vessels of interest have been filled.

Before the start of the procedure, two syringes (appropriately sized based on the number of animals) must be prepared and fitted with 27-gauge needles. One syringe is filled with PBS (4 ml per mouse) and the other with the fluorescent bead suspension. We have used 0.2 μm carboxylate-modified FluoSpheres (Molecular Probes, Eugene, OR) at a 1 : 6 dilution with PBS from the supplied stock (6 ml of final working suspension per mouse): the manufacturer concentration is 2% solids in water. These beads are available in many different fluorescent colors. Because of the brightness of the fluorescence in these beads and their relatively wide excitation/emission spectra, it is best to use fluorescence with a wavelength at one extreme or the other to avoid crosstalk with other fluorescent dyes used for costaining. If conventional fluorescence microscopy is to be used, the "blue" FluoSpheres (365 nm excitation max., 415 nm emission max.) can be used with a standard Hoechst/DAPI filter set in conjunction with rhodamine or Texas Red conjugated secondary antibodies for double staining. For conventional or confocal microscopy with the ability to view far red dyes such as Cy-5, the "dark red" FluoSpheres (660 nm excitation max., 680 nm emission max.) can be used in conjunction with both red and green secondary antibodies.

Mice must be terminally anesthetized such that invasive chest surgery can be performed before they are dead. This is accomplished with a 0.8 ml intraperitoneal injection containing 3 mg/ml sodium pentobarbital and 50 μg/ml nitroglycerin (Tridil, Abbott Laboratories). The nitroglycerin allows maximal vasodilatation of the peripheral vasculature. The fully anesthetized animal (not responding to toe or tail pinch) is taped to a dissection board lying on its back, but without taping down the legs, as such deformation may interfere with the perfusion. The chest is opened through a midline sternotomy, carefully cutting through the diaphragm without damaging the heart. A small hole is produced in the right ventricle to allow blood to drain from the general circulation as it is replaced by the perfusion mixture. The needle of the PBS syringe is carefully inserted into the left ventricular cavity through the apex, with care taken to avoid puncturing the far end of the ventricle or the valve. Roughly 4 ml of PBS is slowly injected over approximately 30–60 sec, making sure to avoid air bubbles. The needle is removed from the heart, where the wound will close but still be visible. The microsphere-filled syringe is then carefully inserted into the wound, again being sure not to go too far and puncture the other end, and the bead suspension is injected as described for the PBS.

After perfusion is complete, muscle tissue is harvested, embedded in OCT compound, and snap frozen as described above. Serial sections can be cut with a cryostat at 10 μm of thickness. If the sections are fixed with formaldehyde for subsequent immunofluorescent staining, the microspheres are not themselves fixed, but they do not tend to diffuse even during a multi-hour staining procedure with many changes of buffer. Other fixation methods have not been tested. Freshly cut sections can be stored frozen at $-20°$ and thawed once without appreciable movement of beads. However, after sections have been stained and mounted in 50% glycerol in PBS, it may be safer to store them at $4°$ rather than in the freezer.

Conclusions

These procedures have been proved useful by numerous investigators as a means of studying the function of specific molecules in the adult organism. This efficient method of "adult transgenesis" is particularly suited to be applied both to gene therapy approaches to neovascularization and as a tool to investigate the basic biology of the angiogenic process.

[10] Use of Phage Display to Identify Novel Peptides for Targeted Gene Therapy

By LORRAINE M. WORK, STUART A. NICKLIN, STEVE J. WHITE, and ANDREW H. BAKER

Introduction

Conceptually and practically the reality of gene therapy as a means of disease intervention has gained much credence in past years as traditional pharmacological interventions have failed to prevent the onset and progression of many disease states.[1,2] While some disease states, such as post-angioplasty restenosis and late vein graft failure, are suited to local gene delivery, clinical trials and research findings have highlighted the obligate need for targeted gene delivery in order to overcome the potentially fatal side effects associated with systemic exposure to high doses of viral vectors.[3] Clinical application of gene therapy requires the identification of suitable targeting molecules thus allowing systemic administration to achieve target tissue-restricted gene delivery. This would require not only lower doses of vector, but also more efficient gene expression at the desired site.[4] Many

---

[1] S. Ylä-Herttuala and J. F. Martin, *Lancet* **355**, 213 (2000).

[2] P. Sinnaeve, O. Varenne, D. Collen, and S. Janssens, *Cardiovasc. Res.* **44**, 498 (1999).

[3] T. Valere, *in* "Understanding Gene Therapy" (N. R. Lemoine, ed.), p. 141. BIOS Scientific Publishers Limited, 1999.

[4] L.-A. Martin, *in* "Understanding Gene Therapy" (N. R. Lemoine, ed.), p. 125. BIOS Scientific Publishers Limited, 1999.

viral and nonviral gene delivery vectors have been studied with the necessity for targeted gene delivery becoming ever more apparent.[5–9]

Of those vectors studied, some have a natural tropism which is advantageous in certain clinical scenarios. For example, persistent infection of cells of the nervous system is achievable with herpes simplex virus (HSV).[7] In the case of adenoviral vectors, however, transduction of target cells is dependent on the level of expression of the Coxsackie-adenovirus receptor (CAR)[10] and α_v integrins.[11] Systemic administration of adenovirus primarily leads to high-level infection in hepatocytes of the liver[12,13] limiting its application purely to pathologies of the hepatic system or where the liver is used as a factory for the production of therapeutic soluble proteins. Furthermore, a number of applications, including delivery of adenovirus using custom-built catheters or by direct infection into the myocardium, aim to limit systemic adenoviral dissemination, thus limiting toxicity. However, a study by Hiltunen et al. (2000)[14] has demonstrated reporter gene expression in nontarget tissue including white blood cells, testis, liver, and lymph node following local gene delivery by intravascular and periadventitial methodology. It is therefore clear that targeted vectors must be developed before successful gene delivery can be achieved in vivo.

Targeted adenoviral systems have been developed by manipulating the mechanism by which adenovirus enters cells—the 2-stage CAR- and integrin-mediated internalization[10,11] (Fig. 1). Directing entry by blocking normal adenoviral tropism through CAR and retargeting to alternate cell surface receptors, specific gene delivery to a number of cell types has been achieved.[15–21] This method has largely utilized candidate cell surface receptors or molecules expressed on the chosen

[5] T. J. Wickham, Gene Ther. 7, 110 (2000).

[6] A. D. Miller, in "Understanding Gene Therapy" (N. R. Lemoine, ed.), p. 43. BIOS Scientific Publishers Limited, 1999.

[7] S. J. Murphy, in "Understanding Gene Therapy" (N. R. Lemoine, ed.), p. 21. BIOS Scientific Publishers Limited, 1999.

[8] K. Peng and S. J. Russell, Curr. Opin. Biotechnol. 10, 454 (1999).

[9] N. Miller and J. Whelan, Hum. Gene Ther. 8, 803 (1997).

[10] J. M. Bergelson, J. A. Cunningham, G. Droguett, E. A. Kurt-Jones, A. Krithivas, J. S. Hong, M. S. Horwitz, R. L. Crowell, and R. W. Finburg, Science 275, 1320 (1997).

[11] T. J. Wickham, P. Mathias, D. A. Cheresh, and G. R. Nemerow, Cell 73, 309 (1993).

[12] A. Kass-Eisler, E. Falck-Pedersen, D. H. Elfenbein, M. Alvira, P. M. Buttrick, and L. A. Leinwand, Gene Ther. 1, 395 (1994).

[13] J. Huard, H. Lochmuller, G. Acsadi, A. Jani, B. Massie, and G. Karpati, Gene Ther. 2, 107 (1995).

[14] M. O. Hiltunen, M. P. Turunen, A.-M. Turunen, T. T. Rissanen, M. Laitinen, V.-M. Kosma, and S. Ylä-Herttuala, FASEB J. 14, 2230 (2000).

[15] P. N. Reynolds, K. R. Zinn, V. D. Gavrilyuk, I. V. Balyasnikova, B. E. Rogers, D. J. Buchsbaum, M. H. Wang, D. J. Miletich, W. E. Grizzle, J. T. Douglas, S. M. Danilov, and D. T. Curiel, Mol. Ther. 2, 562 (2000).

[16] J. Doukas, D. K. Hoganson, M. Ong, W. Ying, D. L. Lacey, A. Baird, G. F. Pierce, and B. A. Sosnowski, FASEB J. 13, 1459 (1999).

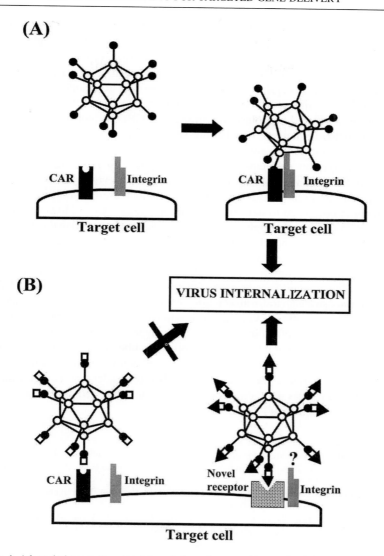

Fɪɢ. 1. Adenoviral retargeting. (A) Adenoviral attachment and internalization is mediated through the knob protein (●) binding to CAR, followed by interaction of the penton base (○) with α_v integrins on the cell surface. Following internalization the virus is localised within cellular endosomes which, on acidification, allows the virions to escape and traffic to the nucleus. (B) Wild-type adenovirus tropism can be modified by blocking the knob protein–CAR interaction (□) and redirecting infection with peptides (▲) mediating binding through a novel or candidate cell-specific receptor. Dependent on the nature of the target receptor and binding interaction the adenovirus–receptor complex may or may not require the secondary integrin interaction for internalization.

target cell. Commonly this approach uses a bispecific antibody whereby the normal adenovirus–receptor interaction is inhibited and tropism is instead redirected via a known cell specific ligand (Fig. 1). Wickham *et al.* (1996)[21] was one of the first studies to highlight the potential of this approach. Using a bispecific antibody directed to α_v integrins, adenovirus transduction of endothelial and smooth muscle cells was increased 7- to 9-fold compared to nontargeted virus.[21] Similarly, E-selectin[18] and fibroblast growth factor (FGF) receptor have been used as the target to drive endothelial- and tumor-specific gene delivery.[16,17,19,20] The first *in vivo* retargeting of adenovirus has been reported with targeting of adenoviral binding through the angiotensin-converting enzyme (ACE), which is highly expressed in the lung, using a bispecific antibody.[15] On *in vivo* systemic delivery of the targeted virus reporter gene expression was increased 20-fold in pulmonary endothelium compared to delivery with nontargeted vector.[15] However, the chosen target molecules, although proving largely more selective than the nontargeted vector, do not lead to entirely cell-specific gene delivery as the target receptor may be expressed on other cell types.

An alternative adenoviral targeting strategy whereby the natural viral tropism is ablated by genetic modification of the fiber protein has been described.[22–24] These studies demonstrated viral transduction of those cells expressing β_2 integrin.[23] Furthermore, retargeting of viral infection could be achieved by replacement of the fiber protein with that of another adenoviral serotype[22] or by the incorporation of an Arg-Gly-Asp (RGD) peptide in the HI loop of the fiber knob domain.[24] In similar studies, Roelvink *et al.* (1999)[25] outlined an approach whereby mutant fiber knob proteins were generated. These were then used to ablate adenoviral binding through CAR and binding was redirected to a novel receptor—the hemagglutination (HA) protein of influenza virus—on 293-HA cells.[25]

[17] D. Gu, A. M. Gonzalez, M. A. Printz, J. Doukas, W. Ying, M. D'Andrea, D. K. Hoganson, D. T. Curiel, J. T. Douglas, B. A. Sosnowski, A. Baird, S. L. Aukerman, and G. F. Pierce, *Cancer Res.* **59,** 2608 (1999).

[18] O. A. Harari, T. J. Wickham, C. J. Stocker, I. Kovesdi, D. M. Segal, T. Y. Huehns, C. Sarraf, and D. O. Haskard, *Gene Ther.* **6,** 801 (1999).

[19] B. E. Rogers, J. T. Douglas, B. A. Sosnowski, W. Ying, G. Pierce, D. J. Buchsbaum, D. D. Manna, A. Baird, and D. T. Curiel, *Tumor Targeting* **3,** 25 (1998).

[20] C. K. Goldman, B. E. Rogers, J. T. Douglas, B. A. Sosnowski, W. Ying, G. P. Siegal, A. Baird, J. A. Campain, and D. T. Curiel, *Cancer Res.* **57,** 1447 (1997).

[21] T. J. Wickham, D. M. Segal, P. W. Roelink, M. E. Carrion, A. Lizonova, G. M. Lee, and I. Kovesdi, *J. Virol.* **70,** 6831 (1996).

[22] D. J. Von Seggern, S. Huang, S. K. Fleck, S. C. Stevenson, and G. R. Nemerov, *J. Virol.* **74,** 354 (2000).

[23] D. J. Von Seggern, C. Y. Chiu, S. K. Fleck, P. L. Stewart, and G. R. Nemerov, *J. Virol.* **73,** 1601 (1999).

[24] I. Dmitriev, V. Krasnykh, C. R. Miller, M. Wang, E. Kashentseva, G. Mikheeva, N. Belousova, and D. T. Curiel, *J. Virol.* **72,** 9706 (1998).

[25] P. W. Roelvink, G. M. Lee, D. A. Einfield, I. Kovesdi, and T. J. Wickham, *Science* **286,** 568.

In addition to adenoviral vector targeting strategies, there have also been studies demonstrating targeted delivery by liposomes[26] and adeno-associated virus (AAV).[27,28] AAV retargeting was achieved using a bispecific antibody for a receptor expressed on megakaryocytes. This allowed a 70-fold increase in AAV infection in the usually nonpermissive megakaryocyte cell lines, DAMI and MO7e.[27] Alternatively, by introduction of a peptide into the viral capsid AAV infection could be redirected to other cell receptors such as integrins.[28] Indeed, Girod et al. (1999)[28] have shown that insertion of a nonviral ligand L14 (a 14 amino acid peptide, QAGTFALRGDNPQG) into the capsid of AAV2 did not affect virus packaging and, thus, points towards a potential application for targeted gene delivery.

The identification of novel peptides that bind to candidate or novel proteins/receptors expressed exclusively on target tissues/organs and subsequent incorporation into gene delivery systems would theoretically result in cell- or organ-specific gene delivery. With this in mind, the ability to screen a large array of small peptide sequences (10^{10}–10^{14}) for their affinity for a particular receptor, cell, or organ is particularly appealing. Phage display of peptide libraries makes this possible.

Phage Display

A bacteriophage or phage is a virus whose host is a bacterium. Phage display of peptide libraries is based on the concept that insertion of a random oligonucleotide encoding a peptide within a structural gene of the phage will lead to the display of the peptide on the surface of the phage.[29–32] Derivatives of Ff (F-pilus dependent) filamentous *Escherichia coli* bacteriophages, such as M13 or fd, are most commonly used as vectors for phage display.[32] Using filamentous bacteriophage, Smith first demonstrated that exogenous peptides could be displayed on the phage surface as fusions with the coat proteins pIII or pVIII.[33]

Construction of phage display libraries whereby 10^{12}–10^{13} peptide sequences are displayed allows the potential to isolate and sequence phage particles displaying peptide sequences which drive high-affinity binding specificity to a chosen target cell or organ. This is achieved by "biopanning." In its simplest form biopanning consists of four stages (Fig. 2). First, a chosen phage library is biopanned on a specific target—either a chosen cell type or immobilized target protein. Weakly

[26] S. L. Hart, R. P. Harbottle, R. Cooper, A. Miller, R. Williamson, and C. Coutelle, *Gene Ther.* **2**, 552 (1995).

[27] J. S. Bartlett, J. Kleinschmidt, R. C. Boucher, and R. J. Samulski, *Nature Biotechnol.* **17**, 181 (1999).

[28] A. Girod, M. Reid, C. Wobus, H. Lahm, K. Leike, J. Kleinschmidt, G. Deléage, and M. Hallek, *Nature Med.* **5**, 1052 (1999).

[29] F. Nilsson, L. Tarli, F. Viti, and D. Neri, *Adv. Drug Deliv. Rev.* **43**, 165 (2000).

[30] E. Koivunen, W. Arap, D. Rajotte, J. Lahdenranta, and R. Pasqualini, *J. Nucl. Med.* **40**, 883 (1999).

[31] D. J. Rodi and L. Makowski, *Curr. Opin. Biotechnol.* **10**, 87 (1999).

[32] H. B. Lowman, *Ann. Rev. Biophys. Biomol. Struct.* **26**, 401 (1997).

[33] G. P. Smith, *Science* **228**, 1315 (1985).

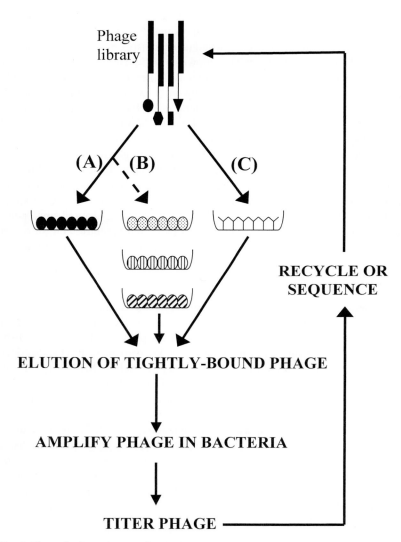

FIG. 2. Phage display on target cells *in vitro*. A phage library is biopanned against (A) target cell, (C) an immobilized protein or cells ectopically expressing a candidate receptor. To deplete the pool of phage which bind ubiquitous receptors, the second round of panning can be performed against nontarget cells (B) before enriching for further rounds on the target (A) or (C) once again. The process is known as preclearing. Following elution of bound phage, they are amplified in pilus-positive bacteria and titered prior to subsequent rounds of biopanning, or plaque selection for sequencing.

associated or unbound phage are removed and tightly bound phage are harvested by either pH disruption or addition of excess known ligand (for strategies based on biopanning against candidate receptors with suitable ligands). Phage are then amplified in pilus-positive bacteria (which are not lysed by the phages but multiple copies of phage displaying a particular peptide insert) whereby the pIII phage coat protein mediates infectivity through the F-pilus of the recipient bacteria. The amplified phage stock is then taken through further rounds of biopanning/amplification to positively enrich the pool in favor of phage containing peptide sequences which bind with high affinity to the chosen target.

Phage display technology has been used for a number of different applications, including mapping antibody–antigen binding epitopes[34] and protein–protein interactions.[30] Of greater interest with regard to targeted gene delivery, however, are recent studies where phages have been biopanned *in vitro* against immobilized purified receptors, proteins, or intact cells to identify novel binding peptides or motifs.[35–46]

Biopanning *in vitro* against candidate proteins has identified a number of motifs or peptides which specifically bind to $\alpha_5\beta_1$ integrin,[46] $\alpha_5\beta_3$ integrin,[45] hyaluronan,[36] vascular endothelial growth factor (VEGF),[42] the kinase domain receptor (KDR),[35] and human transferrin receptor.[38] Biopanning against mouse fibroblasts,[44] mouse thymic epithelial cells,[43] human neutrophils,[40] human umbilical vein endothelial cells (HUVECs),[37] and human epithelial cells[41] has also resulted in identification of peptide ligands which can mediate high-level cell-specific binding.

[34] M. N. Fukuda, C. Ohyama, K. Lowitz, O. Matsuo, R. Pasqualini, E. Ruoslahti, and M. Fukuda, *Cancer Res.* **60,** 450 (2000).

[35] R. Binétruy-Tournaire, C. Demangel, B. Malavaud, R. Vassy, S. Rouyre, M. Kraemer, J. Plouët, C. Derbin, G. Perret, and J. C. Mazié, *EMBO J.* **19,** 1525 (2000).

[36] M. E. Mummert, M. Mohamadzadeh, D. I. Mummert, N. Mizumoto, and A. Takashima, *J. Exp. Med.* **192,** 769 (2000).

[37] S. A. Nicklin, S. J. White, S. J. Watkins, R. E. Hawkins, and A. H. Baker, *Circulation* **102,** 231 (2000).

[38] H. Xia, B. Anderson, Q. Mao, and B. L. Davidson, *J. Virol.* **74,** 11359 (2000).

[39] M. A. Burg, R. Pasqualini, W. Arap, E. Ruoslahti, and W. B. Stallcup, *Cancer Res.* **59,** 2869 (1999).

[40] L. Mazzucchelli, J. B. Burritt, A. J. Jesaitis, A. Nusrat, T. W. Liang, A. T. Gewirtz, F. J. Schnell, and C. A. Parkos, *Blood* **93,** 1738 (1999).

[41] H. Romanczuk, C. E. Galer, J. Zabner, G. Barsomian, S. C. Wadsworth, and C. R. O'Riordan, *Hum. Gene Ther.* **10,** 2615 (1999).

[42] W. J. Fairbrother, H. W. Christinger, A. G. Cochran, G. Fuh, C. J. Keenan, C. Quan, S. K. Shriver, J. Y. K. Tom, J. A. Wells, and B. C. Cunningham, *Biochemistry* **37,** 17754 (1998).

[43] D. B. Palmer, A. J. T. George, and M. A. Ritter, *Immunology* **91,** 473 (1997).

[44] M. A. Barry, W. J. Dower, and S. A. Johnston, *Nature Med.* **2,** 299 (1996).

[45] J. B. Healy, O. Murayama, T. Maeda, K. Yoshino, K. Sekiguchi, and M. Kikuchi, *Biochemistry* **34,** 3948 (1995).

[46] E. Koivunen, B. Wang, and E. Ruoslahti, *J. Cell Biol.* **124,** 373 (1994).

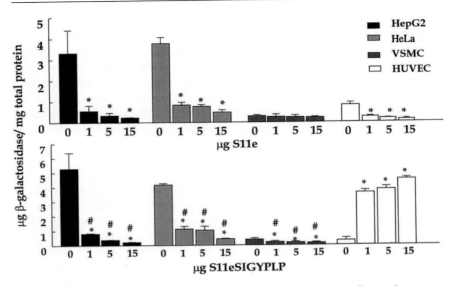

FIG. 3. Retargeting adenoviral infection to vascular endothelial cells *in vitro*. Bar graphs represent the ability of adenoviruses to infect HepG2 hepatocytes, HeLa cancer cells, human vascular smooth muscle cells (VSMCs) and human umbilical vein endothelial cells (HUVECs) in the presence of increasing concentrations of the CAR-binding neutralizing antibody alone (S11e) or neutralizing antibody engineered to display the endothelial cell-homing peptide SIGYPLP (S11eSIGYPLP). The *y*-axis represents the level of β-galactosidase reporter gene expression. *$p < 0.05$ vs nonmodified adenovirus, #$p < 0.05$ vs equivalent data for HUVECs. Readers are encouraged to see Ref. 37 for further details.

With respect to the use of these peptides for retargeting gene delivery, many of the peptides identified *in vitro* have subsequently been used to retarget gene delivery *in vitro* and *in vivo*. For example, within our own group Nicklin *et al.* (2000)[37] have demonstrated a 15.5-fold increase in adenoviral-mediated gene delivery selectively to human umbilical vein endothelial cells (HUVECs) *in vitro* when the heptapeptide SIGYPLP was used to retarget adenoviral tropism (Fig. 3). This is highly significant given the low level of virus uptake by endothelium basally[47,48] and the potential for gene delivery to endothelial cells in a number of disease states. Targeted delivery to human epithelial cells has been achieved by linking polyethylene glycol (PEG)-coupled dodecapeptides identified from *in vitro* biopanning to

[47] A. F. Merrick, L. D. Shewring, and G. J. Sawyer, *Transplantation* **62**, 1085 (1996).
[48] P. Lemarchand, H. A. Jaffe, C. Danel, M. C. Cid, H. K. Kleinman, L. D. Stratford-Perricaudet, M. Perricaudet, A. Pavirani, J.-P. Lecocq, and R. G. Crystal, *Proc. Natl. Acad. Sci. U.S.A.* **89**, 6482 (1992).

the adenoviral surface.[41] Furthermore, Rajotte *et al.*[49,50] identified a peptide GFE-1, capable of binding to lung endothelium through membrane dipeptidase. Subsequently, a bispecific adenoviral conjugate was developed and administration led to selective infection of membrane dipeptidase expressing cells which were previously resistant to infection with nontargeted virus.[51] Rather than attempting cell-selective gene delivery, a number of studies have focused on enhancement of gene delivery using targets expressed at high levels on many cell types. For example, in an attempt to enhance viral transduction of brain microcapillary endothelium (BME), a nonamer phage display library was biopanned against the human transferrin receptor.[38] Representative peptides of the identified motifs were subsequently cloned into the adenovirus fiber and demonstrated a 2- to 34-fold increase in gene transfer to cell lines expressing the receptor and human BME cells *in vitro*.[38]

In vitro biopanning has also been used to identify peptides subsequently shown to be capable of directing organ- and tissue-specific binding *in vivo*.[35,39,52] In one such study, the rat homolog of human melanoma proteoglycan, NG2, was used to screen random phage libraries *in vitro* and two specific NG2-binding peptides were identified. Intravenous injection of homogenous phages containing each peptide resulted in selective homing to tumor vasculature.[39] Koivunen *et al.* (1999)[52] isolated peptides targeted to matrix metalloproteinase-2 and -9, proteolytic enzymes associated with cancer cell invasion and metastasis. These peptides were able to inhibit endothelial and tumor cell migration *in vitro* and, more importantly, prevent tumor growth and invasion *in vivo* in mice models of human carcinoma. This study, therefore, demonstrated isolation of phage that encode peptides with distinct and selective biological activity. As a further example using VEGF-induced angiogenesis as the biological target, peptides which blocked the VEGF–KDR interaction were identified *in vitro* and found to inhibit VEGF-induced endothelial cell proliferation *in vitro*. When administered *in vivo* in a rabbit corneal angiogenesis model VEGF-mediated angiogenesis was selectively blocked.[35] It remains to be defined whether these peptides can also mediate selective gene delivery for applications in human disease.

Of particular interest with regard to gene therapy are studies demonstrating *in vivo* homing of individual peptides isolated using phage display *in vivo* after

[49] D. Rajotte, W. Arap, M. Hagedorn, E. Koivunen, R. Pasqualini, and E. Ruoslahti, *J. Clin. Invest.* **102,** 430 (1998).

[50] D. Rajotte and E. Ruoslahti, *J. Biol. Chem.* **274,** 11593 (1999).

[51] M. Trepel, M. Grifman, M. D. Weitzman, and R. Pasqualini, *Hum. Gene Ther.* **11,** 1971 (2000).

[52] E. Koivunen, W. Arap, H. Valtanen, A. Rainisalo, O. P. Medina, P. Heikkilä, C. Kantor, C. G. Gahmberg, T. Salo, Y. T. Konttinen, T. Sorsa, E. Ruoslahti, and R. Pasqualini, *Nature Biotechnol.* **17,** 768 (1999).

systemic delivery.[49,53–55] In early landmark studies phages displaying peptides which selectively homed to brain and kidney endothelium in Balb/c mice were identified.[55] Intravenous injection of phage carrying an α_v-directed RGD peptide resulted in phage binding selectively to tumor vasculature in tumor-bearing mice.[54] Additionally, aminopeptidase N has since been identified as the receptor for a tumor-homing phage peptide identified *in vivo* containing the Asn-Gly-Arg (NGR) motif.[54,56] Given the role of aminopeptidase N in angiogenesis this highlights a potential new target for both inhibiting angiogenesis and tumor targeting.[56] Subsequent studies *in vivo* identified peptide sequences capable of driving organ- or tissue-specific phage homing in mice.[49] These studies demonstrated the heterogeneity of the endothelium across a number of chosen target organs and provides a unique "address system" in the vasculature which lends itself to targeting exploitation.[57,58] Whether the identified peptides can be used to direct organ- or tissue-specific gene delivery *in vivo* is unknown, but results of such experiments are eagerly awaited. One such peptide containing an α_v-binding RGD motif has, however, been used to target chemotherapy to human breast cancer xenografts in mice.[53]

Clearly, as the field of targeted gene delivery progresses the search for novel, specific, and safe targeting moieties continues. The small nature of the peptides isolated from random phage display peptide libraries permits their integration into existing viral vectors both pre- and post-viral packaging. In addition, immunogenicity of these peptides will be minimal because of their size and thus, sequestration of the retargeted vectors prior to gene expression should be avoided. In light of the recent increase in use of phage display to identify novel targeting peptides it is our aim, in this chapter, to describe the methods currently used to pan phages on cell cultures and identify peptide sequences which can mediate cell-specific gene delivery.

Methods Overview

There are many different types of phage display library available commercially. They can be simply divided into two groups, linear and constrained. Linear libraries are those in which the peptide is constrained at its C terminus to the coat protein of the phage via a short spacer sequence, with the N terminus free. In constrained libraries, the peptide sequence is expressed on the coat protein of

[53] W. Arap, R. Pasqualini, and E. Ruoslahti, *Science* **279,** 377 (1998).
[54] R. Pasqualini, E. Koivunen, and E. Ruoslahti, *Nature Biotechnol.* **15,** 542 (1997).
[55] R. Pasqualini and E. Ruoslahti, *Nature* **380,** 364 (1996).
[56] R. Pasqualini, E. Koivunen, R. Kain, J. Lahdenranta, M. Sakamoto, A. Stryhn, R. A. Ashmun, L. H. Shapiro, W. Arap, and E. Ruoslahti, *Cancer Res.* **60,** 722 (2000).
[57] E. Ruoslahti and D. Rajotte, *Ann. Rev. Immunol.* **18,** 813 (2000).
[58] M. Trepel, W. Arap, and R. Pasqualini, *Gene Therapy.* **7,** 2059 (2000).

the phage flanked by two cysteine residues. Under nondenaturing conditions, the cysteine residues spontaneously dimerize via disulfide bridges giving the peptide secondary structure.

The length of the peptide expressed by the phage is the other variable in available phage display libraries. Peptides can be typically 7-mer, 12-mer, and up to 20-mer. Because of the upper limit of construction of primary phage display libraries, those expressing peptides greater than 7 amino acids risk not expressing the full complement of random peptides. However, libraries expressing longer peptides may have advantages for certain applications. For example, if a hexapeptide library is adequate for a biopanning strategy, a 20-mer library may increase the effective library diversity as each 20-mer contains 15 hexapeptides with different flanking sequences; hence each peptide is constrained in different structural contexts. This strategy may allow the generation of amino acid motifs with a greater affinity for the target. Hence, the choice of phage display library is important and may vary depending on the downstream application.

Within our group we have had success utilizing a commercially available linear 7-mer library (New England Biolabs) to isolate peptides capable of mediating high-level and restricted adenoviral infection into vascular endothelial cells,[37] and also for identifying novel peptides that bind to ectopically expressed candidate receptors.[59] This library expresses peptides on the pIII coat protein of the phage providing 1–5 copies of each peptide per virion. This low-valency peptide display allows the selection of high-affinity interactions in comparison to phage libraries that express the peptide on the pVIII protein (200 peptides per virion). Recombinant M13 phage-expressing peptides have a growth disadvantage in comparison to wild-type M13 phage; this is even more marked when using constrained libraries, and recombinant phage are sensitive to contamination with even very low levels of wild-type phage through multiple amplification steps. Therefore, in the first instance linear libraries are less technically demanding to work with and therefore we would recommend using these for any initial investigations.

Materials

All the microbiological reagent compositions were obtained from *Molecular Cloning A Laboratory Manual.*[60]

PhD Phage Display Peptide Library Kit (New England Biolabs, Beverley, MA).

Luria-Bertani (LB) medium. Per liter: 10 g bacto-tryptone, 5 g yeast extract, 5 g NaCl, pH 7.5. Autoclave and store at room temperature.

[59] S. J. White, S. A. Nicklin, T. Sawamura, and A. H. Baker, *Hypertension* **37**, 449 (2001).

[60] J. Sambrook, E. F. Fritsch, and T. Maniatis, "Molecular Cloning: A Laboratory Manual," 2nd Ed. Cold Spring Harbor Laboratory Press, Cold Spring Harbor, NY, 1989.

LB agar. As for LB medium, +15 g bacto-agar. Autoclave, cool to <70°. Store plates at 4°.

Agarose top. Per liter: 10 g bacto-tryptone, 5 g yeast extract, 5 g NaCl, 1 g MgCl · 6H$_2$O, 7 g agarose. Autoclave, dispense into 25–50 ml aliquots. Store solid at room temperature and melt in microwave as required.

2× M9 salts. Per liter: 12 g Na$_2$HPO$_4$, 6 g KH$_2$PO$_4$, 1 g NaCl, 2 g NH$_4$Cl. Autoclave and store at room temperature.

Minimal plates. Per liter: 500 ml 2× M9 salts, 500 ml 3% bacto-agar, 20 ml 20% glucose, 2 ml 1 M MgSO$_4$, 0.1 ml CaCl$_2$, 1 ml thiamin (10 mg/ml). Autoclave all components separately, except thiamin and glucose, which should be filter-sterilized, cool to <70° before combining. Store plates at 4°.

Polyethylene glycol/NaCl. 20% (w/v) polyethylene glycol-8000 (PEG), 2.5 M NaCl. Autoclave and store at room temperature.

Isopropyl thiogalactoside (IPTG): 200 mg/ml in deionized H$_2$O. Filter-sterilize. Store at 4° protected from light.

5-Bromo,4-chloro-3-indolyl-β-D-galactopyranoside (X-Gal): 20 mg/ml in dimethyl formamide. Filter-sterilize. Store at −20° protected from light.

Tris-buffered saline. 150 mM NaCl, 50 mM Tris-HCl, pH 7.6. Autoclave and store at room temperature.

Phenol : choloroform : isoamyl alcohol (ratio 50 : 49 : 1).

Methods

1.1. Titration of Phage Display Libraries

It is important to determine an accurate titer for phage libraries prior to carrying out any biopanning experiments as variations in phage titer in later rounds can lead to spurious results. In this chapter the bacterial host which will be referred to is *Escherichia coli* ER2537 or ER2738 which are supplied with the New England Biolab's PhD Phage Display Library Peptide Kit (http://www.neb.com). This is a F+ strain of bacteria, where the F' pilus is required to mediate M13 phage infection into the cells. It is important to follow the recommendations for maintaining bacterial hosts, e.g., streak plates onto the correct selective medium to ensure the F' episome is maintained in all progeny. We have also observed that as the plate ages the efficiency with which M13 phage infect the bacteria in liquid culture decreases; therefore we would recommend streaking a fresh plate every 2 weeks. Stocks of bacteria can be archived in glycerol using standard microbiological procedures. The method outlined for titering below is used to determine the titer of the library initially, but also for titration of phage recovered after biopanning before and after amplification. Hence, there are several dilution ranges suggested in point (ii).

(i) Inoculate 15 ml of LB media with a single colony of *E. coli* ER2537 and incubate with shaking at 37° until mid-log phase ($OD_{600} = 0.5$–1.0) for approximately 3–5 hr depending on the age of the streak plate. During this time place 5 LB plates per phage to be titered at 37° to prewarm.

(ii) Make 100-fold dilutions of the phage library in LB in the ranges of 10^1–10^4 for unamplified biopanning eluates or 10^4–10^8 for amplified and purified phage stocks and stock phage libraries.

(iii) Just prior to bacteria being ready, melt agarose top by gentle heating in a microwave, dispense 3-ml aliquots into sterile tubes (one for each phage dilution to be plated out), and equilibrate the tubes in a water bath at 45° until use. Add 4 μl IPTG and 40 μl X-Gal to the agarose top. (N.B. Only melt the agarose top just prior to use as prolonged incubation at 45° is detrimental and makes plaque counting difficult.)

(iv) Once the *E. coli* host culture has reached mid-log phase, dispense 200 μl into Eppendorfs (one per phage dilution to be plated). Add 10 μl of each phage dilution to the bacteria, vortex briefly, and incubate for 1–5 min at room temperature. (N.B. If the phage are added to the tubes first, followed by the bacteria, there is no need to vortex the tube.)

(v) Immediately add the infected bacteria into an aliquot tube of agarose top, quickly pour onto the surface of an LB plate, spread out evenly before it sets, and incubate inverted overnight at 37°. (N.B. As the recombinant phage carry the *LacZα* gene, α-complementing bacteria such as ER2537 plaques will appear as fuzzy blue circles in a lawn of bacteria in the presence of IPTG and X-Gal. As well as making it easier to visualize and count plaques, it also provides a means to distinguish recombinant library phage from any contaminating wild-type M13, which appear as clear plaques.)

(vi) Calculate the titer of the phage by counting plates with approximately 100 plaques. Multiply this number of plaques by the phage dilution factor for that particular plate to give the phage titer per 10 μl (the original phage input). Finally, multiply this figure by 100 to give phage titer as pfu/ml.

1.2. First Round Biopanning on Target Cell Cultures

We have used two strategies for biopanning on cell cultures. They both follow similar steps, except that the second strategy utilizes preclearing steps on nontarget cell types as well as panning on target cells. We found that depletion of the primary library for peptides capable of interacting with ubiquitous cell receptors markedly improves the efficiency with which effective target peptides are selected.[37] However, for depletion of the primary library by biopanning on nontarget cells it has been recommended to carry this clearing step out following a first round of biopanning on target cells. This avoids the likelihood of completely losing those peptides that associate with nontarget cells via a charge rather than specific interaction. The following protocols are for a protocol including preclearing at round 2, but many variations for individual applications can be attempted.

(i) Prepare target cells in 6-well plates. (N.B. Suspension cells can be biopanned by modifying this protocol to carry out centrifuging steps in between each washing or phage incubation step.) It is important to ensure the cells have achieved 100% confluence prior to initiation of phage incubations to eliminate the possibility of selecting phage which bind the tissue culture plastic (peptides rich in aromatic residues, e.g., tryptophan, have previously been demonstrated to target tissue culture plastic[61]). Also, biopanning against postconfluent cells decreases the chance of targeting receptors involved in proliferative phenotypes, unless this is the primary objective.

(ii) Take the initial input titer of phage library (for a library with a complexity of 2×10^9 peptides, we use 2×10^{11} pfu, providing 100 copies of each different peptide). (N.B. However, for biopanning on primary HUVECs, we have found that these cells are very sensitive to the phage incubation step and therefore we had to lower the initial input phage to 1×10^9 pfu. This is a factor that is cell-specific as we have not observed this problem with other cell types.) Remove the medium from the cells and replace with 1 ml cold DMEM containing 0.1–1% bovine serum albumin [(BSA); the concentration of the BSA can be varied to change the stringency of the biopanning]. Place the cells at 4° for 15 min (to prevent endocytosis and subsequent degradation of phage). Add the phage to the media on the cells, swirl to mix, and place the cells back at 4° for 60 min.

(iii) After 60 min, remove the medium from the cells to waste and wash the cells 5 times for 5 min each time with ice-cold phosphate-buffered saline (PBS) (0.132 g/liter calcium, 0.1 g/liter magnesium) and 0.1–1% BSA.

(iv) After the final wash, remove all PBS and add 1 ml of 0.2 M glycine (in H_2O, pH 2.2) to the cells for 10 min on ice. Add 200 μl 1 M Tris-HCl (pH 8.0) to neutralize the glycine and remove the eluate to a sterile Eppendorf. This eluate contains phage which are weak cell-surface binders.

(v) Add 250 μl 30 mM Tris-HCl/10 mM EDTA (pH 8.0) to the cells and incubate for 1 hr on ice. Collect the cell debris, pipette up and down two or three times, and centrifuge at 10,000g for 5 min. Collect the supernatant into a fresh Eppendorf and discard the pellet. The supernatant contains phage which were tightly associated with the cells. Both populations of phage (i.e., weakly and tightly associated) can be used for subsequent rounds of panning, but should be treated as independent populations of phage for all subsequent rounds. In our own experience, tight binders

---

[61] N. B. Adey, A. H. Mataragnon, J. E. Rider, J. M. Carter, and B. K. Kay, *Gene* **156,** 27 (1995).

represent peptides with properties more appropriate for downstream applications (e.g., targeted gene delivery).

(vi) Keep 10 μl of this crude phage stock and titer as described in Section 1.1. This should be carried out after each round as it defines selective phage enrichment after each successive round of biopanning on target cells.[37]

1.3. Phage Amplification

Amplification of the phage stocks between each round of biopanning is important to ensure that the initial input phage titer is maintained throughout all rounds of biopanning.

(i) Inoculate 20 ml of LB medium with a single colony of *E. coli* ER2537 and incubate overnight at 37° with vigorous shaking (200 rpm).

(ii) Dilute this overnight culture 1:100 in fresh LB medium and add the all phage supernatant from the previous round of biopanning.

(iii) Incubate the phage infected cultures at 37° with shaking at 180 rpm for exactly 5 hr and proceed to Section 1.4.

1.4. Phage Purification

(i) Take the bacterial culture containing amplified phage and centrifuge at 6000*g* for 10 min, at 4°. Transfer the supernatant to a fresh tube and repeat the centrifugation step.

(ii) Remove the upper 80% of the supernatant to a fresh tube (only taking the upper 80% ensures no bacteria or bacterial debris is carried through the subsequent purification steps) and add 1/6th volume of PEG/NaCl. Mix the tube gently and place the sample at 4° overnight to precipitate the phage.

(iii) Centrifuge the precipitated phage at 6000*g* for 30 min at 4°C. At this point a white pellet should be observed. Discard the supernatant. Resuspend the pellet in 1 ml of TBS and centrifuge at 10,000*g* at RT to pellet any residual bacteria. Transfer the resulting supernatant to a fresh tube and reprecipitate with 1/6th volume of PEG/NaCl for 60 min on ice.

(iv) Centrifuge the precipitated phage at 10,000*g* at 4° for 10 min, discard the supernatant, and resuspend the phage pellet in 200 μl of TBS containing 0.02% NaN$_3$. Stocks of phage can be stored at 4° for several weeks without loss of titer, but for long term they should be stored in 50% glycerol at −20°. Titer the phage as described in Section 1.1 ready for the next round of biopanning.

1.5. Second Round Biopanning on Nontarget Cells

(i) Prepare 6-well plates containing 2 wells of each of the nontarget cell types you wish to use. The cell types used at this stage will obviously depend to some extent on the target cell, but it is advisable to avoid cell lines if possible as these are likely to express receptors other than those of the primary tissue from which they are derived. For targeting to endothelial cells we precleared on vascular smooth muscle cells (VSMCs), hepatocytes, and peripheral blood mononuclear cells (PBMCs). We rationalized that VSMCs and PBMCs were the cell types in most close contact to endothelial cells. Also, our downstream application was retargeted adenovirus gene transfer; therefore we wished to remove phage capable of mimicking the viruses' natural tropism, i.e., hepatocytes. Again ensure that the cells are 100% confluent.

(ii) Calculate the volume of amplified phage stock required to maintain an identical input titer to that used in round 1.

(iii) Exchange the medium on the first well for 1 ml cold DMEM with 0.1–1% BSA and place at 4° for 15 min as described in Section 1.2.

(iv) Add the phage stock to the first well and place at 4° for 60 min.

(v) Remove the media from the second well and then place the supernatant containing unbound phage from the first well onto these cells.

(vi) Continue this process for all the wells of nontarget cells. Finally, collect the supernatant to a fresh Eppendorf. Save 10 μl of phage to determine the titer obtained after the clearing steps and amplify and titer the phage as described in Sections 1.1, 1.3, and 1.4.

(vii) Once the titer of the freshly amplified phage has been obtained, this is the stock phage. Using the same input titer as originally used this phage can be biopanned on the target cells as described in Section 1.2. Typically four rounds in total of biopanning on target cells are required to obtain consensus peptide sequences, but sequence the DNA insert from 20 plaques after round 3. If no consensus (i.e., any repeated identical peptide sequences or motifs) is observed then carry out the fourth round as described.

1.6. Isolation of Homogenous Populations of Phage and Purification of Single-Stranded Phage DNA

(i) Following completion of the final round of biopanning on target cells, do not proceed to amplification, purification and titration steps. Instead prepare serial dilutions of the eluted phage as recommended in Section 1.1 (i.e., 10^1–10^4), plate in agarose top, invert the plates, and incubate overnight at 37°.

(ii) Identify plates containing approximately 100 plaques for isolating individual phage. Add 3 ml of fresh LB to a universal container [(U.C.), one per plaque to be picked] and inoculate with 30 μl of an overnight culture of E. coli ER2537.

(iii) Pick single well-defined plaques from the agar plates by stabbing them using a sterile glass pipette and transfer the plug to bacteria. (N.B. It is extremely important to ensure that each plaque is well defined and separate from any others to avoid the possibility of obtaining mixed clones.)

(iv) Incubate the cultures containing single plaques at 37° with shaking for 5 hr.

(v) Following the incubation centrifuge the cultures at 10,000g for 5 min at room temperature and transfer the phage-containing supernatant to a fresh tube. Repeat the centrifugation step.

(vi) Transfer the upper 80% (approximately 2.5 ml) of the supernatant to a fresh tube and then transfer 1.5 ml of this supernatant to a second tube for the isolation of ssDNA.

(vii) To the first tube containing 1 ml of phage containing supernatant add an equal volume of glycerol. This is the stable stock of plaque pure phage and should be stored at $-20°$.

(viii) For preparation of ssDNA, add 200 μl of PEG/NaCl to the second tube containing 1.5 ml of phage-containing supernatant, mix the tube by inversion, and incubate at room temperature for 15 minutes.

(ix) Centrifuge the tube for 5 min at 10,000g at room temperature and discard the supernatant, leaving the white pellet. Resuspend the pellet in 100 μl of 1× TE (10 mM Tris, 1 mM EDTA, pH 7.5) and add 50 μl phenol : choloroform : isoamyl alcohol (ratio 50 : 49 : 1).

(x) Vortex the tube vigorously for 20 sec, and incubate at room temperature for 15 min. Repeat the vortexing step and then centrifuge at 10,000g for 3 min at room temperature.

(xi) Transfer the upper aqueous layer containing the DNA to a fresh tube and add 10 μl 3 M sodium acetate (pH 5.5) and 250 μl 100% ethanol followed by incubation at 4° for 10 min.

(xii) Centrifuge at 10,000g for 10 min at room temperature and wash the DNA-containing pellet with 1 ml of ice cold 70% ethanol. Leave the pellets to air dry for 15 min and then redissolve them in 50 μl 1× TE. The integrity and yield of phage ssDNA can be assessed on 1% agarose gels and ethidium bromide staining. Phage DNA appears as a smear as the genome is supercoiled.

1.7. Sequencing of Individual Phage DNA

Phage ssDNA can be sequenced either by standard Sanger chain terminating sequencing[62] or by automated sequencing following standard protocols.

1.8. Characterization of the Specificity of Selected Phage

(i) Amplify, purify, and titer each homogenous phage by adding 20 μl of the glycerol stock to a 1:100 dilution of overnight culture of bacteria as described in Sections 1.1, 1.3, and 1.4.

(ii) Plate target cells into 6-well plates and culture until 100% confluent. Exchange the media for cold DMEM (0.1–1% BSA) and place at 4° for 15 min.

---

[62] F. Sanger, S. Nicklen, and A. R. Coulson, *Proc. Natl. Acad. Sci. U.S.A.* **74**, 5463 (1977).

(iii) Add 1×10^7 pfu of each homogenous phage population to a well of target cells in triplicate and incubate at 4° for 30 min. (N.B. We use a reduced titer and reduced biopanning time for phage characterisation as we have observed cellular toxicity on primary HUVEC, presumably associated with using high titers of homogenous phage which strongly associate with the target cell.)

(iv) Carry out post-biopanning washing steps as described in Section 1.2.

(v) Without amplifying the phage, carry out titering as described in Section 1.1 and the percentage recovery can be calculated for each cell type. (N.B. We use a control phage, without a peptide insert, to normalize for background phage binding. However, care should be taken when using peptideless phage with recombinant phage as wild-type M13 have a selective growth advantage over library phage.)

Another method of quantifying homogenous phage binding to target cells is to use the M13 enzyme-linked immunosorbent assay (ELISA) available from Amersham-Pharmacia Biotech. Once you have selected the phage which express the peptides that are most selective for binding the target cell, the peptides themselves can be used for further applications.

1.9. Biopanning against Ectopically Expressed Receptors

If a candidate receptor is available for a specific application, ectopic expression of the protein on a cell type which does not endogenously express it is a viable means by which to identify novel peptides specific for that receptor. By expressing the protein ectopically on the cell surface, the conformation of the parent protein may be retained more appropriately than if purified protein is dried onto tissue culture plastic. In a recent study[59] we used the cell line HepG2 to overexpress the protein lectin-like oxidized low density lipoprotein-1 (LOX-1) and identified peptides with high affinity for this receptor. HepG2 cells which ectopically expressed LOX-1 on the cell surface after transfection with a eukaryotic expression vector containing LOX-1 or control HepG2 cells transfected with a vector expressing β-galactosidase were generated. We were then able to demonstrate selective binding of LOX-1 enriched phage to the LOX-1 transfected HepG2 cells versus the control transfected HepG2 cells. The protocols detailed below should enable the reader to carry out the same procedures for a chosen protein

(i) Phage libraries should be amplified, purified, titered, and sequenced according to the protocols described in Sections 1.1–1.7.

(ii) A suitable eukaryotic cell type should be selected for ectopic expression of the candidate receptor. Suitable characteristics include: no endogenous expression of the receptor [this should be determined by initial reverse-transcriptase polymerase chain reaction (RT-PCR) and Northern blotting] and a cell type that is readily transfectable so that large populations of receptor-positive cells can be generated.

(iii) Clone the candidate receptor cDNA into an appropriate eukaryotic expression vector such as Lazarus (LZRS).[63] The LZRS vector contains a copy of the Moloney murine leukemia virus long terminal repeat (LTR) linked to the β-galactosidase reporter gene, EBNA-1/oriP maintenance factor/origin of replication, as well as the puromycin resistance gene and is efficiently and episomally maintained in eukaryotic cells placed under puromycin selection. It is also necessary to construct a suitable control vector (i.e., with the same backbone but a control transgene). (N.B. It is appropriate to exchange a control transgene for an empty expression cassette as the important step is to have a control cell line maintained under the same selection conditions as the candidate receptor expressing population. This will ensure that while only one cell type expresses the candidate receptor, the control cells will have an identical phenotype for preclearing steps. However, do not use another cell-surface protein as a control as this could produce a conflict between the phage selected at either biopanning or preclearing steps.)

(iv) Transfect your chosen cell type with the appropriate vector using standard transfection techniques such as calcium phosphate or liposomes. Generate two separate populations, one expressing the candidate receptor, and one the control. Select for cells expressing each plasmid using the antibiotic marker and culture until a large enough population for biopanning is available. Passage these cells into 6-well plates and maintain them under positive selection until they are confluent.

(v) Perform preclearing by incubating the 2×10^{11} pfu phage in 1 ml of biopanning medium [DMEM containing 1% BSA for 1 hr at 4°, on 4 successive wells of control transfected cells (total of 4 hr)]. (N.B Preclearing rounds for candidate receptor biopanning can also be attempted at round 2 as described in Section 1.5. It may be that different methods are viable for different applications.)

(vi) Immediately recover the precleared phage and biopan in duplicate on cultures of candidate receptor transfected cells (in 1 ml biopanning media at 4° for 1 hr).

(vii) Wash the cells 5 times in PBS (0.132 g/liter calcium, 0.1 g/liter magnesium, 1% BSA) for 5 min per wash.

(viii) Elute weakly associated phage in 1 ml 0.2 M glycine (pH 2.2) for 10 min at 4°, followed by neutralization with 200 μl Tris-HCl (pH 8.0).

(ix) Isolate tightly bound phage by lysing the cells in 1 ml of 30 mM Tris/1 mM EDTA (pH 8.0) for 1 hr at 4°.

(x) Remove the cell debris by centrifugation at 10,000g for 5 min at room temperature and remove the phage containing supernatant to a fresh Eppendorf.

(xi) Phage obtained at each step should be amplified and titered as described in Sections 1.1, 1.3, and 1.4.

(xii) Repeat the preclearing step with 1×10^9 pfu of phage library on one well of control transfected cells for 1 hr at 4° and then immediately collect the unbound

[63] T. D. Kinsella and G. P. Nolan, *Hum. Gene Ther.* **7,** 614 (1996).

phage and incubate them with 1 confluent well of candidate receptor expressing cells for 1 hr at 4°. (N.B. We performed a preclearing step before each round of target receptor biopanning to further decrease the opportunity for phage binding nontarget cell receptors to be carried through the procedure. However, depending on the cell type you are using and the receptor being expressed, one round of preclearing may be sufficient. Also, we reduced the titer of biopanned phage at each round in an attempt to bias the biopanning procedure toward those phage with the highest affinity for the target receptor, i.e., create experimental conditions which gave selective pressure toward diminishing the diversity of the library. However, again this may be application specific and therefore keeping the phage titer identical throughout each round of biopanning may be more appropriate for some target receptors.)

(xiii) Repeat steps vi–xii for 4–5 rounds, then plate and pick single phage plaques for extraction of DNA and sequencing as described in Sections 1.6 and 1.7.

(xiv) Finally, for characterisation of phage binding the same basic methodology as described in Section 1.8 can be performed. However, as performed during the initial selection of phage, percentage recovery of candidate peptides can be determined by carrying out homogenous phage panning onto both candidate receptor expressing and control transfected cells. An insertless phage can also be used as a control for background binding, and the phage ELISA mentioned in Section 1.8 can also be used as an alternative method.

Summary

The field of gene therapy has developed at an astonishing pace over the past decade, perhaps too quickly. Clinical studies have highlighted major flaws in the ability of current vectors to deliver genes safely and effectively to patients; hence the further development of vectors is a prerequisite for future success. In this chapter we have discussed advances in development of targeted vectors through isolation of targeting moieties using phage display. The field of gene therapy will benefit considerably by the isolation and use of peptides that are effective for targeting *in vivo,* particularly for diseases affecting individual organs. Only when truly selective and highly efficient vectors are constructed will the tremendous potential of gene therapy be realized.

[11] Helper-Dependent Adenoviral Vectors

By HESHAN ZHOU, LUCIO PASTORE, and ARTHUR L. BEAUDET

Introduction

The human adenovirus genome is a double-stranded, 36-kb linear DNA. The viral genes are divided into early (E) and late (L) genes relative to the onset of viral DNA replication. The E1 and E4 regions encode proteins for regulation of viral and cellular genes; the E3 region is primarily directed toward modification of the host immune response; and the E2 genes encode three proteins directly involved in viral DNA replication. The viral late genes, L1–L5, are expressed from a common major late promoter and encode the structural proteins for packaging viral DNA into infectious particles. All viral proteins can be provided in trans. The only elements required in cis are the two inverted terminal repeats (ITRs, 103 bp for each) that are essential for viral DNA replication, and a 170-bp packaging region that is required for packaging the DNA into virion particles; the packaging region is located at bp 194 to 358 near the left ITR. For more biological details of adenoviruses, an excellent review is available from Shenk.[1]

Adenoviral (Ad) vectors are particularly attractive for *in vivo* and *ex vivo* gene delivery, primarily because they are highly efficient for gene transfer to a variety of cell types, and it is relatively easy to produce stable vectors with high titer. Ad vectors are extensively used in gene therapy, as recombinant viral vaccines and for basic science studies.[2,3]

First-generation Ad vectors utilize expression cassettes of therapeutic or reporter genes replacing the E1 region. E1-deleted Ad vectors are propagated in E1 complementing 293 cells or other more recently developed cell lines. To increase the carrying capacity, the E3 region in some first-generation vectors may also be removed, since this region is not essential for growth of adenovirus in cell culture. First-generation vectors generally produce transient expression of therapeutic genes. In cell culture, this is controlled by the fact that these vectors do not replicate or integrate into the cellular genome with any significant frequency. *In vivo,* the transient expression is primarily due to leaky expression of viral genes leading to immune responses against viral-transduced cells.[4–6] The vector particles also

[1] T. Shenk, *in* "Fields Virology" (B. N. Fields, D. M. Knipe, and P. M. Howley, eds.), p. 2111. Raven Press, Ltd., Philadelphia, 1996.

[2] L. D. Stratford-Perricaudet, M. Levrero, J. F. Chasse, M. Perricaudet, and P. Briand, *Hum. Gene Ther.* **1**, 241 (1990).

[3] B. Quantin, L. D. Perricaudet, S. Tajbakhsh, and J. L. Mandel, *Proc. Natl. Acad. Sci. U.S.A.* **89**, 2581 (1992).

[4] N. Mittereder, S. Yei, C. Bachurski, J. Cuppoletti, J. A. Whitsett, P. Tolstoshev, and B. Trapnell, *Hum. Gene Ther.* **5**, 717 (1994).

can elicit innate immune and inflammatory responses,[7,8] and expression of the transgene carried by the vector may induce host immune responses as well.[9] The viral proteins and the product of the transgene each may act in an adjuvant-like manner to increase the immune response to the other.

One strategy to overcome the limitations of first-generation vectors is to develop vectors with additional deletions of other essential viral genes. Second-generation Ad vectors with deleted E1 and E2 or E1 and E4 genes can be propagated in cell lines that complement both E1 and E2 or E1 and E4, respectively.[10–14] These vectors may provide some degree of improved safety,[11,13,15] but their ability to produce stable transgene expression has not yet been established.[16,9,17] A human fatality occurred using intravascular delivery of an E1/E4 deleted vector,[18] indicating that these vectors have serious limitations.

Another strategy to reduce the immunogenicity and improve the safety of Ad vectors is to delete all viral coding sequences, so that leaky expression of viral proteins is completely eliminated, although the proteins comprising the particle are still delivered. This type of vector depends on a helper virus to provide viral proteins in trans, leading to the terminology of helper-dependent adenoviral (HD-Ad) vectors. These vectors are sometimes called gutless or gutted vectors. One of the earliest HD-Ad vectors removed only a part of the viral coding sequences.[19] Introduction of a mutated packaging region into a first-generation virus resulted in a 90-fold reduction in the packaging of the virus. The combination of

[5] Y. Dai, E. M. Schwarz, D. Gu, W.-W. Zhang, and N. Sarvetnick, *Proc. Natl. Acad. Sci. U.S.A.* **92,** 1401 (1995).

[6] Y. Yang, F. A. Nunes, K. Berencsi, E. E. Furth, E. Gonczol, and J. M. Wilson, *Proc. Natl. Acad. Sci. U.S.A.* **91,** 4407 (1994).

[7] G. Wolff, S. Worgall, N. van Rooijen, W. R. Song, B. G. Harvey, and R. G. Crystal, *J. Virol.* **71,** 624 (1997).

[8] S. Worgall, G. Wolff, E. Falck-Pedersen, and R. G. Crystal, *Hum. Gene Ther.* **8,** 37 (1997).

[9] N. Morral, W. O'Neal, H. Zhou, C. Langston, and A. Beaudet, *Hum. Gene Ther.* **8,** 1275 (1997).

[10] H. Zhou, W. O'Neal, N. Morral, and A. L. Beaudet, *J. Virol.* **70,** 7030 (1996).

[11] M. Gorziglia, M. Kadan, S. Yei, J. Lim, G. M. Lee, R. Luthra, and B. Trapnell, *J. Virol.* **70,** 4173 (1996).

[12] A. Amalfitano, M. A. Hauser, H. Hu, D. Serra, C. R. Begy, and J. S. Chamberlain, *J. Virol.* **72,** 926 (1998).

[13] M. Lusky, M. Christ, K. Rittner, A. Dieterle, D. Dreyer, B. Mourot, H. Schultz, F. Stoeckel, A. Pavirani, and M. Mehtali, *J. Virol.* **72,** 2022 (1998).

[14] H. Zhou and A. Beaudet, *Virology* **275,** 348 (2000).

[15] K. Rittner, H. Schultz, A. Pavirani, and M. Mehtali, *J. Virol.* **71,** 3307 (1997).

[16] B. Fang, H. Wang, G. Gordon, D. A. Bellinger, M. S. Read, K. M. Brinkhous, S. L. Woo, and R. C. Eisensmith, *Gene Ther.* **3,** 217 (1996).

[17] A. Amalfitano, *Gene Ther.* **6,** 1643 (1999).

[18] E. Marshall, *Science* **286,** 2244 (1999).

[19] K. Mitani, F. L. Graham, C. T. Caskey, and S. Kochanek, *Proc. Natl. Acad. Sci. U.S.A.* **92,** 3854 (1995).

a 35-kb helper virus with a crippled packaging signal and purification by CsCl centrifugation yielded an HD-Ad vector (28 kb) with only 1% contamination with helper virus.[20] An important improvement in the HD-Ad system was the use of a 293-derived cell line, designated 293Cre4, that stably expresses the bacterial phage P1 Cre recombinase[21] in combination with a helper virus (AdLC8cluc) with two *loxP* sites flanking the packaging signal.[22] On infection of 293Cre4, the packaging signal in the helper virus DNA is excised through the Cre/*loxP* interaction, so that the helper virus DNA can still replicate and express viral genes, but cannot be packaged. This system further decreased helper virus contamination to as low as 0.01% and is the best strategy for preparation of HD-Ad vectors to date.

HD-Ad vectors are substantially less toxic than other Ad vectors when high doses are administered intravascularly in mice.[23-25] These vectors resulted in remarkably long-term expression of transgenes for 1 to 2 years after a single intravascular dose for delivery to the liver in mice and baboons.[24-28] These initial results suggested that the HD-Ad vector system is one of the most promising viral vector systems for gene therapy, particularly for achieving high-level and long-term expression in the liver. There is a review[29] of the methodology for use of the Cre/*loxP*-based helper-dependent system, and our review will attempt to complement that presentation.

Cre/*loxP* HD-Ad Vector System

The steps in implementing the HD-Ad vector system based on the Cre/*loxP* strategy as developed primarily by the laboratories of S. Kochanek and F. Graham[19-22] are depicted in Fig. 1.

[20] S. Kochanek, P. Clemens, K. Mitani, H.-H. Chen, S. Chan, and C. T. Caskey, *Proc. Natl. Acad. Sci. U.S.A.* **93**, 5731 (1996).

[21] L. Chen, M. Anton, and F. L. Graham, *Somat. Cell Mol. Genet.* **22**, 477 (1996).

[22] R. J. Parks, L. Chen, M. Anton, U. Sankar, M. A. Rudnicki, and F. L. Graham, *Proc. Natl. Acad. Sci. U.S.A.* **93**, 13565 (1996).

[23] N. Morral, R. Park, H. Zhou, C. Langston, G. Schiedner, J. Quinones, F. Graham, S. Kochanek, and A. Beaudet, *Hum. Gene Ther.* **9**, 2709 (1998).

[24] G. Schiedner, M. Nuria, R. J. Parks, Y. Wu, C. Suzanne, C. Langston, F. L. Graham, A. L. Beaudet, and S. Kochanek, *Nature Genet.* **18**, 180 (1998).

[25] M. A. Morsy, M. Gu, S. Motzel, J. Zhao, J. Lin, Q. Su, H. Allen, L. Franlin, R. J. Parks, F. L. Graham, S. Kochanek, A. J. Bett, and C. T. Caskey, *Proc. Natl. Acad. Sci. U.S.A.* **95**, 7866 (1998).

[26] H. H. Chen, L. M. Mack, R. Kelly, M. Ontell, S. Kochanek, and P. R. Clemens, *Proc. Natl. Acad. Sci. U.S.A.* **94**, 1645 (1997).

[27] R. Kumar-Singh and J. S. Chamberlain, *Hum. Mol. Genet.* **5**, 913 (1996).

[28] M. M. Burcin, G. Schiedner, S. Kochanek, S. Y. Tsai, and B. W. O'Malley, *Proc. Natl. Acad. Sci. U.S.A.* **96**, 355 (1999).

[29] P. Ng, R. Park, and F. Graham, in "Gene Therapy Protocols" (J. R. Morgan, ed.), p. 371. Humana Press, 2001.

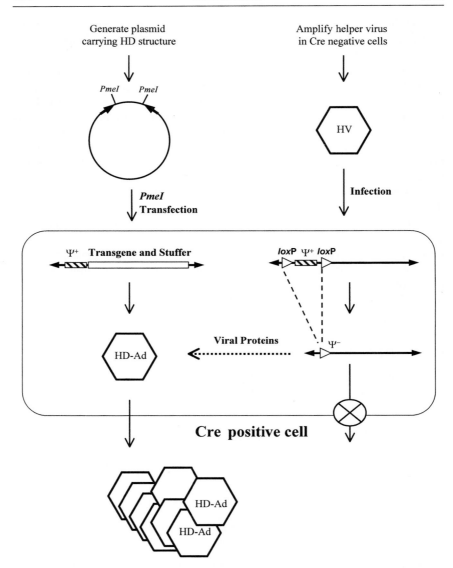

FIG. 1. The Cre/*loxP* helper-dependent adenoviral (HD-Ad) vector system. A plasmid containing the entire HD-Ad vector structure can be generated *in vitro*. The HD-Ad DNA then is released from the circular plasmid with a restriction enzyme and transfected into cells that complement for E1 and provide expression of the Cre recombinase. A helper virus bearing a packaging signal flanked by *loxP* sites is first amplified in E1-complementing cells that do not express Cre. When the helper virus is introduced into the Cre positive cells, the Cre/*loxP*-mediated excision of the packaging signal (Ψ) in the helper virus renders its DNA unable to be packaged. However, the helper DNA with deleted packaging signal is replicated and provides transcripts for all of the necessary proteins for propagation of the HD-Ad vector. Most virions produced in Cre positive cells are the HD-Ad vector.

1. Construction of a plasmid carrying an HD-Ad backbone. A therapeutic gene or reporter gene is inserted into a plasmid that contains the Ad ITRs and the packaging signal.

2. Preparation of helper virus stocks. The helper virus is propagated in E1 complementing cells that do not express Cre such as the 293 cell line.

3. Rescuing the HD-Ad vector in Cre-expressing cells. Linearized plasmid DNA is transfected into E1-complementing cells that also express the Cre recombinase. The helper virus provides viral proteins to amplify and package the HD-Ad DNA. The helper virus is not packaged to any significant extent because the packaging signal is deleted by the action of Cre.

4. The rescued HD-Ad vector and the helper virus are used to coinfect the Cre positive cells for several rounds to increase the titer of the HD-Ad vector.

Cre-Expressing Cell Lines and Cre Activity

293Cre4 Cell Line

The 293 cell line was transfected with a plasmid expressing Cre to develop the 293Cre4 cell line as reported by Chen *et al.*[21] These cells should be maintained under G418 selection. Because it is important to minimize the helper virus contamination of HD-Ad vector preparations and because efficient removal of the packaging signal in the helper virus is essential to achieve this goal, it is important to check the Cre activity of the cell culture. After multiple passages of 293Cre4 cells, the Cre activity in the cultured cells may be lower than in the original cell line. The 293Cre415 cell line is a subclone of 293Cre4 cells, and the cell line was utilized to optimize the preparation of HD-Ad vectors;[30] this subclone of 293Cre4 is available from Merck Research Laboratories.

Monitoring Cre Activity in Cultured Cells

We have developed a convenient method for monitoring Cre activity in cultured cells. The ploxZ plasmid was constructed by inserting a stop codon, such that the open reading frame for expression of *Escherichia coli* β-galactosidase (*lacZ*) would be disrupted.[31] Thus there is no expression of *lacZ* when the plasmid is transfected into Cre negative cells. Excision of the DNA segment containing the stop codon through Cre/*loxP* interaction activates the expression of *lacZ* (Fig. 2). The cells to be tested are first transfected with the ploxZ plasmid DNA using calcium phosphate precipitation[32] or other reagents, such as Lipofectamine Plus

[30] V. Sandig, R. Youil, A. J. Bett, L. L. Franlin, M. Oshima, D. Maione, F. Wang, M. L. Metzker, R. Savino, and C. T. Caskey, *Proc. Natl. Acad. Sci. U.S.A.* **97,** 1002 (2000).

[31] H. Zhou, T. Zhao, L. Pastore, M. Nageh, W. Zheng, X. M. Rao, and A. L. Beaudet, *Mol. Ther.* **3,** 613 (2001).

[32] F. L. Graham, *in* "Methods in Molecular Biology" (E. J. Murray, ed.), Vol. 7, p. 109. The Humana Press Inc., Clifton, NJ, 1991.

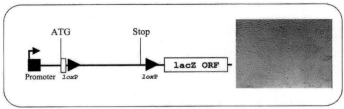

Cre negative cells

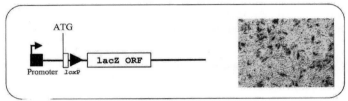

Cre positive cells

FIG. 2. Screening of cell clones for expression of Cre. The ploxZ plasmid contains two *loxP* sites that flank a DNA segment with a stop codon. Expression from the *lacZ* gene in Cre negative cells is blocked because of premature termination of translation. Excision of the segment between the *loxP* sites in Cre expressing cells activates the expression of the *lacZ* gene. Tissue culture dishes with E2T/Cre6 cells at 90% confluence and were transfected with the ploxZ plasmid using calcium phosphate precipitation.

(Life Technologies). The 293 and 293Cre4 are efficiently transfected by the calcium phosphate precipitation method, while other cells may be better transfected with Lipofectamine Plus. Two days after transfection, the medium is removed and the cell monolayer is fixed by adding a solution of 0.2% glutaraldehyde and 2% *p*-formaldehyde in phosphate-buffered saline (PBS). After 5 min, the fixative solution is removed, and an X-Gal solution (0.4 mg/ml X-Gal, 5 mM K$_4$Fe(CN)$_6$, 5 mM K$_3$Fe(N)$_6$, 2 mM MgCl$_2$ in PBS) is added. The ploxZ plasmid is particularly convenient for screening large numbers of clones in a semiquantitative manner. The frequency of positive staining cells approximately correlates with the level of Cre expression in the cells tested.

E2T/Cre6 Cell Line

With the published HD-Ad vector system, contamination with helper virus in the final preparations has been reported to be as low as one plaque-forming unit (pfu) in 10^4 infectious units of HD-Ad vector (0.01% helper contamination).[22,33] We have developed an additional Cre-expressing cell line (E2T/Cre6) in an attempt

[33] R. Parks, C. Evelegh, and F. Graham, *Gene Ther.* **6,** 1565 (1999).

TABLE I

HELPER VIRUSES

Helper viruses	Packaging signal	E3 region	RCA	Reference
AdLC8cluc	wild-type	Luc insert	oversize	22
AdΔE1-lox192	wild-type	wild type	able to be packaged[a]	This report
AdhvΔE1E2	wild-type	wild type	no, ΔE2a	31
H14	modified	human intron insert	oversize	30

[a] Preferably used with PER.C6 or N52E6 or similar cell line.

to further decrease the level of helper virus contamination and improve the vector safety.[31] This cell line is based on the E2T cell line, which complements both E1 and E2a.[14] The E2T cells were co-transfected with a Cre-expressing plasmid and a hygromycin resistance plasmid. Several hundred drug-resistant clones were screened for Cre expression using the ploxZ plasmid. The colonies with higher Cre activity were picked for further analysis by Southern blotting (see below). One of the cell lines, designated E2T/Cre6, derived from E2T cells can efficiently excise the packaging region of the helper virus and results in levels of helper virus contamination as low as 0.001%. We have also developed an E1 and E2a double deleted helper virus, which can be used with the new cell line to produce HD-Ad vectors (Table I). The deletion of both E1 and E2a in the helper may further improve the safety of the HD-Ad system.[31]

Southern Blot Analysis of Cre Activity

The efficiency of Cre/*loxP* mediated deletion can also be evaluated using Southern blot hybridization. The cell lines or clones to be tested are grown in 10 cm dishes and infected with the AdLc8cluc[22] at a moi of 1. Two days later, the cells are collected, washed with PBS, and digested overnight with 0.5 ml of proteinase K (0.04% proteinase K in 10 mM Tris-HCl, pH 7.5/10 mM EDTA/0.5% SDS) at 50°. The total DNA is then extracted once with buffer-saturated phenol : chloroform and precipitated with ethanol. The DNA is then digested with restriction enzyme *Pst*I, which yields a 0.53-kb fragment from the left terminus and a 2.06-kb fragment from the right end of the AdLc8cluc helper (Fig. 3). The hybridization probe is a 0.14-kb fragment isolated from pXCJL.2[32] after digestion with *Eco*RI and *Afl*III. This probe covering the 103-bp ITR hybridizes to both the 2.06-kb and 0.53-kb fragments. Deletion of the packaging region (Ψ) in the helper virus mediated by the Cre/*loxP* interaction decreases the 0.53-kb fragment to 0.26 kb (Fig. 3). The 2.06-kb band representing the right terminus of the helper virus acts as an internal control for equal loading. Hence, the ratio of the 0.26-kb band to the 0.53-kb band represents the excision efficiency of the packaging region by Cre recombinase in

A

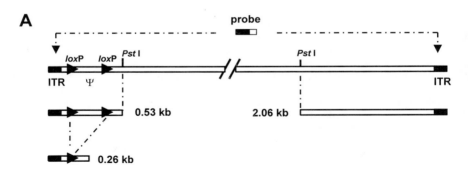

B

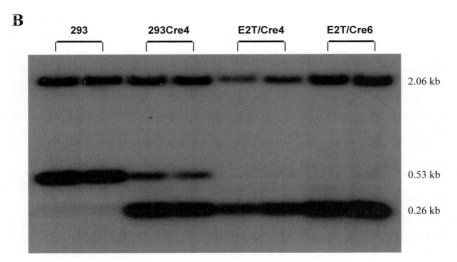

FIG. 3. Southern blot assay for deletion of the packaging region in the helper virus DNA. Panel A: The AdLC8cluc helper virus was used to infect Cre negative 293 cells and various Cre positive cells. Total DNA was isolated from each cell culture and cleaved with *Pst*I. The probe containing the Ad ITR hybridizes with both Ad terminal fragments. Panel B: Deletion of the packaging region in the helper virus decreased the left terminal fragment from 0.53 to 0.26 kb. The fragments cleaved from the right end provide internal controls. The efficiency of the Cre/*loxP*-mediated deletion is measured as the ratio of 0.26-kb to 0.53-kb fragments.

the cells. For 293 cells without Cre expression, there is no deletion of the packaging region (Fig. 3). In 293Cre4 cells, the packaging region is deleted for most copies of the Ad DNA, but about 10% of left end remains as a 0.53-kb fragment under these conditions, although recombination is more efficient in other experiments. For the two E2T/Cre cell lines, the 0.53-kb fragment could barely be detected as most of the DNA was converted to the 0.26-kb fragment (Fig. 3). Excision of the packaging region from the helper virus in E2T/Cre6 cells is highly efficient, and the cell line is also very stable.

Helper Viruses

AdLC8cluc Helper Virus

The AdLC8cluc helper virus is widely used for preparation of HD-Ad vectors. This helper contains two *loxP* sites flanking the sequence essential for packaging the virus. A luciferase expression cassette was inserted in the E3 region of pBHG10 to increase the size of the AdLC8cluc helper to 35.4 kb.[22] Insertion of a reporter gene in the deleted E3 region reduces the potential for generation of replication competent adenovirus (RCA). Recombination between the helper and the Ad sequences in the 293 or 293Cre4 cells, resulting in the transfer of E1 sequences from cellular chromosome to the virus genome, would increase the virus size over 38 kb, thus exceeding the packaging limit for adenovirus. The helper virus must be first amplified in Cre negative cells. The 293 or other E1 complementing cells such as PER.C6[34] or N52E6[35] can be used to prepare large stocks of a helper virus.

Modifications of Helper Viruses

When Sandig *et al.*[30] compared the AdLC8cluc helper virus to a helper with an E3 wild-type region, the AdLC8cluc with luciferase substituted for the E3 region yielded titers one order of magnitude lower than for the virus with wild-type E3. They reported a series of modifications of the helper virus. Because recombination between the packaging signal of the helper and that of the HD-Ad vector can lead to overgrowth of a mutated virus with only a single *loxP* site and an intact packaging signal, they modified the sequence in the packaging signal of the helper virus to minimize such recombination. They found that this modification "did not allow outgrowth of the helper in more than 30 independent HD-rescues."[30] They also inserted human intronic sequence as stuffer in the E3 region rather than the luciferase cassette and reported that this modified helper provided substantially improved yields.[30] This helper, designated H14, is available from Merck Research Laboratories.

More recently, we have constructed a helper AdΔE1-*lox*192 with one *loxP* site inserted at nucleotide 192 of the Ad genome and the other *loxP* at the deleted E1 region (H. Zhou, previously unpublished). This helper contains a wild-type E3 region and the packaging sequence itself is unmodified. The titer of AdΔE1-lox192

[34] F. J. Fallaux, A. Bout, I. Van Der Velde, D. J. M. Van Den Wollenberg, K. M. Hehir, J. Keegan, C. Auger, S. J. Cramer, H. V. Ormondt, A. J. Van Der Eb, D. Valerio, and R. C. Hoeben, *Hum. Gene Ther.* **9**, 1900 (1998).

[35] G. Schiedner, S. Hertel, and S. Kochanek, *Hum. Gene Ther.* **11**, 2105 (2000).

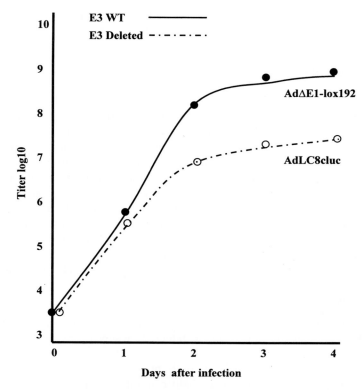

FIG. 4. Growth of helper viruses with wild-type E3 or deleted E3 in 293 cells. Tissue culture plates with 80% confluent 293 cells were infected at a moi of 1, and vector in the medium was measured as infective units at 10 days after infection.

is about 50-fold higher than for the AdLC8cluc when both propagate in 293 cells (Fig. 4). The AdΔE1-*lox*192 helper also forms large plaques on 293 cells as do most first generation vectors, while the plaques of AdLC8cluc are about 1/10 the size of those for AdΔE1-*lox*192. The disadvantage of using a helper virus with wild-type E3 is the potential generation of replication competent adenovirus (RCA). A further improvement would be to develop Cre-expressing derivatives of cell lines such as PER.C6 or N52E6 that complement for E1 but do not include homologous sequences with the helper.

HD-Ad Vectors

For one early form of HD-Ad vectors developed by Kochanek et al.,[20] the propagation and the amplification of the vectors were based on the use of a helper virus with a mutated and inefficient packaging signal. With this system, the vectors

TABLE II
PLASMIDS FOR PREPARATION OF HD-AD VECTORS

HD-Ad plasmids	HD-Ad backbone	Stuffer	Carrying range	References
pSTK51[a]	10 kb	HPRT 9 kb	18 to 26 kb	24
pΔSTK120	16 kb	HPRT5 kb C246 9 kb	12 to 20 kb	25
pSTK120	27 kb	HPRT 16 kb C346 9 kb	1 to 9 kb	25
pΔ21	21 kb	HPRT 16 kb C346 5 kb	7 to 15 kb	This report
pΔ28	27 kb	HPRT 16 kb C346 11 kb	1 to 11 kb	37
C4HSU	28 kb	19 kb C346	1 to 8 kb	30

[a] Backbone for STK109.

were contaminated with helper virus at a level of about 1%, and the yield was relatively low making *in vivo* experiments in large animals impractical.

The development of a Cre/*loxP*-based system allowed the growth and purification of larger amounts of vector with decreased helper virus yielding preparations suitable for *in vivo* testing.[22] One early vector generated to test the Cre/*loxP* system (AdRP1001) contained 5789 bp of the left end and 6143 bp of the right end of the adenovirus genome, and thus it was not totally devoid of adenoviral ORFs. This vector contained a cassette for the expression of lacZ and DNA from phage lambda as stuffer. Subsequently, Parks *et al.*[36] analyzed the effect of the nature of the stuffer DNA in HD-Ad vectors observing that recipient mice can generate a cytotoxic T lymphocyte (CTL) response against vectors containing lambda DNA as stuffer, whereas this response was not observed in vectors containing an intron from the human hypoxanthine-guanine phosphoribosyltransferase (HPRT) gene.

Schiedner *et al.*[24] developed an HD-Ad vector that contained the entire human α_1-antitrypsin gene including all introns. This vector, AdSTK109, contained only 440 bp from the left end and 117 bp from the right end of the adenoviral genome, so it was totally deleted for adenoviral ORFs. The DNA stuffer used in this vector was derived from an intron of the human HPRT gene that had been previously sequenced.

A further series of alternative plasmids for preparation of HD-Ad vectors is available as listed in Table II. The pΔSTK120 plasmid has 5 kb of human intronic sequence from the HPRT locus and 9 kb from the human genomic region HUMDS455A (cosmid C346, GenBank L31948) as stuffer. Because it is reported that Ad vectors should be at least 27 kb in size for efficient packaging, the pΔSTK120 plasmid is suitable for accommodating DNA inserts of 12 to 20 kb. For insertion of smaller cDNA sequences, the pSTK120 plasmid is more suitable

[36] R. J. Parks, J. L. Bramson, Y. Wan, C. L. Addison, and F. L. Graham, *J. Virol.* **73,** 8027 (1999).

as it contains fragments of 16 kb from HPRT and 9 kb from cosmid C346.[25] The pSTK120 plasmid is suitable for inserts of 1 kb to 9 kb. Similar plasmids designated pΔ21 and pΔ28 contain the Ad ITRs, the packaging signal, a 16-kb fragment from HPRT, and fragments from C346 of 5 kb for pΔ21 and 11 kb for pΔ28.[37] A unique *Asc*I site near the right ITR facilitates cloning into pΔ21 and pΔ28, and together they are appropriate for accepting inserts ranging from 1 to 15 kb.

Sandig *et al.*[30] also redesigned the vector backbone by including 400 bp of the right Ad5 terminus and a number of different segments of noncoding DNA of human origin as "stuffer." Comparison of these vectors showed that 9 kb of DNA from human locus HSU71148 and the 400-bp fragment from the right end of the adenovirus genome stabilized the vector C4HSU. The plasmid C4HSU also contains 19 kb of sequence from C346. This sequence was interrupted into four segments of 4.5 kb each and cloned in reversed orientation compared with their original position in the genome to decrease the potential for homologous recombination. This vector is available through Merck Research Laboratories.

Preparation of HD-Ad Vectors

Construction of Plasmids for Conversion to HD-Ad Vectors

We have extensive experience with the pΔ28 plasmid that has been used for construction of HD-Ad vectors expressing a variety of cDNA constructs. This vector contains a bacterial replication origin and drug resistance marker for amplification of the plasmid in bacteria. A cassette with a therapeutic gene or a reporter gene can be inserted into the unique *Asc*I restriction site. The final construct of the plasmid is usually at least 31 kb (3kb of plasmid backbone and 28 to 36 kb for the total HD-Ad vector constructs). Such large plasmids are difficult to manipulate, and in some cases insertion of a transgene may decrease the plasmid stability in a bacterial strain. The HD-Ad vector plasmid DNA used for transfection should be purified either by CsCl gradient centrifugation or Qiagen plasmid column. The plasmid backbone and the HD-Ad vector structure in the pΔ28 are separated by two *Pme*I sites immediately adjacent to the ITRs (Fig. 1). The HD-Ad plasmid is digested with *Pme*I to release the HD-Ad vector DNA in a linear form for rescuing the HD-Ad vector.

Preparation of Helper Virus Stocks

The AdLc8Cluc helper is used in most work published up to the present. In order to minimize formation of RCA, a stuffer sequence was inserted into the E3 region to render any recombinants with the E1 sequence in the chromosome

[37] K. Oka, L. Pastore, I.-H. Kim, A. Merched, S. Nomura, H.-J. Lee, M. Merched-Sauvage, C. Arden-Riley, B. Lee, M. Finegold, A. Beaudet, and L. Chan, *Circulation* **103**, 1274 (2001).

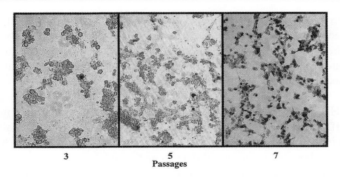

3   5   7
                          Passages

FIG. 5. Increasing titer of the HD-AdΔβ-geo with repeated passages. The plasmid for the HD-Ad vector was transfected into E2T/Cre6 cells with simultaneous infection with helper virus as described in the text. The output from this infection was subjected to several rounds of coinfection in E2T/Cre6 cells. The titer for HD-AdΔβ-geo increased from passages 3 through 7.

of the 293 cells too large to be packaged into virions as discussed above. Since there may be up to 1000 copies per cell for both the helper and HD-Ad vector genomes, there is substantial risk of homologous recombination with even short overlapping sequences. It is important that helper virus be prepared from purified plaque isolates. A large stock should be prepared, analyzed, and stored in aliquots for repeated use. As the conditions for preparation of a helper virus are essentially identical to those used for first-generation Ad vectors, the reader is referred to well-established methods for production of first-generation Ad vectors.[38]

Rescuing HD-Ad Vectors in Cre-Expressing Cell

Rescuing an HD-Ad vector from the plasmid form is initiated by transfection of 293Cre4 or other Cre positive cells with the linearized DNA from HD-Ad plasmid and concurrent infection with the helper virus (Fig. 1). The helper virus provides viral proteins to amplify and package the HD-Ad DNA. However, the helper virus itself cannot be efficiently packaged because of the deletion of the packaging signal. Subsequently, serial passages involve coinfection of the Cre positive cells with the helper virus and the HD-Ad vector to increase the titer of the HD-Ad vector. In rescuing first- and second-generation Ad vectors, cytopathic effect (CPE) can only be induced when an infectious vector is formed through homologous recombination between a shuttle plasmid and an Ad genomic plasmid. In contrast with the HD-Ad vector system, the cells always will develop CPE by virtue of the infectious nature of the added helper virus whether the HD-Ad vector has been generated or not (Fig. 5). Thus rescuing an HD-Ad vector cannot be easily assessed

[38] F. L. Graham and L. Prevec, *Mol. Biotechnol.* **3,** 207 (1995).

by inspection for the presence of CPE as in other Ad vector systems. In addition, the transgenes in an HD-Ad vector cannot be easily monitored in most cases, except when a reporter such as *lacZ* or GFP gene is used. Therefore, monitoring the amplification of the HD-Ad vector presents a major difficulty. Several methods can be used to monitor amplification of the HD-Ad vector.

1. An HD-Ad plasmid containing a *lacZ* transgene can be amplified in parallel with an HD-Ad vector plasmid containing a transgene of interest. X-Gal staining as shown in Fig. 5 can be used to monitor the amplification of the HD-Ad vector with *lacZ*. Although the success of the rescue of the HD-Ad *lacZ* vector does not necessarily assure the rescue of the vector of interest, this is nevertheless a useful control to ensure that the system is functioning properly.

2. A PCR method can be designed to monitor amplification of an HD-Ad vector. A good strategy is to place one primer in the backbone of the HD-Ad vector and another primer in the transgene. These two primers can also be used to confirm the structure of the starting plasmid. After initial transfection, the HD-Ad plasmid DNA will be present in the Cre positive cells. In the following cycles of coinfection, the plasmid DNA will gradually be lost. When the plasmid DNA is converted to replicating HD-Ad DNA with progressive increase in titer, typically over five to seven rounds of coinfection, the amplification can be monitored by PCR. The amount of HD-Ad vector DNA relative to the helper virus can also measured using TaqMan real-time PCR.[30]

3. Southern blot analysis can be used to measure the relative amount of the HD-Ad vector and the helper virus DNA isolated from infected cells.[22]

4. Analysis of the protein or biological function of the transgene carried in an HD-Ad vector is the optimal method, because expressing the proper biological function of a transgene from an HD-Ad vector is the final goal in vector development. The method must be individualized for each vector. For example, we have experience measuring secreted proteins such as α_1-antitrypsin or α-fetoprotein using ELISAs, expressing in hepatocyte cell lines for liver-specific promoters, or using specialized recipient cells such as Chinese hamster ovary (CHO) cells lacking the LDL receptor to measure expression as uptake of LDL.

Not every HD-Ad plasmid containing a transgene can be rescued into an HD-Ad vector form. Some transgene products may be toxic to the cells at high levels, and the Cre positive cells may be killed before the HD-Ad vector is amplified. Since a single cell may contain over 1000 copies of the helper virus and the HD-Ad vector DNA, homologous sequences between the HD-Ad vector and the helper must be avoided. The AdLC8cluc helper virus contains a CMV promoter driving the luciferase gene in the E3 region[22]; therefore, use of the CMV promoter in HD-Ad vectors should be avoided with the AdLC8cluc helper.

Large-Scale Preparation and Purification of HD-Ad Vectors

To produce HD-Ad vectors on a large scale, 16 triple-layer flasks (500 cm^2 per flask) of Cre positive cells are seeded at a density to reach 90% confluence in 1 to 2 days. Alternatively, the cell cube system or other high capacity approach to cell culture can be used.[29] The cells are then coinfected with both an HD-Ad vector (Ad$\Delta\beta$-geo, for example) at a moi of about 10 blue-forming units (BFU), and helper virus AdLC8cluc at a moi of 1–5 infectious units (IFU). The infected cells are harvested when over 90% show CPE (about 2.5 to 3 days). The HD-Ad vector is purified with the CsCl banding method as for purification of first-generation Ad vectors. We have found that each preparation of the HD-Ad Ad$\Delta\beta$-geo from 16 triple flasks generally yields 3 ml of HD-Ad vector with a total of about 10^{12} particles and a particle to BFU ratio of 20 to 100.

Analysis of HD-Ad Vector Preparations

The titer of the purified HD-Ad vector cannot be easily assayed with the methods used for a first-generation Ad vector, and the HD-Ad vector stocks are always contaminated with the helper virus. In addition, the repeated cycles of amplification can result in rearrangement of either the vector or the helper. Analysis of the HD-Ad vector preparation therefore requires specific efforts. The total virus particles and the infectious titers for both the HD-Ad vector and the helper virus should be determined, and their DNA structures should be characterized.

Titer of HD-Ad Vectors

The titers of first-generation Ad vectors are analyzed by several methods: plaque-forming units, infectious units (such as blue-forming units), and total virion particles. There is no standard method that satisfies the needs with different Ad vectors. For the plaque-forming assay, the titer can vary depending on vector size, deletions in the Ad genes, transgene location, and the nature of the transgene. The titer also depends on the cell line and the cell growth conditions. The plaque-forming method cannot be used to determine the titer of HD-Ad vectors, because they cannot form plaques. The widely used method for titering of HD-Ad vectors is the total virion particle count.

1. Total concentration of virion particles. When the helper virus contamination is very low (0.1% or less), the total virion particle count approximates that for the HD-Ad vector. The concentration of total virion particles in the purified preparation is determined with a spectrophotometric method, assuming that 1 OD_{260} unit equals

1.1×10^{12} particles of Ad2 or Ad5.[39] The purified virus is first diluted in PBS with 0.1% sodium dodecyl sulfate (SDS) and then OD_{260} is determined.

2. Measurement of the HD-Ad vector active unit. The infectious titer of an HD-Ad vector can be estimated by the level of transgene expression compared to that obtained with a first-generation vector carrying the same expression cassette,[24,30] but this may be subject to some error. This method requires a first-generation vector which may be not available. When an HD-Ad vector is carrying an easily detectable reporter gene such as *lacZ* gene, an infectious unit can be measured, in this case as BFU. Other strategies are theoretically possible such as measuring expression in recipient cells using immunohistochemistry or quantitating vector genomes reaching the nucleus with fluorescence *in situ* hybridization (FISH).

Level of Helper Virus Contamination

The titer of helper virus can be determined plaque-forming units (PFU)[38] or as infectious units (IFU) using $TCIG_{50}$ end-point dilution.[30] The titer of the AdLC8cluc helper can easily be underestimated, since plaques are small and slow to form and may be obscured by overgrowth of cells. We prefer use of the end point dilution method with modification. In brief, 96-well plates are seeded with 3×10^3 293 or other complementing cells per well, and cells are immediately infected with three-fold serial dilutions of virus. In this way, the virus can efficiently infect the cells while still suspended. Infected cells are incubated for 10 days, and wells are scored visually for CPE.

Southern Blot Analysis of Structure of HD-Ad Vector and Vector:Helper Ratio

Viral DNA can be extracted directly from cells or after purification of the vector to confirm the proper structure and the ratio of the HD-Ad vector to the helper virus. For isolation of the vector DNA from the CsCl-purified preparation, as little as 1 μl of the purified vector is mixed with 2 μg of salmon DNA (as a carrier) and incubated in 100 μl of proteinase K solution (0.04% proteinase K in 10 mM Tris-HCl at pH 7.5/10 mM EDTA/0.5% SDS) overnight at 50°. The total DNA is then extracted once with buffer-saturated phenol : chloroform and precipitated with ethanol. When DNA is isolated from 100 μl of the purified vector stock, the carrier DNA is not necessary, and the proteinase K solution should be 0.5 ml in this case. The isolated DNA samples are then cleaved with the restriction endonuclease, separated on agarose gels, and transferred to Hybond-N positively charged nylon membranes (Amersham International plc, Amersham, UK) by capillary transfer[40] for hybridization with a radioactively labeled DNA probe.

[39] J. V. Maizel, Jr., D. O. White, and M. D. Scharff, *Virology* **36,** 115 (1968).
[40] E. M. Southern, *J. Mol. Biol.* **98,** 503 (1975).

HD-AdΔβ-geo

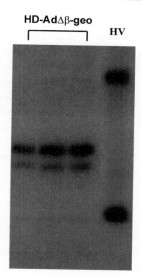

FIG. 6. Southern blot analysis of purified HD-Ad vector DNA. The ITR probe hybridizes with the terminal fragments of helper virus (HV) and HD-Ad vector (HD-AdΔβ-geo) DNA digested with PstI. The helper virus DNA shows hybridization with fragments of 2.06 and 0.53 kb as in Fig. 3. Only a very faint hybridization for these helper virus fragments (not visible in this reproduction) was seen in three different preparations of HD-AdΔβ-geo vector.

We routinely use *Pst*I restriction enzyme to digest the purified DNA, The AdLC8cluc virion DNA produces two terminal fragments of 2.06 and 0.53 kb after digestion, and both are detected by hybridization with the ITR probe (Figs. 3 and 6). Any HD-Ad vector also contains two terminal fragments of differing size. Figure 6 depicts the analysis of a preparation of the HD-Ad vector AdΔβ-geo, and intense bands representing the left and right ends of the vector are seen, whereas fragments from the helper virus are barely visible. Other probes can also be used in this analysis.

In Vivo Testing of HD-Ad Vectors

After the development of the Cre/*loxP*-based system, HD-Ad vectors have been tested extensively in mice and to a limited extent in primates. An HD-Ad vector designated AdSTK109 expressing hAAT from the entire human gene named was used to analyze duration and level of expression and the toxicity of these vectors.[24] This vector is particularly suitable for this type of analysis: it expresses a secreted transgene (hAAT) that is easily monitored in mouse serum with a specific ELISA, and the hAAT does not induce an immune response in many mouse strains. The first report by Schiedner et al.[24] showed an extremely long duration of hAAT expression (over 1 year) and very low hepatic toxicity after intravenous administration of 2×10^{10} particle/mouse. A more detailed analysis performed by Morral et al.[23] demonstrated that even after intravenous administration of extremely high doses of 3.2×10^{11} particle/mouse, there was no elevation of liver enzymes in serum and no alterations in liver morphology. Furthermore, with this dose it was possible to achieve supraphysiological levels of expression of hAAT (5 mg/ml, Fig. 7).

The ability to grow these vectors to very high titer with the Cre/loxP-based system also allowed the intravenous administration of AdSTK109 to juvenile baboons. The treated animals expressed detectable levels of hAAT for more than 2 years, and blood counts and liver function tests were normal for the duration of the experiment.[41]

Parks et al.[33] demonstrated the ability to readminister HD-Ad vectors by switching the adenoviral serotype. One of the advantages of HD-Ad vectors is that they can be packaged with either Ad5 or Ad2 viral proteins using the same ITRs and packaging signals. For this reason the same vector can be packaged into Ad2 or Ad5 virions simply by using helper viruses of different serotypes.

The possibility of regulating transgene expression has been analyzed by Burcin et al.[28] An HD-Ad vector expressing human growth hormone (hGH) under the control of an RU486-responsive system was constructed. High levels of hGH were induced in the presence of RU486, while the transgene expression was undetectable in the absence of the ligand. The induction could be performed several times, and no immune response against hGH or the vector was observed.

Direct administration of an HD-Ad vector to the central nervous system (CNS) has been reported by three groups.[42,43,44] All the results are consistent with reduced inflammation and long-term expression of the transgene, making HD-Ad vector an interesting choice for CNS-directed gene therapy.

The reduced toxicity and the long duration of transgene expression observed with marker genes has led to the use of HD-Ad vectors expressing therapeutic transgenes to correct phenotypes in animal models of human diseases. Morsy et al.[25] used HD-Ad vectors to correct the phenotype of the ob/ob mice, a strain that develop obesity secondary to a deficiency in the protein leptin. Intravenous

[41] N. Morral, W. O'Neal, K. Rice, M. Leland, J. Kaplan, P. A. Piedra, H. Zhou, R. J. Parks, R. Velji, E. Aguilar-Cordova, S. Wadsworth, F. L. Graham, S. Kochanek, K. D. Carey, and A. L. Beaudet, Proc. Natl. Acad. Sci. U.S.A. 96, 12816 (1999).

[42] S. P. Cregan, J. MacLaurin, T. F. Gendron, S. M. Callaghan, D. S. Park, R. J. Parks, F. L. Graham, P. Morley, and R. S. Slack, Gene Ther. 7, 1200 (2000).

[43] C. E. Thomas, G. Schiedner, S. Kochanek, M. G. Castro, and P. R. Lowenstein, Proc. Natl. Acad. Sci. U.S.A. 97, 7482 (2000).

[44] L. Zou, H. Zhou, L. Pastore, and K. Yang, Mol. Ther. 2, 105 (2000).

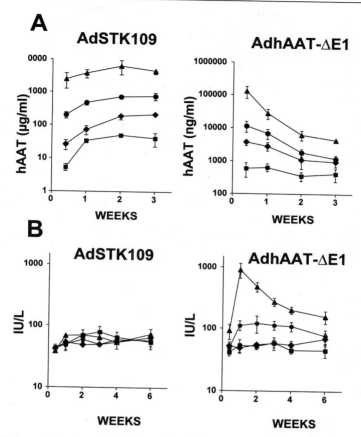

FIG. 7. Comparison of HD-Ad (AdSTK109) and first-generation (AdhAAT-ΔE1) vectors express-ing human α_1-antitrypsin (hAAT). Four groups of mice were injected intravenously with 1.2×10^{10} (■), 3.6×10^{10} (◆), 1.1×10^{11} (●) and 3.2×10^{11} (▲) viral particles per mouse of either vector. Panel A: Levels of hAAT after administration of HD-Ad (left, AdSTK109) or first-generation (right, AdhAAT-ΔE1) vector. Panel B: Levels of alanine aminotransferase (ALT) after administration of HD-Ad (left, AdSTK109) or first-generation (right, AdhAAT-ΔE1) vector. Reproduced from Morral et al.[23] with permission.

administration of HD-Ad vector expressing leptin produced a prolonged weight reduction in the treated *ob/ob* mice with negligible liver toxicity.

Maione et al.[45] showed prolonged expression of mouse erythropoietin (more than 6 months) after intravenous administration of an HD-Ad vector with an associated increase of the hematocrit with no associated liver toxicity. The high bioactivity of erythropoietin can provide a therapeutic increase of the hemat-ocrit after the administration of low doses of vector. Interestingly, intravenous administration of low doses of vectors does not trigger production of neutralizing antibodies against the adenovirus sufficient to prevent subsequent readministration of the same HD-Ad vector. Aurisicchio et al. [46] applied HD-Ad vectors to gene therapy for liver protection against hepatitis by expressing α_1-interferon in a liver restricted manner. Moreover, protection of the liver in the treated mice was extended to noninfectious inflammation agents.

As previously mentioned, one of the main advantages of HD-Ad vectors is a very efficient hepatic expression with the ability to achieve supraphysiological levels of the transgene. Alteration of lipoprotein metabolism to reduce plasma levels of low-density lipoprotein cholesterol (LDL-C) and/or increasing plasma levels of high-density lipoprotein cholesterol (HDL-C) is possible only with the expression of very high levels of transgene, which might be achievable with HD-Ad vectors. Our collaborators[37] have expressed the mouse very low density lipoprotein receptor (VLDLR) in LDL receptor-deficient mice, a model for familial hypercholesterolemia. A long-lasting reduction of plasma LDL-C and a decrease in aortic atherosclerotic lesions was observed.

In conclusion HD-Ad vectors show very prolonged expression of high levels of transgene with minimal hepatic toxicity in murine models of diseases. Further analysis of toxicity in large animals including nonhuman primates will determine the applicability of HD-Ad vectors in clinical settings.

Problems with Current HD-Ad Vector System and Future Development

The major difficulties with the HD-Ad vectors to date are (1) the inability to produce high-quality vectors in large quantity and (2) the lack of data regarding safety and efficacy when high doses of vector are administered intravascularly in large animals or humans. The current system is relatively adequate for administration of modest doses of vector intramuscularly or intracerebrally. We believe that it is very clear that these vectors are substantially superior to any other available system for achieving high-level and long-term expression in hepatocytes in mice based on data such as shown in Fig. 7. These results indicate that very high levels of expression can be achieved in mice without major toxicity. These results also demonstrate that a major component of the acute toxicity in mice is eliminated by the use of vectors with no viral open reading frames, and therefore much of this toxicity must be mediated by leaky expression of the viral genes very

[45] D. Maione, M. Wiznerowicz, P. Delmastro, R. Cortese, G. Ciliberto, N. La Monica, and R. Savino, *Hum. Gene Ther.* **11,** 859 (2000).

[46] L. Aurisicchio, P. Delmastro, V. Salucci, O. G. Paz, P. Rovere, G. Ciliberto, N. La Monica, and F. Palombo, *J. Virol.* **74,** 4816 (2000).

early after cellular entry. Another less likely interpretation is that differences in the protein or other composition of the virions of the HD-Ad vectors compared to first-generation vectors make the HD-AD vectors less toxic. In either case the HD-Ad vectors demonstrate a substantially superior therapeutic index in mice. What is not clear is whether the same is true in large animals and humans. Given the death of a patient due to acute toxicity of a second-generation vector[18] with doses that did not yield substantial hepatic expression, there is great concern that the therapeutic index for hepatic expression of Ad vectors in humans may be much less favorable than in mice. This could be the case if Ad vectors elicit a greater inflammatory and cytokine response in humans than in mice or if a much lower fraction of vector delivered to the vascular compartment enters hepatocytes in humans. There is a great need for additional data using intravascular delivery of high doses of HD-Ad vectors in large animals and humans, but limitations in the ability to produce high-quality preparations in large quantity have delayed these studies.

Refinements of Existing Reagents

Even for the current system, reagents with ideal properties are not available. For example, there is as yet no published work with a Cre-expressing derivative of the PER. C6 or N52E6 cells to permit growth of the HD-Ad vectors in cells that minimize the potential for formation of RCA. A simplified system similar to that developed for preparation of first-generation Ad vectors[47] could be developed for HD-Ad vectors. Some of the reagents such as the H14 helper and C4HSU plasmid with improved stuffer are available for animal work, but material transfer agreements prohibit use in humans. Perhaps further data in large animals, particularly primates, would justify alteration of policies regarding the material transfer agreements. Other relatively routine steps such as production of vector in suspension culture or bioreactors may facilitate the production of vector on a larger scale.

Significant difficulties arise with the existing reagents because of rearrangements involving the helper or the vector. To some extent, these appear to be improved by the use of the H14 helper with reduced homology in the packaging sequence and by the modification of the stuffer segments to minimize repetitive sequences in the C4HSU plasmid.

Another problem is the potential for mutation of a *loxP* site in the helper resulting in overgrowth and high levels of helper contamination. Preliminary data have been presented at meetings suggesting that use of FLP-mediated recombination in cultured cells might address this concern[48] (P. Ng *et al.* and P. Umana *et al.*,

[47] T.-C. He, S. Zhou, L. T. Da Costa, J. Yu, K. W. Kinzler, and B. Vogelstein, *Proc. Natl. Acad. Sci. U.S.A.* **95,** 2509 (1998).
[48] P. Ng, C. Beauchamp, C. M. Evelegh, R. Parks, and F. L. Graham, *Mol. Ther.* **3,** 809 (2001).

Keystone Symposia Gene therapy, 2001). The availability of PER.C6 or N52E6 or similar cells expressing one or more of these recombination systems would represent a progressive improvement of the existing system. The PER.C6 or N52E6 are currently suitable for growth of helper stocks.

Novel Systems

A major weakness of the current approaches is the requirement for a three-component system to produce vector: a specialized cell line, a helper virus, and the HD-Ad vector. Preparation of a packaging cell line with inducible production of viral proteins could eliminate the need for helper virus, but it is likely to be difficult to produce viral proteins with the natural orchestration necessary to yield a high titer of vector. Steinwaerder *et al.*[49] have reported a two-component method for preparation of Ad vectors devoid of all viral genes by recombination between inverted repeats, but this resulted in smaller genomes that were unstable in recipient cells. This is in sharp contrast to HD-Ad vectors that demonstrate remarkably stable expression in mice and baboons. If a system could be developed to provide high levels of stable expression in hepatocytes after a single intravenous injection, many therapeutic avenues would be opened. For example, overexpression of the LDL receptor to decrease LDL cholesterol and/or overexpression of apolipoprotein A-I to raise HDL cholesterol might be highly effective in preventing atherosclerosis of diverse etiologies. Further developments with the HD-Ad system or other strategies are likely to make this possible in the future.

Acknowledgments

We thank F. Graham and Merck Research Laboratories for 293Cre4 cells and the AdLC8cluc helper virus. This work was supported by grants from the NIH (HL5914) to A. L. Beaudet and from the Cystic Fibrosis Foundation (F984) and Texas Higher Education Coordination Board (ATP 004949-0119) to H. Zhou. We thank numerous collaborators who currently work with us on the HD-AD vector system including Brendan H. Lee, Lawrence Chan, Gabriele Toietta, and Kazuhiro Oka.

[49] D. S. Steinwaerder, C. A. Carlson, and A. Lieber, *J. Virol.* **73,** 9303 (1999).

[12] Gene Transfer Methods for Transplantation

By J. S. BROMBERG, P. BOROS, Y. DING, S. FU, T. KU, L. QIN, and R. SUNG

Introduction

Organ transplantation has emerged as the only effective treatment for many end-stage diseases. The overall success of clinical transplantation is, however, limited by donor organ availability. In addition, both early and long-term results may be compromised because of ischemic damage related to procurement, acute and chronic rejection, and the deleterious effects of immunosuppression. Somatic gene therapy offers different approaches to overcome these fundamental problems. Donor shortage could be dramatically improved by the introduction of grafts obtained from transgenic animals, and genetically altered stem cells may also be used to replace organ functions. Furthermore, gene transfer applied to the donor organ could result in improved graft quality by diminishing the secondary inflammation and damage associated with both ischemia and alloimmune responses. Long-term graft function could also be protected by the induction of tolerance via genetic manipulation, eliminating the need for toxic immunosuppression.

This chapter is divided into several parts. The first three parts describe important criteria for selection of vectors, transgenes, and promoters in the context of allotransplantation. The fourth section describes issues of innate and adaptive immune response that influence gene expression and cellular viability. The last two sections describe specific methods for gene transduction of islet, hepatic, and cardiac allografts.

Selection Criteria for Vectors

The selection of a suitable vector is an important aspect of gene therapy in transplantation. Many factors must be considered when choosing a vector, including: (1) the type and dividing state of target cells; (2) size capacity for transgenes; (3) transient or stable expression; (4) transfection efficiency; (5) host immunity; and (6) safety. A variety of gene transfer vectors are described in detail elsewhere in this volume. Each vector has its own advantages and disadvantages when considered in the context of transplantation. Characteristics of the major vectors are listed in Table I.

Viral Vectors

Current gene transfer techniques are mainly focused on viral vectors, which provide high transfection efficiency and high expression level. The commonly used viral vectors in gene therapy include retrovirus, lentivirus, adenovirus,

METHODS IN ENZYMOLOGY, VOL. 346

TABLE I
CHARACTERISTICS OF MAJOR VECTORS IN GENE THERAPY

Vector	Transient/ stable	Capacity of transgene	Advantage	Disadvantage
Viral vectors				
Retrovirus	Stable	10 kb	• Integrate into host chromosome • No immune response	• Only infects dividing cells • Low titers (10^6–10^7 cfu/ml) • Insertional mutagenesis
Lentivirus	Stable	10 kb	• Integrate into host chromosome • Infect nondividing cells, including stem cells and parenchymal cells	• Insertional mutagenesis • Safety concerns of HIV
Adenovirus	Transient	7–8 kb	• High titers (10^{12} pfu/ml) • Infect nondividing cells	• Does not integrate into host chromosome • Host immune and inflammatory response
Adeno-associated virus (AAV)	Stable	4–5 kb	• Infect nondividing cells • Specific integration at chromosome 19 • Limited immune response	• Low titers (10^7–10^9 cfu/ml) • Small capacity (5 kb)
Herpes simplex virus (HSV)	Transient	20–30 kb	• Large capacity (20–30 kb) • Infect nondividing cells, including neuronal cells	• Virus titers 10^4–10^8 cfu/ml • Does not integrate into host chromosome • Host immune response
Nonviral vectors	Transient	Unrestricted	• Large capacity • High yield • Not infectious	• Low transfection efficiency • Transient gene expression • Unmethylated CpG cause host immune response

adeno-associated virus, and herpes simplex virus. A hybrid adenoviral/retroviral vector system has also been developed, combining the advantages of adenovirus and retrovirus in one vector system.

MMLV-Retrovirus. MMLV-retrovirus is RNA virus, based on Moloney murine leukemia virus (MMLV). Retrovirus can integrate into the host chromosome and express the transgenes stably. Nonetheless, gene silencing does occur over time. The capacity for transgene insertion is about 10kb, and the virus titer is relatively low (10^6–10^7cfu/ml).[1] Retrovirus can only infect dividing cells, limiting its use of gene therapy to differentiated cells.[2] Random insertion of viral genomes may lead to mutagenesis. These viruses are useful for transfection of bone marrow stem cells in transplantation, but of limited utility in solid organ allografting.

[1] T. M. Shinnick, R. A. Lerner, and J. G. Sutcliffe, *Nature* **293,** 543 (1981).
[2] D. G. Miller, M. A. Adam, and A. D. Miller, *Mol. Cell Biol.* **10,** 4239 (1990).

Lentivirus. Lentivirus is a member of the retrovirus family and is based on HIV or FIV.[3,4] The virus can also integrate into the host chromosome and express the transgene stably. Lentivirus can infect nondividing cells, including stem cells and parenchymal cells.[5,6] But insertional mutagenesis may occur, and concerns exist about safety problems of HIV-1. These vectors are likely of greater utility in solid organ allografting.

Adenovirus. Adenovirus is a double-stranded DNA virus. Adenovirus can be produced at very high titers (10^{10}–10^{12} pfu/ml) and can infect nondividing cells. The capacity of the transgene is 7–8 kb. But the adenoviral genome is extrachromosomal in target cells, so the transgenes are only transiently expressed. The adenovirus may cause immediate toxicity and inflammatory reactions, secondary to host innate immune responses, in addition to inducing specific, adaptive, anti-adenoviral responses.[7,8] These immune and inflammatory activities likely worsen the immune injury and responses directed toward alloantigen.

Adeno-Associated Virus (AAV). AAV is human parvovirus. It can infect a wide range of host cells, including nondividing cells.[9] AAV integrates into the host cell genome at a specific site (chromosome 19), and transgenes are stably expressed.[10] AAV is nonpathogenic, nontoxic, and induces limited immune responses in the host. The capacity for transgenes is limited (only 4–5 kb) and high-titer pure virus is not easy to obtain. This virus may be preferred for some transplantation applications.

Herpes Simple Virus (HSV). HSV is a double-stranded DNA virus. It can carry large transgenes (20–30 kb), and the virus titer is between 10^4 and 10^8 cfu/ml. HSV can also infect nondividing cells, especially used to transfect genes in neuronal cells. It does not integrate into host chromosome and may cause host immune responses and inflammatory and toxic reactions similar to adenovirus.[11]

Hybrid Adenoviral/Retroviral Vector. A hybrid adenoviral/retroviral vector system has been developed.[12] This system combines the advantages of adenovirus and retrovirus in one vector system, with high transfection efficiency and stable

[3] V. N. Kim, K. Mitrophanous, S. M. Kingsman, and A. J. Kingsman, *J. Virol.* **72,** 811 (1998).

[4] J. C. Johnston, M. Gasmi, L. E. Lim, J. E. Elder, J. K. Yee, D. J. Jolly, K. P. Campbell, B. L. Davidson, and S. L. Sauter, *J. Virol.* **73,** 4991 (1999).

[5] L. Naldini, U. Blomer, P. Gallay, D. Ory, R. Mulligan, F. H. Gage, I. M. Verma, and D. Trono, *Science* **272,** 263 (1996).

[6] E. M. Poeschla, F. Wong-Staal, and D. J. Looney, *Nat. Med.* **4,** 354 (1998).

[7] R. G. Crystal, N. G. McElvaney, M. A. Rosenfeld, C. S. Chu, A. Mastrangeli, J. G. Hay, S. L. Brody, H. A. Jaffe, N. T. Eissa, and C. Danel, *Nat. Genet.* **8,** 42 (1994).

[8] S. Yei, N. Mittereder, K. Tang, C. O'Sullivan, and B. C. Trapnell, *Gene. Ther.* **1,** 192 (1994).

[9] G. Podsakoff, K. K. Wong, Jr., and S. Chatterjee, *J. Virol.* **68,** 5656 (1994).

[10] R. M. Kotin, M. Siniscalco, R. J. Samulski, X. D. Zhu, L. Hunter, C. A. Laughlin, S. McLaughlin, N. Muzyczka, M. Rocchi, and K. I. Berns, *Proc. Natl. Acad. Sci. U.S.A.* **87,** 2211 (1990).

[11] D. S. Latchman, *Mol. Biotechnol.* **2,** 179 (1994).

[12] M. Feng, W. H. Jackson, Jr., C. K. Goldman, C. Rancourt, M. Wang, S. K. Dusing, G. Siegal, and D. T. Curiel, *Nat. Biotechnol.* **15,** 866 (1997).

gene expression simultaneously. Its use in transplantation settings is not currently known.

Nonviral Vectors

Nonviral vectors for gene transfer include various formulations, such as naked DNA, cationic liposomes, and dendrimers. Compared with viral vectors, nonviral vectors have several advantages. First, the capacity of transgenes is unrestricted. Second, the yield is high, and the purification is easy. Third, nonviral vectors are nonpathogenic and have low toxicity. However, the transfection efficiency is low, the transgene expression is transient, and the unmethylated CpG sequences of bacterial DNA cause strong host inflammatory responses.[13] Nonviral vectors may be suitable for oligonucleotide delivery, for example as antisense oligonucleotides, to prevent rejection in allograft transplantation.

Naked DNA. Genes can be transferred into skeletal muscle and cardiac muscle by injection of naked DNA and can be expressed for 2 weeks to several months.[14,15] Further research show that naked DNA can be expressed in liver, lung, kidney, and skin, but gene expression level is relatively low, which limits its clinical application.

Electroporation has been used to transfer DNA into different cells and to improve the transfection efficiency in muscle, liver, brain, skin, and other tissues *in vivo.* Electroporation can increase gene expression 100-fold in muscle over naked DNA injection alone, making electroporation an efficient method of nonviral gene delivery.[16,17]

Cationic Liposomes. Cationic liposomes bind negatively charged DNA, form liposome–DNA complexes, absorb onto cell membranes, and deliver DNA into the cytoplasm. Lipofectin was the first commercially available cationic liposome in 1987,[18] and now a large variety of cationic liposomes are commercially available, including LipofectAMINE, cellfectine, DC-Chol, DOPE, DOTMA, and DOTAP. Cationic liposomes can be used to improve DNA delivery *in vitro* and *in vivo,* but the *in vivo* transfection efficiency is still low, and these compounds are often toxic to vascular endothelial cells, causing thrombosis and complement activation *in vivo* in vascularized allografts.

[13] A. M. Krieg, A. K. Yi, S. Matson, T. J. Waldschmidt, G. A. Bishop, R. Teasdale, G. A. Koretzky, and D. M. Klinman, *Nature* **374,** 546 (1995).

[14] J. A. Wolff, R. W. Malone, P. Williams, W. Chong, G. Acsadi, A. Jani, and P. L. Felgner, *Science* **247,** 1465 (1990).

[15] G. Acsadi, S. S. Jiao, A. Jani, D. Duke, P. Williams, W. Chong, and J. A. Wolff, *New Biol.* **3,** 71 (1991).

[16] G. Rizzuto, M. Cappelletti, D. Maione, R. Savino, D. Lazzaro, P. Costa, I. Mathiesen, R. Cortese, G. Ciliberto, R. Laufer, N. La Monica, and E. Fattori, *Proc. Natl. Acad. Sci. U.S.A.* **96,** 6417 (1999).

[17] H. Aihara and J. Miyazaki, *Nat. Biotechnol.* **16,** 867 (1998).

[18] P. L. Felgner, T. R. Gadek, M. Holm, R. Roman, H. W. Chan, M. Wenz, J. P. Northrop, G. M. Ringold, and M. Danielsen, *Proc. Natl. Acad. Sci. U.S.A.* **84,** 7413 (1987).

Dendrimers. Dendrimers are synthetic spherical polymers.[19] Positive charge densities can be restricted to the polymer surface, so that dendrimers can interact with negatively charged phospholipids and DNA at the same time. Compared with naked DNA alone or liposome-mediated transfection, dendrimers have higher transfection efficiency. Previous data from our laboratory show that dendrimers can mediate efficient DNA expression in murine cardiac allograft transplantation; the efficiency of gene transfer can be increased about 1000-fold.[20]

Selection Criteria for Transgenes

Various mechanisms are involved in graft rejection, and the host immune response plays a critical role during this process. Therefore, central to the application of gene therapy in transplantation is the delivery of immune modulating molecules that intervene in immune responses during and after transplantation and promote survival of the tissue or organ. There are many molecules that can contribute to protecting the graft from the immune response. However, it is important to characterize their cellular distribution as (1) soluble; (2) cell surface; or (3) intracellular. The advantage of choosing soluble over cell surface or intracellular molecules is that successful gene transfer or transduction does not need to be accomplished for all cells in order to confer immunologic protection, since even a few transfected cells could potentially produce enough soluble proteins to affect or "bathe" surrounding tissue in the gene product. Molecules limited to intracellular or cell membrane compartments will function generally only at the single-cell level, so that complete gene transfer must be achieved for this to be a viable approach for immunosuppression. The most commonly used transgenes in transplantation are outlined in Table II. In each case, transduced exogenous nucleic acids are targeted to host immune responses in the cognitive phase, activation phase, or effective phase of the immune responses. However, because of the complexity of the immune response, it is likely that combinations of transgenes will have to be employed to block multiple pathways of immune responsiveness to achieve prolonged graft survival and immunologic tolerance.

Inhibitory Cytokines

Cytokines are a family of soluble protein mediators of both natural and acquired immunity. By delivery of immune-suppressive cytokine genes to the site of the graft, or to the graft itself, the local immune response against graft could be inhibited. Since cytokine molecules are soluble and their activities potent, a small percentage of infected or transduced cells may produce enough functional

[19] J. M. Frechet, *Science* **263,** 1710 (1994).
[20] L. Qin, D. R. Pahud, Y. Ding, A. U. Bielinska, J. F. Kukowska-Latallo, J. R. Baker, and J. S. Bromberg, *Hum. Gene Ther.* **9,** 553 (1998).

TABLE II
COMMONLY USED TRANSGENES

Transgenes	Mechanism of suppression	Prolongation of survival
vIL-10, TGFβ	Inhibit cytokine synthesis, negative regulation of costimulation by APC	Murine cardiac allograft, rat cardiac allograft
sIL-2 receptor	Disrupt IL-2 : IL-2R interaction	Murine cardiac allograft, rat cardiac allograft
sICAM1-Ig, LFA3-Ig	Inhibit cell adhesion, costimulation	Murine cardiac, small bowel, and renal allografts, primate kidney transplants
CTLA4Ig, sCD80, sCD86	Affect CTLA4 : CD80/CD86 interaction	Murine cardiac, renal, and liver Small bowel allografts, murine islet xenografts
sCD40Ig	Block CD40–CD40L interaction	Murine cardiac, skin xenografts murine cardiac, skin xenografts
Donor MHC	Match MHC to induce tolerance	Mouse, rat cardiac, liver allografts
FasL/CD95L, Bcl-2 family	Protect tissue cells from apoptosis	Islet and rat liver, renal allograft
NF-κB/IκB	Affect inflammatory signals	Murine cardiac allograft
α-(1,2)-Fucosyltransferase	Reduce xenoantigen	Xenotransplantation
Ribozymes	Target specific mRNAs	Xenotransplantation, GvHD

protein to modulate local immune responses and protect the entire graft from rejection. Viral IL-10 (vIL-10) and transforming growth factor beta (TGFβ) have been shown to prolong graft survival in various models.[21,22] vIL-10 is a viral form of interleukin 10 (IL-10) encoded by the Epstein–Barr virus. Although it shares many immune regulatory activities with cellular IL-10, it does not posses the stimulatory properties that IL-10 does. TGFβ antagonizes many responses of lymphocytes, including inhibition of T cell activation, CTL maturation, and macrophage activation. Another approach is the delivery of soluble cytokine receptor molecules that act as dominant negative molecules to prevent normal ligand–receptor interactions to modulate local immune responses. These soluble receptors require high affinity and higher expression levels compared to the use of cytokine genes. These approaches to transfer immune suppressive cytokines achieve local immunosuppression, but not necessarily robust immunological tolerance.

Inhibition of Leukocyte Adhesion

Leukocytes—including macrophages, NK cells, neutrophils, and dendritic cells—and activated endothelial cells in the graft or surrounding sites are capable of

[21] L. Qin, K. D. Chavin, Y. Ding, H. Tahara, J. P. Favarro, J. E. Woodward, J. Lin, P. Robbins, M. T. Lotze, and J. S. Bromberg, *J. Immunol.* **156,** 2316 (1996).
[22] L. Qin, Y. Ding, and J. S. Bromberg, *Hum. Gene Ther.* **7,** 1981 (1996).

initiating antigen nonspecific inflammatory responses that recruit more immune responses cells to induce antigen-specific responses. Adhesion molecule interactions including LFA-1/ICAM-1, LFA-3–CD2, and VLA-4/VCAM-1 promote interactions between leukocytes and other cells. LFA-1–ICAM-1 and LFA-3–CD2 interactions are suggested to be among the most important in graft rejection. Soluble molecules, including sICAM-1 or sLFA-3 that retained the extracellular ligand-binding domain, have been developed to block adhesion interactions.[23,24] Since truncated soluble proteins may not be very stable, carboxy-terminal immunoglobulin constant region domains are fused in frame with those molecules to increase stability and half-life *in vivo,* while also increasing valency and thus avidity.

MHC Genes

MHC mismatch between donor and host plays a critical role in graft rejection. Efforts have been made to transfer donor MHC-encoding cDNAs to host-derived cells which are then transferred back to the recipient to induce antigen-specific tolerance before organ transplantation. Both MHC class I and MHC class II genes can prolong murine cardiac allograft survival in these models.[25]

Blockade of Costimulation

Besides TCR–MHC interaction (first signal), antigen-specific T lymphocytes require costimulatory signals (second signal) to be activated fully. The absence of costimulation can induce T lymphocyte anergy or apoptosis. The second signal can be provided by the interaction of CD28 and B7-2 (CD80 or CD86, respectively), or CD40 and CD40L (CD154). In addition, cytotoxic T-lymphocyte antigen 4 (CTLA4) (CD152) is expressed on the T cell surface as an alternative ligand for CD80 and CD86 and delivers negative signals to the T cells. Delivering soluble CTLA4Ig by gene transfer into cardiac allografts caused significant prolongation of graft survival.[26] Genes encoding a single chain anti-CD40 mAb have also been used to block CD40–CD40L interactions. Furthermore, the combination of blockade of CD40–CD40L and B7–CD28 is a particularly potent regimen in transplant models.

Death Ligands and Signals

Lymphocyte apoptosis is triggered by the engagement of cell surface Fas by its ligand FasL (CD95L). Methods to protect allogeneic grafts against immune cell infiltration and destruction by overexpressing FasL have generated

[23] M. Isobe, H. Yagita, K. Okumura, and A. Ihara, *Science* **255,** 1125 (1992).
[24] R. J. Kaplon, P. S. Hochman, R. E. Michler, P. A. Kwiatkowski, N. M. Edwards, C. L. Berger, H. Xu, W. Meier, B. P. Wallner, P. Chisholm, and C. C. Marboe, *Transplantation* **61,** 356 (1996).
[25] E. K. Geissler, W. J. Korzum, and C. Graeb, *Transplantation* **64,** 782 (1997).
[26] M. G. Levisetti, P. A. Padrid, G. L. Szot, N. Mittal, S. M. Meehan, C. L. Wardrip, G. S. Gray, D. S. Bruce, J. R. Thistlethwaite, Jr., and J. A. Bluestone, *J. Immunol.* **159,** 5187 (1997).

conflicting results. The survival of islet allografts cotransplanted with myoblasts overexpressing FasL was significantly prolonged,[27] whereas mice transplanted with FasL-expressing islet allografts showed accelerated islet destruction.[28] Recent results suggests that soluble FasL may provide protection, while membrane bound molecules may confer susceptibility to neutrophilic inflammation.[29,30] Prolongation of allogeneic islet survival has been achieved by overexpressing the soluble form of FasL. Protective Bc12 family members (bcl-2 and Bcl-XL) have demonstrated protective effects in bone marrow transplantation.[31]

Xenotransplantation

Antibody-mediated hyperacute rejection response against xenoantigens plays a significant role in xenotransplant (the transplantation of cells or tissues between species) graft rejection. The Galα(1,3)-gal-β(1,4)-GlcNac, also known as the α-galactosyl, epitope is a major xenoantigen. Efforts have been made to overexpress α-(1,2)-fucosyltransferase which can covert the xenoantigen to nonantigentic epitope.[32]

Ribozymes to Target mRNA

Ribozymes are RNA molecules that act as enzymes capable of catalyzing cellular reactions. Delivery of modified ribozymes to target specific mRNAs may also be used in transplantation gene therapy. An adenoviral vector expressing a ribozyme has been used to target α(1,3)-galactosyltransferase which, as noted above, is involved in the production of xenoantigenic determinants to prolong xenograft survival.[33] FasL mRNAs have been targeted to prevent graft host disease (GvHD).[34]

Selection Criteria for Promoters

Viral Promoters

Viral promoters, such as human cytomegalovirus immediate early promoter (hCMVie), Rous sarcoma virus long terminal repeat (RSV-LTR), Moloney murine leukemia virus long terminal repeat (MMLV-LTR), SV40 promoter, and SR-alpha

[27] H. T. Lau, M. Yu, A. Fontana, and C. J. Stoeckert, Jr., *Science* **273**, 109 (1996).

[28] A. V. Chervonsky, Y. Wang, F. S. Wong, I. Visintin, R. A. Flavell, C. A. Janeway, Jr., and L. A. Matis, *Cell* **89**, 17 (1997).

[29] K. M. Swenson, B. Ke, T. Wang, J. S. Markowitz, M. A. Maggard, G. S. Spear, D. K. Imagawa, J. A. Gos, R. W. Busuttil, and P. Seu, *Transplantation* **65**, 155 (1998).

[30] T. Suda, H. Hashimoto, M. Tanaka, T. Ochi, and S. Nagata, *J. Exp. Med* **186**, 2045 (1997).

[31] S. Kondo, D. Yin, T. Morimura, Y. Oda, H. Kikuchi, and J. Takeuchi, *Cancer Res.* **54**, 2928 (1994).

[32] N. Osman, I. F. McKenzie, K. Ostenried, Y. A. Ioannou, R. J. Desnick, and M. S. Sandrin, *Proc. Natl. Acad. Sci. U.S.A.* **94**, 14677 (1997).

[33] S. Hayashi, T. Nagasaka, C. Koike, T. Kobayashi, H. Hamada, I. Yokoyama, I. Saito, and H. Takagi, *Transplant Proc.* **29**, 893 (1997).

[34] Z. Du, C. Ricordi, E. Podack, and R. L. Pastori, *Transplant Proc.* **29**, 2224 (1997).

promoter, are widely used promoters in gene transfer and gene therapy. The advantage of using viral promoters is their strong activity in many different cell types *in vitro* and *in vivo*. However, studies have demonstrated that transgene expression under viral promoters *in vivo* is limited to a short period of time.[35-37]

Several viral promoters, including CMVie, SV40, MMLV-LTR, and RSV-LTR, demonstrated promoter attenuation phenomenon *in vivo* and *in vitro* when exposed to certain cytokines, such as IFNγ and TNFα.[38-40] These anti-viral cytokines, produced as results of specific or nonspecific immune responses, represent basic cellular defenses against viral nucleic acids, a feature desirable for host anti-viral immunity, but undesirable for gene transfer utilizing viral promoters. In particular for transplantation, gene transfer vectors are expected to be exposed to a cytokine-rich environment, as a result of antigen-specific immune responses (such as allograft rejection) or antigen nonspecific response (such as ischemia, trauma, and infection). Therefore, cytokine-sensitive viral promoters are less desirable for gene therapy for allografting. In contrast to using viral promoters, gene therapy vectors based on mammalian promoters offer the potential for increased cell specificity and less susceptibility than viral promoters to transcriptional attenuation by host cytokines.

Cell- and Tissue Type-Specific Promoters

One of the biggest challenges facing transplantation is to generate localized, donor-specific unresponsiveness or tolerance, while maintaining systemic immune responses against pathogens or malignant cells. Gene therapy can approach this goal by using a tissue-specific delivery system.

Many cell- and tissue type-specific promoters have been described. In the setting of liver transplantation, hepatocyte-specific albumin,[41,42] apolipoprotein E,[43] human alpha-1 anti-trypsin,[43] elongation factor-1 alpha,[44] transthyretin,[45] and human factor IX[46] promoters can be considered. For heart transplantation, cardiac myocyte specific promoters, such as atrial natriuretic factor,[47] ventricle-specific

[35] A. Kass-Eisler, E. Falck-Pedersen, M. Alvira, J. Rivera, P. M. Buttrick, B. A. Wittenberg, L. Ciproani, and L. A. Leinwand, *Proc. Natl. Acad. Sci. U.S.A.* **90,** 11498 (1993).

[36] S. D. Rettinger, S. C. Kennedy, X. Wu, R. L. Saylors, D. G. Hafenrichter, M. W. Flye, and K. P. Ponder, *Proc. Natl. Acad. Sci. U.S.A.* **91,** 1460 (1994).

[37] P. M. Challita and D. B. Kohn, *Proc. Natl. Acad. Sci. U.S.A.* **91,** 2567 (1994).

[38] J. S. Harms and G. A. Splitter, *Hum. Gene Ther.* **6,** 1291 (1995).

[39] G. Gribaudo, S. Ravagalia, M. Gaboli, M. Gariglio, R. Cavallo, and S. Landolfo, *Virology* **221,** 251 (1995).

[40] L. Qin, Y. Ding, D. R. Pahud, E. Chang, M. J. Imperiale, and J. S. Bromberg, *Hum. Gene Ther.* **8,** 2019 (1997).

[41] S. Kuriyama, T. Sakamoto, M. Kikukawa, T. Nakatani, Y. Toyokawa, H. Tsujinoue, K. Ikenaka, H. Fukui, and T. Tsujii, *Gene Ther.* **5,** 1299 (1998).

[42] S. I. Miyatake, S. Tani, F. Feigenbaum, P. Sundaresan, H. Toda, O. Narumi, H. Kikuchi, N. Hashimoto, M. Hangai, R. L. Martuza, and S. D. Rabkin, *Gene Ther.* **6,** 564 (1999).

[43] C. H. Miao, K. Ohashi, G. A. Patijn, L. Meuse, X. Ye, A. R. Thompson, and M. A. Kay, *Mol. Ther* **1,** 522 (2000).

myosin light chain-2, or atrial-and ventricular-specific alpha-myosin heavy chain[48] promoters are the choices. For lung transplantation, lung epithelium specific cytokeratin 18[49,50] and surfactant protein B[51] or C[52] promoters are available. In addition, endothelial cell specific promoters E-selectin,[53–55] and prepro-endothelin-1[56] or smooth muscle specific promoters[57–59] can be used for targeting graft vasculature. Transplantation also offers the advantage that since tissues or organs can be exposed to vectors *ex vivo,* the promoters do not necessary have to be organ specific; specificity is determined by *ex vivo* targeting.

Chimeric and Synthetic Promoter Elements

Selective gene targeting using cell- or tissue-specific promoters plays an important role in the development of site-selective vectors for gene therapy. However, the expression of these highly selective promoters is usually insufficient, and their applications are limited. The promoter activities could be enhanced when coupled with other viral, cellular, or synthetic enhancer elements, while maintaining specificity. For example, the activities of the lung-specific K18 promoter could be enhanced by SV40 3' untranslated region,[49] skeletal muscle actin promoter by CMV immediate early enhancer,[60] and vascular smooth muscle alpha-actin promoter by SV40 early promoter.[59] It was reported that tumor-specific promoters

[44] F. Park, K. Ohashi, and M. A. Kay, *Blood* **96,** 1173 (2000).

[45] L. Aurisicchio, P. Delmastro, V. Salucci, O. G. Paz, P. Rovere, G. Ciliberto, N. La Monica, and F. Palombo, *J. Virol.* **74,** 4816 (2000).

[46] H. Hoag, J. Gore, D. Barry, and C. Mueller, *Gene Ther.* **6,** 1584 (1999).

[47] K. Eizema, H. Fechner, K. Bezstarosti, S. Schneider-Rasp, A. van der Laarse, H. Wang, H. P. Schultheiss, W. C. Poller, and J. M. Lamers, *Circulation* **101,** 2193 (2000).

[48] W. M. Franz, T. Rothmann, N. Frey, and H. A. Katus, *Cardiovasc. Res.* **35,** 560 (1997).

[49] D. R. Koehler, Y. H. Chow, J. Plumb, Y. Wen, B. Rafii, R. Belcastro, M. Haardt, G. L. Lukacs, M. Post, A. K. Tanswell, and J. Hu, *Pediatr. Res.* **48,** 184 (2000).

[50] Y. H. Chow, H. O'Brodovich, J. Plumb, Y. Wen, K. J. Sohn, Z. Lu, F. Zhang, G. L. Lukacs, A. K. Tanswell, C. C. Hui, M. Buchwald, and J. Hu, *Proc. Natl. Acad. Sci. U.S.A.* **94,** 14695 (1997).

[51] M. S. Strayer, S. H. Guttentag, and P. L. Ballard, *Am. J. Respir. Cell. Mol. Biol.* **18,** 1 (1998).

[52] R. Dhami, K. Zay, B. Gilks, S. Porter, J. L. Wright, and A. Churg, *J. Mol. Med.* **77,** 377 (1999).

[53] U. Modlich, C. W. Pugh, and R. Bicknell, *Gene Ther.* **7,** 896 (2000).

[54] T. Walton, J. L. Wang, A. Ribas, S. H. Barsky, J. Economou, and M. Nguyen, *Anticancer Res.* **18,** 1357 (1998).

[55] R. T. Jaggar, H. Y. Chan, A. L. Harris, and R. Bicknell, *Hum. Gene Ther.* **8,** 2239 (1997).
F. Moorman, G. Brem, and H. A. Katus, *Cardiovasc. Res.* **43,** 1040 (1999).

[56] U. Jager, Y. Zhao, and C. D. Porter, *J. Virol.* **73,** 9702 (1999).

[57] K. Aoki, L. M. Akyurek, H. San, K. Leung, M. S. Parmacek, E. G. Nabel, and G. J. Nabel, *Mol. Ther.* **1,** 555 (2000).

[58] W. M. Franz, O. J. Mueller, M. Fleischmann, P. Babij, N. Frey, M. Mueller, U. Besenfelder, A. F. Moorman, G. Brem, and H. A. Katus, *Cardiovasc. Res.* **43,** 1040 (1999).

[59] M. C. Keogh, D. Chen, J. F. Schmitt, U. Dennehy, V. V. Kakkar, and N. R. Lemoine, *Gene Ther.* **6,** 616 (1999).

[60] J. N. Hagstrom, L. B. Couto, C. Scallan, M. Burton, M. L. McCleland, P. A. Fields, V. R. Arruda, R. W. Herzog, and K. A. High, *Blood* **95,** 2536 (2000).

TABLE III
REGULATORY ELEMENTS

Regulatory elements	Regulators	References
Glucose-responsive elements (GRE)	Glucose	76,77
Insulin-sensitive promoter	Insulin	77
Tetracycline transcription activator	Tetracycline	78–80
FK506-inducible system	Rapmycin	65,66
RU486-inducible transactivator	RU486	81
CAMP-regulatable promoter	Forskolin	82
Glucocorticoid-responsive elements	Dexamethasone	83
Radiation-inducible promoter WAF1	Radiation	84
Radiation-responsive Egr-1 gene	Radiation	85
Heat shock protein 70	Heat	86
MHC class I	IFN-gamma	38
STAT-binding sequence	IFN-gamma	67
NFkB tandem-binding site	TNF-alpha	53
Hypoxia-responsive elements (HREs)	Hypoxia	53,68–70
Stress-inducible promoter grp78	Glucose deprivation	71,72

such as CEA promoter could be enhanced by the Cre/loxP system,[61] and AFP promoter by SV40 or retroviral LTR.[62] These enhancer elements should be able to enhance other cellular promoters as well. However, the cytokine effect on these viral enhancer elements was not investigated.

A synthetic promoter construction method (SPCM)[63] has been developed. Following DNA database searches, SPCM constructs synthetic promoters containing multiple active cis motifs that dramatically enhance promoter activity. Synthetic muscle promoters,[64] constructed by random assembly of several myogenic regulatory elements followed by screening for transcriptional activity, exhibit activities that greatly exceed those of natural myogenic and viral promoters.

Regulatable Promoters

Regulation of transgene expression is desirable for gene therapy in transplantation applications. It could be achieved by incorporating regulatory elements into cell- or tissue-specific promoters. Transgene expression can be regulated by pharmacological agents, hormones, radiation, heat, hypoxia, stress, and cytokines (Table III). All these regulatory elements can be considered for gene

---

[61] K. Ueda, M. Iwahashi, M. Nakamori, M. Nakamura, H. Yamaue, and H. Tanimura, *Oncology* **59**, 255 (2000).

[62] G. Cao, S. Kuriyama, H. Tsujinoue, Q. Chen, A. Mitoro, and Z. Qi, *Int. J. Cancer* **87**, 247 (2000).

[63] G. M. Edelman, R. Meech, G. C. Owens, and F. S. Jones, *Proc. Natl. Acad. Sci. U.S.A.* **97**, 3038 (2000).

[64] X Li, E. M. Eastman, R. J. Schwartz, and R. Draghia-Akli, *Nat. Biotechnol.* **17**, 241 (1999).

therapy for allografting. In particular, an FK506-inducible system[65,66] may be desirable since its regulator, rapmycin, is an immune suppressant routinely used for posttransplantation patients. Utilizing the FK506-inducible system, the level of immunosuppression could be regulated by the doses of rapmycin, through both the direct drug effect and the amount of transgene expressed, when clinically indicated. Cytokine-inducible promoters,[38,53,67] especially IFNγ and TNFα inducible promoters, will be particularly useful, as these two cytokines increase during graft rejection. Hypoxia-responsive elements[53,68–70] and stress-inducible promoter grp78[71,72] can also be considered, since graft rejection may be associated with ischemia and glucose deprivation.

In addition to promoter activity, specificity, and regulatory features, promoter size is also an important issue, especially when viral vectors with low packaging capacity (e.g., AAV and retroviral vectors) are used and large transgenes are required. Efficient cytoplasmic expression of target genes using a T7 RNA polymerase autogene[73] can be considered when nonviral vectors, with inefficient nuclear transport, are used. If the expression of two transgenes in the same cell is required, the use of internal ribosome entry site (IRES) is a potential approach.[74] If controlled switching between the expression of the two genes is desired, a regulatory system based on the tetracycline-controlled transactivators and reverse tTA is applicable.[75]

[65] F. M. Rossi and H. M. Blau, *Curr. Opin. Biotechnol.* **9,** 451 (1998).

[66] X. Ye, V. M. Rivera, P. Zoltick, F. Cerasoli, Jr., M. A. Schnell, G. Gao, J. V. Hughes, M. Gilman, and J. M. Wilson, *Science* **283,** 88 (1999).

[67] R. L. Saylors, K. C. Stine, and J. Derrick, *Gene Ther.* **6,** 944 (1999).

[68] H. Prentice, N. H. Bishopric, M. N. Hicks, D. J. Discher, X. Wu, A. A. Wylie, and K. A. Webster, *Cardiovasc. Res.* **35,** 567 (1997).

[69] K. Eizema, H. Fechner, K. Bezstarosti, S. Schneider-Rasp, A. van der Laarse, H. Wang, H. P. Schultheiss, W. C. Poller, and J. M. Lamers, *Circulation* **101,** 2193 (2000).

[70] T. Shibata, A. J. Giaccia, and J. M. Brown, *Gene Ther.* **7,** 493 (2000).

[71] G. Gazit, G. Hung, X. Chen, W. F. Anderson, and A. S. Lee, *Cancer Res* **59,** 3100 (1999).

[72] X. Chen, D. Zhang, G. Dennert, G. Hung, and A. S. Lee, *Breast Cancer Res. Treat.* **59,** 81 (2000).

[73] M. Brisson, Y. He, S. Li, J. P. Yang, and L. Huang, *Gene Ther.* **6,** 263 (1999).

[74] M. J. Wagstaff, C. E. Lilley, J. Smith, M. J. Robinson, R. S. Coffin, and D. S. Latchman, *Gene Ther.* **5,** 1566 (1998).

[75] U. Baron, D. Schnappinger, V. Helbl, M. Gossen, W. Hillen, and H. Bujard, *Proc. Natl. Acad. Sci. U.S.A.* **96,** 1013 (1999).

[76] P. M. Thule and J. M. Liu, *Gene Ther.* **7,** 1744 (2000).

[77] P. M. Thule, J. M. Liu, and L. S. Phillips, *Gene Ther.* **7,** 205 (2000).

[78] T. Kafri, H. van Praag, F. H. Gage, and I. M. Verma, *Mol. Ther.* **1,** 516 (2000).

[79] B. Massie, F. Couture, L. Lamoureux, D. D. Mosser, C. Guilbault, P. Jolicoeur, F. Belanger, and Y. Langelier, *J. Virol.* **72,** 2289 (1998).

[80] S. A-Mohammadi and R. E. Hawkins, *Gene Ther.* **5,** 76 (1998).

[81] T. Oligino, P. L. Poliani, Y. Wang, S. Y. Tsai, B. M. O'Malley, D. J. Fink, and J. C. Glorioso, *Gene Ther.* **5,** 491 (1998).

[82] M. Suzuki, S. N. Singh, and R. G. Crystal, *Gene Ther.* **4,** 1195 (1997).

Immunologic Considerations

While immunity to gene therapy vectors has been well characterized in a variety of applications,[87–89] very little attention has been given to the inhibitory role of other facets of immunity on vector transgene expression in the transplant setting. The ischemia–reperfusion injury that occurs immediately following transplantation, and the subsequent adaptive alloimmune response, both induce additional cascades of immune activation and cytokine release in addition to those elaborated by vector. Both inhibitory cytokines produced by Th1 lymphocytes and cytodestructive CD8-mediated effector responses against the allograft may lead to a further limitation of transgene expression and a more rapid clearance of vector. Conversely, because of the dual stimuli of alloantigen and gene therapy vector, adaptive immune responses against the allograft may be greater than that which would be induced in the absence of vector. This may lead to accelerated graft rejection.

Innate Immunity

While the major mechanism for the attenuation of gene expression is the specific cellular immune response mediated by cytotoxic T lymphocytes and immunoglobulins to either viral proteins and to the encoded transgene product,[87–89] loss of transgene expression frequently precedes the development of specific immune responses by a significant period of time.[90] This suggests the importance of nonspecific, innate immune mechanisms mediated by macrophages, neutrophils, natural killer (NK) cells, or eosinophils in limiting vector efficiency and persistence, independent of adaptive immunity.[91,92] Resident macrophages clear a vast majority of adenovirus vector genomes within 24 hr of administration,[93,94] and

[83] K. Narumi, M. Suzuki, W. Song, M. A. Moore, and R. G. Crystal, *Blood* **92**, 822 (1998).

[84] J. Worthington, T. Robson, M. Murray, M. O'Rourke, G. Keilty, and D. G. Hirst, *Gene Ther.* **7**, 1126 (2000).

[85] Y. Manome, T. Kunieda, P. Y. Wen, T. Koga, D. W. Kufe, and T. Ohno, *Hum. Gene Ther.* **9**, 1409 (1998).

[86] A. Vekris, C. Maurange, C. Moonen, F. Mazurier, H. De Verneuil, P. Canioni, and P. Voisin, *J. Gene Med.* **2**, 89 (2000).

[87] Y. Yang, H. D. Ertl, and J. M. Wilson, *Immunity* **1**, 33 (1994).

[88] Y. Yand, Q. Su, and J. M. Wilson, *J. Virol.* **70**, 7209 (1996).

[89] S. K. Tripathy, H. B. Black, E. Goldwasser, and J. M. Leiden, *Nat. Med.* **2**, 545 (1996).

[90] L. Qin, Y. Ding, D. R. Pahud, M. D. Robson, A. Shaked, and J. S. Bromberg, *Hum. Gene Ther.* **8**, 1365 (1997).

[91] S. Yei, N. Mittereder, S. Wert, J. A. Whitsett, R. W. Wilmott, and B. C. Trapnell, *Hum. Gene Ther.* **5**, 731 (1994).

[92] M. R. Adesanya, R. S. Redman, B. J. Baum, and B. C. O'Connell, *Hum. Gene Ther.* **7**, 1085 (1996).

[93] S. Worgall, P. L. Leopold, G. Wolff, B. Ferris, N. Van Roijen, and R. G. Crystal, *Hum. Gene Ther.* **8**, 1675 (1997).

[94] S. Worgall, G. Wolff, E. Falck-Pedersen, and R. G. Crystal, *Hum Gene Ther.* **8**, 37 (1997).

Kupfer cell depletion by gadolinium chloride treatment substantially reduces this vector clearance in the liver.[95] The vectors may cause the release of cytokines (TNFα, IL-6, IFNγ, IL-1β) and chemokines (MCP-1, RANTES, IP-10, MIP-1α) which induce a profound inflammatory response.[95,96]

This inflammatory response is often not dependent on viral or transgene expression, as such responses occur very early or immediately after infection[97] and adenoviral vectors progressively deleted of viral genes incite similar inflammatory responses and interference with vector persistence.[98] Furthermore, specific antigen recognition may not be required,[99] suggesting that infection of parenchymal cells, which are not part of the leukocytic immune system, induces an innate stereotyped response in these cells that inhibits vector function and gene expression.[100–102] This may occur by cytotoxicity of vector-transduced cells, or by inhibition of vector promoter expression by cytokines.[40]

Innate immune responses are characterized by pattern-recognition receptors such as mannan-binding lectin, which focus on highly conserved sequences common to many pathogens. Effector responses are rapid and do not require effector cell proliferation. Signaling receptors, such as those of the toll family, induce expression of a variety of genes involved in innate and adaptive immunity such as cytokines and chemokines.[103]

Assessment of innate immune responses has included the following: (1) The detection of the expression of relevant cytokines and chemokines and their receptors by RT-PCR or Northern hybridization. Primer sequences are published in the literature for many genes encoding these proteins. In addition, RNase protection assay kits are commercially available for detection of a number of these molecules. (2) Analysis of expression of the toll receptors TLR2 and TLR4 by RT-PCR or Northern hybridization. (3) Analysis of graft or organ-infiltrating cells (identification of NK or monocyte infiltration) by immunohistochemistry or FACS. (4) Utilization of monoclonal antibodies (such as exist against many cytokines and several chemokines/chemokine receptors, including CXCR3, RANTES, MCP-1, and MIP 3α) or knockout mice (e.g., CCR5 KO, TNFα-R1 KO, CXCR2 KO, CCR1 KO, MIP-1α), all of which have impaired components of innate immunity.

[95] A. Lieber, C. Y. He, L. Meuse, D. Schowalter, I. Kirillova, B. Winther, and M. A. Kay, *J. Virol.* **71**, 8798 (1997).

[96] D. A. Muruve, M. J. Barnes, I. E. Stillman, and T. A. Liebermann, *Hum. Gene Ther.* **10**, 965 (1999).

[97] K. Otake, D. L. Ennist, K. Harrod, and B. C. Trapnett, *Hum. Gene Ther.* **9**, 2207 (1998).

[98] M. Lusky, M. Christ, K. Rittner, A. Dieterle, D. Dreyer, B. Mourot, H. Schultz, F. Stoeckel, A. Pavirani, and M. Mehtali, *J. Virol.* **72**, 2022 (1998).

[99] S. Gangappa, J. S. Babu, J. Thomas, M. Daheshia, and B. T. Rouse, *J. Immunol.* **161**, 4289 (1998).

[100] L. G. Guidott, P. Borrow, A. Brown, H. McClary, R. Koch, and F. V. Chisari, *J. Exp. Med.* **189**, 1555 (1999).

[101] A. Heim, S. Zeuke, S. Weiss, W. Ruschewski, and I. M. Grumbach, *Circ. Res.* **86**, 753 (2000).

[102] R. Amin, R. Wilmott, Y. Schwarz, B. Trapnell, and J. Stark, *Hum. Gene Ther.* **6**, 145 (1995).

[103] R. Medzhitov and C. Janeway, Jr., *N. Engl. J. Med.* **343**, 338 (2000).

Anti-Vector Adaptive Immunity

Adenovirus vectors elicit the most vigorous immune responses, whereas more recently utilized vectors such as lentivirus and adeno-associated virus vectors are less immunogenic.[87,104,105] Nonviral vectors may generate adaptive immune responses to the encoded gene product.[89] The short-lived transgene expression that characterizes adenoviral gene therapy has been attributed to these antiviral adaptive immune responses. These processes, which involve CD8+ CTL, CD4+ TH1 cells, IFNγ, and IL-2, lead to destructive inflammation and loss of vector.[87,106,107] Adenoviral vectors also elicit an antiviral humoral immune response that limits the effectiveness of secondary vector administration.[108] Adenoviral vectors also enhance costimulatory signals for lymphocyte or neutrophil activation and homing by up-regulation of adhesion molecules such as ICAM-1 and VCAM.[109,110] All of these responses may be specific to adenovirus proteins[88] or to the transgene product itself.[3] The significance of adaptive immunity in limiting transgene expression can be illustrated by the demonstration of prolonged transgene expression in immune deficient mice[107] or mice depleted of CD4+ cells,[111] among others.

Anti-adenoviral adaptive immunity can be detected by standard protocols commonly employed in immunology research. Many of these protocols can be found in Current Protocols in Immunology.[112] These include proliferation assay of primed splenic lymphocytes; cytotoxic T-lymphocyte assay against vector-infected targets; neutralizing antibody assay; and Th1 cytokine production by primed splenic lymphocytes by ELISA.

Alloantigen Specific Adaptive Immunity

Transplantation evokes an array of antigen-specific immunological events directed against alloantigen. Alloantigen is derived from foreign major histocompatibility complex (MHC) proteins and recognized by host T cell receptors (TCR). Following T cell activation, immune responses to alloantigen are generally similar

[104] F. Park, K. Ohashi, and M. A. Kay, *Blood* **96**, 1173 (2000).

[105] K. Jooss, Y. Yand, K. J. Fisher, and J. M. Wilson, *J. Virol.* **72**, 4212 (1998).

[106] Y. Yang, F. A. Nunex, K. Berencsi, E. E. Furth, E. Gonczol, and J. M. Wilson, *Proc. Natl. Acad. Sci. U.S.A.* **91**, 4407 (1994).

[107] Y. Dai, E. M. Schwarz, D. Gu, W. W. Zhang, N. Sarvetnick, and I. M. Verma, *Proc. Natl. Acad. Sci. U.S.A.* **92**, 1401 (1995).

[108] Y. Yand, Q. Li, H. C. Ertle, and J. M. Wilson, *J. Virol.* **69**, 2004 (1995).

[109] B. C. Trapnell and M. Gorziglia, *Curr. Opin. Biotechnol.* **5**, 617 (1994).

[110] J. M. Pilewski, D. J. Sott, J. M. Wilson, and S. M. Albelda, *Am. J. Respir. Cell. Mol. Biol.* **12**, 142 (1995).

[111] R. P. DeMatteo, J. F. Markmann, K. F. Kozarsky, C. F. Barker, and S. E. Raper, *Gene Ther.* **3**, 4 (1996).

[112] F. M. Ausubel, R. Brent, R. E. Kingston, D. D. Moore, J. G. Seidman, J. A. Smith, and K. Struhl (eds.), "Current Protocols in Immunology." Wiley, New York, 2000.

to those against other peptide antigens. Although both cellular immunity and humoral immunity may contribute to rejection of the allograft, acute cellular rejection is the dominant mechanism in the absence of immunosuppression. Specific antibody generated against alloantigen following transplantation may lead to long-term graft failure.[113]

The interaction between anti-vector immunity and anti-alloantigen immunity has been largely unexplored. We have demonstrated anti-vector immune responses in allotransplant models, and have shown that adenoviral gene transfer of vIL-10 inhibits immune responses to both alloantigen and adenoviral antigen.[90]

Vector Administration in Cellular Transplants

Currently, the cell types most frequently transplanted for cellular transplantation are bone marrow or hematopoietic stem cells.[114,115] Pancreatic islets have also been the focus of much preclinical and, more recently, clinical transplantation efforts.[116–118] Over the past decade, there has been increasing interest in the manipulation of stem cells *ex vivo*. Stem cells can self-renew and differentiate into multiple lineages within a suitable microenvironment. Thus, stem cells from various sources [e.g., bone marrow, parenchymal organs, embryonic stem (ES) cells] have become important targets for gene therapy.

Our lab has focused on ES cells and pancreatic islets. The protocols in the following section cover our current methods for manipulation and differentiation of ES cells *in vitro,* and procuring islets from adult mouse pancreas. Methods for transducing ES cells and islets and for the *in vivo* transplantation of single cells underneath the renal capsule or into the portal vein of streptozotocin-induced diabetic mice are included.

In Vitro Differentiation of Endocrine Type Cells from ES Cells

For the growth and maintenance of ES cells, and the generation of embryoid bodies (EBs) from ES cells, follow the detailed protocol of Dr. Gordon Keller and his colleagues.[119] In order to differentiate EB cells into pancreatic lineages, plate single cells from EBs onto 24- or 6-well tissue culture plates for further differentiation. This step is required for maturation of committed cells. It is known that providing a suitable environment containing desirable growth factors is the key to support lineage-specific cell differentiation and maturation.

[113] S. Yilmaz, E. Taskinen, T. Paavonen, A. Mennander, and P. Hayry, *Transpl. Int.* **5,** 85 (1992).

[114] R. J. O'Reilly, *Blood* **62,** 941 (1983).

[115] I. L. Weissman, *Science* **287,** 1442 (2000).

[116] J. Oberholzer, F. Triponez, J. Lou, and P. More, *Ann. N. Y. Acad. Sci.* **875,** 189 (1999).

[117] R. P. Robertson, C. Davis, J. Larsen, R. Stratta, and D. E. Sutherland, *Diabetes Care* **23,** 112 (2000).

[118] T. Berney and C. Ricordi, *Cell Transplant.* **8,** 461 (1999).

[119] G. M. Keller, S. Webb, and M. Kennedy, *Meth. Mol. Med.,* in press.

Isolation of Islets from Adult Mouse Pancreas

Materials and Reagents

Dulbecco's phosphate-buffered saline (DPBS):

> Sigma # D-1408, 10×
> Make 1× solution by adding H_2O and filter sterilize.
> Bovine serum albumin (BSA), Fraction V, 7.5% solution (Sigma)

Dithizone solution:

Dithizone (Sigma)	10 mg
Alcohol, 100%	3 ml
Ammonium hydroxide	3 drops
DPBS/0.1% BSA	300 ml

> Dissolve dithizone powder in 100% alcohol. Add 3 drops of ammonium hydroxide in a fume hood, using 1000-ml Gilson pipette. The solution will turn red. Add to 300 ml DPBS/0.1% BSA and filter sterilize. Store solution at 4°. This can last about 2 weeks.

Collagenase P (Boehringer Manniheim)

> Make stock solution in DPBS as 100 mg/ml and store at $-20°$. Use final concentration of 1 mg/ml. Because there is tremendous variability in collagenase P preparation, it is necessary to test optimal concentrations of a particular batch for the yield of islets.

Density solution, $\rho = 1.080$ g/ml:

OptiPrep (Nycomed Pharma)	2.33 ml
DMEM	7.67 ml

> Do not add penicillin/streptomycin or protein in DMEM, as this will change the density of the solution.

Methods

Anesthetize the mice by intraperitoneal injection of 0.4 mg/kg chloral hydrate. Restrain the mice on a Styrofoam surgical platform. Sterilize abdominal hair using 70% alcohol. Use scissors and forceps to incise abdominal wall along the ventral midline from sternum to pubis. Open the abdominal cavity using a retractor on the abdominal wall. Further open the upper abdominal cavity by pulling the tip of the sternum away using a hemostatic forceps. Find the distal pancreatic duct as it enters the duodenum and occlude it using a hemastatic forceps. Turn the animal around to let the head face the surgeon. Find the pancreatic duct, and carefully inject 1.5 ml of 1 mg/ml collagenase P solution using 30-gauge needles and 3 ml syringes. Pancreatic lobes should be distended as a result. If it is difficult to see the pancreatic duct, use an anatomic microscope to facilitate.

Cut off the distended lobes, place in a 50 ml conical tube, and incubate at 37° for 12 min. Euthanize the mouse. Stop collagenase P reaction by addition of 40 ml ice-cold DPBS/0.1% BSA. Shake the tube by hand. At this time, islets should be released from the surrounding tissue. Avoid shaking too vigorously, as this will break up islets into single cells.

After collection of pancreata from all mice, wash cells, twice with ice-cold DPBS/0.1% BSA. Resuspend cells in 10 ml room temperature DPBS (without BSA), and carefully layer over one volume of 1.080 g/ml Opti-Prep density solution. An interface should be formed. Centrifuge at 2500 rpm at room temperature for 15 min on a Sorvall RT7, or equivalent, desktop centrifuge with swing rotor. Care should be taken to make rate of acceleration and deceleration minimal; high rate will disturb the interface.

Collect the cells at the interface, which should be enriched for islets. Approximately 80 to 85% of the loaded islets are retained in the interface. Wash cells once with dithizone solution. Resuspend cells in 3 ml DPBS/0.1%BSA, and transfer to a 60 × 15 mm petri dish. Observe cells under an inverted microscope. Islets should appear red. Hand pick individual islets using a mechanical pipette with 20 μl tip. Suspend islets at desired concentration.

To check the total recovery of the islets, one could determine the total number of islets in one pancreas by infusing dithizone solution into the whole pancreas of age-controlled mice. Cut the pancreas out, trim into smaller pieces, and mount on slides. Determine the total number of red clusters (islets) under the inverted microscope. By comparing the final number of recovered islets to the total number of islets in whole pancreas, one can determine the yields of islets in the procedure.

Gene Transduction of ES Cells

Whereas transduction efficiency for whole organ transplants is limited, individual cells manipulated *ex vivo* have the potential to be 100% transduced. Because it is possible to select and purify specific types of cells *in vitro,* the problem of targeting genes to unwanted cell types when administered *in vivo* is eliminated. Several methods can be used to transduce genes in cells *in vitro.* As discussed in the earlier section, these include viral and nonviral vectors.

ES cells are in general difficult to transduce. In our laboratory, we routinely use murine stem cell virus (MSCV) and electroporation for gene transfer. MSCV retroviral vector, which is MSCV vector, can be found on the Clontech Web site at www.clontech.com. For producer cell lines, Clontech has the RetroPack PT67 cell line available. The Phoenix line is specifically designed to target ES cells and has been shown to effectively transduce ES cells.[120] MSCV vector can be purchased from Clontech. Detailed protocols have been compiled by Drs. Mike Rothenberg and Garry P. Nolan at Stanford University and are available at www.stanford.edu/group/nolan/NL-retopt.html.

[120] R. G. Hawley, F. H. Lieu, A. Z. Fong, and T. S. Hawley, *Gene Ther.* **1,** 136 (1994).

Electroporation for gene transfer in ES cells has been useful to generate both heterozygous and homozygous mutant cell lines. A good protocol can be found in *Current Protocols in Molecular Biology.*[112]

Gene Transduction of Islets

Isolated islets are washed twice in serum-free RPMI/0.1% BSA prior to infection. Serum can interfere with infection by adenoviral and other viral vectors. Use a control adenovirus expressing a reporter gene; infect islets at multiplicities of infection (MOI) of 1–1000 to determine the optimal conditions for a particular virus preparation. MOI is defined as the number of infectious plaque-forming units divided by the number of cells to infect. Incubate islets with the virus at 37° for 2 hr. Following infection, islets are washed twice in serum-free RPMI/0.1% BSA and then cultured for 48 hr in RPMI medium containing 10% FCS. Now the islets are ready for further manipulation, such as transplantation or gene expression study.

Transplantation of Single Cells underneath Renal Capsule or into Portal Vein of Streptozotocin-Induced Diabetic Mice

The renal capsule is one of several sites to which cells can be transplanted in a diabetic mouse. Diabetes is induced in mice after treatment with streptozotocin (STZ). The effects of STZ are dependent on mouse strain and age. Therefore, a dose-response for STZ induction of hyperglycemia should be established. For 10- to 15-week-old female C57BL/6 mice, tail vein injection of 200 mg/kg STZ gives a consistent result. Check blood glucose level with Glucometer Elite XL monitoring system (Bayer Company). Mice become hyperglycemic (>300 mg/dl) 4 days after injection. Intraperitoneal injection of STZ induces hyperglycemia in some, but not all, mice.

Material and Reagents

Streptozotocin (STZ) solution, 20 mg/ml

STZ (Sigma)	100 mg
50 m*M* citric acid (Sigma C-2404) in H_2O, pH 4.5	5 ml

Inject STZ solution slowly into the tail vein of mice. The acidity of the solution
   may cause spasm when injected too quickly.
Chloral hydrate, 4%:

Chloral hydrate powder (Sigma)	2 g
DPBS	50 mL

This solution could be kept at room temperature for a few months.
Gel loading pipet tips, size 1–200 μl, beveled (Fisher):
Use a sharp knife to bevel the tips. Select those ones that are sharp enough to
   penetrate the renal capsule. Sterilize by autoclaving.

Surgical instruments
Silk suture, 4-0, Ethicon K952
Nylon suture, 9-0, Ethicon 2829
Cotton-tipped applicators
Cotton gauze
Serrated dressing forceps
Accu-Temp cautery (Opto-Systems)

Methods

Renal subcapsular injection. Anesthetize the mice by intraperitoneal injection of 0.4 mg/kg chloral hydrate. Shave the abdominal hair. Restrain the mice on a Styrofoam surgical platform. Sterilize the abdominal skin using 70% alcohol. Use scissors and forceps to incise abdominal wall along the ventral midline, from sternum to 0.5 cm above the pelvis. Moisten cotton gauze and spread it over the thorax of the mouse. Eviscerate the intestine, using moistened cotton-tipped applicators, and place onto the moistened gauze. This will expose the left kidney.

Turn around the surgical platform to let the head of the mouse face toward the surgeon. Select a sharp, beveled, gel-loading pipet tip, and pipet up to 10 μl of the cell solution. While applying an opposite force to stretch the capsule using a serrated dressing forceps, insert the pipet tip into the kidney just underneath the capsule. Inject the cell solution, which should cause the capsule to bulge. Remove the tip from the capsule, and immediately cauterize the opening.

Turn around the surgical platform again and suture the abdominal wall and skin with a 4-0 silk suture. Place the recipient in a recovery pan that is warmed with a heater and allow about 1 hr for recovery.

Portal vein injection. Prepare mouse as stated for renal capsule transplantation. Open the abdominal cavity using a retractor on the abdominal wall. Further open the upper abdominal cavity by pulling the tip of the sternum away using a hemostatic forceps. Find the portal vein under a surgical microscope. Temporarily ligate the proximal part of the vein to stop the blood flow with 9-0 suture. Inject cell solution to the portal vein using a 30-gauge needle and 1 ml syringe. Remove needle and observe the position of the needle opening. Suture the opening with simple 9-0 nylon suture. Care should be taken not to reduce the diameter of the portal vein. Untie the earlier tie on the vein, restoring portal vein flow, and ensure normal portal blood flow. Suture the abdominal wall and skin with a 4-0 silk suture. Place the recipient in a recovery pan that is warmed with a heater and allow about 1 hr for recovery.

Vector Administration to Parenchymal Organ Transplants

The major challenge of gene therapy for whole organs is to achieve sufficient transgene expression and functional protein levels without compromising graft function or survival. Gene transfer can be performed in different settings *in vivo* by (1) injection into the recipient; (2) donor pretreatment; or (3) *ex vivo* by manipulating the graft. Here we describe several experimental procedures successfully used in various solid organs.

Gene Transfer by Treatment of the Recipient

Alloantigen pretreatment by blood transfusions, a widely used method to modulate the allospecific immune response, can be avoided by recipient gene therapy. This approach allows antigen delivery before or at the time of transplantation, and is primarily used to achieve tolerance or unresponsiveness. Recipient-type bone marrow cells are transduced to express donor-specific MHC genes using retroviral delivery systems, and then reinfused. The injection of transduced cells—with or without agents modulating T cell function (i.e., anti-CD4 monoclonal antibody)—has been demonstrated to induce tolerance in skin, heart, and kidney grafts in both rodent and pig models.[121–123] In addition to systemic injection, delivery of the transduced cells to the thymus has also been shown to induce tolerance to donor organs subsequently transplanted at peripheral sites.[124,125]

Gene Transfer into Prospective, Grafts via Donor Pretreatment

Modifying organ grafts prior to transplantation is an attractive approach to modulate posttransplant events. Using this method, effective functional protein levels in the target organ can be detected very early after surgery compared to the other methods. Thus donor pretreatment could be of particular importance in preventing or ameliorating ischemia–reperfusion injury (IRI) and the early events of acute cellular rejection.[126,127] The use of adenoviral or other vectors for gene delivery to the liver is particularly relevant here since systemic administration of vectors often results in preferential hepatic delivery and/or expression. Studies have indeed demonstrated that donor pretreatment with adenoviral vectors encoding the superoxide dismutase gene, a free-radical scavenger molecule, decrease IRI to preserved liver grafts, and decrease the incidence of IRI-associated graft loss after orthotopic liver transplantation in syngeneic rats.[128] Similarly, rat livers transduced in the

[121] M. Sykes, D. H. Sachs, A. W. Nienhuis, D. Pearson, A. D. Moulton, and D. M. Bodine, *Transplantation* **55,** 197 (1993).

[122] W. Wong, S. Stranford, P. Morris, and K. Wood, *Transplant. Proc.* **29,** 1130 (1997).

[123] A. Yasamoto, K. Yamada, T. Sablinska, C. LeGuern, M. Sykes, and D. H. Sachs, *Transplant. Proc.* **29,** 1132 (1997).

[124] A. M. Posselt, C. F. Barker, J. E. Tomaszewski, J. F. Markman, M. A. Choti, and A. Naji, *Science* **249,** 1293 (1990).

[125] R. DeMatteo, R. Raper, M. Ahn, K. Fisher, C. Burke, A. Radu, G. Widera, B. Claytor, C. Barker, and J. Markmann, *Ann. Surg.* **222,** 229 (1995).

[126] Y. Takahashi, D. A. Geller, A. Gambatto, S. C. Watkins, J. J. Fung, and N. Murase, *Surgery* **128,** 345 (2000).

[127] H. Tashiro, K. Shinozaki, H. Yahata, K. Hayamzu, T. Okimoto, H. Tanji, Y. Fudara, H. Yamamoto, X. Fan, H. Ito, and T. Asahara, *Transplantation* **70,** 336 (2000).

[128] T. G. Lehman, M. D. Wheeler, R. Schoonhoven, H. Bunzendahl, R. J. Samulski, and R. G. Thurman, *Transplantation* **69,** 1051 (2000).

donor with an adenovirus vector expressing human IL-10 survived significantly longer in allogeneic hosts. Additional genes encoding other important mediators related to IRI (Bcl-2, heme oxygenase-1) have also been successfully transfected into livers prior to harvesting the organ.[129,130] In addition to data obtained with liver grafts, the most significant advantage of this method—the higher levels of transgene expression at the critical time of reperfusion and early postoperative period—has also been demonstrated by transtracheal gene transfection to donor lungs.[131]

Protocol for Transfer of Adenoviral Vectors into Liver Grafts with a Single Intravenous Injection to the Donor

Vector delivery. Adenoviral vectors are the best vechicle for pretransplant gene transfer into liver grafts. For rats weighing 200–300g, the recommended titer of the vector is between 1–3×10^9 and 1×10^{10} pfu/animal. The vector should be diluted to 1 ml, and injected into a peripheral (penile, femoral, or ileocecal) vein under anesthesia.

Organ harvest and transplantation. To achieve sufficient expression at the time of surgery, the timing of organ harvest is crucial. Livers should be harvested between 24 and 72 hr following gene delivery, and transplanted according to the standard liver transplantation procedure.[132]

Transfection efficiency. Intravenous injection to the donor results in dispersed expression throughout the liver. The number of hepatocytes transfected reaches 80% per section 48 hr after liver transplantation.

Expression pattern. Using enhanced green fluorescence protein encoded by an adenoviral vector (AdEGFP) as a marker gene, expression can be detected as early as 1–3 hr after liver transplantation. Expression peaks at 48 h and is maintained up to 14 days after surgery. If the time period between harvest and donor pretreatment is either less than 12 hr or more then 1 week, the critical early phase of expression is completely absent.

Toxicity. With the recommended doses of viral vector, the levels of transaminases during the early postreperfusion period, a sensitive index of liver injury, do not differ significantly from levels observed in animals transplanted with control grafts. Gene transfer by donor pretreatment does not result in additional damage to liver grafts.

[129] G. Bilbao, J. L. Contreras, L. Gomez-Novarro, D. E. Eckoff, G. Mickhleva, V. Krasnykh, T. Hynes, F. T. Thoma, J. M. Thoma, and D. T. Curiel, *Transplantation* **67,** 775 (1999).

[130] F. Amersi, R. Buelow, H. Kato, B. Ke, A. J. Coito, X. D. Shen, D. Zhao, J. Zaky, J. Melnak, C. R. Lassman, J. K. Kollino, J. Alam, T. Ritter, H. P. Volk, D. G. Farmer, R. W. Busuttil, and J. W. Kupiec-Weglisnki, *J. Clin. Invest.* **104,** 1631 (1999).

[131] S. D. Cassivi, J. A. Cardella, S. Fischer, M. Liu, A. S. Slutsky, and S. Keshavjee, *Heart Lung Transpl.* **18,** 1181 (1999).

[132] N. Kamada and R. Y. Calne, *Surgery* **93,** 64 (1983).

Ex Vivo Gene Delivery Methods to Whole Organs

Delivery of biologically active molecules to a graft can be achieved either by cotransplantation of genetically modified cells along with the graft or by direct genetic modification of the graft itself. To be efficient, the carrier cells must be easily transfectable, and the transgene expression stable and sustained. Despite obvious limitations, this approach has been shown to be effective in some cases. Myoblasts of recipient origin transfected with cytotoxic T lymphocyte antigen (CTLA)4-Ig or Fas ligand prevent the rejection of allogeneic islets provided they are transplanted together under the capsule of recipient mice.[133,134]

Ex vivo genetic modification of the transplanted organ has a particularly significant potential as an approach to modulate posttransplant events. The strategy selected for gene transfer is of crucial importance, and depends on the susceptibility of the organ to transduction. Early studies compared various viral and nonviral methods including retroviral, herpes simplex virus, and adenovirus vectors and naked DNA, and have demonstrated that transgene expression varies according to the type of vector used.[21,135] Different transfer techniques for the same vector and organ have been developed and compared (i.e., continuous perfusion and clamp method for the liver; direct injection and intracoronary perfusion for the vascularized heart grafts) to achieve optimal transduction. Adenoviral delivery systems have been used successfully to transduce liver,[136,137] lung,[138] heart,[139–141] and corneal grafts.[142] In general, transgene expression peaks between 3 and 6 days after injection, and varying grades of inflammatory response have been reported. Nonviral strategies have also been used for whole organs to overcome the adverse consequences of inflammation associated with the use of viral vector delivery systems. Despite relatively low transfection efficiency, both plasmid- and oligonucleotide (ODN)–liposome complexes are capable of transferring genes to

---

[133] A. Chahine, M. Yu, M. McKernan, C. Stoeckert, and H. Lau, *Transplantation* **59,** 1313 (1995).

[134] H. Lau, A. Fontana, and C. Stoeckert, *Science* **273,** 109 (1996).

[135] L. Qin, K. Chavin, Y. Ding, J. Favaro, J. Woodward, J. Lin, H. Tahara, P. Robbins, A. Shaked, and J. Bromberg, *Transplantation* **59,** 809 (1995).

[136] K. Drazan, L. Wu, X. Shen, D. Bullington, O. Jurin, R. Busuttil, and A. Shaked, *Transplantation* **59,** 70 (1995).

[137] S. H. Chia, D. A. Geller, M. R. Kibbe, S. C. Watkins, J. J. Fung, T. E. Starzl, and N. Murase, *Transplantation* **66,** 1545 (1998).

[138] A. Chapelier, C. Daniel, M. Mazmanian, E. Bacha, H. Sellak, M. Gilbert, P. Serve, and P. Lemarchand, *Hum. Gene Ther.* **7,** 1837 (1996).

[139] J. Lee, H. Laks, A. Dinkwater, A. Blitz, L. Lan, Y. Shitaishi, P. Chang, T. Drake, and A. Ardehlali, *J. Thorac. Cardiovasc. Surg.* **111,** 246 (1996).

[140] B. Asfour, B. J. Bryrne, H. A. Hammel, R. H. Hruban, M. Weyand, M. Dend, and H. H. Scheld, *Thorac. Cardiovasc. Surg.* **47,** 311 (1999).

[141] A. David, J. Chetrit, L. Tesson, J. M. Heslan, M. C. Cuturi, J. P. Soulillou, and I. Anegon, *Gene Ther.* **7,** 505 (2000).

[142] D. Larkin, H. Oral, C. Ring, N. Lemoine, and A. George, *Transplantation* **61,** 363 (1996).

liver,[143] heart,[144–147] and kidney grafts,[148] as suggested by the apparent biological effects of the respective recombinant proteins.

Nondistending pressure has also been successfully introduced to transduce whole organs *ex vivo:* hyperbaric pressure transfects antisense (AS)-ODNs to ICAM-1 into rodent hearts with consistently high efficiency and minimal toxicity.[149,150] Dendrimer polymers have been applied to facilitate gene transfer into both vascularized and nonvascularized murine heart grafts.[20,151] These synthetic, highly branched, spherical polyamidoamine molecules are nonimmunogenic and highly soluble, and they significantly enhance DNA gene transfer, thus offering an additional effective and nontoxic transfection method.[152]

Protocols to Transfer Adenovirus Vectors into Rat Livers

Vector Delivery

Perfusion methods. Using this approach, transduction is achieved through continuous perfusion of the organ. Donor livers are flushed with cold lactated Ringer's (LR) solution, harvested, and placed in a perfusion circuit connected with a nonpulsatile pump (Cole Parmer, Chicago, IL). Gene delivery is carried out via the portal vein alone or simultaneously perfusing the portal vein and the hepatic artery in recirculating mode. The viral vector (1×10^8–1×10^9) is dissolved in a total volume of 10–15 ml of LR. Perfusion is performed for 30 min at $4°$ with flow rates of 2 ml/min (portal vein) and 6 ml/min (hepatic artery).

Clamp technique. Sufficient gene transfer is achieved by trapping the vector in the vasculature of the liver. The donor liver is perfused *in situ* with 15–20 ml of University of Wisconsin (UW) solution through the aorta and dissected. The isolated graft is slowly perfused with 4–6 ml UW containing the adenovirus vector

[143] K. Yamabe, W. Kamike, S. Shimizu, S. Waguri, J. Hasegawa, S. Okuno, Y. Yohsikoa, Y. Sawa, Y. Uchiyama, and H. Matsuda, *Transpl. Proc.* **29,** 384 (1997).

[144] J. Dalesandro, H. Akimoto, C. Gorman, T. McDonald, R. Thomas, H. Liggit, and M. Allen, *J. Heart Lung Transplant.* **15,** 857 (1996).

[145] L. A. Bruyne, K. Li, S. Y. Chan, L. Qin, D. K. Bishop, and J. S. Bromberg, *Gene Ther.* **5,** 1079 (1998).

[146] Y. Sawa, Y. Kaneda, H. Z. Bai, K. Suzuki, J. Fujimoto, R. Morisjita, and H. Matsuda, *Gene Ther.* **5,** 1472 (1998).

[147] M. Isoba, J. Suzuki, R. Morishita, Y. Kaneda, and J. Amano, *Ann. N. Y. Acad. Sci.* **902,** 77 (2000).

[148] M. Isoba, J. Suzuki, R. Morishita, Y. Kaneda, and J. Amano, *Ann. N. Y. Acad. Sci.* **902,** 77 (2000).

[149] R. S. Poston, M. J. Mann, E. G. Hoyt, M. Ennen, V. J. Dzau, and R. C. Robbins, *Transplantation* **68,** 825 (1999).

[150] B. T. Feeley, R. P. Poston, A. K. Park, M. P. Ennen, E. G. Hoyt, P. W. H. E. Vriens, and R. C. Robbins, *Transplantation* **69,** 1067 (2000).

[151] Y. Wang, P. Boros, J. Liu, L. Qin, Y. Bai, A. U. Bielinska, J. F. Kukowska-Latallo, J. R. Baker, and J. S. Bromberg, *Mol. Ther.* **2,** 602 (2001).

[152] J. F. Kukowska-Latallo, A. U. Bielinska, J. Johnson, R. Spindler, D. A. Tomalia, and J. R. Baker, *Proc. Nat. Acad. Sci. U.S.A.* **93,** 4897 (1996).

(1×10^9 pfu/liver) in two stages. After an initial flush-out with 1–2 ml through the portal vein and the hepatic artery, vascular clamps are placed on the suprahepatic and infrahepatic inferior vena cava. Additional perfusion is performed through the portal vein and the hepatic artery with 2 ml. After visible expansion of the organ, the remaining open vessels (portal vein and hepatic artery) are clamped and ligated, respectively.

Liver transplantation. Using any *ex vivo* gene transfer method, the liver can be transplanted either immediately or after preservation in the actual perfusion buffer according to standard procedures. LR allows for several hours of cold storage, while livers preserved in UW may be stored up to 18 hr.[137]

Transfection efficiency and expression pattern. Both viral titer and incubation time are positively correlated with transduction rate using the perfusion methods. For the liver, however, the clamp technique provides a more efficient gene delivery and requires relatively low viral titers. Both the number of positive cells for the reporter gene and the amount of recombinant protein per gram of liver protein are significantly higher compared with the perfusion methods. Transgene expression remains stable for 2 weeks and declines after 28 days. Additionally, the clamp technique results in selective transduction of hepatocytes with minimal extrahepatic transgene expression.

Toxicity. Transaminase levels are increased at 12 hr after liver transplantation, but quickly normalize thereafter, and graft quality and survival are not significantly compromised by either of these methods.

Protocol for Viral, DNA/Liposome-, and DNA/Dendrimer Complex Mediated Gene Transfer into Murine Hearts by Intracoronary Infusion

Vectors. Optimal doses for the different vectors may vary depending both on the actual reagent and the transgene. For adenoviral vectors, the recommended titer is $1–3 \times 10^9$ pfu/heart. With the cationic lipid N-(3-aminopropyl)-N,N-dimethyl-2,3-bis(dodecyloxy)-1-propanimium bromide/dioleoyl phosphatidylethanolamine (γAP DLRIE/DOPE, VICAL, San Diego, CA) efficient gene transfer is attainable with 1500 μg/ml DNA combined with 500 μg/ml lipid. For G5 EDA Starburst PAMAM polyamidoamine dendrimers (Aldrich, Milwaukee, WI) the effective DNA : dendrimer charge ratio for transfecting mouse hearts is around 1:10.

Vector Delivery and Organ Harvest. Donor animals are anesthetized, and the heart is arrested by injecting 0.5 ml of cold LR or UW solution into the inferior vena cava. The aortic arch is ligated proximal to the right subclavian artery. To expel intracoronary blood, the aortic root is flushed with 0.1 ml of preservation solution. The vector is dissolved in the actual cold preservation buffer, and 250 μl is injected into the aortic root. The heart is removed from the chest and placed into a bath of corresponding preservation solution for 1 hr, then transplanted heterotopically according to standard procedures without further perfusion or washout.

Transfection Efficiency and Toxicity. Lipid-mediated transfer of plasmid DNA or OD into murine hearts via intracoronary infusion results in efficient and diffuse transfection 3 days after surgery lasting for 14 days. Under optimal conditions, 50–70% of myocytes are transduced. Liposome-induced myocardial transduction is significantly less efficient compared with viral vectors. DNA/dendrimer complexes offer a very efficient transfection modality for heart grafts.

Transgene expression and protein production are stable between 7 and 14 days after transplant and decline after 4 weeks, rendering this method suitable for transferring immunosuppressive genes. Dendrimers appear to be more effective compared with liposomes using the same DNA concentration. Increased DNA : dendrimer ratios and higher dendrimer concentration, however, may be cardiotoxic by causing particle aggregation and thrombosis.

[13] Generation and Growth of Gutted Adenoviral Vectors

By Dennis Hartigan-O'Connor, Catherine Barjot, Giovanni Salvatori, and Jeffrey S. Chamberlain

Introduction

Gene transfer vectors derived from adenovirus (Ad) have been the subject of enormous study in recent years as they provide highly efficient gene transfer to many tissues. Ad vectors can be rapidly manipulated *in vitro,* have a moderately large cloning capacity, and can be grown to extremely high titers. However, conventional Ad vectors contain a large number of viral genes that contribute to immunological reactions by the host against infected tissues. These immunological problems, together with a need for even greater cloning capacities, have spurred the development of a new class of vectors known as gutted adenoviruses. Gutted Ad vectors are also known as "helper-dependent," "gutless," or "high-capacity" Ad vectors. Gutted vectors carry fewer viral sequences than conventional Ad vectors; usually, all viral coding sequences are removed. As a consequence, gutted vectors (1) have a greater carrying capacity for foreign DNA; (2) do not direct expression of viral proteins, which may be advantageous for gene therapy; and (3) are more difficult to propagate and purify than conventional Ad vectors.

Conventional Ad vectors are prepared by replacement of viral coding regions with exogenous DNA; these vectors must be grown in a complementing cell line, which supplies the missing gene product(s). This strategy is effective when the deleted viral genes code for nontoxic proteins that are normally expressed at a relatively low level. Deletion of many genes from conventional Ad vectors has not been possible because expression of the missing gene products at sufficiently high levels has been toxic to the packaging cell lines. Since gutted vectors lack all viral coding regions, their deficiencies must be complemented using a helper virus, rather than a stable cell line.[1-3] The helper virus replicates in tandem with the gutted vector, supplying viral proteins in *trans* for replication and packaging of gutted vector genomes.

The requirement for parallel growth of gutted and helper virus introduces complexity not encountered in preparation of conventional Ad vectors. Contamination of gutted vector stocks with helper virus is an obvious concern, but robust replication of the helper virus is required for efficient production of viral particles. In addition, gutted vectors are true defective viruses, whose presence in a cell inhibits the replication of helper virus. Efficient gutted vector production requires that growth of gutted and helper viruses be balanced. At the final stage of production, producer cells should be infected with an optimal number of gutted and helper viral particles. When this is achieved, high-titer production of gutted vectors results.

General Principles

Gutted vectors are first prepared as plasmids. Conversion of the plasmid into purified gutted virus is accomplished in several steps (Fig. 1): (a) "rescue" of the gutted viral genome from plasmid form into linear viral form, which is covalently linked to terminal protein; (2) amplification of the rescued vector by serial passage; (3) removal of helper; virus; and (4) further purification away from cellular components and helper virus. Of these steps, amplification of rescued virus is subject to the greatest variability. During this stage, a very small amount of gutted virus must be expanded considerably, until it becomes the predominant virus produced during cell lysis. Two technological advances have facilitated the amplification of gutted virus. First, it is now possible to convert plasmid vectors to gutted virus with very high efficiency, which reduces the number of passages and the degree of expansion required. Second, use of Cre-expressing cell lines to remove helper virus now allows the experimenter to control the ratio between gutted and helper virus, which can be very important to obtain high yields of gutted viruses that do not grow well (Fig. 2).[4,5]

[1] K. J. Fisher, H. Choi, J. Burda, S. J. Chen, and J. M. Wilson, *Virology* **217,** 11 (1996).

[2] S. Kochanek, P. R. Clemens, K. Mitani, H. H. Chen, S. Chan, and C. T. Caskey, *Proc. Nat. Acad. Sci. U.S.A.* **93,** 5731 (1996).

[3] R. Kumar-Singh and J. S. Chamberlain, *Hum. Mol. Genet.* **5,** 913 (1996).

[4] S. Hardy, M. Kitamura, T. Harris-Stansil, Y. Dai, and M. L. Phipps, *J. Virol.* **71,** 1842 (1997).

[5] R. J. Parks, L. Chen, M. Anton, U. Sankar, M. A. Rudnicki, and F. L. Graham, *Proc. Nat. Acad. Sci. U.S.A.* **93,** 13565 (1996).

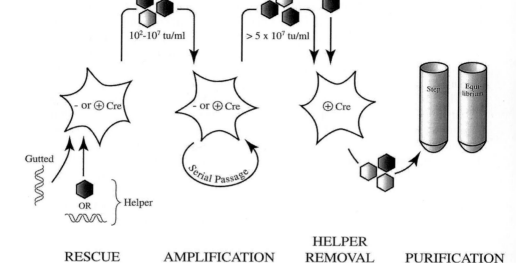

RESCUE   AMPLIFICATION   HELPER REMOVAL   PURIFICATION

FIG. 1. Preparation of gutted vectors can be divided into rescue, amplification, helper removal, and purification stages.

Gutted Vector Design

Gutted vectors are designed to act as defective adenovirus genomes[6]; therefore, they must contain the critical *cis*-acting elements of all Ad genomes and they must be of an appropriate size. It is questionable whether all the *cis*-acting elements that contribute to efficient genome replication have been identified; however, the minimal requirements are known. Exogenous, nonviral sequences in the vector can also influence its replication; unfortunately, the impact of a given sequence must be determined empirically. Generally, the following factors must be considered in gutted vector design: origin of replication, packaging signals, size, reporter genes, and other sequence elements.

Origin of Replication. The natural adenovirus origin consists of sequences from the inverted terminal repeat together with terminal protein, to which the DNA strand is covalently attached. The plasmid form of a gutted virus does not contain bound terminal protein, and as a result initiation of replication is inefficient. However, the sequence context of the Ad origin can greatly improve replication of these plasmid origins. The more functional the origin is, the more frequently replication will be initiated by the viral machinery and the more gutted virus will be produced. Two types of functional Ad origins lacking terminal protein are known (Fig. 3).

---

[6] S. Mak, *J. Virol.* **7,** 426 (1971).

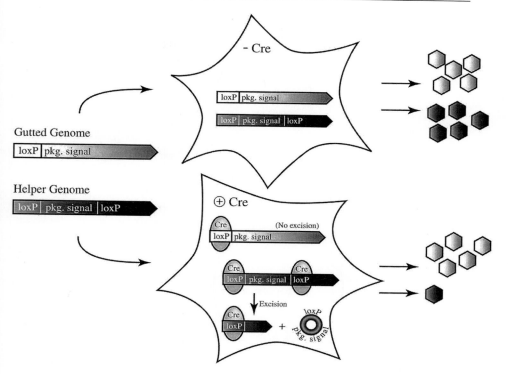

FIG. 2. The Cre/loxP system for selection against helper virus packaging. Left: The packaging signal of the helper virus is flanked by loxP sites, each with the same orientation. The gutted viral genome contains a single loxP site so that recombination between gutted and helper within the packaging signal does not yield an unselectable helper. Center, top: In cells that do not express Cre, both viruses are packaged. Center, bottom: In cells that express Cre, the gutted virus is packaged, because its packaging signal remains intact. The helper virus packaging signal is excised, however, so helper genomes cannot be packaged. As a result, the number of viral particles that contain helper genomes is reduced.

First, ITR sequences are functional in the absence of terminal protein when they are found near the terminus of a DNA strand.[7,8] One design strategy therefore incorporates two ITRs, separated from each other by vector sequences and flanked on each side by an infrequently cutting restriction enzyme, such as *Pac*I or *Fse*I (Fig. 3, bottom). The gutted viral plasmid is digested to release the ITRs, which can then serve as functional origins after cotransfection into packaging cells with helper viral DNA. The advantages of this design, as compared to the "fused" design

[7] F. Tamanoi and B. W. Stillman, *Proc. Nat. Acad. Sci. U.S.A.* **79**, 2221 (1982).
[8] B. G. van Bergen, P. A. van der Ley, W. van Driel, A. D. van Mansfeld, and P. C. van der Vliet, *Nucl. Acids Res.* **11**, 1975 (1983).

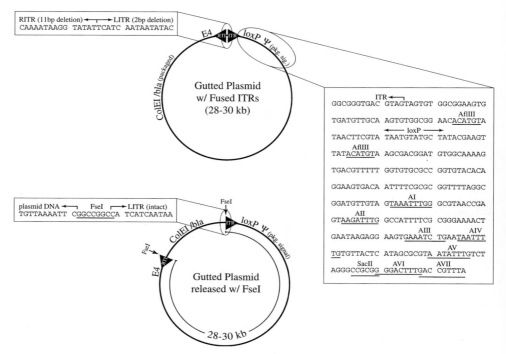

RITR (11bp deletion) ←—┬—→ LITR (2bp deletion)
CAAAATAAGG TATATTCATC AATAATATAC

Gutted Plasmid
w/ Fused ITRs
(28-30 kb)

ColEI/bla (packaged)

E4 RITR-LITR loxP Ψ (pkg. site)

ITR ←—┐
GGCGGGTGAC GTAGTAGTGT GGCGGAAGTG
                              AflIII
TGATGTTGCA AGTGTGGCGG AACACATGTA
         ←— loxP —→
TAACTTCGTA TAATGTATGC TATACGAAGT
AflIII
TATACATGTA AGCGACGGAT GTGGCAAAAG

TGACGTTTTT GGTGTGCGCC GGTGTACACA

GGAAGTGACA ATTTTCGCGC GGTTTTAGGC
                    AI
GGATGTTGTA GTAAATTTGG GCGTAACCGA
AII
GTAAGATTTG GCCATTTTCG CGGGAAAACT
                  AIII   AIV
GAATAAGAGG AAGTGAAATC TGAATAATTT
                              AV
TGTGTTACTC ATAGCGCGTA ATATTTGTCT
   SacII   AVI   AVII
AGGGCCGCGG GGACTTTGAC CGTTTA

plasmid DNA ←—┐ FseI ┌—→ LITR (intact)
TGTTAAAATT CGGCCGGCCA TCATCAATAA

FseI

FseI

ColEI/bla

E4 ITR loxP Ψ (pkg. signal)

Gutted Plasmid
released w/ FseI

28-30 kb

FIG. 3. Two designs for gutted vectors. Top: Adenovirus ITRs are active when fused in head-to-head orientation. As shown to the left, one or both ITRs may suffer a small deletion when cloned in this configuration. Replication proceeds outward from the fusion point, so the plasmid replication origin and antibiotic resistance gene will be replicated and packaged. In addition to the ITRs, the Ad packaging signal must be included in the vector. As shown to the right, the packaging signal comprises seven A/T-rich elements; the first five are included in most gutted vectors. A single loxP site is usually included between the left ITR and the packaging signal. Bottom: Ad ITRs are also active when located close to the end of a DNA fragment. In this design, the gutted genome can be released by digestion with EseI.

described below, include a much higher rescue efficiency, easier construction of the vector due to more efficient replication of the plasmid in bacterial cells, and the ability to remove prokaryotic vector sequences before rescue. One disadvantage is the requirement that the gutted vector lack *Pac*I or *Fse*I sites.

Second, ITR sequences also serve as active origins when fused in head-to-head orientation (Fig. 3, top).[9] Since perfect palindromes are often unstable in bacterial cells, at least one of the two ITRs normally suffers a minor deletion at the fusion point. Often, the fused ITRs are subcloned from a plasmid such as pFG140, which contains an origin known to be functional.[3,9] Replication proceeds from the fusion

[9] F. L. Graham, *EMBO J.* **3**, 2917 (1984).

point outward, so in this design the entire plasmid, including the prokaryotic origin of replication and antibiotic resistance gene, is incorporated into the gutted vector. Plasmids constructed in this way contain an imperfect palindrome, so they are difficult to clone into and to propagate in *Escherichia coli*. Other disadvantages of this design include a lower efficiency of rescue and the inclusion of prokaryotic vector sequences in the rescued vector. One advantage is the ability to go directly from supercoiled plasmid to virus production without the need for restriction digestion and DNA purification prior to transfection into packaging cell lines.

Packaging Signals. Cis-acting sequences necessary for Ad packaging are located adjacent to the left ITR. Seven A/T-rich pseudo-repeats, known to be important for packaging, are found between base pairs 240 and 375 of the Ad5 sequence (Fig. 3, right).[10–12] Many Ad vectors in common use, including gutted vectors, contain only the first five pseudo-repeats, truncating the packaging signal at the SacII site found at approximately position 353 of the Ad5 sequence. Deletion of elements VI and VII has no effect on packaging in a wild-type virus background.

Gutted vectors are often grown in conjunction with a "floxed" helper virus, whose packaging signal is flanked by loxP sites, to allow for removal of the packaging signal in Cre-expressing cells. When using the Cre/loxP system, recombination between the packaging signals of the gutted and helper viruses can be deleterious, since this leads to an unselectable helper virus that contains only one loxP site (to the right of the packaging signal).[4] To avoid this problem, it can be helpful to include a single a loxP site in the gutted virus, to the left of the packaging signal, so that the leftmost sequences of the gutted and the helper virus are identical.[4] After recombination within the packaging signal, therefore, the sequence of the floxed helper virus remains unchanged and Cre selection is still effective. One common and convenient location for insertion of the loxP site is position 154 of the Ad5 sequence, in the AflIII site (Fig. 3, right).

Finally, Sanding *et al.*[13] have reported that the E4 promoter region of adenovirus (base pairs 118–400 from the right end of the virus) contains a sequence that facilitates growth of the gutted vector. Since these authors detected no effect on genome replication, they suggest that the sequence may facilitate genome packaging. Therefore, inclusion of 400 base pairs from the right end of the virus, rather than only the right ITR, may be advantageous. Indeed, vectors carrying these sequences have been shown to grow to particularly high titers.[3]

Size. Gutted Vectors usually carry much less DNA than a wild-type adenovirus, which contains about 36kb. As a result, gutted vector particles are less dense than helper particles and can be separated on equilibrium cesium gradients. In

[10] P. Hearing, R. J. Samulski, W. L. Wishart, and T. Shenk, *J. Virol.* **61,** 2555 (1987).
[11] M. Grable and P. Hearing, *J. Virol.* **64,** 2047 (1990).
[12] M. Grable and P. Hearing, *J. Virol.* **66,** 723 (1992).
[13] V. Sandig, R. Youil, A. J. Bett, L. L. Franlin, M. Oshima, D. Maione, F. Wang, M. L. Metzker, R. Savino, and C. T. Caskey, *Proc. Nat. Acad. Sci. U.S.A.* **97,** 1002 (2000).

addition, small genome size should provide an advantage to the gutted virus during successive cycles of replication.

Graham and his co-workers[14,15] demonstrated that Ad genomes between 27.7 kb and 37.6 kb can be packaged and serially passaged with fair stability. Genomes of 25.7 kb and below, or 37.9 kb and above, were unstable. Since a short genome size is advantageous for purification and possibly for replication of gutted vectors, we recommend that vectors be designed to yield a virus of 27–30 kb. Theoretically, gutted vectors could be designed to carry up to 37.6 kb, the approximate packaging limit. In this case, purification of the vector on equilibrium cesium gradients would have to be forgone or a very short helper virus would have to be designed, so as to establish a density difference between the two viruses.

Reporter Genes. Inclusion of a reporter gene in a gutted vector is a great convenience but can be problematic for gene therapy applications. During growth of a gutted virus, some optimization is usually required, and the presence of a reporter gene allows for rapid analysis of different growth conditions. However, it is known that exogenous proteins can trigger a strong immune response in test animals, which is a drawback for *in vivo* experiments.

When possible, we recommend inclusion of an inducible reporter gene cassette, especially during early attempts to grow gutted viruses. In our laboratory, the ecdysone promoter has been used to drive expression of β-galactosidase. This strategy allows easy assay of the gutted virus in crude cell lysates but greatly reduces gene expression *in vivo*. Use of a strong, constituitive promoter is never recommended, since the resulting high protein levels can result in pseudo-transduction, which complicates vector titering when very low levels of gutted virus are present.[16]

For *in vivo* experiments that may be sensitive to an immune response, at least one version of the vector should lack a reporter gene. It may be useful to prepare two versions in parallel, one containing and one lacking the reporter cassette. Growth of the viruses can then be optimized using the reporter-containing vector, but the reporter-negative vector can be used for experiments *in vivo*.

Other Sequence Elements. Because of the size constraints on gutted genome size, stuffer sequences must usually be included in the vector. There are reports that the choice of stuffer sequence can be affect transgene expression, generation of a cytotoxic T-lymphocyte response, the efficiency of DNA replication, and genome stability.[13,17] Unfortunately, there are few rules that allow prediction as to whether a stuffer will affect these parameters in a positive or a negative way. The best results to date have been obtained with mammalian DNA sequences lacking protein coding regions, such as might be found in introns or intergenic regions.

[14] R. J. Parks and F. L. Graham, *J. Virol.* **71**, 3293 (1997).

[15] A. J. Bett, L. Prevec, and F. L. Graham, *J. Virol.* **67**, 5911 (1993).

[16] I. E. Alexander, D. W. Russell, and A. D. Miller, *Hum. Gene Ther.* **8**, 1911 (1997).

[17] R. J. Parks, J. L. Bramson, Y. Wan, C. L. Addison, and F. L. Graham, *J. Virol.* **73**, 8027 (1999).

Some groups have selected against repetetive sequences and for high G/C content, although the importance of the latter criterion is unclear. In our laboratory, we have used long stretches of intronic DNA from within the human dystrophin gene.

Finally, it is very important to be aware of potential overlaps between gutted and helper virus sequences. Ad genomes are known to recombine frequently during replication. Every shared sequence element presents a potential problem, especially when stringent Cre selection is used to select against helper virus packaging. We have observed escape from Cre selection through transfer of the left and of the gutted virus to the *right* end of the helper virus: in one case, this transfer was mediated by homologous recombination at a shared, 200-bp polyadenylation signal.

Rescue of Gutted Virus from Plasmid DNA

The goal of rescue is to prepare for serial amplification by producing as many gutted viral particles as possible, at the highest possible concentration. The higher the number of gutted viral particles produced in rescue, the lower the overall amplification required before purification and use of the virus in experiments. Also, the higher the concentration of gutted viral particles produced, the sooner it becomes possible to infect producer cells with a multiplicity of infection of one (for the gutted virus). We have found that gutted virus is produced most efficiently when every producer cell is infected with at least one gutted viral particle. Rescue of gutted virus at low titer (less than about 3×10^6 tu/ml) results in suboptimal serial passages.

The earliest methods for gutted virus rescue yielded less than 100 transducing units per milliliter of medium (tu/ml).[3,5,14] Efficient production of high-titer stocks from such small starting quantities can be extremely difficult. A variety of improvements have facilitated this process to the point that gutted vectors are now routinely rescued at about 10^7 tu/ml, a 100,000-fold improvement over the original methods.

There are two basic methods for cointroduction of gutted and helper viral DNA into cells, which involve either transfection of helper viral DNA, or infection with helper viruses (Fig. 4). When the gutted viral genome can be excised from a plasmid by restriction enzyme digest, *cotransfection* yields approximately 10 times more gutted virus after rescue than *transfection/infection* (i.e., transfection of gutted vector DNA followed by infection with helper virus). Ideally, helper viral genomic DNA is also released from a plasmid by restriction digestion and the two forms of DNA, identical at their origins of replication, are transfected together (Fig. 4, first bar). Several such plasmid-embedded helper viruses have been made available since introduction of a simple method for their preparation.[18] If such a

[18] C. Chartier, E. Degryse, M. Gantzer, A. Dieterle, A. Pavirani, and M. Mehtali, *J. Virol.* **70**, 4805 (1996).

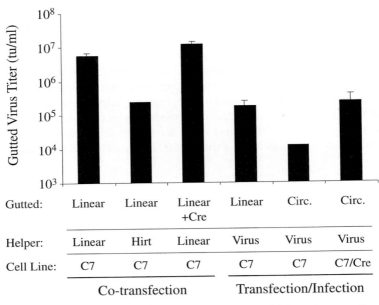

Fig. 4. Rescue of gutted virus from plasmid DNA. Bar height indicates the average titer of gutted virus observed after rescue under the indicated conditions. For example, the first bar indicates the average titer observed after cotransfection of linear gutted viral DNA (released from plasmid by restriction digest) and linear helper viral DNA (also released from plasmid) into C7 cells. The error bars indicate standard error. Linear, genomes released from plasmid by restriction digest; Circ., intact plasmid with fused ITRs; Hirt, protease-treated viral DNA prepared by the Hirt procedure; Virus, intact virus was used.

helper is not available, protease-treated viral DNA may be used, but the titer of gutted virus produced will be substaintially less, probably similar to what could be achieved by transfection/infection. When the gutted and helper DNAs have identical plasmid-derived origins of replication, we find that the yield of gutted virus is greater, presumably because the helper virus initially has no advantage over gutted virus in terms of its affinity for replication factors.[19] Cotransfection of two such substrates into the C7 cell line (discussed below) yields approximately 6×10^6 gutted tu/ml, or around 3 gutted vector transducing units per cell.

The titer of helper virus after lysis is typically around 5×10^8 tu/ml, or 100 times higher than the titer of gutted virus. Therefore, if the first passage is performed at a gutted vector MOI of one, the helper MOI will be less than 100, which is acceptable. As the gutted virus titer rises, the helper MOI moderately decreases. Since it is desirable to remove most of the helper virus before purification, the last two

---

[19] D. Hartigan-O'Connor, A. Amalfitano, and J. S. Chamberlain, *J. Virol.* **73,** 7835 (1999).

passages are normally performed on Cre-expressing cells. For the final passage, helper virus must be added to achieve the desired MOI (5 to 40, depending on the helper virus).

If the helper virus packaging signal is floxed, cells that constitutively express Cre should not be used for cotransfection. The cotransfection protocol introduces helper virus DNA, not live virus, and the amount of Cre expressed by stable cell lines is usually sufficient to prevent accumulation of helper virus to levels necessary for lysis of the cells. One method to increase the yield of gutted virus is to transiently express moderate amounts of Cre recombinase in the packaging cell lines. Transient expression can be achieved by inclusion of a small quantity of a Cre expression plasmid in the initial cotransfection of gutted and helper genomes (see Methods, below). The amount of Cre produced is sufficient to give the gutted virus a packaging advantage as lysis proceeds, but not enough to prevent lysis. As a result, the yield of gutted virus can be approximately doubled, to over 10^7 tu/ml (Fig. 4, third bar).

The *transfection/infection* protocol consists of transfecting gutted viral DNA and subsequently infecting with intact helper virus. For linear gutted viruses, this protocol is inferior. For circular gutted viruses with fused ITRs, when using a floxed helper virus, transfection/infection is preferred to cotransfection. Primarily, this preference is due to the fact that Cre-expressing cells can be used with the transfection/infection protocol (infection with a helper virus leads to such efficient replication and packaging that robust helper function and cell lysis is obtained even in the presence of a stably expressed Cre recombinase). Use of Cre cells both increases the titer of gutted vector recovered and decreases the level of contamination with helper virus. Since the gutted vector titer recovered is low—about 10^5 tu/ml or 100 times lower than can be achieved using linear gutted genomes—the level of helper virus contamination is important (Fig. 4, last bar). On serial passage, the ratio of gutted to helper virus changes slowly. Without Cre cells, many passages would be required before a reasonable proportion of the particles produced contained gutted viral genomes. Using Cre cells at each passage, the amount of helper virus is reduced by at least 100-fold at each step and the proportion of gutted virus is correspondingly increased.

We have found that the cell line used for rescue can influence the recovery of gutted virus. C7 cells, which express adenoviral DNA polymerase and preterminal protein, allow for recovery of at least 5–10 times more gutted virus than unmodified 293 cells, in a wide variety of protocols.[19] The viral replication factors expressed by C7 cells facilitate initiation of DNA replication on suboptimal Ad templates, like those carried by gutted viral plasmids. In cotransfection protocols, when the helper viral DNA also carries an abnormal origin, this feature is especially important because C7 cells also facilitate helper replication, which allows for rapid lysis of the plate, saving time and contributing to efficient rescue. As mentioned above, addition of a small amount of Cre expression plasmid to a cotransfection mixture can increase the yield of gutted virus. This is true when rescue is performed in C7 cells or their derivatives; in 293 cells, addition of Cre expression plasmid tends to prevent lysis and yields are usually depressed.

In general, the cell line used for rescue should robustly support replication of adenovirus after transfection of genomic DNA. We have found that some lines of 293 cells, though capable of supporting the replication of intact virus, do not support the outgrowth of adenovirus from plasmid DNA and are not suitable for gutted virus rescue. One simple method for testing a cell line is to count the number of plaques produced after transfection of plasmid-derived Ad DNA, in comparison to low-passage 293 cells obtained from Microbix, which are known to support robust production of gutted Ad vectors.

Serial Passage of Gutted Virus

The goal of serial passage is to prepare for purification by attaining, at the final passage, optimal conditions for producing large numbers of viral particles containing gutted genomes and minimal numbers containing helper genomes. Maximal production of gutted virus requires that each cell be infected by at least one gutted particle and sufficient helper particles to support its replication, but not by so many helper particles as to overwhelm the Cre recombinase available for negative selection. These conditions can be easily defined in pilot experiments, although the conditions may vary for different combinations of gutted and helper viruses. For example, using purified gutted viral and helper viral particles, we found that a gutted MOI of 5–10 and a gutted : helper ratio of 1 : 5 were optimal. Using a gutted : helper ratio of 1 : 1 produced slow lysis and lower gutted viral titers. This most likely occurs because gutted viruses can act as interfering particles, slowing the replication of helper virus and lysis of the plate. At high gutted : helper ratios, lysis of the plate is prevented.

Note that the optimal gutted : helper ratio can be different when the gutted viral particles are contained in a crude lysate and helper particles are added from purified stocks, as is normally the case during serial passage. Under these conditions, we find that a gutted : helper ratio of 1 : 20 is ideal (Fig. 5). Again, lower ratios lead to slow lysis of the plate and low gutted virus production. It is unclear why the optimal ratio is different under these circumstances; possibly, mixture of purified helper with a crude cell lysate reduces its infectivity. For most gutted/helper combinations, a gutted MOI of 2 and a helper MOI of 40 at the last passage should produce good results.

When the gutted vector is rescued at around 10^7 tu/ml, these conditions are easily achieved (Fig. 6A). The lysate may be passaged at a gutted MOI of 1–2, in the absence of negative selection against helper, until a sufficient volume of lysate is obtained. During these passages, the gutted titer may rise slightly to around 5×10^7 tu/ml. After this expansion, the gutted virus is fairly concentrated but still represents at most 10% of the viral particles in the lysate: the

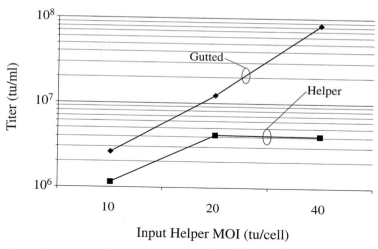

FIG. 5. Sample experiment for determination of the optimal helper MOI for gutted virus production. A gutted MOI of 2.0 tu/cell was achieved by addition of a crude lysate containing gutted virus particles to C7-Cre cells in a 60-mm plate. The lysate contained approximately three times more gutted than helper virus. Additional helper virus was then added from a purified stock to achieve helper MOIs of 10, 20, or 40 tu/cell. When complete CPE was observed, the cells and medium were collected together and frozen at $-80°$. The gutted and helper viruses were titered by colorimetric assay for β-galactosidase and alkaline phosphatase production, respectively.

bulk of the virus is helper. A Cre-expressing cell line can then be used to re-move helper virus. We recommend one or two passages on Cre cells at a gutted MOI of 2. If the gutted : helper ratio is sufficiently high, a single passage on Cre cells may be satisfactory to remove helper. If the gutted : helper ratio is too low, the helper MOI during this first passage may be excessive and negative selec-tion against helper virus suboptimal. In this case, a final passage on Cre cells can be performed after addition of helper virus to achieve a gutted : helper ratio of approximately 1 : 20.

When the vector is rescued at low titer, serial passage can be challenging and the results variable. Gutted vectors vary in the ease with which they can be expanded. In many cases, if the viral lysates are passaged on cells that do not express Cre, the gutted vector titer rises only slowly. It can be difficult to predict how many passages a given vector will take to reach an acceptable titer. If negative selection against helper virus packaging is employed, however, most gutted vectors can be expanded to high titer (Fig. 6B). At each passage, Cre cells remove the majority of the helper virus from the preparation, thereby increasing the gutted : helper ratio. Fresh helper virus is added at every step to assure infection of each cell. When a sufficient volume of a high-titer stock is obtained, the gutted virus can be purified.

Use of Cre cells for amplification of low-titer rescues is effective as long as no recombination occurs that allows the helper virus to escape selection.

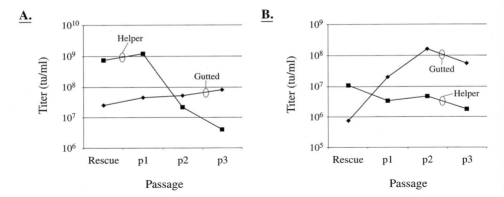

FIG. 6. Serial passage of gutted virus. (A) When the gutted virus is rescued at very high titer (2.5×10^7 tu/ml in this example), the lysate may be passaged without negative selection. The helper virus titer remains high during early passages. To remove helper virus, the last two passages are carried out on Cre-expressing cells. In the final passage, intact helper virus must be added to achieve synchronous lysis of the producer cells. (B) When the gutted virus is rescued at low titer, use of Cre selection ensures a rapid rise in titer. Intact helper virus must be added in every passage. Since this process requires a larger number of passages on Cre-expressing cells, unselectable helper virus is more likely to emerge.

Unfortunately, expansion for a large experiment may require 5–6 cycles of stringent selection. Under these conditions, even byzantine recombinations can occur, leading to generation of helper viruses with a selective advantage. All of the recombinations that we have observed transfer the unselectable packaging signal from the gutted to the helper virus. In some cases, the transfer occurs in a rather unpredictable way resulting in the generation of a defective, but packageable, helper. In other words, not all recombinations result in production of a virus that can grow in the absence of *trans*-complementation. For example, one recombination we observed transferred the left and of the gutted vector to the E3 region of the helper virus, resulting in deletion of the essential E4 region at the right end of the helper. Nevertheless, the recombinant quickly grew to high concentration, presumably because it provides most late proteins, and, therefore, its own capsids, so long as it receives complementation for the missing E4 region from residual intact helper virus. The best strategy for avoiding such unforeseen rearrangements is to avoid any inclusion of similar sequences in both gutted and helper viruses.

Helper Virus Contamination

All preparations of gutted virus contain a small amount of helper virus. The best values reported in the literature are approximately 0.2% after equilibrium gradient centrifugation.[13] It is unclear how great an influence this level of contamination might have on the therapeutic usefulness of gutted vectors.

Some consideration suggests that 0.2% contamination is admirable given the reagents available today. Our laboratory's most efficient Cre-expressing cell line reduces helper virus packaging by approximately 250-fold. During a production run, the maximum gutted virus titers observed are approximately 5-fold lower than those obtained with helper virus in the absence of Cre. Therefore, a gutted : helper ratio of 50 : 1, or 2% contamination, is a good result before purification. After a single step gradient, this level of contamination is reduced; after equilibrium centrifugation, between 0.2 and 1% contamination can be expected. The current techniques might be improved through development of new Cre-expressing cell lines and gutted vectors that grow to higher titers; work of just this kind allowed Sandig *et al.* to obtain levels of helper contamination below 0.2%.[13] Unfortunately, further improvement will probably require the development of a new (or additional) selection strategy.

The level of helper virus contamination can be measured using a variety of different assays. Some assays give numbers that can be used as an internal benchmark for comparison of different preparations, but do not reflect the true level of contamination. For example, when growing a gutted virus carrying a reporter gene with the aid of a helper virus that does not, it is common to measure gutted concentration with a reporter gene assay and helper concentration with a plaque-forming unit (pfu) assay. Because the plaque assay is less sensitive than most reporter gene assays, the amount of helper virus is underestimated by such a procedure and the calculated percent contamination is always impressively low.[20] In addition, as discussed above, the helper virus can rearrange so as to yield particles that carry viral genes and may elicit an immune response but cannot form plaques. In our opinion, the best method for estimation of helper contamination in purified preps is a real-time PCR assay, such as the TaqMan assay, which measures genome numbers directly. Unfortunately, the TaqMan assay is inconvenient for crude lysates, which contain unpackaged viral genomes in addition to viral particles. To use the TaqMan assay for crude lysates, an estimate of the number of intact particles produced per genome must be made. For crude lysates, whenever possible, we favor use of two reporter gene assays, one for the gutted and one for the helper.

Methods

Rescue of Gutted Viruses from Digested Plasmid DNA by Cotransfection

This method is used when the genomes of both the gutted and helper viruses can be released from their plasmid backbones by restriction enzyme digestion. For large-scale production, or when very high yields are required, addition of terminal protein to the linearized gutted viral genomes *in vitro* can be advantageous (Fig. 7). A procedure that accomplishes this modification is at the end of the chapter.

[20] N. Mittereder, K. L. March, and B. C. Trapnell, *J. Virol.* **70,** 7498 (1996).

A.

B.

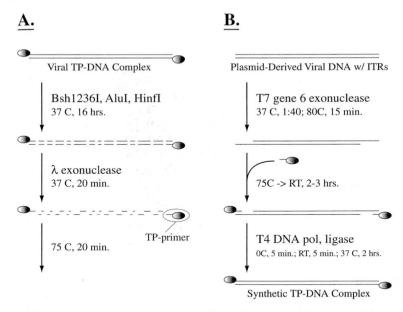

FIG. 7. Protocol for conversion of plasmid-derived Ad origins to a natural structure. (A) Preparation of TP-primer reagent. (B) Conversion of cloned origins to natural form.

1. The day before transfection, plate approximately 10^6 Cre-negative cells (293, C7, or equivalent) in 60 mm plates using 5 ml of medium per plate. At transfection, cells should be approximately 80% confluent. It is critical that the cells not be confluent at the time of transfection; however, if the cells are too sparse the transfection will be toxic and virus yield will be reduced.

2. Digest 5 μg each of gutted and helper plasmid DNA to completion. The plasmids can be digested together in a single reaction if the same enzyme is used for each.

3. Extract once with phenol/chloroform; ethanol precipitate. During these procedures, do not vortex. Vortexing can shear long DNA, such as viral genomes. (Note: If C7 cells are used for the transfection and the helper virus contains a floxed packaging signal, add 15 ng of a CMV-Cre expression plasmid to the DNA mixture before precipitation.)

4. Centrifuge the precipitated plasmids for 15 min at 4°. Remove the ethanol from the pellet under sterile conditions, using a syringe fitted with a 27.5-gauge needle. It is not necessary to wash the pellet with 70% ethanol.

5. Completely resuspend the pellet in 220 μl of 0.1× TE, pH 8 (10 mM tris-HCL, 0.1 mM EDTA). Resuspension can be difficult but is important for successful transfection. Do not vortex instead, pipette up and down using a 1-ml tip that has been truncated with a razor blade, to increase the size of the opening. Incubation for 10–15 min at 37° may be helpful.

6. Add 250 μl of 2× HEPES-buffered Saline (HBS), prepared and optimized for transfection efficiency as described by Sambrook *et al.*[21] Mix gently by pipetting with a truncated tip. Commercial preparations of 2× HBS often yield poor results.

7. Slowly add 31 μl of 2 M CaCl$_2$, mixing gently and continuously. Allow the solution to stand for 20 min at room temperature.

8. Add the transfection mixture dropwise to the cell culture medium. Gently mix by sliding the plate back and forth.

9. Add 5.5 μl of 100 mM chloroquine and gently mix. Return the cells to the incubator for 4.5 hr. (Note: These conditions are appropriate for 293 cells and their derivatives. Other cell lines may require different conditions for optimal transfection.)

10. Remove the medium from the cells. Add 1.5 ml of 15% glycerol in 1× HBS. Incubate for 40 sec, then remove the glycerol by aspiration and add complete culture medium. Apirate the culture medium, which still contains a small amount of glycerol, and add fresh medium for a second time.

11. Viral cytopathic effect (CPE) will first become evident after approximately 5 days.

12. Harvest the cells when CPE is complete, usually after 8 to 12 days. CPE can be considered complete when virtually all cells are rounded and most cells are detached from the bottom of the plate. To harvest, simply collect all the medium and cells together and freeze at −70°. Be sure to collect all cells; some may need to be dislodged from the substrate using a stream of culture medium.

13. If the gutted virus contains a reporter gene, measure its concentration in the lysate by assaying the reporter gene following infection of an appropriate indicator cell line. If the virus does not carry a reporter, estimate its concentration, to provide a starting point for serial passage, using Table I. The results of a real-time PCR assay are not reliable at this stage, since the lysate is contaminated with plasmid DNA.

Rescue of Gutted Viruses from Circular Plasmids by Transfection/Infection

This method is used to rescue gutted virus from a plasmid that has fused ITRs in a head-to-head orientation. The procedure requires a helper virus with a floxed packaging signal and a Cre-expressing cell line.

1. Plate Cre-expressing cells for transfection as described in step one of the previous method.

2. Dilute 9 μg of your gutted viral plasmid into 220 μl of 0.1× TE, pH 8. Ethanol precipitation is not necessary. (Note: Plasmids containing fused ITRs can

---

[21] J. Sambrook, E. F. Fritsch, and T. Maniatis (eds.), "Molecular Cloning: A Laboratory Manual." Cold Spring Harbor Laboratory Press, New York, 1989.

TABLE I
EXPECTED TITERS AFTER RESCUE OF GUTTED VIRUS FROM
DIGESTED PLASMID DNA BY COTRANSFECTION

Cell line	Cre plasmid?	Average titer
LP293	No	1.1×10^6
C7	No	5.6×10^6
C7	Yes	1.2×10^7

be difficult to grow and purify. We recommend the following procedure. Electroporate bacteria, choose a single fresh colony, and inoculate into 500 ml of 2 × YT broth (16 mg/mL bactotryptone, 10 mg/mL yeast extract, 5 mg/mL NaCL) containing antibiotic. 20 hr later, harvest the cells and purify DNA using Qiagen's protocol for very low copy plasmids. Use 35 ml of buffers P1, P2, and P3; purify on a tip 500.)

3. Transfect as described in steps 6 through 10 of the previous method.

4. Five to 16 hours later, add 5 transducing units of helper virus per transfected cell. (Note: It is critical that the helper virus not be contaminated with any unselectable virus. Any such virus will outgrow the selectable helper quickly. If possible, the helper virus should be derived from a plasmid clone and purified using only disposable supplies. For example, the purification method of Gerard and Meidell[22] avoids the use of a sonicator or homogenizer.)

5. When CPE is complete, usually after 2 to 3 days, harvest as described in step 12 of the previous method.

6. Measurement of the gutted virus concentration is not necessary at this stage. Typically, rescue of a circular gutted virus in low-passage 293-Cre cells yields fewer than 10^3 tu/ml; rescue in C7 cells yields an average of about 10^5 tu/ml.

Serial Passage of Gutted Viruses Rescued at Low Titer

When a gutted virus is rescued at low titer, it is not possible to infect all producer cells during every passage. Those cells that can be infected produce a variable amount of gutted virus; many other cells lyse more quickly and produce only helper virus. Depending on how successfully the gutted virus competes for replication and packaging factors, the proportion of virus that contains gutted genomes may or may not increase. Some gutted viruses, including the first published examples, compete very effectively and grow to a high titer without selection against helper virus. Others do not compete efficiently and may never rise to a concentration that is acceptable for purification. Propagation of both types of virus is greatly

[22] R. D. Gerard and R. S. Meidell, *in* "DNA Cloning: A Practical Approach." IRL Press, New York, (1995).

aided by selection against the packaging of helper virus. Selection ensures that the proportion of virus containing gutted genomes rises at every passage. In this way, each successive lysate contains a higher concentration of gutted virus. Eventually, the concentration reaches a level that allows uniform infection of producer cells in the final passage.

1. Estimate the gutted virus titer in the crude rescue lysate. This can be accomplished through direct measurement of reporter gene transduction (after one rapid freeze/thaw cycle to release intracellular virions) or an estimate based on previous experience. TaqMan assays are not useful at this stage, since transfected DNA, as well as viral DNA, will be detected.

2. Determine the volume of crude lysate that would be required to infect new plates of producer cells at a gutted MOI of 1.0. If the volume is greater than about 2 ml per 10 cm petri dish, proceed to step 3. If the volume is less than 2 ml per 10 cm dish, proceed to the next protocol for serial passage of high titer lysates.

3. Add 0.2 volumes of crude rescue lysate to the culture medium of Cre-expressing producer cells and then quickly rock back and forth. For example, add 2 ml of crude lysate to 10 ml of medium over the cells in a 10-cm dish.

4. If the rescue was carried out by transfection/infection in Cre-expressing cells, add 10 transducing units of helper virus per cell. Collect the cells after complete CPE has appeared, which usually takes 3 days. (Note: It is critical that the helper virus not be contaminated with any unselectable virus. Any such virus will outgrow the selectable helper quickly. Contamination is likely to come from impure vector stocks or tools used in purification that are not well cleaned. If possible, the helper virus should be derived from a plasmid clone and purified using only disposable supplies. For example, the purification method of Gerard and Meidell[22] avoids the use of a sonicator or homogenizer.)

5. Titer the gutted virus in the new lysate. It is important to determine the correct titer at each passage so that you will know when the concentration is high enough to warrant purification. For a given combination of gutted and helper virus, the procedure is reproducible enough to allow estimating the correct titer from several previous runs. Otherwise, titering may be accomplished using a transduction assay or the TaqMan assay. The TaqMan assay measures total numbers of genomes, rather than only packaged genomes, so the genome concentration must be adjusted downward to yield an estimate of gutted virus titer. The adjustment factor may vary from virus to virus. In our experience, the transducing unit concentration is approximately 20-fold lower than the genome concentration.

6. Determine the volume of crude lysate that will be required to infect new plates of producer cells at a gutted virus MOI of 1.0. If this volume is more than can be added to a plate of producer cells, than use 0.1 volumes of crude lysate per plate. Otherwise, infect at a gutted MOI of 1.0. (Note: If this is the final passage, it may be advantageous to infect at a gutted MOI of 2.0, depending on whether the optimal

conditions for a production run have been defined. Note that determination of the optimal conditions for production is a simple experiment: infect single plates of producer cells at a gutted MOI of 1–2, and helper virus at varying concentrations, and measure gutted virus production by titer or TaqMan assay.)

7. Add 10 transducing units of helper virus per cell. Collect the cells after complete CPE has appeared, which usually takes 3–4 days. Take note of the time required for complete lysis. If more than 4 days are required, repeat the passage using more helper virus.

8. Titer the gutted virus in the new lysate.

9. Repeat steps 6–8 until a sufficient number of producer cells have been infected at a gutted MOI of 1.0. At that point, the lysate should contain large amounts of gutted virus and relatively small amounts of helper virus. Purify the gutted virus from the crude lysate as described below.

Serial Passage of Gutted Viruses Rescued at High Titer

When producer cells can be infected with gutted virus at an MOI of at least 1.0, there is little danger of a large drop in titer during a given passage. Since all cells will contain gutted viral genomes, the helper virus has no safe haven in which to replicate without interference by the gutted virus. As a result, gutted titers tend to remain stable or rise slowly to a plateau. In these circumstances, crude lysates can be passaged without selection against the helper virus. By reserving Cre selection until the final passages, the chance that an unselectable helper virus will emerge is minimized. Since high concentrations of crude cell lysates can be toxic to producer cells, the gutted virus must be fairly concentrated if an MOI of 1.0 is to be achieved. We have found that 3×10^6 tu/ml is a practical minimum rescue titer for passage in the absence of Cre selection.

1. Estimate the gutted virus titer in the crude rescue lysate. This can be accomplished through direct measurement of reporter gene transduction (after freeze/thaw) or an estimate based on previous experience. TaqMan assays are not useful at this stage, since transfected DNA, as well as viral DNA, will be detected.

2. Determine the volume of crude lysate that will be required to infect new plates of producer cells at a gutted virus MOI of 1.0. You should now be able to calculate the approximate factor by which the number of infected cells will increase at each passage. With this information, decide how many passages to carry out. (Note: We typically perform a viral purification from approximately 100 150-mm plates, or approximately 1.75×10^9 cells. Usually, two passages on Cre-negative cells, followed by two passages on Cre-positive cells, are sufficient for preparations of this size. If the rescue titer is relatively high, such a procedure may yield a sufficient volume of intermediate lysates for many production runs from a single rescue.)

3. Infect new Cre-negative cells at a gutted MOI of 1.0 by adding the appropriate volume of crude lysate to the culture medium, and then quickly rocking the plate back and forth. Collect the cells after complete CPE has appeared, which usually takes 2 days.

4. If possible, titer the gutted virus in the new lysate. Titering may be accomplished using a transduction assay or the TaqMan assay. The TaqMan assay measures total numbers of genomes, rather than only packaged genomes, so the genome concentration must be adjusted downward to yield an estimate of gutted virus titer. The adjustment factor may vary from virus to virus. In our experience, the transducing unit concentration is approximately 20-fold lower than the genome concentration.

5. Repeat steps 3 and 4 until you have carried out all but the last two passages. For the penultimate passage, infect Cre-expressing cells at a gutted virus MOI of 1.0. Collect the cells after complete CPE has appeared, which usually takes 2 days.

6. If possible, titer the gutted virus in the new lysate. Note that the titer of gutted virus does not spike after a single passage on Cre cells. In other words, the titer of gutted virus may be higher now than at the last passage, but the percentage rise is no higher than could have been expected in Cre-negative cells. The goal of using Cre cells at this stage is removal of helper virus, to allow setting the optimal gutted : helper ratio during the last passage. The gutted titer is usually higher than at rescue. It is not necessary to measure the helper virus titer at this stage.

7. For the final passage, infect Cre-expressing cells at a gutted MOI of 2.0. Add purified helper virus to achieve the optimal helper MOI; in our experience, a helper MOI of 40 often provides excellent results. Note that determination of the optimal helper MOI for each gutted virus is a simple experiment: infect single plates of producer cells at a gutted MOI of 1–2, add helper virus at varying concentrations, and measure gutted virus production by titer or TaqMan assay.

8. Purify gutted virus from the crude lysate as described below.

Purification of Gutted Virus from Lysates

Purification of gutted virus is very similar to purification of conventional Ad vectors, except that additional centrifugation steps increase the purity of the virus.

1. Prepare step gradients according to any standard method. We find that the methods of Graham[23] and Gerard[22] both work well. Before centrifugation of the step gradient, mark the interface well, since the viral band will usually be fainter than is the case with conventional Ad vectors.

[23] F. L. Graham and L. Prevec, in "Gene Transfer and Expression Protocols" (E. J. Murray, ed.), p. 109. The Humana Press Inc., Clifton, NJ, 1991.

2. After centrifugation of the step gradient, the gutted and helper viruses comigrate at the interface. A black background and strong, direct overhead lighting will facilitate visualization of the band. We use a fiber-optic light source in a dark room. Depending on the degree of helper virus contamination, it may be apparent that the viral band is really composed of two closely apposed subbands.

3. Collect the viral band by piercing the side of the tube with an 18.5-gauge needle. Do not attempt to exclude helper virus at this point.

4. Load this band onto an equilibrium CsCl gradient (1.34 g/ml) in a Beckman heat-sealable tube or equivalent.

5. Centrifuge at approximately 320,000g for 12 hr and 73,000g for at least 12 hr. The first spin at high speed sets up the gradient while the second spin shallows it, allowing the two viral bands to move apart. Increased distance between the bands makes it possible to collect one band with a needle and exclude the other.

6. Arrange the tube so that the viral bands are clearly visible. With good lighting, the lower band, containing helper virus, is usually visible. The dominant upper band, containing gutted virus, will found approximately 4 mm higher in the tube. Sometimes a faint third band is visible above the band that contains gutted virus. (Note: If you did not use Cre recombinase to select against helper virus packaging, the helper band, (which is lowest in the tube, will be the most prominent band.)

7. Pierce the top of the sealed tube with a 16-gauge needle. This allows removal of the gutted viral band without creating a vacuum.

8. Attach an 18.5-gauge needle to a syringe; loosen the syringe by working the plunger back and forth a few times. Collect the gutted viral band by piercing the side of the tube above the helper virus band, angling the needle upward until it lies just below the gutted band, and slowly drawing the virus into the syringe.

9. If the helper contamination was low at the start of purification (less than 5%), you may want to dialyze and freeze the virus for storage, depending on the experiments that have been planned. Additional equilibrium gradients increase the purity of the preparation but result in loss of virus. To carry out a second or third round of purification, return to step 4.

10. Dialyze the virus against 20 mM HEPES, pH 7.4, containing 5% sucrose. We use Pierce Slide-a-Lyzer cassettes to minimize loss of volume. Aliquot and freeze quickly in a dry ice/ethanol bath.

TaqMan Assay for Viral Genomes

TaqMan is a convenient method for measurement of DNA copy number. It is the method of choice for measurement of helper virus contamination in gutted virus preparations. TaqMan can also be used to measure the number of viral genomes in crude lysates. The number of genomes in a lysate is much higher than the number of packaged genomes, so TaqMan does not provide a reliable estimate of titer in lysates. A rough estimate of titer may be obtained after determination of the average factor by which the genome count exceeds the titer. In purified preparations, the genome count as determined by TaqMan is usually very close to the particle count as determined by OD_{260} readings.

1. Choose primer pairs and probes for measurement of gutted and helper virus (using the PrimerExpress program from ABI). Be sure that the target sequences are not found in packaging cell lines. 293 cells contain a poorly defined stretch of DNA from the left end of the Ad genome, including the packaging signal. For a gutted virus target, choose a sequence that will be present in many different viruses constructed in the laboratory, perhaps a common promoter or stuffer sequence not found in the producer cell line. (Note: For helper virus quantitation, we use a sequence in the L2 region of the virus defined by the following primers: forward, 5'-CGCAACGAAGCTATGTCCAA-3'; reverse, 5'-GCTTGTAATCCTGCTCT-TCCTTCTT-3'; and probe, 5'-VIC-CAGGTCATCGCGCCGGAGATCTA-TAM RA-3'.)

2. Crude lysates must be diluted at least 1000-fold in order to eliminate quenching of the fluorescent TaqMan signal by tissue culture medium. Concentrated lysates may be diluted 10,000 times and the number of gutted virus genomes will still be within the range of measurement. No special effort to open viral capsids is required before measurement.

3. For assays of purified virus, disrupt the capsids before measurement. Dilute the purified stock fivefold into virion lysis solution (0.1% SDS, 10 mM Tris-Cl pH 7.4, and 1 mM EDTA) and incubate for 10 min at 56°. Perform further dilutions in water before carrying out the TaqMan assay.

Addition of Terminal Protein to Gutted Viral Genomes

We have developed a method for addition of terminal protein to plasmid DNA (Fig. 7). Using this protocol, synthetic terminal protein–DNA complex from the gutted virus can be created and used for rescue. This procedure requires extra time but yields gutted virus at titers as high as 4×10^7 ml^{-1}. Such high titers are an asset when consistent, high-level production is required.

1. Purify terminal protein–DNA complex from the helper virus. We recommend the method of Miyake et al.[24]

2. For rescue from a single 60-mm plate, digest 8 μg of terminal protein–DNA complex for at least 16 hr at 37° with 20 U *Bsh*1236I, 10 U *Alu*I, and 5.5 U *Hinf*I (in New England Biolabs buffer 3). *Bsh*1236I cuts between base pairs

[24] S. Miyake, M. Makimura, Y. Kanegae, S. Harada, Y. Sato, K. Takamori, C. Tokuda, and I. Saito, *Proc. Natl. Acad. Sci. U.S.A.* **93**, 1320 (1996).

73 and 74 of the Ad5 ITR, so this digestion results in terminal protein linked to a 73-bp, double-stranded DNA molecule.

3. Add 20 U of lambda exonuclease (New England Biolabs) and incubate for 20 min at 37°. This enzyme catalyzes the removal of 5′ mononucleotides from duplex DNA. Since the enzyme acts in a 5′ to 3′ direction, strands linked to terminal protein are not degraded; all other strands are degraded until a single-stranded region is reached. The products of this digestion, therefore, include terminal protein linked to 73 unpaired bases of the nontemplate strand of the Ad5 ITR (TP-primer). Inactivate the enzymes by incubation at 75° for 20 min.

4. Digest 4 μg of a plasmid containing gutted viral genomic DNA to release the genome.

5. Carry out very limited digestion with T7 gene 6 exonuclease to expose single-stranded regions at the viral termini. This can be accomplished by adding the exonuclease directly to the restriction digest. Cool the digest on ice. Add 3 units of exonuclease, mix quickly using a wide-bore pipette, and incubate at 37° for 1 min, 40 sec. Inactive the exonuclease by incubation at 80° for 15 min.

6. Add the TP-primer reagent, prepared in step 3, to the exonuclease-treated restriction digest. Raise the temperature of the mixture to 75° and allow the temperature to fall slowly (over 2–3 hr) to room temperature.

7. Extend hybridized TP-primer molecules using T4 DNA polymerase and repair nicks using T4 DNA ligase as follows. The reaction should be carried out in a final volume of 100 μl; use New England Biolabs buffer 2 to adjust the volume as necessary. Add 0.5 mM each dNTP, 1 mM ATP, 10 units T4 polymerase, and 8 Weiss units T4 ligase. Incubate the reaction for 5 min at 0°, 5 min at room temperature, and 2 hr at 37°.

8. Add EDTA to a final concentration of 15 mM and store the reaction on ice.

9. Dialyze the reaction products against 1.06× HBS. This is prepared by mixture of 2× HBS (prepared as described by Sambrook et al.[21]) and 0.1× TE in a 25 : 22 ratio.

10. After dialysis, the volume of the reaction may have changed. Note the new volume. Add 4 μg of terminal protein–DNA complex from the helper virus and 0.9 μg of CMV-Cre expression plasmid. Adjust the volume to 470 μl using 1.06× HBS. Slowly add 31 μl of 2 M CaCl$_2$, mixing gently and continuously as you do so. Allow the solution to stand for 20 min at room temperature.

11. This mixture can be transfected into producer cells as described in steps 8 to 13 of the first method in this chapter.

[14] Adenovirus-Mediated Gene Transfer for Cardiovascular and Renal Diseases

By JULIE CHAO, CINDY WANG, and LEE CHAO

Introduction

Tissue kallikrein (E.C. 3.4.21.35) belongs to a subgroup of serine proteinases. It processes kininogen substrates and releases vasoactive kinin peptides.[1] Kinins activate second messengers and mediate a broad spectrum of biological effects at target organs via binding to the bradykinin B_1 and B_2 receptor.[2] Reduced urinary kallikrein excretion has been reported in hypertensive animal models and essential hypertensive patients.[3,4] A study aimed at identifying the genetic factors associated with cardiovascular risks using highly informative family pedigrees suggested that a dominant gene expressed as renal or urinary kallikrein may be associated with a reduced risk of hypertension.[5] The finding suggests that high urinary kallikrein may have a protective effect against the development of high blood pressure. Oral administration of purified pig pancreatic kallikrein has been shown to temporarily reduce blood pressure of hypertensive patients.[6,7] In our studies, we have shown that transgenic mice overexpressing human tissue kallikrein are hypotensive throughout their life span.[8,9] These findings suggest that genetic and/or pharmacological manipulation of the vasodilating kallikrein–kinin system could be beneficial for the treatment of high blood pressure and cardiovascular and renal diseases.

Methods

Adenovirus Preparation

Adenovirus can infect a wide variety of cell types and tissues, including both dividing and nondividing cells. Recombinant adenoviruses (serotype 5, $\Delta E1 \Delta E3$) are constructed with a simplified approach as follows[10]:

[1] K. D. Bhoola, C. D. Figueroa, and K. Worthy, *Pharmacol. Rev.* **44**, 1 (1992).

[2] D. Regoli, N. E. Rhaleb, G. Drapeau, and D. Dion, *J. Cardiovasc. Pharmacol.* **15**, S30 (1990).

[3] S. H. Zinner, H. S. Margolius, B. Rosner, and E. H. Kass, *Circulation* **58**, 908 (1978).

[4] S. Favaro, B. Baggio, A. Antonello, A. Zen, G. Cannella, S. Todesco, and A. Borsatti, *Clin. Sci. Mol. Med.* **49**, 69 (1975).

[5] T. D. Berry, S. J. Hasstedt, S. C. Hunt, L. L. Wu, J. B. Smith, K. O. Ash, H. Kuida, and R. R. Williams, *Hypertension* **13**, 3 (1989).

[6] A. Overlack, K. O. Stumpe, C. Ressel, R. Kolloch, W. Zywzok, and F. Kruck, *Klin. Wochenschr.* **58**, 37 (1980).

[7] A. Overlack, K. O. Stumpe, R. Kolloch, C. Ressel, and F. Krueck, *Hypertension* **3**, I18 (1981).

[8] J. Wang, W. Xiong, Z. Yang, T. Davis, M. J. Dewey, J. Chao, and L. Chao, *Hypertension* **23**, 236 (1994).

METHODS IN ENZYMOLOGY, VOL. 346

1. Insert the human tissue kallikrein or β-galactosidase cDNA into multiple cloning site of pAdTrack-CMV shuttle vector. The expression of transgene is under the control of cytomegalovirus promoter/enhancer (CMV) and SV40 poly(A) signal provided by the shuttle vector.

2. Linearize the shuttle vector containing the gene of interest with *Pme*I (100–500 ng) and cotransform it with an adenoviral backbone plasmid pAdEasy-1 (100 ng) by electroporation of electrocompetent *Escherichia coli* BJ5183 cells. Recombinant adenoviral plasmids, pAd.CMV-cHK and pAd.CMV-LacZ, are generated via homologous recombination and selected by kanamycin.

3. Culture human embryonic kidney 293 packaging cells in six-well or 35-mm tissue culture plate until 50–70% confluence. Transfect 293 cells with *Pac*I-linearized pAd.CMV-cHK or pAd.CMV-LacZ with Effectene (Qiagen, Valencia, CA) to generate homogeneous adenoviruses.

4. Scrape off 293 cells with a rubber policeman at 7 to 10 days after transfection. Release adenoviruses from 293 cells by three cycles of freeze in dry ice/ethanol and thaw at 37°.

5. To propagate adenoviruses, infect 293 cells (30 × 150-mm dishes) at 70% confluence with viral lysate at 2–5 plaque-forming unit (pfu)/cell.

6. At 36 to 48 hr after infection, scrape off 293 cells with a rubber policeman. Release adenoviruses into 18 ml phosphate-buffered saline (PBS) by three cycles of freeze (dry ice/ethanol) and thaw (37°). Collect crude viral lysate after centrifugation at 2000 rpm for 15 min in Beckman GS-6R centrifuge.

7. Load the crude viral lysate on top of a discontinuous gradient of cesium chloride (10 ml of 1.43 mg/ml and 10 ml of 1.25 mg/ml in 10 mM Tris-HCl, pH 8.0) in a SW-28 centrifuge tube. After centrifugation at 20,000 rpm for 2 hr at 5°, adenoviruses (lower band) are separated from defective viral particles (upper band). Remove the upper band and collect adenoviruses by puncturing the wall of the centrifuge tube with a 23-gauge needle attached to a 5 ml syringe.

8. Dilute the adenovirus samples by adding equal volume of 10 mM Tris-HCl, pH 8.0, and load it on top of a discontinuous gradient of cesium chloride (5 ml of 1.43 mg/ml and 5 ml of 1.25 mg/ml) in a SW-41 centrifuge tube. After centrifugation at 20,500 rpm for 18 hr at 5°, collect adenoviruses as described in step 7.

9. Dialyze the adenovirus samples against prechilled 10 mM HEPES, pH 7.2, 140 mM NaCl, 1 mM MgCl$_2$ at 5° for 6 hr by using preframed dialysis membranes (Pierce, Rockford, IL).

---

[9] Q. Song, J. Chao, and L. Chao, *Clin. Exp. Hypertens.* **18**, 975 (1996).

[10] T.-C. He, S. Zhou, L. T. da Costa, J. Yu, K. W. Kinzler, and B. Vogelstein, *Proc. Natl. Acad. Sci. U.S.A* **95**, 2509 (1998).

10. Determine the concentration of the viral particles by measuring OD at 260 and 280 nm.[11]

$$\text{Adenovirus (particles/ml)} = OD_{260\,nm} \times \text{dilution fold} \times (1 \times 10^{12})$$

11. Dilute adenovirus to 5×10^{12} particles/ml with storage buffer containing 2 mM Tris-HCl, pH 8.0, 20 mM NaCl, 0.02% BSA, 10% glycerol. Make 0.5 to 1 ml aliquots and store at $-80°$.

The adenovirus preparation system is commercially available from Qbiogene (AdEasy system) and Stratagene (AdEasy Adenoviral System).

Adenovirus-Mediated Human Tissue Kallikrein Gene Transfer

Systemic Gene Delivery

Intravenous injection. Adenoviruses containing the human tissue kallikrein (Ad.CMV-cHK) or β-galactosidase gene (Ad.CMV-LacZ) are administered into various animal models. Rats are warmed in a 38° incubator for 15 to 20 min prior to injection. Adenoviruses are injected *via* lateral tail vein at a dosage of $1.2–4.0 \times 10^{10}$ pfu per rat using a 27-gauge needle.

Tissue-Targeted Gene Delivery

Intramuscular gene delivery. Spontaneously hypertensive rats (SHRs) are anesthetized by intraperitoneal injection of pentobarbital (50 mg/kg) and an incision is made to expose quadriceps. Adenoviruses carrying the human tissue kallikrein gene (2×10^{10} pfu) are injected into quadriceps at 20 sites with a 27-gauge needle. The depth of the injection is 2–3 mm. Control rats receive Ad. CMV-LacZ at the same dose.

Carotid artery gene delivery. Balloon angioplasty is performed on male Sprague-Dawley rats (400–450 g) with a 2F embolectomy balloon catheter (Baxter Health). Blood flow is temporarily stopped by placing artery clips on the left common and internal carotid artery. A small incision is made using microscissors on the left external carotid artery, only large enough to allow the catheter to be forced through. A balloon catheter is inserted into the common carotid artery via the external carotid artery with the help of a catheter introducer (Harvard Apparatus). The balloon catheter is inflated with saline and pulled back three times. After balloon injury, the injured segment is rinsed with sterile saline. The adenoviral particles of Ad.CMV-cHK or Ad.CMV-LacZ (2×10^9 pfu in 20 μl) are infused and incubated for 15 min.[12] After incubation, the cannula is removed and blood flow to the common carotid artery is restored.

---

[11] N. Mittereder, K. L. March, and B. C. Trapnell, *J. Virol.* **70**, 7498 (1996).

[12] H. Murakami, K. Yayama, R. Q. Miao, C. Wang, L. Chao, and J. Chao, *Hypertension* **34**, 164 (1999).

Intracerebroventricular (ICV) gene delivery. Rats are anesthetized by intraperitoneal injection of pentobarbital (50 mg/kg). A 25-gauge stainless steel cannula fitted into a 3×4 mm membrane-valve plastic block (Umberto Danuso, Milan, Italy) is implanted sterotaxically into the left lateral cerebral ventricle (1.5 mm lateral, 1.0 mm posterior to the bregma, and 4.5 mm into the skull surface).[13] The plastic block is anchored to the skull with screws embedded in dental acrylic cement. ICV cannula is flushed with 5 μl of saline at 24 hr after surgery. Adenoviruses Ad.CMV-cHK (1×10^9 pfu in 10 μl) are injected via ICV cannula into spontaneously hypertensive rats at 2 days after surgery. Control rats receive the same dose of control virus Ad.CMV-LacZ.

Intra-salivary gland gene delivery. Rats are anesthetized by intraperitoneal injection of pentobarbital (50 mg/kg). Salivary glands are exposed by a central, longitudinal cut at the neck. Adenoviruses (Ad.CMV-cHK, 4×10^9 pfu in 200 μl) are introduced into the salivary glands of male Sprague-Dawley rats (280 g) via a direct intracapsular injection using a 30-gauge needle.[14] Control rats receive the same amount of control adenoviral particles (Ad.CMV-LacZ). The incision is closed with skin clips.

Measurement of Cardiomyocyte Diameter

1. Anesthetize rats by intraperitoneal injection of pentobarbital (50 mg/kg).
2. Remove hearts and kidneys, rinse in saline, blot, and weigh.
3. Preserve heart slices in 4% PBS-buffered formaldehyde and embed the tissues in paraffin.
4. Cut 4 μm-thick sections with a microtome (Cut 4055, Olympus) and stain with Gordon and Sweet's silver staining.
5. Calibrate an ocular micrometer with an engraved measuring scale against a stage micrometer, and calculate conversion factor for low ($4 \times$ objective) and high ($45 \times$ objective) magnifications.
6. Measure cardiomyocyte diameters in two perpendicular directions using an ocular micrometer.[15] Cardiomyocytes with nuclear profile are judged to be cut in cross section when the difference of the shortest and longest perpendicular measurements is less than 2 μm. The average of the two measurements is then recorded as the cross-sectional diameter. The mean diameter of 300 cardiomyocytes in each group is measured at a magnification of $450 \times$.

[13] P. Madeddu, N. Glorioso, A. Soro, G. Tonolo, P. Manunta, C. Troffa, M. P. Demontis, M. V. Varoni, and V. Anania, *Hypertension* **15,** 407 (1990).

[14] C. Wang, C. Chao, L. Chao, and J. Chao, *Immunopharmacology* **36,** 221 (1997).

[15] J. Chao, J. J. Zhang, K. F. Lin, and L. Chao, *Hum. Gene Ther.* **9,** 21 (1998).

Gordon and Sweet's Silver Staining

1. Deparaffinize and rehydrate the tissue sections in the following solutions:
   (a) Xylene: 5 min × 2
   (b) 100% ethanol: 1 min × 2
   (c) 95% ethanol: 1 min
   (d) 70% ethanol: 1 min
   (e) 50% ethanol: 1 min
   (f) Water
2. Oxidize in acidified potassium permanganate for 10 min.
3. Rinse three times in distilled water. Tissues should be dark brown.
4. Bleach in 1% oxalic acid for 3 min. Rinse three times in distilled water. Tissues should be opaque now.
5. Incubate in 2.5% ferric ammonium sulfate for 20 min. Rinse quickly three times in double-distilled water. Over-rinse may wash out the mordant.
6. Treat with silver solution for 10 sec, then rinse well five times in distilled water.
7. Reduce in 10% aqueous formalin. Rinse three times in distilled water.
8. Add 1% gold chloride and incubate for 5 min. Rinse three times in distilled water.
9. Treat in 5% sodium thiosulfate for 3 min. Rinse three times in distilled water.
10. If desired, counterstain in Erlich's hematoxylin for 2 min, rinse in tap water, dip in 1% acidic alcohol, rinse in tap water, dip in 1% ammonia alcohol, rinse.
11. Dehydrate and mount as follows:
    (a) 95% ethanol: 5 dips
    (b) 95% ethanol: 10 dips
    (c) 100% ethanol: 10 dips
    (d) 100% ethanol: 10 dips
    (e) Xylene: 10 dips × 2
    (f) Coverslip

Solutions

Acidified potassium permanganate. Mix 95 ml of 0.5% potassium permanganate [0.5% potassium permanganate: dissolve 0.5 g potassium permanganate (Sigma) in 100 ml of dH_2O] with 5 ml 3% sulfuric acid (3% Sulfuric acid: Mix 3.0 g sulfuric acid with 100 ml dH_2O).

1% oxalic acid. Dissolve 1 g oxalic acid (Sigma) in 100 ml distilled water.

2.5% ferric ammonium sulfate (iron alum). Dissolve 2.5 g ferric ammonium sulfate (Sigma) in 100 ml distilled water.

Silver solution

1. To 5 ml of 10.2% silver nitrate [10.2% silver nitrate: dissolve 10.2 g silver nitrate (Sigma) in 100 ml dH$_2$O], add ammonium hydroxide drop by drop until the formed precipitate barely dissolves.

2. Add 5 ml 3.1% sodium hydroxide (3.1% sodium hydroxide: 3.1 g sodium hydroxide in 100 ml dH$_2$O).

3. Add ammonium hydroxide drop by drop until the resulting precipitate barely dissolves (the solution should not be completely clear).

4. Make up the solution to 50 ml with distilled water.

10% Formalin. Dissolve 10 ml formalin (37% formaldehyde) in 90 ml distilled water.

1% Gold chloride. Dissolve 1 g gold chloride in 100 ml distilled water.

5% Sodium thiosulfate. Dissolve 5 g sodium thiosulfate (Sigma) in 100 ml distilled water.

Picosirius Red: Phosphomolybdic Acid Staining

1. De-paraffinize and rehydrate.
2. Incubate with 0.2% phosphomolybdic acid for 5 min.
3. Rinse briefly in distilled water.
4. Stain in picosirius red solution for 90 min.
5. Rinse in 0.01 *N* HCl and change solutions 3 or 4 times.
6. Dehydrate and mount.

Solutions. 0.2% Phosphomolybdic acid (Sigma): dissolve 0.2 g phosphomolybdic acid in 100 ml distilled H$_2$O.

Picosirius red solution

1. Prepare a 1% Sirius red F3BA (direct red 80, Aldrich) aqueous solution.
2. Mix 10 ml of this dye solution with 90 ml of the saturated picric acid (Sigma).
3. Adjust the pH to 2.0 with 10 *N* NaOH.
4. Stand at room temperature for at least 24 hr before use.

Periodic Acid Schiff (PAS) Staining

1. Deparaffinize and rehydrate.
2. Incubate in 1% periodic acid (freshly prepared) for 5 min.
3. Wash 3 times with dH$_2$O.
4. Incubate for 15 min in Schiff reagent (bring Schiff reagent to room temperature prior to use).
5. Wash for 15 min with running tap water.
6. Place slides in hematoxylin for 2–4 min.
7. Wash with running tap water.

8. Dip in 1% acidic alcohol.
9. Rinse with ddH$_2$O.
10. Dip in 1% ammonia alcohol.
11. Rinse with ddH$_2$O.
12. Dehydrate and mount.

Solutions

1% Periodic acid. Dissolve 5 g Periodic acid (Sigma) in 500 ml dH$_2$O.
Ehrlich's hematoxylin
1. Dissolve 10.6 g hematoxylin (Sigma) in 533 ml ethanol (solution 1).
2. Dissolve 16 g aluminum potassium sulfate in 467 ml distilled water, and stir 1 hr or until dissolved (solution 2).
3. Dissolve 1.6 g sodium iodate in 100 ml distilled water (solution 3).
4. Add 533 ml glycerin to solution 2, and stir until well mixed (solution 4).
5. Add solution 1 to solution 2, and stir for a few min (solution 5).
6. Add solution 5 to solution 3, and stir for a few min (solution 6).
7. Add 53 ml glacial acetic acid to solution 6, and stir for 30 min.

1% Acidic alcohol: mix 350 ml 95% ethanol with 145 ml water and 5 ml concentrated hydrochloric acid
1% ammonia alcohol: mix 425 ml 95% ethanol with 70 ml water and 5 ml ammonia hydroxide

Measurement of Glomerular Filtration Rate and Renal Blood Flow

Renal hemodynamics is determined as polyfructosan (Inutest) and *p*-aminohippuric acid (PAH) clearance.[16]

1. Anesthetize rats by intraperitoneal injection of pentobarbital (50 mg/kg). Place the rats on a 37° heating pad.
2. After tracheotomy, cannulate jugular vein for infusion of fluids and drugs. Cannulate the right femoral artery for the measurement of blood pressure and for blood sampling.
3. Cannulate the bladder to allow urine collection.
4. Prepare intravenous infusion mixture containing 2.5% polyfructosan (Inulin; Laevosan, Linz, Austria) and 2% *p*-aminohippuric acid (PAH; Merck, Sharp & Dohme, West Point, PA) in 0.9% NaCl. Infuse via the jugular vein using syringe pump at 1.2 ml/hr for 45 min to reach equilibrium.
5. Collect two timed urine samples (20 min for each collection) from cannulated bladder. Measure urine volume. Collect blood (0.6 ml) from femoral artery immediately after each urine collection. Collect renal vein blood after the

[16] W. R. Fitzgibbon, A. A. Jaffa, R. K. Mayfield, and D. W. Ploth, *Hypertension* **27**, 235 (1996).

second urine collection. Blood samples are collected in Winthrop tubes (Becton Dickinson).

6. Excise kidneys, blot, and weigh at the end of each experiment.

7. Determine inulin and PAH concentrations by modified anthrone and colorimetric methods.

8. Calculate glomerular filtration rate (GFR) and renal plasma flow (RPF) from the clearance of inulin and PAH. Calculate renal blood flow (RBF) from RPF and hematocrit. Normalize clearance data with kidney weight.

$$\text{GFR (ml/min/g kidney wt)} = \frac{U_{inulin}(mg/ml) \cdot UV(ml)}{P_{inulin}(mg/ml) \cdot 20\ min \cdot kidney\ wt\,(g)}$$

$$\text{RBF (ml/min/g kidney wt)} = \frac{RPF(ml/min/g\ kidney\ wt)}{1 - Ht}$$

where

$$\text{RPF (ml/min/g kidney wt)} = \frac{C_{PAH}}{ER}$$

$$C_{PAH} = \frac{U_{PAH}(mg/ml) \cdot UV(ml)}{P_{PAH}(mg/ml) \cdot 20\ min \cdot kidney\ wt\,(g)}$$

$$ER = \frac{P_{PAH}(mg/ml) - V_{PAH}(mg/ml)}{P_{PAH}(mg/ml)}$$

Ht: hematocrit

Measurement of Regional Blood Flow with Fluorescent Microspheres

Fluorescence in the tissues is determined by sedimentation method. The following protocol is adapted from Klemm and Moody[17] and manufacturer's instructions (Molecular Probes Inc., Eugene, OR):

1. Anesthetize rats by intraperitoneal injection of pentobarbital (50 mg/kg).

2. Cannulate the right carotid artery and right femoral artery. Advance polyethylene 50 tubing into the aortic arch and monitor the left ventricular pressure (LVP).

3. Sonicate the microspheres (15 μm) and vortex thoroughly.

4. Inject 0.2 ml microsphere solution ($\sim 2 \times 10^5$ microspheres in 0.9% saline containing 0.02% Tween 20) in 10 sec via carotid artery, followed by 0.5 ml 0.9% saline to flush.

5. Withdraw 1 ml of blood at 0.68 ml/min (Q_r) from the femoral artery using infusion/withdrawal syringe pump.

6. Dissect and weigh tissues (Wt_{Tiss}).

7. Mince and digest the tissues and blood in 10× volumes of 2 M ethanolic KOH containing 0.5% Tween 80 for 48 hr at 56°.

---

[17] K. Klemm and F. G. Moody, *Ann. Surgery* **227**, 126 (1998).

8. Centrifuge the digested mixture at 2000g for 20 min in Beckman GS-6R centrifuge.

9. Remove the supernatant until less than 1 ml of the solution remains.

10. Rinse microspheres in the pellet with 9 ml 0.25% Tween 80 in deionized water.

11. Vortex and repeat steps 9 and 10.

12. Rinse the pellet with 9 ml deionized water.

13. Centrifuge at 2000g for 20 min and remove as much supernatant as possible.

14. Add 3 ml 2-ethoxyethyl acetate (Sigma) to the pelleted microspheres, vortex, and let stand at room temperature for more than 4 hr to ensure complete dissolution of the microsphere and fluorescent dyes.

15. Votex and centrifuge at 2000g for 10 min.

16. Total fluorescence in blood samples (reference fluorescence, $Fluor_r$) and fluorescence in tissues ($Fluor_{Tiss}$) is measured in a luminescence spectrometer (PerkinElmer LS50B) at excitation of 570 nm and emission of 598 nm.

Solutions. For 2 M ethanolic KOH containing 0.5% Tween 80; dissolve 56 g KOH and 2.5 ml Tween 80 in 500 ml 95% ethanol.

Calculations

$$\text{Regional blood flow (ml/min/100g)} = \frac{Fluor_{Tiss}}{Wt_{Tiss}} \times \frac{Q_r}{Fluor_r} \times 100$$

where

$$Q_r = \text{reference blood flow withdrawal rate (m/min)}$$
$$Fluor_r = \text{reference fluorescence}$$
$$Fluor_{Tiss} = \text{fluorescence in tissues}$$
$$Wt_{Tiss} = \text{tissue weight}$$

Measurement of Cardiac Function with Fluorescent Microspheres

Fluorescence is determined according to the following protocol adapted from Klemm and Moody[17] and manufacturer's instruction (Molecular Probes Inc., Eugene, OR):

1. Anesthetize rats by intraperitoneal injection of pentobarbital (50 mg/kg).

2. Cannulate the right carotid artery and right femoral artery. Advance polyethelene 50 tubing into the aortic arch and measure left ventricular end-diastolic pressure (LVEDP).

3. Sonicate the microspheres (15 μm) and vortex thoroughly.

4. Inject 0.2 ml microsphere solution ($\sim$2 $\times$ 10^5 microspheres in 0.9% saline containing 0.02% Tween 20) in 10 sec via carotid artery, followed by 0.5 ml 0.9% saline to flush.

5. Withdraw 1 ml of blood at 0.68 ml/min (Q_r) from femoral artery using infusion/withdrawal syringe pump.

6. Digest blood in 10× volumes of 2 M ethanolic KOH containing 0.5% Tween 80 for 48 hr at 56°.

7. Centrifuge the digested mixture at 2000× g for 20 min in Beckman GS-6R centrifuge.

8. Remove the supernatant until less than 1 ml of solution remains.

9. Rinse microspheres in the pellet with 9 ml 0.25% Tween 80 in deionized water.

10. Vortex and repeat steps 8 and 9.

11. Rinse the pellet with 9 ml of deionized water.

12. Centrifuge at 2000g for 20 min and remove as much supernatant as possible.

13. Add 3 ml 2-ethoxyethyl acetate to the pelleted microspheres, vortex and let stand at room temperature for more than 4 hr to ensure complete dissolution of the microsphere and fluorescent dyes.

14. Votex and centrifuge at 2000g for 10 min.

15. Total injected fluorescence (Fluor$_i$) and fluorescence in blood samples (refe-rence fluorescence, Fluor$_r$) are measured in a luminescence spectrometer (PerkinElmer LS50B) at excitation of 570 nm and emission of 598 nm.

Solution. For 2 M ethanolic KOH containing 0.5% Tween 80; dissolve 56 g KOH and 2.5 ml Tween 80 in 500 ml 95% ethanol.

Calculations

$$\text{Cardiac output (CO)} = Q_r \frac{\text{Fluor}_i}{\text{Fluor}_r}$$

$$\text{Cardiac index (CI)} = \frac{\text{CO}}{\text{body weight}}$$

$$\text{Total peripheral resistance index (TPRI)} = \frac{\text{MAP}}{\text{CI}}$$

where

$$Q_r = \text{reference blood flow withdrawal rate (ml/min)}$$

$$\text{Fluor}_i = \text{total injected fluorescence}$$

$$\text{Fluor}_r = \text{reference fluorescence}$$

$$\text{MAP} = \text{mean arterial pressure}$$

*Terminal Deoxynucleotidyltransferase-Mediated Nick
End Labeling (TUNEL) Assay*

Apoptosis was detected by using *In Situ* Cell Death Detection Kit, POD (Roche Molecular Biochemicals) according to manufacturer's instructions.

1. Deparaffinize and rehydrate.

2. Incubate the tissue section with proteinase K (20 μg/ml in 10 mM Tris-HCl, pH 7.4) for 15 min at 21–37°.

3. Rinse the slides twice with PBS.

4. Incubate the slides with blocking solution (0.3% H_2O_2 in methanol) for 1 hr at room temperature.

5. Rinse slides twice with PBS and dry the area around sample.

6. Add 50 μl TUNEL reaction mixture (containing terminal deoxynucleotidyl-transferase) on sample and cover the slide with Parafilm. Incubate slides in a humidified chamber at 37° for 60 min. Fluorescein is incorporated to free 3'-OH DNA ends.

7. Rinse slides 3 times with PBS. Samples can be analyzed under a fluorescence microscope at this state.

8. Use PBS containing the 3% BSA to block nonspecific binding for anti-fluorescein-POD (horseradish peroxidase).

9. Rinse slides 3 times with PBS.

10. Dry the area around the sample.

11. Add 50 μl Converter-POD (25 μl converter + 25 μl PBS) on the sample and cover the sample with Parafilm. Anti-fluorescein antibody Fab fragments conjugated with POD in this solution bind specifically to the incorporated fluorescein.

12. Incubate the slides in a humidified chamber for 30 min at 37°.

13. Rinse slides 3 times with PBS.

14. Add 150 μl DAB-substrate solution (0.025% DAB and 0.03% H_2O_2 in 50 mM Tris-HCl).

15. Incubate the slides for 5 min at room temperature (monitor the reaction under microscope.)

16. Rinse the slides 2 times with PBS and 3 times with dH_2O.

17. Mount the slides under glass coverslip and analyze under light microscope.

18. Negative control: Incubate fixed and permeabilized sample in 50 μl label solution (without terminal transferase).

19. Positive controls: Incubate fixed and permeabilized samples with DNase I (100 ng/ml DNAase 1 in 50 mM Tris-HCl containing 1 mM $MgCl_2$ and 1 mg/ml BSA) for 10 min at room temperature.

Detection of Superoxide in Frozen Tissues with 3,3'-Diaminobenzidine-Manganese (DAB/Mn) Cytochemical Method

1. Embed the tissues of interest in OCT compound (Fisher Scientific) and freeze them in liquid nitrogen.

2. Cut 5–8 μm sections with a microtome and keep the sections on ice. Store the sections at −20°.

3. Rinse the frozen sections twice in PBS for 5 min each.

    4. Incubate the sections in prewarmed equilibrating media in Coplin staining jar at 37° for 5 min.

    5. Incubate the sections in staining medium for 30 min.

    6. Rinse briefly with equilibrating medium.

    7. Rinse twice with PBS.

    8. Fix the slides in 10% formalin for 5 min. If desired, counterstain 20 seconds in eosin.

    9. Dehydrate in graded alcohols and clear in xylene.

    10. Cover with coverslip.

Equilibrating Medium. HBSS containing 10 mM Hepes and 1 mM sodium azide.

Staining Medium. Dissolve 10 mg 3,3′-diaminobezidine (DAB) in 100 ml equilibrating medium. Add 100 μl 50 mM MnCl$_2$ (0.5 mM final). Adjust the pH to 7.2.

Results

Tissue Distribution of Human Tissue Kallikrein Transgene

    At 3 days post intravenous injection of Ad.CMV-cHK, human tissue kallikrein mRNA can be detected by RT-PCR Southern blot analysis in the liver, kidney, heart, lung, spleen, adrenal gland, and aorta of the injected rats, but not in control rats receiving Ad.CMV-LacZ.[18–22] After intravenous gene transfer immunoreactive human tissue kallikrein is present in the circulation and urine as measured by ELISA.[18–22]

    The expression of human tissue kallikrein and its mRNA can be identified solely in injected muscle after intramuscular gene delivery.[23] Immunoreactive human tissue kallikrein is present in the circulation and urine after intramuscular gene transfer,[23] which provides direct evidence that circulatory kallikrein can be secreted into the urine.

    Following carotid artery gene transfer, human tissue kallikrein mRNA can only be identified in the injured left carotid artery but not in the control right carotid artery, aorta, heart, liver, or kidney by RT-PCR Southern blot analysis.[12] Kininogenase activity in the blood vessels increase 4.2-fold after kallikrein gene transfer as compared to injury plus infection with control adenovirus.[12]

[18] J. Chao, L. Jin, L. M. Chen, V. C. Chen, and L. Chao, *Hum. Gene Ther.* **7,** 901 (1996).

[19] L. Jin, J. J. Zhang, L. Chao, and J. Chao, *Hum. Gene Ther.* **8,** 1753 (1997).

[20] K. Yayama, C. Wang, L. Chao, and J. Chao, *Hypertension* **31,** 1104 (1998).

[21] E. Dobrzynski, H. Yoshida, J. Chao, and L. Chao, *Immunopharmacology* **44,** 57 (1999).

[22] W. C. Wolf, H. Yoshida, J. Agata, L. Chao, and J. Chao, *Kidney Int.* **58,** 730 (2000).

[23] J. J. Zhang, C. Wang, K. F. Lin, L. Chao, and J. Chao, *Clin. Exp. Hypertens.* **21,** 1145 (1999).

The expression of human tissue kallikrein and its mRNA can be detected in the cortex, cerebellum, brain stem, hippocampus, and hypothalamus after ICV injection of Ad.CMV-cHK.[24] Cellular localization of β-galactosidase can be identified in the thalamus, hypothalamus, and third ventricle of rats injected with Ad.CMV-LacZ by X-Gal staining.[24]

A single injection of Ad.CMV-cHK results in a sustained expression of human tissue kallikrein in rat salivary glands.[14] β-Galactosidase activity is localized in the granular convoluted tubular and striated duct cells of rat submandibular gland.[25] Immunoreactive human tissue kallikrein is present in rat sera and saliva after intrasalivary gene delivery. Targeted gene delivery to the salivary gland may provide systemic route to deliver therapeutic proteins in saliva and the systemic circulation.

Expression Level and Time Course

Expression levels of immunoreactive human tissue kallikrein in rats are assessed by ELISA. The highest level of immunoreactive human tissue kallikrein is detected in rat plasma (from 200 to 1000 ng/ml) at 3 days post systemic gene transfer of Ad.CMV-cHK.[18–22] Expression of human kallikrein transgene in rat plasma decreases gradually after day 3 of injection and lasts for 5–6 weeks. Urinary excretion of immunoreactive human kallikrein gradually declines after day 3 of intravenous injection and lasts for 4 weeks. The decline of human kallikrein levels in rats receiving Ad.CMV-cHK is attributed to immunogenicity and inactivation or clearance of the adenovirus by the host. The generation of neutralizing antibody also blocks effective readministration of adenoviral vectors. Immunoreactive human tissue kallikrein cannot be detected in the circulation of control rats injected with Ad.CMV-LacZ, indicating that the rabbit anti-human tissue kallikrein antibody is specific for human kallikrein and does not cross-react with rat tissue kallikrein.

Blood Pressure-Lowering Effect

To explore the potential role of kallikrein gene in cardiovascular and renal function, we deliver adenovirus carrying the human tissue kallikrein cDNA into genetically and experimentally induced hypertensive animal models. These include Dahl salt-sensitive (Dahl-SS), spontaneously hypertensive rats (SHR), deoxycorticosterone acetate (DOCA)-salt, two-kidney, one clip (2K1C) hypertensive rats, and rats with five-sixths reduction of renal mass. A single intravenous injection of Ad.CMV-cHK causes a sustained delay in blood pressure increase from day 2 to 5 weeks post injection, as compared to control rats receiving adenovirus Ad.CMV-LacZ.[18–22] Adenovirus-mediated kallikrein gene delivery has no effect on the blood pressure of normotensive Wistar-Kyoto rats.[19] A maximal blood

[24] C. Wang, C. Chao, P. Madeddu, L. Chao, and J. Chao, *Biochem. Biophys. Res. Commun.* **244,** 449 (1998).
[25] W. Xiong, J. Chao, and L. Chao, *Biochem. Biophys. Res. Commun.* **231,** 494 (1997).

pressure reduction of 30–40 mm Hg is observed in rats receiving kallikrein gene delivery as compared with control rats. These findings show that the expression of human tissue kallikrein via gene delivery attenuates blood pressure increase in hypertensive rats.

A single intramuscular injection of Ad.CMV-cHK causes a significant delay in blood pressure increase for 5 weeks,[23] which shows that a continuous supply of human tissue kallikrein in the circulation is sufficient to reduce blood pressure. A single injection of the human tissue kallikrein gene via ICV cannula causes a rapid and prolonged blood pressure-lowering effect, which suggests that the tissue kallikrein–kinin system may function in the central control of blood pressure homeostasis.

Following Ad.CMV-cHK injection, urine excretion and urinary sodium output significantly increases as compared to the control rats receiving Ad.CMV-LacZ.[18,20] Kallikrein gene delivery causes increases in urinary kinin excretion, which is accompanied by significant increases of urinary cAMP, cGMP, and nitrite/nitrate levels.[18–22] These results indicate that the protective effects of kallikrein gene delivery are mediated via activation of bradykinin receptors through cGMP-NO and cAMP-PGE$_2$ dependent signal transduction pathways.

Cardiovascular Effect

The tissue kallikrein–kinin system is present in the heart. Kinin has been shown to have cardioprotective effects. In Dahl-SS, 2K1C hypertensive rats, and rats with gentamicin-induced nephrotoxicity or five-sixths reduction of renal mass, a single injection of Ad.CMV-cHK can attenuate cardiac hypertrophy and improve cardiac function.[20,22,26,27] Adenovirus-mediated kallikrein gene delivery causes a significant reduction in the left ventricular mass and cardiomyocyte size. Extracellular matrix production was quantified on Sirius red stained sections. Cardiomyocytes stain yellow while collagen stains red with Sirius red staining. Myocardial fibrosis is attenuated in hypertensive rats receiving kallikrein gene delivery.[22,26] In addition, cardiac output and regional blood flow is increased and peripheral vascular resistance is decreased after human kallikrein gene transfer as compared to the control rats receiving Ad.CMV-LacZ.

We also investigated the potential beneficial effects of human kallikrein gene transfer in myocardial ischemia/reperfusion injury.[28] One week after systemic kallikrein gene delivery, rats are subjected to a 30 min coronary occlusion followed by a 2-hour reperfusion. Kallikrein gene delivery causes a significant decrease in the ratio of infarct size to ischemic area at risk and in the incidence of ventricular fibrillation compared to the control rats. Kallikrein gene delivery also attenuates

[26] J. Chao, J. J. Zhang, K. F. Lin, and L. Chao, *Kidney Int.* **54,** 1250 (1998).
[27] H. Murakami, K. Yayama, L. Chao, and J. Chao, *Kidney Int.* **53,** 1305 (1998).
[28] H. Yoshida, J. J. Zhang, L. Chao, and J. Chao, *Hypertension* **35,** 25 (2000).

programmed cell death in the ischemic area as assessed by TUNEL assay. Icatibant, a specific bradykinin B_2 receptor antagonist, abolishes these kallikrein-mediated beneficial effects in myocardial ischemia/reperfusion injury.[28] Superoxide production in the ischemic heart can be determined in forzen tissue sections by using DAB/Mn cytochemical method. These findings show that the expression of human tissue kallikrein via gene delivery exerts protective effects against cardiovascular dysfunction in experimental animal models.

Renal Effect

In Dahl-SS, SHR, DOCA-salt, 2K1C hypertensive rats, gentamicin-induced nephrotoxicity, or five-sixths reduction of renal mass, a single injection of Ad.CMV-cHK attenuates renal lesions and enhances renal function. Kallikrein gene transfer reduces glomerulosclerosis index, tubulointerstitial lesion, lumenal protein cast accumulation, and interstitial inflammation.[20,22,26,27] Total urinary protein and albumin excretion are significantly lower in rats receiving kallikrein gene as compared to control rats receiving Ad.CMV-LacZ. Adenovirus-mediated kallikrein gene delivery significantly increases renal blood flow, glomerular filtration, and urine flow rates, indicating enhanced renal function. In addition, kallikrein gene transfer also significantly reduces cellular necrosis and blood urea nitrogen levels.[22,27] Kallikrein gene delivery not only exhibits protection against renal lesions in various animal models, but also reverses salt-induced renal injury in Dahl-SS rats.[26] Previous studies reported that long-term infusion of purified rat tissue kallikrein *via* a minipump attenuates glomerulosclerosis without affecting the blood pressure of Dahl-SS rats fed on a high-salt diet.[29] Furthermore, the renal protective effect is abolished by a bradykinin B_2 receptor antagonist.[29] Therefore, it is possible that tissue kallikrein exerts a renoprotection, which is independent of its blood pressure-lowering effect. These combined results raise the potential for kallikrein gene therapy to treat renal diseases.

Inhibition of Hyperplasia in Blood Vessels after Balloon Angioplasty

Rat primary VSMC is infected with Ad.CMV-cHK and proliferation of rat primary VSMC is evaluated by measurement of ^{3}H-thymidine incorporation. Expression of human tissue kallikrein causes a significant inhibition in the proliferation of VSMC.[12] Adenovirus-mediated human kallikrein gene delivery causes a significant reduction in intima/media ratio at the balloon-injured rat carotid artery as compared with that of control rats. Administration of the bradykinin B_1 and B_2 receptor antagonists blocks the protective effect of kallikrein and partially reverses the intima/media ratio toward the control ratio.[12,30] Furthermore, kallikrein

[29] Y. Uehara, N. Hirawa, Y. Kawabata, T. Suzuki, N. Ohshima, K. Oka, T. Ikeda, A. Goto, T. Toyo-oka, K. Kizuki, and M. Omata, *Hypertension* **24**, 770 (1994).
[30] J. Agata, R. Q. Miao, K. Yayama, L. Chao, and J. Chao, *Hypertension* **36**, 364 (2000).

gene delivery results in the regeneration of endothelium. In addition, the effect of adenovirus-mediated human tissue kallikrein gene delivery is evaluated in a mouse model of arterial remodeling induced by permanent alteration in sheer stress conditions.[31] Mice undergoing ligature of the left common carotid artery are injected intravenously with 1.8×10^9 pfu of Ad.CMV-cHK or Ad.CMV-LacZ. Human kallikrein gene delivery significantly reduces neointima formation as compared to the control rats. The protective action of Ad.CMV-cHK on neointima formation is significantly reduced in knockout mice lack of the bradykinin B_2 receptor gene as compared to the wild-type control mice. In contrast, the effect of Ad.CMV-cHK is amplified in transgenic mice overexpressing human bradykinin B_2 receptor as compared to the wild-type control mice.[31] Kinin, nitrite/nitrate, cGMP, and cAMP levels in balloon-injured arteries and urinary excretion also significantly increase after kallikrein gene delivery.[12,30,31,32] These results indicate that both B_1 and B_2 receptors contribute to the reduction of neointima formation via the promotion of re-endothelialization and inhibition of VSMC proliferation and migration.

Concluding Remarks

Gene transfer is a novel means of providing sustained and localized delivery of the required therapeutic protein. Currently, viral vectors are the vehicles of choice for gene delivery because of their high transfer efficiency. The most commonly used viral vectors are adenovirus, adeno-associated virus, and retrovirus. Retroviral vector has severe limitations as it can only infect dividing cells. Adeno-associated viral vector is attractive for long-term gene expression but the protocol for the production of high-titer viral stocks is a rate-limiting step. Replication-deficient adenovirus is also an excellent vector for gene delivery because of its high transfection efficiency in infecting cells and its ability to target nondividing cells. We have used the first-generation adenoviral vector to demonstrate that the gene transfer approach can attenuate hypertension and organ damage in experimental animal models. However, long-term benefits of gene transfer cannot be achieved because of the cytopathic effects of the viruses and host immune response. Gutless adenoviral vector devoid of all virus proteins has been developed to remedy these problems.[33] Moreover, the routes of gene delivery could be by local or systemic injection. Systemic delivery is less invasive, but may result in pathological effects in remote tissues, thus producing unwanted side effects such as diabetic retinopathy or tumor growth. To allow gene delivery of therapeutic proteins to the heart

[31] C. Emanueli, M. B. Salis, J. Chao, L. Chao, J. Agata, K. F. Lin, A. Munao, S. Straino, A. Minasi, M. C. Capogrossi, and P. Madeddu, *Arteriosc. Thromb. Vasc. Biol.* **20,** 1459 (2000).

[32] H. Murakami, R. Q. Miao, L. Chao, and J. Chao, *Immunopharmacology* **44,** 137 (1999).

[33] G. Schiedner, N. Morral, R. J. Parks, Y. Wu, S. C. Koopmans, C. Langston, F. L. Graham, A. L. Beaudet, and S. Kochanek, *Nature Genetics* **18,** 180 (1998).

and kidney by intravenous injection, it is essential to develop highly efficient expression vectors carrying heart- or kidney-specific promoters capable of directing specific gene expression to the target organs. Our studies indicate that adenovirus-mediated kallikrein gene transfer attenuates blood pressure increase and cardiac and renal damage in various animal models. These results are crucial for future clinical applications in gene therapy for cardiovascular and renal diseases.

Acknowledgments

This work was supported by the National Institutes of Health Grant HL29397.

[15] Gene Transfer to Blood Vessels Using Adenoviral Vectors

By Yi Chu and Donald D. Heistad

Introduction

Gene transfer to blood vessels is a useful tool for functional genomics and vascular biology and is a promising therapy for vascular diseases. Replication-deficient adenoviruses are widely used as vectors. Advantages of the adenoviral vector are the ability to transduce nondividing as well as dividing cells, relatively high efficiency for transduction, a relatively large cloning capacity to accommodate a variety of transgene inserts, and ease of production of high-concentration viral stocks. Disadvantages include propensity of the virus to elicit a strong host immune response, which leads to transient expression of the transgene: killing of the transduced cells by cell-mediated immunity, elimination of the transgene product by antibody-mediated mechanisms, and nullification of repeated administration of the virus by neutralizing antibodies. Transient expression, however, is an advantage when the risk of the condition being treated, such as vasospasm after subarachnoid hemorrhage,[1] is transient.

Immune responses to adenoviral vectors could be dramatically reduced, if not eliminated, by reduction of dose.[2,3] On the other hand, "second-generation" adenoviral vectors do not reduce inflammatory responses in blood vessels.[3] The

---

[1] K. Toyoda, F. M. Faraci, Y. Watanabe, T. Ueda, J. J. Andresen, Y. Chu, S. Otake, and D. D. Heistad, *Circ. Res.* **87,** 818 (2000).

[2] C. A. Gerdes, M. G. Castro, and P. R. Löwenstein, *Mol. Ther.* **2,** 330 (2000).

[3] S. Wen, D. B. Schneider, R. M. Driscoll, G. Vassalli, A. B. Sassani, and D. A. Dichek, *Arterioscler. Thromb. Vasc. Biol.* **20,** 1452 (2000).

helper-dependent or "gutted" vector (which eliminates all viral sequences except ITRs and the packaging sequence which are required in cis for replication and packaging of the vector) may not reduce the immune response to the primary (initial) load of virus, but may greatly reduce the secondary immune response[4] (also see articles 11 and 13, this volume). We will limit our discussion to the "first-generation" adenoviral vector (E1-deleted and partially E3-deleted) which has been used most widely in gene transfer to blood vessels, for research but not large-scale clinical applications.

Construction of Adenoviral Vectors

The easiest way to obtain an adenoviral vector is to request it from an existing source. Hundreds of adenoviral vectors, containing different genes and promoters, have been constructed and published. Sources for vectors include many vector cores/facilities listed on the Web.

Traditional construction of a vector consists of three steps: (1) cloning of the transgene into a shuttle vector, (2) cotransfection of the shuttle vector and the adenovirus backbone DNA (E1 deleted and E3 partially deleted) to a packaging cell line (most commonly human embryonic kidney 293 cells, which complement viral E1 genes in trans), and (3) identification of positive viral plaques that occur from homologous recombination, using a functional assay for the transgene product. Many improvements in construction have been made to increase the efficiency of ligation between the shuttle vector and viral backbone, and to reduce the possibility of formation of a wild-type virus. For example, the highly efficient Cre/loxP recombination method has been used to replace the less efficient homologous recombination[5]; direct ligation of the transgene to plasmids containing E1-deleted full-length viral DNA has been used to replace homologous recombination[6]; the use of the plasmid form of viral backbone (E1-deleted and partially E3-deleted full-length viral DNA), instead of the backbone that is derived from an intact viral DNA, drastically reduces wild-type virus[7]; and cells expressing E1 genes with limited surrounding adenoviral sequence also helps to reduce generation of wild-type virus.[8]

Materials for construction of a recombinant adenovirus may be obtained commercially (e.g., Clontech, PanVera, and Stratagene) or from vector cores/facilities

---

[4] S. Kochanek, *Hum. Gene Ther.* **10**, 2451 (1999).

[5] P. Ng, R. J. Parks, D. T. Cummings, C. M. Evelegh, U. Sankar, and F. L. Graham, *Hum. Gene Ther.* **10**, 2667 (1999).

[6] T. C. He, S. Zhou, L. T. Da Costa, J. Yu, K. W. Kinzler, and B. Vogelstein, *Proc. Natl. Acad. Sci. U.S.A.* **95**, 2509 (1998).

[7] R. D. Anderson, R. E. Haskell, H. Xia, B. J. Roessler, and B. L. Davidson, *Gene Ther.* **7**, 1034 (2000).

[8] F. J. Fallaux, A. Bout, I. Van Der Velde, D. J. Van Den Wollenberg, K. M. Hehir, J. Keegan, C. Auger, S. J. Cramer, H. Van Ormondt, A. J. Van Der Eb, D. Valerio, and R. C. Hoeben, *Hum. Gene Ther.* **9**, 1909 (1998).

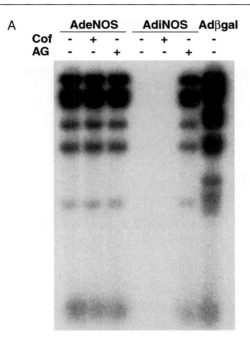

FIG. 1. An inhibitor of inducible nitric oxide synthase (iNOS) is necessary for replication of AdiNOS in 293 cells. (A) Viral DNA replication detected by Southern blotting.[9] Cells were infected with adenoviral vectors (AdeNOS, AdiNOS, and Adβgal) at 10 PFU/cell, in the absence or presence (Cof) of cofactors for NOS (FAD 5 μM, FMN 5 μM, NADPH 10 μM, and tetrahydrobiopterin 50 μM; plus substrate L-arginine 600 μM) or aminoguanidine (AG, 1 mM), an inhibitor of iNOS. Viral DNA was isolated 20 hr postinfection, cut by *Hind*III, electrophoresed, and transferred to a membrane by Southern blotting. The blot was hybridized with a probe of Adβgal DNA labeled by random-primer extension, and exposed to X-ray film. The bands shown are *Hind*III fragments of viral DNA of the three viruses. No viral DNA is detected with AdiNOS in the absence of aminoguanidine, whereas viral DNA replicates after AdiNOS in the presence of aminoguanidine. Adβgal serves as a positive control. (B) Production of nitric oxide measured indirectly as concentration of nitrite in medium 16 hr postinfection. Viral DNA replication appears to be prevented by greater production of NO by AdiNOS when aminoguanidine is absent.

(e.g., University of Iowa Gene Transfer Vector Core). Importantly, if the transgene product is potentially inhibitory to viral replication in 293 cells, an inhibitor of the transgene product may need to be included in the medium. For example, we found that an inhibitor of inducible nitric oxide synthase (iNOS) is necessary for AdiNOS to replicate in 293 cells (Fig. 1).

[9] Y. Chu, K. Sperber, L. Mayer, and M. T. Hsu, *Virology* **188**, 793 (1992).

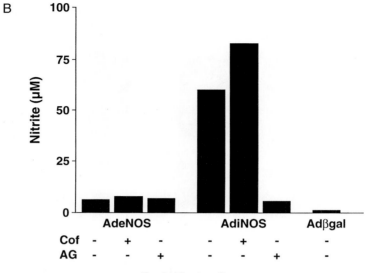

FIG. 1. (*Continued*)

Propagation of Adenoviral Vectors

Adenoviral vectors may be amplified using a standard method.[10] Briefly, 293 cells are seeded on >20 15-cm tissue culture plates, infected at a confluence of 80–90% with the vector at a multiplicity of infection (MOI) of 5 plaque-forming unit (PFU) per cell. Within a day, cytopathic effects are observed (i.e., cells becoming enlarged and rounding up). Cells are then harvested (around 30 hr postinfection), lysed by freezing and thawing, loaded onto a cesium chloride (CsCl) cushion, and centrifuged, and the virus band in the CsCl gradient is harvested. The virus may be further purified using a second round of CsCl centrifugation. Purified virus is dialyzed against a buffer (e.g., 3% sucrose in phosphate-buffered saline, PBS) to remove CsCl. Physical quantification of the virus is usually by measurement of optical density (1 OD/260 nm $= 1 \times 10^{12}$ particles/ml). Biological quantification usually is performed by plaque assay on 293 cells to determine PFU/ml. The ratio of particles to PFU generally ranges from 10 to 100 : 1. Two papers,[11,12] using different assay conditions and mathematical modeling, argued that the ratio between particles and infectious unit (IU) should approach 1 : 1. Thus, it is important for authors to provide a detailed description of methods used to titer the vector.

[10] F. L. Graham and L. Prevec, *Methods Mol. Biol.* **7,** 109 (1991).

[11] N. Mittenreder, K. L. March, and B. C. Trapnell, *J. Virol.* **70,** 7498 (1996).

[12] C. Nyberg-Hoffman, P. Shabram, W. Li, D. Giroux, and E. Aguilar-Cordova, *Nat. Med.* **3,** 808 (1997).

Virus stocks should be aliquotted and stored in a −80° freezer; some of the stocks should be saved for future comparisons, to examine reproducibility of effects. After several rounds of amplification, or upon receipt of an adenoviral stock about which there is no information on wild-type adenovirus, the viral stock should be examined for wild-type adenovirus. Wild-type adenovirus may be generated during propagation by recombination between an adenoviral vector and the adenoviral sequence that is integrated in the genome of 293 cells, or by contamination, and has replication advantage over an adenoviral vector. Wild-type adenovirus can be detected using primers for E1 genes in a regular PCR with a maximal sensitivity of 1 in a million viral genomes. A sensitive real-time PCR method is capable of detecting 1 wild-type in a billion viral genomes.[7] If wild-type adenovirus is detected, plaque purification or end-point dilution of the viral stock onto 293 cells should be performed to isolate the adenoviral vector, and a new stock amplified.

Optimal Expression of Transgene

The promoter/enhancer is an important determinant of the level and duration of expression of the transgene. For instance, human cytomegalovirus (hCMV) immediate early gene promoter/enhancer (IE P/E) produces a high level of expression of the transgene for a short duration (1–3 days), whereas Rous sarcoma virus (RSV) long-terminal repeat promoter/enhancer produces lower level but longer lasting expression of the transgene after gene transfer to blood vessels.[13,14] Murine CMV (mCMV) IE P/E appears to be 1000-fold stronger than the human counterpart in driving expression of *Escherichia coli* β-galactosidase (β-Gal) after *in vivo* adenovirus-mediated gene transfer to the brain of rats.[2] However, a longer sequence of hCMV IE P/E is about twofold stronger than mCMV IE P/E in driving expression of TGF-β after *in vivo* adenovirus-mediated gene transfer to the lungs of rats.[15] Another study indicated that mCMV IE P/E is stronger in several species, whereas hCMV IE P/E is stronger in human than other cells.[16] Thus, promoter/enhancer must be optimized for different tissues (such as blood vessels in different vascular beds) and species.

A cellular promoter/enhancer (such as human elongation factor-1α) may be used to avoid possible silencing of a viral promoter/enhancer.[17,18] In addition, cis sequences that augment expression through posttranscriptional (e.g., intron and

[13] S. D. Christenson, K. D. Lake, H. Ooboshi, F. M. Faraci, B. L. Davidson, and D. D. Heistad, *Stroke* **29**, 1411 (1998).

[14] H. C. Champion, T. J. Bivalacqua, F. M. D'Souza, L. A. Ortiz, J. R. Jeter, K. Toyoda, D. D. Heistad, A. L. Hyman, and P. J. Kadowitz, *Circ. Res.* **84**, 1422 (1999).

[15] P. J. Sime, Z. Xing, R. Foley, F. L. Graham, and J. Gauldie, *CHEST* **111**, 89S (1997).

[16] C. L. Addison, M. Hitt, D. Knunsken, and F. L. Graham, *J. Gen. Virol.* **78**, 1653 (1997).

[17] G. J. Clesham, H. Browne, S. Efstathiou, and P . L. Weisberg, *Circ. Res.* **79**, 1188 (1996).

[18] Z. Ye, P. Qiu, J. K. Burkholder, J. Turner, J. Culp, T. Roberts, N. Shahidi, and N. Yang, *Hum. Gene Ther.* **9**, 2197 (1998).

splicing sequences) and translation (e.g., adenovirus tripartite leader sequence) mechanisms may be used in a vector to increase expression. Importantly, the level of expression of the transgene contributes to the sensitivity that is required for determination of efficiency of transduction.[2] Efficiency of transduction can only be underestimated but not overestimated.[2,11,12]

Gene Transfer to Vascular Cells *in Vitro*

Endothelial cells (EC), smooth muscle cells (SMC), and fibroblast cells, derived from vascular intima, media, and adventitia, respectively, may be obtained from primary culture and/or transformed cell lines from a variety of animals, and studied for efficiency of gene transfer by adenoviral vector.[19–21] All of these cells can be transduced by adenoviral vectors, albeit at a lower efficiency than transformed epithelial cells in tissue culture. Thus, a high MOI (100 PFU/cell or higher) and long incubation time (>1 hr) often are needed for efficient transduction. The underlying mechanism for low efficiency of gene transfer may be that vascular cells are relatively deficient in expression of a receptor for adenovirus, Coxsackievirus–adenovirus receptor (CAR), which is responsible for high-affinity binding of adenovirus.[22,23] Vascular cells also may be transduced by adenoviral vector by non-CAR dependent mechanisms through other cellular molecules including integrins[24] and heparan sulfate glycosaminoglycans.[25] Adenoviral vectors with a modification to bind to heparan sulfate or integrin transduce both EC and SMC at a higher efficiency than unmodified vector.[26] With more specific targeting (see Chapter 10, this volume), each type of vascular cell may be transduced specifically.

Protocol for Gene Transfer to Vascular Cells in Vitro

1. Primary or transformed cells are grown on tissue culture plates in growth medium containing 10–20% fetal bovine serum (FBS), supplements of vitamins and growth factors, and antibiotics, in a 37° incubator (95% air and 5% CO_2).

---

[19] P. Lemarchand, H. A. Jaffe, C. Danel, M. C. Cid, H. K. Kleinman, L. D. Stratford-Perricaudet, M. Pericaudet, A. Pavirani, J. P. Lecocq, and R. G. Crystal, *Proc. Natl. Acad. Sci. U.S.A.* **89,** 6482 (1992).

[20] R. J. Guzman, P. Lemarchand, R. G. Crystal, S. E. Epstein, and T. Finkel, *Circulation* **88,** 2838 (1993).

[21] P. J. Pagano, J. K. Clark, E. Cifuentes-Pagano, S. M. Clark, G. M. Callis, and M. T. Quinn, *Proc. Natl. Acad. Sci. U.S.A.* **94,** 14483 (1997).

[22] J. M. Bergelson, J. A. Cunningham, G. Droguett, E. A. Kurt-Jones, A. Krithivas, J. S. Hong, M. S. Horwitz, R. L. Crowell, and R. W. Finberg, *Science* **275,** 1320 (1997).

[23] R. P. Tomko, R. Xu, and L. Phillipson, *Proc. Natl. Acad. Sci. U.S.A.* **94,** 3352 (1997).

[24] G. R. Nemerow, *Virology* **274,** 1 (2000).

[25] M. C. Dechecchi, A. Tamanini, A. Bonizzato, and G. Cabrini, *Virology* **268,** 382 (2000).

[26] T. J. Wickham, E. Tzeng, L. L. Shears II, P. W. Roelvink, Y. Li, G. M. Lee, D. E. Brough, A. Lizonova, and I. Kovesdi, *J. Virol.* **71,** 8221 (1997).

2. Cells, at 70–80% of confluence, are incubated with an adenoviral vector at an MOI of 50–200 PFU/cell in incubation medium containing 2% FBS and antibiotics for 2–5 hr. This low percentage of FBS stabilizes adenoviral vector, whereas higher percentage of FBS may neutralize adenoviral vector, if the serum contains antibody specific for adenovirus.

3. After the viral suspension is aspirated, cells are rinsed with incubation medium, fresh growth medium is added, and cells are grown for 24–48 hr.

4. Cells and/or medium are harvested for assay for the transgene product and/or the end product catalyzed by the transgene product (an enzyme).

Gene Transfer to Blood Vessels *ex Vivo*

Gene transfer to segments of blood vessels *ex vivo* provides an initial step to examine effects of the transgene on vessels, with the ultimate goal to introduce the vector to vessels *in vivo*. Vessels are excised and cut into several segments to allow gene transfer in the presence of inhibitors, substrate, after denudation of endothelium, or in the presence of a variety of stimulants. The aorta from rabbits and other animals can be cut into many rings, and muscular arteries (such as the carotid) can be cut into several rings. Augmentation of nitric oxide-mediated vascular relaxation has been demonstrated after *ex vivo* gene transfer of adenoviral vector expressing endothelial nitric oxide synthase (AdeNOS) to vessels including rabbit carotid artery,[27] canine basilar artery,[28] and human pial artery.[29] Because the adventitia is an important target for gene transfer as well as an integral part of the vessel, one should not attempt to remove the adventitia.

Protocol for Transduction of Arteries ex Vivo

1. After removal, arteries are placed in Krebs bicarbonate solution of the following composition (mmol/liter): 118.3 NaCl, 4.7 KCl, 2.54 CaCl$_2$, 1.2 MgSO$_4$, 1.2 KH$_2$PO$_4$, 25 NaHCO$_3$, and 11.1 D-glucose. Loose connective tissue is removed gently without disruption of adherent adventitia. Vessels are cut into rings 3 mm in length, and intraluminal blood is gently rinsed out of the rings.

2. Each ring is incubated with 100 μl of vehicle or viral suspension that contains 10^8–10^{10} PFU/ml, in a 96-well tissue culture plate. After 2 hr rings are rinsed in PBS and incubated with minimum essential medium (MEM) containing 100 μg/ml penicillin, 100 U/ml streptomycin, and 5% fetal bovine serum, in a 37° incubator aerated with 95% O$_2$ and 5% CO$_2$.

[27] H. Ooboshi, Y. Chu, C. D. Rios, F. M. Faraci, B. L. Davidson, and D. D. Heistad, *Am. J. Physiol.* **273**, H265 (1997).

[28] A. F. Y. Chen, T. O'Brien, M. Tsutsui, H. Kinoshita, V. J. Pompili, T. B. Crotty, D. J. Spector, and Z. S. Katusic, *Circ. Res.* **80**, 327 (1997).

[29] V. Khurana, L. A. Smith, D. A. Weiler, M. J. Springett, J. E. Parisi, F. B. Meyer, W. R. Marsh, T. O'Brien, and Z. S. Katusic, *J. Cereb. Blood Flow Metab.* **20**, 1360 (2000).

3. After 24 hr, rings are analyzed for transduction efficiency (e.g., histochemical staining for β-Gal, or antibody detection for the transgene product) and/or used for functional studies (e.g., NO-mediated vasodilation of the rings after gene transfer of AdeNOS). Because several rings can be obtained from vessels of most species, it is often possible to examine several parameters in one vessel. Quantitative measurements usually are performed in duplicate (on two vascular rings).

Protocol for Detection of β-Galactosidase

Although histochemistry is less sensitive than antibody detection or fluorescence assay for β-gal, it is the easiest and most useful method to detect and localize cells expressing β-gal following gene transfer to blood vessels. Determination of transduction efficiency based on this method will underestimate the level of β-gal, especially if expression is low. Two protocols (A and B, modified from Refs. 30 and 31, respectively) are given below, each with essentially 100% success rate of the assay. We use the two protocols interchangeably, and find no advantage of one over the other.

1. After infection with Adβgal, rings of vessels are rinsed in PBS twice, each with fresh solution. Vessels are fixed with solution A (2% paraformaldehyde and 0.025% glutaraldehyde in PBS) or solution B (0.2% glutaraldehyde, 0.1 M phosphate buffer, pH 7.3, 5 mM EGTA, 2 mM MgCl$_2$) for 10 min. Repeat the fixing once using fresh solution.
2. Vessels are rinsed with solution A (PBS) or solution B (0.1 M phosphate buffer, pH 7.3, 2 mM MgCl$_2$, 0.01% sodium deoxycholate, 0.02% NP-40) for 10 min. Repeat the rinsing once using fresh solution.
3. Vessels are incubated in stain solution A (1 mg/ml X-Gal, 5-bromo-4-chloro-3-indolyl-β-D-galactopyranoside, 5 mM each of potassium ferri- or ferrocyanide, 2 mM MgCl$_2$ in PBS) or stain solution B (1 mg/ml X-Gal, 5 mM each of potassium ferri- or ferrocyanide, in the rinse solution B above) at 37° for 2–4 hr. It is important that vessels or other tissues not be incubated in X-gal solution for longer than 4 hr, because endogenous β-Gal may be stained after a longer incubation.
4. Vessels are rinsed in PBS, embedded in paraffin, sectioned, and counterstained with fast nuclear red. β-Gal positive cells can be counted, using a microscope, as blue cells.

Gene Transfer to Blood Vessels *in Vivo*

In this section, we will describe briefly some approaches used by others, and then describe in detail the approach we have used.

[30] J. R. Sanes, J. L. R. Rubenstein, and J. F. Nicolas, *EMBO J.* **5,** 3133 (1986).
[31] B. Hogan, R. Beddington, F. Costantini, and E. Lacy, *in* "Manipulating the Mouse Embryo," 2nd Ed., p. 373. Cold Spring Harbor Laboratory Press, Cold Spring Harbor, NY, 1994.

A. Intraarterial Delivery

The first gene transfer to blood vessels *in vivo* delivered endothelial cells, which were transduced *ex vivo* by a retroviral vector expressing β-gal, to femoral and iliac arteries of pig.[32] Blood flow was stopped for 30 min to allow the endothelial cells to seed onto denuded vessel. Subsequently, the same retrovirus was used as a vector, instead of using cells transduced with the retrovirus.[33] The retrovirus was directly injected into iliofemoral arteries of pig using the a double-balloon catheter.[33] These early studies established the feasibility of local intraarterial gene transfer.

A detailed study was performed on adenovirus-mediated gene transfer to uninjured rat carotid arteries *in vivo*.[34] A 1-cm segment of the common carotid artery was isolated. A carotid artery was exposed, and an arteriotomy was made on the external carotid artery. A 24-gauge polytetrafluoroethylene catheter was introduced through the external carotid arteriotomy, and the isolated segment of vessel was flushed with 1 ml medium 199. Fifty μl of adenoviral or control solution was drawn into a 24-gauge catheter mounted on a 1 ml syringe. The catheter was inserted into the external carotid arteriotomy and secured with a silk tie, and the solution was infused into the isolated carotid segment, resulting in distension (presumably mild, to avoid damage from a large increase in intravascular pressure) of the common carotid artery. The solution was allowed to dwell in the vessel segment for 20 min, and then withdrawn into the catheter.

The catheter was removed, and the external carotid artery was ligated. Blood flow was reestablished through the common and internal carotid arteries. Three days later, the vessel segment was harvested for analysis of reporter transgene expression. In summary, a viral concentration of 1×10^{10}–1×10^{11} PFU/ml (5×10^8–5×10^9 PFU, total) was needed for transgene expression in $\sim$35% of luminal endothelial cells (EC). Under this range of virus, no toxicity to EC and no immune infiltrates were found. Lower concentrations of adenovirus did not give detectable transgene expression, and higher concentration produced EC denudation and neointima formation. A similar study using rabbits confirmed an optimal concentration of adenovirus, and suggested that acute host-mediated EC injury is an important determinant for the outcome of adenovirus-mediated gene transfer to vessels *in vivo*.[35]

One limitation of this intraarterial gene transfer is that blood flow must be stopped, which makes transduction of peripheral vessels much easier than transduction of cerebral or coronary circulation which do not tolerate prolonged interruption of blood flow. A second limitation is that transduction is limited to the endothelium and sometimes also to adventitia, presumably through vasa vasorum

[32] E. G. Nabel, G. Plautz, F. M. Boyce, J. C. Stanley, and G. J. Nabel, *Science* **244**, 1342 (1989).

[33] E. G. Nabel, G. Plautz, and G. J. Nabel, *Science* **249**, 1285 (1990).

[34] A. H. Schulick, G. Dong, K. D. Newman, R. Virmani, and D. A. Dichek, *Circ. Res.* **77**, 475 (1995).

[35] K. M. Channon, H. Qian, S. A. Youngblood, E. Olmez, G. A. Shetty, V. Neplioueva, M. A. Blazing, and S. E. George, *Circ. Res.* **82**, 1253 (1998).

from branch arteries, and the efficiency may be low, due in part to the short period of contact between the virus and cells.

Protocol for Intraluminal Delivery to Carotid Artery

1. Rabbits are anesthetized with an intramuscular injection of xylazine/ketamine (10/50 mg/kg), and the anterior neck is shaved and disinfected with betadine. Midline incision is made, and the common carotid artery is exposed running beneath the sternohyoid muscle. A 3 cm segment is isolated, and any small branches of the common carotid artery are ligated. Typically, the thyroid artery requires ligation.

2. Proximal and distal endpoints are marked and secured with vessel clamps to stop blood flow. An incision is made in the distal end of the common carotid artery, and a polyethylene catheter (PE-10) is introduced through the carotid arteriotomy.

3. Blood in the lumen is flushed once with 150 μl of PBS using a 1 ml syringe. The injectate is removed by aspiration into the syringe.

4. A fresh 1 ml syringe containing an aliquot of adenovirus (100–150 μl in 3% sucrose in PBS) is attached to the catheter, and the virus is instilled into the carotid artery until the artery is filled. Care is taken to avoid overdistention of the artery (i.e., not too much volume or pressure). The catheter is then secured, with the syringe in place. There is no backflow.

5. After 20 min, the viral solution is withdrawn into the syringe, and the catheter is removed.

6. The arteriotomy is repaired by 9-0 silk purse string closure technique.

7. The vessel clamps are removed, blood flow is reestablished, and the wound is closed.

B. Adventitial Delivery

To circumvent the need to stop blood flow, we have used an adventitial gene transfer strategy in vivo (Fig. 2A). This strategy is especially useful for transgenes whose products are diffusible to vascular media. Compared with intraluminal delivery, this approach provides prolonged exposure of virus, which may increase the efficiency of transduction. Another important advantage of this approach is that the host immune response is greatly reduced compared with intraluminal delivery. A 1- to 2-cm segment of carotid or femoral artery is surgically exposed. Adβgal (0.3–1.0 ml of 10^{10} PFU/ml stock) is injected within the carotid or femoral sheath with a 26-gauge needle. One day after gene transfer, vessel segments are analyzed for expression of β-gal. About 24% of adventitial cells are stained positive by X-gal histochemistry, whereas no endothelial cells or smooth muscle cells are stained.[37]

[36] H. Ooboshi, C. D. Rios, and D. D. Heistad, Mol. Cell. Biochem. **172,** 37 (1997).

[37] C. D. Rios, H. Ooboshi, D. Piegors, B. L. Davidson, and D. D. Heistad, Arterioscler. Thromb. Vasc. Biol. **15,** 2241 (1995).

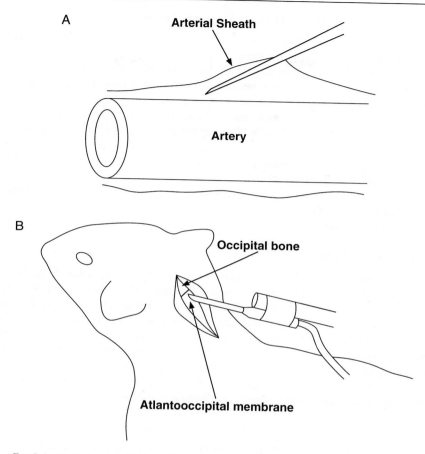

FIG. 2. Methods for gene delivery. (A) Perivascular injection of adenoviral vector in peripheral vessels. Gene transfer is limited to cells in the outer two-thirds of the adventitia, in a symmetrical pattern. No inflammation is observed in the adventitia. (B) Injection of adenoviral vector into the cisterna magna. Gene transfer occurs in cerebral blood vessels and meninges. [Modified with permission from H. Ooboshi, C. D. Rios, and D. D. Heistad, *Mol. Cell Biochem.* **172,** 37 (1997).]

Gene transfer of eNOS by this strategy increases NO-mediated relaxation, even in the absence of endothelium,[38] which demonstrates that NO, generated from overexpression of eNOS in adventitial fibroblast, can diffuse to media to produce relaxation.

---

[38] I. J. Kullo, G. Mozes, R. S. Schwartz, P. Gloviczki, T. B. Crotty, D. A. Barber, Z. S. Katusic, and T. O'Brien, *Circulation* **96,** 2254 (1997).

Protocol for Adventitial Delivery to Carotid Artery

1. Rabbits are anesthetized with xylazine/ketamine and the anterior neck shaved and disinfected with betadine.

2. The rabbit is placed on its back on a surgical table that is set at a 45° angle, with its head elevated. This position facilitates injection of virus during withdrawal of the needle by minimizing leakage of the injectate (step 6).

3. A midline incision is made and both carotid arteries are exposed.

4. Using a 22-gauge, 1.5 inch, blunt tip needle, 150 μl of virus is drawn into a 1 ml syringe.

5. A small incision is made in the rostral end of the carotid sheath and the needle is slid down into the sheath.

6. As the needle is pulled back in the sheath, the virus is injected into the space occupied by the needle.

7. The wound is closed.

C. Intracranial Delivery

The method was first developed in rats[39] and has been used in dogs,[40] rabbits,[1] and mice[13] (Fig. 2B). The virus is injected in a solution of 3% sucrose and, because the solution is heavier than cerebrospinal fluid, the virus settles to the perivascular space around blood vessels on the ventral surface of the brain. The transgene is expressed on the ventral surface of the cerebrum, especially along major blood vessels (Fig. 3). When the position of the head is altered during and after injection of virus, gene transfer is achieved mainly in the most dependent part of the brain, resulting in position-dependent targeting of gene transfer.[39] A major limitation is transient inflammation and leukocyte infiltration in the cerebrospinal fluid.

Protocol for Gene Transfer to Cerebral Spinal Fluid (CSF)

1. Rabbits (2.5–3.0 kg) are anesthetized with xylazine/ketamine, and the dorsal neck is shaved and disinfected with betadine.

2. With the animal lying on its side, the occipital protuberance (the lower margin of the occipital bone) is located by palpation. The neck is flexed with care not to obstruct the airway.

3. No incision is made in the skin. A 5/8 to 7/8 inch, 25-gauge needle attached to a 1 ml syringe is inserted at a point 1–2 cm below the occipital protuberance. The needle is advanced until CSF can be withdrawn from the cisterna magna. It is crucial not to insert the needle too deep, and to hold the needle at the shallowest level that permits CSF to be withdrawn, to avoid traumatic injury to the brain stem by the needle.

---

[39] H. Ooboshi, M. J. Welsh, C. D. Rios, B. L. Davidson, and Donald D. Heistad, *Circ. Res.* **77**, 7 (1995).

[40] A. F. Y. Chen, S. W. Jiang, T. B. Crotty, M. Tsutsui, L. A. Smith, T. O'Brien, and Z. S. Katusic, *Proc. Natl. Acad. Sci. U.S.A.* **94**, 12568 (1997).

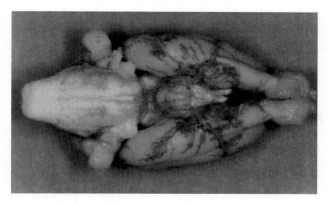

FIG. 3. Expression of transgene after gene transfer to cerebrospinal fluid. Adβgal was injected into the cisterna magna with the rat in the nose-down position. The brain was stained with X-gal 1 day after gene transfer, with positive staining on the ventral surface shown as black. Transduction is seen especially along major cerebral arteries. [Reprinted with permission from H. Ooboshi, M. J. Welsh, C. D. Rios, B. L. Davidson, and Donald D. Heistad, *Circ. Res.* **77,** 7 (1995).]

4. Withdraw 250–300 μl of CSF and remove the syringe from the needle.

5. With the needle in place, attach syringe containing 250–300 μl of adenovirus (in 3% sucrose in PBS) to the needle, and inject virus into the cisterna magna.

6. Withdraw the needle. No bleeding or leak of CSF is observed, and thus pressure does not need to be applied.

7. The animal is turned and propped in an upright position (resting on its stomach) with its head tilted forward by 30° for 30 min. Because the viral solution in sucrose is heavier than CSF, virus settles to the dependent portion of the intracranial cavity. By tilting the head for 30 min, preferential expression can be obtained in either cerebral hemisphere, on the brain stem, or around cerebral vessels of anterior and posterior circulation.[39]

Augmentation of Adenovirus-Mediated Gene Transfer to Blood Vessels

Vascular cells have relatively fewer receptors for adenovirus, Coxsackie- and adenovirus receptor (CAR), and thus transduction efficiency is suboptimal using adenoviral vector. On the other hand, a major concern, the host immune response, is strictly proportional to the dose of adenoviral vector. Thus, augmentation of adenovirus-mediated gene transfer with a reduced viral dose is highly desired and may be achieved by (1) construction of a potent transgene expression cassette (by using a strong promoter/enhancer and including sequences that augment expression), (2) increase of contact time between virus and vessels, as in gene transfer to perivascular sheath, and (3) increase of binding of virus to blood vessels through CAR-independent mechanisms. Here we describe a simple method which uses coprecipitates of adenoviral vector and calcium phosphate (CaPi) to augment gene transfer to blood vessels *ex vivo* and *in vivo,* through a CAR-independent binding mechanism.[41,42]

Protocol for Preparation of Adenovirus/CaPi Coprecipitates

1. $5-40 \times 10^{10}$ particles of an adenoviral vector are added to 1 ml of a vehicle solution in a sterile microcentrifuge tube (1.5 ml). Different vehicles are used depending on the subsequent usage (for cells, blood vessels *ex vivo* or *in vivo*): Dulbecco's modified Eagle medium (DMEM), minimum essential medium (MEM), or Medium 199 (M 199, all purchased from Life Technologies). Each vehicle contains 1.8 mM Ca^{2+} and 0.86 mM P_i, and Krebs' bicarbonate solution contains 2.54 mM Ca^{2+} and 1.2 mM P_i.

2. Add Ca^{2+} from the stock solution of 2 M $CaCl_2$ to the above viral suspension to make a final concentration of total Ca^{2+} of 8 mM.

3. Mix by gentle pipette tip aspiration, and incubate at room temperature for 30 min, to allow the coprecipitate of adenovirus and CaPi to form.

4. Immediately apply the mixture to cells, vessels *ex vivo,* or vessels *in vivo.* A delay in using the adenovirus/CaPi may reduce the potency of the virus, presumably due to suboptimal coprecipitate at delayed time or instability of the virus under the condition of the coprecipitate.

We use MEM for *ex vivo* gene transfer to rabbit carotid artery and aorta. The adenovirus/CaPi coprecipitate augments expression of β-gal $\sim$6-fold compared with the virus alone, with increased transduction of both endothelium and adventitia.[42] We use Krebs' bicarbonate solution for *in vivo* gene transfer to CSF. The coprecipitate augments expression of β-gal $\sim$15-fold in basilar artery compared with the virus alone, and is equivalent to a 10 times higher dose of the virus alone. Augmented expression results solely from greater transduction of the adventitia.[42] Importantly, no adverse effect on the animals has been observed with the use of the coprecipitate, which produces a slight elevation in Ca^{2+} concentration in the CSF. This approach has not been tested in other vascular beds, using intraluminal or perivascular gene transfer.

Acknowledgments

We thank Drs. Donald Lund and Yoshimasa Watanabe for contributions to the protocols for *in vivo* gene transfer. Original studies by the authors are supported by NIH Grants HL62984, HL16066, NS24621, HL14388, DK54759, and funds provided by the VA Medical Service.

[41] A. Fasbender, J. H. Lee, R. W. Walters, T. O. Moninger, J. Zabner, and M. J. Welsh, *J. Clin. Invest.* **102,** 184 (1998).
[42] K. Toyoda, J. J. Andresen, J. Zabner, F. M. Faraci, and D. D. Heistad, *Gene Ther.* **7,** 1284 (2000).

[16] Rearrangements in Adenoviral Genomes Mediated by Inverted Repeats

By CHERYL A. CARLSON, DIRK S. STEINWAERDER, HARTMUT STECHER, DMITRY M. SHAYAKHMETOV, and ANDRÉ LIEBER

Introduction

Adenoviruses (Ads) are now commonly adapted for use as gene transfer vectors. Ad genomes are double-stranded, linear DNA molecules, approximately 35 kb in length. Ad replicates its genome based on a unique strategy that involves stabilized single-stranded replication intermediates.[1] Single-stranded DNA is known to be a preferred substrate for homologous recombination between repetitive sequences.[2,3] The fact that wild-type Ads undergo efficient genetic recombination has been known since the early 1970s.[4–7] Investigators first utilized this trait to form a genetic map of Ad mutants (based on recombination frequencies) and delineate their complementation groups.[8] Efficient recombination appeared to be independent of E3 or E4 gene products but dependent on viral replication.[7] Recombined genomes arose only after initiation of viral DNA replication and continued to form throughout the replication cycle.[9] More recently, investigators have tried to harness Ad homologous recombination for gene targeting purposes.[10–12] Although integration resulting from homologous recombination was more efficient with Ad vectors as compared to electroporation or transfection of plasmid DNA, the overall efficiency was too low to be therapeutically useful.

We have demonstrated that homologous recombination between inverted repeats can be utilized to manipulate the genomes of E1/E3 deleted, first-generation Ad vectors.[13] First-generation Ad vectors possess a number of features which

[1] P. C. van der Vliet, *in* "The Molecular Repertoire of Adenoviruses" (P. B. W. Doerfler, ed.), Vol. 2, p. 1. Springer Verlag, Berlin, 1995.

[2] M. S. Meselson and C. M. Radding, *Proc. Natl. Acad. Sci. U.S.A.* **72,** 358 (1975).

[3] D. Dressler and H. Potter, *Annu. Rev. Biochem.* **51,** 727 (1982).

[4] J. F. Williams and S. Ustacelebi, *J. Gen. Virol.* **13,** 345 (1971).

[5] N. Takemori, *Virology* **47,** 157 (1972).

[6] M. J. Ensinger and H. S. Ginsberg, *J. Virol.* **10,** 328 (1972).

[7] C. S. Young, *Curr. Top. Microbiol. Immunol.* **199,** 89 (1995).

[8] H. S. Ginsberg and C. S. Young, *Adv. Cancer Res.* **23,** 91 (1976).

[9] C. S. Young, G. Cachianes, P. Munz, and S. Silverstein, *J. Virol.* **51,** 571 (1984).

[10] Q. Wang and M. W. Taylor, *Mol. Cell Biol.* **13,** 918 (1993).

[11] K. Mitani, M. Wakamiya, P. Hasty, F. L. Graham, A. Bradley, and C. T. Caskey, *Somat. Cell Mol. Genet.* **21,** 221 (1995).

[12] A. Fujita, K. Sakagami, Y. Kanegae, I. Saito, and I. Kobayashi, *J. Virol.* **69,** 6180 (1995).

[13] D. S. Steinwaerder, C. A. Carlson, and A. Lieber, *J. Virol.* **73,** 9303 (1999).

METHODS IN ENZYMOLOGY, VOL. 346

render them attractive gene transfer vehicles *in vitro* and *in vivo*.[14,15] These include their ability to transduce a large variety of cell types including nondividing cells and the ease at which new recombinant vectors can be created, produced, and purified to high titers. However, they are not without disadvantages, especially short-term expression due to their episomal nature, cytotoxicity from expressed viral proteins, and elicitation of immune responses.

In order to combat these problems, new recombinant Ad vectors devoid of all viral genes have been developed.[16] These "gutless" vectors demonstrate significantly less cytotoxicity and decreased immunogenicity. However, production of these vectors is labor-intensive involving multiple viral passages in the presence of helper virus.[17]

In this chapter we provide a new method for creating Ad vectors devoid of all viral genes (ΔAd) through directed homologous recombination. We also demonstrate that homologous recombination between first-generation Ad vectors can be used to achieve replication-dependent, tumor-specific gene expression.

Principle

Homologous elements can be designed and inserted into the E1 region of first-generation Ad vectors in order to mediate recombination and thereby generate specifically rearranged vector genomes. There does not seem to be any specific sequence requirements for either the homologous elements or the flanking regions; however, the efficient formation of the genomic derivatives is dependent on viral DNA replication. The homologous elements can be provided by either one or two parental vectors depending on the desired application (Fig. 1). Note that because the 5' end of a first-generation vector is used as the recombination site, the progeny genomes of interest are devoid of all viral genes and contain duplicated packaging signals which allow them to be packaged into capsids derived from their parental first-generation vectors.

One of the main applications of this system is the generation of Ad vectors devoid of all viral genes. The recombined progeny vectors have a lighter buoyant density than their precursors. Therefore, they can be purified and separated from their parental vectors in CsCl gradients and subsequently employed. This system can also be used to create replication activated transgene expression, since Ad homologous recombination is dependent on Ad DNA replication (see Applications).

As expected, this system follows the standard rules of homologous recombination.[2,3] Providing two homologous elements by one parental vector mediates more efficient recombination than splitting the homologous elements between two

[14] K. Benihoud, P. Yeh, and M. Perricaudet, *Curr. Opin. Biotechnol.* **10**, 440 (1999).

[15] M. M. Hitt, C. L. Addison, and F. L. Graham, *Adv. Pharmacol.* **40**, 137 (1997).

[16] S. Kochanek, *Hum. Gene Ther.* **10**, 2451 (1999).

[17] V. Sandig, R. Youil, A. J. Bett, L. L. Franlin, M. Oshima, D. Maione, F. Wang, M. L. Metzker, R. Savino, and C. T. Caskey, *Proc. Natl. Acad. Sci. U.S.A.* **97**, 1002 (2000).

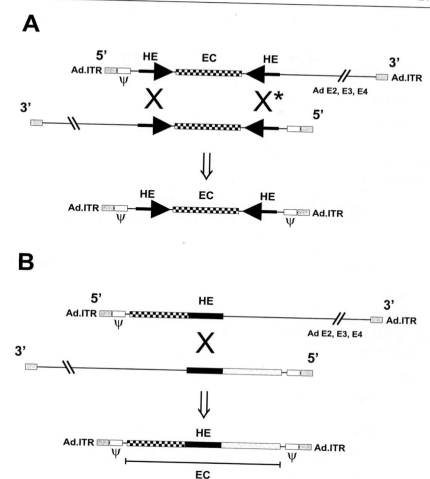

FIG. 1. The principle of harnessing homologous recombination to efficiently generate specifically rearranged vector genomes from first generation parental Ad vectors on viral DNA replication. (A) One parental vector can be designed containing two inverted homologous elements (HEs) flanking an expression cassette (EC) in place of the Ad E1 region. After infection of cells that are permissive for Ad DNA replication, recombination between two of these parental Ad genomes at either HE (depicted by "X" or "X*") forms a deleted progeny vector with duplicated Ad packaging signals (Ψs) and Ad inverted terminal repeats (Ad.ITRs) flanking the EC. No viral genes remain in this progeny vector. Theoretically, a second recombined genome should also be formed containing duplicated Ad E2, E3, and E4 regions flanking the EC. The length of this genome and its lack of packaging signals would prevent it from being packaged. We have not yet observed the formation of this product. (B) Two parental vectors can be designed, each containing one HE (in inverted orientations) within the EC. Similar to above, recombination between the two different parental vectors at the HEs (depicted by "X") forms a deleted progeny vector with duplicated Ψs and Ad.ITRs flanking the EC.

parental vectors. Likewise, increasing the length of homology increases the yield of recombined genomes. The choice of what sequence to use as the homologous element is obviously dependent upon the intended application. So far, we have successfully employed a variety of sequences between 500 and 3000 bp long.

Materials

Cell Line

293 cells,[18] human embryonic kidney cells expressing the left 11% of the Ad type 5 genome, are used for Ad vector generation and amplification. These cells can be purchased from Microbix (Toronto, Ontario, Canada). 293 cells are grown in Dulbecco's modified Eagle's medium supplemented with 10% fetal bovine serum, 2 mM L-glutamine, 100 units/ml penicillin G, and 100 μg/ml streptomycin sulfate (GIBCO, Grand Island, NY). Overlay media for 293 cells is made by mixing 37° 2× Minimum Essential Medium (supplemented with 25% fetal bovine serum, 4 mM L-glutamine, 200 units/ml penicillin G, and 200 μg/ml streptomycin sulfate) with an equal volume of 60° 1% UltraPURE agarose (GIBCO, Grand Island, NY).

Plasmids

For Ad vector generation, we recommend using the pΔE1sp1A/B[19] plasmids with pJM17[20] or pBHG10/11,[19] which are all available from Microbix (Toronto, Ontario, Canada).

Buffers

2× HBS: 274 mM NaCl, 9.9 mM KCl, 2.8 mM Na$_2$HPO$_4$ • 2H$_2$O, and 42.0 mM HEPES with the pH adjusted to 6.67 (at 25°) by addition of NaOH. Store at room temperature during use but −20° for long-term purposes. (Note: the pH of this buffer is critical for transfection efficiency and may need to be adjusted slightly higher or lower depending on the quality of the DNA preparation used.)

Pronase stock: 5 mg/ml pronase (Roche, Indianapolis, IN) in 10 mM Tris-Cl (pH 7.5) needs to be preincubated at 56° for 15 min. followed by 37° for 1 hr. Store at −20°.

Pronase buffer: 10 mM Tris-Cl (pH 7.4), 10 mM EDTA (pH 8.0), and 1% sodium dodecyl sulfate. Store at room temperature.

TE pH 8.0: 10 mM Tris-Cl (pH 8.0) and 1 mM EDTA (pH 8.0). Store at room temperature.

Dialysis buffer: 10 mM Tris-Cl (pH 7.5), 10% glycerol, and 10 mM MgCl$_2$. Store at 4°.

[18] F. L. Graham, J. Smiley, W. C. Russell, and R. Nairn, *J. Gen. Virol.* **36**, 59 (1977).
[19] A. J. Bett, W. Haddara, L. Prevec, and F. L. Graham, *Proc. Natl. Acad. Sci. U.S.A.* **91**, 8802 (1994).
[20] W. J. McGrory, D. S. Bautista, and F. L. Graham, *Virology* **163**, 614 (1988).

Equipment

QIAGEN Plasmid Maxi kit (QIAGEN, Valencia, CA)
Limulus Amebocyte Lysate PYROTELL test kit (Cape Cod, Falmouth, MA)
Ultra-Clear centrifuge tubes 14 × 89mm (Beckman, Palo Alto, CA)
Ultracentrifuge with SW-41 Ti rotor (Beckman, Palo Alto, CA)
Thermocycler (any model)
SP PhosphoImager (Molecular Dynamics, Sunnyvale, CA)

General Methods

Production of Parental First-Generation Vectors

Cotransfection. In order to create first-generation parental Ad vectors, we recommend employing the system developed in Frank Graham's laboratory consisting of cotransfection of pΔE1sp1A/B[19] as the left-end shuttle plasmid with pJM17[20] or pBHG10/11[19] as the right end in 293 cells. First, the desired combination of homologous regions, transgenes, and any regulatory elements should be cloned into pΔE1sp1A/B using standard molecular biology techniques creating the left-end shuttle plasmid. Next, 10 μg of Qiagen column purified left-end shuttle plasmid should be linearized within the plasmid backbone with an appropriate restriction enzyme in a total volume of 40 μl and subsequently heat inactivated. (We have found that although linearizing is not necessary, plaques appear earlier, and the plaque yield is greater.) For cotransfection, we use a modification of the standard calcium phosphate method. The 40 μl of linearized shuttle plasmid is mixed with 14 μg right-end shuttle plasmid (pJM17 or pBHG10/11) and water to a final volume of 450 μl. Then 50 μl of 2.5 *M* CaCl$_2$ is mixed in. While the DNA/CaCl$_2$ solution is continuously vortexed, 500 μl of 2× HBS is added dropwise. After ~2 min incubation, the precipitate is added dropwise to 90% confluent low passage 293 cells (less than passage 40) in a 10 cm dish with 10 ml medium (freshly changed 2 hr prior to transfection). Six hr after transfection, the medium is changed again.

Overlays. Twenty-four hr after transfection, cells are carefully overlaid with 15 ml overlay medium. After 4 days, the cells are overlaid again with 10 ml. After 3 more days, 10 ml overlay medium is added, and if necessary, 5 ml more is added 6 days later. Plaques generally appear between 6 and 14 days after transfection.

Plaque Analysis. For each virus, 10 plaques are generally chosen to be analyzed. Plaques are picked as agarose plugs, mixed with 1 ml medium, subjected to 4 freeze/thaw cycles (in order to lyse cells and release virus), and added to 3 × 10^4 low-passage 293 cells seeded the day before in a 24-well plate. After the development of cytopathic effect (CPE) (usually 2–4 days), the virus is amplified by collecting the cells plus medium, freezing and thawing four times, and adding them to 3 × 10^5 low-passage 293 cells along with 0.5 ml fresh medium. After the development of CPE (usually 48 hr), one-third of the cells and media are frozen at −80° for later amplification and two-thirds are used for DNA analysis.

For DNA analysis, the cells are pelleted by centrifugation at 14K rpm for 2 min, and all but 200 μl of the supernatant is discarded. The pellet should be resuspended by pipetting and slowly added to 200 μl pronase stock diluted 1 : 5 in pronase buffer (final concentration of 0.5 mg/ml pronase). The sample should be well mixed by pipetting and vortexing. It should then be incubated at 37° for 2–6 hr. Next, the sample is extracted with an equal amount of phenol/chloroform/isoamyl alcohol (400 μl), saving the aqueous phase. The aqueous phase is extracted again with 400 μl chloroform. DNA is precipitated from the aqueous phase with 1/10 volume 3 M NaOAc (pH 5.3) plus 3 volumes ethanol and incubation of at least 1 hr at −80°. The DNA is pelleted by centrifuging 12 min at 14K rpm and 4°, and the supernatant is discarded. The pellet is washed with 500 μl 70% ethanol and repelleted 5 min at 14K rpm. The pellet should then be briefly air dried and dissolved in 50 μl TE. Twenty μl of the sample can then be digested with a restriction enzyme appropriate for analysis (HindIII is usually a good choice) and run on an agarose gel. Those with the correct restriction pattern can be further amplified and purified.

Virus Amplification and Purification. For a large-scale preparation, the correct frozen one-third viral stock from above should be subjected to four freeze/thaw cycles to release virus and added to 4×10^5 high-passage 293 cells in one well of a 12-well plate with a final volume of 1.5 ml medium. After the development of CPE, cells and medium must once again be collected, frozen/thawed four times, and this time added to 2×10^6 high-passage 293 cells in a 6 cm dish with a final medium volume of 5 ml. The next amplification step should be the infection of an 80% confluent 10 cm dish of high-passage 293 cells in 10 ml medium. Then, a 95% confluent 15 cm dish is infected with 25 ml medium. From one 15 cm dish, six 15 cm dishes are infected and then 30 15 cm dishes (the last amplification step). After the development of CPE this time, the cells are pelleted by centrifuging at 2000 rpm for 10 min, and the supernatant is discarded. The cells are then resuspended in 1 ml Dulbecco's phosphate buffered saline (GIBCO, Grand Island,) and 10 mM $MgCl_2$ per 15 cm dish. The cells are lysed by four freeze/thaw cycles. The lysates are centrifuged at 2000 rpm for 10 min to remove cellular debris, and the supernatants are digested for 30 min at 37° with 1 mg DNase I and 0.5 mg RNase A per ml final concentration (Roche, Indianapolis, IN). Next, the lysates are run on CsCl step gradients for purification. The gradients are formed by layering 3.5 ml 1.25 g/ml CsCl on top of 3.5 ml 1.32 g/ml CsCl in an ultracentrifuge tube and adding 5 ml lysate on top. The gradients are ultracentrifuged for 4 hr at 35,000 rpm (SW-41 rotor) and 14°. The band containing parental full-length virus is carefully collected from each tube (shown in Fig. 2), combined, and subjected to ultracentrifugation in an equilibrium gradient derived from 1.35 g/ml CsCl for 18 hr at 14° and 35,000 rpm. The full-length band is collected again and dialyzed twice against 1 liter dialysis buffer at 4° in the dark for 6–8 hr each. The virus is aliquotted and stored at −80°. The viral DNA should be checked again by restriction analysis as described above using 50 μl virus diluted with 150 μl serum-free medium.

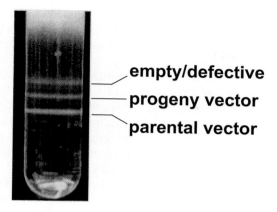

empty/defective

progeny vector

parental vector

FIG. 2. Picture of a parental full-length Ad vector and its progeny ΔAd vector banded by ultracentrifugation in the first CsCl step gradient for purification. Also depicted is the position of banded empty or defective Ad particles.

Titer. The titer can be determined by three different methods. First, a photospectrometer can be used by lysing dilutions of virus (we generally use a 1 : 20 dilution) in TE with 0.1% sodium dodecyl sulfate and measuring their absorbance at 260 nm, where one optical density unit equals 10^{12} particles/ml. Second, a quantitative Southern blot can be used as described below for ΔAd vectors. Ultimately, the titer should be determined by counting plaques from overlaid low-passage 293 cells infected with different dilutions of virus, yielding plaque-forming units (pfu) per ml. Usually, the particle/genome to pfu ratio is approximately 20.[21]

Quality Control. All vector preparations should be tested for contamination with replication competent Ad (E1$^+$) as well as bacterial endotoxin. In order to assess replication competent Ad contamination, we recommend employing the polymerase chain reaction assay described by Zhang *et al.*[22] which can detect one E1$^+$ genome in 10^9 pfu of recombinant virus. To detect the presence of endotoxin, the limulus amebocyte lysate PYROTELL test kit can be used which detects as little as 0.03 endotoxin units per ml by a gel-clot technique.

Production of Progeny ΔAd Vectors

Production and Purification. For the generation of ΔAd vectors from one parental vector (Fig. 1A), 30 100% confluent 15 cm dishes of high-passage 293 cells should be infected at a multiplicity of infection (MOI) of 20 pfu/cell. For the generation of ΔAd vectors from two parental vectors (Fig. 1B), 30 100% confluent 15 cm dishes should be coinfected at an MOI of 15 pfu/cell with each parental

[21] N. Mittereder, K. L. March, and B. C. Trapnell, *J. Virol.* **70**, 7498 (1996).
[22] W. W. Zhang, P. E. Koch, and J. A. Roth, *Biotechniques* **18**, 444 (1995).

vector. In order to boost viral DNA replication, the medium is changed 16 to 24 hr after infection. After the development of CPE (usually between 36 and 48 hr), the cells are harvested and processed as described above for parental vectors and run on the same first CsCl step gradient (Fig. 2). The collected band containing deleted vectors is also subjected to ultracentrifugation in an equilibrium gradient except employing 1.32 g/ml CsCl (instead of 1.35 g/ml CsCl) for 24 hr at 14° and 35,000 rpm. The ΔAd vector band is collected, dialyzed twice against 1 liter dialysis buffer at 4° in the dark for 6–8 hr each, aliquotted, and stored at −80°. These purified deleted vectors should also be tested for endotoxin plus replication competent Ad contamination and subjected to restriction analysis (extracting DNA from 50 μl) as described above for the parental vectors.

Titer. Since these progeny vectors are devoid of all viral genes, plaque titering is not an option. Instead, quantitative Southern blot should be employed to determine the number of genomes/ml. We recommend extracting DNA from 25 μl vector preparation mixed with 175 μl medium plus 2×10^5 293 cells (for a source of carrier DNA) as described under plaque analysis. Next, 5, 2.5, and 1.25μl of the 50 μl extracted DNA should be run on an agarose gel against a standard curve from 0 to 15 ng and analyzed by Southern blot with a phosphoimager. The following formula can then be used to determine the titer:

$$\left(\frac{\# \text{ng}}{\text{ml virus prep}} \right) \left(\frac{1 \text{ mol bp}}{6.50 \times 10^{11} \text{ng}} \right) \left(\frac{6.022 \times 10^{23} \text{bp}}{1 \text{ mol bp}} \right) \left(\frac{1 \text{ genome}}{\# \text{bp}} \right)$$

(average MW   (Avogadro's   (conversion
of DNA bp)   number)   to genomes)

= titer (genomes/ml)

where the number of nanograms of DNA per milliliter of virus preparation is determined from the standard curve and corrected for the volume lost during extraction. As an example, the Southern blot for an 11.6 kb ΔAd vector with a calculated titer of 2.2×10^{11} genomes/ml is shown in Fig. 3. The Southern blot also allows for the contamination with parental vectors to be estimated, which can then

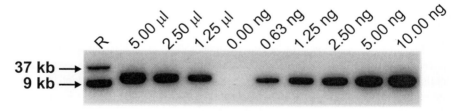

FIG. 3. Southern blot used for titering an 11.6 kb purified progeny deleted vector, derived from two 37 kb parental vectors. 5.00, 2.50, and 1.25 μl of the 50 μl extracted vector plus cellular DNA was run against a standard curve of 0.00 to 10.00 ng. The calculated titer was 2.2×10^{11} genomes/ml. As a reference (R), 37 kb and 9 kb cross-hybridizing DNA was also included. Note that contamination with parental vectors was not detectable.

be confirmed by dilution plaque titering on low-passage 293 cells. Plaque titering of this vector preparation determined that the contamination was less than 0.001%.

Depending on the length of homology and number of parental vectors, we have observed yields from ∼50 packaged progeny genomes per cell [2 parental vectors with 500 bp of homology (1.1×10^{11} genomes/ml)] to ∼1×10^{4} progeny per cell [1 parental vector with 1200 bp of homology (2.1×10^{13} genomes/ml)]. ΔAd vector preparations are generally contaminated with less than <0.01% parental vectors.

Applications

Replication Activated Transgene Expression

Since efficient Ad recombination is dependent on Ad DNA replication, replication activated transgene expression can be achieved by designing parental vectors (either 1 or 2) to bring a promoter into conjunction with a transgene only upon homologous recombination. As an example of this, we have constructed the single parental vector shown in Fig. 4A, which employs two copies of a rabbit β-globin intron (640 bp) as the homologous elements.[23] The utility of this system is underscored by the recent findings that first-generation Ad vectors replicate efficiently in a variety of tumor cell lines,[24] whereas they do not in primary human cells[23] and mouse liver in vivo.[25] In a mouse tumor model with liver metastases derived from human cervical carcinoma cells, a single systemic administration of this parental vector achieved transgene expression in every metastasis while no extra tumoral transgene induction was observed (Fig. 4B). Therefore, tumor-specific expression of cytotoxic or proapoptotic transgenes using this system holds promise for the treatment of a variety of cancers.

The idea that genomic rearrangements (mediated by homologous recombination between inverted repeats) can be used to generate a functional promoter/gene constellation only upon viral DNA replication represents a new principle of selective transcriptional activation. This principle can be applied to any type of conditionally replicating Ad vector, including the E1B55k deleted ONYX015-based vectors,[26] adenovirus E1A mutants,[27,28] or first-generation vectors that express wild-type E1a under the control of a heterologous promoter.[29]

[23] D. S. Steinwaerder, C. A. Carlson, D. L. Otto, Z. Y. Li, S. Ni, and A. Lieber, *Nat. Med.* **7**, 240 (2001).

[24] D. S. Steinwaerder, C. A. Carlson, and A. Lieber, *Hum. Gene Ther.* **11**, 1933 (2000).

[25] J. E. Nelson and M. A. Kay, *J. Virol.* **71**, 8902 (1997).

[26] C. Heise, A. Sampson-Johannes, A. Williams, F. McCormick, D. D. Von Hoff, and D. H. Kirn, *Nat. Med.* **3**, 639 (1997).

[27] K. Doronin, K. Toth, M. Kuppuswamy, P. Ward, A. E. Tollefson, and W. S. Wold, *J. Virol.* **74**, 6147 (2000).

[28] C. Heise, T. Hermiston, L. Johnson, G. Brooks, A. Sampson-Johannes, A. Williams, L. Hawkins, and D. Kirn, *Nat. Med.* **6**, 1134 (2000).

[29] P. L. Hallenbeck, Y. N. Chang, C. Hay, D. Golightly, D. Stewart, J. Lin, S. Phipps, and Y. L. Chiang, *Hum. Gene Ther.* **10**, 1721 (1999).

A

Parental Vector

Progeny Vector

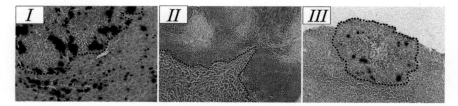

B

FIG. 4. (A) Schematics of the parental and progeny vectors used to achieve replication-activated transgene expression. A promoter (P) is brought into conjunction with the transgene (TG) only upon recombination, which in turn is dependent upon replication. We used a RSV promoter in this system to drive expression of the reporter gene, β-galactosidase (β-gal). As homology elements, two copies of a 640 bp rabbit β-globin intron (I) were used that do not contain any transcription stop sites and are spliced out upon transcription. A bidirectional SV40 polyadenylation signal (pA) was used to terminate transcription and prevent the formation of β-gal antisense RNA. Note that the parental vector should not express the transgene. The Avr II and Blp I restriction sites shown for the parental vector can be used to change the transgene in the corresponding left-end shuttle plasmid. (B) A mouse tumor model mimicking hematogenic liver metastasis was created by injecting 2×10^6 HeLa cells (cervical carcinoma cells [ATCC CCL-2]) into the portal vein of immunodeficient NIH-III mice. Metastases in the three panels are encircled by a dotted line. Panel I shows a liver section from a mouse injected through the tail vein (14 days posttransplantation) with 5×10^9 pfu of a standard first-generation Ad vector constitutively expressing β-gal from an RSV promoter. $13.3 \pm 2.9\%$ of tumor cells were transduced and $16.5 \pm 6.8\%$ of hepatocytes. (Section was stained with X-gal and counter stained with neutral red. 200× original magnification.) Panel II is a liver section from a mouse injected through the tail vein (14 days posttransplantation) with two doses of 5×10^9 pfu of a control vector similar to the replication activated parental vector described in (A), except lacking the second HE. Therefore, this vector cannot undergo recombination to form a progeny vector. No X-gal staining was observed. (Section was stained with X-gal and counterstained with hematoxilin. 100× original magnification.) Panel III shows a liver section from a mouse injected through the tail vein (14 days posttransplantation) with 5×10^9 pfu of the replication activated parental vector described in (A). β-Gal expression was seen in $5.4 \pm 1.4\%$ of tumor cells, whereas no β-gal expression was observed in the normal liver tissue. Importantly, all metastases seen demonstrated β-gal staining. (Section was stained with X-gal and counterstained with hematoxilin. 100 × original magnification.)

Ad Vectors Devoid of All Viral Genes

Homologous recombination in 293 cells provides an easy and efficient method to generate ΔAd vectors.[13] A typical example of a vector that can be generated in this fashion is shown in Fig. 5A. Here, an insulator element (1.2 kb) derived from the HS-4 region of the chicken β-globin locus is used as the homologous element. In this example, there is only room for an expression cassette of 5.6 kb or less. However, by decreasing the homologous regions to 500 bp and using two parental vectors, the capacity can be increased to 15 kb.

Because these progeny vectors are devoid of all viral genes, they demonstrate considerably less toxicity than their parental vectors (Fig. 5B). However, their transduction efficiency and stability seem to be cell line dependent. For example, in SKHEP-1 cells[30] (human endothelial cells), we found that a ΔAd progeny vector was able to transduce cells with the same efficiency as its parental vector, but transgene expression declined more rapidly because of vector DNA loss (Fig. 5C). On the other end of the spectrum, the transduction efficiency of MO7e cells[31] (human leukemic cells) with a ΔAd vector was 100 times less efficient than its parental vector (data not shown). Therefore, each new target cell line must be tested for transduction efficiency and stability of ΔAd vectors. We have not yet tested these ΔAd vectors *in vivo*.

Transient transgene expression from ΔAd vectors has potential applications in cell biology or cell cycle studies, which require the absence of toxicity or side effects associated with the expression of adenoviral proteins.

ΔAd Hybrid Vectors

Elements from other viruses or bacteria can be added to this ΔAd system in order to change the vectors' properties. For example, we have added two AAV ITRs as homologous elements to generate an integrating Δ Ad vector (Fig. 6A).[32] When compared to a recombinant AAV vector with the same expression cassette, our hybrid AAV ΔAd vector was about half as efficient at mediating integration under G418 selection in SKHEP-1 cells (Fig. 6B). However, 100% colony formation efficiency was reached with the ΔAd vector at an MOI of 10^6 genomes/cell, which could not be achieved with the recombinant AAV vector since its titer was 500 times lower. Note that the parental vector was also capable of mediating long-term G418 resistance, but its efficiency was clearly affected by dose-dependent toxicity.

[30] S. C. Heffelfinger, H. H. Hawkins, J. Barrish, L. Taylor, and G. J. Darlington, *In Vitro Cell Dev. Biol.* **28A,** 136 (1992).
[31] P. C. Hendrie, K. Miyazawa, Y. C. Yang, C. D. Langefeld, and H. E. Broxmeyer, *Exp. Hematol.* **19,** 1031 (1991).
[32] A. Lieber, D. S. Steinwaerder, C. A. Carlson, and M. A. Kay, *J. Virol.* **73,** 9314 (1999).

A

Parental Vector

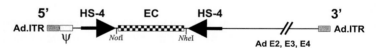

Progeny Vector

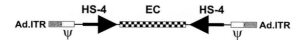

B

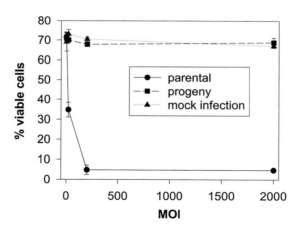

FIG. 5. (A) Schematic of a parental vector that efficiently generates ΔAd vectors through homologous recombination and its progeny vector. Two copies of an insulator element derived from the HS-4 region of the chicken β-globin locus (HS-4) are used as the homologous elements flanking the expression cassette (EC). The *Not*I and *Nhe*I restriction sites depicted for the parental vector can be used to change the expression cassette in the corresponding left-end shuttle plasmid. (B) Toxicity study in MO7e cells comparing the parental (first-generation) and progeny (ΔAd) vectors described in (A) along with mock infected cells at different MOIs (genomes/cell) 48 hr after infection. (C) Transduction efficiency and vector stability studies in SKHEP-1 cells comparing parental and progeny vectors (described in A) over time at on MOI of 2000 genomes/cell. Transgene (human α1-antitrypsin) expression was examined by ELISA and vector DNA persistence was detected by Southern blot.

C

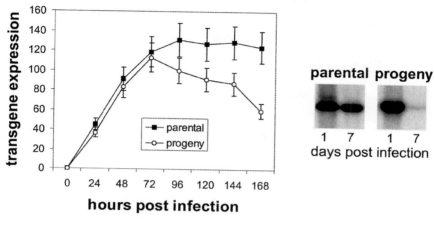

FIG. 5. (*Continued*)

One problem arose with this system. Over time with increased passaging, the AAV ITR/s within the parental vector tended to become rearranged and/or deleted. In order to overcome this, we have developed the system shown in Fig. 6C where only one AAV ITR is located in the parental vector and another sequence is used for homology. One AAV ITR per parental vector appears to be stable (data not shown).

Toxic Transgenes

First-generation Ad vectors cannot be created to express transgenes that inhibit Ad replication or are toxic to 293 cells since their gene products interfere with the production and amplification of these vectors. We hypothesize that the recombination system can be used to circumvent this problem by dividing toxic or problematic transgenes between two parental vectors such that neither encodes a functional product (Fig. 7). Only upon coinfection and after the onset of replication will the toxic product be produced. We are currently in the process of testing this system.

Concluding Remarks

Homologous recombination is a flexible and powerful technique for manipulating and creating Ad vectors for use in gene transfer. The production of high-titer progeny vectors devoid of all viral genes is less labor intensive than

A

Parental Vector

Progeny Vector

B

% G418 resistant colonies (SEM)			
MOI	rAAV	Progeny	Parental
10^1	<0.001	<0.001	<0.001
10^2	0.01	0.01	0.06
10^3	2.7 (1.6)	1.3 (1)	5.4 (3.0)
10^4	90.8 (7.0)	48.0 (8.9)	12.9 (7.2)
10^5	N/A	93.1 (5.4)	3.8 (2.1)
10^6	N/A	100	<0.001

FIG. 6. (A) Schematics of a hybrid parental vector and the derived progeny vector where AAV ITRs are employed as homology elements. (B) The formation of SKHEP-1 cell G418-resisant colonies after infection with the progeny and parental vectors described in (A) and a recombinant AAV vector, all containing the same neomycin expression cassette. Different MOIs (genomes/cell) were employed, and infected SKHEP-1 cells were subjected to 4 weeks of G418 selection. (C) Schematics of a hybrid parental vector and the derived progeny vector where HS-4 insulator elements are used as homologous elements and recombination duplicates the AAV ITR. The *Not*I and *Nhe*I restriction sites indicated for the parental vector can be used to change the expression cassette in the corresponding left-end shuttle plasmid.

helper-dependent production of "gutless" vectors[33] or packaged Ad mini-chromosomes,[34] both of which require multiple passages of virus. Furthermore, this system's dependence on viral DNA replication can be exploited to generate replication-activated transgene expression for tumor gene therapy as well as to produce ΔAd vectors expressing toxic transgenes.

[33] G. Schiedner, N. Morral, R. J. Parks, Y. Wu, S. C. Koopmans, C. Langston, F. L. Graham, A. L. Beaudet, and S. Kochanek, *Nat. Genet.* **18,** 180 (1998).
[34] R. Kumar-Singh and D. B. Farber, *Hum. Mol Genet.* **7,** 1893 (1998).

C

Parental Vector

Progeny Vector

FIG. 6. (*Continued*)

Parental Vectors

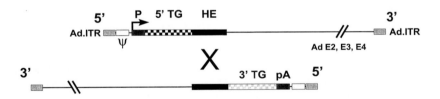

Progeny Vector

FIG. 7. Schematics of proposed parental and progeny vectors designed to express toxic transgenes. A promoter (P) followed by a 5′ portion of a toxic gene (5′ TG) is cloned into the E1 region of an Ad vector. A second parental vector is created containing a polyadenylation signal (pA) followed by an overlapping 3′ portion of the toxic gene (3′ TG). Note that the second parental vector is represented in the opposite orientation of the first. The overlapping regions of the gene form the necessary homologous elements (HEs). Therefore, the toxic gene is reconstituted in the progeny vector.

However, this system is not without problems. Every progeny vector created thus far has been less stable than traditional first-generation Ad vectors. Also, the progeny vectors' transduction efficiencies as compared to parental vectors have ranged from being equal to 100 times less efficient depending on the target cell type. With these disadvantages in mind, we forsee a potential application for these new ΔAd vectors in the *ex vivo* transduction of hematopoietic stem cells with an integrating hybrid AAV ΔAd vector containing a retargeted capsid. All Ad capsid retargeting strategies can be applied to these progeny vectors since they are packaged within the same capsids as their first-generation Ad progenitors. A recent study has demonstrated that exchanging the Ad 5 fiber with an Ad 35 fiber conferred a new vector tropism. This feature allowed for efficient transduction of human CD34+ cells, particularly subsets with potential stem cell activity.[35,36]

In summary, homologous elements can be used to create predictable genetic rearrangements within the framework of Ad replication. Harnessing this homologous recombination allows for the reliable and efficient generation of progeny vectors devoid of all viral genes that can be employed in a large variety of applications.

[35] P. Yotnda, H. Onishi, H. Heslop, D. Shayakhmetov, A. Lieber, M. Brenner, and A. Davis, *Gene Ther.* **8,** 930 (2001).
[36] D. M. Shayakhmetov, T. Papayannopoulou, G. Stamatoyannopoulos, and A. Lieber, *J. Virol.* **74,** 2567 (2000).

[17] High-Capacity, Helper-Dependent, "Gutless" Adenoviral Vectors for Gene Transfer into Brain

By P. R. LOWENSTEIN, C. E. THOMAS, P. UMANA, C. A. GERDES,
T. VERAKIS, O. BOYER, S. TONDEUR, D. KLATZMANN, and M. G. CASTRO

Introduction

Current Status of Adenovirus-Mediated Gene Transfer into CNS

Adenoviral-mediated gene transfer into brain cells *in vitro* or the CNS *in vivo* is very efficient and can be effective in experimental models of brain disease.[1–8]

[1] S. Akli, C. Caillaud, E. Vigne, L. D. Stratford-Perricaudet, L. Poenaru, M. Perricaudet, A. Kahn, and M. R. Peschanski, *Nat. Genet.* **3,** 224 (1993).
[2] D. L. Choi-Lundberg, Q. Lin, Y. N. Chang, V. L. Chaing, C. M. Hay, J. Mohajero, B. L. Davidson, and M. C. Bohn, *Science* **275,** 838 (1997).

To date most vectors utilized have been first-generation, partially deleted vectors. The most commonly used have been E1/E3-deleted vectors, or vectors with further deletions in E2 or E4.[9–15]

First generation adenoviral vectors injected into the brain parenchyma cause acute cellular[10–14,16–20] and cytokine-mediated inflammatory responses,[21–23] and do not allow stable transduction in the presence of anti-adenoviral immune responses.[17–19,21,24–30] Direct virion toxicity, inflammation, and/or anti-adenoviral T-cell responses to capsid and other viral proteins limit long-term transgene

[3] B. L. Davidson, E. D. Allen, K. F. Kozarsky, J. M. Wilson, and B. J. Roessler, *Nat. Genet.* **3**, 219 (1993).

[4] B. J. Geddes, T. C. Harding, S. L. Lightman, and J. B. Uney, *Nat. Med.* **3**, 1402 (1997).

[5] A. Ghodsi, C. Stein, T. Derksen, G. Yang, R. D. Anderson, and B. L. Davidson, *Hum. Gene Ther.* **9**, 2331 (1998).

[6] G. Le Gal La Salle, J. J. Robert, S. Berrard, V. Ridoux, L. D. Stratford-Perricaudet, M. Perricaudet, and J. Mallet, *Science* **259**, 988 (1993).

[7] A. F. Shering, D. Bain, K. Stewart, A. L. Epstein, M. G. Castro, G. W. G. Wilkinson, and P. R. Lowenstein, *J. Gen. Virol.* **78**, 445 (1997).

[8] S. Windeatt, T. D. Southgate, R. Dewey, F. Bolognani, M. J. Perone, A. T. Larregina, T. Maleniak, I. Morris, R. Goya, D. Klatzman, P. R. Lowenstein, and M. G. Castro, *J. Clin. Endocrinol. Metab.* **85**, 1296 (2000).

[9] K. Benihoud, P. Yeh, and M. Perricaudet, *Curr. Opin. Biotechnol.* **10**, 440 (1999).

[10] R. A. Dewey, G. Morrissey, C. M. Cowsill, D. Stone, F. Bolognani, N. J. F. Dodd, T. D. Southgate, D. Katzmann, H. Lassmann, M. G. Castro, and P. R. Lowenstein, *Nat. Med.* **5**, 1256 (1999).

[11] C. A. Gerdes, M. G. Castro, and P. R. Lowenstein, *Mol. Ther.* **2**, 330 (2000).

[12] T. Kafri, D. Morgan, T. Krahl, N. Sarvetnick, L. Sherman, and I. Verma, *Proc. Natl. Acad. Sci. U.S.A.* **95**, 11377 (1998).

[13] D. Maione, M. Wiznerowicz, P. Delmastro, R. Cortese, G. Ciliberto, N. La Monica, and R. Savino, *Hum. Gene Ther.* **11**, 859 (2000).

[14] W. K. O'Neal, H. Zhou, N. Morral, C. Langston, R. J. Parks, F. L. Graham, S. Kochanek, and A. L. Beaudet, *Mol. Med.* **6**, 179 (2000).

[15] T. W. Trask, R. P. Trask, E. Aguilar-Cordova, H. D. Shine, P. R. Wyde, J. C. Goodman, W. J. Hamilton, A. Rojas-Martinez, S. H. Chen, S. L. Wood, and R. G. Grossman, *Mol. Ther.* **1**, 195 (2000).

[16] A. P. Byrnes, J. E. Rusby, M. J. A. Wood, and H. M. Charlton, *Neuroscience* **66**, 1015 (1995).

[17] A. P. Byrnes, R. E. MacLaren, and H. M. Charlton, *J. Neurosci.* **16**, 3045 (1996).

[18] A. P. Byrnes, M. J. A. Wood, and H. M. Charlton, *Gene Ther.* **3**, 644 (1996).

[19] C. E. Thomas, G. Schiedner, S. Kochanek, M. G. Castro, and P. R. Lowenstein, *Proc. Natl. Acad. Sci. U.S.A.* **97**, 7482 (2000).

[20] C. E. Thomas, D. Birkett, I. Anozie, M. G. Castro, and P. R. Lowenstein, *Mol. Ther.* **3**, 36 (2001).

[21] M. J. A. Wood, H. M. Charlton, K. J. Wood, K. Kajiwara, and A. P. Byrnes, *Trends Neurosci.* **19**, 497 (1996).

[22] M. J. A. Wood, A. P. Byrnes, M. McMenamin, K. Kajiwara, A. Vine, I. Gordon, J. Lang, K. J. Wood, and H. M. Charlton, "Protocols for Gene Transfer in Neuroscience" (P. R. Lowenstein and L.W. Enquist, eds.). John Wiley and Sons Ltd., 1996, p. 365.

[23] T. Cartmell, T. Southgate, G. S. Rees, M. G. Castro, P. R. Lowenstein, and G. N. Luheshi, *J. Neurosci.* **19**, 1517 (1999).

[24] K. Brand, R. Klocke, A. Possling, D. Paul, and M. Strauss, *Gene Ther.* **6**, 1054 (1999).

expression. Significantly, many of these side effects of gene transfer are due to the high vector doses used to transduce cells.[20,24,31–37] It has thus become clear that two objectives need to be achieved to render adenovirus-mediated gene expression safe: (i) reduction of the dose of viral vectors required for optimal transgene expression, and (ii) elimination of adenoviral protein encoding sequences from vector genomes.

Vector dose could be reduced if the amount of protein produced from each infectious adenoviral genome could be significantly increased. In search of higher levels of transgene expression, we explored the activity of various promoter systems in brain cells *in vitro* and *in vivo*.[7,11,38] Highest levels of transgene expression were obtained with the major immediate early murine cytomegalovirus (mCMV) promoter, which in the brain was 1000 times more potent than the major immediate early human cytomegalovirus (hCMV) promoter. By using the mCMV promoter to drive transgene expression it is possible to reduce the dose of vector needed to transduce equal rat brain areas by 2–3 logs, compared to a vector containing the hCMV promoter.[11]

Immune-mediated and toxic side effects resulting from the expression of adenoviral genes are avoided by helper-dependent, high-capacity, "gutless" adenoviral vectors (HC-Adv). HC-Adv are efficient gene transfer vectors devoid of all viral coding sequences (Fig. 1).[39–42] Consequently, they display reduced *in vivo*

[25] W. T. Hermens and J. Verhaagen, *Hum. Gene Ther.* **8**, 1049 (1997).

[26] Y. Ohmoto, M. J. Wood, H. M. Charlton, K. Kajiwara, V. H. Perry, and K. J. Wood, *Gene Ther.* **6**, 471 (1999).

[27] C. E. Thomas, G. Schiedner, S. Kochanek, M. G. Castro, and P. R. Lowenstein, *Hum. Gene Ther.* **12**, 839 (2001).

[28] Y. Yang, F. A. Nunes, K. Berencsi, E. E. Furth, E. Gonczol, and J. M. Wilson, *Proc. Natl. Acad. Sci. U.S.A.* **91**, 4407 (1994).

[29] Y. Yang, H. C. J. Ertl, and J. M. Wilson, *Immunity* **1**, 433 (1994).

[30] Y. Yang, Q. Li, H. C. J. Ertl, and J. M. Wilson, *J. Virol.* **69**, 2004 (1995).

[31] N. Morral, O'Neal, H. Zhou, C. Langston, and A. Beaudet, *Hum. Gene Ther.* **8**, 1275 (1997).

[32] M. T. O'Leary and H. M. Charlton, *Gene Ther.* **6**, 1351 (1999).

[33] J. G. Smith, S. E. Raper, E. B. Wheeldon, D. Hackney, K. Judy, J. M. Wilson, and S. L. Eck, *Hum. Gene Ther.* **8**, 943 (1997).

[34] R. M. Easton, E. M. Johnson, and D. J. Creedon, *Mol. Cell. Neurosci.* **11**, 334 (1998).

[35] C. Cowsill, T. D. Southgate, G. Morrissey, R. A. Dewey, A. E. Morelli, T. C. Maleniak, Z. Forrest, D. Klatzmann, G. W. G. Wilkinson, P. R. Lowenstein, and M. G. Castro, *Gene Ther.* **7**, 679 (2000).

[36] M. C. Bohn, D. L. Choi-Lundberg, B. L. Davidson, C. Leranth, D. A. Kozlowski, J. C. Smith, M. K. O'Banion, and D. E. Redmond, Jr., *Hum. Gene Ther.* **10**, 1175 (1999).

[37] M. S. Lawrence, H. G. Foellmer, J. D. Elsworth, J. H. Kim, C. Leranth, D. A. Kozlowski, A. L. Bothwell, B. L. Davidson, M. C. Bohn, and D. E. Redmond, Jr., *Gene Ther.* **6**, 1368 (1999).

[38] A. E. Morelli, A. T. Larregina, J. Smith-Arica, R. A. Dewey, T. D. Southgate, B. Ambar, A. Fontana, M. G. Castro, and P. R. Lowenstein, *J. Gen. Virol.* **80**, 571 (1999).

[39] K. Mitani, F. L. Graham, C. T. Caskey, and S. Kochanek, *Proc. Natl. Acad. Sci. U.S.A.* **92**, 3854 (1995).

[40] S. Kochanek, P. R. Clemens, K. Mitani, H. H. Chen, S. Chan, and C. T. Caskey, *Proc. Natl. Acad. Sci. U.S.A.* **93**, 5731 (1996).

and *in vitro* toxicity; have a large cloning capacity (up to a theoretical value of approximately 28–32 kbp), close to the potential capacity of HSV-1 derived amplicons (~50 kbp),[43] and provide long-term transgene expression.[13,19,44–50] Further, HC-Adv direct transgene expression *in vivo* with unsurpassed efficiency among gene therapy vectors.[13]

Work from our group has shown that HC-Adv sustain transgene expression even in the presence of immune responses against E1/E3-deleted adenovirus type 5. We established that immunization either following[19] or preceding[27] injection of adenovirus into the brain does not abolish transgene expression from HC-Adv, whereas expression from E1/E3-deleted adenoviral vectors is eliminated rapidly following either immunization schedule, when doses of 10^3 infectious units (IU) and above are used (Gerdes, Castro, and Lowenstein, unpublished observations). These two paradigms mimic conditions which may occur in patients undergoing injections of adenovirus into the brain as part of gene therapy clinical trials. Immunization following intracerebral injection resembles the condition of patients naïve to adenovirus who undergo gene therapy and then subsequently suffer an upper respiratory infection due to wild-type adenovirus. Immunization preceding intracerebral gene delivery reproduces the situation occurring in patients with preexisting anti-adenovirus immunity, who may then become subjects in a clinical trial. Our experiments have confirmed that injections of adenovirus into the brain do not prime an anti-adenovirus immune response.[19,27] Priming of the immune response needs to occur in lymph nodes,[51–55] and in order to prime an

[41] R. J. Parks, L. Chen, M. Anton, U. Sankar, M. A. Rudnicki, and F. L. Graham, *Proc. Natl. Acad. Sci. U.S.A.* **93,** 13565 (1996).

[42] S. Hardy, M. Kitamura, T. Harris-Stansil, Y. Dai, and M. L. Philips, *J. Virol.* **71,** 1842 (1997).

[43] X. Wang, G. R. Zhang, T. Yang, W. Zhang, and A. I. Geller, *Biotechniques* **1,** 102 (2000).

[44] H. H. Chen, L. M. Mack, R. Kelly, M. Ontell, S. Kochanek, and P. R. Clemens, *Proc. Natl. Acad. Sci. U.S.A.* **94,** 1645 (1997).

[45] G. Schiedner, N. Morral, R. J. Parks, Y. Wu, S. C. Koopmans, C. Langston, F. L. Graham, A. L. Beaudet, and S. Kochanek, *Nat. Genet.* **18,** 180 (1998).

[46] M. A. Morsy, M. Gu, S. Motzel, J. Zhao, J. Lin, Q. Su, H. Allen, L. Franlin, R. J. Parks, F. L. Graham, S. Kochanek, A. J. Bett, and C. T. Caskey, *Proc. Natl. Acad. Sci. U.S.A.* **95,** 7866 (1998).

[47] N. Morral, R. J. Parks, H. Zhou, C. Langston, G. Schiedner, J. Quinones, F. L. Graham, S. Kochanek, and A. L. Beadet, *Hum. Gene Ther.* **9,** 2709 (1998).

[48] N. Morral, W. O'Neal, K. Rice, M. Leland, J. Kaplan, P. A. Piedra, H. Zhou, R. J. Parks, I. R. Velji, E. Aguilar-Cordova, S. Wadworth, F. L. Graham, S. Kochanek, K. D. Carey, *Proc. Natl. Acad. Sci. U.S.A.* **96,** 12816 (1999).

[49] H. H. Chen, L. M. Mack, R. Lelley, M. Ontell, and S. Kochanek, *Proc. Natl. Acad. Sci. U.S.A.* **94,** 1645 (1997).

[50] H. H. Chen, L. M. Mack, S. Y. Choi, M. Ontell, S. Kochanek, and P. R. Clemens, *Hum. Gene Ther.* **10,** 365 (1999).

[51] M. K. Matyszak, *Prog. Neurobiol.* **56,** 19 (1998).

[52] P. G. Stevenson, S. Hawke, D. J. Sloan, and C. R. M. Bangham, *J. Virol.* **71,** 145 (1997).

[53] M. J. Carson and J. G. Sutcliffe, *J. Neurosci. Res.* **55,** 1 (1999).

[54] H. F. Cserr and P. M. Knopf, *Immunol. Today* **13,** 507 (1992).

[55] P. R. Lowenstein, *Trends Immunol.* **23,** in press.

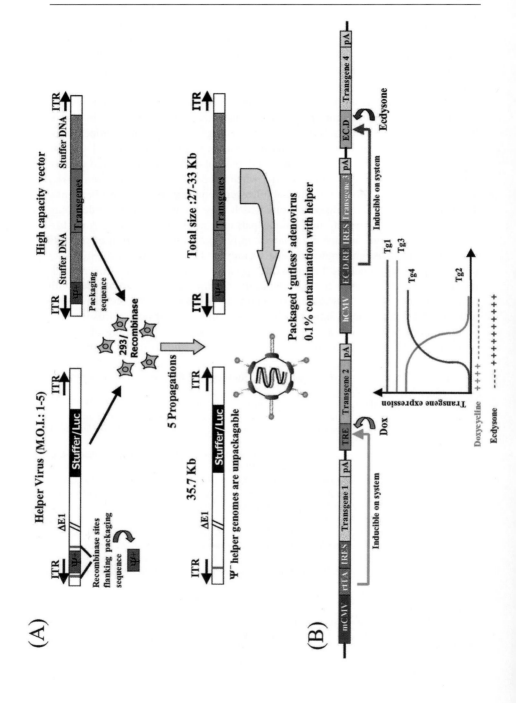

anti-adenoviral immune response, adenovirus needs to be injected peripherally, outside of the brain, e.g., intradermally.[19,27,56] Adjuvant is not needed to immunize against adenovirus.[57]

Although HC-Adv genomes can evade brain-infiltrating anti-adenovirus specific T cells, injection of HC-Adv into the brain, similarly to injection of E1/E3-deleted vectors, causes an acute and transitory brain inflammatory response.[19,20,27] To summarize, over the past few years we have characterized transgene expression and inflammatory responses elicited in the brain following the intrastriatal injection of E1/E3-deleted or HC-Adv vectors in four different paradigms[11,19,20,27]: (i) short term in naïve rats (e.g., 3–5 days postinjection); (ii) long-term in naïve animals (e.g., 3–12 months postinjection); (iii) long-term in naïve animals, followed by a peripheral immunization against adenovirus type 5; and (iv) medium-term in animals immunized against adenovirus type 5 prior to the intracranial vector injection. These studies have shown the following: (i) Injection of 1×10^7 IU of either E1/E3-deleted vectors or HC-Adv into the brain causes an acute, transitory, inflammatory response, with activation of microglia and astrocytes, and up-regulation of ICAM-1, MHC-I, and MHC-II expression. Acute inflammation per se does not appear to affect directly long-term transgene expression[20] (see Fig. 2). (ii) Within 12 months the expression of a marker transgene from first-generation vectors, but not from HC-Adv, decreases[19] (unpublished observations). (iii) Following subsequent peripheral immunization against adenovirus, transgene expression from an

[56] C. E. Thomas, E. Abordo-Adesida, T. C. Maleniak, D. Stone, C. A. Gerdes, and P. R. Lowenstein, "Current Protocols in Neuroscience" (J. N. Crawley, C. R. Gerfen, R. McKay, M. A. Rogawski, D. R. Sibley, and P. Skolnick, eds.), p. 4.23.1. John Wiley and Sons, New York, 2000.
[57] B. A. Rubin and L. B. Rorke, in "Vaccines," 2nd Ed. (S. A. Plotkin and E. A. Mortimer, Jr., eds.), p. 475. W. B. Saunders Company, Philadelphia, 1994.

FIG. 1. (A) Schematic description of the construction of a high-capacity, helper-dependent adenovirus. Producer cells, e.g., Ela expressing 293 cells encoding a recombinase, are infected with helper virus and a high-capacity vector. The helper virus has been engineered to contain a packaging signal (Ψ) flanked by two sequences recognised by the recombinase. During replication, the genomes of both the helper virus and the vector will be replicated. The recombinase will delete the packaging signal from the genome of the helper virus, leaving mainly vector genomes to be packaged into virions. (B) Idealized vector genome of a helper-dependent adenovirus. This figure shows a complex construct that would allow the expression of at least 4 different transgenes (Tg). Two of these, Tg 1 and 3, would be under the constitutive control of either the mCMV or the hCMV promoters, while Tg 2 would be under the inducible doxycycline-dependent transcriptional activator system (rtTA), and Tg 4 would be under the control of the inducible ecdysone-dependent transcriptional activator (ECD). Currently, such large constructs cannot be inserted into any of other currently available vectors. As indicated below the scheme of the vector genome, Tg 2 would be expressed only in the presence of doxycycline, and Tg 4 would only be expressed in the presence of ecdysone. Further, Tg 1 would be expressed at higher levels than Tg 3, because in the brain the mCMV promoter allows higher levels of expression than the hCMV promoter. Thus, ideally, such a complex construct would allow a complex array of transgene expression to be achieved from a single viral vector.

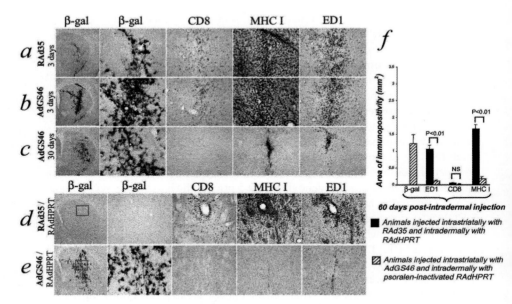

FIG. 2. Transgene expression (β-galactosidase) and inflammation in the brains of animals injected with either the E1/E3-deleted vector RAd35 or the HC-Adv AdGS46. Rows a and b show β-galactosidase expression (columns 1 and 2), CD8+ cell infiltration (column 3), MHC class I upregulation (column 4) and microglial cell activation (ED1, column 5), 3 days after injection of 1×10^7 IU of RAd35 (row a) or AdGS46 (row b) into the brains of naïve animals. Levels of transgene expression and inflammation mediated by the E1/E3-deleted, or the HC-Adv vector, were indistinguishable in naïve animals. Transgene expression from 1×10^7 IU of both RAd35 and AdGS46 is stable for at least 30 days in naïve animals and inflammation resolves within this time frame (shown in row c only for AdGS46). However, transgene expression from E1/E3-deleted vectors in the CNS is rapidly eliminated and accompanied by severe brain inflammation if animals receive a subsequent *peripheral* infection with adenovirus (row d; RAd35 injected in the brain at day 0, RAdHPRT[62] injected in the skin at day 60; see also the quantitative analysis of all markers, in panel f). In contrast, transgene expression from HC-Adv remains stable, and no brain inflammation is elicited when animals are subsequently injected with RAdHPRT in the skin (row e; AdGS46 injected in the brain at day 0, RAdHPRT injected in the skin at day 60). (Modified from Ref. 19.)

E1/E3-deleted adenovirus vector previously injected into the brain is abolished within 60 days, while expression from a HC-Adv remains unaffected[19] (Fig. 2); down-regulation of transgene expression appears to occur at the level of transcription (specific reduction of transgene mRNA levels), rather than through immune-mediated cell killing (unpublished data). (iv) Expression from an E1/E3-deleted adenovirus in brains of animals previously immunized against adenovirus is eliminated by >90% by 14 days postinjection, while expression from a HC-Adv is only reduced to 50% and remains stable for at least 2 months[27] (Fig. 3).

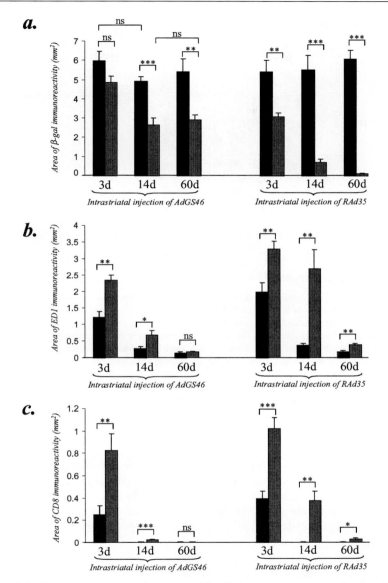

FIG. 3. Preimmunization against adenovirus rapidly eliminates transgene expression from E1/E3 deleted vectors, but not from high-capacity vectors. Quantification of the area within 40 μm thick brain sections occupied by β-galactosidase (*a*), ED1 (*b*), or CD8 (*c*). Error bars show the SEM value from the 5 animals in each experimental group. Student's *t* test was used to calculate the degree of significance of differences between levels of transgene expression and inflammation in the brains of nonimmunized animals (black bars) and immunized animals (gray bars) after intrastriatal injection of RAd35 or AdGS46. (Modified from Ref. 27.)

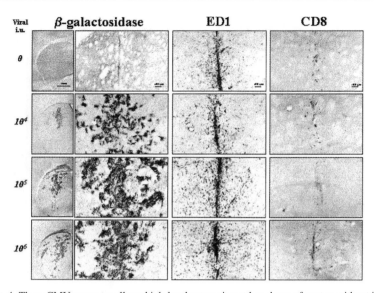

FIG. 4. The mCMV promoter allows high-level expression at low doses of vector, without inflammation. Increasing doses of RAd36 (mCMV-βgal) (10^4–10^6) were injected into the striatum of adult Sprague-Dawley rats in a total volume of 3 μl, and animals were perfused 5 days later. Brains were processed for immunohistochemistry, to detect β-galactosidase (transgene expression; first two left-hand columns), ED1 (macrophages-microglial cells), and CD8 (infiltrating lymphocytes and NK cells). Low power view of the striatum is shown in the left-hand column, and a higher power view is shown in the column next to it. Scale bar, shown in the upper left = 1 mm; all others = 100 μm. Note that RAd36 allows good transduction of the striatum, in the absence of recruitment of inflammatory cells, compared to animals injected with saline. Also note that we have previously shown quantitatively[11] that transduction measured as expression of β-galactosidase at the dose of 10^4–10^5 IU is comparable to that seen with RAd35 (hCMV βgal) at a dose of 10^7 IU, which induces much higher inflammation. (Modified from Ref. 11.)

To improve the use of adenoviral vectors in both experimental and clinical neuroscience applications, safe, nontoxic, and long-term, regulated transgene expression needs to be achieved. In this chapter we will describe our developments toward achieving these aims. Achieving safe and nontoxic expression will depend on being able to lower the vector doses needed (Fig. 4) and using novel viral vectors which cannot be recognized by the immune system.

The factors determining long-term expression from viral vectors in the brain remain to be explored. Although it has been proposed that long-term expression would be stable when using vectors that integrate their genomes into the chromosomes of host cells, this has not always been the case. Especially in the brain, it has been shown that long-term expression can be achieved in the absence of integration into host chromosomes by vectors derived from HSV-1, HC-Adv, and possibly AAV vectors (although AAV vector genomes can integrate into the host chromosomes they may also express from episomes, and their precise fate in the nuclei of brain cells remains to be determined). The mechanisms and disposition of vector genomes in the nuclei of brain cells during long-term transgene expression are important areas of future work, which will need in-depth analysis. Nevertheless, lack of integration may confer additional safety features on the vectors, since they will be less likely to cause irreversible chromosomal alterations. Importantly, the large cloning capacity of HC-Adv and HSV-1 derived vectors allows the incorporation of regulatory sequences in addition to the transgene. This should further improve the safety of vectors by allowing transgene expression to be selectively turned "on" or "off".[58–61]

Methods

Stereotactic Surgery Procedures

Adult Sprague-Dawley rats of 250 g body weight (Charles River Breeding Laboratories) are anesthetized with halothane and placed in a stereotactic apparatus that is modified for use with inhalational anesthetic (described by us in detail[56]). Animals are injected with HC-Adv into the left striatum (coordinates; 0.6 mm forward, and 3.4 mm lateral from bregma, and 5.0 mm from the dura) with 1×10^7 IU of virus vector using a 10 μl Hamilton syringe with a removable 26-gauge needle. Virus is administered in a volume of 2–3 μl, and each injection is performed over a period of 3 min, with the needle being left in place for a further 5 min before it is slowly withdrawn from the brain. Anesthesia is maintained throughout surgery with 1% halothane in 67% medical O_2 and 33% medical N_2O. We have tested doses from 1×10^1 to 1×10^9 IU, depending on the vector used. At doses above 1×10^8 IU we have detected severe cytotoxicity and chronic inflammatory responses. In absolute terms, we believe 1×10^7 infectious units to be the highest safe dose administered for any adenovirus vector, but this upper limit will vary with virus prepared, titered, and purified in different laboratories. Ideally, the highest dose that causes no major long-term side effects should be determined by each laboratory. A high degree of quality control (e.g., low levels of lipopolysaccharide, absence of replication competent adenovirus, low viral particle : IU ratio, and careful surgical technique)[56] is crucial for obtaining low levels of inflammation. The dose needed will depend on the levels of transgene product needed, and the levels

[58] M. M. Burcin, G. Schiedner, S. Kochanek, S. Y. Tsai, and B. W. O'Malley, *Proc. Natl. Acad. Sci. U.S.A.* **96,** 355 (1999).
[59] T. C. Harding, B. J. Geddes, D. Murphy, D. Knight, and J. B. Uney, *Nat. Biotechnol.* **16,** 553 (1998).
[60] G. S. Ralph, A. Bienemann, T. C. Harding, M. Hopton, J. Henley, and J. B. Uney, *NeuroReport* **11,** 2051 (2000).
[61] J. Smith-Arica, A. E. Morelli, A. T. Larregina, J. Smith, P. R. Lowenstein, and M. G. Castro, *Mol. Ther.* **2,** 579 (2000).

of transgene expression obtained by different promoters. Careful dose-response curves need to be performed for any new vector. At the end of the experiment animals are anesthetized with an overdose of anesthetic and perfused–fixed. We perfuse with oxygenated Tyrode solution, followed by 4% paraformaldehyde in 0.1 M phosphate-buffered saline (PBS), pH 7.4.

Quantification of β-Galactosidase Enzyme Activity

Following the removal of the brain from the cranium, striata injected with vectors are rapidly dissected on ice, homogenized in PBS, pH 7.4, and centrifuged in a microfuge. Supernatants can be stored at $-80°$ until assayed for β-galactosidase activity using a spectrophotometric assay, as described by us,[63] using the substrate O-nitrophenol-β-D-galactopyranoside. Enzyme units are defined as $380 \times A_{420nm}$/min. Enzyme units need to be standardized against protein concentration assayed with, e.g., the BCA Protein Assay Kit.

Immunohistochemistry

Coronal sections (30–50 μm thick) are cut throughout the whole anterior–posterior extent of the striatum using a vibratome. Free-floating immunohistochemistry is performed to detect either the particular transgene of interest or markers of inflammatory or immune cells. Endogenous peroxidase is first inactivated by incubating the section in 0.3% hydrogen peroxide in PBS, and nonspecific epitopes are blocked with 10% normal horse serum before incubating overnight with primary antibody diluted in PBS and containing 1% normal horse serum and 0.5% Triton X-100. Three to five washes in PBS separate incubations in antibodies. The presence of antigen is finally revealed either through the use of secondary antibodies labeled with fluorescent markers, or through the use of biotinylated secondary antibodies and avidin–biotin–peroxidase.

The following set of primary antibodies allows a detailed visualization of vector-mediated transgene expression and virus-induced inflammation in the brain. The different antibodies are used to monitor transgene expression, the appearance and CNS distribution of immunological and inflammatory markers, or cellular integrity.[56] Commercial antibody names are underlined, and recommended dilutions are given in parentheses.

Transgene Expression

anti-β-galactosidase (1 : 1000); Promega; mouse monoclonal

[62] T. D. Southgate, D. Bain, L. D. Fairbanks, A. E. Morelli, A. T. Larregina, H. A. Simmonds, M. G. Castro, and P. R. Lowenstein, *Metab. Brain. Dis.* **14,** 207 (1999).
[63] T. D. Southgate, S. Windeatt, J. Smith-Arica, C. A. Gerdes, M. J. Perone, I. Morris, J. R. Davis, D. Klaztmann, P. R. Lowenstein, and M. G. Castro, *Endocrinology* **141,** 3493 (2000).

Immunological and Inflammatory Markers

anti-CD8, recognizing cytotoxic T lymphocytes and NK cells (1 : 500); Serotec; mouse monoclonal

anti-CD8β, recognizing exclusively T lymphocytes (1 : 200); Serotec; mouse monoclonal

anti-CD4, recognizing helper T-cells (and microglia in rat brain) (1 : 200); Serotec; mouse monoclonal

anti-CD43, recognizing all lymphoid and myeloid-derived cells, but not B-cells (1 : 500); Serotec; mouse monoclonal

anti-CD161, recognizing natural killer (NK) cells (1 : 2000); Serotec; mouse monoclonal

anti-CD45RA, recognizing B cells (1 : 2000); Pharmingen; mouse monoclonal

OX-6, recognizing rat MHC-II (1 : 200); Serotec; mouse monoclonal

OX-18, recognizing rat MHC-I (1 : 200); Serotec; mouse monoclonal

OX-62, recognizing dendritic cells and $\gamma\delta$-T cells (1 : 20); Serotec; mouse monoclonal

ICAM-I, perivascular microglia and activated endothelial cells (1 : 100); Serotec; mouse monoclonal

anti-ED1, recognizing activated macrophages/microglial cells (1 : 1000); Serotec; mouse monoclonal

Markers of Axonal, Neuronal, and Glial Integrity

anti-myelin basic protein (MBP) (1 : 2000); Dako; rabbit polyclonal anti-human

anti-glial fibrillary acidic protein (GFAP), Sigma, astrocytes (1 : 200); Roche; mouse monoclonal

anti-neuN, neuronal nuclei; Chemicon (1 : 50); mouse monoclonal

Secondary Antibodies

biotinylated rabbit anti-mouse or biotinylated swine anti-rabbit (Dako), used, detected, and examined as described by us elsewhere[11,19,20,27,56]

Quantification of Area of Brain Tissue Transduced

Quantitative analysis to determine the area occupied by cells immunoreactive with antibodies against β-galactosidase and immune markers within brain sections is performed using a Leica Quantimet 600 Image Analysis System controlled by QWIN software (Leica Microsystems, Cambridge, UK) using a Leica RMDB microscope, as described by us before.[19,20,27,56] Statistical analysis should be performed to determine the degree of statistical significance.

Quantitative Real-Time PCR for mRNA Encoding HSV-1
Thymidine Kinase or lacZ

Ideally, when assessing gene transfer to the brain, three parameters should be measured: transgene expression, levels of mRNA, and the presence of the adenoviral genome. Usually, most studies only analyze transgene expression. For many purposes this may be sufficient. When assessing the persistence of vectors in the brain, it is useful to determine the presence or absence of genomic vector sequences. In our studies, we have used quantitative PCR to detect and quantify levels of both mRNA coding for transgenes of interest, and the corresponding vector genome sequences. As examples, we describe the protocols for β-galactosidase and HSV1-thymidine kinase. Assays could be developed to detect other genome sequences, and, in the future, it will be of paramount importance to develop *in situ* PCR to determine simultaneously the genomic integrity and anatomical localization of vector sequences.

DNA is obtained from homogenized brain samples of animals injected with adenovirus vectors. Final DNA is obtained using spin columns (Qiagen). RNA samples are obtained from separate brain samples by lysis in Trizol buffer and reverse transcribed into cDNA by using random hexamers (Boehringer Mannheim, Germany), and Moloney murine leukemia virus reverse transcriptase (RT) (Life Technologies, UK). The cDNA is then submitted to standard phenol/chloroform extraction and ethanol precipitation and resuspended in water. Two hundred and fifty ng of DNA or, alternatively, a quantity of cDNA corresponding to 100 ng of total RNA is amplified by PCR in a 50 μl reaction using the TKf′-CGA GCC GAT GAC TTA CTG CG-3′ and TKr 5′-CCC CGG CCG ATA TCT CAC-3′ primers, and the 6-carboxyfluorescein-labeled TKp 5′-TAC ACC ACA CAA CAC CGC CTC GAC C-3′ TaqMan probe (PerkinElmer, UK). RNA control is performed using 2 primers and a VIC^{TM}-labeled 18S ribosomal probe (PerkinElmer). The final concentration is 0.2 μM for each primer, probe 0.2 μM, dNTP 0.2 mM (except for dUTP 0.4 mM), MgCl$_2$ 5 mM, UNG 0.5 U in PCR buffer in the presence of 1.25 units of AmpliTaq Gold polymerase (PCR core reagents, PerkinElmer) with an initial step at 50° for 2 min, than at 95° for 10 min, followed by 40 of the following cycles: denaturation at 95° for 15 sec, annealing extension at 60° for 1 min. Raw data are analyzed using the SDS v1.63 software (PerkinElmer).

The measurement of *lacZ* mRNA and adenoviral genome encoding the *lacZ* gene is performed as described in Ref. 64 . The *lacZ* gene sequence is amplified by TaqMan PCR[65] using the following primers: 5′-TAC TGT CGT CGT CCC CTC AAA-3′ (246F) and 5′-TAA CAA CCC GTC GGA TTC TCC-3′ (368R), and the

[64] M. Senoo, Y. Matsubara, K. Fujii, Y. Nagasaki, M. Hiratsuka, S. Kure, S. Uehara, K. Okamura, A. Yajima, and K. Narisawa, *Mol. Genet. Metab.* **69,** 269 (2000).

[65] C. A. Heid, J. Stevens, K. J. Livak, and P. M. Williams, *Genome Res.* **10,** 986 (1996).

internal fluorogenic hybridization probe 5'-TAT CCC ATT ACG GTC AAT CCG CCG-3' (313T). The composition of the PCR mixture, units of AmpliTaq Gold DNA polymerase (PerkinElmer), and template DNA is as described in Ref. 64. Amplification is performed using an ABI PRISM 7700 sequence detection system (PerkinElmer). The reaction mixture is incubated at 50° for 2 min to degrade possible carryover PCR products from previous reactions and then heated at 95° for 10 min to inactivate uracil-N-glycosylase and to activate the modified DNA polymerase. The subsequent thermoprofile consists of 40 cycles of denaturation at 95° for 30 s, annealing at 59° for 30 s, and extension at 72° for 30 s.

For both genes, fluorescent emission signal from the reporter dye is monitored in real time to determine the "threshold cycle" at which the level of fluorescent emission exceeds the baseline. Copy numbers are determined using a plasmid-based standard curve.

Immunization against Adenovirus Vectors

Two weeks preceding the intracerebral injection of HC-Adv vectors, rats are briefly anesthetized with halothane and injected intradermally in the back with either sterile saline solution or 5×10^8 IU of wild-type (wt) adenovirus type 5. Intradermal injections are performed in a volume of 100 μl of PBS.[19,27,56] Immunization against adenovirus does not require the injection of adjuvants such as Freud's.[57]

Adenovirus-Neutralizing Serum Antibody Assays

To verify that animals have been immunized against adenovirus type 5, titers of neutralizing antibodies in response to the intradermal injection of adenovirus are measured in serum samples using an *in vitro* adenovirus neutralization assay.[19,27,56,66a,66b] The neutralizing antibody titer is determined by infecting 293 cells with a test vector expressing β-galactosidase, which has been previously incubated with serial 1 : 2 dilutions of serum from immunized animals. The neutralizing titer for each sample is described as the reciprocal of the highest dilution of serum that inhibits RAd35-mediated transduction of 293 cells by 50%.[19,27,66a,66b,67]

Assessing Quality of Recombinant Adenoviral Vectors

Three main parameters need to be assessed in any adenoviral preparation which is to be used either for *in vitro* or for *in vivo* experiments: the level of contaminating lipopolysaccharide, the levels of potentially contaminating replication-competent

---

[66a] T. D. Southgate, D. Stone, J. C. Williams, P. R. Lowenstein, and M. G. Castro, *Endocrinology* **142,** 464 (2001).

[66b] R. Parks, C. Eevelegh, and F. Graham, *Gene Ther.* **6,** 1565 (1999).

[67] K. Kajiwara, A. P. Byrnes, Y. Ohmoto, H. M. Charlton, M. J. Wood, and K. J. Wood, *J. Neuroimmunol.* **103,** 8 (2000).

adenovirus, and the purity of the adenovirus preparation (in terms of the ratio of infectious particles to total particles). Failure to perform stringent quality control assays (described in detail in Ref. 68) can lead to spurious experimental results, and subsequent difficulties in interpreting experimental data. The levels of contaminating lipopolysaccharide, a very powerful inflammatory stimulus, can be assessed using the Limulus Amebocyte lysate pyrogen kit (from BioWhittaker), or the E-toxate assay (from Sigma), following manufacturer's instructions. The levels of contaminating lipopolysaccharide should be below 2 endotoxin units/ml.[69] The levels of replication-competent adenovirus contamination should be tested using the technique described in detail in Ref. 70. Although viral vector titers are determined using biological assays, which indicate the number of infectious units, plaque-forming units, blue-forming units (in the case of a β-galactosidase expressing vector), etc., it is likely that in the future, titrations will be done by directly assessing the numbers of genomes present using quantitative molecular determinations.[71,72]

Transport of Adenovirus on Dry Ice

Clinical trials and collaborations between research groups using recombinant adenoviral vectors require the use of dry ice transport between laboratories in order to keep the virus stocks frozen. Transport on dry ice is essential when the virus is to be used immediately (without retitration), as the titer used for the experiments will be the one determined by the laboratory which manufactured and shipped the virus. It is recommended to retitrate the virus after shipment, as titration protocols (and therefore results) vary from lab to lab. This is especially important when comparing the effects of two or more viruses against one another.

It has been observed that significant loss of viral titer occurs during dry ice transport.[73] Thawing of the dry ice over 24 hr results in increased production of CO_2 gas into the transport container, which can penetrate the tubes containing the viruses, thereby decreasing the pH of the virus solution. The viral capsid is unstable at acid pH, and we have found that virus aliquots in small Eppendorf tubes placed directly into dry ice for 48 hr can lose up to 6 log units of infectious titer. A common method of virus transport currently utilized is placing the Eppendorf tubes containing the virus aliquots into 50 ml polypropylene centrifuge tubes,

[68] T. Southgate, P. Kingston, and M. G. Castro, in "Current Protocols in Neuroscience" (J. N. Crawley, C. R. Gerfen, R. McKay, M. A. Rogawski, D. R. Sibley, and P. Skolnick, eds.), p. 4.23.1. John Wiley and Sons, New York, 2000.

[69] M. Cotten, A. Baker, M. Saltik, E. Wagner, and M. Buschle, *Gene Ther.* **1,** 239 (1994).

[70] D. L. Dion, J. Fang, and R. I. Garver, Jr., *J. Virol. Meth.* **56,** 99 (1996).

[71] W. J. Bowers, D. F. Howard, and H. J. Federoff, *Mol. Ther.* **1,** 294 (2000).

[72] P. Umana, C. A. Gerdes, J. R. E. Davis, M. G. Castro, and P. R. Lowenstein, *Nat. Biotechnol.* **19,** 582 (2001).

[73] C. Nyberg-Hoffman and E. Aguilar Cordova, *Nat. Med.* **5,** 955 (1999).

before placing in dry ice. We determined that this precaution is still not sufficient for maintaining titer; a 2-log decrease in titer was observed when virus was stored in this manner in dry ice for 48 hr. We have found that a combination of sealing the Eppendorf tubes in Parafilm and placing this into a 50 ml centrifuge tube, then subsequently into two polyethylene bags maintains the viral titer after storage in dry ice for 48 hr. Alternative methods are described in the original manuscript by Nyberg-Hoffman et al.[73]

Construction of High-Capacity, Helper-Dependent (HC-Adv) Adenoviruses Using FLPe Recombinase

Experimental data using HC-Adv indicate that these vectors are able to transfer larger constructs (>20 kbp) into target brain cells, and are able to do so, even in the presence of anti-adenoviral immune responses.[19,27] The large cloning capacity of HC-Adv makes these vectors ideal for expressing large transgenic constructs under the control of complex regulatory elements.[58,61]

Completely deleted vector need to be grown in the presence of a helper virus to provide in trans all necessary elements for replication and packaging. Helper virus contamination is then significantly reduced through recombinase-mediated excision of its packaging signal. To date, this has been done using *Cre* recombinase expressed in 293 cells. Engineered helper viruses contain a packaging site flanked by *loxP* sites. Growth of such helper virus in 293-*Cre* cells removes the packaging signal from the majority of helper genomes. HC-Adv genomes retain the packaging signal and can be packaged into adenoviral virions. The HC-Adv is initially generated by transfection of producer cells with HC-Adv DNA, followed by infection with helper virus. The resulting HC-Adv is subsequently amplified in a series of helper-plus-HC-Adv coinfection passages in *Cre* recombinase-expressing cells. Following CsCl density-based separation of helper and HC-Adv, the levels of helper virus remaining in the final HC-Adv preparation are approximately 0.1% (ratio of titers of helper virus [expressed as plaque-forming units (PFU)/ml) to titer of HC-Adv (e.g., blue-forming units/ml)].[13,19,41,42,44–48,74]

We describe here the development of a HC-Adv system in which the helper virus packaging site is eliminated through recombination using the yeast FLP recombinase. FLP mediates maximum levels of excision close to 100%, in contrast to 80% for Cre.[75] As wild-type enzyme is not very active at 37°, we utilized an FLP mutant (FLPe) developed by *in vitro* evolution, which is active at 37°.[76] In this system, helper virus contamination of the final preparation of HC-Adv is <0.1% prior to density-based separation by ultracentrifugation,[72] allowing viral purification by chromatographic methods.

[74] V. Sandig, R. Youil, A. J. Bett, L. L. Franlin, M. Oshima, D. Maione, F. Wang, M. L. Metzker, R. Savino, and C. T. Caskey, *Proc. Natl. Acad. Sci. U.S.A.* **97,** 1002 (2000).
[75] L. Ringrose, V. Lounnas, L. Ehrlich, F. Buchholz, R. Wade, and A. F. Stewart, *J. Mol. Biol.* **284,** 363 (1998).

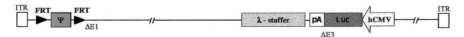

FIG. 5. An E1/E3-deleted helper adenovirus (FL helper) with a packaging signal sensitive to FLP-mediated excision. FL helper was generated by homologous recombination in 293 cells after cotransfection with plasmids pΔE1sp1A-2xFRT and pBHG10-CMVluc-λ. pΔE1sp1A-2xFRT carries left-end adenoviral sequences including the left inverted terminal repeat (ITR) and a packaging signal (Ψ) consisting of domains A1–A5, flanked by parallel, minimal FLP-recombinase target (FRT) sites. It also carries a deletion in the E1 adenoviral sequences. pBHG10-CMVluc-λ is an adenoviral backbone plasmid carrying most of the viral genome except for deletions in the left end (including the packaging signal, ΔΨ) and in the E3 region. It carries a luciferase expression cassette, under the control of the human CMV promoter, plus a stuffer fragment of bacteriophage-λ DNA, subcloned into the *Pac*I restriction site situated in the ΔE3 region. (Figure modified from Ref. 72.)

Introduction of FLP-Recombinase Target (FRT) Sites Flanking Adenoviral Packaging Signal

Minimal (34 bp) FRT sites were introduced by PCR in parallel orientation flanking the adenoviral packaging signal. The PCR product was then subcloned into the left end adenoviral sequences present in plasmid pΔE1sp1A (Microbix, Toronto, Canada). The resulting plasmid was designated pΔE1sp1A-2xFRT (Fig. 5).

Subcloning of Luciferase Expression Cassette and Stuffer DNA in Adenoviral ΔE3 Region

A luciferase expression cassette under the control of the human CMV promoter and including an SV40 polyadenylation sequence, a stuffer DNA fragment, and a bacterial kanamycin resistance cassette were then subcloned into the adenoviral-backbone plasmid pBHG10 (Microbix). A clone with the luciferase expression cassette in antiparallel orientation to the E3 expression unit (pBHG10-CMVluc-λ) was selected, and the kanamycin resistance cassette removed.

Generation of Helper Virus with Packaging Signal Sensitive to Flpe-Mediated Excision

pΔE1sp1A-2xFRT and pBHG10-CMVlucλ DNA were cotransfected into low-passage 293 cells. A first generation, replication-defective, E1/E3-deleted, helper adenovirus, with its viral packaging site (Ψ) flanked by FRT sites and with a luciferase expression cassette plus stuffer DNA subcloned into the ΔE3 region (total DNA size of 35.7 kbp), was generated by homologous recombination in cotransfected 293 cells as described.[68] The resulting virus (FL helper virus) was isolated and amplified by standard methods and titered by an end-point dilution, using the cytopathic end-point assay.[68]

---

[76] F. Buchholz, P. O. Angrand, and A. F. Stewart, *Nat. Biotech.* **16,** 657

Generation of 293-Cell Line Expressing High Levels of FLPe Recombinase

Low-passage 293 cells were transfected with a linearized plasmid pCAGG-SFLPeIRESpuro using the calcium phosphate method. The plasmid pCAGG-SFLPeIRESpuro, containing an *in vitro* evolved FLP recombinase which displays higher activity at 37° compared to wild-type FLP,[76] under the control of a chicken *β*-actin/hCMV promoter, followed by an IRES linked to a puromycin resistance gene, was kindly provided by F. Stewart from EMBL in Heidelberg, Germany.[76] Individual clones were isolated, amplified, and FLPe recombinase activity was determined. A highly active clone, 293Flpe6, was selected for further experiments.

HC-Adv Virus Rescue and Amplification

To rescue HC-Adv virus from HC-Adv plasmid, 293-FLPe6 cells (grown in 1 T25 flask) were transfected with linearized, *β*-galactosidase-expressing HC-Adv plasmid pGS46[19] (kindly provided by Stefan Kochanek, University of Cologne, Germany) using the calcium phosphate transfection method, as described in detail in Ref. 75. Following 16 hr of incubation, cells were infected with FL helper virus using an MOI of 5. Upon full cytopathic effect, virus was harvested by lysing the cells by 3 freeze–thaw cycles. The lysate from this initial rescue was used to amplify the HC-Adv in 3–4 serial steps by coinfecting 293-Flpe cells (at approximately 80% confluency) with FL helper virus plus lysate from the previous step, as described by us in detail (Fig. 6).[72]

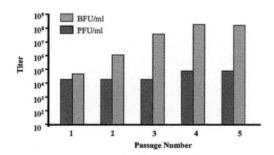

FIG. 6. Rescue and amplification of a *β*-galactosidase-expressing HC-Adv using 293-FLPe cells and FL helper virus. 293-FLPe cells were transfected with an HC-Adv plasmid for *β*-galactosidase expression (pGS46) and HC-Adv virus was rescued by infection with FL helper virus (passage number 1). HC-Adv virus was amplified in a series of 5 passages by coinfecting 293-FLPe cells with lysate from the previous passage plus FL helper virus (added at an MOI of 5). In each passage 293-FLPe cells were infected when cells were approximately 80% confluent and virus was harvested upon full cyto-pathic effect. GS46/Flpe HC-Adv virus titer was measured as blue-forming-units (BFU)/ml, whereas contaminating FL helper virus titer was measured as plaque-forming-units (pfu)/ml. Note the low level of contamination of helper virus. (Figure modified from Ref. 72.)

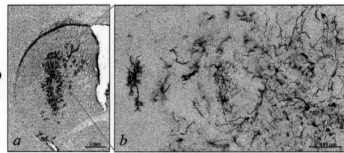

FIG. 7. *In vivo* β-galactosidase expression in rat brain infected with either GS46-FLP HC-Adv vector. 7×10^7 BFU of virus (in 3 μl) was injected into the striatum of male Sprague-Dawley rats. Six days after virus inoculation, β-galactosidase expression was analyzed by immunohistochemistry. (a,b) GS46-FLP HD virus. β-Galactosidase is expressed under the control of the hCMV promoter. (Figure modified from Ref. 72.)

Titration of HC-Adv and FL Helper Virus

HC-Adv virus was titrated by quantifying blue-forming units (BFU).[72] FL helper virus was titrated by quantifying plaque-forming units (pfu) in 96-well plates as described.[68]

Stereotactic Injections into Brain

GS46-Flpe HC-Adv virus was injected into rat brain as described above. Two to three μl of virus containing 7×10^7 BFU was injected into the striatum of male Sprague-Dawley rats (200–250 g), and animals were perfused 5 days later. We detected the distribution of transgene β-galactosidase throughout the striatum (Fig. 7). Staining detected was comparable to that achieved through the injection of a similar dose of an E1/E3-deleted vector.

Conclusions

Novel generations of high-capacity, helper-dependent adenovirus vectors should provide further tools for exploiting the highly efficient gene delivery capacity of adenoviral vectors. The absence of any adenoviral coding sequences allows cloning of large transgenic sequences and reduces the presentation of anti-genic adenoviral epitopes to the immune system. Our data demonstrate the stability of adenovirus genomes in the brains of injected animals, even in the presence of a preexisting anti-adenoviral immune response (a worst-case scenario in potential clinical trials of gene therapy for neurological diseases). Further developments in adenovirus vectorology are likely to involve engineering HC-Ad vector genomes to provide increased long-term genome stability, and combining "retargeted" adenovirus capsids with HC-Ad genomes to provide safe and stable transgene expression in a selected subpopulation of target cells or tissues.

Acknowledgments

Neurological gene therapy projects in our laboratory are funded by The Wellcome Trust, the Parkinson's Disease Society (UK), BBSRC (UK), MRC (UK), EU Biomed II Grants BMH-4-CT98-3277 and BIO-CT98-0297, EU Fifth Framework Grant QLK-CT1999-00365, and the Royal Society. P.R.L. was a Research Fellow of The Lister Institute of Preventive Medicine. We are grateful to R. Poulton for secretarial assistance, and T. Maleniak, J. Podesta, M. C. Burland, and M. Ackyroyd for expert technical assistance.

[18] Gene Therapy Methods in Cardiovascular Diseases

By MIKKO O. HILTUNEN, MIKKO P. TURUNEN, and SEPPO YLÄ-HERTTUALA

Introduction

Gene therapy can be defined as transfer of genetic material into specific cells in order to obtain a therapeutic effect.[1] Techniques for gene transfer have rapidly emerged as an alternative approach for the treatment of various genetic and acquired diseases. Blood vessels are among the easiest targets for gene therapy because of ease of access. The target cells in arteries are endothelial cells, smooth muscle cells (SMC), macrophages, and fibroblasts. Sufficient expression of a gene of interest in the vessel wall can be achieved using either extravascular or intravascular gene delivery approaches. We have demonstrated that also using *ex vivo* delivery of transduced SMCs it is possible to obtain very high transgene expression in target vessel wall.[2] Intramuscular gene transfer can be used for the treatment of ischemic heart disease or peripheral ischemia.

Intravascular gene transfer is easily performed during angioplasty, stenting, and other intravascular manipulations.[3] Limitations of intravascular gene transfer are the presence of anatomical barriers, such as the internal elastic lamina and atherosclerotic lesions[1,4,5] and the presence of the blood complement system

---

[1] S. Ylä-Herttuala, *Curr. Opin. Lipidol.* **8**, 72 (1997).

[2] P. Leppänen, J. Koponen, M. P. Turunen, T. Pakkanen, and S. Ylä-Herttuala, *J. Gene Med.* **3**, 173 (2001).

[3] S. Ylä-Herttuala and J. F. Martin, *Lancet* **355**, 213 (2000).

which efficiently inactivates many gene transfer vectors.[6] Several different types of catheters are commercially available, which can be used for gene transfer.[7] The double balloon catheter is made of two latex balloons which, when inflated into the target arterial segment, delineate a "transfection chamber" of varying length, into which gene transfer solution can be infused. This catheter was the first to be used for catheter-based arterial gene transfer.[8] The limitations of this catheter type are stopped blood flow and likely leakage through arterial side branches. The Dispatch catheter is a balloon catheter that forms separate compartments adjacent to the target vessel wall when the catheter is inflated. Prolonged gene transfer vector infusion can be performed since blood flow occurs through the central core of the catheter. This system has been successfully used to achieve substantial gene delivery into the endothelium and superficial medial layers of atherosclerotic human arteries.[9] Porous and microporous catheters have also been used for arterial delivery of marker genes. The channeled balloon catheter has 24 longitudinal channels, each containing a 100 μm pore, and also allows continuous blood flow into peripheral tissues.[10] Iontophoretic catheters are catheters that use an electroporation technique in combination with a balloon catheter. New catheters have also been developed for intravascular injections, such as the transport catheter, stented porous balloon catheter, and nipple infusion catheter.

When a gene transfer vector is administered on the adventitial surface of the vessel wall, it can stay in close contact with arterial cells for a long time.[11,12] Adventitial gene transfer can be used for the delivery of therapeutic genes into the arterial wall during bypass operations, prosthesis and anastomosis surgery, and endarterectomies.[3] Adventitial gene delivery can be performed with silastic[11] or

[4] J. J. Rome, V. Shayani, M. Y. Flugelman, K. D. Newman, A. Farb, R. Virmani, and D. A. Dichek, *Arterioscler. Thromb.* **14,** 148 (1994).

[5] L. J. Feldman, P. G. Steg, L. P. Zheng, D. Chen, M. Kearney, S. E. McGarr, J. J. Barry, J. F. Dedieu, M. Perricaudet, and J. M. Isner, *J. Clin. Invest.* **95,** 2662 (1995).

[6] C. Plank, K. Mechtler, F. C. Szoka, Jr., and E. Wagner, *Hum. Gene Ther.* **7,** 1437 (1996).

[7] J. E. Willard, C. Landau, D. B. Glamann, D. Burns, M. E. Jessen, M. J. Pirwitz, R. D. Gerard, and R. S. Meidell, *Circulation* **89,** 2190 (1994).

[8] E. G. Nabel, G. Plautz, and G. J. Nabel, *Science* **249,** 1285 (1990).

[9] M. Laitinen, K. Mäkinen, H. Manninen, P. Matsi, M. Kossila, R. S. Agrawal, T. Pakkanen, J. S. Luoma, H. Viita, J. Hartikainen, E. Alhava, M. Laakso, and S. Ylä-Herttuala, *Hum. Gene Ther.* **9,** 1481 (1998).

[10] M. K. Hong, S. C. Wong, A. Farb, M. D. Mehlman, R. Virmani, J. J. Barry, and M. B. Leon, *Coron. Artery. Dis.* **4,** 1023 (1993).

[11] M. Laitinen, T. Pakkanen, E. Donetti, R. Baetta, J. Luoma, P. Lehtolainen, H. Viita, R. Agrawal, A. Miyanohara, T. Friedmann, W. Risau, J. F. Martin, M. Soma, and S. Ylä-Herttuala, *Hum. Gene Ther.* **8,** 1645 (1997).

[12] M. Simons, E. R. Edelman, J. L. DeKeyser, R. Langer, and R. D. Rosenberg, *Nature* **359,** 67 (1992).

biodegradable collar,[13] biodegradable gel,[14] or direct injection into adventitia. The limitation of this technique is that the gene transfer vector or secreted/diffusible gene product has to reach the target cells, which in the case of intimal cells is difficult to accomplish with most vectors.

Ex vivo gene delivery with transduced SMC and endothelial cells has also been tested in cardiovascular systems.[2] This method has the major advantage that the transduction is made in cell culture and no viruses need be transferred into the human body. The cells can be well characterized and implantation of the transduced cells results in localized expression of the treatment gene. Transferred SMC survive several weeks around the arteries and the expression has a long duration. Problems associated with the *ex vivo* gene delivery include a complicated nature of the procedure and difficulty in obtaining autologous cells.

In Vivo Gene Transfer Models

New Zealand White rabbits weighing 2.1–3.5 kg are used in this study. Fentanyl fluanisone (0.3 ml/kg, s.c.; Janssen Pharmaceutica, Beerse, Belgium) and Midazolam (1.5 mg/kg, i.m.; Roche, Basel, Switzerland) were used for anesthesia. Local anesthesia is achieved with intracutaneous injection of lidocaine. For adventitial operations the initial anesthesia is usually enough for the whole operation, but in the gene transfer protocol to the aorta rabbits need additional anesthetics because of the lengthy operation. The most convenient way to keep rabbits anesthetized through the whole intravascular operation is to inject a 0.05 ml bolus of fentanyl fluanosine when needed through an iv cannula placed in the ear vein. Intravenous injection of 0.05 ml naloxon (0.4 mg/ml) can be used for stopping the anesthesia after the operations. To avoid inflammation, intramuscular administration of 125 mg cefuroxim should be used in every setting.

Replication-deficient E1-E3 deleted adenoviruses for the gene transfers were produced in 293 cells and concentrated by ultracentrifugation. Adenoviruses were manufactured in a GMP clinical grade facility and preparations were analyzed for the absence of helper viruses and bacteriological contaminants.[9] An adenovirus titer of 1–2×10^{10} plaque-forming units (pfu) was used for gene transfer.

Animals were bred in the National Experimental Animal Center, Kuopio, Finland. Rabbits were fed with Lactamin K5 (Sverige) and water ad libitum. All animal procedures were approved by the Animal Care and Use Committee, University of Kuopio, Finland.

[13] T. M. Pakkanen, M. Laitinen, M. Hippelainen, M. O. Hiltunen, E. Alhava, and S. Ylä-Herttuala, *J. Gene Med.* **2,** 52 (2000).
[14] C. C. Stephan and T. A. Brock, *P. R. Health Sci. J.* **15,** 169 (1996).

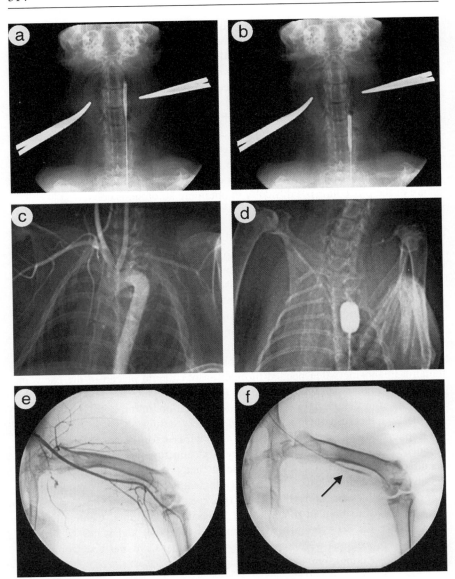

FIG. 1. Denudation of carotid arteries, aorta, and femoral arteries of New Zealand White rabbits. (a,b) Common carotid artery denudation (Sorin Biomedical, CA); (c) angiography of thoracic aorta; (d) inflated embolectomy catheter inside thoracic aorta; (e) angiography of femoral artery; (f) denudation of femoral artery using PTCA catheter (arrow).

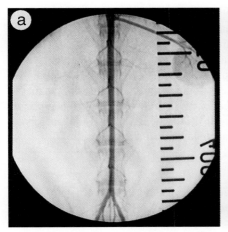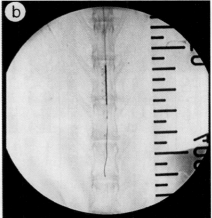

FIG. 2. Gene transfer to the New Zealand White rabbit abdominal aorta using a channeled balloon local drug delivery catheter (Boston Scientific Corp). (a) Angiography of aorta for determining gene transfer sections with no side branches. (b) Balloon fully inflated. These angiographs are needed at the time of gene transfer and sacrifice to localize the target area. The position of the catheter can be easily determined from either renal artery or iliac bifurcation.

Intravascular Gene Transfer

In the model for intravascular gene transfer to aorta, the intimal proliferation is induced by endothelial denudation (Fig. 1).[15] The whole aorta is denuded twice by using an arterial embolectomy catheter. A similar denudation operation can also be done in carotid or femoral arteries of the rabbit as shown in Fig. 1.

Through a midline neck incision the right common carotid artery is exposed and prepared from surrounding tissues and vagus nerve for a length of 4 cm. The carotid artery is then ligated at the distal end and the proximal end is closed with a surgical clip. An introducer sheath (Arrow International, Reading, PA) is positioned inside the artery through incision near the distal closure. Gene transfer is done to the abdominal aorta using a channeled balloon local drug delivery catheter (Boston Scientific Corp., Maple Grove, MA). Under fluoroscopical control, the balloon catheter is positioned caudal to the left renal artery in a segment free of side branches and inflated at 6 atm with a mixture of contrast medium and saline (Fig. 2).[16] Adenovirus is infused in the final volume of 2 ml of 0.9% NaCl at

[15] M. O. Hiltunen, M. Laitinen, M. P. Turunen, M. Jeltsch, J. Hartikainen, T. T. Rissanen, J. Laukkanen, M. Niemi, M. Kossila, T. P. Hakkinen, A. Kivela, B. Enholm, H. Mansukoski, A. M. Turunen, K. Alitalo, and S. Ylä-Herttuala, *Circulation* **102,** 2262 (2000).
[16] M. O. Hiltunen, M. P. Turunen, A. M. Turunen, T. T. Rissanen, M. Laitinen, V. M. Kosma, and S. Ylä-Herttuala, *FASEB J.* **14,** 2230 (2000).

6 atm pressure for 10 min. Side branches will cause leakage and decrease the gene transfer efficiency. The anatomical location of the catheter is determined by measuring its distance from the aortic orifice of the left renal artery (Fig. 2). The right common carotid artery is ligated at the proximal end after removal of the introducer sheath. Intraarterial manipulation during gene transfer to aorta may cause severe thrombus formation and neurological complications. Subcutaneous injections of heparin (200 U) before and after the operation prevents this problem. To prevent parasympathetic effects caused by the operation 0.04 mg of glyco-pyrrone bromide can be injected subcutaneously before the operation.

Periadventitial Gene Transfer

For studying periadventitial gene transfer we use a local periadventitial gene delivery model.[11] Through a midline neck incision of 5 cm the common carotid arteries are exposed between sternohyoid and sternocleidomastoid muscles. The arteries are gently prepared from surrounding tissues and vagus nerve for a length of 3.5 cm and inert silastic collars (Ark Therapeutics Ltd, Kuopio, Finland) are installed around the arteries (Fig. 3). Preparation of both carotid arteries from

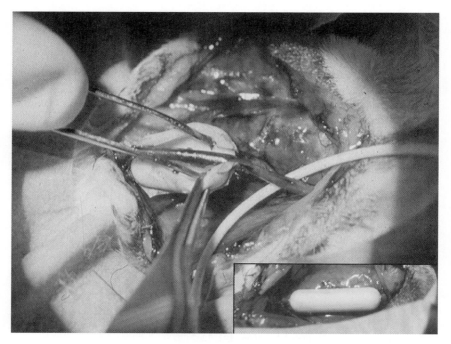

FIG. 3. Installation of a silastic collar to the common carotid artery of a New Zealand White rabbit. The artery is gently prepared from surrounding tissue and collar is wrapped around the artery. (*Inset*) Collar fully installed. A silicone glue or sutures can be used to ensure the closure of the collar.

surrounding tissues can cause damage to the vagus nerve, leading to paralysis of vocal chords and severe breathing difficulties. This can be avoided by using surgical tape around arteries during the procedure. Gene transfer, by opening the collar and filling it with gene transfer solution, can be done immediately after the collar installation or at varying time points depending on the function of the gene of interest. For a gene transfer at a later time point the collar is exposed and opened through the same approach as in the collar installation.

Ex Vivo Gene Transfer Model

Rabbit vascular SMCs are obtained from the central ear artery.[2] The artery is flushed with saline, the medial part is cut into 5 mm pieces, and the explants are cultured in DMEM supplemented with 20% (vol/vol) fetal bovine serum, 100 IU/ml penicillin, 100 μg/ml streptomycin, and 2.5 μg/ml Fungizone (Gibco BRL, Rockville, MD). Cell identity is verified as SMCs with α/γ-actin immunostaining (HHF-35 antibody, Enzo Diagnostics, Farmingdale, NY). Medium is changed every 3 days.

Four hours prior to transfection SMCs are stimulated with a growth supplement (SMGS, Cascade Biologics Inc., Portland, OR). The cells are transfected with 6.6×10^6 retrovirus particles (MOI 2) and polybrene (8 μg/ml). pBAG retrovirus containing Escherichia coli β-galactosidase (lacZ) cDNA is pseudotyped with the G glycoprotein of vesicular stomatitis virus (VSV-G) as described.[17] The collected virus supernatants are filtered (0.45 μm) and concentrated by ultracentrifugation in a Beckman SW-28 rotor at 50,000g (25,000 rpm) at 4° for 2 hr. The concentrated virus is resuspended in 200 μl 0.1 × Hanks/1% sucrose in PBS and titered on NIH 3T3 cells. The titer is 1.0×10^8 cfu/ml. Virus preparations are shown to be free of helper viruses, mycoplasma, other microbiological contaminants, and lipopolysaccharide.[9]

After 1 hr incubation 250 μl FBS is added. The medium is changed the next day, and selection is started by adding G418 (0.5 mg/ml, Gibco BRL). After 10 days, the selection is completed and approximately 4×10^5 cells are transferred to 300 μl of autologous rabbit serum. Cells at the third passage are used. The β-galactosidase activity is determined on Chamber slides with X-Gal staining for 3 hr.

Rabbits are anesthetized as described above. Carotid arteries are exposed using a midline neck incision. Two different methods can be used for the implantation of SMCs. For the first method, a 3 cm silastic collar (Ark Therapeutics Ltd, Kuopio, Finland) is positioned around the artery,[18] where it serves as a reservoir for the

[17] J. C. Burns, T. Friedmann, W. Driever, M. Burrascano, and J. K. Yee, Proc. Natl. Acad. Sci. U.S.A. **90,** 8033 (1993).

[18] M. P. Turunen, M. O. Hiltunen, M. Ruponen, L. Virkamaki, F. C. Szoka, A. Urtti, and S. Ylä-Herttuala, Gene Ther. **6,** 6 (1999).

SMC/serum suspension. In the second method, a surgical collagen sheet (Braun, Melsungen, Germany) is placed around the carotid artery. The cell suspension is applied between the collagen sheet and the artery. The collagen sheet is then wrapped around the artery.[13]

Collection of Tissue Specimens

General anesthesia is achieved as described above followed by a 0.4 ml intravenous bolus of fentanyl fluanosine. Arteries are then removed, flushed gently with saline, and divided into three equal parts. The proximal third is immersion-fixed in 4% paraformaldehyde/15% sucrose (pH 7.4) for 4 hr, rinsed in 15% sucrose (pH 7.4) overnight, embedded in paraffin, and used for immunohistochemistry.[19] The medial third is snap frozen in liquid nitrogen and stored at $-70°$ for RT-PCR analyses. The distal third is immersion-fixed in 4% paraformaldehyde/phosphate buffered saline (pH 7.4) for 10 min, rinsed 24 hr in phosphate-buffered saline (pH 7.2), and embedded in OCT compound and used for X-Gal staining. Various methods for tissue fixation have been reported. We have found 10 min fixation in the above-mentioned buffer optimal for arterial samples and 30 min fixation for the other tissues. Longer fixation, e.g., in the liver, abolishes the background from the low expression of endogenous β-galactosidase activity.

β-Galactosidase Assay and Histology

Evaluation of the gene transfer efficiency is done using X-Gal staining of OCT embedded tissue sections. Ten-μm sections are cut on a cryostat. X-Gal reagent (100 mg/ml) is prepared by dissolving 1 g X-Gal (5-bromo-4-chloro-3-indolyl-β-D-galactoside, MBI Fermentas) to 10 ml dimethyl formamide. X-Gal reagent is diluted dropwise 1 : 100 with X-Gal solution [5 mM K$_3$Fe(CN)$_6$, 5 mM K$_4$Fe(CN)$_6$, 2 mM MgCl$_2$] by slowly mixing to avoid precipitation. Of this solution, 100 μl is pipetted directly on the frozen sections and incubated in dark at $+37°$ for 8 hr. Sections are rinsed with 1 × PBS, dehydrated with rising ethanol concentrations, counterstained with Mayer's Carmalum, and embedded with Permount (Fischer Scientific). Gene transfer efficiency is calculated as a percentage of X-Gal positive cells of all arterial cells. Using plasmid/liposome complexes the gene expression time in the vessel wall lasts up to 2–3 weeks.[18] Using adenoviruses the expression time is longer, lasting up to 4 weeks. False positive results may be obtained because of the endogenous expression of β-galactosidase. Too long incubation times in X-Gal often result in background staining from endogenous β-galactosidase activity. The use of positive and negative controls in X-Gal staining is essential.

---

[19] S. Ylä-Herttuala, M. E. Rosenfeld, S. Parthasarathy, C. K. Glass, E. Sigal, J. L. Witztum, and D. Steinberg, *Proc. Natl. Acad. Sci. U.S.A.* **87,** 6959 (1990).

TABLE I
ANTIBODIES USEFUL FOR IMMUNOCYTOCHEMISTRY OF RABBIT AND
HUMAN TISSUES

Antibody	Target	Dilution	Producer
RAM-11	Rabbit macrophages	1 : 50	Dako
CD68	Human macrophages	1 : 100	Dako
HHF-35	SMC α-actin	1 : 500	Enzo Diagnostics
Anti-β-galactosidase	β-Galactosidase protein	1 : 100	Sigma
CD-31	Endothelial cells	1 : 50	Dako
MCA-805	T cells	1 : 30	Dako

Background staining can also be easily recognized by using nuclear targeted *lacZ* constructs.[11] The number of cells positive for transfected gene varies between sections. It is essential to count cells from at least 10 different randomly selected sections and give the average count. Analysis of a large number of samples can be automated using image analysis systems.

Paraffin sections are used for immunocytochemical detection of SMCs (HHF35; DAKO, dilution 1 : 50), macrophages (RAM-11; DAKO, dilution 1 : 50), endothelium (CD31; DAKO, dilution 1 : 50), and T-cells (MCA 805; D AKO, dilution 1 : 100) (Table 1). Controls for immunostainings include incubations with class and species matched immunoglobulins and incubations where primary antibodies are omitted.[19]

Reverse Transcriptase Polymerase Chain Reaction

RT-PCR is used to detect expression of the transfected marker gene in target tissues.[16] Total RNA is extracted using Trizol Reagent (Gibco BRL) according to the manufacturer's instructions, and cDNA synthesis is performed with 4 μg of total RNA using random hexamer primers (Promega). The primers should be designed to distinguish between endogenous *lacZ* and transduced *lacZ* by selecting the 5′ primer from the transcribed promoter region and the 3′ primer from *lacZ* gene.[16]

Summary

Local gene transfer into the vascular wall offers a promising alternative to treat atherosclerosis-related diseases. Blood vessels are among the easiest targets for gene therapy because of percutaneous, catheter-based treatment methods. On the other hand, gene transfer to the artery wall can also be accomplished from adventitia either by *ex vivo* gene transfer and implantation of transfected cells or by direct *in vivo* gene transfer methods. In the future, as the pathological processes in arteries are better understood, several therapeutic genes could be combined and these "gene cocktails" are expected to produce enhanced therapeutic effects in vascular gene therapy.

We have developed a new, efficient technique for performing *ex vivo* gene transfer to rabbit arterial wall using autologous SMC.[2] The cells were harvested from rabbit ear artery, transfected *in vitro* with VSV-G pseudotyped *lacZ* retrovirus, and returned back to the adventitial surface of the carotid artery using a silicone collar or collagen sheet placed around the artery. The transduced SMCs implanted with a high efficiency and expressed β-galactosidase marker gene at a very high level 7 days and 14 days after the operation. The level of *lacZ* expression decreased thereafter, but was still easily detectable for at least 6 months and was exclusively localized to the site of cell implantation inside the collar.

Development of new vectors, such as baculovirus,[20] for gene transfer will provide targeted, efficient, and safer methods for gene delivery. Plasmids and viruses coding for more than one protein, and bearing regulatory elements, would be useful for future gene therapy applications. Also, constructing second-generation viruses that contain fewer endogenous genes in their genome may reduce immunological reactions caused by the first-generation adenoviruses. In conditions where stable expression of therapeutic proteins is needed, it is necessary to develop better *ex vivo* and *in vivo* gene transfer strategies. Also, production of viruses that can efficiently transfect nondividing cells will be important for future applications of vascular gene therapy. However, current knowledge from vascular gene transfer experiments strongly suggests that vascular gene transfer is a promising new alternative for the treatment of cardiovascular diseases.[3]

[20] K. J. Airenne, M. O. Hiltunen, M. P. Turunen, A. M. Turunen, O. H. Laitinen, M. S. Kulomaa, and S. Ylä-Herttuala, *Gene Ther.* **7,** 1499 (2000).

Section III

Adeno-Associated Virus

[19] Gene Delivery to Cardiac Muscle

By Nathalie Neyroud, H. Bradley Nuss, Michelle K. Leppo,
Eduardo Marbán, and J. Kevin Donahue

Introduction

Gene transfer into the heart has a variety of potential uses. From a traditional scientific perspective, the introduction of defined genes into the heart facilitates the determination of the function of that gene within the proper context of its physiological function. Much insight can be gained from overexpression of wild-type genes or suppression of existing gene products by dominant-negative or antisense constructs. Another set of insights, more relevant to gene therapy, is the use of somatic gene transfer to define and to refine genetic approaches to modification of the substrate toward salutary ends. Recent work indicates the promise of gene therapy for the treatment of common cardiac disorders such as heart failure[1] and arrhythmias.[2] Such work will never come to full fruition without substantial pre-clinical groundwork defining the best genes, vectors, and delivery strategies.

Our focus has been on the use of adenoviral vectors to manipulate cardiac excitability. Studying the role of ion channels in cardiac electrophysiology is complicated by culture-induced changes in the control action potential.[3] To circumvent culture-related effects on the action potential waveform, we choose to deliver ion channel constructs by direct intramyocardial injection of recombinant adenovirus vectors *in vivo* and to study the electrophysiology of cardiomyocytes freshly isolated from such hearts. If a sufficiently large region of the heart is injected, end points can also include global *in vivo* measures such as the electrocardiogram. This approach has helped us to define a number of promising candidate genes for use in antiarrhythmic gene therapy. Meanwhile, we have been developing improved means to deliver genes using clinically realistic intravascular, catheter-mediated approaches.

This article describes these two different methods of *in vivo* gene delivery into the heart. Although our focus has been on manipulation of excitability, the methods described are generally applicable.

Intramyocardial Injection of Adenoviruses

Intramyocardial injection is a method of gene delivery that allows transfer of genes from any vector. Here, in order to study the electrophysiological effects of

---

[1] R. J. Hajjar, F. Del Monte, T. Matsui, and A. Rosenzweig, *Circ. Res.* **86,** 616 (2000).
[2] J. K. Donahue, A. W. Heldman, H. Fraser, A. McDonald, J. M. Miller, J. J. Rade, T. Eschenhagen, and E. Marban, *Nature Med.* **6,** 1395 (2000).

the overexpression of ion channels in heart cells, we use recombinant adenovirus vectors to transfer ion channel genes into adult rat cardiomyocytes.

Adenoviral Constructs

As emphasized above, the delivery methods are generalizable to a variety of vectors, given that they are capable of introducing genes into heart cells. We have used adenoviral constructs, and we describe our usual application as an example. To control the expression of ion channel genes *in vivo*, we have used an ecdysone-inducible promoter system to regulate transgene expression. The ecdysone-inducible promoter enables graded, reversible activation of transgene expression.[4] It also facilitates the elaboration of viral constructs in which the transgene may have a negative selective pressure, as the transgene remains silent during the processes of recombination, amplification, and purification. Since the expression of most genes is tightly controlled under physiological conditions, precise and predictable regulation of transgene expression in target cells is an important tool. In the case of overexpression of a protein or a part of a protein that can be toxic for cells, it is also highly useful that the gene remain silent until its expression is induced in target cells.

The coding sequence of the ion channel gene of interest is cloned into the multiple cloning site of a plasmid containing an ecdysone-inducible promoter. In order to visualize the expression of the ion channel protein in infected cells, a gene encoding the green fluorescent protein (GFP) is also cloned into the same plasmid behind an internal ribosome entry site (IRES).[5,6] This system allows the expression of two genes in the same cells from one plasmid.

The adenovirus AdDBEcR provides the hybrid ecdysone receptor DBEcR[7] to infected cells. This receptor combines the advantages of the *Drosophila* (DmEcR) and the *Bombyx* ecdysone receptor (BmEcR) and preserves the ability to bind to the ecdysone promoter without exogenous retinoid X receptor (RxR).

Adenovirus vectors are generated by Cre-lox recombination of purified $\Psi 5$ viral DNA and shuttle vector DNAs in CRE8 cells expressing the CRE recombinase.[4,8] The recombinant products are plaque-purified and expanded and purified on cesium chloride gradients yielding concentrations of the order of 10^{10} plaque-forming units (pfu)/ml. These are recombinant, replication-deficient adenoviruses which lack the E1 and E3 viral genes required for viral replication.

[3] J. S. Mitcheson, J. C. Hancox, and A. J. Levi, *Cardiovasc. Res.* **39**, 280 (1998).

[4] D. C. Johns, R. Marx, R. E. Mains, B. O'Rourke, and E. Marban, *J. Neurosci.* **19**, 1691 (1999).

[5] S. K. Jang, M. V. Davies, R. J. Kaufman, and E. Wimmer, *J. Virol.* **63**, 1651 (1989).

[6] J. Pelletier and N. Sonenberg, *Nature* **334**, 320 (1988).

[7] U. C. Hoppe, E. Marban, and D. C. Johns, *Mol. Ther.* **1**, 159 (2000).

[8] S. Hardy, M. Kitamura, T. Harris-Stansil, Y. Dai, and M. L. Phipps, *J. Virol.* **71**, 1842 (1997).

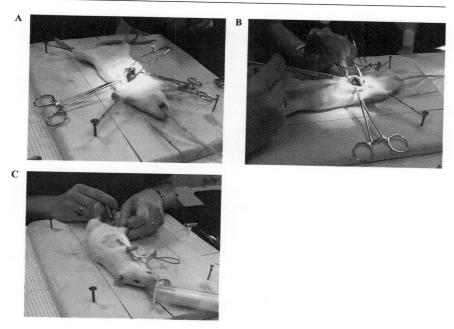

FIG. 1. Intramyocardial adenovirus injection. After the chest is opened (A), adult rats undergo a direct intramyocardial adenovirus injection. The anterior wall of the left ventricle is infiltrated under direct viewing 3 to 5 times with a total volume of 150 μl of the adenovirus mixture (B). After the chest is closed, the rats are injected intraperitoneally with 40 mg of GS-E (C).

Intramyocardial Injection

The animal procedures were approved by the institutional animal care committee and conform to AALAC guidelines.

Normal female rats of the Sprague-Dawley strain from 200 to 250 g are anesthetized with methoxyflurane. During the onset of anesthesia, animals are shaved around the surgical area. An aseptic procedure is followed: instruments are autoclaved in advance and kept sterile until the time of surgery. As shown in Fig. 1, a left thoracotomy and pericardiotomy are performed in the fifth or sixth intercostal space and intermittent positive pressure ventilation with 100% oxygen is given. Under direct visualization, the anterior wall of the left ventricular myocardium is infiltrated three to five times with a total volume of 150 μl of an adenovirus mixture containing approximately 5×10^8 pfu of each adenovirus, using a 30-gauge needle.

After intramyocardial injection, the muscle layer is closed with a single pursestring stitch. The cutaneous layers are approximated with three staples. After the chest is closed, rats are injected intraperitoneally with 40 mg of the non-steroidal ecdysone receptor agonist, GS-E [N-(3-methoxy-2-ethylbenzoyl)-N'-(3,5-dimethylbenzoyl)-N'-*tert*-butylhydrazine] kindly provided by Rohm and Haas Co., Spring House, PA, USA), dissolved in 90 μl of DMSO and 360 μl of sesame oil. GS-E is a member of the bisacylhydrazine chemical family that has been shown to have no adverse effects over a broad dosage range in mammals.[9] This dose of GS-E produces robust and possibly maximal transgene induction in the rat or guinea pig; we have yet to perform dose-response curves to establish the full range of induction by the ecdysone analog.

Animals are then placed in a recovery cage and closely watched until the effects of anesthesia wear off and they regain consciousness and have control of their posture. An intramuscular injection of morphine sulfate (2 mg/kg) is given for postsurgical discomfort, as required to manage pain.

Cell Isolation

Seventy-two hours after intramyocardial injection, left ventricular myocytes are isolated. Animals are subjected to euthanasia by an intraperitoneal injection of pentobarbital (50 mg/kg). After opening of the chest cavity by a large thoracotomy, all the arteries and veins are cut and the heart is rapidly removed and transferred to a beaker containing oxygenated Tyrode's solution with 1 mM Ca^{2+} (composition: 1 mM CaCl$_2$, 140 mM NaCl, 5 mM KCl, 1 mM MgCl$_2$, 10 mM HEPES, and 10 mM glucose; pH 7.4). The heart is mounted on a Langendorff perfusion apparatus by cannulating the aorta, and is retrogradely perfused at a constant flow (6 ml/min) with Tyrode's solution containing 1 mM Ca^{2+} at 37°. This retrograde perfusion by the aorta allows the perfusion of the coronary arteries.

Once the preparation appears stable and the coronary vessels have cleared of blood, the perfusion is switched to a Ca^{2+}-free Tyrode's solution for 5 min (composition: 140 mM NaCl, 5 mM KCl, 1 mM MgCl$_2$, 10 mM HEPES, and 10 mM glucose; pH 7.4). The perfusate is then switched to a Ca^{2+}-free Tyrode's solution containing 1 mg/ml collagenase type II (305 U/mg, Worthington Biochemical Corp., Lakewood, NJ) and 0.1 mg/ml protease type XIV (4.4 U/mg, Sigma, St. Louis, MO). This solution is recirculated to give a total exposure to the enzymes of 10 min. The heart is then perfused again with a Ca^{2+}-free Tyrode's solution for 5 min.

At the end of the perfusion, the heart is cut down and transferred to a bath of high-K$^+$ Tyrode's solution (composition: 120 mM potassium glutamate, 25 mM KCl, 1 mM MgCl$_2$, 10 mM HEPES, 1 mM EGTA, and 10 mM glucose; pH 7.4). The left ventricle is cut open and the injected area is dissected and chopped. Isolated cells are harvested by filtration on a nylon mesh (Lab Pak). The image of a single cardiac myocyte expressing GFP is shown in Fig. 2.

[9] T. S. Dhadialla, G. R. Carlson, and D. P. Le, *Annu. Rev. Entomol.* **43,** 545 (1998).

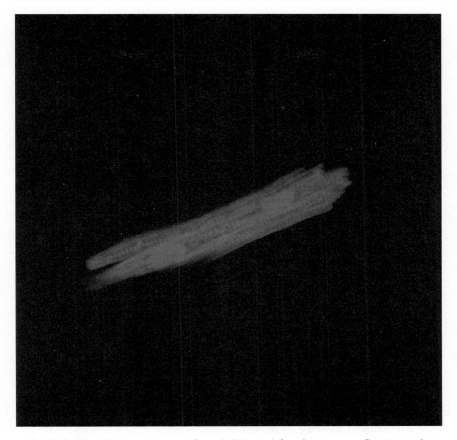

FIG. 2. Confocal fluorescence image of a typical *in vivo* infected rat myocyte. Seventy-two hours after adenovirus injection and stimulation with GS-E, myocytes are isolated and imaged with a confocal microscope. The presence of GFP indicates induced expression of the ion channel protein from the adenovirus.

Confocal Imaging

The image (Fig. 2) was taken on a laser confocal inverted microscope (Zeiss LSM 410; Carl Zeiss Inc., Thornwood, NY, USA) with 40× water immersion objective lens. After isolation, a drop of cell suspension containing left-ventricle cells in high-K^+ Tyrode's solution is placed on a coverslip on the stage of the confocal microscope. GFP is imaged with an argon laser at an excitation wavelength of 488 nm.

In the injected area, the yield of infected myocytes is typically 1 to 3%. Infected myocytes can be readily identified by epifluorescence microscopy and used for single-cell studies.[10] Figure 2 shows a typical freshly isolated cardiac myocyte, which can be seen to have been infected by inspection under direct epifluorescent illumination. Given that most of the constructs express the reporter as part of a bicistronic construct, expression of the ion channel protein of interest can be predicted by the presence of the coexpressed GFP.

Intramyocardial injection of adenoviruses can also be performed to study the effects of ion channel overexpression on global cardiac function. For example, for electrocardiogram (ECG) recordings, the left ventricle is injected in a more widespread manner at multiple sites (about 10 times) from the base to the apex of the anterior, lateral, and posterior wall with a total volume of 220 μl of the adenovirus mixture.[10] Surface ECGs are recorded immediately after operation and 72 hr after intramyocardial injection. Rats are anesthetized with methoxyflurane and needle electrodes are placed under the skin. Electrode positions are optimized to obtain maximal amplitude recordings, enabling accurate measurements of QT intervals. ECGs are simultaneously recorded from standard lead II, modified lead I with arm electrode placed at the base of the sternum, and modified lead III with the arm electrode placed to the back of the left shoulder. Needle electrode positions are marked postoperatively on the rats' skin to ensure exactly the same electrode position for 72-hr controls.[10] Such recordings have enabled us to create the first animal models of genetically induced long QT syndrome without germline manipulation.[10,11]

Although direct injection via thoracotomy can be used to modify cellular or global function, its traumatic and invasive nature preclude its use as a favored method of clinical gene delivery. The next section focuses on intravascular delivery as a promising alternative method.

Intravascular Delivery to Myocardium

Current gene therapy models are limited by inadequate vector delivery. The initial delivery methods utilized injection of adenovirus directly into the myocardium and achieved high-density gene transfer, but only within close proximity to the needle track.[12] Intracoronary catheterization models succeeded in broadening the distribution of virus delivery, but the percentage of infected cells within the target area was low. The best result was 30% of ventricular myocytes in a rabbit model.[13] A subsequent study, however, delivered a similar concentration of virus after visualization and needle puncture of porcine coronary arteries and fewer than 1%

[10] U. C. Hoppe, E. Marban, and D. C. Johns, *J. Clin. Invest.* **105,** 1077 (2000).

[11] U. C. Hoppe, E. Marban, and D. C. Johns, *Proc. Natl. Acad. Sci. U.S.A.* **98,** 5335 (2001).

[12] R. J. Guzman, P. Lemarchand, R. G. Crystal, S. E. Epstein, and T. Finkel, *Circ. Res.* **73,** 1202 (1993).

[13] E. Barr, J. Carroll, A. M. Kalynych, S. K. Tripathy, K. Kozarsky, J. M. Wilson, and J. M. Leiden, *Gene Ther.* **1,** 51 (1994).

of myocytes were infected.[14] Coronary infusion protocols performed under cold, *ex vivo* conditions prior to cardiac transplantation have been disappointing, with gene transfer to less than 1% of myocytes.[15,16] A novel delivery system infused virus into rat left ventricles during aortic and pulmonary artery cross-clamping and achieved gene transfer in approximately 50% of ventricular myocytes in a patchy, heterogeneous distribution.[17] To date, no delivery method has successfully reached the entire myocardium in a homogeneous fashion.

Biological Parameters Important for Efficient Delivery to Solid Organs

In order to enhance the efficiency of adenovirus-mediated gene transfer, we investigated the characteristics of virus–cell interactions in isolated rabbit cardiac myocytes and *ex vivo*-perfused rabbit hearts.[18] Infection was evaluated by using a recombinant adenovirus encoding the *Escherichia coli* β-galactosidase gene and by staining cells with 5-bromo,4-chloro,3-indolyl-β-D-galactopyranoside (X-Gal) to quantify gene transfer. In isolated myocytes, we found that virus attachment to cells typifies a receptor–ligand interaction. The reaction has two components, virus diffusion to the cell surface and attachment of the virus to the receptor. The final percentage of infected cells (and thus gene transfer) depends on the starting amounts of virus particles and cells as well as on the temperature, exposure time, and culture medium (Fig. 3). If the reaction is not allowed to go to completion, the percentage of infected cells is dependent on the concentration of the reactants. When sufficient time is given, the final percentage of infected cells is related to the starting virus amount in a sigmoidal fashion. Temperature affected both the diffusion and the attachment steps. At $37°$, both steps proceeded efficiently. At $24°$, diffusion was slowed but attachment was unaffected, so the reaction rate was decreased but the final percentage of infected cells was unaffected. At $4°$, both diffusion and attachment were hindered. The culture medium was found to affect virus attachment to myocytes when inhibitors to infection were present (red blood cells, antibodies, intravenous contrast agents).

In the *ex vivo* model, a relationship was found between infection efficiency and coronary flow rate, virus concentration, and infection time (Fig. 4). The effect of increasing coronary flow demonstrated threshold behavior, i.e., increasing coronary flow rate improved infection up to 30 ml/min. Above 30 ml/min, the infection

[14] J. Muhlhauser, M. Jones, I. Yamada, C. Cirielli, P. Lemarchand, T. R. Gloe, B. Bewig, S. Signoretti, R. G. Crystal, and M. C. Capogrossi, *Gene Ther.* **3**, 145 (1996).

[15] S. Gojo, S. Kitamura, K. Niwaya, Y. Yoshida, H. Sakaguchi, and K. Kawachi, *Transplant Proc.* **28**, 1818 (1996).

[16] J. Lee, H. Laks, D. C. Drinkwater, A. Blitz, L. Lam, Y. Shiraishi, P. Chang, T. A. Drake, and A. Ardehali, *J. Thorac. Cardiovasc. Surg.* **111**, 246 (1996).

[17] R. J. Hajjar, U. Schmidt, T. Matsui, L. Guerrero, K. Lee, J. K. Gwathmey, G. Dec, M. Semigran, and A. Rosenzweig, *Proc. Natl. Acad. Sci. U.S.A.* **95**, 5251 (1998).

[18] J. K. Donahue, K. Kikkawa, D. C. Johns, E. Marban, and J. H. Lawrence, *Proc. Natl. Acad. Sci. U.S.A.* **94**, 4664 (1997).

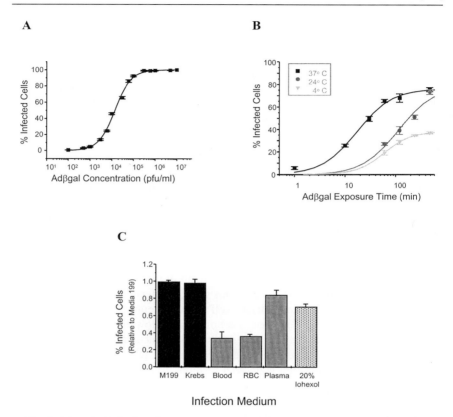

FIG. 3. Adenovirus infection of myocytes *in vitro*. (A) Dose-response curve of adenovirus infection of rabbit ventricular myocytes plated at a density of 10^5 cells/plate. (B) Response of adenoviral infection efficiency to variance of time and temperature at a virus concentration of 10^5 pfu/ml and a cell density of 10^5 myocytes/plate. (C) Comparison of infection efficiency with different infection media. All results are normalized to M199 culture media for 2 hr infection of 10^5 myocytes with 10^5 pfu of adenovirus.

rate plateaued. Virus concentration and exposure time were complementary variables, so that increases in virus concentration could be used to compensate for decreases in exposure time.

Improving Vector Delivery by Increasing Microvascular Permeability

After investigating the fundamental characteristics of adenovirus infection of the myocardium, we sought conditions in the *ex vivo* perfusion model that would reduce the exposure time required for gene transfer (Fig. 5A).[19] A potential

[19] J. K. Donahue, K. Kikkawa, A. Thomas, E. Marban, and J. H. Lawrence, *Gene Ther.* **5**, 630 (1998).

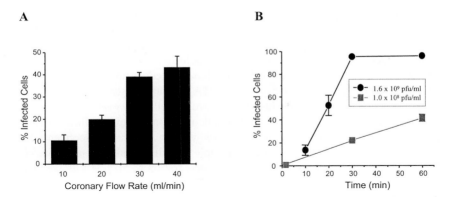

FIG. 4. Adenovirus infection of *ex vivo* perfused rabbit hearts. (A) Improvement of infection effi-ciency with increasing coronary flow rate. Rabbit hearts were perfused with 10^8 pfu/ml of adenovirus for 2 hr. (B) Variation of the percentage of infected myocytes with virus exposure time and concentration. Coronary flow rate for these experiments was 30 ml/min.

solution to this problem was to break down the vascular and extravascular barriers to virus diffusion. Since inflammatory mediators have been shown to increase microvascular permeability,[20] we evaluated gene transfer efficacy after transient exposure to serotonin or bradykinin. Treatment with 10 μM serotonin or bradykinin

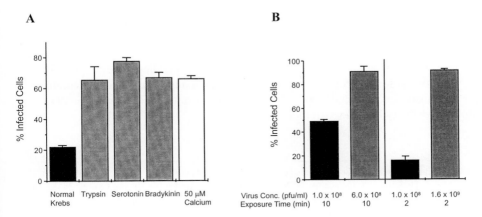

FIG. 5. Modulation of vascular permeability in *ex vivo* perfused rabbit hearts. (A) Perfusion with 1 mg/ml trypsin for 30 sec, or 50 μmol/liter calcium for 2 min, or 10 μmol/liter of either serotonin or bradykinin for 15 min prior to exposure to 10^8 pfu/ml of adenovirus for 30 min. Experiments using low calcium, serotonin, or bradykinin pretreatment continued the same agent during infection. (B) Achievement of gene transfer to >90% of cardiac myocytes in the *ex vivo* perfused heart after 2- or 10-min infection. Hearts were treated with 10 μmol/liter serotonin for 15 min and 50 μmol/liter calcium for 2 min before and during infection.

for 15 min prior to recombinant adenoviral infection significantly improved the efficiency of gene transfer.

Following a similar line of reasoning, we postulated that transient reduction in the extracellular calcium concentration would disrupt endothelial tight junctions and possibly extracellular matrix attachment. Reduction in the extracellular calcium concentration by chelation with EDTA was previously shown to increase adenoviral gene transfer to hepatocytes.[21] Cardiac cells, however, do not tolerate similar treatment, since the elimination of calcium from the extracellular environment is toxic when calcium is reintroduced to myocytes.[22] A calcium concentration of 50 μM is sufficient, however, to avoid this so-called "calcium paradox." Indeed, a 2-min wash-down of the calcium concentration to 50 μM followed by a 30-min adenovirus exposure caused an increase in infection similar to that seen with serotonin or bradykinin. A synergistic effect was found between serotonin and low calcium. Ultimately, gene transfer to more than 90% of cardiac myocytes was achieved with 10 μM of serotonin treatment and reduction of extracellular calcium to 50 μM prior to adenovirus exposure (Fig. 5B).

The permeabilizing effects of serotonin and bradykinin, as well as those from histamine, vascular endothelial growth factor (VEGF), substance P, and a variety of other compounds, result from activation of a series of intracellular enzymes including phospholipase Cγ, protein kinase Cα, constitutive nitric oxide synthase, guanylate cyclase, and cGMP-dependent protein kinases.[23–27] Various elements of this intracellular signaling pathway can be exploited to potentiate gene transfer to cardiac myocytes.

In the *ex vivo* perfusion model, VEGF exposure improved gene transfer in a dose-dependent fashion (Fig. 6A). Likewise, exogenous administration of either nitric oxide or cGMP from nitroglycerin or 8-Br-cGMP perfusion improved gene delivery. Prevention of cGMP metabolism by phosphodiesterase 5 (PDE-5) inhibition caused an increase in gene transfer when administered alone; combined administration of VEGF and a PDE-5 inhibitor resulted in a dramatic increase in gene delivery in the *ex vivo* model (Fig. 6B).

[20] I. Breil, T. Koch, M. Belz, K. Vanackern, and H. Neuhof, *Acta Physiol. Scand.* **159,** 189 (1997).

[21] W. K. De Roos, F. J. Fallaux, A. W. Marinelli, A. Lazaris-Karatzas, A. B. von Geusau, M. M. van der Eb, S. J. Cramer, O. T. Terpstra, and R. C. Hoeben, *Gene Ther.* **4,** 55 (1997).

[22] R. A. Chapman, *Biomed. Biochem. Acta* **46,** S512 (1987).

[23] K. Nagata, E. Marban, J. H. Lawrence, and J. K. Donahue, *J. Mol. Cell. Cardiol.* **33,** 575 (2001).

[24] H. M. Wu, Q. Huang, Y. Yuan, and H. J. Granger, *Am. J. Physiol.* **271,** H2735 (1996).

[25] P. Kubes and D. N. Granger, *Am. J. Physiol.* **262,** H611 (1992).

[26] P. Xia, L. P. Aiello, H. Ishii, Z. Y. Jiang, D. J. Park, G. S. Robinson, H. Takagi, W. P. Newsome, M. R. Jirousek, and G. L. King, *J. Clin. Invest.* **98,** 2018 (1996).

[27] H. M. Wu, Y. Yuan, D. C. Zawieja, J. Tinsley, and H. J. Granger, *Am. J. Physiol.* **276,** H535 (1999).

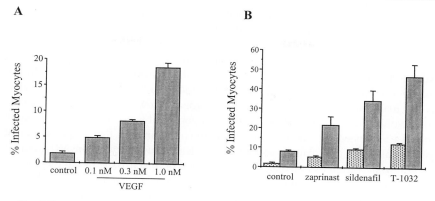

FIG. 6. Improved gene transfer with VEGF and phosphodiesterase 5 inhibitor (PDE-5) exposure to *ex vivo* perfused rabbit hearts. (A) Dose-response curve for VEGF exposure 2 min before and during a 2-min infection with 10^8 pfu/ml of adenovirus. (B) Improvement in gene transfer with exposure to 10 μmol/liter of the indicated PDE-5 inhibitor. Dotted bars are from isolated exposure to the PDE-5 inhibitor prior to 2-min infection with 10^8 pfu of adenovirus. Solid bars are from combined exposure to 10 μmol/liter of the indicated PDE-5 inhibitor and 10 μmol/liter of VEGF before and during a 2-min infection with 10^8 pfu/ml of adenovirus.

Putting It All Together: *In Vivo* Delivery

These data demonstrate that adenovirus-mediated gene transfer is essentially a diffusion-limited process, so optimization of parameters relevant to diffusion, such as virus and receptor concentration, particle size, and temperature, will increase the efficiency of gene transfer. Likewise, breaking down structural barriers by inflammatory mediators or reduced calcium concentration should improve results. The poor efficiency of adenovirus-mediated gene transfer in previously published *in vivo* models can be readily explained by these data. For the previously reported transplant models,[15,16] low temperature and slow flow rate would limit infection. In the coronary catheterization models,[13,14] the presence of blood and contrast dye would have an inhibitory effect, and poor flow rate, short exposure time, and low virus concentration would decrease infection efficiency. Pericardial delivery[28] and direct myocardial injection[12] would have the advantages of maintaining local virus concentration, but both would be limited from achieving generalized delivery by diffusional barriers.

More recently, two *in vivo* models have validated the concept of improving gene transfer by increasing vascular permeability. A rabbit catheterization model reported 50% gene transfer efficiency by perfusing the left circumflex coronary

[28] K. G. Lamping, C. D. Rios, J. A. Chun, H. Ooboshi, B. L. Davidson, and D. D. Heistad, *Am. J. Physiol.* **272**, H310 (1997).

artery with serotonin for 2 min and then 10^{10} pfu/ml of adenovirus at 0.75 ml/min for 2 min.[29] A porcine model achieved gene transfer in 45% of AV nodal myocytes with a protocol that perfused the AV nodal artery with VEGF for 3 min before a 30 sec infusion of 7.5×10^9 pfu/ml adenovirus at 2 ml/min.[2] There has not yet been a report of *in vivo* gene transfer to all, or nearly all, myocytes. In our experience, an impediment to the transfer of the *ex vivo* results to *in vivo* models is the inability to achieve the necessary coronary flow rate and virus exposure time without vascular volume overload or myocardial ischemia. Further research in this area is needed to identify a combination of parameters that allows sufficient gene transfer but also overcomes these obstacles. In summary, the investigation of several basic variables in isolated cells and then in intact tissue has allowed us to identify parameters important for successful adenovirus-mediated gene transfer. The current challenge is to transfer this information to usable *in vivo* models and eventually into human subjects for gene therapy of a variety of congenital and acquired cardiac diseases.

Acknowledgments

We warmly thank Dr. M. Akao for his help in cardiac cell isolation. N.N. was supported by the Fondation Bettencourt and the Fondation Simone et Cino Del Duca.

[29] H. J. Weig, K. L. Laugwitz, A. Moretti, K. Kronsbein, C. Stadele, S. Bruning, M. Seyfarth, T. Brill, A. Schomig, and M. Ungerer, *Circulation* **101**, 1578 (2000).

[20] Recombinant AAV-Mediated Gene Delivery Using Dual Vector Heterodimerization

By Ziying Yan, Teresa C. Ritchie, Dongsheng Duan, and John F. Engelhardt

Introduction

The development of dual vector strategies for adeno-associated virus-2 (AAV-2)-mediated gene therapy has enhanced the potential use of this vector for a broad range of genetic diseases.[1-3] Although already an attractive vector for clinical applications because of its nonpathogenicity, AAV was previously restricted to

[1] R. J. Samulski, *Nat. Biotechnol.* **18**, 497 (2000).
[2] P. Tattersall, *Proc. Natl. Acad. Sci. U.S.A.* **97**, 6239 (2000).
[3] T. R. Flotte, *Respir. Res.* **1**, 16 (2000).

the delivery of only relative small disease genes due to its 5.0 kb packaging size.[4] Research breakthroughs, however, have now effectively doubled the size of genes that can be expressed by this vector.[5-8] This technology involves dividing a gene expression cassette into two parts, each encoded within separate AAV virions. When coadministered to cells, the linear viral genome in each of the vectors forms heterodimers and concatamers through a mechanism(s) involving intermolecular recombination between linear and/or circular forms of the AAV genomes.[9-11] This mechanism of concatamers occurs in a time-dependent fashion. By constructing paired rAAV vectors containing the two halves of a gene with appropriate splicing signal sequences, this feature of rAAV cell biology can be harnessed for the expression of intact functional proteins.

In this chapter, we will address issues relevant to the design, construction, and analysis of dual, "trans-splicing" AAV vectors for the expression of intact transgenes in cells and tissues. AAV dual vector strategies also include a "cis-activation" approach for augmenting transgene expression though enhancer/promoter elements encoded in a second virion.[6] However, "cis-activation" approaches, while mechanistically similar to "trans-splicing," have a distinct set of considerations for vector design, and the discussion here will be limited to methods for split gene, "trans-splicing" vectors. Similarly, a comprehensive presentation of AAV cell biology is beyond the scope of this chapter, and the reader is referred to several review articles.[12-14]

Mechanisms of Circularization and Heterodimerization of the rAAV Genome

AAV-2 is a single-stranded DNA parvovirus with a 4680-nucleotide genome. This virus is replication defective and requires coinfection with a helper virus, such as adenovirus or herpes virus, for lytic phase productive replication. In the absence of a helper virus, wild-type AAV establishes a latent, nonproductive

[4] J. Y. Dong, P. D. Fan, and R. A. Frizzell, *Hum. Gene Ther.* **7,** 2101 (1996).

[5] Z. Yan, Y. Zhang, D. Duan, and J. F. Engelhardt, *Proc. Natl. Acad. Sci. U.S.A.* **97,** 6716 (2000).

[6] D. Duan, Y. Yue, Z. Yan, and J. F. Engelhardt, *Nat. Med.* **6,** 595 (2000).

[7] L. Sun, J. Li, and X. Xiao, *Nat. Med.* **6,** 599 (2000).

[8] H. Nakai, T. A. Storm, and M. A. Kay, *Nat. Biotechnol.* **18,** 527 (2000).

[9] C. H. Miao, H. Nakai, A. R. Thompson, T. A. Storm, W. Chiu, R. O. Snyder, and M. A. Kay, *J. Virol.* **74,** 3793 (2000).

[10] J. Yang, W. Zhou, Y. Zhang, T. Zidon, T. Ritchie, and J. F. Engelhardt, *J. Virol.* **73,** 9468 (1999).

[11] D. Duan, Z. Yan, Y. Yue, and J. F. Engelhardt, *Virology* **261,** 8 (1999).

[12] K. I. Berns and C. Giraud, "Adeno-Associated Virus (AAV) Vectors in Gene Therapy." Springer, New York, 1996.

[13] P. J. Carter and R. J. Samulski, *Int. J. Mol. Med.* **6,** 17 (2000).

[14] R. M. Linden and K. I. Berns, *Gene Ther.* **4,** 4 (1997).

infection withlong-term persistence primarily by integrating into the host chromosomal DNA.[15] With high frequency, wild-type AAV integrates into a specific locus, AAVS1, on human chromosome 19.[16–18] In contrast to wild-type AAV, the mechanism for latent phase persistence of recombinant AAV-2 (rAAV-2) was until recently unknown. The wild-type AAV genome consists of two families of genes, Rep (encoding regulatory proteins) and Cap (encoding capsid proteins). These genes are flanked by inverted terminal repeats (ITRs), whose palindromic sequences form a "T"-shaped hairpin structure. In the construction of rAAV vectors, the Rep and Cap genes are deleted and replaced by a foreign gene between the ITRs. These ITRs encode the minimal sequence required for recombinant AAV DNA replication, provirus integration, and packaging of AAV DNA into virus particles.[19,20]

Without helper virus or genotoxic stimuli, rAAV will enter its latent cycle and the viral genome either integrates randomly into a chromosome, because of the absence of the Rep protein, or persists as an episomal provirus.[21,22] Although the existence of rAAV episomal genomes has been suspected for several years, confirming evidence came from recent studies. With an rAAV shuttle vector (AV.GFP3ori) containing a bacterial plasmid replicon and the ampicillin resistance gene, Duan *et al.*[23] successfully retrieved circularized AAV molecules from rAAV infected cell lines and muscle tissue *in vivo.*[23] Southern blot analysis demonstrated that head-to-tail circularization of the rAAV genome forms the majority of the transduction intermediates. Chemical sequencing of the ITR junction revealed that the "double D" structure is the predominant type of ITR array contained in the AAV circular transduction intermediates.[11] The Double D sequence consists of a single ITR flanked by "D" sequences on both ends (D-A'-B'-B-C'-C-A-D').[24] These findings suggested that the formation of these circular molecules was mediated by the ITRs. In addition, these studies also revealed larger circular forms containing two or more copies of the viral genome.

[15] K. I. Berns, *Microbiol. Rev.* **54,** 316 (1990).

[16] R. M. Kotin, M. Siniscalco, R. J. Samulski, X. D. Zhu, L. Hunter, C. A. Laughlin, S. McLaughlin, N. Muzyczka, M. Rocchi, and K. I. Berns, *Proc. Natl. Acad. Sci. U.S.A.* **87,** 2211 (1990).

[17] R. J. Samulski, X. Zhu, X. Xiao, J. D. Brook, D. E. Housman, N. Epstein, and L. A. Hunter, *EMBO J.* **10,** 3941 (1991).

[18] R. M. Kotin, R. M. Linden, and K. I. Berns, *EMBO J.* **11,** 5071 (1992).

[19] R. J. Samulski, L. S. Chang, and T. Shenk, *J. Virol.* **61,** 3096 (1987).

[20] R. J. Samulski, L. S. Chang, and T. Shenk, *J. Virol.* **63,** 3822 (1989).

[21] T. R. Flotte, S. A. Afione, and P. L. Zeitlin, *Am. J. Respir. Cell. Mol. Biol.* **11,** 517 (1994).

[22] S. A. Afione, C. K. Conrad, W. G. Kearns, S. Chunduru, R. Adams, T. C. Reynolds, W. B. Guggino, G. R. Cutting, B. J. Carter, and T. R. Flotte, *J. Virol.* **70,** 3235 (1996).

[23] D. Duan, P. Sharma, J. Yang, Y. Yue, L. Dudus, Y. Zhang, K. J. Fisher, and J. F. Engelhardt, *J. Virol.* **72,** 8568 (1998).

[24] X. Xiao, W. Xiao, J. Li, and R. J. Samulski, *J. Virol.* **71,** 941 (1997).

Furthermore, the percentage of high molecular weight dimer or multiple viral genome concatamers increased with the time course after rAAV infection, correlating with stable, long-term transgene expression in muscle.[23]

To address whether the mechanism of formation of these circular concatamers involved rolling circle replication from a single viral genome or intermolecular recombination, mice were coinfected with two independent rAAV vectors harboring either EGFP (enhanced green fluorescent protein) or alkaline phosphatase reporter genes in the same muscle (tibialis anterior). In AAV circular intermediates rescued from the Hirt DNA purified from tissue samples, the abundance of heterodimers and concatamers expressing both reporter genes reached 33% at 120 days post coinfection.[10] Molecular analysis of these concatamers confirmed that ITR-mediated intermolecular recombination resulted in the formation of single circular molecules containing two distinct AAV genomes. After coinfection of the same cells, the two different AAV genomes recombined into a single molecule. These groundbreaking studies led to the initial hypothesis that the 5 kb packaging limitation of rAAV vector could be bypassed by an approach based on heterodimerization of two independent vectors. Subsequent research confirmed this hypothesis and led to the development of two separate technologies using dual AAV vectors. The first utilizes "split-gene, trans-splicing" vectors for large transgenes exceeding the single vector packaging limit.[5,7,8] The second technology, termed "cis-activation," is useful for augmenting expression from a gene that will just fit into one AAV vectors with no room remaining for promoter or enhancer elements.[6]

Design and Construction of Dual Trans-Splicing AAV Vectors

Principles of Vector Design

Based on the principle that two independent AAV genomes can form a chimeric molecule after coinfection of the same cell, the premise of the trans-splicing approach is that a large transgene can be divided between two different rAAV vectors and successfully reconstituted through trans splicing. The first vector harbors a promoter/enhancer driving the 5' half of the transgene flanked by a donor splice site. The second vector encodes the remaining half of the gene and a polyA signal preceded by an acceptor splice site. As described below, the splicing sequences can either be from an endogenous intron or consist of artificially engineered heterologous splice sites. RNA transcription from a correctly oriented head-to-tail heterodimer proceeds through the ITR junction. During post-transcriptional processing, the cell splicing machinery recognizes the splicing signal sequences and removes the introns in the primary transcript (hnRNA). An intact mRNA for the target protein containing all exon segments originally divided between separate AAV genomes is generated. The methods given below first present information applicable to general dual trans-splicing vector design, followed by issues and examples specific for using either a genomic locus or a cDNA for the target protein.

Choosing the Site for Splitting a Gene. The trans-splicing approach effectively doubles the package capacity of rAAV, and in principle, a transgene expression cassette [with the promoter and a poly(A) signal] larger than approximately 5 kb but less than 10 kb is an appropriate candidate for this strategy. From a practical viewpoint, the choice of a site for splitting a gene should take into consideration the optimal packaging limit of 4.7 kb for each individual vector. Although individual virions up to 5 kb in size have been successfully generated, above 110% of the wild-type virus size (4675 nucleotides), the packaging efficiency is significantly decreased.[4] Additionally, the ITRs, the only AAV sequences required for vector construction, are each 145 nucleotides in length. This brings the optimal size of a transgene expression cassette in an individual virion to 4385 nucleotides. The inclusion of auxiliary elements in one or both of the vectors to augment or track gene expression or to facilitate molecular analysis of heterodimers should also be considered when choosing where to divide the gene (see AAV Vector section below). Sequence considerations for a division point are based on the consensus splice donor and acceptor sequences.

Consensus Splice Signal Sequence. The consensus splicing sequences at exon–intron boundaries and the interior of introns are conserved in eukaryotes (Fig. 1A). At either end of most introns, the most highly conserved nucleotides are (5') GT and (3') AG. In addition, approximately 30 nucleotides at the 3' end of the intron are necessary for correct splicing. This region includes the 5-nucleotide consensus sequence C-T-A/G-A-C/T around the branch point adenine (A) and a 15-base pyrimidine-rich region. Factors to consider include the size of the intron, and most importantly, the existence of restriction endonuclease sites flanking the division site and in the interior of intron for cloning. Within a cDNA, a potential splice junction between different exons can be identified by the following sequence: N-A/C-A-G‖G-N-N.[25] It is mandatory to choose a site in the cDNA with sequences identical to exon junctional boundaries in the central portion of the cDNA for inserting the engineered intron. In constructing trans-splicing vectors for a large cDNA, it is usually not feasible to put an endogenous intron back into the sequence, since most introns are relatively large. A short artificially engineered endogenous intron or a heterologous chimeric intron such as that from the commercially available pCI-neo vector (Promega; Fig. 1C) can be used instead. Since the availability of precisely placed and unique restriction sites can be problematic, in many cases a restriction site can be introduced by means of a silent point mutation technique. With a cDNA, PCR primers can be designed to include the required intron consensus

---

[25] P. Senapathy, M. B. Shapiro, and N. L. Harris, *Methods Enzymol.* **183,** 252 (1990).

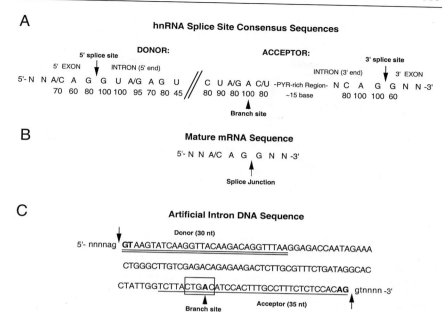

FIG. 1. Splice consensus sequences. (A) Sequences at 5′ and 3′ exon/intron boundaries in a genomic locus. The numbers below the sequence indicate the percentage occurrence of specific nucleotides in the splice signaling regions. (B) The sequence at the splice junction between adjacent exons in mature mRNA. (C) The DNA sequence of an artificial chimeric intron (pCI-neo, Promega) that can be used in PCR strategies for adding splice signal sequences to the segments of a large cDNA. Underlined regions indicate the suggested sequences for donor (double underline) and acceptor (single underline) sequences to be included in PCR primers during cloning. Bolded letters indicate 100% conserved nucleotides and the boxed region indicates the branch site consensus of the acceptor splice signal.

sequences for the donor or acceptor vectors, together with a convenient restriction site for cloning.

AAV Vectors for Cloning a Therapeutic or Reporter Gene

Two AAV proviral plasmids containing multiple cloning sites between the AAV ITR sequences for inserting foreign genes, as well as other elements useful for propagation or analysis, are illustrated in Fig. 2. The basis of the design of these vectors is that sequences inserted in the correct orientation between ITRs will be packaged into virions after transfection of packaging cell lines infected with helper adenovirus (see methods, below). The two ITRs are mirror images of each other, and the orientation of the ITRs with respect to inserted sequences is critical for correct packaging of the transgene into the virion. The "D" sequence

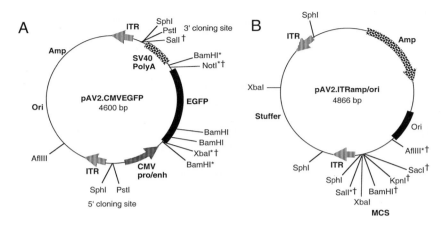

FIG. 2. rAAV cloning vectors. (A) Structure of a general use rAAV proviral plasmid with 5′ and 3′ multiple cloning regions for inserting a transgene expression cassette between the AAV ITRs. (B) An rAAV shuttle vector that contains the bacterial replicon and ampicillin resistant gene and a multiple cloning site (MCS) for inserting a transgene segment between the ITRs. Restriction enzyme sites marked by asterisks are suggested for cloning of the transgene. Sites marked by † have unique single cutting sites within the plasmid.

of each ITR should be directed toward the inserted transgene. Sequences lying outside the ITRs serve functions such as stabilizing the plasmids, or providing elements for amplifying the proviral plasmid DNA in bacteria. Note that these external sequences do not contribute to the packaging size of the virus. The two vectors shown are a general AAV cloning vector and an AAV shuttle vector.

pAV2.CMVEGFP is a general use AAV vector (Fig. 2A). This vector contains AAV-2 ITR sequences derived from pSub201[19] on either end, an EGFP reporter gene driven by the human Cytomegalovirus (hCMV) promoter/enhancer, and an SV40 poly(A) signal. By digestion with restriction endonucleases at the 5′ and 3′ multiple cloning sites (*Sph*I or *Pst*I), the entire EGFP expression cassette can be removed in order to clone the foreign gene of interest. The nucleotide length cloned between the ITRs should be no more than 4.5 kb (see above). As warranted by the requirements, the EGFP gene can be excised (using *Bam*HI or *Xba*I/*Not*I digestion) and replaced by another reporter or therapeutic gene, while leaving the hCMV and SV40-poly(A) intact. This vector can be used for producing the donor and/or acceptor vectors for the trans-splicing technique.

Although not a requirement for trans-splicing vectors, a second AAV cloning vector (AAV shuttle vector; Fig. 2B) was engineered to facilitate molecular analyses of circular AAV genomes (methods and examples are given below). The shuttle vector contains elements that allow for the retrieval of circular viral genomes from infected cells and provide a mechanism for their amplification in bacteria. This will enable the generation of DNA for restriction mapping, Southern blotting, or other molecular techniques. A revised version of the general use AAV vector, the shuttle vector is termed pAV2.ITRamp/ori (Fig. 2B). The pUC18 plasmid backbone containing the bacterial replicon and the ampicillin-resistant gene has been inserted between the "D" sequences of the two ITRs. These sequences will also be packaged in rAAV virions along with the inserted transgene. A stabilizing "stuffer" sequence consisting of a 1.6 kb segment from the luciferase gene is positioned externally to the ITRs. When the plasmid is cut with *Afl*III along with a MCS enzyme (*Sal*I, *Bam*HI, *Kpn*I, or *Sac*I), a 2.0–2.3 kb foreign sequence can be incorporated and packaged in the virus together with the plasmid backbone.

Example Strategy for Cloning Gene from Genomic Locus into Trans-Splicing AAV Vector

The human erythropoietin (Epo) gene contains 5 exons separated by 4 introns.[26] A human genomic DNA fragment encoding the Epo gene was split into two parts at a unique *Bcl*I site located in intron 3. The two fragments were then cloned into two independent rAAV vectors, AV.Epo1 and AV.Epo2 (Fig. 3; from Ref. 5). The 5′ portion of the Epo gene contains exons 1–3, introns 1–2, and the splicing donor signal from intron 3. This 5′ fragment (1468 bp) and the hCMV immediate early promoter/enhancer were cloned into the rAAV shuttle vector (Fig. 2B), which contains sequences for the bacterial replicon and the ampicillin resistance gene, to produce AV.Epo1. The remainder of the Epo sequence, including exons 4 and 5, intron 4, and the splicing acceptor from intron 3 (914 bp), was cloned into a second vector, AV.Epo2. A ribosome internal entry site (IRES) element, followed by an enhanced green fluorescent protein (EGFP) gene, was incorporated at the immediate 3′ end of the Epo gene-coding region. With this design, Epo and EGFP proteins will only be translated from the same bicistronic mRNA derived from a correctly orientated head-to-tail heterodimer of these two rAAV vectors. The expression of the EGFP gene provides a useful indicator for continually monitoring successful trans-splicing in cultured cells coinfected with AV.Epo1 and AV.Epo2. Last, a prokaryotic selective marker, the kanamycin resistance gene, was also encoded within AV.Epo2. Since the plasmid replicon and the ampicillin resistance gene were incorporated in AV.Epo1, this allows a simple method for retrieval of chimeric viral genomes from the coinfected cells or tissues by double antibiotic selection (see methods below). Figures 3B and 3C illustrate the sequence of intracellular processing events that result in the expression of functional proteins from the two trans-splicing vectors.

---

[26] F. K. Lin, S. Suggs, C. H. Lin, J. K. Browne, R. Smalling, J. C. Egrie, K. K. Chen, G. M. Fox, F. Martin, and Z. Stabinsky, *Proc. Natl. Acad. Sci. U.S.A.* **82**, 7580 (1985).

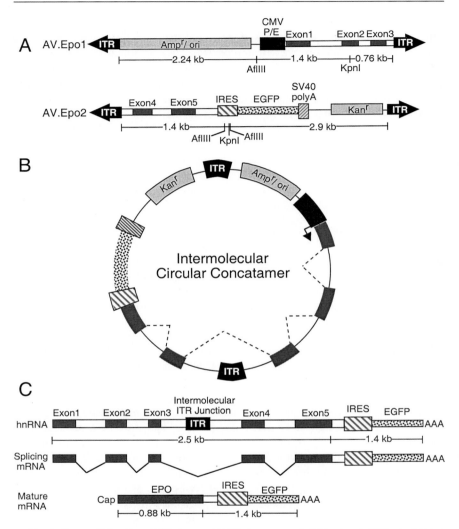

FIG. 3. Trans-splicing Epo rAAV vectors. (A) Viral structure of AV.Epo1 and AV.Epo2. (B) Predicted functional head-to-tail heterodimer resulting from intermolecular concatamerization between AV.Epo1 and AV.Epo2. Splicing of Epo exons is indicated by the dashed line. (C) Maturation of hnRNA Epo transcripts resulting from intermolecular head-to-tail concatamerization. The functional mature transcript expresses both Epo and EGFP via an internal IRES. This figure is reprinted with permission from Z. Yan, Y. Zhang, D. Duan, and J. F. Engelhardt, *Proc. Natl. Acad. Sci. U.S.A.* **97,** 6716 (2000). Copyright (2000) National Academy of Sciences, U.S.A.

Strategy for Cloning Gene from a Large cDNA into Trans-Splicing AAV Vectors

For generating trans-splicing vectors from a large cDNA, an endogenous intron can be inserted in the appropriate position of the sequence. However, because of the large size of most introns, this approach is usually not feasible. Alternatively, a 132 bp engineered heterologous chimeric intron (Fig. 1C) can be used in place of an endogenous intron. This chimeric intron is from a commercially available plasmid (pCI-neo, Promega), and it is composed of the splice donor from the first intron of the human beta-globin gene and the branch and the acceptor site from the intron of an immunoglobin gene.[27] Based on the sequence of this intron, a PCR strategy for splitting a large cDNA expression cassette and inserting artificial splicing signals to generate the trans-splicing vectors was designed (Fig. 4). The approach for cloning utilizes blunt ligation to avoid problems associated with locating appropriate restriction sites common between the transgene and the AAV vector. Of note in this regard, AAV is a single-stranded DNA virus, and both + and − strands are packaged into virions at equal frequency. Also, the two ITRs have identical sequences (they are mirror images) so orientation of the transgene segments between the ITRs in the proviral plasmids is not a consideration. Thus, directional cloning of the transgene expression cassette into the AAV vectors would provide no advantage.

Cloning Procedure (Fig. 4)

1. The cDNA of interest is cloned into a eukaryotic expression vector to produce a plasmid, pTransgene, containing all elements of an expression cassette [promoter, transgene, and poly(A) tail]. Unique restriction endonuclease cutting sites in the plasmid are determined by sequence analysis. Potential exon–exon boundary sites with the consensus splice site motif of NAG‖GN within the central portion of the cDNA are located.

2. The division site for splitting the gene is chosen. It should be taken into consideration that two independent unique restriction sites (RE), one on either side of the division site, are required for the PCR cloning approach as illustrated in Fig. 4. The binding site for RE1 can be several hundred base pairs upstream from the division site NAG‖GN, while that for RE2 lies downstream from the division site. Shorter fragments generated by PCR are preferred as there is less chance for Taq errors in the cloning.

3. Splice donor PCR primer design: For the 5' segment of the transgene, the splice donor sequence must be inserted into the sequence at the division site. The forward primer for this reaction (5F) contains 25 nucleotides of the sequence around the unique restriction site for RE1. The 65-mer reverse primer

[27] A. L. Bothwell, M. Paskind, M. Reth, T. Imanishi-Kari, K. Rajewsky, and D. Baltimore, *Cell* **24,** 625 (1981).

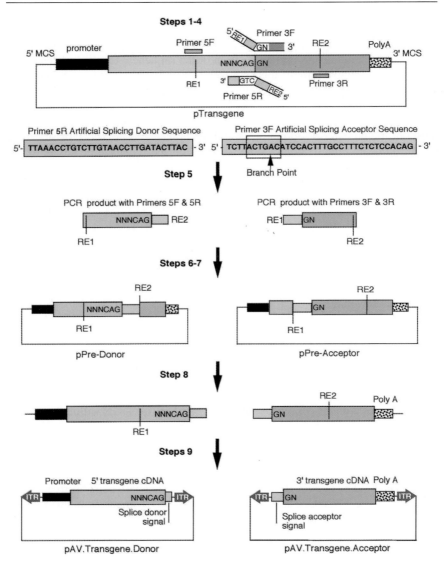

FIG. 4. PCR cloning strategy for generating trans-splicing rAAV vectors. This schematic figure outlines a PCR cloning strategy for dividing a large cDNA and inserting splice donor and acceptor sequences derived from an artificial intron. See text for details.

(5R) comprises 25 nucleotides complementary to the region of the transgene ending with the potential exon boundary sequence NNNCAG, and a 5' overhang of 30 nucleotides complementary to the 5' end of the chimeric intron (Fig. 1C). The final 5' portion of the primer provides the recognition sequence for the RE2 restriction site.

4. Splice acceptor PCR primer design: For the 3' segment of the transgene, the 70-mer forward primer (3F) contains first the RE1 recognition site followed by 35 nucleotides from the 3' portion of the chimeric intron (Fig. 1C). This is followed with 25 nucleotides of homology to the transgene beginning with the potential exon boundary GN. The acceptor-splicing region of the artificial intron includes the conserved branch point adenine and the pyrimidine-rich region. The reverse primer (3R) is complementary to the transgene sequence around and including the restriction site RE2.

5. Two independent PCR reactions are performed to introduce the overhanging splice donor and acceptor sequences into the appropriate transgene segments. With primers 5F and 5R, the PCR product (5' donor) contains several hundred base pairs of upstream sequence from the division site, ending with the splicing donor sequence and the RE2 restriction site. Similarly, PCR with the primer set 3F and 3R produces a product (3' acceptor) containing the downstream portion of the transgene with the splice branch site and acceptor, as well as restriction site RE1 at its 5' end.

6. Both PCR products (5' donor and 3' acceptor) are digested with restriction enzymes RE1 and RE2 at 37° for 2 hr, respectively. The bands are resolved by agarose gel electrophoresis and are recovered using a Geneclean Kit (Bio 101) or other similar method. At the same time, the plasmid pTransgene (from step 1) is also digested with RE1 and RE2 to release the region containing the division site.

7. Two independent ligation reactions are performed to separately clone each of the purified PCR bands (5' donor and 3' acceptor) between the RE1 and RE2 sites of the plasmid, pTransgene. One plasmid (pPre-Donor) contains the promoter, the 5' segment of the transgene, and the artificial splice donor. pPre-Acceptor contains the artificial splice acceptor, the 3' segment of the transgene, and a poly(A) signal. Bacteria are transformed and the resulting colonies are screened by restriction analysis to confirm the identity of two plasmids. It is also advisable to sequence the plasmids to determine whether mutations were introduced during PCR. Any mutations located in the sequence can be repaired using site-directed mutagenesis.

8. pPre-Donor is digested with a unique enzyme from the 5' multiple cloning site (MCS) region of the plasmid vector, and with RE2. This digestion releases the 5' portion of the expression cassette containing the promoter, the 5' segment of the transgene, and the donor splicing sequence. Similarly, pPre-Acceptor is cut by RE1 and a unique enzyme within the 3' MCS of the plasmid to release a fragment containing the acceptor splicing sequence, the 3' segment of the transgene,

and the poly(A) signal. The ends are blunted with T4 DNA polymerase, and then both the fragments of the expression cassette are purified after agarose gel electrophoresis.

9. Preparation of the AAV proviral plasmid vector for cloning: The rAAV general vector (Fig. 2A) can be digested with *Pst*I or *Sph*I for replacing the entire EGFP expression cassette with a segment of the foreign gene cassette. The fragments are separated by electrophoresis on a 1% gel and two bands are resolved (2.0 and 2.6 kb). The larger 2.6 kb band is excised and purified, and the ends are blunted with T4 DNA polymerase. In the hypothetical case presented in Fig. 4, both segments of transgene cassette are cloned into the prepared general proviral vector to produce two rAAV proviral plasmids, pTransgene.Donor and pTransgene.Acceptor.

Notes

1. As indicated above, if size constraints permit, it is useful to clone one segment of the transgene expression cassette into the rAAV shuttle vector (Fig. 2B) to facilitate subsequent molecular analyses. In preparation for cloning, the shuttle vector is either linearized by digestion with one of the enzymes in the multiple cloning site (e.g., *Sal*I), or is cut with both *Afl*11I and *Sal*I to release a 300 bp fragment. This will increase the size of an insert to be cloned into this vector from 2.0 to 2.3 kb. After digestion and agarose electrophoresis, the vector band is either 4.5 or 4.8 kb, depending on the digestion chosen.

2. The reader is referred to *Molecular Cloning*[28] for common molecular methods such as PCR, blunt ligation, and agarose gel electrophoresis.

LacZ as Example of Split Gene Trans-Splicing Strategy for Large cDNAs. To test the efficiency of the artificial splicing signals, the β-galactosidase transgene was divided by means of the PCR cloning strategy described above. The junction of a potential exon-exon boundary site (NAG∥GN) was selected as the division point in the LacZ cDNA. Two independent PCR reactions were performed to split the entire LacZ expression cassette (including the RSV promoter/enhancer, LacZ cDNA, and the SV40 pA) into two parts, while introducing the splice donor or branch and splice acceptor signals. These DNA fragments were then cloned into separate AAV vectors to produce AV.LacZ.Donor and AV.LacZ.Acceptor (Fig. 5A). Figure 5B illustrates the two possible heterodimers that will be formed after coinfection of eukaryotic cells, such as primary fibroblasts, with the two virus vectors. In this case, the 3' segment of the LacZ gene was cloned into the rAAV shuttle vector. Thus, rescue and molecular analyses could be used to confirm the structures of the rAAV circular genomes. Following intermolecular recombination in coinfected

[28] J. Sambrook, T. Maniatis, and E. F. Fritsch, "Molecular Cloning: A Laboratory Manual." Cold Spring Harbor Laboratory, Cold Spring Harbor, NY, 1989.

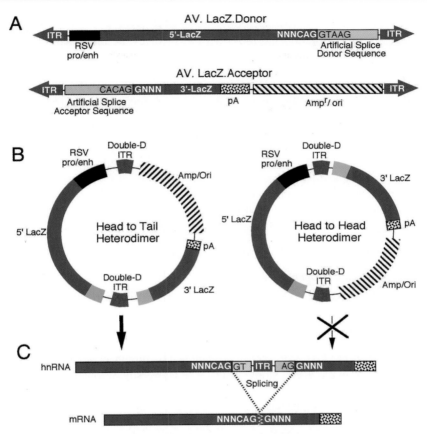

FIG. 5. LacZ trans-splicing rAAV vectors. (A) Structure of the LacZ trans-splicing vectors. (B) Structure of the head-to-tail and head-to-head heterodimers rescued from coinfected cells. (C) Processing of the primary hnRNA transcript from head-to-tail heterodimers.

cells, both head-to-tail (correct orientation) and head-to-head heterodimers are formed. (Note that homodimers in both head-to-tail and head-to-head orientations will also be formed but are not shown.) Processing and splicing of the primary transcript (hnRNA) from the head-to-tail heterodimer will produce an intact mRNA for the target transgene β-galactosidase (Fig. 5C). Functional integrity of the protein product can be confirmed by histological staining of coinfected cells with X-Gal.[29]

[29] G. R. MacGregor, A. E. Mogg, J. F. Burke, and C. T. Caskey, *Somat. Cell. Mol. Genet.* **13,** 253 (1987).

Production and Purification of Recombinant AAV Virus

The strategy for producing rAAV virions involves the introduction of both proviral and auxiliary DNA into host cells preinfected by a helper adenovirus. Two plasmids are cotransfected, an AAV proviral plasmid containing the DNA fragment of interest and an auxiliary plasmid (pTrans) containing DNA sequences for the AAV regulatory (Rep) and capsid (Cap) genes that are deleted from recombinant AAV vectors. The helper adenovirus donates other required proteins for the production of rAAV, with the exception that the host cell line supplies adenoviral E1 gene product. rAAV virions are released by lysing the cells and can be purified and concentrated by one of two methods. The first method, which is rather time-consuming and laborious, consists of three rounds of CsCl density gradient centrifugation.[30–32] In this chapter, we describe an adaptation of a newer more rapid protocol for producing highly purified rAAV with HPLC methodology.[33] Last, methods are presented for titering and analyzing the purity of rAAV preparations.

Cotransfection of Host Cells and Pretreatment of Cell Lysates

Materials and Reagents

Human 293 cells (ATCC #CRL-1573, Adenovirus-transformed cells that constitutively express the E1 gene product)
DMEM (Dulbecco's modified Eagle's medium)
Fetal bovine serum (FBS)
Helper virus: Ad.CMVlacZ, an E1-deficient recombinant adenovirus (available from the Vector Core Facility, The University of Iowa)
rAAV proviral plasmid
Auxiliary plasmid (pTrans)
2.5 M CaCl$_2$
2× HBS Buffer: 0.3 M NaCl, 1.5 mM Na$_2$PO$_4$, and 40 mM HEPES, pH 7.05
DNase I (Sigma D4513, 11 mg protein/vial, total 33K kuniz units)
0.5% Trypsin
10% Sodium deoxycholate
Tissue culture disposables

Protocol

1. Human 293 cells are split 1 : 5 and plated on 40 150-mm dishes in DMEM medium containing 10% fetal bovine serum (FBS). Incubation at 37° with 5% CO$_2$ is continued for 2 days prior to viral production.

[30] L. M. de la Maza and B. J. Carter, *J. Biol. Chem.* **255**, 3194 (1980).
[31] L. M. de la Maza and B. J. Carter, *J. Virol.* **33**, 1129 (1980).
[32] C. A. Laughlin, C. B. Cardellichio, and H. C. Coon, *J. Virol.* **60**, 515 (1986).
[33] K. R. Clark, X. Liu, J. P. McGrath, and P. R. Johnson, *Hum. Gene Ther.* **10**, 1031 (1999).

2. Cells are infected with helper virus when they are approximately 70% confluent. Then, 5×10^{11} particles of Ad.CMVlacZ are diluted in 800 ml DMEM/2% FBS. For infection, the culture medium is removed from the cells and they are refed immediately with 20 ml of the virus-containing medium for each dish. The dishes are then placed back in the 37° incubator.

3. A transfection cocktail is prepared 1 hr after infection with the helper virus. rAAV proviral plasmid (0.5 mg) is mixed with 1.5 mg pTrans plasmid in a volume of 40.5 ml H$_2$O using a 50 ml conical tube. Next, 4.5 ml of 2.5 M CaCl$_2$ is added for a final concentration of 0.25 M. To initiate the formation of a DNA–CaPO$_4$ precipitate, this DNA–Ca solution is added drop by drop with swirling into 45 ml 2× HBS.

4. The CaPO$_4$/DNA precipitate is added to the 293 cells (2.25 ml/dish) with gentle swirling, and the cells are incubated at 37° for an additional 40 hr, after which time cytopathic effects should be clearly evident.

5. Cells are dislodged by tapping the plate and harvested after 40 hr by centrifugation at 3500 rpm for 15 min at 4°. The medium, which contains dilute virus, is discarded using methods consistent with handling biohazardous material.

6. The cell pellet is resuspended in 12 ml 10 mM Tris-Cl, pH 8.1, followed by three freeze–thaw cycles using dry ice/ethanol and a 40° water bath.

7. The crude cell lysate is forced through a 25-gauge needle 3 times.

8. DNA is digested with by adding 1 vial of DNAse I (approximately 33 units) to the cell lysates.

9. 0.5% Trypsin (1.5 ml) and 1.5 ml of 10% sodium deoxycholate are added to the cell lysate to aid in the release of virus from nuclei. The mixture is incubated at 37° for 30 min.

Heparin Affinity Purification of rAAV by HPLC

HPLC affinity purification of AAV-2 is based on findings that heparan sulfate proteoglycan mediates cell surface binding of this virus.[34] The protocol uses a heparin chromatographic matrix, as previously described.[33] A support protocol for packing and maintenance of the affinity column is also provided.

Materials and Reagents

Beckman Biosys 2000 Workstation (semipreparative HPLC system)
Poros Self-Pack packing device (PerSeptive, Applied Biosystems)
4.6 mm D/100 mm L Peek column (1.7 ml bed volume)
Poros HE/20 medium (PerSeptive, Applied Biosystems)
Dilution Buffer: 20 mM Tris-Cl, pH 8.0, 150 mM NaCl, 0.5% sodium deoxycholate

---

[34] C. Summerford and R. J. Samulski, *J. Virol.* **72**, 1438 (1998).

Buffer A: 20 mM Tris-Cl, pH 8.0/100 mM NaCl
Buffer B: 20 mM Tris-Cl, pH 8.0/1 M NaCl
HEPES dialysis buffer: 20 mM HEPES, 150 mM NaCl, pH 7.8
Dialysis tubing (1/4 inch diameter, 12,000 molecular weight cutoff, Gibco BRL)

Protocol

1. The pretreated cell lysate is diluted with 400 ml of Dilution Buffer and heated at 60° for 60 min to inactivate contaminating helper adenovirus.

2. The mixture is centrifuged at 3000 rpm for 15 min, and the clarified supernatant is filtered through a 2 μm pore size disk filter.

3. The Poros HE column is preequilibrated with 10 bed-volumes ($\cong$17 ml) buffer A. The clarified AAV containing supernatant ($\sim$410 ml) is applied to the column using the pump-loaded mode at a rate of 4 ml/min. After loading, the pump and the column are washed thoroughly with 30 bed volumes of equilibration buffer A.

4. AAV is eluted from the column by application of 10 bed volumes of a linear NaCl gradient (0.1 M to 1 M NaCl) at a flow rate of 3 ml/min, and 1 ml gradient fractions are collected. Gradients are produced using buffer A and B. The elution is monitored at A_{280}.

5. Typically, the dominant protein peak on the chromatographic profile correlates to rAAV particles eluting at 0.4 M NaCl. The virus containing fractions (usually about 2–3 ml) are pooled and dialyzed against HEPES dialysis buffer, and stored in aliquots at −80° in 5% glycerol.

Support Protocol for HE Column Packing and Maintenance. [Note: A prepacked column, Poros HE/M (PerSeptive, Applied Biosystems), is also available.]

1. Suspend 0.8 g of Poros HE medium in 10 ml 0.5 M NaCl and wash twice by centrifugation at 1000 rpm for 3 min. Then suspend the beads in 12 ml buffer A.

2. Install the Self-Pack unit with the Peek column in the HPLC system. Apply 12 ml of the medium slurry into the device. Seal the unit and pack the column at a flow rate of 30ml/min with 35 ml buffer A. The backpressure of the HPLC system should not exceed 2000 psi during packing.

3. A well-packed column can be reused at least 60 times. If the back pressure of the column reaches 600 psi, it should be repacked after soaking the beads in 5 M NaCl overnight.

4. For long-term storage, the column should be flushed with buffer A containing 0.05% sodium azide as a preservative and kept at 4°. The HPLC system should be flushed with distilled water to prevent the precipitation of buffer salts. However, for shorter intervals between runs (days to 1 week), buffer A can remain in the system.

Titering and Analysis of Purity of rAAV Preparations

Each preparation of rAAV vectors should be titered, especially for dual vector strategies such as trans-splicing, which require that the two recombinant viruses be administered in equal doses. Additionally, the purity of the viral stocks should be checked to rule out contamination with helper adenovirus or wild-type AAV. Following are commonly used methods for these procedures.

Materials and Reagents

Alkaline Digestion Buffer: 0.4 M NaOH, 20 mM EDTA
Hybond-N plus membrane (Amersham Pharmacia)
Bio-Dot SF manifold microfitration apparatus (Bio-Rad)
20× SSC (3 M NaCl, 0.3 M sodium citrate, pH 7.0)
Hybridization Solution (5× SSC, 5× Denhardt's Solution, 1% SDS, 50% formamide; add 100 μg/ml denatured salmon sperm DNA just before use)
X-ray film
Anti-rep antibody (American Research Products, Belmont, MA)

Protocol

1. The physical titer (particles/ml) of the viral stocks is determined by alkaline digestion/and slot-blot hybridization with a ^{32}P-labeled probe against the gene of interest. As copy number standards, the proviral plasmid DNA at 10^7, 10^8, 10^9, 10^{10}, 10^{11} molecules/μl is utilized. One μl virus or 1 μl standard DNA is added to 50 μl of digestion buffer. After heating at 90°, samples are diluted in 500 ml digestion buffer and loaded onto Hybond-N plus membrane with the Bio-Dot SF manifold microfitration apparatus. The membrane is prehybridized, hybridized with the probe, and washed with 1 × SSC/1% SDS at 50° for 15 min, twice. The membrane is then exposed to X-ray film and developed. The particle titer of the virus is determined by comparison with the standard DNA. Typical viral yields are in the range of 1–5 × 10^{12} DNA particles/ml.

2. Contamination with helper adenovirus is evaluated by infection of 293 cells and histochemical staining for β-galactosidase activity, which is encoded by the helper virus (Ad.CMVlacZ). For these studies a 100 mm plate of subconfluent 293 cells is infected with 10^{10} DNA particles of rAAV in 5ml of serum free DMEM for 2 hr. Five ml of 20% FBS/DMEM is then added and the cells are cultured for an additional 72 hr prior to staining. Contaminating adenovirus will be noted as small X-Gal positive foci. Typically, helper virus contamination is less than 1 in 10^{10} DNA particles.

3. Wild-type AAV contamination can be evaluated by coinfection of 293 cells with the purified rAAV virus and Ad.CMVlacZ, with subsequent immuno-cytochemical staining with anti-rep antibody. For these studies a 100 mm plate of subconfluent 293 cells is coinfected with 10^{10} DNA particles of rAAV and

Ad.CMVlacZ (at an MOI of 20 particles/cell) in 5 ml of serum free DMEM for 2 hr. Five ml of 20% FBS/DMEM is then added and the cells are cultured for an additional 16 hr prior to immunofluorescent staining for Rep. Note that it is important to perform the immunofluorescent analysis prior to the onset of cytopathic effects induced by adenoviral replication at 20–24 hr. Transfection with pTrans is used a positive control for immunostaining. Typically, no anti-rep staining is observed in 293 cells infected with 10^{10} rAAV particles.[23] The limit of sensitivity of this assay is 1 wt AAV particle per 10^{10} rAAV particles.

Notes

1. A functional titer for recombinant AAV carrying segments of a transgene cannot easily be determined since the individual viruses express no transgene. Methods for measuring the functional (infectious) titer of rAAV vectors involve an adaptation of the "replication center" assay.[35] The modification is the use of a transformed cell line expressing the AAV rep and cap proteins.[36,37] rAAV DNA is quantified after coinfection of these cells with serially diluted rAAV together with helper adenovirus.

2. The replication center assay can also be used as an alternative method for detecting wild-type AAV contamination in a recombinant AAV stock. For details see Bartlett and Samulski, "Methods for the Construction and Propagation of Recombinant Adeno-Associated Virus Vectors," in Ref. 38.

Molecular Characterization of Heterodimerization between Independent rAAV Genomes

Since the trans-splicing rAAV vector strategy is based on the formation of circular AAV genomes (monomers, dimers, and concatamers) as nonintegrated episomes, molecular characterization of correctly oriented heterodimers can be performed using methods for the isolation of low molecular weight (Hirt) DNA.[39] The inclusion of a bacterial origin of replication, and antibiotic resistance genes in one of the pair of vectors (as illustrated for the Epo and LacZ dual vectors above), allows for the rescue, selection, and amplification of circular rAAV forms after transformation of bacteria with the Hirt DNA (Fig. 6; Table I). The DNA can be then be analyzed by methods such as restriction mapping, Southern blotting and/or DNA sequencing.

[35] W. W. Hauswirth, A. S. Lewin, S. Zolotukhin, and N. Muzyczka, *Methods Enzymol.* **316,** 743 (2000).
[36] K. R. Clark, F. Voulgaropoulou, D. M. Fraley, and P. R. Johnson, *Hum. Gene Ther.* **6,** 1329 (1995).
[37] K. R. Clark, F. Voulgaropoulou, and P. R. Johnson, *Gene Ther.* **3,** 1124 (1996).
[38] P. D. Robbins, "Gene Therapy Protocols." Humana Press, Totowa, NJ, 1997.
[39] B. Hirt, *J. Mol. Biol.* **26,** 365 (1967).

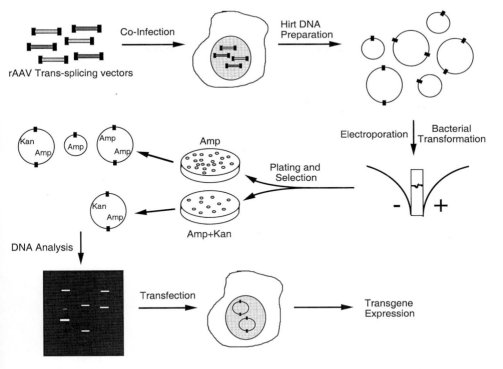

FIG. 6. Rescue of rAAV circular genomes. Methods for retrieval of circular genomes from rAAV infected cells are schematically outlined. See text for details.

Previous analyses have shown that protein expression in skeletal muscle from trans-splicing vectors is readily detectable by 10 days after the first injection of virus, and reaches peak levels by about 60 days.[5] Peak protein expression is maintained for at least 110 days, the longest time point evaluated. For molecular characterization, it has been shown that circular rAAV concatamers increase steadily

TABLE I
NUMBER OF ANTIBIOTIC-RESISTANT COLONIES EXPECTED FOLLOWING TRANSFORMATION OF E. coli
WITH HIRT DNA FROM DUAL VECTOR INFECTED CELLS

Cell type	Viral dose (each vector)	# Cells	Days after infection	Amp^R colonies	$[Amp^R/Kan^R]$ colonies	$[Amp^R/Kan^R]/$ Amp (%)
Fibroblasts	5×10^9	5×10^8	10	10,600	920	9
Tibialis anterior	2×10^{11}	1/2 muscle	150	27,160	11,320	42

from about 30 days through at least 120 days.[10] This section describes methods for infecting primary fibroblasts and skeletal muscle tissue with dual trans-splicing rAAV vectors *in vivo,* as well as techniques for Hirt DNA purification and rescue of circular intermediates by transformation of *Escherichia coli.* Analysis of protein expression can involve enzyme assays, Western blots, histological techniques, or function assays, and are characteristic for each transgene studied. These methods will not be addressed here.

Infection of Fibroblasts in Vitro and Purification of Hirt DNA

Materials and Reagents

Primary fetal fibroblasts (or other eukaryotic cells)
DMEM
Fetal bovine serum (FBS)
Hirt Extraction Buffer: 10 mM Tris-Cl, pH 8.0, 1% SDS, and 10 mM EDTA
Proteinase K (10 mg/ml)
Pronase (10 mg/ml)
5 M NaCl
Phenol : chloroform : isoamyl alcohol (24 : 24 : 1)
Absolute ethanol
Ice-cold 70% ethanol

Protocol

1. Fetal fibroblasts are seeded on six-well plates the day before infection at a density of 8×10^5 cells per well in DMEM supplemented with 10% FBS and 1% penicillin–streptomycin. The plates are incubated at 37% overnight.

2. Before infection, the confluent monolayers are washed twice with DMEM without serum. They are then infected with AV.Epo1 and /or AV.Epo2 (multiplicity of infection $= 10^4$ particles per cell) in 1 ml DMEM without FBS for 2 hr. After this incubation, the level of FBS is increased to 10% by adding an equal volume of 20% FBS/DMEM.

3. Medium is harvested from the cells for analysis of protein expression or reporter gene expression. Generally, the cells are refed with fresh medium the day before the medium is harvested.

4. For Hirt DNA purification, fibroblasts are washed with ice-cold PBS twice and lysed by adding 0.5 ml of Hirt extraction buffer.

5. Cells are then incubated at 37° for 30 min.

6. Add 50 μl proteinase K (10 mg/ml) and 25 μl pronase (10 mg/ml) to each well. Thoroughly mix by gently swirling and continue the digestion at 37° for 2 hr.

7. Add 140 μl of 5 M NaCl to each well for a final concentration of 1.1 M. Mix by gently swirling the plate several times. Incubate overnight at 4°.

8. Transfer the mixture to microcentrifuge tubes and pellet insoluble debris by centrifugation at 15,000g at 4° for 30 min.

9. Carefully remove the supernatant to a new tube and extract with an equal volume of phenol : chloroform : isoamyl alcohol (24 : 24 : 1), followed with an equal volume of chloroform.

10. Precipitate nucleic acids with 2.5 volumes absolute ethanol. Pellet Hirt DNA by centrifugation at 15,000g for 15 min at 4°. Discard the supernatant and wash the DNA with ice-cold 70% ethanol. Air dry the pellet and resuspend in 20 μl of TE buffer.

Infection of Murine Muscle Tissue in Vivo and Purification of Hirt DNA from Tissue

This section provides a model system for testing dual rAAV vectors *in vivo*. Skeletal muscle tissue has been previously demonstrated to be highly transducible with AAV-2[40–42] and has become, in effect, a model system for investigations of AAV cell biology. This model was employed for the studies providing the initial identification of AAV circular intermediate genomes[23] and was also used in the development and evaluation of the dual rAAV vector approaches for both trans-splicing[5] and cis-activation.[6] Although AAV has been shown to transduce a variety of tissues and organs in addition to muscle,[43–46] tropism of a particular tissue for AAV should be evaluated empirically. In addition, evidence suggests that intracellular trafficking of AAV can vary among different cell types.[47] Because these issues could profoundly influence the success of trans-splicing rAAV approaches for a given disease entity, preliminary evaluations of the cell biology of rAAV transduction in the relevant tissue should be performed.

Materials and Reagents

4 to 5-week-old C57BL6 mice
Anesthetic
Surgical instruments
Titered rAAV dual vectors
Hirt Extraction Buffer: 10 mM Tris-Cl, pH 8.0, 1% SDS, and 10 mM EDTA
HEPES-buffered saline

[40] X. Xiao, J. Li, and R. J. Samulski, *J. Virol.* **70**, 8098 (1996).

[41] P. D. Kessler, G. M. Podsakoff, X. Chen, S. A. McQuiston, P. C. Colosi, L. A. Matelis, G. J. Kurtzman, and B. J. Byrne, *Proc. Natl. Acad. Sci. U.S.A.* **93**, 14082 (1996).

[42] K. J. Fisher, K. Jooss, J. Alston, Y. Yang, S. E. Haecker, K. High, R. Pathak, S. E. Raper, and J. M. Wilson, *Nat. Med.* **3**, 306 (1997).

[43] T. R. Flotte, S. A. Afione, C. Conrad, S. A. McGrath, R. Solow, H. Oka, P. L. Zeitlin, W. B. Guggino, and B. J. Carter, *Proc. Natl. Acad. Sci. U.S.A.* **90**, 10613 (1993).

Protocol

1. Four- to 5-week-old C57BL6 mice are anesthetized with anesthetic compatible with a brief surgical procedure according to NIH and institutional guidelines.

2. The skin overlying the tibialis anterior muscle is reflected and the muscle is exposed. To infect the muscle, 4×10^{11} total viral particles (a 1 : 1 mixture of AV.donor and AV.receptor vectors) are suspended in 30 μl HEPES-buffered saline and injected. The animals are monitored while they recover from anesthesia and returned to their cages.

3. Three days after the first injection, the injection procedure is repeated.

4. Following injection of virus, mice are housed for 1 week to 120 days or more prior to harvest of the muscle tissue for analysis.

Hirt DNA Extraction from Infected Muscle Tissue

1. Harvest the infected muscle (one half of the tibialis anterior is sufficient for Hirt DNA extraction and the other half can be used for assessing transgene expression). The muscle sample is chopped into small pieces and put into a mortar. Immediately fill the mortar with liquid nitrogen, and grind the muscle pieces into fine powder.

2. Resuspend the powder with 600 μl Hirt extraction buffer and transfer the mixture to a 1.5 ml tube. Incubate at 37° for 90 min with continuous shaking.

3. Add 60 μl proteinase K (10 mg/ml) and incubate at 55° until the entire sample is dissolved. Lower the temperature to 37° and add 30 μl pronase (10 mg/ml) into the digestion. Incubate for an additional 2 hr to complete the digestion.

4. Add 168 μl of 5 M NaCl to each tube and incubate overnight at 4°.

5. Perform phenol/chloroform extraction and ethanol precipitation as described above for cell line Hirt preparation, and resuspend the DNA pellet in 20 μl of TE buffer.

Retrieval of Circular AAV Genomes from Dual Vector Infected Cells and Tissues

Circular viral genomes containing the plasmid replicon can be rescued from Hirt DNA by transformation of competent bacteria (Fig. 6). This protocol requires a highly efficient transformation method using electroporation, and super-competent *E. coli* cells. As illustrated for the dual vectors constructed for expression of Epo (Fig. 3), different antibiotic resistance genes were included in each virus of the

[44] R. O. Snyder, C. H. Miao, G. A. Patijn, S. K. Spratt, O. Danos, D. Nagy, A. M. Gown, B. Winther, L. Meuse, L. K. Cohen, A. R. Thompson, and M. A. Kay, *Nat. Genet.* **16,** 270 (1997).

[45] A. S. Lewin, K. A. Drenser, W. W. Hauswirth, S. Nishikawa, D. Yasumura, J. G. Flannery, and M. M. LaVail, *Nat. Med.* **4,** 1081 (1998), *Nat. Med.* **4,** 967 (1996).

[46] R. L. Klein, E. M. Meyer, A. L. Peel, S. Zolotukhin, C. Meyers, N. Muzyczka, and M. A. King, *Exp. Neurol.* **150,** 183 (1998).

[47] J. Hansen, K. Qing, and A. Srivastava, *J. Virol.* **75,** 4080 (2001).

pair. In this case, double antibiotic selection (e.g., ampicillin and kanamycin) can be used to differentiate and select bacterial colonies with heterodimers from those with monomer or homodimer rAAV genomes. Table I gives the approximate numbers of colonies that can be expected from Hirt DNA prepared from coinfected cultured cells (such as primary fibroblasts) or muscle tissue.

Materials and Reagents

E. coli competent cells (Electro Max DH10B, #18290015, Gibco Life Technologies)
E. coli Pulser (Bio-Rad)
Gene Pulser Cuvette, 0.1 cm gap (Bio Rad; #165-2089)
14 ml polypropylene round-bottom tube, 17 × 100 mm (Falcon 4059)
Amp selection LB agar plates (100 μg/ml, Ampicillin)
Amp+Kan selection LB agar plates (100 μg/ml Ampicillin, 33 μg/ml Kanamycin)

Protocol

1. Four μl Hirt DNA (1/5 of the preparation) is mixed with 50 μl electrocompetent cells on ice and then placed into a Gene Pulser cuvette.

2. Electroporation is performed with an *E. coli* Pulser at a setting of 1.7 kV.

3. After electroporation, add 950 μl SOC medium to the cuvette, and transfer the mixture to a Falcon tube. Incubate cells for 1 hr at 37° and 220 rpm.

4. Spread 200 μl culture on antibiotic selection plates and incubate overnight at 37°.

Acknowledgments

This work was supported by NIH Grant 2RO1 HL 58340 (JFE) and the Center for Gene Therapy funded by NIH (P30 DK54759) (J.F.E) and the Cystic Fibrosis Foundation.

[21] Designing and Characterizing Hammerhead Ribozymes for Use in AAV Vector-Mediated Retinal Gene Therapies

By JASON J. FRITZ, D. ALAN WHITE, ALFRED S. LEWIN,
and WILLIAM W. HAUSWIRTH

Introduction

Ribozyme is the term used to describe RNA molecules that possess the ability to catalyze a variety of biochemical reactions. The processes promoted by these RNA enzymes include sequence-specific cleavage, ligation and polymerization of nucleotides, phosphoryl exchange, and the formation of amide and carbon–carbon bonds.[1–4] Historically, naturally occurring ribozymes have been grouped into several classes: (1) self-cleaving viral agents, (2) self-splicing introns, (3) RNase P, (4) *Neurospora* vs ribozyme, (5) spliceosome, and (6) ribosome.[5,6] The self-cleaving viral agents include the hepatitis delta virus and components of plant viruses that perform site-specific cleavage of their RNA genomes during rolling-circle replication. The self-splicing introns include the group I and group II introns found in bacteria, as well as those found in the mitochondria and chloroplasts of eukaryotes. The RNA component of RNase P is required for the 5′ processing of tRNAs. In addition to RNase P, it has been suggested that RNA catalysis occurs in other RNA–protein complexes, such as the eukaryotic spliceosome and ribosome.[7–9]

Although protein enzymes have had medical applications for many years, the therapeutic use of ribozymes is comparatively recent and remains largely untested. The self-cleaving group I intron ribozymes[10,11] and the ribozymes based on the RNA component of RNase P[12,13] hold clear promise for gene therapy. However, the self-cleaving viral agents, such as the hairpin and hammerhead ribozymes,

[1] T. R. Cech and B. L. Bass, *Annu. Rev. Biochem.* **55,** 599 (1986).
[2] B. Zhang and T. R. Cech, *Nature* **390,** 96 (1997).
[3] B. Zhang and T. R. Cech, *Chem. Biol.* **5,** 539 (1998).
[4] B. Seelig and A. Jaschke, *Chem. Biol.* **6,** 167 (1999).
[5] T. R. Cech, *Biochem. Int.* **18,** 7 (1989).
[6] T. R. Cech, *Science* **289,** 878 (2000).
[7] T. Tuschl, P. A. Sharp, and D. P. Bartel, *EMBO J.* **17,** 2637 (1998).
[8] P. Khaitovich, T. Tenson, P. Kloss, and A. S. Mankin, *Biochem.* **38,** 1780 (1999).
[9] R. Green and H. F. Noller, *Biochemistry* **38,** 1772 (1999).
[10] B. A. Sullenger and T. R. Cech, *Nature (London)* **371,** 619 (1994).
[11] J. T. Jones and B. A. Sullenger, *Nature Biotechnol.* **15,** 902 (1997).
[12] F. Liu and S. Altman, *Genes Dev.* **9,** 471 (1995).
[13] C. Guerrier-Takada, R. Salavati, and S. Altman, *Proc. Natl. Acad. Sci. U.S.A.* **94,** 8468 (1997).

because of their substrate specificity and small size currently offer the greatest potential for ribozyme-mediated gene therapies. Because these small ribozymes can carry out sequence-specific cleavage of RNA molecules in *trans,* researchers have used them as inhibitors of viral replication[14,15] and cell proliferation[16–18] and in the treatment of diseases caused by dominant mutations, such as autosomal dominant retinitis pigmentosa (ADRP).[19,20] Currently, Phase I clinical trials are underway to test the safety of ribozyme mediated therapy for human immunodeficiency virus type I infections (HIV-1).[21–23]

Hairpin and hammerhead ribozymes were initially characterized from the satellite RNA of tobacco ringspot virus, and their names reflect the schematic secondary structure each conforms to when bound to a substrate molecule (Fig. 1). Both ribozymes catalyze sequence-specific cleavage resulting in the formation of two products, one with a $5'$-hydroxyl and the other with a $2',3'$-cyclic phosphate. The hairpin ribozyme consists of a 34 nucleotide catalytic core, composed of loops A and B and four helices numbered 1–4, of which helices 1 and 2 are responsible for substrate recognition (Fig. 1a). Helix 1 may be of variable length, but helix 2 is confined to four base pairs.[24] The hammerhead ribozyme has a conserved catalytic core of just 11 nucleotides.[25] This core is tethered to a stable hairpin structure (helix II) and flanked by two arms that are responsible for target recognition, helices I and III (Fig. 1b)

Hairpin ribozymes recognize substrates containing the sequence $5'$YNGUC$3'$, where N is any nucleotide and Y is a pyrimidine. The $5'$YNGUC$3'$ sequence exists in many gene transcripts, but it is not prevalent enough to ensure that it will frequently occur within close proximity of a disease-causing point mutation. The

[14] N. Sarver, E. M. Cantin, P. S. Chang, J. A. Zaia, P. A. Lande, D. A. Stephens, and J. J. Rossi, *Science* **247,** 1222 (1990).

[15] A. Gervix, L. Scharz, A. D. Ho, D. Looney, T. Lane, and F. Wong-Staal, *Hum. Gen. Ther.* **8,** 2229 (1997).

[16] M. Koizumi, H. Kamiya, and E. Ohtsuka, *Biol. Pharm. Bull.* **16,** 879 (1993).

[17] M. Kashani Sabet, T. Funato, V. A. Florenes, O. Fodstad, and K. J. Scanlon, *Cancer Res.* **54,** 900 (1994).

[18] F. Czubayko, A. M. Schulte, G. J. Berchem, and A. Wellstein, *Proc. Natl. Acad. Sci. U.S.A.* **93,** 14753 (1996).

[19] A. S. Lewin, K. A. Drenser, W. W. Hauswirth, S. Nishikawa, D. Yasumura, J. G. Flannery, and M. M. LaVail, *Nature Med.* **4,** 967 (1998).

[20] M. M. LaVail, D. Yasumura, M. T. Matthes, K. A. Drenser, J. G. Flannery, A. S. Lewin, and W. W. Hauswirth, *Proc. Natl. Acad. Sci. U.S.A.* **97,** 11488 (2000).

[21] F. Wong-Staal, E. M. Poeschla, and D. J. Looney, *Hum. Gene Ther.* **9,** 2407 (1998).

[22] R. G. Amado, R. T. Mitsuyasu, G. Symonds, J. D. Rosenblatt, J. Zack, L. Q. Sun, M. Miller, J. Ely, and W. Gerlach, *Hum. Gene Ther.* **10,** 2255 (1999).

[23] D. Cooper, R. Penny, G. Symonds, A. Carr, W. Gerlach, L. Q. Sun, and J. Ely, *Hum. Gene Ther.* **10,** 1401 (1999).

[24] A. Hampel, R. Tritz, M. Hicks, and P. Cruz, *Nucleic Acids Res.* **18,** 299 (1990).

[25] M. J. McCall, P. Hendry, and P. A. Jennings, *Proc. Natl. Acad. Sci. U.S.A.* **89,** 5710 (1992).

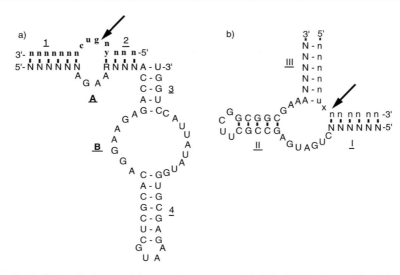

FIG. 1. Schematic diagram of the secondary structures of the hairpin (a) and hammerhead ribozymes (b). Ribozyme sequences are indicated by capital letters and target RNA sequences are shown in lowercase. The arrows designate the cleavage sites in target RNAs. "N" or "n" is used to represent any nucleotide, "x" represents any nucleotide except guanosine, "y" indicates a pyrimidine nucleotide, and "R" indicates a purine nucleotide. The helices in the hairpin ribozyme have been numbered using Arabic numerals and the loops designated A and B. The helices in the hammerhead ribozyme are designated by Roman numerals.

hammerhead ribozyme has a less restricted target sequence requirement than the hairpin ribozyme and can be designed to target any substrate RNA containing an $^{5'}NUX^{3'}$ triplet, where N is any nucleotide and X is any nucleotide except guanosine. Its relaxed target sequence requirement enables the hammerhead ribozyme to be used for the specific down-regulation of many mutant alleles, because variations of the NUX triplet occur frequently throughout all RNA messages, and it is usually possible to identify a potential hammerhead ribozyme cleavage site near the targeted missense mutation. It should be noted, however, that not all NUX triplets are cleaved with the same efficiency. Initially the GUC triplet was reported to be the most efficiently cleaved followed by CUC, UUC, and then the remaining NUX combinations.[26] Requirements for the cleavage triplet have since been extended to include any NHH sequence, where N is any nucleotide and H is any nucleotide except guanosine.[27] In addition to its minimal target sequence requirements, the hammerhead ribozyme is also capable of discriminating *in vivo* between

[26] T. Shimayama, S. Nishikawa, and K. Taira, *Biochemistry* **34**, 3649 (1995).
[27] A. R. Kore, N. K. Vaish, U. Kutzke, and F. Eckstein, *Nucleic Acids Res.* **26**, 4116 (1998).

two target molecules that differ by only one nucleotide, for example, cleaving the message from a mutated H-ras target but not the wild-type H-ras message.[28] This ability to discriminate between two mRNA molecules based on a single distinguishing nucleotide gives the hammerhead ribozyme a significant advantage over traditional antisense RNA approaches to inactivation of gene expression. In 1997 Hormes et al.[29] performed a head-to-head comparison in which traditional antisense RNA molecules and hammerhead ribozymes were directed against identical HIV-1 target sequences and assayed for their abilities to inhibit replication. Hammerhead ribozymes inhibited viral replication 2- to 10-fold better than antisense RNA molecules recognizing the same sequences.

Dominant diseases are of two types, those that lead to the accumulation of proteins that are either toxic or unable to form a functional complex (dominant negative mutations) and those that lead to insufficient production of essential proteins (haploinsufficiency). ADRP is an example of the former type, in which the production of mutated proteins (e.g., rhodopsin or peripherin/rds) ultimately results in the apoptotic death of photoreceptor cells.[30,31] There are more than 100 point mutations in the rhodopsin gene alone that cause ADRP[32] and hence much of our research has focused on producing a ribozyme protocol for testing the potential of ribozymes for gene therapy of rhodopsin-linked ADRP. Our hypothesis has been that ribozymes can reduce the production of aberrant rhodopsin protein by selectively cleaving mRNA molecules encoding the mutated forms of the proteins while the level of mRNA molecules encoding the wild-type protein remains unaffected. If the targeted mRNA molecules are cleaved between their 5′, 7-methylguanosine cap and 3′ poly (A) tail, we expect that the two RNA cleavage products will be quickly and efficiently degraded by cellular nucleases.[33,34]

In this work we will confine our discussion to in vitro methods and the rationale we have developed to assess the therapeutic potential of hammerhead and hairpin ribozymes targeted against retinal mRNAs associated with autosomal dominant retinitis pigmentosa (ADRP). The strategies employed are directed towards delivery by recombinant adeno-associated virus (rAAV). However, specific protocols for production of rAAV particles are discussed in detail elsewhere,[35] as well as in

[28] T. Funato, T. Shitara, T. Tone, L. Jiao, M. Kashani-Sabet, and K. J. Scanlon, *Biochem. Pharmacol.* **48**, 1471 (1994).

[29] R. Hormes, M. Homann, I. Oelze, P. Marschall, M. Tabler, F. Eckstein, and G. Sczakiel, *Nucleic Acids Res.* **25**, 769 (1997).

[30] R. Adler, *Arch. Ophthamol.* **114**, 79 (1996).

[31] J. K. Phelan and D. Bok, *Mol. Vision* **6**, 116 (2000).

[32] S. Van Soest, A. Westerveld, P. T. V. M. Dejong, E. M. Bleeker-Wagemakers, and A. A. Bergen, *Surv. Ophthalmol.* **43**, 321 (1999).

[33] A. Beelman and R. Parker, *Cell* **81**, 179 (1995).

[34] P. Mitchell and D. Tollervy, *Curr. Opin. Gen. Dev.* **10**, 193 (2000).

[35] W. W. Hauswirth, A. S. Lewin, S. Zolotukhin, and N. Muzycka, *Methods Enzymol.* **316**, 743 (2000).

this volume. Additionally, for a review of the uses of ribozymes in retinal gene therapy consult Hauswirth and Lewin.[36]

Materials

Enzymes and Reagent Chemicals

T7 RNA polymerase, DNA polymerase I, restriction endonucleases, and polynucleotide kinase are available from several manufacturers including Invitrogen (Carlsbad, CA) and New England BioLabs (NEB) (Beverly, MA). Recombinant placental ribonuclease (RNase) inhibitor can be obtained as RNasin from Promega (Madison, WI). Purified, ribonuclease free water is critical for RNA experiments. All water is distilled and deionized, and purified by a filtration and ion-exchange chromatography system such as Barnstead NanoPureII (Dubuque, IA). We steam-sterilize such water and use it directly without diethyl polycarbonate (DEPC) treatment. Henceforth, water so treated will be referred to simply as "H_2O." Should RNase contamination prove a problem, water can be treated for 1–15 hr with DEPC at a 0.02% final concentration prior to autoclaving.

Plasmids, Nucleotides, and Oligonucleotides

Transcription vectors such as pT7T3-19 (Invitrogen) and pBluescriptII KS+ (Stratagene, La Jolla, CA) contain promoter sequences for bacteriophage RNA polymerases such as T7 and T3 RNA polymerase. TA PCR cloning kits are available from Invitrogen and include T7 and T3 promoters in the plasmid DNA provided. Unlabeled nucleoside triphosphates should be obtained from a high-quality vendor such as Pharmacia (Piscataway, NJ) and purchased as a buffered stock solution (typically 100 mM). Labeled nucleotides for ribozyme assays can be purchased from Amersham (Arlington Heights, IL), ICN (Irvine, CA), or DuPont/NEN (Boston, MA). DNA oligonucleotides encoding ribozyme sequences with appropriate flanking restriction sites can be ordered from a variety of sources; we have had success with GIBCO/Life Technologies. RNA oligonucleotides to serve as targets for ribozyme cleavage assays can be ordered from fewer companies; we have used Dharmacon (Boulder, CO). RNA oligonucleotides are protected on their 2′ residues and must be deprotected according to the manufacturer's instructions (see below).

Chromatography and Electrophoresis

Sephadex G-25 fine or G-50 fine resins are purchased directly from Pharmacia (Piscataway, NJ). Reagents for denaturing electrophoresis [acrylamide,

---

[36] W. W. Hauswirth and A. S. Lewin, *Prog. Retinal Eye Res.* **19,** 689 (2000).

methylene(bis)acrylamide, urea] should be highly purified and are available from a wide variety of companies. Silica gel thin-layer chromatography (TLC) plates for UV-shadowing should contain fluorescent (F254) indicator and are available through Fisher Scientific (Pittsburgh, PA) and other vendors.

Methods

Identification of Target Sites

For a ribozyme to be effective at down-regulating gene expression, the point mutation must be within four bases of the scissile phosphodiester bond or else the ribozyme will fail to distinguish between wild-type and mutant transcripts.[37] This consideration limits ribozyme targeting to a fixed region immediately upstream and downstream of the point mutation (e.g., P23H opsin). Most importantly, the region of mRNA containing the targeted cleavage triplet needs to be accessible for base pairing with the flanking arms of the ribozyme in order to allow helix II to form and sequence-specific cleavage to take place. Consequently, the single largest barrier to be overcome in a ribozyme-mediated therapy is the reliable prediction of accessible target regions within full-length mRNA molecules. Because of intramolecular base pairing, long mRNAs characteristically exhibit complex secondary and tertiary structures.[38] Several strategies exist to predict the accessibility of a target site, including the use of computer RNA folding programs,[39] the functional screening of random ribozyme libraries,[40] and accessibility mapping of RNA using enzymes, chemical reagents,[41] or RNase H.[42] However, such in vitro assays cannot account for RNA binding proteins that appear to alter mRNA structure in vivo.[43] Thus, the ultimate test of a ribozyme's catalytic efficiency and substrate specificity is experimental, preferably in an in vivo system.

Ribozyme Design

Hammerhead Design. There are no hard-and-fast rules for designing a hammerhead ribozyme to ensure absolute discrimination between mutant and wild-type mRNA sequences. However, we normally include a few alterations to the *trans* cleaving hammerhead ribozymes first introduced by Haseloff and Gerlach

[37] M. Werner and O. C. Uhlenbeck, *Nucleic Acids Res.* **23,** 2092 (1995).

[38] O. C. Uhlenbeck, A. Pardi, and J. Feigon, *Cell* **90,** 833 (1997).

[39] M. Zuker, *Science* **244,** 48 (1989).

[40] A. Lieber and M. Strauss, *Mol. Cell Biol.* **15,** 540 (1995).

[41] T. B. Campbell, C. K. McDonald, and M. Hagen, *Nucleic Acids Res.* **25,** 4985 (1997).

[42] K. R. Birikh, Y. A. Berlin, H. Soreq, and F. Eckstein, *RNA* **3,** 429 (1997).

[43] O. Heidenreich, S. H. Kang, D. A. Brown, X. Xu, P. Swiderski, J. J. Rossi, F. Eckstein, and M. Nerenberg, *Nucleic Acids Res.* **23,** 2223 (1995).

in 1988.[44] Generally, it is recommended to keep the ribozyme's hybridizing arms (helices I and III) relatively short. The combined length of the sequence targeted by helices I and III can be up to 12 nucleotides and still discriminate a single base mismatch in either of the arms.[45] Consequently, after identifying a potential NUX sequence that lies within close proximity of a desired point mutation, hybridizing arms (helices I and III) are designed to base pair with a stretch of 12 nucleotides surrounding the "X" nucleotide in the NUX triplet site (Fig. 1b). This base separates helices I and III and does not form a conventional base pair. While the use of symmetric hammerhead ribozymes (i.e., six base helices I and III) has been the usual approach, asymmetric hammerheads may allow faster turnover while still providing the desired target specificity.[46] Ribozymes having hybridizing arms more than six bases long can be designed to distinguish between wild-type and mutant sequences if the mutation generates a unique NUX site. Helix II should be four or five base pairs closed by an RNA tetraloop (GNRA or UUCG) for increased stability.

Hairpin Design. Many of the same considerations given above apply to the design of hairpin ribozymes as well. However, the hairpin ribozyme recognition sequence is more constrained than that of the hammerhead and, as mentioned before, the length the hybridizing arm forming helix 2 is limited to four base pairs (Fig. 1a). It is important, therefore, to keep helix 1 short in order to promote selective cleavage, because hairpin ribozymes are able to cut mismatched targets with reduced efficiency if the mismatches are located distal to the cleavage site.[47] *In vitro* selection and site-specific mutagenesis have determined the nucleotides essential for activity and have expanded the target range of these ribozymes.[47–52] Although fewer targets for hairpin ribozymes are typically present in a target mRNA, hairpin activity is increased as a result of their greater inherent stability under physiologic temperature and magnesium conditions relative to hammerhead ribozymes.[53] Their activity can be increased further by replacing the naturally

[44] J. Haseloff and W. L. Gerlach, *Nature* **334**, 585 (1988).

[45] K. J. Hertel, D. Herschlag, and O. C. Uhlenbeck, *EMBO J.* **15**, 3751 (1996).

[46] P. Hendry and M. McCall, *Nucleic Acids Res.* **24**, 2679 (1996).

[47] J. M. Burke and A. Berzal-Herranz, *FASEB J.* **7**, 106 (1993).

[48] A. Berzal-Herranz, S. Joseph, and J. M. Burke, *Genes Dev.* **6**, 129 (1992).

[49] A. Berzal-Herranz, S. Joseph, B. M. Chowrira, S. E. Butcher, and J. M. Burke, *EMBO J.* **12**, 2567 (1993).

[50] P. Anderson, J. Monforte, R. Tritz, S. Nesbitt, J. Hearst, and A. Hampel, *Nucleic Acids Res.* **22**, 1096 (1994).

[51] A. Siwkowski, R. Shippy, and A. Hampel, *Biochemistry* **36**, 3930 (1997).

[52] N. K. Vaish, P. A. Heaton, O. Fedorova, and F. Eckstein, *Proc. Natl. Acad. Sci. U.S.A.* **95**, 2158 (1998).

[53] A. M. Chowira, A. Berzal-Herranz, and J. M. Burke, *Biochemistry* **32**, 1088 (1993).

a)

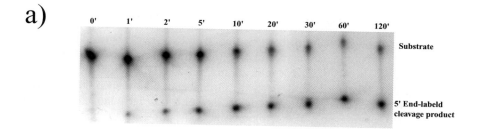

b)

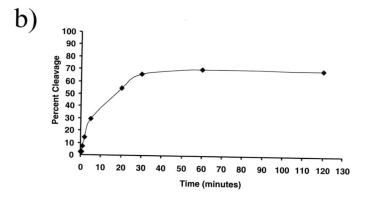

Fig. 2. Autoradiograph of the cleavage time course analysis for hammerhead ribozyme 656 targeted against a synthetic 13-nucleotide substrate corresponding to the sequence of the human P23H rod opsin mRNA associated with ADRP (a). The cleavage data have been plotted in (b) as function of time.

occurring three-base loop at the terminus of helix 4 with a tetraloop,[54,55] and by changing the uridine at position 39 to a cytosine.[54]

Cloning of Ribozymes and Targets

Synthetic DNA oligonucleotides are used to clone ribozymes in packaging vectors to generate both expression vectors for *in vitro* transcription and recombinant adeno-associated viruses (rAAV). Sense and antisense DNA oligonucleotides coding for the ribozyme sequence are designed with flanking restriction sites. Cloning efficiency can be increased by polyacrylamide gel purification of the oligonucleotides before using them in a ligation reaction.

[54] B. Sargueil, D. B. Pecchia, and J. M. Burke, *Biochemistry* **34**, 7739 (1995).

[55] M. Yu, E. Poeschla, O. Yamada, P. Degrandis, M. C. Leavitt, J. Heusch, J. K. Yees, F. Wong-Stall, and A. Hampel, *Virology* **206**, 381 (1995).

Gel Purification of Oligonucleotides

Protocol. Of each oligonucleotide, 100 nmol suspended at 20 nmol/μl is mixed with an equal volume of formamide loading dye [90% (w/v) formamide, 50 mM EDTA, 0.4% (w/v) xylene cyanol, and 0.4% (w/v) bromphenol blue] and separated on a 20% (w/v) acrylamide, 8 M urea gel run in 0.5× TBE buffer [45 mM Tris-borate (pH 8.3), 1 mM EDTA]. After the bromphenol blue tracking dye has run about two-thirds the length of a 40 cm gel, the gel is transferred to a 20 × 20 cm silica gel TLC plate containing F254 fluorescent indicator. The TLC plate should be covered with plastic wrap to prevent transfer of the chromatography powder to the gel. In a dark room, a short-wavelength UV hand light is used to create a "shadow" of the RNA band against the fluorescent background. The bands are then excised with a sterile scalpel and eluted overnight in 1 M ammonium acetate, 50 mM Tris HCl (pH 7.5), 20 mM EDTA, 0.5% (w/v) SDS at 37°. The elution solution is then removed from the gel slices and the DNA is ethanol precipitated, resuspended in 100 μl H$_2$O, and its concentration is determined by absorbance at 260 nm.

Polynucleotide Kinase Reaction for DNA Oligonucleotides

Protocol. A 10 μl reaction contains the following:

    5 μl Oligonucleotide (approximately 5 nmol)
    1 μl 10 × PNK Buffer [700 mM Tris-HCl (pH 7.6), 100 mM MgCl$_2$, 5 mM dithiothreitol (DTT)] (NEB)
    1 μl 10 mM ATP
    2 μl H$_2$O
    1 μl (10 units/μl) T4 polynucleotide kinase (NEB)

The reactions are then incubated at 37° for 1 hr. Phosphorylated oligonucleotides can be stored at −20°, but it is generally best to use them immediately in the following DNA ligation reaction. It is also possible to avoid this step by purchasing 5′-phosphorylated oligonucleotides.

Gel Purification of Vector DNA

Protocol. Before ligation, agarose gel purification of the linearized AAV packaging vector or other expression plasmid is performed by running the digested vector on a 1% agarose gel, staining with ethidium bromide, and excising the proper band after visualization with low-intensity UV light. The excised band is then mixed with an equal volume of phenol and crushed in a microcentrifuge tube. The contents are then incubated at −70° for at least 30 min and spun for 15 min at 14,000 rpm in a microcentrifuge. The aqueous phase is next extracted with an equal volume of phenol : chloroform : isoamyl alcohol (50 : 50 : 1, v/v/v). The DNA is precipitated by the addition of two volumes of ethanol, and, finally, the sample is resuspended at a concentration of 0.5 μg/μl in H$_2$O.

Annealing and Ligating Phosphorylated Oligonucleotides. Before beginning the reaction, oligonucleotides should be resuspended at a final concentration of approximately 2 pmol/μl. The following reaction has been scaled to work with 5 μg of an 8 kb plasmid DNA. For different sized plasmid DNAs, the amount of oligonucleotide and linearized plasmid DNA should be adjusted to maintain a 10 : 1 molar ratio of oligonucleotide : vector ends. For reactions using synthetic DNA oligonucleotides that have already had a 5' phosphate added during synthesis, one may resuspend directly at a 1 : 1 oligonucleotide : vector molar ratio.

Protocol. To initiate the reaction, 5 μl, or 10 pmol, of each oligonucleotide (sense and antisense) are mixed and heated to 90° for 3 min on a heating block. The reaction is then slow-cooled by removing the block from the heating element and waiting until the block containing the reaction tubes has cooled to room temperature. Add 2 μl each of 10 × ligase buffer [500 mM Tris-HCl (pH 7.5), 100 mM MgCl$_2$], 10 mM ATP, 10 mM DTT, 40% PEG 8000, 1 μl of packaging plasmid (0.5 μg/μl) that has been linearized with the proper enzymes and resuspended in H$_2$O, and 1 μl (10 units) of T4 DNA ligase. The reactions are incubated at room temperature for 2 hr or 16° overnight. Two μl of this reaction are then used to transform chemically competent or electro-competent *E. coli* cells according to Sambrook and Russel.[56]

Bacterial Transformations. We use one of several recombination deficient strains of *Escherichia coli,* such as JC 8111 or Sure (Stratagene), in order to prevent loss of the AAV inverted terminal repeat (ITR) sequences present in the AAV packaging vector. ITRs are critical for packaging plasmid inserts as recombinant AAV vectors.[35] Transformed cells are grown at 30° to a low density to help preserve ITRs. All clones, of course, are sequenced for proper ribozyme orientation and sequence and also screened for ITR retention before attempting to package the plasmids as recombinant AAV.

Special Considerations. For longer sequences such as hairpin ribozymes, we have used mutually primed synthesis of overlapping DNA oligonucleotides containing the ribozyme sequences and flanked by restriction sites necessary for subsequent cloning. These oligonucleotides are annealed as above, and the large fragment of DNA polymerase I is used to make completely duplex DNAs. The annealed, extended oligonucleotides are then digested with the proper restriction enzyme(s) and ligated to the packaging vector. A more detailed description of this method appears in Volume 316 of this series.[57]

Full-length cDNA targets are cloned in transcription vectors to allow synthesis of long target RNAs for assessing cleavage site accessibility. We use the TA and

[56] J. Sambrook and D. W. Russell, *in* "Molecular Cloning: A Laboratory Manual," 3rd Ed., p. 1.84. Cold Spring Harbor Laboratory Press, New York, 2001.
[57] L. C. Shaw, P. O. Whalen, K. A. Drenser, W. S. Yan, W. H. Hauswirth, and A. S. Lewin, *Methods Enzymol.* **316,** 761 (2000).

TA TOPO cloning vectors (Invitrogen), which allow direct cloning of PCR or RT-PCR products. These vectors contain T7 and T3 RNA polymerase promoters that can be used to generate transcripts of the cloned genes *in vitro* (discussed below).

In Vitro Assay of Short Synthetic RNA Targets

Initial cleavage assays use short synthetic RNA oligonucleotide targets ranging in length from 13 to 16 bases and synthetic or transcribed ribozymes. Such short RNA targets lack the secondary and tertiary structures that are likely to exist in the full-length mRNA of *in vivo* target molecules, and synthetic ribozymes lack the incorporation of vector RNA sequence that is unavoidable following cloning. These extra nucleotides could interfere with target recognition.

Deprotection of 2'-Orthoester Protected RNA Oligonucleotides. Ribozymes and short synthetic RNA oligonucleotides substrates containing 6–8 nucleotides flanking the cleavage site are ordered from Dharmacon Research, Inc. (Boulder, CO). RNA oligonucleotides are 2'-orthoester protected and are usually ordered on a 0.2 μmol synthesis scale. The RNA synthesis process employed by Dharmacon has an average coupling yield of 99% per step. Therefore the yield of full-length product is given by $(0.99)^{(oligo\ length)} \times$ micromoles ordered. The 2'-orthoester RNA oligonucleotides are deprotected by incubation in 100 mM acetic acid (adjusted to pH 3.8 with TEMED) for 30 min at 60° according to the manufacturer's instructions. Deprotected RNA oligonucleotides are dried under vacuum and resuspended in H$_2$O to yield a 300 pmol/μl working stock and stored at −80°. We have not found it necessary in most instances to gel purify synthetic ribozyme or targets purchased from Dharmacon Research, Inc., but they may be purified as described for transcribed ribozymes and targets (see below).

5' End-Labeling of Deprotected RNAs

Protocol. Combine the following reagents in a sterile 1.5 ml microcentrifuge tube for a 10 μl end-labeling reaction:

2 μl RNA oligonucleotide (20 pmol)
1 μl 10 × Kinase Reaction Buffer [700 mM Tris-HCl (pH 7.6), 100 mM MgCl$_2$, 5 mM dithiothreitol (DTT)] (NEB)
1μl RNasin [diluted 1 : 10 in 0.1 M dithiothreitol (DTT) from 40 units/μl stock] (Promega)
4 μl H$_2$O
1 μl [γ^{32}-P ATP (167 μCi) (ICN)
1 μl T4 Polynucleotide kinase (10 units/μl) (NEB)

Polynucleotide kinase may be obtained from any reputable vendor, but we use either Invitrogen or NEB. Place the tube at 37° for 30 min. After incubation, add

90 μl water and heat inactivate the kinase at 65° for 3 min. Extract for 60 sec with 100 μl phenol/chloroform and once with 100 μl chloroform. Purify the aqueous layer by filtration through a 1 ml Sephadex G-25 (Pharmacia) spin column. A stock solution will be 0.2 pmol/μl and should be stored at 4° or −80°.

Hammerhead Ribozyme Cleavage Time Course Analysis. Cleavage time course reactions are performed under substrate excess conditions, usually at a molar ratio of target to ribozyme of 20 : 1. Under such conditions, it is possible to determine the time interval during which the substrate cleavage is linear with time and no more than 10–20% of the substrate has been converted to its cleavage products. Under these criteria, multiple turnover kinetic analysis as an estimate of the initial velocity (V_o) can be carried out. The precise number and time interval of reaction aliquots needed to make this determination must be ascertained experimentally to ensure that the relationship between the accumulation of cleavage products and time maintains linearity during the initial phases of the reaction.

The possibility exists that a ribozyme designed to perform sequence-specific cleavage of a mutant allele may also recognize and cleave the wild-type substrate as well. As a result, cleavage time course analysis should also be performed using substrates corresponding to the wild-type mRNA sequence. Candidate ribozymes that do not cleave their intended targets selectively may not be good candidates for therapeutics.

Initial cleavage time course conditions usually consist of 50 nM ribozyme and 1μM substrate incubated in a 100 μl reaction mix containing a final concentration of 20 mM MgCl$_2$ and 40 mM Tris-HCl, pH 7.5, for varying amounts of time at 37°. Experimental conditions such as time of incubation, magnesium concentration, ribozyme concentration, and substrate concentration should be varied independently until the initial levels of cleavage are linear with time for up to 10–20% target cleavage. Ribozymes that cleave substrate molecules only under conditions of ribozyme excess or MgCl$_2$ concentrations greater than 20 mM should be eliminated from further *in vitro* analysis, since it is unlikely that they would perform well under physiological conditions.

Protocol. Combine the following reagents in sterile 1.5 ml microcentrifuge tubes:

Ribozyme solution	Target solution
25 μl H$_2$O	38 μl H$_2$O
10 μl 400 mM Tris-HCl, pH 7.5 at 37°	4 μl 25 pmol/μl unlabeled target
5 pmol ribozyme	2 μl 0.2 pmol/μl ^{32}P-end-labeled target
Add H$_2$O to 36 μl	

Heat target solution to 65° for 2 min, quick cool on ice, and then equilibrate to 37°. Heat the ribozyme solution at 65° for 2 min, then remove it from the heat source and allow the tube to cool at room temperature for 10 min. Add 10 μl

200 mM MgCl$_2$ and 10 μl RNasin (diluted 1 : 10 from stock in 0.1 M DTT) to the ribozyme solution, mix gently, and equilibrate the tube at 37° for 10 min in a 37° water bath. Following the incubation, add the target solution to the ribozyme solution, mix gently, and immediately remove a 10 μl aliquot from the tube. Add this aliquot to a prelabeled sterile 1.5 ml microcentrifuge tube containing 10 μl of formamide loading dye, and place the tube on ice for 1 min. Continue to remove six more 10 μl aliquots at various time points thereafter (e.g., $t = 1, 5, 10, 30,$ 60, and 120 min). Always prepare at least 25% more reaction volume than needed to account for pipetting/transfer errors. Store the reaction aliquots at $-20°$ until ready to analyze by denaturing polyacrlyamide gel electrophoresis.

Cleavage products for short synthetic RNA oligonucleotide substrates are analyzed by electrophoresis on 8 M urea, 10% acrylamide (w/v) sequencing gels run in 1 × TBE buffer. Once gels have been poured and allowed to polymerize, they are prerun to warm the gel to approximately 45°. Denature reaction aliquots at 85° for 2 min and quick cool on ice. Then load 10 μl of each cleavage time course reaction aliquot onto the gel and run at high voltage (1500–2000 V) until the bromphenol blue dye has run two-thirds of the way to the bottom of a 40 cm gel, in order to separate the 5′-end-labeled cleavage product from the substrate. The gel is then fixed in 40% methanol, 10% acetic acid, and 3% glycerol for 30–45 min at room temperature. The gel can then be dried and imaged by autoradiography with X-ray film (Fig. 2a). Better quantitation of the gel bands is achieved by exposing the dried gel to a radioanalytic phosphorescent screen and scanning the screen on a Molecular Dynamics PhosphoImager. The percentage of substrate cleaved can then easily be determined from the ratio of radioactivity in the 5′-end-labeled cleavage product band (P) to the sum of the radioactivity in the 5′-end-labeled cleavage product band (P) and the substrate band (S):

$$\text{Percentage cleaved}(\%C) = P/P + S$$

Percentage substrate cleaved can then be plotted against the time of the reaction to generate a graphical representation of the cleavage time course (Fig. 2b). The reaction velocity is simply calculated as the concentration of substrate cleaved per minute.

Multiple Turnover Kinetic Analysis. The hammerhead ribozyme catalytic reaction can be fitted to the Michaelis–Menten equation as long as the following conditions are fulfilled: (1) the substrate concentration is in molar excess of the ribozyme concentration, (2) the formation of a ribozyme–substrate complex is rapid and reversible, and (3) the rate-limiting step is the catalytic step.[58] To fulfill the first requirement, we perform the multiple turnover kinetic analysis by holding

[58] T. McConnell, *in* "Methods in Molecular Biology," Vol. 74, "Ribozyme Protocols" (P. C. Turner, ed.), p. 187. Humana Press Inc., Totowa, NJ, 1997.

the concentration of ribozyme constant at 15 nM and increasing the substrate concentration over a range from 150 nM to 150 μM. Increasing ratios of ribozyme to target may be necessary to reach saturating conditions for the ribozyme. Multiple turnover kinetic analysis should always be performed in triplicate and replicates should yield reproducible cleavage levels.

Protocol. To perform these experiments in triplicate, prepare the following amounts of each stock solution:

Ribozyme stock: 36 μl of 0.3 pmol/μl (1 : 1000 dilution of 300 pmol/μl stock)
Target stocks: 93 μl of 30 pmol/μl[a] (1 : 10 dilution of 300 pmol/μl)
             61 μl of 10 pmol/μl (1 : 3 dilution of 30 pmol/μl)
             66 μl of 3 pmol/μl (1 : 10 dilution of 30 pmol/μl)
             42 μl of 1 pmol/μl (1 : 10 dilution of 10 pmol/μl)

To make 200 μl of 30 pmol/μl stock add:
   20 μl of 5′-end-labeled RNA oligonucleotide (0.2 pmol/μl)
   20 μl of 300 pmol/μl stock
   160 μl of H$_2$O

Always prepare at least 25% more reaction volume than needed to account for pipetting/transfer errors. Equilibrate target stock solutions in a 37° water bath.

Following Table I, combine the indicated amounts of H$_2$O, 400 mM Tris-HCl, and ribozyme in appropriately prelabeled sterile 1.5 ml microcentrifuge tubes. Incubate the tubes at 65° for 2 min to denature the ribozyme and allow the tubes to

TABLE I
MULTIPLE TURNOVER KINETIC ANALYSIS

Tube #	H$_2$O (μl)	400 mM Tris-HCl pH 7.5 (μl)	Ribozyme 0.3 pmol/μl	200 mM MgCl$_2$ (μl)	RNasin : 0.1M DTT (1 : 10) (μl)	Target 3 pmol/μl[a]	Target 30 pmol/μl[a]	Rz : target ratio
1, 13, 25	14	2	0	2	1	1	n/a	0 to 10
2, 14, 26	13	2	1	2	1	1	n/a	1 to 10
3, 15, 27	12	2	1	2	1	2	n/a	1 to 20
4, 16, 28	10	2	1	2	1	4	n/a	1 to 40
5, 17, 29	8	2	1	2	1	6	n/a	1 to 60
6, 18, 30	6	2	1	2	1	8	n/a	1 to 80
7, 19, 31	13	2	1	2	1	n/a	1	1 to 100
8, 20, 32	12	2	1	2	1	n/a	2	1 to 200
9, 21, 33	10	2	1	2	1	n/a	4	1 to 400
10, 22, 34	8	2	1	2	1	n/a	6	1 to 600
11, 23, 35	6	2	1	2	1	n/a	8	1 to 800
12, 24, 36	4	2	1	2	1	n/a	10	1 to 1000

[a] n/a = not applicable.

TABLE II

PREPARATION OF SOLUTIONS CONTAINING KNOWN AMOUNTS OF 5′-END-LABELED
OLIGONUCLEOTIDE TARGET

Tube # (in duplicate)	H₂O (μl)	Target (μl)	Target solution used (pmol/μl)	Picomoles of target
1,13	100	0		0
2,14	99	1	1	1
3,15	98	2	1	2
4,16	96	4	1	4
5,17	94	6	1	6
6,18	92	8	1	8
7,19	99	1	10	10
8,20	98	2	10	20
9,21	96	4	10	40
10,22	94	6	10	60
11,23	92	8	10	80
12,24	90	10	10	100

cool at room temperature for 10 min. Then add the 200 mM MgCl$_2$ and RNasin, and equilibrate the tubes at 37° for an additional 10 min. Stagger the addition of the target stocks to the ribozyme by 30 to 60 sec and incubate reactions at 37° for the time period estimated to reach 10–20% of full cleavage as determined by the cleavage time-course reaction. Terminate the reactions after the appropriate time interval by adding 20 μl of ice-cold formamide RNA loading dye to the reaction tube. Place terminated reaction tubes on ice, then store reactions at −20° until the entire reaction set is ready to be analyzed by polyacrylamide gel electrophoresis.

To prepare a calibration curve for the multiple turnover kinetic reaction use Table II to prepare solutions containing known amounts of 5′-end-labeled oligonucleotide target. Using a slot blot apparatus, filter each target solution onto a HybondN$^+$ membrane (Amersham Pharmacia Biotech, Piscataway, NJ) that has been presoaked with H$_2$O. After scanning the calibration slot-blot, average the pixel density values for each set of duplicate solutions and plot these values vs the known amount of 5′-end-labeled oligonucleotide substrate to generate the calibration curve.

Analyze each multiple turnover kinetic reaction by electrophoresis on 8 M urea, 10% acrylamide (w/v) sequencing gels as above. Load 5 μl of each sample on the gel, run, fix, and dry gel as described above. The dried gel is then exposed to a radioanalytic phosphorescent screen along with the above calibration slot-blot. The screen is then scanned using a Molecular Dynamics PhosphoImager, and the labeled cleavage product of each multiple turnover reaction quantitated. Using the calibration curve, the pixel density of each 5′-end-labeled cleavage product can be converted into picomoles of cleavage product present in each reaction. The values for V_{max} and K_m can be determined using double-reciprocal plots of velocity vs

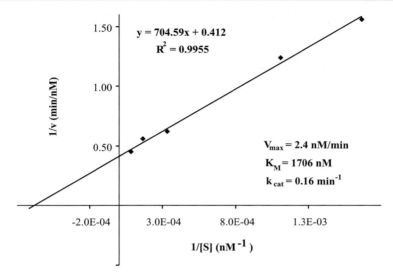

$y = 704.59x + 0.412$
$R^2 = 0.9955$

$V_{max} = 2.4$ nM/min
$K_M = 1706$ nM
$k_{cat} = 0.16$ min^{-1}

1/v (min/nM)

1/[S] (nM^{-1})

FIG. 3. Lineweaver–Burk plot for the multiple turnover kinetic analysis of a P347S-specific hammerhead ribozyme cleaving a synthetic 14-nucleotide substrate. Kinetic constants based on this double-reciprocal plot are shown.

substrate concentration (Lineweaver–Burk plots) or by plots of reaction velocity vs the ratio of velocity to substrate concentration (Eadie–Hofstee plots). First, plot observed reaction rates (in moles/minute) against the substrate concentrations to determine if the ribozyme has reached saturation. It is important that the reaction rate has reached saturation in order to estimate V_{max} and k_{cat}. Finally, use a Lineweaver–Burk plot (Fig. 3) or Eadie–Hofstee plot to determine the kinetic constants. For Lineweaver–Burk plots:

V_{max} = the absolute value of $1/Y$ where $X = 0$, and the units will be nM/min

K_m = the absolute value of $1/X$ where $Y = 0$, and the units will be nM

k_{cat} = $V_{max}/$[Ribozyme], and the units will be min^{-1}

In Vitro Transcription

For the kinetics experiments detailed above, synthetic RNA oligonucleotide targets are used. Ribozymes may be purchased as RNA oligonucleotides as well, but the yield is relatively low for longer transcripts ($0.99^{35} = 70\%$ for the average hammerhead, and $0.99^{50} = 60\%$ for the average hairpin). Consequently, it is usually more economical to produce ribozymes by T7 RNA polymerase transcription of plasmid DNA. We also routinely clone full-length target cDNAs into TA cloning vectors, as mentioned above, for long-target cleavage assays to determine target site accessibility.

For runoff transcription, plasmid DNAs containing the target or ribozyme sequences are linearized with restriction endonucleases that leave blunt ends or $5'$ overhangs downstream of the sequence to be transcribed. Full-length cDNA clones can be linearized with any enzyme that has a unique recognition site at the $3'$ end of the gene, or preferably with one that cuts just downstream of the cloned gene. Several rare sites (i.e., *Not*I) in the TA vectors (Invitrogen) are available for this purpose. Linearized plasmids are then extracted with phenol : chloroform : isoamyl alcohol (50 : 50 : 1, v/v/v), ethanol precipitated, and resuspended in water so that their final concentration is around 100 ng/μl. Transcripts are subsequently generated with T7 or the appropriate RNA polymerase and labeled by incorporation of [α-^{32}P]UTP (ICN) using the general protocol of Grodberg and Dunn.[59]

Protocol. A 25 μl transcription reaction contains:

1 μl of digested plasmid (100 ng/μl)
5 μl 5 × phosphate buffer (100mM sodium phosphate, pH 7.7)
5 μl Magnesium/spermidine solution (40mM MgCl$_2$, 16mM spermidine)
2 μl NTP solution (20 mM each of ATP, CTP, GTP)
1 μl 1 M DTT
1 μl RNasin (Promega)
1 μl [α-^{32}P]UTP (10 μCi per μl)
1 μl (5 units) T7 RNA polymerase
H$_2$O to 25 μl

For generating ribozymes, a high concentration of unlabeled UTP (800 μM) is included in the reaction to produce a transcript of low specific activity in which the radioactivity serves only as a tracer to allow for quantitation.

For full-length target RNA transcripts, a high specific activity is desired and the level of unlabeled UTP in the reaction is reduced fourfold (to 200 μM). However, dropping the concentration of UTP further severely reduces the yield of full-length transcripts. Reactions are incubated at 37° for 2 hr, and then 70 μl of sterile water and 5 μl of 10% (w/v) SDS (sodium dodecyl sulfate) are added and the labeled transcripts are extracted with 100 μl of phenol–chloroform–isoamyl alcohol (50 : 50 : 1, v/v/v). After extraction, fractionate ribozyme transcripts on 1 ml Sephadex G-25 fine (Pharmacia) spin columns and store the RNA in the flow-through at 4°. Long target transcripts may instead be purified by high-salt ethanol precipitation: add 100 μl of 5 M ammonium acetate to the 100 μl aqueous phase from the phenol–chloroform–isoamyl alcohol extraction and then precipitate the labeled transcripts by addition of 400 μl of absolute ethanol. Resuspend the sample and repeat the precipitation. Wash the final pellets twice with 70% ethanol,

[59] J. Grodberg and J. J. Dunn, *J. Bacteriol.* **170,** 1245 (1988).

and resuspend the transcripts in 50 μl of H_2O. Determine the concentration of the transcripts by spotting 1 μl on filter paper and measuring incorporation of ^{32}P in a liquid scintillation counter. Concentration is determined as follows:

$$\left(\frac{1 \times 10^{12} \text{ pmol}}{\text{mol}}\right) \left[\frac{\left(\dfrac{\text{cold U (mol)}}{\mu\text{Ci of labeled U}}\right)\left(\dfrac{1 \mu\text{Ci}}{2.22 \times 10^{11} \text{ cpm}}\right)}{\text{number of U's in transcript}}\right.$$

$$\left. \times (\text{cpm in sample per } \mu\text{l}) = \text{pmol}/\mu\text{l} \right]$$

This assumes that there is a 100% counting efficiency for ^{32}P in scintillation fluid and that correction has been made for decay of the radioactive label if it is used after its reference date. Hairpin ribozymes may occasionally undergo aggregation or denaturation after being ethanol precipitated. However, if the concentration is kept below 1 nM, reannealing should restore activity.

Cleavage Time Courses for Long Targets

Ribozyme cleavage of long RNA targets cannot be used to determine the kinetic parameters of a ribozyme in question because the rate of cleavage is often limited by the secondary structure of the target. However, long targets (>100 nt) provide an estimate of how well the ribozyme cleaves relative to alternative ribozymes directed against the same mRNA and gauge the accessibility of the chosen cleavage site. Cleavage time courses are preformed basically as described for short RNA targets. The molar ratio of ribozyme to target is kept constant (generally 1 : 10) and the reaction is stopped and analyzed after increasing intervals.

Protocol. Perform the time course reactions as follows: in a 1.5 ml microcentrifuge tube mix 13 μl of 400 mM Tris-HCl (pH 7.4–7.5) with 10 pmol of ribozyme, 100 pmol of labeled full-length target, and a volume of sterile water sufficient to give a final reaction volume of 130 μl. Heat the ribozyme to 65° for 5 min and allow the tube to cool to room temperature (this usually takes 30–45 min). Add 13 μl of a 1 : 10 RNasin (Promega) to 0.1 M DTT mixture and start the reaction by adding 13 μl of 200 mM MgCl$_2$. Incubate the reactions at 37° and immediately remove 10 μl aliquots after 0, 1, 5, 10, 30, 60, 120, 240, and 360 min. Add the aliquots to 20 μl of formamide dye mix to stop the cleavage reactions and place the reactions on ice for 1 min. Store the reactions at $-20°$ until they are ready to be analyzed. To analyze, heat denature the samples at 95° for 10 min and then run 6 μl aliquots on 4% (w/v) polyacrylamide 8 M urea gels run in 1 $\times$ TBE. Include control lanes of ribozyme and target alone. Fix and dry the gels as described above. Visualization of the cleavage time course by autoradiography and quantification of the cleavage products using a PhosphorImager should be performed as described above.

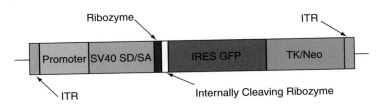

FIG. 4. Schematic of rAAV packaging plasmid including the IRES GFP element and neomycin resistance genes downstream of the ribozyme cassette. Plasmids for use in animal models are essentially the same but may lack the IRES GFP and neomycin resistance elements.

Packaging of Ribozymes in AAV Vectors

Adeno-associated viruses (AAV) are human parvoviruses composed of a 4.7 kb single-stranded linear DNA genome packaged in a capsid composed of three structural proteins. They are also known as dependoviruses because they require coinfection with either adenovirus or herpes virus to yield a productive infection. The AAV replication cycle requires the products of its two open reading frames, termed *rep* (for replication associated proteins) and *cap* (for capsid associated proteins). These genes are located between two 145-base inverted terminal repeats (ITRs) which contain the viral origin of replication and are necessary for viral packaging and integration of the virus into the host genome. To make recombinant AAV vectors, therapeutic sequences are inserted into a packaging plasmid containing a cloning site and regulatory sequences flanked by the ITR sequences. This construct is then cotransfected into HEK 293 cells along with a helper plasmid that encodes both the *rep* and *cap* AAV genes and the adenovirus early gene products essential for AAV replication.[60] This generates replication deficient, recombinant AAV containing the therapeutic DNA of interest.[61] Such recombinant AAV vectors infect a broad range of target cell types and, with the inclusion of strong cis-acting promoter sequences, generate high levels of expression of the passenger gene product.[62] Since these recombinant vectors are missing the *rep* and *cap* genes, they are not able to achieve productive infection of their target cells; however, as they do contain ITR structures, they are able to replicate in infected cells. They are also able to integrate into the genome of infected cells.

Several of the features of AAV vectors are essential for achieving good ribozyme expression in target cells (Fig. 4). The most important is the promoter. For example, vectors based on pTR-UF2[62] and employing either the 472-bp murine rod opsin promoter or the equivalent 691-bp fragment of the proximal bovine rod

[60] A. Kern, K. Rittner, and J. A. Kleinschmidt, *Hum. Gene Ther.* **9,** 2745 (1998).

[61] R. M. Kotin, *Hum. Gene Ther.* **5,** 793 (1994).

[62] J. G. Flannery, S. Zolotukhin, M. I. Vaquero, M. M. LaVail, N. Muzyczka, and W. W. Hauswirth, *Proc. Natl. Acad. Sci. U.S.A.* **94,** 6916 (1997).

opsin promoter were sufficient to efficiently direct expression of reporter genes in photoreceptor cells.[63] In contrast, previous efforts with vectors containing the lacZ reporter gene driven by the less specific immediate early CMV promoter resulted in only moderate levels of expression in many retinal cell types, predominantly in cells of the retinal pigment epithelium. Cell-type specific promoters also provide a level of safety that may be critical when designing therapeutic strategies for human use. Further cell-type and temporal specificity can be achieved through the use of inducible promoters. Studies with transgenic mice have shown that tetracycline inducible transactivators can be used to induce photoreceptor specific expression of a lacZ gene when under the control of the tetracycline operator cassette.[64]

A potentially important feature of AAV vectors expressing a ribozyme is the presence of an internally cleaving hairpin ribozyme cassette located downstream of the site at which the therapeutic ribozyme cassette is cloned. Self-cleavage of the primary vector transcript by this ribozyme serves to create uniform 3′ ends on mature ribozymes[65] that may be necessary to increase activity *in vivo*. Ribozyme cassettes contain both therapeutic and internal processing ribozymes preceded by an SV40 early intron and followed by a polyadenylation signal to promote nuclear export of the ribozymes.[66] For testing AAV-delivered ribozymes in tissue culture, a neomycin resistance gene driven by a thymidine kinase promoter and/or humanized GFP protein driven by a second promoter or an internal ribosomal entry site (IRES) are often included downstream of the ribozyme expression cassette. These features permit selection and/or visualization of cells expressing the ribozymes. Further information about development of ribozymes for treatment of retinal disease and AAV-mediated ribozyme delivery to retinal cells can be found in Lewin *et al.* (1998),[19] Shaw *et al.* (1999),[67] and LaVail *et al.* (2000).[20]

Acknowledgments

The authors thank the National Institutes of Health, the March of Dimes Foundation, the Foundation Fighting Blindness, Macular Vision Research Foundation, and Research to Prevent Blindness Inc., for supporting our ongoing gene therapy research. We also thank Dr. Lynn C. Shaw for his work in developing many of these methods and for allowing us to present some of his results as examples.

[63] S. Zolotukhin, M. Potter, W. W. Hauswirth, J. Guy, and N. Muzyczka, *J. Virol.* **70,** 4646 (1996).

[64] M. A. Chang, J. W. Horner, B. R. Conklin, R. A. DePinho, D. Bok, and D. J. Zack, *IOVS* **41,** 4281 (2000).

[65] M. Altschuler, R. Tritz, and A. Hampel, *Gene* **122,** 85 (1992).

[66] E. Bertrand, D. Castanotto, C. Zhou, C. Carbonelle, N. S. Lee, P. Good, S. Chatterjee, T. Grange, R. Pictet, D. Kohn, D. Engelke, and J. J. Rossi, *RNA* **3,** 75 (1997).

[67] L. C. Shaw, P. O. Whalen, K. A. Dresner, W. Yan, W. W. Hauswirth, and A. S. Lewin, *in* "Retinal Degenerative Diseases and Experimental Therapy" (J. G. Hollyfield, R. E. Anderson, and M. M. LaVail, eds.), p. 267. Kluwer Academic/Plenum Publishers, New York, 1999.

[22] Adeno-Associated Viral Vector-Mediated Gene Therapy of Ischemia-Induced Neuronal Death

By TAKASHI OKADA, KUNIKO SHIMAZAKI, TATSUYA NOMOTO,
TAKASHI MATSUSHITA, HIROAKI MIZUKAMI, MASASHI URABE,
YUTAKA HANAZONO, AKIHIRO KUME, KIYOTAKE TOBITA,
KEIYA OZAWA, and NOBUFUMI KAWAI

Introduction

Stroke remains a major brain disorder that often renders patients severely impaired and permanently disabled. Despite multiple clinical trials as a treatment for stroke, neuroprotective strategies have yet to be proved effective. Reperfusion injury is defined as the enhancement of the damage that occurs in ischemic cells during the reperfusion period. Transient occlusion of the carotid arteries in the rodent causes cell death in CA1 pyramidal neurons with a delay of a few days after reperfusion. The mechanisms underlying neuronal death after ischemia involve an excess of the excitatory neurotransmitter glutamate in the synapse. This excess causes a mobilization of free cytosolic calcium in the postsynaptic neuron, leading to calcium-dependent oxygen radical formation, cytoskeletal damage, and protein malfolding.[1] These degenerative events ultimately result in necrotic or apoptotic death in a subset of neurons. Because such neuronal death ensues over days, it is plausible to use protective gene therapy techniques during that interval.

Considerable interest has focused on the possibility of using viral vectors to deliver genes to the central nervous system for the purpose of decreasing neuronal injury. Adeno-associated virus (AAV) is a potentially useful gene transfer vehicle for neurologic gene therapies.[2–4] The advantages of AAV vector include the lack of any associated disease with a wild-type virus, the ability to transduce nondividing cells, possible integration of the gene into the host genome, and long-term expression of transgenes. The development of novel therapeutic strategies for central nervous system (CNS) diseases by using AAV vector has an increasing impact on

[1] J. M. Lee, G. J. Zipfel, and D. W. Choi, *Nature* **399**, A7 (1999).

[2] K. Ozawa, D. S. Fan, Y. Shen, S. Muramatsu, K. Fujimoto, K. Ikeguchi, M. Ogawa, M. Urabe, A. Kume, and I. Nakano, *J. Neural. Transm. Suppl.* **58**, 181 (2000).

[3] Y. Shen, S. I. Muramatsu, K. Ikeguchi, K. I. Fujimoto, D. S. Fan, M. Ogawa, H. Mizukami, M. Urabe, A. Kume, I. Nagatsu, F. Urano, T. Suzuki, H. Ichinose, T. Nagatsu, J. Monahan, I. Nakano, and K. Ozawa, *Hum. Gene Ther.* **11**, 1509 (2000).

[4] D. S. Fan, M. Ogawa, K. I. Fujimoto, K. Ikeguchi, Y. Ogasawara, M. Urabe, M. Nishizawa, I. Nakano, M. Yoshida, I. Nagatsu, H. Ichinose, T. Nagatsu, G. J. Kurtzman, and K. Ozawa, *Hum. Gene Ther.* **9**, 2527 (1998).

gene therapy research. Among the apoptosis repressor genes studied in mammalian cells, the protooncogene *bcl-2* has attracted attention as a potential regulator of neuronal survival. We have previously shown that postischemic injection of an AAV vector is effective for *bcl-2* gene delivery conferring neuroprotection in the gerbil hippocampus.[5] Here we present protocols for adenovirus-free preparation of an AAV vector and methods of gene transfer into gerbil brain for the gene therapy of ischemia-induced neuronal death.

Epitope Tagging to Target Gene by PCR

Principle of Molecular Tagging

To fuse the coding sequences of the epitope and the target protein, a tag DNA sequence is added by oligonucleotide-mediated mutagenesis with PCR to either the N terminus or the C terminus of a known protein coding sequence. The resulting DNA will encode a fusion product that contains both the target protein and the epitope tag. We introduce a strategy to construct the pWF*bcl-2* vector plasmid harboring FLAG-tagged *bcl-2* driven by the RSV promoter and flanked by ITRs (inverted terminal repeats) as an example.

Factors for Designing PCR Primers

1. For easy insertion of the PCR tagged gene into convenient vector cloning sites, add unique restriction sites to the 5' ends of PCR primers. Add a few extra bases before the 5' end of each restriction site to ensure that restriction enzyme cleavage will be efficient.

2. When adding the tag to the amino-terminal coding region of the target gene, you should also add an ATG start codon before the 5' end of the tag DNA sequence. The amino-terminal restriction site may contain an internal ATG start codon as in *Nco*I (CC<u>ATG</u>G) or *Nde*I (CAT<u>ATG</u>) sites. The Kozak consensus translation initiation sequence, 5'-CCACC<u>ATG</u>(G)-3',[6] may be used in place of the ATG start codon to maximize the efficiency of translation in eukaryotic cells.

3. To ensure accurate and efficient PCR amplification, optimize the portion of the primers by using commercially available primer analysis software. Generally, design each PCR primer so it will have a T_m around 60°. During the PCR reaction, primer-template annealing temperature is set 5° below the calculated T_m of the primer pair. Avoid making primers that can form primer dimer artifacts, significant secondary structures, and hairpin loops.

[5] K. Shimazaki, M. Urabe, J. Monahan, K. Ozawa, and N. Kawai, *Gene Ther.* **7,** 1244 (2000).
[6] M. Kozak, *Nucleic Acids Res.* **15,** 8125 (1987).

A **Upstream primer**
5'-GCCGCC <u>ATG</u> <u>GACTACAAGGACGATGATGACAAA</u> <u>GGC</u> <u>GCGCACGCTGGGAGAAC</u>-3'
           Start   FLAG   Gly   Target protein

Downstream primer
5'-ATCCT<u>GGATCC</u>AGGTGTGCAGG-3'
       Bam HI

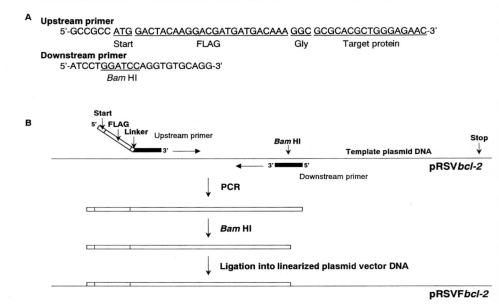

FIG. 1. Epitope tagging to a target gene by PCR. (A) The upstream primer includes extra bases, the ATG start codon, the FLAG epitope tag sequence, 1-amino acid linker, and the DNA sequence for the amino terminus of the target protein. The downstream primer includes a restriction enzyme site (*Bam*HI; GGATCC) and the neighboring DNA sequence including the restriction site of the target protein. (B) The upstream primer constructed to add a tag sequence partially anneal to the template DNA during PCR. The epitope tag sequence and the start codon will not anneal to the template. The downstream primer includes a restriction enzyme site (*Bam*HI) in the DNA sequence of the target protein. The PCR product was digested with *Bam*HI and 5'-phosphorylated with polynucleotide kinase. The purified restriction product was ligated into the blunt-*Bam*HI site of plasmid vector DNA.

Procedure

To construct pWF*bcl-2* harboring FLAG-tagged *bcl-2* driven by the RSV promoter and flanked by ITRs (inverted terminal repeats), human *bcl-2* fragment was excised from pB4 (a gift from Dr. Y. Tsujimoto) by double digestion with *Eco*RI and *Eco*T221. The resulting 846 bp fragment was cloned into the *Eco*RI-*Pst*I site of an expression vector pSV, harboring the RSV promoter and the SV40 early polyadenylation signal (pRSV*bcl-2*). To generate FLAG-tagged *bcl-2*, PCR was performed by using *Pfu* DNA polymerase (Stratagene, La Jolla, CA) with pB4 as a template. The 8-amino acid FLAG tag contains enterokinase cleavage site (DYKDDDDK).[7] For attaching a tag sequence to the N-terminal coding region of the target gene, the upstream primer includes the ATG start codon, the FLAG epitope tag sequence (GACTACAAGGACGATGATGACAAA), 1-amino acid linker glycine (GGC), and the DNA sequence for the amino terminus of the target protein (GCGCACGCTGGGAGAAC) (Fig. 1). For insertion of other epitopes, the

TABLE I
COMBINATIONS OF COMMERCIALLY AVAILABLE EPITOPE AND ANTIBODY
USED FOR MOLECULAR TAGGING

Epitope	Sequence	Antibody	Immunogen
FLAG	DYKDDDDK	M1, M2, M5	Enterokinase cleavage site
c-myc	EQKLISEEDL	9E10	Residues 410–419 of human c-myc protein
HA	YPYDVPDYA	3F10	Human influenza virus hemagglutinin HA
His6	HHHHHH	6-His, 6× His	Recombinant His-tagged fusion protein
VSV-G	MNRLGK	P5D4	C-terminal 15 residues of VSV-G

appropriate tag sequence[7] from Table I in place of the FLAG sequence can be substituted. The downstream primer includes a restriction enzyme site (*Bam*HI; GGATCC) and the neighboring DNA sequence including the restriction site of the target protein (ATCCTGGATCCAGGTGTGCAGG).

Procedures for epitope tagging by PCR must be optimized empirically for each individual experimental system. Conduct PCR following the manufacturers guidelines included with the proof reading polymerase(such as *Pwo, Pfu,* or *Pfx* polymerase) to set up the amplification profile. The 610 bp purified PCR product was digested with *Bam*HI and 5′-phosphorylated with polynucleotide kinase. The purified restriction product was ligated into the *Cla*I (blunt)-*Bam*HI site of pRSV*bcl*-2 (pRSVF*bcl*-2). Isolated transformants can be screened for the presence of the epitope-tagged gene by doing direct PCR with the primers used to produce the original epitope-tagging. Plasmid minipreparations for restriction mapping and DNA sequencing for both strands are also required. Finally, the entire *bcl*-2 expression cassette was inserted between the ITR of a pUC-based plasmid pW1 (pWF*bcl*-2).

Protocol for AAV Vector Production

Principle of Triple Plasmid Transfection System

The minimum sets of regions in helper adenovirus that mediate AAV vector replication are the E1, E2A, E4, and VA.[8] A human embryonic kidney cell line 293 encodes the E1 region of the Ad5 genome.[9] When the helper plasmid assembling E2A, E4, and VA regions (Ad-helper plasmid) is cotransfected into the 293 cells, along with plasmids encoding the AAV vector genome (vector plasmid) as well as

[7] J. W. Jarvik and C. A. Telmer, *Annu. Rev. Genet.* **32,** 601 (1998).

[8] T. Matsushita, S. Elliger, C. Elliger, G. Podsakoff, L. Villarreal, G. J. Kurtzman, Y. Iwaki, and P. Colosi, *Gene Ther.* **5,** 938 (1998).

[9] F. L. Graham, J. Smiley, W. C. Russell, and R. Nairn, *J. Gen. Virol.* **36,** 59 (1977).

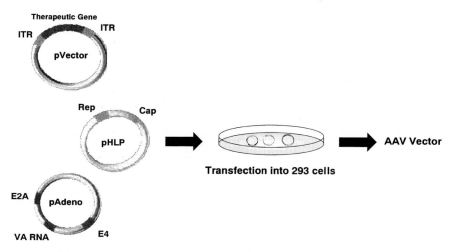

FIG. 2. Schematic representation of triple plasmid transduction for rAAV production. For rAAV vector production, 293 cells are transfected with vector plasmid (pVector), AAV-helper plasmid (pHLP), and Ad-helper plasmid (pAdeno) in the absence of helper adenovirus.

rep and *cap* genes (AAV-helper plasmid), AAV vector is produced as efficiently as when using adenovirus infection as a helper virus (Fig. 2). Elimination of the heat-inactivation for the contaminated adenovirus can improve yield of the virus. Furthermore, contamination of most adenovirus proteins can be avoided in AAV vector stock made by this helper virus-free method.

Reagents

Helper plasmid DNA (pHLP, pAdeno)
Vector plasmid harboring gene of interest flanked with ITRs
290 mM NaCl, 50 mM HEPES buffer, 1.5 mM Na$_2$HPO$_4$, pH 7.10 (2× HBS)
300 mM CaCl$_2$
293 HEK cells
DME/F12 culture medium, 10% fetal calf serum
Phosphate buffered saline (PBS)
1 M HEPES buffer, pH 7.4
100 mM Tris HCl, pH 8.0, 150 mM NaCl (TBS)
0.5 M EDTA, pH 8.0
40% sucrose, 0.01% BSA in TBS
50 mM HEPES, pH 7.6, 0.15 M NaCl, 10 mM MgCl$_2$ (DNase buffer)
50 mM HEPES, pH 7.4, 0.15 M NaCl, 25 mM EDTA (HNE)

CsCl in HNE (1.25 g/ml)
CsCl in HNE (1.50 g/ml)

Plasmids

The AAV vector plasmid, pAAVlacZ, harbors a β-galactosidase expression cassette flanked by inverted terminal repeats (ITRs). pWF*bcl-2* is composed of FLAG-tagged *bcl-2* driven by the RSV promoter and flanked by ITRs.[3] AAV-helper plasmid pHLP harboring Rep and Cap was previously reported as pHLP19.[10] Ad-helper plasmid pAdeno identical to pVAE2AE4-5 encodes the entire E2A and E4 regions, and VA RNA I and II genes.[8]

Transfection and Viral Extraction

This protocol is for the transfection of one 225 cm^2 flask. For additional flasks or different capacity, increase volumes on a linear basis. Trypsinize 293 cells and plate 5×10^6 cells per 225 cm^2 flask to achieve a monolayer confluency of 20 to 40% when cells initially attach to surface of the flask. The volume of medium per flask is 40 ml. Try to avoid plating clumps of cells and make sure that cells are distributed evenly across the plate. Even cell density in all areas of the plate is essential for high yield and can be attained by moving the plates of newly plated cells gently in a cross pattern before the cells attach. Transfer plates to a 5% CO_2 incubator and grow to a confluency of 80% over the next 24 to 48 hr.

At 1 hr before the transfection, exchange half of the medium in tissue culture flasks with fresh DME/F12 culture medium containing 10% FCS. Add 23 μg each of vector and helper plasmids to 4 ml of 300 mM CaCl$_2$. Gently add this solution to 4 ml of 2× HBS and immediately mix by gentle inversion 3 times. Immediately pipette this mixture into a 225 cm^2 flask of 293 cells containing 40 ml of DME/F12 medium plus 10% FCS and swirl to produce a homogeneous solution. Immediately transfer plates to a 5% CO_2 incubator and incubate at 37° for 4 to 6 hr. Do not disturb plates during this period. At the end of the incubation, replace medium with prewarmed DME/F12 culture medium containing 2% FCS. Three days after the transfection, add 1 ml of 0.5 M EDTA to each flask and incubate for 3 min at room temperature. Collect cell suspension and centrifuge at 300g for 10 min. Remove supernatant and resuspend cell pellet in 2 ml of TBS.

Freeze and thaw cells suspended in TBS three times by placing them alternately in a dry ice/ethanol bath until completely frozen and in a 37° water bath until

[10] M. Urabe, K. Shimazaki, Y. Saga, T. Okada, A. Kume, K. Tobita, and K. Ozawa, *Biochem. Biophys. Res. Commun.* **276,** 559 (2000).

completely thawed. Immediately transfer sample back to ice bath when completely thawed. Remove tissue debris by centrifuging at 10,000g for 10 min and collect supernatant.

AAV Vector Purification

For the material obtained from 24 flasks, add 11 ml of 40% sucrose/0.01% BSA in TBS to a sterile ultracentrifugation tube (Ultrabottle; Nalge Nunc, Rochester, NY). Carefully overlay 48 ml of the freeze/thaw supernatant on the solution. Pellet the crude virus particles by centrifuging at 100,000g for 16 hr at 4°. Resuspend the pellet vigorously in 5 ml DNase buffer. Add 1000 U of DNase I and incubate for 1 hr at 37°. Add 250 μl of 0.5 M EDTA and then remove debris by centrifuging at 10,000g for 2 min followed by filtration with low protein binding 5 μm syringe filter (Sterile Acrodisc Syringe Filter; Pall Gelman Laboratory, Ann Arbor, MI). Load the material onto a two-tier CsCl gradient (1.25 and 1.50 g/ml) in HNE buffer. Spin the gradient at 35,000 rpm for 2 hr at 16° in a SW-40 rotor (Beckman Instruments, Palo Alto, CA). Collect the virus-rich fraction by measuring refractive index (RI; 1.371–1.380) and load again onto a two-tier CsCl gradient (1.25 and 1.50 g/ml) in HNE buffer. Spin the gradient at 65,000 rpm for 2 hr at 16° in a VTi 65.2 rotor (Beckman Instruments, Palo Alto, CA). Collect 0.5 ml each of fraction and select the virus-rich fraction by measuring RI (1.371–1.380), semiquantitative PCR analysis, Western blotting analysis for Cap, or quantitative DNA dot-blot hybridization. By using dialysis cassette Slide-A-Lyzer (Pierce, Rockford, IL), desalt the virus-rich fraction by three cycles of dilution with 300 ml of HNE buffer. Concentrate the material to 50 μl with Ultrafree-4 (Millipore, Bedford, MA) according to the manufacturer's instruction. The final titer is usually $1–5 \times 10^{13}$ particles from 5×10^8 293 cells as determined by quantitative DNA dot-blot hybridization or Southern blotting. In the case of AAV vector expressing lacZ, vector production can be functionally assessed by titration with 293 cells in the presence of a wild-type adenovirus type 2 at an MOI of 50. The transduced cultures are incubated for 24 hr at 37° before fixation and X-gal staining. The stained cells are counted under light microscopy. The genome particle to functional ratio is determined as the level of total genomes divided by that of function units (X-gal staining) to verify infectivity.

Brain Ischemia Procedure

Procedure

The Mongolian gerbil (*Meriones unguiculatus*) has been used as a model for cerebral ischemia. Since this animal lacks an interconnection between the carotid

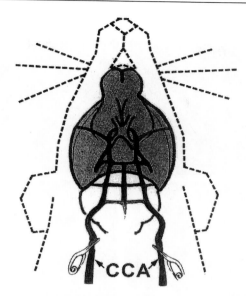

FIG. 3. Procedure for experimental gerbil brain ischemia. Three-month-old male Mongolian gerbils were used for the experiments. The gerbils were anesthetized with 3.0% halothane in a mixture of 30% of O_2/70% N_2O. Both common carotid arteries (CCA) were clamped with surgical clips thereby blocking cerebral blood supply completely. After 3 min occlusion, the clips were removed and the skin was sutured.

and vertebro-basilar circulation, one can easily produce forebrain ischemia by occlusion of the common carotid arteries at the neck.[11,12] Transient bilateral carotid occlusion between 3 min and 5 min can produce an apoptosis-like neuronal cell death in the CA1 area of the hippocampus.[11,13,14]

Three-month-old male Mongolian gerbils (70–90 g) are anesthetized with 3.0% halothane in a mixture of 30% O_2/70% N_2O. The rectal temperature is measured by inserting a thermocouple probe (Unique Medical, Tokyo, Japan) into the anus, and the rectal temperature is monitored and maintained at 37–38° throughout the operation by means of a feedback-controlled infrared heating lamp. Both common carotid arteries are clamped with surgical clips thereby blocking cerebral blood supply (Fig. 3). After 3 min occlusion, the clips are removed and the skin is sutured.

[11] T. Kirino, *Brain Res.* **239,** 57 (1982).
[12] W. A. Pulsinelli, J. B. Brierley, and F. Plum, *Ann. Neurol.* **11,** 491 (1982).
[13] K. Shimazaki, A. Ishida, and N. Kawai, *Neurosci. Res.* **20,** 95 (1994).
[14] C. Charriaut-Marlangue, D. Aggoun-Zouaoui, A. Represa, and Y. Ben-Ari, *Trends Neurosci.* **19,** 109 (1996).

(A)

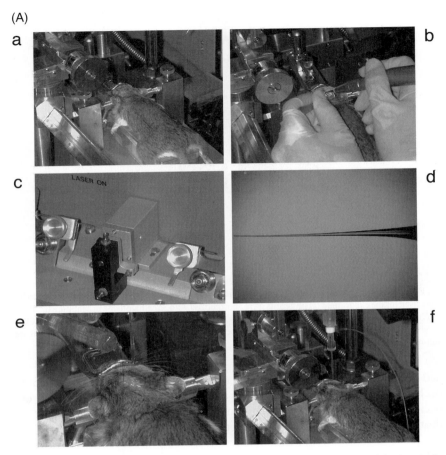

Fig. 4. Injection of AAV vectors into gerbil brain. (A) Prior to AAV vector injection, male Mongolian gerbils are anesthetized with 2.5% chloral hydrate (400 mg/kg, i.p.) and placed in a stereotaxic apparatus. (a) Following induction of anesthesia, gerbils are placed in a stereotaxic frame (Type SR-50 NARISHIGE, Tokyo, Japan). (b) A linear scalp incision is made and a burr hole is drilled to the pericranium. (c, d) Glass micropipettes (30–50 μm in diameter) are fabricated by CO_2 laser based micropipette puller P-2000 (Sutter Instrument, Novato, CA) according to the manufacturer's instructions. Injection of the AAV vectors into the hippocampus through a glass micropipette is provided with electrical stimuli and recording. (e) For injection into the CA1/CA2 of hippocampus, a microelectrode is placed 1.5 mm below the pial surface via a burr-hole, 3.0 mm to the right, and 2.5 mm posterior to point bregma. Electrical repetitive stimuli (square-wave pulses of 0.3 ms duration, 0.1/s frequency) are delivered to the hippocampal commissure through bipolar tungsten microelectrodes. The electrodes are placed within the hippocampus ipsilaterally to the recording site. For the ventricles lateralis, a micropipette is placed 2.2 mm below the pial surface, 1.2 mm to the right and 0.4 mm posterior to point bregma. (f) The vectors are injected through a glass micropipette at a rate of 1.0 μl/min by using a microinfusion pump. Following injection, the micropipette is left in place for 3 min and withdrawn slowly to prevent reflux. The muscles are sutured together and the scalp incision is closed with sutures.

(B)

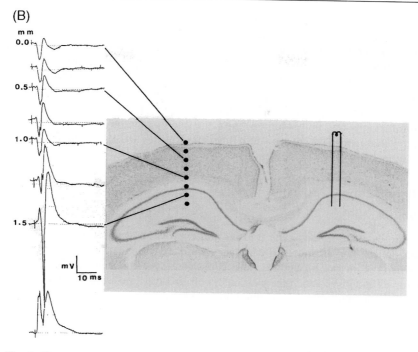

FIG. 4. (*Continued*) (B) CA1 commissural responses from gerbil. Figure shows responses from successive depths by a recording electrode penetrating the cortex. The stimulating electrode is fixed in CA1 of the right hippocampus while the recording electrode is placed at contralateral sites.

Protocol for Injection of AAV Vector into Gerbil Brain

Procedure

Prior to AAV vector injection, gerbils are anesthetized with chloral hydrate and placed in a stereotaxic apparatus. Male Mongolian gerbils are anesthetized with 2.5% chloral hydrate (400 mg/kg, ip) and placed in a stereotaxic frame (Type SR-50 NARISHIGE, Tokyo, Japan), and a burr hole is drilled to the pericranium. The procedure is shown in Fig. 4A.

Glass micropipettes (30–50 μm in diameter) are fabricated by CO_2 laser based micropipette puller P-2000 (Sutter Instrument, Novato, CA) according to the manufacturer's instructions. Injection of AAV vectors into the hippocampus through a glass micropipette is provided along with electrical stimuli and recording. For injection into the CA1/CA2 of hippocampus, a microelectrode is placed 1.5 mm below the pial surface via a burr-hole, 3.0 mm to the right, and 2.5 mm posterior to point bregma. Electrical repetitive stimuli (square-wave pulses of 0.3 ms duration, 0.1/s frequency) are delivered to the hippocampal commissure through bipolar tungsten microelectrodes. The electrodes are placed within the hippocampus ipsilateral to the recording site. Extracellular field potentials are recorded by microelectrodes stereotaxically positioned. The typical patterns are shown in Fig. 4B. For the ventricles lateralis, a micropipette is placed 2.2 mm below the pial surface, 1.2 mm to the right, and 0.4 mm posterior to point bregma. The vectors are injected through a glass micropipette at a rate of 1.0 μl/min by using a microinfusion pump. Following injection, the micropipette is left in place for 3 min and withdrawn slowly to prevent reflux. The muscles are sutured together and the scalp incision is closed with sutures.

Histochemical Analysis

Principle

To evaluate *in vivo* gene transfer and expression at the histologic level, histochemical studies are required. Histochemical studies on brain tissue are often carried out on frozen sections. This is the gentlest method for the preparation of samples and gives good preservation of cell structure and antigens. In this chapter we describe a free-floating method on frozen sections.

Reagents

Reagents for Fixation

Nembutal (50 mg pentobarbital/ml; Abbott Laboratories, North Chicago, IL)
Normal saline (0.9% NaCl)
0.1 M Phosphate buffer, pH7.4 (PB)
7.5% Sucrose in PB
Fixative: 4% paraformaldehyde in 0.1 M PB containing sucrose at 7.5%

Reagents for Immunohistochemistry for Anti-FLAG Antibody

0.1 M phosphate-buffered saline, pH 7.4 (0.9% NaCl in PB)
Anti-FLAG M5 monoclonal antibody (Kodak, New Haven, CT)
Normal horse serum (Vector Laboratories, Burlingame, CA)
ABC Elite Kit: Biotinylated anti-mouse/rabbit IgG (H+L) made in horse
    Avidin DH and biotinylated horseradish peroxidase (Vector Laboratories, Burlingame, CA)
Diaminobenzidine tetrahydrochloride (DAB)
Hydrogen peroxide (H_2O_2)

In Situ Staining of DNA Fragmentation

Proteinase K
Hydrogen peroxide (H_2O_2)

Terminal deoxynucleotidyltransferase (Boehringer Mannheim, Indianapolis, MD)

Biotinylated dUTP (Travigen Inc., Gaithersburg, MD)

Labeling buffer: 140 mM sodium cacodylate, 1 mM cobalt chloride in 30 mM Tris-HCl (pH 7.2)

Stopping buffer (2× SSC): 0.3 M NaCl, 30 mM sodium citrate (pH 7.0)

Streptavidin-peroxidase (Travigen, Inc., Gaithersburg, MD)

Diaminobenzidine tetrahydrochloride (DAB)

Commercial kits: Travigen, Inc. (Gaithersburg, MD) provide kits (TACS *In Situ* Apoptosis Detection Kit) to identify apoptotic cells on the basis of terminal deoxynucleotidyltransferase reaction.

Reagents for Counterstaining

1% Methyl green in 0.1 M sodium acetate buffer, pH 4.0
95% Ethanol
100% Ethanol
100% Xylene

Fixation of Brain for Histochemical Analysis

Gerbils are anesthetized by intraperitoneal injection of pentobarbital (30 mg/kg body weight) and perfused with 50 ml of normal saline (0.9% NaCl) followed by 100 ml of fixative. Fixation is performed by transcardiac perfusion. Brains are removed and preserved overnight in fixative at 4°. The brains are then transferred to 0.1 M PB containing 15% sucrose and kept in a refrigerator.

Detection of Gene Expression with Anti-FLAG Antibody

In order to evaluate the *in vivo* expression with immunohistochemistry, avidin–biotin-complex (ABC) methods are used (Fig. 5). Coronal sections of the brain are made by a freezing microtome at 50 μm. Following 1 hr of incubation in block solution (3% normal horse serum in PBS), sections are incubated with anti-FLAG M5 antibody (1–10 μg/ml in blocking solution) overnight at 4°. Sections are washed 2 times for 15 min each in PBS and incubated for 1 hr with biotin-ylated anti-mouse/rabbit IgG. Sections are washed in PBS and further incubated for 1 hr with an avidin–biotin complex (Vector Laboratories, ABC Elite Kit) applied according to the instructions of the manufacture. Following a final wash in PBS, the peroxidase reaction is carried out with diaminobenzidine (0.5 mg/ml) in the presence of hydrogen peroxide (0.01%). Staining is terminated by washing 2 times for 5 min each in PB. Sections are then mounted on gelatin-coated glass slides, dehydrated and, cleared in xylene and then sealed with coverslides.

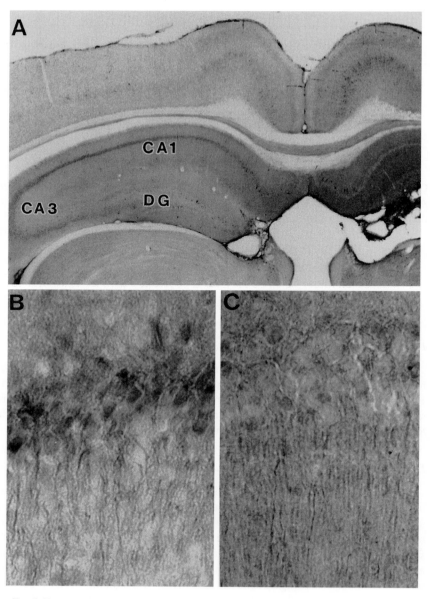

Fig. 5. Transgene expression after the injection of AAV *bcl-2* in gerbil brain. 5 days following 5 μl ($4.25 \times 10^9 - 6.0 \times 10^{10}$ particles/5 μl) of AAV *bcl-2* injection in gerbil brain. The transgene product was detected with anti-FLAG antibody. (A) Labeling pattern of the injection site (left) and contralateral site (right). (B) Higher magnification of the injection site of CA1 region. The pyramidal neurons were extensively positive and the apical dendrites are clearly visible. (C) Higher magnification of the contralateral site. CA1, CA3; CA1 and CA3 field of Ammon's horn, respectively. DG, Dentate gyrus.

Caution

1. Development times may differ depending on the level of antigen. DAB generally should be developed for 2–10 min.

2. In the presence of nickel ions, the precipitate formed by DAB is gray/black rather than brown. This may enhance the sensitivity of the staining procedure.

Detection of Apoptosis

The DNA fragmentation in the hippocampus of the experimental ischemic model is described as a key phenomenon for the apoptotic cell death. DNA nick end labeling of the sections is performed according to the methods of Gavrieli *et al.*[15] with partial modifications. Nuclei of the sections are stripped from proteins by incubation with 20 μg/ml proteinase K in Tris-HCl pH 6.6 for 15 min at room temperature. Endogenous peroxidase is inactivated by 2% H_2O_2 for 5 min at room temperature. Terminal deoxynucleotidyltransferase and biotinylated dUTP are diluted in labeling buffer at a concentration of 0.15 eu/ml and 0.8 nmol, respectively. The sections are immersed in the solution, and then incubated at 37° for 60 min. The reaction is terminated by transferring the sections to the stopping buffer for 5 min at room temperature. The sections are rinsed with distilled water and then treated with streptavidin-peroxidase for 10 min at room temperature. Sections are then developed in diaminobenzidine (0.5 mg/ml) in the presence of hydrogen peroxide (0.01%).

Counterstaining

The sections are counterstained to aid in the morphological characterization (Fig. 6). Staining time varies with cell type and must be empirically determined for optical results. Staining times can vary from 5 seconds to 5 minutes. Standard steps are as follows:

1. Methyl green solution for 5–10 sec
2. 2 times in deionized water, each for 10 sec
3. 2 times in 95% ethanol, each for 30 sec
4. 2 times in 100% ethanol, each for 1 min
5. 2 times in 100% xylene, each for 5 min

Concluding Remarks

In this chapter we describe adenovirus-free system for the production of AAV vectors. The preparation of AAV vector for gene therapy study of neuronal diseases

---

[15] Y. Gavrieli, Y. Sherman, and S. A. Ben-Sasson, *J. Cell. Biol.* **119**, 493 (1992).

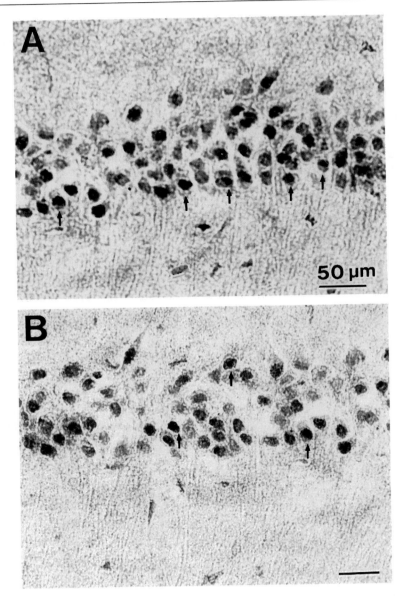

FIG. 6. *In situ* staining of fragmented DNA 2 days after ischemia in hippocampal CA1 area. DNA fragmentation was detected using the TUNEL technique. (A) The DNA fragmented neurons were densely distributed in the CA1 pyramidal cell layer of control gerbil. (B) On AAV bcl-2 injected gerbil, far fewer CA1 neurons were positive for DNA fragmentation. The data strongly suggest that AAV bcl-2 transduction was effective in inhibiting DNA fragmentation in CA1 neurons after ischemia. Arrows indicate typical DNA fragmented neurons. The sections were counterstained with methyl green.

is greatly facilitated. The AAV vectors can be used to identify specific activity of the gene of interest *in vivo* along with the application of the vectors to the gerbil ischemic brain. AAV vectors have been demonstrated to be capable of constitutive expression of Bcl-2 in ischemic brain and protect neuronal cell death when delivered after ischemia.

Acknowledgments

We thank Avigen, Inc. (Alameda, CA) for providing pAAVlacZ, pHLP, and pAdeno. This work was supported in part by grants from the Ministry of Health, Labour and Welfare of Japan, Grants-in-Aid for Scientific Research from the Ministry of Education, Culture, Sports, Science and Technology of Japan.

[23] Recombinant Adeno-Associated Viral Vector Production Using Stable Packaging and Producer Cell Lines

By Lydia C. Mathews, John T. Gray, Mark R. Gallagher, and Richard O. Snyder

Introduction

Recombinant AAV vectors (rAAV) are capable of long-term gene expression with low toxicity *in vivo* (reviewed in Snyder[1]). Reports are now common of therapeutic and sustained gene expression resulting in correction of animal models of disease, and human clinical trials are in progress for cystic fibrosis, hemophilia, and muscular dystrophy. The therapeutic dose of vector will depend on the tissue and disease targets, ranging from 1×10^8 vector genomes for retinal gene therapy to 1×10^{14} vector genomes for muscle and liver targets. Methods are being developed to meet the needs of maturing clinical trials involving more patients and higher doses, and ultimately, FDA-licensed rAAV products.

The production of rAAV vectors requires four genetic elements[2,3] (1) mammalian tissue culture cells, (2) vector sequences containing a transgene flanked by AAV inverted terminal repeats, (3) AAV helper sequences comprising the AAV open reading frames (ORFs), and (4) helper virus (i.e., adenovirus or herpes virus) genes which act throughout the AAV lytic cycle, including activation of AAV gene

[1] R. O. Snyder, *J. Gene Med.* **1**, 166 (1999).
[2] P. L. Hermonat and N. Muzyczka, *Proc. Natl. Acad. Sci. U.S.A.* **81**, 6466 (1984).
[3] R. J. Samulski, L. S. Chang, and T. Shenk, *J. Virol.* **63**, 3822 (1989).

expression. A standard method for producing rAAV virions is by transient transfection of rAAV vector and helper sequences into tissue culture cells and infection with adenovirus[4] resulting in yields of 200–300 infectious units (IU) per cell. However, transient transfection is inefficient, is difficult to scale up, and requires the certification of several components (plasmid DNA, transfection reagents, and cell lines) for clinical grade production. In addition, homologous and illegitimate recombination between vector and AAV helper plasmids may result in the generation of recombinant wild-type AAV (rcAAV), albeit at very low levels.[5–7] Stable cell lines harboring the AAV2 helper sequences alone or together with vector sequences facilitate efficient large-scale rAAV production.[8–13]

The establishment of stable cell lines has been problematic, as overexpression or even stable expression of the AAV Rep proteins in tissue culture cells is toxic or lethal[14] and is likely related to the apoptotic effects of the proteins.[15] This property of the nonstructural proteins has been observed for other parvoviruses as well.[16] Further development of rAAV packaging and producer cell lines with yields approaching wild-type levels of 1×10^4 infectious virions/cell (1×10^6 vector genomes/cell)[17] requires innovative methods to control the temporal expression of the toxic *rep* gene along with sufficient *cap* gene expression.[18–20] Adaptation of the cell lines from flat stock to suspension culture allows further scale-up; however, the agent used to induce AAV gene expression needs to be compatible with this format.

[4] R. O. Snyder, X. Xiao, and R. J. Samulski, *in* "Current Protocols in Human Genetics" (N. Dracopoli, J. Haines, B. Krof, D. Moir, C. Morton, C. Seidman, J. Seidman, and D. Smith, eds.), pp. 12.1.1. John Wiley and Sons, New York, 1996.

[5] J. M. Allen, D. J. Debelak, T. C. Reynolds, and A. D. Miller, *J. Virol.* **71,** 6816 (1997).

[6] X. S. Wang, B. Khuntirat, K. Qing, S. Ponnazhagan, D. M. Kube, S. Zhou, V. J. Dwarki, and A. Srivastava, *J. Virol.* **72,** 5472 (1998).

[7] A. Salvetti, S. Oreve, G. Chadeuf, D. Favre, Y. Cherel, P. Champion-Arnaud, J. David-Ameline, and P. Moullier, *Hum. Gene Ther.* **9,** 695 (1998).

[8] T. R. Flotte, X. Barraza-Ortiz, R. Solow, S. A. Afione, B. J. Carter, and W. B. Guggino, *Gene Ther.* **2,** 29 (1995).

[9] K. R. Clark, F. Voulgaropoulou, D. M. Fraley, and P. R. Johnson, *Hum. Gene Ther.* **6,** 1329 (1995).

[10] G. P. Gao, G. Qu, L. Z. Faust, R. K. Engdahl, W. Xiao, J. V. Hughes, P. W. Zoltick, and J. M. Wilson, *Hum. Gene Ther.* **9,** 2353 (1998).

[11] N. Inoue and D. W. Russell, *J. Virol.* **72,** 7024 (1998).

[12] G. Chadeuf, D. Favre, J. Tessier, N. Provost, P. Nony, J. Kleinschmidt, P. Moullier, and A. Salvetti, *J. Gene Med.* **2,** 260 (2000).

[13] X. L. Liu, K. R. Clark, and P. R. Johnson, *Gene Ther.* **6,** 293 (1999).

[14] Q. Yang, F. Chen, and J. P. Trempe, *J. Virol.* **68,** 4847 (1994).

[15] M. Schmidt, S. Afione, and R. M. Kotin, *J. Virol.* **74,** 9441 (2000).

[16] D. Legendre and J. Rommelaere, *J. Virol.* **66,** 5705 (1992).

[17] B. J. Carter, *in* "Handbook of Parvoviruses," Vol. 1 (P. Tijssen, ed.) p. 155. CRC Press, Boca Raton, FL, 1990.

[18] J. Li, R. J. Samulski, and X. Xiao, *J. Virol.* **71,** 5236 (1997).

[19] K. A. Vincent, S. T. Piraino, and S. C. Wadsworth, *J. Virol.* **71,** 1897 (1997).

[20] J. M. Allen, C. L. Halbert, and A. D. Miller, *Mol. Ther.* **1,** 88 (2000).

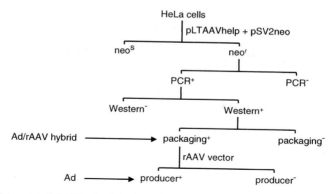

FIG. 1. Summary of rAAV packaging cell line construction and its uses. HeLa cells were transfected with pLTAAVHelp and pSV2neo. G418 resistant clones (neo^r) were isolated under selection and screened for AAV intron sequences by PCR. PCR positive clones were screened for synthesis of AAV proteins by Western blotting. Western positive clones were tested for the packaging of an rAAV-nlacZ vector. The packaging cell lines can be used with a hybrid adenovirus harboring an AAV vector. Alternatively, AAV vector sequences can be introduced to create producer cell lines. Both the packaging and producer cell lines may be adapted to spinner and serum-free culture to facilitate large-scale production.

In this chapter we present protocols for generating rAAV packaging and producer cell lines which stably harbor the AAV2 open reading frames (ORFs). We also describe the characterization and performance of cell lines generated using these protocols. The cell lines do not synthesize detectable amounts of AAV proteins until infected with adenovirus and the induction involves a DNA amplification mechanism that is dependent on adenovirus. The best packaging clone is capable of rAAV vector titers of up to 300 IU per cell when vector sequences are delivered by transient transfection or hybrid adenovirus, and this output has been maintained for more than 6 months. Prototype stable producer cell lines harboring AAV helper and vector sequences were isolated with average outputs of up to 300 IU per cell. A large-scale manufacturing process can be achieved using packaging and producer cell lines together with previously published column chromatographic purification methods[21,22] and those described by Potter *et al.* in Chapter [24], this volume.

Generation of Packaging Cell Lines

Our strategy for generating stable rAAV packaging cell lines is outlined in Fig. 1. Because the large Rep proteins are cytotoxic, cell lines were made using

[21] K. R. Clark, X. Liu, J. P. McGrath, and P. R. Johnson, *Hum. Gene Ther.* **10,** 1031 (1999).
[22] G. Gao, G. Qu, M. S. Burnham, J. Huang, N. Chirmule, B. Joshi, Q. C. Yu, J. A. Marsh, C. M. Conceicao, and J. M. Wilson, *Hum. Gene Ther.* **11,** 2079 (2000).

an AAV helper plasmid harboring all AAV2 ORFs under control of their native promoters, but with attenuated expression of the large Rep 78 and 68 proteins. Decreased large Rep expression was achieved by incorporating a mutated Kozak sequence to reduce translation initiation efficiency[18] (see below). This modification results in greater production of small Rep proteins (Rep52 and Rep40), structural capsid proteins, and vector titers by transient transfection. The AAV helper plasmid pLTAAVHelp was transfected along with a plasmid expressing a neomycin drug-selectable marker, pSV2neo, into HeLa cells. HeLa cells were chosen because the AAV promoters are not active in this background and this cell line can be adapted to suspension culture. Following selection, 149 single-cell clones were expanded (149 of 287 clones picked were viable) and screened for the presence of the AAV intron sequences, an important regulatory element controlling AAV gene expression.[23,24] As a control for DNA quantity and integrity, polymerase chain reaction (PCR) for cellular β-actin sequences was performed on the same samples. Replica plates were made of the 62 AAV PCR-positive clones and AAV protein expression was detected by immunoblotting after infection with adenovirus (Ad). Synthesis of the Rep and Cap proteins was dependent on Ad (Fig. 2). Cap expression was induced to a high level in clone #26 following Ad infection from a basal level seen in the absence of infection (Fig. 2A); significant induction in other clones was not observed. A reduced amount of Rep 78 and 68 proteins and an increased amount of Rep 52 protein was observed in stable clones compared with pAAV/Ad, a plasmid harboring all of the AAV ORFs without any modifications (Fig. 2B). In clone #26, Rep 40 was detected following Ad infection, but was not detected in other clones. Efficient rAAV packaging requires high levels of Cap and Rep 52 and 40 proteins.

The 11 clones expressing the Cap proteins were screened for their ability to support the packaging of a rAAV-nlacZ vector by transient transfection of the vector plasmid and infection with Ad. Five of 11 clones tested were able to support rAAV-nlacZ packaging to varying levels. A summary of the selection and screening process is given in Table I. Two clones (#23 and #26) capable of vector packaging were further characterized for their stability and the status of the engineered AAV *rep* and *cap* sequences. Vector plasmid transfection efficiencies of 5% for clone #26, and 50–70% for clone #23 and naïve HeLa cells were obtained. A titration of Ad and a timecourse of infection were carried out to optimize rAAV vector yield following transient transfection of vector plasmid; an MOI of 25–50 was best for a 72 hr infection. During 6 months of continuous culture, clones #23 and #26 were able to reproducibly generate rAAV-lacZ titers of ~25 and 300 LacZ forming units (LFU) per cell, respectively (approximately 5000 and 60,000 vector genomes/cell given a particle : LFU ratio of 200). Transient cotransfection of control naïve HeLa cells yielded an average of 350 LFU/cell during the 6 months (Fig. 3).

[23] S. P. Becerra, F. Koczot, P. Fabisch, and J. A. Rose, *J. Virol.* **62**, 2745 (1988).
[24] M. B. Mouw and D. J. Pintel, *J. Virol.* **74**, 9878 (2000).

A

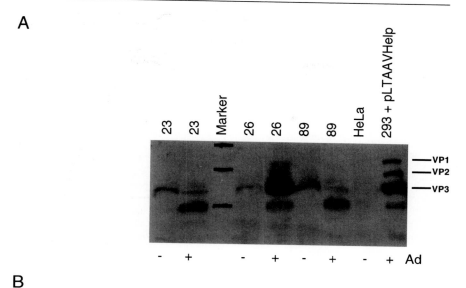

B

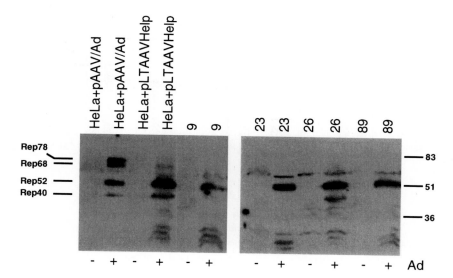

FIG. 2. AAV protein expression of cell line clones. Cell lysates from 2×10^5 cells were separated on 10% (w/v) sodium dodecyl sulfate–polyacrylamide gel electrophoresis (SDS–PAGE) gels and probed with an anti-Cap antibody (A) or anti-Rep antibody (B). Clone numbers are indicated. + or − denotes if the cells were infected with Ad. HeLa: naïve HeLa cells; 293+pLTAAVHelp: 293 cells transfected with pLTAAVHelp; HeLa+pAAV/Ad: HeLa cells transfected with pAAV/Ad; HeLa+pLTAAVHelp: HeLa cells transfected with pLTAAVHelp. AAV proteins and molecular weight markers (kilodaltons) are indicated. In (A) molecular mass markers of 123, 83, and 51 kDa are indicated.

TABLE I
SELECTION AND SCREENING OF AAV PACKAGING
CELL LINES

Selection/screen	Number of clones
G418 resistant	149
PCR	
AAV intron[+]	62/149 (41.9%)
β-Actin[+]	148/149 (99.3%)
Western	
Cap[+]	11/45 (24.4%)[a]
Rep[+]	5/7 (71.4%)[b]
Packaging[+]	5/11 (45.5%)[c]

[a] 45 of the 62 AAV intron[+] clones were assayed for capsid protein expression.

[b] 7 of the 11 Cap[+] clones were assayed for Rep protein expression.

[c] The 11 Cap[+] clones were assayed for their ability to package rAAV-nlacZ virions.

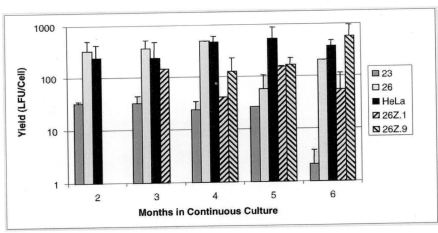

FIG. 3. Stability and yield of rAAV cell line clones. Packaging cell line clones #23 and #26 were continuously cultured and tested bimonthly for vector production by transient transfection of pAAV-nlacZ and infection with Ad. rAAV-nlacZ producer cell line clones 26Z.1 and 26Z.9 were continuously cultured and infected bimonthly with Ad for vector production. Control naïve HeLa cells were transiently cotransfected with rAAV-nlacZ vector and pLTAAVHelp and infected with Ad. Yields are indicated in LFU per cell determined by titering the vector stock and dividing the total LFU by the number of packaging or producer cells staining positive for lacZ. 1 LFU is approximately equal to 200 vector genomes. Averaged bimonthly yields ± SDEV are plotted.

General Procedures and Reagents

Safety Considerations

All cell and viral manipulations are performed in Class II biological safety hoods wearing appropriate protective clothing. Work outside of the hood requires cells and viruses to be contained in sterile leakproof vessels.

Cells and Media

The human cervical carcinoma HeLa cell line harboring the E6/7 genes of HPV18 and the human embryonic kidney 293 cell line harboring the E1 genes of adenovirus were obtained from the ATCC (Manassas, VA). Cells are maintained in DMEM (BioWhittaker) supplemented with 10% heat-inactivated fetal bovine serum (BioWhittaker), 2 mM L-glutamine (BioWhittaker), and 40 μg/ml gentamycin (BioWhittaker). AAV packaging cell line clones are maintained in the above medium (referred to as DMEM-complete) containing 600 μg/ml active G418 (LifeTechnologies). HeLa conditioned medium is made by culturing naïve HeLa cells in DMEM-complete for 24–72 hr, collecting the medium, and centrifuging to remove any cells and cell debris. Cells are split when 80–90% confluent by washing with PBS without Mg^{2+} or Ca^{2+}, adding trypsin, incubating until cells are detached, and diluting the cells in DMEM-complete. A heamocytometer is used to count viable and dead cells after mixing 100ul of cells 1 : 1 with trypan blue (Sigma). Cells are seeded into new plates and incubated at 37°, 5% CO$_2$ in appropriate volumes of media. Cell staining for LacZ activity using 5-bromo-4-chloro-3-indolyl-β-D-galactoside (X-Gal) is done using standard methods.[25]

Viruses

Adenovirus type 5 (Ad) was purchased from the ATCC (VR-5). Ad stocks are grown on 293 or HeLa calls, purified on CsCl gradients, and titered by plaque assay on 293 cells. To make crude rAAV lysates: transfer culture media and infected cells to disposable centrifuge tubes, centrifuge at 500g, and aspirate culture medium, resuspend cell pellet in Lysis Buffer (150 mM NaCl and 50 mM HEPES–NaOH, pH, 7.6), freeze and thaw three times using dry ice–ethanol and 37° water baths, centrifuge at 3000g, and collect the supernatants containing viruses. CsCl gradient purification of rAAV has been described.[4] Store adenovirus and recombinant AAV virions at −20° or −80° and limit the number of freeze/thaws to preserve titer.

Plasmids

The AAV helper plasmid pLTAAVHelp harbors the AAV type 2 *rep* and *cap* genes inserted into pLITMUS29 (New England BioLabs). Expression of Rep 78 and 68 proteins is attenuated by changing the ATG start codon to an inefficient ACG start codon.[18] To construct pLTAAVHelp, the *rep-cap* fragment from pAAV/Ad[3] was inserted into pLITMUS29 in a four-part ligation which included (1) the 5' *Xba*I to *Bsa*I 101bp terminal fragment; (2) a double-stranded synthetic linker formed by annealing oligonucleotides with the following sequences: 5'**ATTTT**GAAGCGGGAGGTTTGAACGCGCAGCC<u>ACCACGG</u>CGGGGTTTTACGAGATTGTGATTAAG3' and 5'**GACC**TTAATCACAATCTCGTAAAACC CGC<u>CCGTG</u> GTGGCTGCGCGTTCAAACCTCCCGCTTCA3', where bold denotes the *Bsa*I and *Ppu*MI compatible overhangs, and the attenuated translation start codon and flanking sequence are underlined; (3) the *Ppu*MI to *Xba*I 3' 4140bp terminal fragment; and (4) the *Xba*I-digested pLITMUS29 backbone. The selectable marker plasmid pSV2neo was obtained from Clontech. The pAAV-nlacZ plasmid harbors a nuclear-localized β-galactosidase gene driven by the CMV promoter and a bovine growth hormone poly(A) flanked by AAV ITRs. Plasmids were sequenced and prepared by the alkaline lysis method using CsCl gradients.

Antibodies and Immunoblotting

259.5, a monoclonal antibody that detects all four AAV Rep proteins, is used at 1 : 1000 (American Research Products). Monoclonal antibody B1 (American Research Products) is used at 1 : 2000 to detect capsid expression. Detection is carried out using HRP-conjugated sheep anti-mouse (Amersham) at 1 : 5000 and Super Signal (Pierce).

Protocol

1. Transfect HeLa cells (2×10^6 in a 10-cm culture dish containing 10 ml DMEM-complete) with AAV helper plasmid and selectable marker plasmid at a ratio of 19 : 1 using a HeLa cell-optimized calcium phosphate method: (a) Replace the culture medium with fresh DMEM-complete 2–3 hours prior to transfection. (b) Mix 20 μg of DNA in 1 ml of 150 mM $CaCl_2$ with 1 ml of $2 \times$ HBS (280 mM NaCl, 50 mM HEPES–NaOH, 1.4 mM Na_2HPO_4, pH 7.05) and incubate for 1.5 min at room temperature. (c) Add the 2 ml of precipitate dropwise to the medium, swirl the plate to disperse evenly, and incubate 6 hr at 37°, 5% CO_2.

2. Select clones after 72 hr by changing the culture medium to DMEM-complete supplemented with 600 μg/ml active G418. Culture cells for 12 days changing the medium every 4 days until isolated clones (10–50 cells/clone) develop. The amount of G418 is determined prior to transfection by performing a killing curve on the parental naïve HeLa cells using a range of active G418 of 100–1100 μg/ml with an end point of 8 days for 80% killing.

3. Pick clones using a Pipetman and seed into 96-well plates containing 200 μl conditioned medium supplemented with L-glutamine and G418. Conditioned medium is used to supply growth factors to promote cell growth and may

---

[25] J. Alam and J. L. Cook, *Anal. Biochem.* **188,** 245 (1990).

overcome apoptotic effects of Rep; at low cell numbers the concentration of these factors is very low. Medium is changed every 5 days to remove dead cells using a multichannel pipetter and barrier tips that are changed between each addition to prevent cross contamination.

4. Expand clones stepwise into larger culture dishes (96-well to 24-well to 6-well plates) until able to seed 10-cm plates ($\sim 5 \times 10^6$ cells). Wean cells from conditioned medium to fresh DMEM-complete medium containing G418 at the 6-well plate stage.

5. Make replica plates of the clones and isolate genomic DNA from one replicate and carry the other. Assay for the AAV intron sequences by PCR using the primers Splice1: 5'TCGTCAAAAAGGCGTATCAG3' and CAP2: 5'TCCCTTGT-CGAGTCCGTTGA3' located in the AAV intron and at the 5' end of the cap ORF, respectively. PCR reactions contain 0.5 μg genomic DNA, 50 pmol of each primer, 0.2 mM dNTPs, 1$\times$ buffer with MgCl$_2$, and 1 unit of Taq polymerase in 50 μl. Reactions are carried out in a thermocycler as follows: 94° for 2 min [94° 20 sec, 55° 20 sec, 72° 20 sec] for 30 cycles; this generates a 323-bp PCR product. To control for genomic DNA quality and quantity, the endogenous cellular human β-actin gene is detected separately using an amplimer set and conditions specified by the manufacturer (Clontech) to generate an 825 bp product.

6. Screen intron PCR-positive clones for the presence of AAV proteins. Make replicas of the clones and infect one of the replicates with Ad at an MOI of 35 for 48–72 hr to induce Rep and Cap protein expression. Prepare whole-cell extracts from infected and uninfected cells, separate protein from 2×10^5 cells per lane on 10% SDS–PAGE gels, transfer to nitrocellulose, and probe with anti Rep and Cap antibodies.

7. Screen immunoblot-positive clones for their ability to package a recombinant AAV vector expressing an easily scored transgene product (i.e., LacZ or GFP). Transfect each of two 10-cm plates of packaging clones (4×10^6 cells) with 20 μg of an rAAV vector plasmid using the conditions stated above and infect with Ad at an MOI of 35. The transfection efficiency is determined by scoring transgene expression in one of the two plates 24–48 hr after transfection and infection. For virus production, incubate cells 72 hr until full cytopathic effect (CPE) develops and make a viral lysate. Titer crude AAV lysates: seed 8×10^5 293 cells (or other appropriate cell type) in 6-well plates 12–24 hr prior to titering, coinfect subconfluent cells with dilutions of lysate together with an E1-deleted adenovirus,[26] incubate for 16–20 hr 37°, 5% CO$_2$, and detect transgene expression.

8. Expand the clones further so enough cells ($1–5 \times 10^8$ cells) are generated at the earliest possible passage and freeze in aliquots of 5×10^6 cells in freezing

[26] F. K. Ferrari, T. Samulski, T. Shenk, and R. J. Samulski, *J. Virol.* **70**, 3227 (1996).

medium [90% FBS and 10% dimethyl sulfoxide (DMSO)] using controlled rate freezing jars placed at −80°. Transfer vials to liquid nitrogen storage to serve as a cell bank.

Use of Adenovirus/AAV Hybrid Vector with Packaging Cell Lines

AAV vector sequences have been incorporated into the E1A region of Ad.[10,13,27,28] In the presence of *rep* and *cap* (supplied in a stable cell line, by transient transfection, or conjugated to the exterior of the hybrid adenovirus), rAAV vectors can be produced at titers similar to traditional transient transfection. In addition, this approach adds flexibility because construction of adenovirus hybrid viruses is less time consuming than generating stable producer cell lines (see below). Incorporating the AAV *rep* and *cap* genes into Ad has been very difficult and attempts have been met with disappointing results. This difficulty is due to an unknown toxicity of the AAV gene products toward adenovirus which may be of similar origin to the toxicity seen in tissue culture cells. Herpes virus constructs harboring the AAV genes have been made for AAV production,[29] indicating that *rep* and *cap* are better tolerated in this background.

An Ad/AAV-nlacZ hybrid vector was constructed by isolating the AAV-nlacZ vector fragment from pAAV-nlacZ and inserting it into the E1 region of an adenovirus shuttle plasmid which was recombined with an adenovirus type 5 viral backbone by homologous recombination in *Escherichia coli*.[30] Linearized plasmid DNA harboring the adenovirus hybrid virus was transfected into 293 cells to generate plaques that were screened for lacZ expression. Using the Ad/AAV-nlacZ hybrid virus together with Ad5 (used to supply the E1 gene), clones #23 and #26 were coinfected following the protocol of Liu *et al.*[13] Cells were infected with an MOI of 10 of Ad5 17 hr prior to infection with the hybrid (MOI 7.5) and incubated an additional 50–60 hr. Following heat treatment (30 min, 56°) to inactivated the hybrid adenovirus, rAAV-nlacZ yields of 30–40 LFU/cell were obtained from clone #26, 0.5–1.0 LFU/cell from clone #23, and 0 LFU for naïve HeLa cells.

Generation of Stable Producer Cell Lines

Two different approaches have been taken for creating stable rAAV producer cell lines. The first relies on transfecting naïve HeLa cells with a plasmid harboring AAV *rep-cap,* vector, and selectable marker sequences.[9] The second makes use of

[27] A. J. Thrasher, M. de Alwis, C. M. Casimir, C. Kinnon, J. Page, J. Lébkowski, A. W. Segal, and R. J. Levinsky, *Gene Ther.* **2**, 481 (1995).

[28] K. J. Fisher, W. M. Kelley, J. F. Burda, and J. M. Wilson, *Hum. Gene Ther.* **7**, 2079 (1996).

[29] J. E. Conway, C. M. Rhys, I. Zolotukhin, S. Zolotukhin, N. Muzyczka, G. S. Hayward, and B. J. Byrne, *Gene Ther.* **6**, 986 (1999).

[30] C. Chartier, E. Degryse, M. Gantzer, A. Dieterle, A. Pavirani, and M. Mehtali, *J. Virol.* **70**, 4805 (1996).

an established AAV packaging cell line transduced with a rAAV vector stock[31]; in this published example a rAAV-neo[r] vector was used followed by selection with G418. We wanted to determine the likelihood of isolating stable proviral producer lines generated using this second approach in the absence of selection, a more likely scenario for generating producer clones harboring vectors with therapeutic transgenes. In addition, this approach adds flexibility because a robust packaging cell line can be transduced with different rAAV vectors to generate different producer lines.

Creating producer cell lines with vector sequences integrated into a defined chromosomal locus is advantageous because rescue efficiency should be consistent between different cell lines. AAV transduction in the presence of residual AAV Rep protein results in the majority of proviral sequences integrating in the AAVS1 locus[32–34] rather than randomly integrating throughout the genome.[35] The AAVS1 locus has been sequenced and wild-type AAV proviruses are rescued efficiently from this locus when present in tandem repeats.[36,37] In the producer cell lines described by Liu et al.[38] they observed integration of their plasmid into AAVS1 and efficient rescue of vector sequences.

We chose to make a prototype producer cell line using a rAAV-nlacZ vector because the transgene is easily scored, which facilitated screening, and the rAAV-nlacZ vector is highly requested by other investigators. A screen of cell clones for the presence of the rAAV-nlacZ vector could have been carried out using FACS following fluorescein di-β-D-galactopyranoside staining. Instead, limiting dilution cloning and manual screening of attached cells was performed because we wanted to assess cell cloning efficiency and the frequency of obtaining producer clones using this manual technique. In addition, our future cell lines will harbor vectors encoding intracellular or secreted proteins, the live screening of which would not be easily accommodated by a FACS-based approach.

Packaging cell lines #23 and #26 were transduced with a purified and characterized rAAV-nlacZ vector stock at a MOI of 100 LFU to increase the likelihood of isolating productive clones and to achieve a high proviral copy number per cell. A portion of each population of producer lines (23ZPOP and 26ZPOP) was directly stained 3 weeks posttransduction and approximately 5% of cells in both

---

[31] K. Tamayose, Y. Hirai, and T. Shimada, *Hum. Gene Ther.* **7,** 507 (1996).

[32] R. M. Kotin, M. Siniscalco, R. J. Samulski, X. D. Zhu, L. Hunter, C. A. Laughlin, S. McLaughlin, N. Muzyczka, M. Rocchi, and K. I. Berns, *Proc. Natl. Acad. Sci. U.S.A.* **87,** 2211 (1990).

[33] M. D. Weitzman, S. R. Kyostio, R. M. Kotin, and R. A. Owens, *Proc. Natl. Acad. Sci. U.S.A.* **91,** 5808 (1994).

[34] R. T. Surosky, M. Urabe, S. G. Godwin, S. A. McQuiston, G. J. Kurtzman, K. Ozawa, and G. Natsoulis, *J. Virol.* **71,** 7951 (1997).

[35] E. A. Rutledge and D. W. Russell, *J. Virol.* **71,** 8429 (1997).

[36] S. K. McLaughlin, P. Collis, P. L. Hermonat, and N. Muzyczka, *J. Virol.* **62,** 1963 (1988).

[37] X. Xiao, W. Xiao, J. Li, and R. J. Samulski, *J. Virol.* **71,** 941 (1997).

[38] X. Liu, F. Voulgaropoulou, R. Chen, P. R. Johnson, and K. R. Clark, *Mol. Ther.* **2,** 394 (2000).

populations were transduced. Another portion of each population was infected with Ad and viral lysates were made: 23ZPOP and 26ZPOP had vector yields of 0.7 LFU/cell and 17.2 LFU/cell, respectively.

Single-cell clones were isolated and expanded, resulting in 59% of 23Z and 17% of 26Z clones being viable. The clones were scored for the presence of rAAV-nlacZ after 6 weeks in culture by making two replica plates, infecting one of the replicates with Ad for 24 hr, and staining both with X-gal. Approximately 15% of 23Z clones and 9% of 26Z clones expressed lacZ, and a twofold increase in the number of lacZ$^+$ 26Z clones was obtained upon Ad infection. It was predicted that a rAAV-nlacZ vector, if present, would replicate in the presence of Ad and Rep (present in the cell lines) to increase a potentially weak lacZ signal. In addition to low copy number, a weak signal could be due to the ability of Rep to suppress heterologous promoters.[39] We observed several different classes of clones: (1) clones without lacZ expression, (2) clones with variegated lacZ expression ranging from 5% to 90% of the cells within the clone, and (3) clones with greater than 90% of the cells being lacZ positive. Even upon subcloning and Ad infection, variegated clones maintained their phenotype.

Clones scoring positive for lacZ staining were infected with adenovirus for 72 hr to package the rAAV-nlacZ vector and vector titers were determined. Several clones with nearly 100% of the cells being LacZ positive did not produce rAAV on infection with Ad; this is likely due to incomplete ITRs generated during transduction,[35] and variegated clones generated lower viral yields than homogeneously staining clones. No 23Z clones were isolated and this may be due to their instability (see below). Cell line clones 26Z.1 and 26Z.9 were chosen for their ability to reproducibly generate the highest titers of rAAV-nlacZ (98 and 287 LFU/cell, respectively) and greater than 90% of cells in both of the clones express lacZ. It may be possible to further subclone these producer cell lines to identify cells with greater yields. A summary of the cloning results is presented in Table II, showing that in the absence of selection, a high rate of obtaining producer clones was achieved in the #26 background, which suggests that this is a viable method for making producer cell lines harboring rAAV vectors with therapeutic transgenes.

Three CsCl purified large-scale rAAV-nlacZ preparations were generated from 26Z.1 and 26Z.9 cells infected with Ad. The purified preparations had infectious titers of 1×10^8–1×10^9 LFU/ml (1.7×10^{11} vector genomes/ml) or 5–30 LFU/cell. Yields of 20% are typical using CsCl gradients, and the particle:pfu ratios were 200–350. Wild-type AAV purified on CsCl gradients typically has a particle:pfu ratio of 100.

Protocol

1. Transduce packaging cell lines (1×10^5 cells in a single well of a 12-well plate) with a well characterized rAAV vector at an MOI of 100. Ten μl of a virus

TABLE II
SCREENING OF AAV-nlacZ PRODUCER CELL LINES

	Packaging cell line	
	#23	#26
Viable clones	480	88
lacZ$^+$ without Ad[a]	74/480 (15.4%)	8/88 (9.1%)
lacZ$^+$ with Ad[b]	74/480 (15.4%)	18/88 (20.5%)
Producing$^+$	0[c]	4/18 (22.2%)[d]

[a] Clones were stained with X-Gal and scored for lacZ.
[b] Clones were infected with Ad for 24 hours, stained with X-Gal, and scored for lacZ.
[c] The 74 lacZ$^+$ clones identified in the presence of Ad were assayed for their ability to produce rAAV-nlacZ virions.
[d] The 18 lacZ$^+$ clones identified in the presence of Ad were assayed for their ability to produce rAAV-nlacZ virions.

stock with an infectious titer of 1×10^9/ml is added to the cells in 1 ml of DMEM-complete and incubated for 4 days without changing the medium.

2. Single-cell clone the transduced cells by trypsinizing, counting, and diluting the cells in conditioned media supplemented with L-glutamine and G418 so that on average 100 μl of medium contains 1 cell. Seed one cell per well in 96-well plates already containing 100 μl conditioned media supplemented with L-glutamine and G418. Medium (200 μl) is changed every 5 days using a multichannel pipetter and barrier tips that are changed between each addition to prevent cross contamination.

3. Following expansion to 6-well plates, wean cells from conditioned medium to fresh DMEM-complete medium supplemented with G418, and make duplicate plates of the clones.

4. Screen one of the replicates using a transgene-specific assay (i.e., ELISA, enzymatic activity, immunohistochemical staining). This may require infecting replicates of the clones with Ad to increase the signal.

5. Expand positive clones into two 10 cm plates and infect one with Ad at an MOI of 35 for 72 hr to produce rAAV virions. Make cell lysates and determine titers as described above by scoring for transgene-specific expression, or by the infectious center assay[4] or real-time PCR.[40]

6. Expand the clones further so enough cells ($1–5 \times 10^8$ cells) are generated at the earliest possible passage and freeze in aliquots of 5×10^6 cells in freezing

[39] M. A. Labow, L. H. Graf, Jr., and K. I. Berns, *Mol. Cell Biol.* **7**, 1320 (1987).
[40] L. Drittanti, C. Rivet, P. Manceau, O. Danos, and M. Vega, *Gene Ther.* **7**, 924 (2000).

medium (90% FBS and 10% DMSO) using controlled rate freezing jars placed at $-80°$. Transfer vials to liquid nitrogen storage to serve as a cell bank.

Characterization of Cell Lines

Reagents used in the manufacturing process of clinical gene therapy vectors need to be characterized in order to ensure process reproducibility and freedom from adventitious agents. The stability of the packaging and producer cell lines was evaluated by periodically testing the clones for vector production, and as shown in Fig. 3, they maintained outputs for 6 months ($\sim$30 passages). Eventually, packaging clone #23 lost packaging capability after 7 months and yields were reduced at 6 months; also, no 23Z producer clones were isolated (Table II), indicating that this clone was unstable. In addition, 23Z clones unable to make rAAV-nlacZ were transfected with pAAV-lacZ plasmid, and no rAAV-nlacZ was produced, further indicating that the rep and/or cap sequences were lost or inactivated. Packaging clone #26 and 26Z producer clones (26Z.1 and 26Z.9) were stable for 11 and 7.5 months, respectively. As a practical consideration for clinical manufacturing, characterized master cell banks (MCBs) of packaging and producer clones should be made from early passage cells. For vector production, a vial of cells from an MCB is thawed, expanded over a few passages, and infected with Ad to make a rAAV vector lot; therefore, long-term propagation is not necessary.

Variable rates of growth were observed for both the packaging and producer clones with growth rates 50–75% of naïve HeLa cells, which double every 20 hr. During the expansion process, packaging cell clones were frozen at the earliest possible passage (passage 3) and then thawed and propagated continuously for over 6 months. Clones were also frozen at later passages (between passages 10 and 20) and thawed again to test stability during freezing and thawing: no significant loss of vector production was observed. Cell viability measured by trypan blue staining is greater than 95% for clones and naïve HeLa cells.

An analysis of clone #23 and #26 rep-cap sequences was carried out to evaluate the status of these engineered sequences. Southern blot analysis of genomic DNA probed with rep and cap sequences indicates that clone #23 harbors two copies of pLTAAVHelp and clone #26 harbors 10 copies in integrated head-to-tail or high molecular weight episomal configurations (Fig. 4). Previously described AAV packaging cell lines harbor 1–10 copies of the AAV rep and cap sequences that amplify as episomes up to 100-fold when infected with adenovirus.[10,12,38,41] To determine if DNA replication is involved here, genomic or low molecular weight (LMW)[42] DNA was isolated from uninfected and Ad infected packaging

[41] J. Tessier, G. Chadeuf, P. Nony, H. Avet-Loiseau, P. Moullier, and A. Salvetti, *J. Virol.* **75**, 375 (2001).

[42] B. Hirt, *J. Mol. Biol.* **26**, 365 (1967).

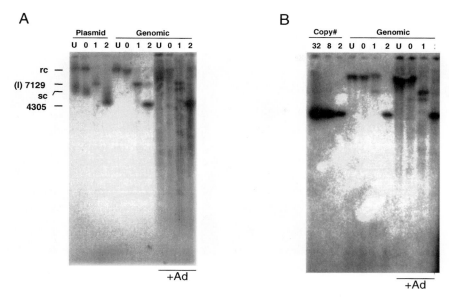

FIG. 4. Structure of the *rep-cap* sequences. Genomic DNA isolated from uninfected or Ad infected clone #26 (A) and clone #23 (B) was digested with restriction enzymes, separated on 1% agarose gels, blotted to nylon membranes, and probed with *rep* and *cap* sequences. Genomic, genomic DNA; copy#, plasmid copy number standards of 32, 8, and 2 copies in (B), and plasmid controls of 8 copies per lane were digested in (A). U, Undigested DNA; 0, *Eag*I does not cut pLTAAVHelp; 1, *Acc*I cuts pAAVLTHelp once in the *rep* gene; and 2, *Xba*I cuts at two sites in pLTAAVHelp flanking the AAV sequences. +Ad, DNA isolated from cells infected with Ad; rc, relaxed circular plasmid DNA; (1), linearized plasmid DNA; sc, supercoiled plasmid DNA. Fragment sizes are indicated.

clones and control HeLa (transiently transfected with pLTAAVHelp) cell lines, digested, and probed with *rep-cap* sequences (Fig. 5). This revealed a 10-25-fold Ad-dependent amplification of the engineered AAV sequences, and digestion of the DNA with methylation-sensitive restriction enzymes indicated that the *rep-cap* sequences had been replicated (Figs. 5A and 5B). This is also true for plasmid sequences transfected into naïve HeLa cells infected with Ad (Fig. 5C), a result contradictory to those of Liu *et al.*[38] Following Ad infection and digestion with *Acc*I or *Eco*NI, a significant portion of the amplified sequences appear to be in a head-to-head inverted orientation (Figs. 5A, 5B, and 5D), a structure found in amplified DNA.[43,44]

To determine if the amplified *rep-cap* sequences in #23 and #26 are episomal, *E. coli* was transformed with genomic and LMW DNA from uninfected and

[43] G. H. Nonet, S. M. Carroll, M. L. DeRose, and G. M. Wahl, *Genomics* **15,** 543 (1993).
[44] B. E. Windle and G. M. Wahl, *Mutat. Res.* **276,** 199 (1992).

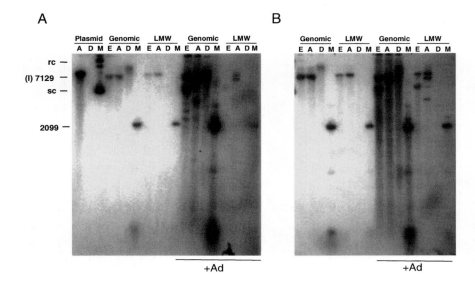

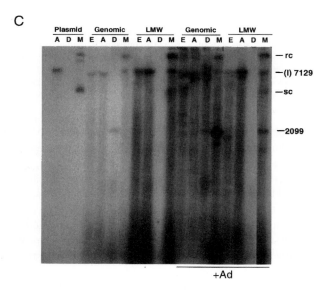

FIG. 5. Amplification of the *rep-cap* sequences. Genomic and low molecular weight DNA isolated from uninfected or Ad infected cells was digested with restriction enzymes, separated on 1% agarose gels, blotted to nylon membranes, and probed with *rep* and *cap* sequences. Analysis of clone #23 (A), clone #26 (B), and DNA from HeLa cells transiently transfected with pLTAAVHelp (C). Genomic, genomic DNA; LMW, low molecular weight DNA. Plasmid controls of 8 copies per lane were digested

D

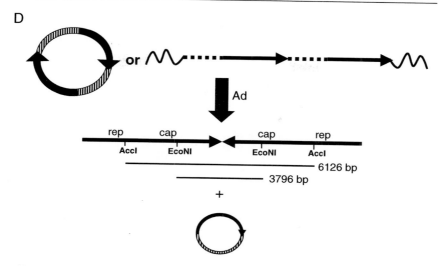

FIG. 5. (*continued*) with the indicated restriction enzymes. E, *Eco*NI cuts pLTAAVHelp once in the *cap* gene; A, *Acc*I cuts pAAVLTHelp once in the *rep* gene; D, *Dpn*I cleaves input methylated plasmid DNA at several sites, the largest fragment being 2099 bp; M, *Mbo*I, an isoschizomer of *Dpn*I, cleaves unmethylated (replicated) DNA. +Ad, DNA isolated from cells infected with Ad; rc, relaxed circular plasmid DNA; (1), linearized plasmid DNA; sc, supercoiled plasmid DNA. Fragment sizes are indicated. A model of the amplification is presented in (D). pLTAAVHelp is maintained either as an episome or integrated into the cellular genome in head-to-tail configuration. AAV sequences and plasmid backbone are indicated by a solid arrow and a dashed line, respectively. On Ad infection the AAV sequences are amplified, and head-to-head structures are generated in addition to free pAAVLTHelp (see Fig. 6).

Ad-infected cells. Few rescuable episomes exist prior to Ad infection, but an increase in rescued plasmid occurs upon Ad infection (Fig. 6A). The size of the supercoiled rescued plasmids is similar to the size of parental plasmid (7129 bp). *Pst*I digests of the plasmids revealed the structure of the majority of the plasmids rescued from Ad-infected packaging cell lines is the same as pLTAAVHelp (Fig. 6B). Since we could rescue intact plasmid sequences from adenovirus infected cell lines and digest them with *Pst*I, the cellular amplified sequences are double-stranded. The head-to-head structure was not rescued and possibly was too large for bacterial transformation or may have lost plasmid backbone sequences.

Wild-Type AAV Contamination

The parental wild-type AAV (wtAAV) virus has not been associated with any known human disease[45]; however, as a potential contaminant of rAAV vector

[45] K. I. Berns and R. M. Linden, *Bioessays* **17**, 237 (1995).

A

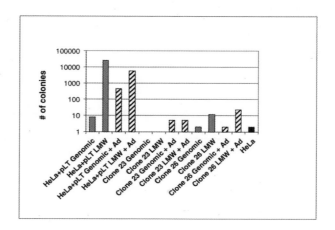

B

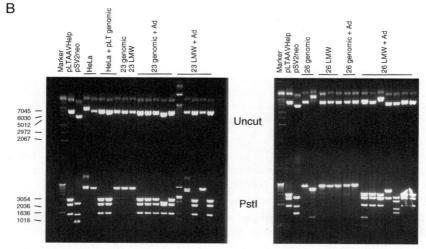

FIG. 6. Rescued plasmids. Genomic and low molecular weight DNA isolated after 13 weeks of culture from 3×10^4 uninfected or Ad infected cells was used to transform *E. coli* DH12S by electroporation. (A) Number of ampillicin-resistant colonies obtained from the different samples. HeLa+LT Genomic, genomic DNA isolated from HeLa cells transfected with pLTAAVHelp; HeLa+LT LMW, low molecular weight DNA isolated from HeLa cells transfected with pLTAAVHelp; clone 23 or 26 genomic, genomic DNA isolated from clone #23 or #26; Clone 23 or 26 LMW, low molecular weight DNA isolated from clone #23 or #26; +Ad, DNA isolated from Ad-infected cells. (B) Size and structure of rescued plasmids. Undigested (top tier) or *Pst*I digested (bottom tier) plasmid DNA isolated from bacterial colonies was separated on 1% agarose gels. Control pLTAAVHelp and pSV2neo plasmid DNA was run undigested or following digestion with PstI, and the sizes of supercoiled (top tier, Life Technologies) and standard (bottom tier, Life Technologies) molecular weight markers are shown in base pairs.

stocks, it should be evaluated because AAV Rep protein may affect transduction frequency and gene expression, and possibly change the integration specificity of the rAAV provirus. As stated above, transient transfection of AAV helper and vector plasmids can generate wild-type-like AAV (rcAAV) by nonhomologous recombination. We tested for the presence of wtAAV/rcAAV in the packaging cell lines or in vector preparations made using the producer cell lines. Chromosomal and low molecular weight DNA isolated from uninfected or adenovirus-infected packaging cell lines was tested by PCR with the primers D1 : 5′ACTCCATCAC TAGGGGTTCC3′ and 2S2: 5′TCAGAATCTGGCGGCAACTCCC3′ under previously published conditions[46] and no wtAAV/rcAAV was detected. Southern blot analysis of pre- and postamplified (Ad infected) genomic and LMW DNA from packaging lines #23 and #26 revealed no bands in the undigested and "no-cutter" lanes consistent with wtAAV/rcAAV genomes, such as monomer (4679 bp) and dimer (9237 bp, sharing a single internal ITR) replicative forms (Fig. 4). Lastly, Southern blot analysis was carried out on LMW DNA isolated from 293 cells coinfected with Ad and 1×10^7 or 1×10^9 LFU of CsCl purified rAAV-nlacZ vector made using the 26Z.1 and 26Z.9 producer cell lines. rcAAV or wtAAV was not detected in 1×10^7 LFU (2×10^9 vector genomes); however, at 1×10^9 LFU replication of the vector in the presence of Ad was seen. Although indirect, replication of rAAV in the presence of Ad indicates a contamination with rcAAV that would supply Rep; however, a direct probe of *rep-cap* sequences resulted in no signal. LMW DNA isolated from cells coinfected with rAAV-nlacZ, wtAAV, and Ad served as a positive control, and replication of the rAAV-nlacZ vector and input wtAAV genomes was detected when probed.

Discussion

We have presented protocols for creating rAAV packaging and producer cell lines, and characterized cell lines generated using these protocols. Reproducible vector titers obtained from the cell lines depended on Ad infection to induce AAV protein expression. The Ad-dependent DNA amplification of engineered *rep-cap* sequences appears to be a common property of these and similar cell lines. The increase in copy number may mimic what occurs during natural infection, and one study[41] identified some of the required elements for amplification. In another approach, SV40 T antigen-mediated amplification of AAV sequences has been shown to be effective for rAAV production.[11,47]

The Ad genes required for AAV vector production can be supplied by transfecting them into cells along with the AAV vector and helper plasmids, without

[46] R. O. Snyder, C. H. Miao, G. A. Patijn, S. K. Spratt, O. Danos, D. Nagy, A. M. Gown, B. Winther, L. Meuse, L. K. Cohen, A. R. Thompson, and M. A. Kay, *Nat. Genet.* **16,** 270 (1997).
[47] J. A. Chiorini, C. M. Wendtner, E. Urcelay, B. Safer, M. Hallek, and R. M. Kotin, *Hum. Gene Ther.* **6,** 1531 (1995).

the generation of infectious Ad.[48,49] The ability to establish producer cell lines harboring rAAV vector, *rep-cap,* and Ad sequences will depend on tight control of both AAV and Ad gene expression because, in addition to the AAV gene products, some of the required Ad proteins are cytotoxic and stimulate AAV gene expression. Furthermore, it may be difficult to isolate productive cell lines harboring the Ad genes if Ad replication centers cannot be established.[50] Transient transfection of Ad DNA into packaging cell lines does not result in replication center establishment, *rep-cap* sequence amplification, or rAAV vector production.[12] Our attempts to use an E1 deleted adenovirus in the producer cell lines did not induce rAAV production even though HPV E6/E7 proteins have similar properties to the Ad E1 proteins.[51–53] Supplying the adenoviral genes through infection is a convenient way to induce AAV gene expression, but the Ad that is generated must be physically removed or inactivated (inactivation by itself does not remove the cytotoxic Ad structural proteins).

From a regulatory point of view, making stable rAAV producer cell lines by transduction may be problematic because vector preparations made by standard transient transfection may be difficult to define and it may be difficult to ensure that vector sequences are intact. In addition, vector stocks made by transient transfection may be contaminated with rcAAV. Introducing sequenced rAAV vector plasmid DNA lends better control to the production process by creating defined cell-based reagents.

Compared to vectors based on other AAV serotypes, AAV2 vectors are being evaluated most extensively in animals and human clinical trials. AAV vectors based on other serotypes can transduce certain tissues more efficiently.[54–59] As the requests grow for these vectors, large-scale production and purification methods will be developed. Establishing stable packaging and producer cell lines harboring sequences from the different AAV serotypes will help to meet this demand.

[48] X. Xiao, J. Li, and R. J. Samulski, *J. Virol.* **72,** 2224 (1998).

[49] T. Matsushita, S. Elliger, C. Elliger, G. Podsakoff, L. Villarreal, G. J. Kurtzman, Y. Iwaki, and P. Colosi, *Gene Ther.* **5,** 938 (1998).

[50] M. D. Weitzman, K. J. Fisher, and J. M. Wilson, *J. Virol.* **70,** 1845 (1996).

[51] W. C. Phelps, C. L. Yee, K. Munger, and P. M. Howley, *Cell* **53,** 539 (1998).

[52] S. A. Sedman, M. S. Barbosa, W. C. Vass, N. L. Hubbert, J. A. Haas, D. R. Lowy, and J. T. Schiller, *J. Virol.* **65,** 4860 (1991).

[53] P. Massimi, D. Pim, A. Storey, and L. Banks, *Oncogene* **12,** 2325 (1996).

[54] B. L. Davidson, C. S. Stein, J. A. Heth, I. Martins, R. M. Kotin, T. A. Derksen, J. Zabner, A. Ghodsi, and J. A. Chiorini, *Proc. Natl. Acad. Sci. U.S.A.* **97,** 3428 (2000).

[55] C. L. Halbert, E. A. Rutledge, J. M. Allen, D. W. Russell, and A. D. Miller, *J. Virol.* **74,** 1524 (2000).

[56] E. A. Rutledge, C. L. Halbert, and D. W. Russell, *J. Virol.* **72,** 309 (1998).

[57] W. Xiao, N. Chirmule, S. C. Berta, B. McCullough, G. Gao, and J. M. Wilson, *J. Virol.* **73,** 3994 (1999).

[58] J. E. Rabinowitz, W. Xiao, and R. J. Samulski, *Virology* **265,** 274 (1999).

[59] H. Chao, Y. Liu, J. Rabinowitz, C. Li, R. J. Samulski, and C. E. Walsh, *Mol. Ther.* **2,** 619 (2000).

Acknowledgments

We thank Dr. Richard C. Mulligan (Harvard Gene Therapy Initiative) and Dr. Olivier Danos (Généthon) for their support. We are appreciative of Kathleen Skarre, Timothy Sheahan, Demetrios Karafilidis, Michael Rutenberg, Melanie Watkins, and Bonnie Ziegler for technical assistance. This work was supported by the Association Française contre les Myopathies.

[24] Streamlined Large-Scale Production of Recombinant Adeno-Associated Virus (rAAV) Vectors

By MARK POTTER, KYE CHESNUT, NICHOLAS MUZYCZKA, TERRY FLOTTE, and SERGEI ZOLOTUKHIN

Recent progress in development of rAAV production protocols in several laboratories has allowed a widespread testing of purified vectors of high titers in a variety of animal models. Many of these trials have resulted in successful therapies derived from long-term expression of transgenes delivered by rAAV. The positive outcome of most rAAV-mediated therapies appears to depend on the quality of the vector reagent, i.e., its purity, infectivity, and titer. The last parameter is a function of manufacturing capacity, specifically, the number of cells included into one production run. The requirements of preclinical *in vivo* studies in most cases reach beyond the regular production formats that rely on cells grown in flasks or dishes. In this report we describe a preindustrial scale-up protocol carried out in a cell factory format (about 10^9 cells per factory), allowing a modest facility to increase vector production at least 10- to 100-fold.

Earlier, we developed a new rAAV purification protocol resulting in higher yield and improved infectivity of viral particles.[1,2] This method utilizes a bulk purification of a crude lysate through an iodixanol step gradient followed by conventional heparin affinity or HPLC ion-exchange chromatography. Although quick and reproducible, this protocol is not readily amenable to large-scale production of a clinical-grade vector because of the limiting capacity of the iodixanol centrifugation step. To incorporate the requirements of an increased starting cell volume we describe further improvements in the production protocol introducing new chromatography purification steps that eliminate the need for any centrifugation methods. We also characterize the purified rAAV in terms of purity, infectivity,

[1] S. Zolotukhin, B. J. Byrne, E. Mason, I. Zolotukhin, M. Potter, K. Chesnut, C. Summerford, R. J. Samulski, and N. Muzyczka, *Gene Ther.* **6,** 973 (1999).

[2] W. W. Hauswirth, A. S. Lewin, S. Zolotukhin, and N. Muzyczka, *Methods Enzymol.* **316,** 743 (2000).

and packaged particle composition. As a case study, we describe the production of a National Reference Standard (NRS) rAAV vector, sponsored by the National Gene Vector Laboratory.

The National Reference Standard effort stems from a multicenter effort to share preclinical data with regard to the long-term potential risks for insertional mutagenesis and/or germline transmission of rAAV. The types of preclinical studies which may be required to adequately address these issues may be very large and may require resources beyond those traditionally available to academic centers involved in the treatment of rare genetic diseases. In an effort to address this issue, a joint FDA/NIH Workshop was held on May 2nd and 3rd, 1999. At that workshop, members of the rAAV gene therapy community from academia, industry, and the federal government discussed the potential for developing a shared platform of preclinical data to address these vector-specific safety issues. It was generally recognized that in order to pool preclinical data in a meaningful way, it would be necessary for a wide range of groups to be able to discuss vector dosage, strength, and potency in equivalent titer units. It was further recommended that to facilitate this goal, a reference standard stock of rAAV with a precisely defined titer should be generated and made generally available to all members of the research community. All users of this reference stock would essentially be able to calibrate their titering assays against a common standard, thus allowing each group to state their titers in units that were precisely understood by all. The goal of the current application is to generate the rAAV reference standard stock, aliquot it into a large number of individual user vials, validate its utility as a reference standard among a handful of rAAV laboratories, and then transfer it to an appropriate distribution service. Within this application, we describe the production process to be used in the generation of this rAAV stock, the basic quality control assays that will be in place to confirm its purity and identity, and the standard protocols (SPs) to be used for physical and biological titering by the producing center, the validating centers, and the eventual individual users.

The rAAV construct chosen for the National Vector Standard was pTR-UF5.[3] It contains a humanized *gfp* gene[4] under the control of a CMV promoter and a *neo* gene under the control of a TK promoter.

Propagation of Cells

Protocol

Low-passage ($P < 40$) HEK 293 cells are used to propagate the rAAV vector. A working cell bank of 293 cells (P30), derived from a batch certified by

[3] R. L. Klein, E. M. Meyer, A. L. Peel, S. Zolotukhin, C. Meyers, N. Muzyczka, and M. A. King, *Exp. Neurol.* **150,** 183 (1998).

[4] S. Zolotukhin, M. Potter, W. W. Hauswirth, J. Guy, and N. Muzyczka, *J. Virol.* **70,** 4646 (1996).

Microbiology Associates for GMP-grade production, is being stored in vials (10^7 cells/aliquot) in liquid N_2.

Seeding Cell Factory with 293 Cells from T255 Flasks

1. Cells from one vial are thawed and seeded into a T225 flask. After cells reach confluence, they are split and seeded into three T225 flasks using regular cell culture techniques, followed by one more split into eight T225 flasks. The initial propagation cycle takes about 10 days.

2. Cells from eight confluent T225 flasks are then seeded into one cell factory (6320 cm^2 of a culture area, NUNC, Roskilde, Denmark). To split cells, 2 liters of medium [Dulbecco's modified Eagle's medium (DMEM) supplemented with 5% fetal bovine serum (FBS), 1× antibiotic–antimycotic] is warmed to 37°. Five ml of 10 × trypsin-EDTA stock solution is diluted in 45 ml 1× phosphate-buffered saline (PBS) in a 50 ml conical tube. Medium from eight flasks is discarded and cells in each flask are washed with 10 ml PBS. Four ml of diluted trypsin–EDTA is added and flasks are rocked until cells start to peel, at which point flasks are knocked against a hard surface to lift cells off the surface. The trypsin–EDTA solution is then neutralized by addition of 16 ml per flask DMEM–5% FBS. The cells are resuspended by triturating with a pipette and collected in a 250 ml disposable conical tube. About 1090 ml of medium is mixed with the pooled cells in an aspirator bottle and the cell factory is then loaded following the manufacturer's instructions. The cell factory is incubated at 37° for about 20 hr until transfection.

Seeding Four Cell Factories from One Confluent Factory. A fresh stock of trypsin–EDTA solution is made by diluting 25 ml of 10 × stock with PBS in a 250 ml conical tube while medium is being warmed at 37°. Medium from the confluent cell factory is poured off and cells are rinsed with ~250 ml PBS. Trypsin/EDTA solution is added to the factory and cells are dislodged after rocking the factory at room temperature for 2 min. About 850 ml of media is added to the aspirator bottle, and the cells are rinsed off with the media and transferred into an empty media bottle (total volume is about 1.1 liter). Cell clumps are dispersed by vigorously shaking the capped bottle and cells are aliquoted into the four factories. The cell factory that was used for seeding is reused and receives 200 ml of cell suspension (along with 1 liter of fresh medium). This factory is then used to seed new factories on subsequent days. One cell factory can be recycled for seeding up to five times. The remaining three factories receive 300 ml of cell suspension and 1 liter of fresh medium each and are transfected the next day.

Transfection

The basic strategy for producing rAAV via cotransfection of two plasmid DNAs has been extensively described.[1,2] The scale-up protocol adapted for one cell factory is described below.

Protocol

Cell confluency is determined by focusing the microscope on the plane of the bottom layer of the cell factory. Transfection is carried out when cells are about 75% confluent.

$2\times$ Hepes-buffered Saline (HBS) buffer is thawed and kept at $37°$ until ready to use. One liter of media and 50 ml of FBS are prewarmed and mixed in an aspirator bottle. The helper plasmid pDG and the rAAV plasmid (1867.5 μg and 622.5 μg, respectively) are mixed in a 250 ml conical tube and the mixture is diluted with H_2O up to 46.7 ml final volume. $CaCl_2$ is added to the same tube (5.2 ml of 2.5 M stock solution) and mixed with the DNA. Old media from the cell factory is discarded and formation of the $CaPO_4$ precipitate is initiated by adding 52 ml of prewarmed $2\times$ HBS buffer. The mixture is swirled for 1 min and transferred into an aspirator bottle containing medium. Medium is then loaded into the cell factory, and the cells are incubated at $37°$ for 48–60 hr without additional media change.

Harvesting Transfected Cells

Protocol

In a dedicated rAAV hood the cell factory is rocked to dislodge nonadherent cells and medium is discarded. Cells are washed with 500 ml of PBS using an aspirator bottle. EDTA is added to 500 ml PBS to a 5 mM final concentration, and the solution is transferred into the factory and spread evenly over all layers. The cells are lifted off the plastic by vigorously shaking the factory, and the cells are poured into two 250 ml conical tubes. The remaining cells are rinsed out with an additional 500 ml PBS and transferred into two more conical tubes. Cells are harvested by low-speed centrifugation at 300g for 10 min at $4°$. The supernatant is discarded and cell pellet is stored at $-20°$ until processing.

Purification of rAAV Vector

Generation of Crude Lysate

As an alternative to repeated freeze–thaw cycles, which produce a crude cell lysate requiring further purification either by filtration or centrifugation prior to any subsequent processing, we sought a method that would generate a lysate that would be immediately suitable for chromatography. One such method utilizes a microfluidizer processor (model M-110, Microfluidics International Corporation). The processor works by introducing low-pressure air (or gas) through a series of sequentially smaller cylinder piston pumps into a product stream generating high-pressure force (1 psi of supply air generates 230 psi liquid pressure). The high-pressure stream enters the interaction chamber into precisely defined microchannels where three primary forces, shear, impact, and cavitation, act on the sample as it passes through the chamber. We have found one pass of a stock material through the microfluidizer to be sufficient for our needs. However, recycling the product through the interaction chamber is possible. To process the cell suspension through the chamber, we routinely pool cell pellets from factories transfected with the same vector. In the case of the National Reference Standard, cell pellets from 10 individual cell factories were pooled.

Protocol

Cells are allowed to thaw at room temperature for approximately 15 min and placed on ice. To each cell pellet 15 ml deoxycholate lysis buffer (0.5% sodium deoxycholate, 20 mM Tris-HCl pH 8.0, 150 mM NaCl) is added and mixed well by triturating, and mixtures are pooled into one 250 ml conical tube. To reduce the viscosity the lysate is treated with benzonase (100 U/ml) for 30 min at 37° on the addition of MgCl$_2$ (1 mM final concentration). The sample is then split into two equal parts, 125 ml each, and the volumes in each tube are adjusted to 150 ml by the addition of an equal volume of fresh deoxycholate lysis buffer. Both aliquots of diluted sample are then homogenized by one pass through a microfluidizer and collected in a single vessel.

Streamline Heparin Affinity Chromatography. All chromatography steps are performed using AKTA FPLC Pharmacia's system. A Pharmacia FPLC XK/26 column containing the Streamline Heparin chromatography matrix (fixed bed format, column bed volume of 55 ml) is equilibrated with deoxycholate lysis buffer (see above). The entire crude sample (~300 ml) is applied to column at a flow rate of 3 ml/min. After loading, the column is washed with 330 ml (about 6 column volumes, CV) of the same buffer at a flow rate of 5 ml/min, and then with 18 CV (about 1 liter) of PBS at a flow rate of 10 ml/min. The sample is then eluted from the column by a single step gradient using PBS containing 0.5 M NaCl at a flow rate of 3 ml/min. Ten ml fractions are collected (Fig. 1A) and positive fractions are determined by fluorescent cell assay (FCA).[1] In the example shown in Fig. 1A fractions 6 through 8 are pooled.

Phenyl-Sepharose Hydrophobic Interaction Chromatography. The pooled heparin column fractions (30 ml) are adjusted to 1 M NaCl by the addition of 5 M stock solution. The sample is then loaded onto a Pharmacia FPLC XK/16 column (column bed of 18 ml) containing hydrophobic interaction chromatography (HIC) medium phenyl Sepharose at a flow rate of 5 ml/min. Prior to loading sample the column is preequilibrated with PBS–1 M NaCl. The column is monitored by UV absorption and the virus elutes in the flow-through (~100 ml) (Fig. 1B).

Heparin Affinity Chromatography. The virus is concentrated by chromatography on heparin. The phenyl Sepharose fraction is diluted to ~150 mM NaCl by the addition of dH$_2$O (usually a 6-fold dilution). The diluted sample (~700 ml)

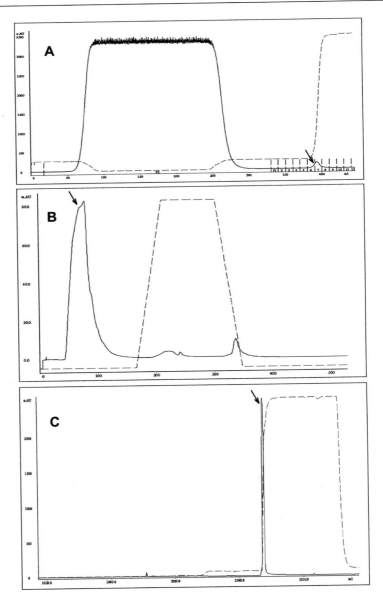

Fig. 1. Chromatography profiles of NRS rAAV purification steps. (A) Streamline heparin affinity chromatography. (B) Phenyl Sepharose hydrophobic interaction chromatography (C). Heparin affinity chromatography. Broken lines show conductivity profiles of the respective processes. Solid lines indicate OD$_{280}$. Arrows mark the pooled peak fractions.

TABLE I
SUMMARY OF rAAV-UF5 PURIFICATION

Run	Crude lysate		Streamline heparin affinity chromatography		HIC phenyl sepharose chromatography		Heparin affinity chromatography	
	Total inf. particles	% yield	Total inf. particles	% yield	Total inf. particles	% yield	Total inf. particles	% yield
1	4.8×10^{12}	100	2.7×10^{12}	56	2.3×10^{12}	48	1.3×10^{12}	27
2	9.6×10^{12}	100	1.8×10^{13}	186	1×10^{13}	104	4.2×10^{12}	43
3	3.6×10^{12}	100	1.2×10^{12}	33	1.3×10^{12}	36	1.2×10^{12}	33

is then loaded onto a heparin affinity column, BioPerceptive's Poros HE/20 column (bed volume 1.7 ml) at a flow rate of 5 ml/min. After loading, the column is washed with 100 ml PBS at the same flow rate and virus is eluted with PBS containing 0.5 M NaCl using a single step gradient. One ml fractions are collected and aliquots analyzed by fluorescence cell assay (FCA). The major peak of UV absorbance contains rAAV-GFP vector (Fig. 1C).

To estimate the yield, the vector obtained after each purification stage is titered by FCA. The yields of three representative vector runs, each consisting of 10 pooled cell factories, are shown in Table I.

Characterization of Purified rAAV

An important index of virus quality is the ratio of physical particles to infectious particles in a given preparation. To characterize the quality of the virus, we used several independent assays to titer both physical and infectious rAAV particles. A conventional dot-blot assay (not shown) and real-time polymerase chain reaction (PCR) (Fig. 2) were used for physical particle titers. Infectious titers were determined by infectious center assay (ICA) and a fluorescence cell assay (FCA), which scored for expression of GFP (not shown). In order to avoid adventitious contamination of rAAV stocks with wtAAV, the use of wtAAV was eliminated in the ICA (as well as all other protocols in the laboratory). This was made possible by the use of the C12 cell line,[5] which contains integrated wtAAV *rep* and *cap* genes, for both the infectious center assay and the fluorescent cell assay. Adenovirus serotype 5 (Ad5), which was used to coinfect C12 along with rAAV, was titered using the same C12 cell line in a serial dilution cytopathic effect (CPE) assay. The amount of Ad producing well-developed CPE in 48 hr on C12 cells was used to provide helper function for both the ICA and FCA assays. Both physical particle

---

[5] K. R. Clark, F. Voulgaropoulou, D. M. Fraley, and P. R. Johnson, *Hum. Gene Ther.* **6,** 1329 (1995).

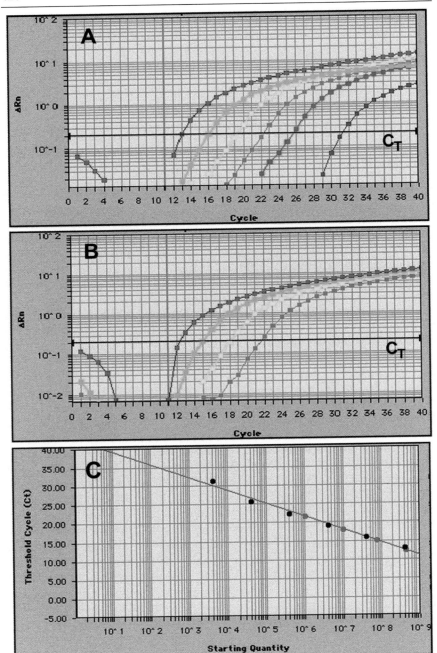

TABLE II
PHYSICAL AND INFECTIOUS TITERS OF NATIONAL REFERENCE STANDARD rAAV
AS DETERMINED BY FOUR ASSAYS

	Dot blot	Real-time PCR	ICA	FCA
NRS rAAV titer	1.12×10^{13} part/ml	1.46×10^{13} part/ml	2.0×10^{12} infect.part/ml	2.16×10^{12} infect.part/ml

titers and infectious titers, each obtained by two independent methods, were generally in agreement, differing in most cases by a factor of 2 or less (Table II). The particle to infectivity ratio (p/i) was approximately 6.

Dot-Blot Assay (DBA)

Protocol

1. Eppendorf tubes are marked 1 through 12 (tubes for $2\times$ dilution series) and 1* through 12* (tubes to denature the diluted DNAs). Two additional tubes to denature viral DNA samples are marked 10^0 and 10^{-1}.

2. To make a standard DNA dilution curve, 50 μl of H_2O is aliquotted into each of the tubes marked 1 through 12. The first dilution is made by adding 50 μl of standard plasmid DNA (5 ng/μl) containing the gene of interest to tube 1. Normally, the same transfer vector plasmid that was originally packaged is being used. Twofold series dilutions are made by vortexing the tube and transferring 50 μl from tube 1 into tube 2 and so on until all dilutions are done. *Caution:* Make sure to change tips after each dilution.

3. To denature DNA, 200 μl of alkaline buffer (0.4 M NaOH–10 mM EDTA) is aliquotted into each tube marked 1* through 12*, including the tubes 10^0 and 10^{-1}. Diluted DNA samples, 10 μl each, are transferred from tubes 1 through 12 into tubes 1* through 12*, respectively, and mixed by vortexing. *Note:* The transfer is usually done in reverse order (starting from tube 12) using one tip.

4. To prepare the viral DNA sample, the purified viral stock is first treated with DNase I to digest any contaminating unpackaged DNA. Four μl of a purified

FIG. 2. Real-time PCR assay for NRS rAAV. (A) Amplification profiles of standard plasmid pTR-UF5 dilution series. Red graph, 1 ng of plasmid DNA per reaction; green graph, 100 pg; yellow, 10 pg; light blue, 1 pg; purple, 100 fg; dark blue, 10 fg. (B) Amplification profiles of NRS viral DNA dilution series. Red graph, 5 μl out of 10^2 dilution (including 10-fold original dilution after DNAse I and Proteinase K treatment) per reaction; green graph, 5 μl out of 10^3 dilution; yellow, 5 μl out of 10^4 dilution; light blue, 5 μl out of 10^5 dilution. (C) Calibration graph of threshold cycle values plotted against starting genome equivalents of the standard plasmid. Black dots represent plasmid DNA values generated at different dilutions (A); red dots represent viral DNA dilutions (B).

virus stock is incubated with 10 U of DNase I (Boehringer) in a 200 μl reaction mixture, containing 50 mM Tris-HCl, pH 7.5, 10 mM MgCl$_2$ for 1 hr at 37°. At the end of the reaction, 20 μl of 10 × proteinase K buffer (10 mM Tris-HCl, pH 8.0, 10 mM EDTA, 1% SDS final concentration) is added, followed by the addition of 2 μl of proteinase K (18.6 mg/ml, Boehringer). The mixture is incubated at 37° for 1 hr. Viral DNA is purified by phenol/chloroform extraction (twice), followed by chloroform extraction and ethanol precipitation using 10 μg of glycogen as a carrier. The DNA pellet is dissolved in 40 μl of water, resulting in a 10-fold dilution of the original sample. To denature viral DNA, 1 μl and 10 μl are transferred into tubes containing prealiquotted alkaline buffer, marked 10^{-1} and 10^{0}, respectively.

5. To apply DNA to the filter, an 11.5 × 4 cm rectangular piece of nylon filter (0.45 μm, MagnaGraph Nylon Transfer Membrane, Osmonics Inc.) is wetted in H$_2$O. The filter is then placed on top of two layers of prewetted Whatman 3 MM filter papers (8 × 11 cm) and the whole stack is assembled in a dot-blot manifold apparatus (Schleicher and Schull Inc.) following the manufacturer's instructions. To equilibrate the membrane alkaline buffer (0.5 M NaOH–1.5 M NaCl) is pipetted into each well (400 μl per well) and house vacuum is applied for slow (about 100 μl/min) suction until wells are empty. Denatured DNA is then transferred into the wells in a reverse order, i.e., starting with tube 12*. Viral DNA samples are then applied, skipping one full row in the minifold. *Caution:* Eliminate inadvertent bubbles by slowly pipetting the solution in a well up and down. After wells are dry, the filter is washed one time with alkaline buffer (400 μl/well). The apparatus is disassembled and the nylon membrane is placed on dry Whatman 3 MM paper for 5 min. The filter is then rinsed in 4× SSC for 10-15 sec; air dried on Whatman 3 MM filter; and DNA is fixed to the nylon membrane in a microwave oven (high setting for 4 min). *Caution:* Be sure to have about 500 ml of water in a separate beaker in the microwave or the membrane may catch on fire.

6. The membrane is prehybridized in 7% (w/v) SDS–0.25 M NaHPO$_4$ (pH 7.2)–1 mM EDTA (pH 8.0) at 65° for at least 1 hr before adding the denatured probe. The probe is a ^{32}P-labeled DNA (any fragment of the DNA in the rAAV cassette being titered). About 20 ng of a random-primed labeled probe (specific activity of 10^{9} cpm/μg) is denatured and added to 5 ml of hybridization buffer (see above). The hybridization is carried in a glass cylinder in a rotational oven at 65° for 12 hr. The membrane is then washed three times (20 min each) in 2× SSC at room temperature, air dried, and exposed in a Phospho-Imager (Storm 860, Molecular Dynamics) cassette.

7. The image is processed using ImageQuant software and subsequently plotted to derive the standard curve using software GraphPhad Prism (GraphPad Software, Inc). The viral physical particles titer is calculated taking into account the dilution factor using the average values of two dilutions (see Table II for the titer of NRS rAAV).

Real-Time PCR Assay (RTPA)

Earlier we described two independent assays for determining rAAV physical titer: dot-blot assay and QC-PCR.[1] Below we describe another PCR-based assay using the PE-Applied Biosystems Prism 7700 sequence detector system similar to the method described by Clark *et al.*[6] To perform the assay, viral DNA is isolated by sequential treatment with DNase I and Proteinase K as described in the preceding section.

PCR primers, designed by PerkinElmer to amplify GFP have the following sequences:

Forward: 5'-TTTCAAAGATGACGGGAACTACAA-3'
Reverse: 5'-TCAATGCCCTTCAGCTCGAT-3'
Probe: 5'-6FAM-CCCGCGCTGAAGTCAAGTTCGAAG-TAMRA-3'

The reaction components used in setting up the reactions and their corresponding volumes are listed below:

TaqMan Universal PCR Master Mix ($2\times$)	$25\ \mu l$
Forward primer ($9\ \mu M$, $10\times$ stock solution)	$5\ \mu l$
Reverse primer ($9\ \mu M$, $10\times$ stock solution)	$5\ \mu l$
TaqMan probe ($2\ \mu M$, $10\times$ stock solution)	$5\ \mu l$
DNA sample	$5\ \mu l$
Water	$5\ \mu l$
Total volume	$50\ \mu l$

The reactions are initiated by incubating samples at 50° for 2 min to activate uracil-*N*-glycosylase (UNG), followed by incubation at 95° for 10 min (AmpliTaq Gold activation step). The cycling parameters are as follows: 15 sec at 95°, 1 min at 60°, 40 cycles total.

Protocol

To quantify the titer, semilog plots of the increase in reporter fluorescence (ΔRn) vs PCR cycle number are derived for dilutions of the standard and test sample DNAs.

1. To derive a standard curve, plasmid DNA pTR-UF5 is serially diluted in H_2O to derive 6 separate dilutions covering a range of 10 fg to 1 ng (4×10^3–4×10^8 genome equivalents of single-stranded viral DNA, respectively) in a volume of

---

[6] K. R. Clark, X. Liu, J. P. McGrath, and P. R. Johnson, *Hum. Gene Ther.* **10**, 1031 (1999).

5 μl. This is added into the PCR reaction and amplified as described above. The amplification profiles of six different dilutions are shown in Fig. 2A.

2. To derive a viral DNA sample amplification curve, four 10-fold serial dilutions are made using 10 μl of purified DNA derived from a viral stock, which had been digested with DNase I and treated with proteinase K (see above). After all dilutions are made, 5 μl of each dilution is added to 45 μl of preassembled mix and the reactions are performed as described above. The amplification profiles of serial dilutions are shown in Fig. 2B.

3. To generate a copy number for the standard curve, the C_T value (an arbitrary set threshold cycle) is plotted against the input molecule copy number (Fig. 2C, black dots, coefficient of linearity, 0.985). The viral titer of 7.3 $\times$ 10^7 is derived using the system's software on plotting viral DNA amplification values (Fig. 2C, red dots). This value is then corrected for the dilution factor to derive a final value of 1.46 $\times$ 10^{13} particles/ml (Table II).

Infectious Center Assay/Fluorescent Cell Assay

Protocol

A modification of the previously published protocol[7] was used to measure the ability of the virus to infect C12 cells, unpackage, and replicate. The same cells are used for both ICA and FCA if rAAV encodes the GFP reporter gene.

1. C12 cells are plated in DMEM in a 96-well dish at about 75% confluence. Serial dilutions of the rAAV to be titered are set up as follows: add 250 μl of medium to the first well and add 225 μl of medium to the adjacent wells. Add 2.5 μl of virus to be titered to the first well and serially dilute (10$\times$ steps) by transferring 25 μl per dilution to the adjacent well, being certain to change tips after each dilution. Add adenovirus (Ad5) to the cells at a multiplicity of infection (MOI) of 20. Leave a few wells as "adenovirus only" controls.

2. At 40 hr postinfection, cells infected with rAAV-UF5 are visually scored using a fluorescence microscope. Green fluorescence is monitored using high sensitivity CHROMA filter 41012 High Q FITC.

3. To perform ICA, set up a 12-port filter manifold built to hold 4-mm diameter filters as follows: first, wet two pieces of Whatman (Clifton, NJ) paper with PBS and apply it to the entire manifold. Next, apply a nylon DNA transfer membrane (MagnaGraph; Micron Separations, Westboro, MA) wetted with PBS to the Whatman paper. Tighten the assembled manifold and fill each well of the manifold with 5 ml of PBS. (Alternatively, individual glass-fritted 4-mm diameter disk filter holders fitted into a vacuum manifold can be used.)

[7] S. K. McLaughlin, P. Collis, P. L. Hermonat, and N. Muzyczka, *J. Virol.* **62,** 1963 (1988).

4. Detach cells from the 96-well dish by pipeting vigorously eight times and apply each infected cell sample to one well of the manifold. Wash the well with 200 μl of PBS and apply this also to the appropriate well. After transferring all of the cells to the filter manifold, allow 5 min to pass before gentle vacuum (about 1 cm of water) is applied.

5. Allow the nylon membrane to air dry for 5 min on Whatman 3MM paper. Denature the DNA on the filter for 5 min in 0.5 M NaOH, 1.5 M NaCl and then blot on Whatman paper. Neutralize the membrane for 5 min in 1.5 M NaCl, 0.5 M Tris-HCl, pH 7.8, and blot on Whatman paper. Rinse in 4$\times$ SSC (1$\times$ SSC is 0.15 M NaCl plus 0.015 M sodium citrate) for 30 sec and air dry for 10 min. Microwave the nylon membrane on a high setting for 4 min to fix the DNA as described above. *Caution:* Be sure to have about 300 ml of water in a separate beaker in the microwave or the membrane may catch on fire.

6. Prehybridize the membrane in 7% (w/v) SDS–0.25 M NaHPO$_4$ (pH 7.2)–1 mM EDTA (pH 8.0) at 65° for at least 1 hr before adding the probe. The probe is a ^{32}P-labeled DNA (any fragment of the DNA in the rAAV cassette being titered). Denature about 50 ng of a random-primed labeled probe (specific activity of 10^9 cpm/μg), add to 15 ml of hybridization buffer (see above) in a glass hybridization cylinder, and hybridize at 65° for 12 hr. Wash the membrane twice for 1 hr each in the same buffer at 60°, dry, and expose the filter to standard X-ray film. An example of such autoradiographs was published earlier.[1] To calculate the infectious titer of the rAAV preparation, the number of positive dots on the filters (optimally 20–200 positive dots per filter) is counted and corrected for the dilution of the stock used on that filter.

Analysis of Vector Purity by SDS–Gel Electrophoresis

To investigate the purity of the final vector stock further, virus was analyzed using a 12% SDS polyacrylamide gel for 5 hr at 200 V under standard buffer conditions and visualized by silver staining (Fig. 3A). Tenfold serial dilutions were loaded into three separate wells. In addition to three major bands representing rAAV capsid proteins V1, VP2 and VP3, some low molecular weight bands are visible in lane 1. These proteins appeared to be products of proteolysis of viral capsid proteins, since the Western blotting analysis performed on a separate segment of the gel revealed that all low-molecular peptides hybridized to anti-capsid antibodies (Fig. 3B). The serial dilution experiment over a 100-fold range illustrates that the vector is 99.9% pure.

Analysis of Vector Composition by Continuous Iodixanol Density Gradient

To examine the density of viral particles a new analytical procedure has been introduced. The virus-containing fractions from the POROS column were pooled

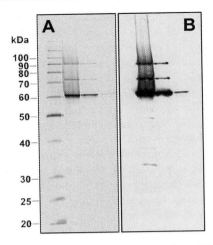

FIG. 3. SDS-protein gel analysis of purified NRS rAAV. Tenfold serial dilutions were loaded into three separate lanes in duplicate. The duplicate portions were stained with either silver to visualize proteins (Panel A) or Western blotted with the monoclonal antibody B1[9] against all three AAV capsid proteins (Panel B). Lane 1 in panel A contains protein standards where molecular masses are shown on the left in kDa.

and analyzed in a continuous gradient of iodixanol. The gradient is formed by mixing virus with 30% iodixanol prepared in PBS containing 0.9 M MgCl$_2$. The sample is transferred into a quick-sealing tube, placed in a Ti-70 rotor, and spun at 70,000 rpm for 8 hr at 15°. After centrifugation the sample is fractionated by puncturing the tube at the bottom and collecting 1 ml fractions.

Fractions are analyzed by FCA (Fig. 4A) and by polyacrylamide gel electrophoresis (Fig. 4B). Two distinct peaks of rAAV are seen upon silver stain analysis: fractions 6 through 8 and fractions 13 through 15. The virus peak that bands in the gradient at a refractive index of about 1.425 η contains most of the infectious virus (Fig. 4A, filled diamonds), whereas the more prominent peak in the middle of the gradient contains mostly empty, noninfectious particles. To compare two viral stocks processed by the current method and the method described earlier[1,2] we subjected an rAAV-GFP vector preparation that had been purified by an iodixanol step gradient and conventional heparin affinity chromatography to the iodixanol continuous gradient described above. As expected, the empty particle peak is no longer seen in the viral stock (Fig. 4C) since it was separated away during the preceding iodixanol density step gradient purification. The only peak that is seen, fractions 6 through 9, consists of fully infectious particles, as judged by FCA (Fig. 4A, open circles). The additional bands seen in fraction 22 are contaminating cellular proteins.

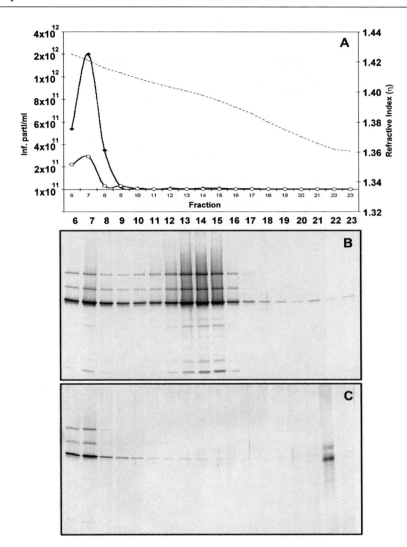

FIG. 4. Continuous iodixanol gradient analysis of purified rAAV. (A) Fluorescent cell assay analysis of gradient fractions. Filled diamonds; infectious profile of NRS rAAV fractions (separated as shown in Panel B); open circles, rAAV-UF5, prepurified as described earlier by iodixanol step gradient/heparin affinity chromatography[1] and subsequently separated as shown in Panel C. A plot of refractive index of gradient fractions is shown by a dotted line. (B) Silver-stained SDS/protein gel analysis of iodixanol fractions of the NRS rAAV separated by continuous gradient. (C) Similar analysis of iodixanol fractions of the prepurified rAAV-UF5 rAAV.

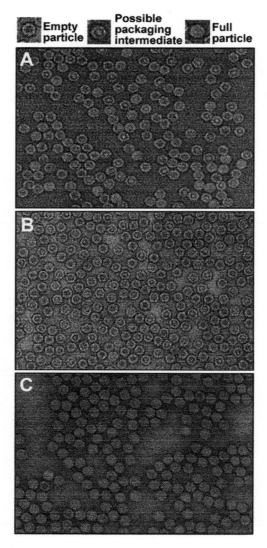

FIG. 5. Electron microscopy analysis of NRS rAAV particles. Electron micrographs were obtained using a Zeiss EM-10A transmission electron microscope operating at 80 kV accelerating voltage. The magnification factor is 49,500. (A) Electron micrograph of rAAV purified by three successive chromatography steps shown in Fig. 1. The preparation consists of a mixture of full particles (uniformly stained), empty particles (filled circle inside a particle), and a possible packaging intermediate (open circle inside a particle). (B) Electron micrograph of virus pooled in fractions 13–15 of iodixanol gradient (Fig. 4B). The preparation essentially consists of empty particles and packaging intermediates. (C) Electron micrograph of virus pooled in fractions 6–8 of iodixanol gradient (Fig. 4B). The preparation consists predominantly of full particles.

It is worth noting that separation of full and empty particles does not improve the physical-to-infectious particle ratio of a given stock, since both titering assays used in this protocol (DBA and RTPA) are based on quantification of packaged genomes, rather than on the assay of assembled particles. Removal of empty particles, however, improves the overall quality of a viral preparation by decreasing the capsid antigen burden of the stock and eliminating a competitor for cell surface receptors.

Analysis of Virus Structure by Electron Microscopy

To confirm the structure of infectious and noninfectious particles, two peaks from the continuous iodixanol gradient are concentrated by POROS HPLC chromatography as described earlier.[1] Following chromatography, EM analysis of concentrated samples is performed as described below and shown in Fig. 5. The sample is prepared by placing 5 μl of purified virus stock on support films of Formvar/Carbon 400 mesh copper grids (Ted Pella, Inc.) for 1 min. Excess sample is removed by blotting with a filter paper. The sample is then stained with 5 μl of 2% uranyl acetate for 10 sec and excess stain is removed as described above.

rAAV purified by three successive chromatography steps consists of a mixture of full (uniformly stained) and empty (filled circle inside a particle) particles, as well as a possible packaging intermediate (open circle inside a particle) (see Fig. 5A and captions above). This particle structure is consistent with the data documented by Grimm et al.[8] After separation in an iodixanol gradient as shown in Fig. 4, the virus pooled in fractions 13–15 consists essentially of empty particles and packaging intermediates (Fig. 5B). The virus pooled in fractions 6–8 of the iodixanol gradient consists predominantly of full particles (Fig. 5C).

Conclusions

Here we describe an improved protocol adapted for large-scale production of a preclinical grade rAAV. This protocol consists of three sequential chromatography purification steps resulting in highly purified (99.9% pure) and infectious (particle-to-infectivity ratios less than 10) vector preparations. In addition, we describe a new centrifugation procedure that allows the separation of full and empty particles. The described protocol was successfully implemented for the production the National Reference Standard rAAV vector, sponsored by the National Gene Vector Laboratory.

[8] D. Grimm, A. Kern, K. Rittner, and J. A. Kleinschmidt, *Hum. Gene Ther.* **9**, 2745 (1998).
[9] C. E. Wobus, B. Hugle-Dorr, A. Girod, G. Petersen, M. Hallek, and J. A. Kleinschmidt, *J. Virol.* **74**, 9281 (2000).

Acknowledgments

The authors express their gratitude to Dr. Verlander-Reed and to Melissa Lewis in the Electron Microscope Core Facility at the University of Florida for their assistance in obtaining the electron micrographs in this publication. This work was supported by a grant from the NCRR/National Gene Vector Laboratory.

Section IV
Lentivirus

[25] Gene Transfer to the Brain Using Feline Immunodeficiency Virus-Based Lentivirus Vectors

By COLLEEN S. STEIN and BEVERLY L. DAVIDSON

Introduction

Neurodegenerative diseases of the central nervous system (CNS) can display restricted or widespread CNS pathology. Examples of disorders demonstrating extensive CNS involvement include Alzheimer's disease,[1] the neuronal ceroid lipofuscinoses (CLN or NCL),[2,3] and the mucopolysaccharidoses (MPS types I to VII).[4] Disorders with relatively restricted neuronal cell loss include Parkinson's disease[5] and motor neuron diseases such as amyotrophic lateral sclerosis (ALS)[6] and the spinocerebellar ataxias (SCA).[7] For many neurodegenerative diseases, gene therapy approaches are being investigated, and may prove viable treatment options. Studies in our own laboratory include investigation of CNS gene transfer strategies for treatment of CLN, MPS, ALS, and the SCA. The CLN and MPS are inherited lysosomal storage diseases characterized by abnormal lysosomal storage deposits. CNS manifestations of these diseases include advancing neuronal dysfunction resulting in progressive cognitive and visual deterioration. ALS is characterized by degeneration of motor neurons in the spinal cord and brain stem, followed by further degeneration of corticospinal neurons in the cerebral cortex. ALS patients manifest increasingly severe muscular weakness and eventual death due to neuromuscular respiratory failure. Approximately 5% of ALS is caused by mutations in superoxide dismutase (SOD1) while 95% are of unknown origin. The SCA are inherited disorders characterized by selective loss of cerebellar neurons leading to motor dysfunction and ataxia.

Gene transfer strategies are often based on delivery of a correct cDNA copy of the affected gene to the relevant cells *in vivo*. For many of the lysosomal storage diseases, the affected gene encodes a soluble lysosomal protein that when overexpressed can be secreted and taken up by neighboring cells via mannose or

[1] E. Braak, K. Griffing, K. Arai, J. Bohl, H. Bratzke, and H. Braak, *Eur. Arch. Psychiatry Clin. Neurosci.* **249,** 14 (1999).

[2] L. Peltonen, M. Savukoski, and J. Vesa, *Curr. Opin. Genet. Dev.* **10,** 299 (2000).

[3] M. J. Bennett and S. L. Hofmann, *J. Inherit. Metab. Dis.* **22,** 535 (1999).

[4] E. F. Neufeld and J. Muenzer, *in* "The Metabolic and Molecular Bases of Inherited Disease" (C. R. Scriver, A. L. Beaudet, W. S. Sly, and D. Valle, eds.), p. 2465. McGraw-Hill, New York, 1995.

[5] M. C. Bohn, *Mol. Ther.* **1,** 494 (2000).

[6] L. J. Martin, A. C. Price, A. Kaiser, A. Y. Shaikh, and Z. Liu, *Int. J. Mol. Med.* **5,** 3 (2000).

[7] H. Y. Zoghbi and H. T. Orr, *Annu. Rev. Neurosci.* **23,** 217 (2000).

mannose 6-phospate receptors. As a result, transduction of a small proportion of CNS cells can potentially mediate widespread correction. The task is more difficult with a membrane-integral lysosomal protein, where prevention and/or reversal of neurodegeneration may necessitate delivery of the gene to most affected CNS cells.

In instances where the defective protein performs a dominant negative function, or deleterious function, provision of a normal version of the gene may be without beneficial effect. Several of the SCA fall into this category; expanded polyglutamine tracts create a toxic gain of function in the encoded protein.[7] Similarly the SOD1 mutation in ALS is a gain of function mutation leading to accumulation of toxic reactive metabolites.[6] Gene transfer approaches for such disorders are aimed at either down-regulating expression of the mutated gene or countering its toxic effects. Neuroprotective strategies such as transfer of genes expressing neurotrophic factors or anti-apoptotic molecules could be undertaken to prevent or delay neuronal cell death, and may find general application to neurodegenerative diseases.[8-13]

Targeting of neurons requires use of a vector that can transduce fully differentiated nondividing cells. *In vivo* rodent studies have found that although recombinant adenoviruses based on serotype 5 can infect neurons *in vivo,* their preference is for astrocytes,[14-16] and the vector-associated inflammatory response often curtails transgene expression.[15,17] Commonly used retroviral vectors based on oncoretroviruses show strict requirements for cell division,[18] and thus are not be useful for targeting neurons. Retroviral vectors based on lentiviruses, such as human immunodeficiency virus (HIV) or feline immunodeficiency virus (FIV), possess nuclear import mechanisms and demonstrate ability to transduce nondividing, fully

[8] D. L. Choi-Lundberg, Q. Lin, Y.-N. Chang, C. M. Hay, H. Mohajeri, and B. L. Davidson, *Science* **275,** 838 (1997).

[9] J. H. Kordower, M. E. Emborg, J. Bloch, S. Y. Ma, Y. Chu, L. Leventhal, J. McBride, E. Y. Chen, S. Palfi, B. Z. Roitberg, W. D. Brown, J. E. Holden, R. Pysalski, M. D. Taylor, P. Carvey, Z. Ling, D. Trono, P. Hantraye, N. Déglon, and P. Aebisher, *Science* **290,** 767 (2000).

[10] D. L. Choi-Lundberg and M. C. Bohn, *in* "Stem Cell Biology and Gene Therapy" (P. J. Quesenberry, G. S. Stein, B. Forget, and S. Weissman, eds.), John Wiley & Sons, New York, 1996.

[11] M. Caleo, M. C. Cenni, E. Menna, M. Costa, L. Zentilin, and M. Giacca, *Proc. Soc. Neurosci.* **1,** 122.11 (abstract) (2000).

[12] R. Dalal, F. E. Samson, and Z. Suo, *Proc. Soc. Neuroscience* **1,** 307.8 (abstract) (2000).

[13] M. Yamada, T. Oligino, M. Mata, J. R. Goss, J. C. Glorioso, and D. J. Fink, *Proc. Natl. Acad. Sci. U.S.A.* **96,** 4078 (1999).

[14] J. M. Alisky, S. M. Hughes, S. L. Sauter, J. M. Alisky, S. M. Hughes, S. L. Sauter, D. J. Jolly, T. W. Dubensky, P. D. Staber, J. A. Chiorini, and B. L. Davidson, *NeuroReport* **11,** 2669 (2000).

[15] U. Blömer, L. Naldini, T. Kafri, D. Trono, I. M. Verma, and F. H. Gage, *J. Virol.* **71,** 6641 (1997).

[16] K. Moriyoshi, L. J. Richards, C. Akazawa, D. D. M. O'Leary, and S. Nakanishi, *Neuron* **16,** 255 (1996).

[17] K. Kajiwara, A. P. Byrnes, H. M. Charlton, M. J. Wood, and K. J. Wood, *Hum. Gene Ther.* **8,** 253 (1997).

[18] P. F. Lewis and M. Emerman, *J. Virol.* **68,** 510 (1994).

differentiated cell types including postmitotic neurons, *in vitro* and *in vivo*.[15,19–21] Lentiviral vector gene transfer to the brain is accompanied by minimal if any vector-associated inflammatory response, and transgene expression is stable.[15] Thus lentivirus-based vectors show promise for neuronal gene transfer. Our recent studies have investigated the use of recombinant FIV-based vectors for direct gene transfer to the MPS VII mouse brain cerebrum[22] as a model for gene therapy application for widespread neurodegenerative diseases. We have also investigated the feasibility of recombinant FIV as a vector for gene transfer to cerebellar neurons,[14] for potential therapeutic use in degenerative diseases of the cerebellum.

Wild-type FIV causes an immunodeficiency disease in cats and, despite prevalent exposure, is not known to cause infection or disease in humans.[23] Like other retroviruses, FIV is an enveloped virus, with a single stranded RNA genome. Cell entry of enveloped viruses is a fusogenic process directed by the virus envelope protein, resulting in delivery of nucleocapsids either directly into the cytoplasm or into intracellular vesicular compartments. Reverse transcription, mediated by the retroviral enzyme reverse transcriptase (RT), then proceeds to generate double-stranded DNA. In oncoretroviruses, such as the Moloney murine leukemia virus (MLV, used commonly in recombinant form as a gene transfer vector), access of the double-stranded DNA preintegration complex to the host cell genome requires dissolution of the nuclear membrane, as occurs during cell mitosis. In contrast, lentiviral preintegration complexes are equipped to mediate nuclear import, thus enabling integration of genetic material into the genomes of nondividing as well as dividing cells.

The native FIV genome contains the basic retroviral *gag, pol,* and *env* open reading frames (Fig. 1A) for production of matrix and nucleocapsid proteins, RT, polymerase and integrase proteins, and envelope proteins, respectively.[24] The FIV genome is simpler than the HIV genome, and encodes only three accessory proteins: vif, orf2, and rev.[25] Reports by Poeschla *et al.,*[20] Johnston *et al.,*[19] and Curran *et al.*[26] describe construction of recombinant FIV gene transfer vectors that can be produced at high particle titer and are able to efficiently transduce dividing and nondividing cells. These protocols utilize a triple plasmid transfection system. The packaging plasmid provides all the necessary viral proteins in

[19] J. C. Johnston, M. Gasmi, L. E. Lim, J. H. Elder, J. K. Yee, D. J. Jolly, K. P. Campbell, B. L. Davidson, and S. L. Sauter, *J. Virol.* **73,** 4991 (1999).

[20] E. M. Poeschla, F. Wong-Staal, and D. J. Looney, *Nat. Med.* **4,** 354 (1998).

[21] G. Wang, V. Slepushkin, and J. Zabner, *J. Clin. Invest.* **104,** R55 (1999).

[22] C. S. Stein, A. I. Brooks, J. A. Heth, T. W. Dubensky, Jr., S. L. Sauter, K. Townsend, D. A. Cory-Slechta, M. A. Howard, H. J. Federoff, and B. L. Davidson, *Proc. Soc. Neurosci.* **1,** 668.7 (2000).

[23] J. K. Yamamoto, H. Hansen, E. W. Ho, T. Y. Morishita, T. Okuda, T. R. Sawa, R. M. Nakamura, and N. C. Pedersen, *J. Am. Vet. Med. Assoc.* **194,** 213 (1989).

[24] T. Miyazawa, K. Tomonaga, Y. Kawaguchi, and T. Mikami, *Arch. Virol.* **134,** 221 (1994).

[25] K. Tomonaga and T. Mikami, *J. Gen. Virol.* **77,** 1611 (1996).

[26] M. A. Curran, S. M. Kaiser, P. L. Achacoso, and G. P. Nolan, *Mol. Ther.* **1,** 31 (2000).

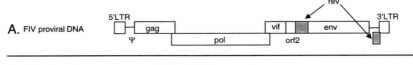

B. Production of recombinant FIV particles

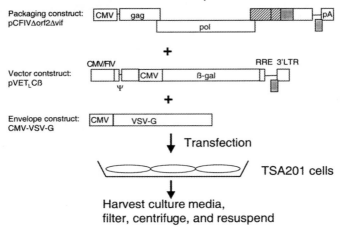

FIG. 1. Recombinant FIV particle production. (A) FIV proviral DNA contains three large open reading frames (*gag, pol,* and *env*), and three small regions encoding accessory proteins (*vif, orf2,* and *rev*). The packaging signal (Ψ) extents from noncoding sequences just downstream of the 5′ LTR to coding sequences within the 5′ portion of *gag*. (B) For particle production, TSA201 cells are transfected with packaging, vector, and envelope plasmids. Full-length transcripts from the vector plasmid contain an intact packaging signal and are packaged into enveloped particles. At 24, 36, and 72 h posttransfection the culture media (containing particles) is harvested and filtered, and the particles are concentrated by centrifugation.

trans (except for the env protein), but it is not packaged into particles. The vector plasmid contains the transgene expression cassette and retains the minimal *cis*-acting viral sequences to allow efficient genome packaging, reverse transcription, and integration. The third plasmid encodes a chosen envelope protein in *trans*. Most lentiviral vector studies to date have used recombinant vectors pseudotyped with the envelope glycoprotein from vescicular stomatitis virus (VSV-G), because it imparts particle stability and mediates widespread cellular tropism.[27] This chapter outlines techniques for (1) production of VSV-G-pseudotyped recombinant FIV particles encoding the reporter protein bacterial β-galactosidase (FIVβgal), (2) vector injection in mouse brain striatum and cerebellum, and (3) immunofluorescent staining for determination of cell types transduced.

[27] J. C. Burns, T. Friedmann, W. Driever, M. Burrascano, and J. K. Yee, *Proc. Natl. Acad. Sci. U.S.A.* **90,** 8033 (1993).

Materials

Recombinant FIV Particle Production

Plasmids: packaging construct, vector construct, and envelope construct (Fig. 1B): The FIV packaging construct (pCFIVΔorf2Δvif)[19] was derived from the FIV molecular clone p34TF10. The packaging construct retains full-length *gag* and *pol*, and *rev*, but contains a deletion in the *env* gene, and mutations in *vif* and *orf2* genes. The vif and orf2 accessory proteins have been determined to be dispensable both for recombinant particle production and for transduction of brain *in vivo*. The third accessory gene, *rev*, is retained since rev is necessary for efficient nuclear export of long and full-length transcripts. The native 5′ LTR has been replaced by the human cytomegalovirus (CMV) immediate early promoter/enhancer, and the 3′ LTR has been replaced with the simian virus 40 polyadenylation signal. The vector construct (pVET$_L$Cβ)[19] carries the transgene of interest driven by an internal promoter. Here the *lacz* gene, encoding the reporter protein bacterial β-galactosidase, is driven by the CMV promoter. The 5′ U3 of FIV has been replaced with the CMV promoter. pVET$_L$Cβ has been deleted of all viral coding regions with the exception of a 5′ portion of *gag* (which contains part of the packaging signal). Last a third plasmid, the envelope construct, provides the envelope protein in trans. Here we use a plasmid encoding the VSV-G directed by the CMV promoter.[28]

TSA201 cells[29] maintained in exponential growth in DMEM-10. TSA201 cells were derived from 293 human embryonic kidney epithelial cells (ATCC CRL-1573) and grow as monolayers. They are easily lifted with trypsin for passage.

Incubator at 37° with 5% CO_2
HEPES buffered saline (HBS):

HEPES	5.0 g
NaCl	8.0 g
KCl	0.37 g
$Na_2HPO_4 \cdot 7H_2O$	0.188 g
Glucose	1.0 g

Bring to 1 liter in ddH$_2$O, pH to 7.1 with concentrated NaOH, filter sterilize and store at 4°
2.5 M CaCl$_2$
DMEM: Dulbecco's modified Eagle's medium
DMEM-10: DMEM with 10% fetal calf serum, 100 U/ml penicillin, and 100 μg/ml streptomycin

[28] J.-K. Yee, T. Friedmann, and J. C. Burns, *In* "Methods in Cell Biology," Vol. 43, p. 99. Academic Press, San Diego, 1994.
[29] E. S. Shen, G. M. Cooke, and R. A. Horlick, *Gene* **156,** 235 (1995).

DMEM-2: DMEM with 2% fetal bovine serum, 100 U/ml penicillin, and 100 μg/ml streptomycin

Bottle-top filters (500 ml capacity, Nalgene PES low protein binding)

Lactose buffer: phosphate-buffered saline (PBS), pH 7.4 (Sigma P-3813) with 40 mg/ml lactose, filter-sterilized

Sorvall centrifuge RC 26 Plus, with SLA 1500 rotor

Sorvall centrifugation bottles (250 ml capacity)

Determination of Transduction Titer by X-Gal Staining

HT-1080 cells (ATCC CRL-121) maintained in exponential growth in DMEM-10. These cells are derived from human fibrosarcoma and grow as monolayers. They are easily lifted with trypsin for passage.

Incubator at 37° with 5% CO_2

Six-well tissue culture dishes

DMEM-2 and DMEM-10 (see above)

Polybrene stock: 8 mg/ml in ddH_2O, filter sterilized

DMEM-2/Polybrene: On the day of use, dilute the Polybrene stock 1/2000 in DMEM-2, for a final Polybrene concentration of 4 μg/ml

Dilution tubes (3.5 ml polystyrene sterile tubes)

1% glutaraldehyde in PBS

KC Mixer:

35 mM K_3 $Fe(CN)_6$	5.74 g
35 mM K_4 $Fe(CN)_6 \cdot 3HO$	7.35 g
2 mM $MgCl_2$	1 ml of 1 M stock
0.01% sodium desoxycholate	0.5 ml of 10% stock
0.02% NP40	1.0 ml of 10% stock

Add to PBS for a final volume of 500 ml. Do not add the $MgCl_2$ until the previous ingredients have dissolved. Filter through a 0.45 μm bottle top filter and store in the dark at 4°.

X-Gal (5′-bromo-4-chloro-3-indolyl-β-D-galactopyranoside) stock: Dissolve at 40 mg/ml in N',N-dimethyl formamide, and store at −20° (does not freeze).

X-Gal solution (staining solution): Dilute X-Gal stock to 1 mg/ml in the KC mixer just prior to use. Prewarming separately the KC mixer and the X-Gal stock at 37° before mixing can help to avoid precipitation that can occur with longer incubation times.

Vector Injection into Adult Mouse Striatum or Cerebellum

Ketamine/xylazine mix: combine 8.9 ml of sterile PBS with 1 ml of 100 mg/ml ketamine and 0.1 ml of 100 mg/ml xylazine for final concentrations of 10 mg/ml ketamine and 1 mg/ml xylazine.

Insulin syringe with attached 28-gauge needle

Underpads (blue chucks)

Eye ointment (bacitracin zinc and polymyxin B sulfate ophthalmic ointment)

Razor to shave mouse head

Iodine tincture

Surgical instruments (autoclaved): scalpel and blades, curette, forceps, scissors, hemostat

Dry sterilizer (Germinator 500, Cellpoint Scientific, Inc.)

Small sterile beaker with 70% ethanol

Surgical suture (4.0 silk) with attached needle (1/2 × 17 mm)

Small animal stereotaxic frame with mouse adaptor (KOPF Instruments)

Microprocessor-controlled pump (World Precision Instruments UltraMicro-Pump) mounted onto the stereotaxic frame

Microprocessor-based controller (World Precision Instruments Micro 1 model)

Small box (we use tip box) taped to the base of the stereotaxic frame, to lay the mouse on

10 μl glass Hamilton syringe with a removable stainless steel blunt-ended 33-gauge needle (Hamilton) (for striatal injection)

5 μl glass Hamilton syringe with a pulled glass microcapillary tube attached (for cerebellar injection): the microcapillary tubes are pulled out on microcapillary needle puller, and cemented to the syringe with Superglue

Drill (Dremel Moto-Tool Model 395 Type 5, or equivalent) with 003 bit

Dissecting microscope (optional)

Lactated Ringer's

3 ml syringe with 25-gauge needle

Recovery mouse cage with clean towel bedding

Lamp with 75 watt bulb, suspended 2.5 feet above the recovery cage

Mouse brain atlas (*The Mouse Brain in Stereotaxic Coordinates,* K. B. I. Franklin and G. Paxinos, Academic Press, 1997)

FIVβgal: concentrated recombinant FIV particles encoding the β-galactosidase transgene, kept on ice

Animal Perfusion/Fixation

Ketamine/xylazine mix (see recipe above)

PBS

2% paraformaldehyde in PBS: in a fume hood, add 2 g paraformaldehyde (powder grade) to near 100 ml of PBS. Cover and heat to 50–60° with stirring. Do not boil. Add a few drops of concentrated NaOH to help dissolution, and bring pH back to 7.4 with concentrated HCl. Bring final volume to 100 ml with PBS. Filter through a 0.45 μm bottle top filter and cool on ice. Store at 4°. Use within 1 week of preparation.

Peristaltic perfusion pump
Flexible rubber hosing 2–3 mm in diameter
Butterfly IV catheter, 23 or 25 gauge
Instruments: scissors, forceps
Styrofoam board
Spray bottle with 70% ethanol

Dissecting Out the Brain

Razor blade or scalpel
Ronguers (Roboz Surgical Instruments)
Curette (spoonlike instrument)
30% Sucrose in PBS
Plastic molds (Peel-a-ways from VWR Scientific Products)
OCT (Tissue-Tek)
Dry ice/95% ethanol bath

Tissue Sectioning

Cryostat
Glass slides (Fisher Superfrost glass plus microscope slides)
Slide boxes
For thick, free-floating sections:
    Forceps
    24-well tissue culture dishes
    PBS
    PBS/azide: PBS with 0.02% sodium azide

Dual Immunofluorescent Staining

Staining of Cryosections on Glass Slides
Humidity chamber (wet paper towels in closed shallow container)
PAP pen (Electron Microscopy Sciences)
PBS
BSA: bovine serum albumin (Sigma, ELISA grade)
Block: PBS with 10% normal (goat) serum (from same species as the secondary antibody), 0.1% Triton X-100, and 0.02% sodium azide (can use 3% BSA in place of or in addition to the serum)
Primary antibody diluent: 3% BSA in PBS with 0.1% Triton X-100 and 0.02% azide
Wash buffer: PBS containing 1% normal (goat) serum, 0.1% Triton X-100, and 0.02% azide
Secondary antibody diluent: PBS containing 1% normal (goat) serum, 0.1% Triton X-100, and 0.02% azide

Primary antibodies:

Polyclonal rabbit anti-β-galactosidase (BioDesign B59136R, 10 mg/ml). Dilute 1/1500.

Mouse mAb to NeuN (Chemicon MAB377, 1 mg/ml). Dilute 1/200.

Mouse mAb to GFAP (Sigma C9205, 1 mg/ml, Cy3 conjugate). Dilute 1/2000.

We purchase this antibody directly conjugated to Cy-3; it works well and eliminates the need for a secondary antibody.

Mouse monoclonal antibody to calbinbin-D-28K (Sigma, ascites). Dilute 1/3000.

Secondary antibodies:

Goat anti-rabbit Alexa 488 (Molecular Probes). Dilute 1/200.

Goat anti-mouse IgG lissamine-rhodamine (Jackson ImmunoResearch). Dilute 1/200.

Glass coverslips

Mounting media: gelmount, or "Vectashield" (from Vector Laboratories, Inc.)

Confocal microscope and associated software

Staining of Free-Floating Thick Sections. The materials are the same as for staining of sections on glass slides except:

The humidity chamber is not needed.

Reagents containing Triton X-100 are prepared with 0.3% Triton X-100 instead of 0.1%.

24-well tissue culture dishes are needed.

A small paint brush is needed.

Glass slides are needed to put sections onto after staining.

Methods

Recombinant FIV Particle Production

We routinely produce recombinant FIV particles utilizing the triple plasmid system and constructs described by Johnston *et al.*[19] (Fig. 1). This system of particle production involves concurrent transfection of TSA201 cells with three plasmids, followed by harvest of particle-containing culture media, and concentration of particles. Below are the steps for preparing 3 ml of concentrated vector particles from an 18 plate (150 mm diameter) transfection.

1. Seed TSA201 cells into 18 150-mm diameter flat-bottom tissue culture dishes at a density of 10^7 cells per dish.

2. The next day, add 34 ml of HBS to two 50-ml conical tubes. The HBS should be at room temperature.

3. Add 225 μg of the packaging plasmid, 337.5 μg of the vector plasmid, and 112.5 μg of the envelope plasmid to each HBS-containing tube and vortex well.

4. Slowly add 1.7 ml of 2.5 M CaCl$_2$ to each tube while slowly vortexing or shaking the HBS-plasmid mixture. CaCl$_2$ should be at room temperature.

5. Let the solution stand for 25 min to allow precipitate formation. The solution should appear slightly translucent or cloudy.

6. Add both tubes of precipitate directly to 200 ml of DMEM. Briefly mix.

7. Aspirate off the medium from the cells (9 plates at a time).

8. Gently pipette the transfection solution onto the cells (15 ml per dish), and return the cells to the incubator.

9. Four to 6 h after transfection aspirate off the medium and provide 15 ml of fresh DMEM-10 per dish.

10. Collect the medium (containing vector particles) at 24, 36, and 72 h, replacing this medium with fresh DMEM-10 at the 24 and 36 h time points. At each collection, filter the medium through a 0.45 μm bottle-top filter, and store short-term at 4°, or long-term in 50 ml aliquots at −80°.

11. Just prior to intended use, concentrate the particles by centrifuging the collected medium at 4° for 16 h at 7400g (7000 rpm in the Sorvall centrifuge with the SLA 1500 rotor). Carefully pour off the supernatants, drain well, and resuspend particles in lactose buffer. We typically resuspend the particles produced from an 18-plate transfection into a total 3 ml volume.

Comments on FIV Vector Particle Production

1. We have found that transfections work best with CsCl-purified plasmids.

2. We routinely suspend our concentrated particles in PBS/lactose as this buffer is physiological and acceptable for *in vivo* use, and the lactose has a stabilizing effect. However, other buffers maintaining similar pH and salt concentrations may be suitable alternatives. These include saline, TNE (50 mM Tris-HCL, pH 7.8, 130 mM NaCl, 1 mM EDTA), and culture medium.

3. Lentivirus vector-containing culture media suffers minimal loss in transduction titer when stored at −80°, while centrifuge-concentrated vector loses approximately one log in titer after freezing. Substantial loss of titer of concentrated preparations is also observed within 24 h of storage at 4°. Thus we routinely concentrate the virus immediately prior to use.

Determination of Particle Concentration and Transduction Titers

Standardized methods of determining the concentration of FIV-based lentivirus particles and the concentration of transduction-competent particles within lentiviral preparations have not yet been established. An ELISA assay can be used to measure FIV p24 nucleocapsid antigen,[19,30] as an indicator of particle concentration. For determination of transduction titers, we transduce HT-1080 cells with serially diluted particles, followed by quantification of transduced cells either by staining and counting the transgene-expressing cells, or by quantitative PCR detection of pro-vector DNA sequences.

Determination of Transduction Titer by X-Gal Staining

1. One day prior to transduction, seed a 6-well flat-bottom plate with 2 million HT-1080 cells per well in DMEM-10.

2. For transduction, make a 10-fold dilution series of concentrated FIV as follows. Place 1.485 ml of DMEM-2/polybrene in first tube and 1.35 ml in tubes 2 through 6. Add 15 μl of virus to the first tube and vortex. Transfer 150 μl from the first to the second tube and vortex, and so on for the remaining tubes.

3. Remove medium from wells. Add 1 ml of each dilution to separate wells. For wells #1 through 6, the dilution factors will thus be 10^2, 10^3, 10^4, 10^5, 10^6, and 10^7. Return the cells to the incubator.

4. Incubate the HT-1080 cells for 72 h, then feed with 1 ml of DMEM-10.

5. Incubate a further 24 h. Rinse monolayers with PBS, and then fix with 1% glutaraldehyde for 5 min at room temperature.

6. Wash once with PBS and add enough X-Gal solution to cover the cells.

7. Incubate 6 h in the dark at 37° without CO_2.

8. Rinse the wells 2× with PBS, and add PBS to cover the cells.

9. Using an inverted microscope, count the number of β-galactosidase-expressing (blue) cells in each well. The first two or three wells will often have too many positive cells to count. Doublets or small clusters of cells are counted as one, as they likely originated by division of a single transduced cell.

10. For each well, multiply the total blue cell count by the dilution volume (1 ml) and by the dilution factor. Determine the mean of all the wells. This number represents the transducing units per ml (TU/ml) of concentrated virus. Using this method, our concentrated FIVβgal preparations typically contain 5×10^7 to 10^9 TU/ml.

Determination of Transduction Titer by Quantitative PCR.
A PCR-based assay system for titering the FIV vector is described elsewhere.[21] Briefly, HT-1080 target cells are transduced with serial dilutions of FIV vector preparations. The transduced cells are collected 48 h posttransfection, and genomic DNA is extracted according to standard techniques. Total genomic DNA is quantified by staining with Hoechst dye H33258 and comparison with calf thymus DNA standards, using the CytoFluor II fluorometer (PerSeptive Biosystems, Framingham, MA). Quantitative PCR is performed on 100 ng of each DNA sample, employing a PE ABI

---

[30] G. K. Tilton, T. P. O'Connor, Jr., C. L. Seymour, K. L. Lawrence, N. D. Cohen, P. R. Anderson, and Q. J. Tonelli, *J. Clin. Microbiol.* **28**, 898 (1990).

Prism 7700 system (Perkin-Elmer Corp., Norwalk, CT) and a synthetic oligonucleotide primer set directed against FIV packaging signal sequences yielding an 80-bp product. The resulting fluorescence is used to determine the provector copy number for each HT-1080 DNA sample, and from this the transducing units per ml (TU/ml) of the original concentrated virus is then calculated. Based on the quantitative PCR method, our concentrated FIV preparations typically contain 10^8 to 5×10^9 TU/ml.

Comments on Titering. Titering of transduction-competent particles by staining and counting of transgene-expressing cells can be applied only when a suitable staining method is available. X-Gal staining as described above is a simple and reliable method for detecting β-galactosidase-expressing cells. For other transgene products, immunostaining methods can be applied, assuming there is a suitable antibody available and the HT-1080 cells have low to no endogenous expression of the antigen. X-Gal staining or immunohistochemistry is useful for titer comparison between preparations of the same FIV vector. However, it is not recommended for comparisons between vector preparations carrying different transgenes, as the sensitivities of staining procedures for different transgene products may vary. Titering by PCR requires access to a quantitative PCR system, and effort to optimize the procedure, but once this system is established it can be applied to FIV vectors encoding any transgene. Both titering methods are based on transduction of HT-1080 cells. HT-1080 cell transduction readily occurs with VSV-G-pseudotyped FIV. However, HT-1080 cell transduction may be less efficient if the FIV vector is pseudotyped with a different envelope. Many envelope proteins exhibit restricted cellular tropism, and it might be necessary to identify more suitable cell lines for determining transduction titers of alternate pseudotypes.

Pseudotyping with Alternative Envelopes

We have described here the production of a lentivirus vector pseudotyped with the VSV-G envelope protein. VSV-G lends enhanced structural stability to vector particles, allowing for concentration by ultracentrifugation with minimal loss of infectivity.[27] In addition, VSV-G reportedly mediates viral entry via interaction with a membrane phospholipid component,[31] rather than with a specific cell surface receptor protein, and this imparts VSV-G-pseudotyped vectors with an extremely broad host-cell range, including cells of nonmammalian species.[27] These features have resulted in the common use of VSV-G for pseudotyping MLV and lentiviral gene transfer vectors.

However, there can be disadvantages to using the VSV-G envelope. Despite the wide tropism, VSV-G pseudotyped vectors do not always mediate efficient transduction. For example, VSV-G pseudotyped FIV was unable to transduce polarized airway epithelial cells when applied to the apical surface.[21] Furthermore,

[31] P. Mastromarino, C. Conti, P. Goldoni, B. Hauttecoeur, and N. Orsi, *J. Gen. Virol.* **68,** 2359 (1987).

the observations that VSV-G can be toxic to cells,[27] and that VSV-G pseudotyped vectors are inactivated upon exposure to human serum,[32] may limit the clinical application of VSV-G-pseudotyped vectors. Lastly, particularly for direct *in vivo* applications, widespread tropism can be a drawback when it is desirable to restrict transduction to specific cell types or tissues.

Several reports describe successful pseudotyping of HIV-based lentiviral vectors with alternative envelope proteins. Early HIV-1-vectors were pseudotyped with the MLV amphotropic envelope,[33,34] and the human T-cell leukemia virus envelope,[35] as well as the VSV-G envelope glycoprotein.[34] HIV vectors incorporating envelope glycoproteins from Marburg or Ebola viruses have been produced and show a wide range of infectivity.[36]

Attempts at pseudotyping can also meet with failure. Efficient incorporation of envelope proteins into particles depends on appropriate interaction of cytoplasmic envelope sequences with encapsidated genomes. Maedi-visna virus envelope constructs were unable to pseudotype MLV- or HIV-derived vector particles.[37] Pseudotyping of an HIV vector with the gibbon ape leukemia virus (GaLV) envelope glycoprotein was similarly unsuccessful.[38] However, use of a chimeric construct substituting the cytoplasmic tail of the GaLV envelope with that of the MLV amphotropic envelope protein resulted in the formation of infective GaLV-pseudotyped HIV vectors.[38]

Since FIV and HIV are related lentiviruses, successful pseudotyping of HIV with the aforementioned envelope proteins may predict similar results with FIV vectors. Preliminary studies in our laboratory indicate that FIV vectors can be pseudotyped with unmodified envelope proteins from amphotropic MLV, Marburg virus, and Ross river virus (unpublished observations), and we are currently testing these FIV pseudotypes for cell tropisms in the CNS. Also worthy of mention are a few alternative envelopes described in the literature that, although not reported in combination with lentivirus vectors, have successfully been used to pseudotype MLV vectors. Lymphocytic choriomeningitis virus envelope-pseudotyped MLV particles were structurally stable, withstanding concentration by ultracentrifugation, and exhibited cross-species tropism.[39] Pseudotyping of MLV with the Sendai

[32] N. J. DePolo, J. D. Reed, P. L. Sheridan, K. Townsend, S. L. Sauter, and D. J. Jolly, *Mol. Ther.* **2**, 218 (2000).

[33] K. A. Page, N. R. Landau, and D. R. Littman, *J. Virol.* **64**, 5270 (1990).

[34] J. Reiser, G. Harmison, S. Kluepfel-Stahl, R. O. Brady, S. Karlsson, and M. Schubert, *Proc. Natl. Acad. Sci. U.S.A.* **93**, 15266 (1996).

[35] N. R. Landau, K. A. Page, and D. R. Littman, *J. Virol.* **65**, 162 (1991).

[36] S. Y. Chan, R. F. Speck, and M. C. Ma, *J. Virol.* **74**, 49233 (2000).

[37] U. Zeilfelder and V. Bosch, *J. Virol.* **75**, 548 (2001).

[38] J. Stitz, C. J. Buchholz, M. Engelstadter, W. Uckeret, U. Bloemer, I. Schmitt, and K. Cichutek, *Virology* **273**, 16 (2000).

[39] H. Miletic, M. Bruns, K. Tsiakas, B. Vogt, R. Rezai, C. Baum, K. Kuhlke, F. L. Cosset, W. Ostertag, H. Lother, and D. van Laer, *J. Virol.* **73**, 6114 (1999).

virus fusion protein imparted a restricted tropism for asialoglycoprotein receptor-bearing cells, indicating promise for hepatocyte-directed gene delivery.[40]

The successful production of FIV-based vectors pseudotyped with an alternative envelope may simply involve following the steps outlined above, with a plasmid encoding high-level expression of the alternative envelope protein used in place of the VSV-G envelope plasmid. High-titer production of particles that display appropriate cell tropism would suggest successful pseudotyping. Low titers may indicate poor incorporation of envelope proteins into the particles, or poor receptor expression on the TSA201 or HT-1080 cells. The former problem of incompatibility may be overcome with selective mutations in the envelope protein or construction of chimeric envelopes, whereas the latter problem would necessitate usage of more appropriate cells lines in place of TSA201 and/or HT-1080 cells.

Vector Injection into Adult Mouse Striatum

We have studied the potential of gene transfer to the CNS for treatment of the neurological aspects of lysosomal storage diseases using the MPS VII mouse model. MPS VII is caused by a deficiency of β-glucuronidase, a soluble lysosomal enzyme involved in the degradation of glycosaminoglycans.[4] Like other lysosomal proteins, β-glucuronidase can be secreted and taken up by neighboring cells.[41,42] In the MPS VII (β-glucuronidase-deficient) mouse a single intra-striatal injection of most vectors encoding β-glucuronidase results in local transduction, yet the enzyme penetrates much of the injected hemisphere to provide widespread correction of pathology in both glia and neurons.[43] Vectors encoding the reporter protein bacterial β-galactosidase (a nonsecreted protein) are useful for discerning the vector transduction volume and cell types transduced (see "dual immunofluorescent staining").

Here we describe the steps involved in intrastriatal injection of FIVβgal. Sterile techniques are adhered to for brain injections. The bench space is covered with clean underpads (blue chucks), the surgical instruments are autoclaved, the mouse is prepped with iodine, and the syringe/needle is flushed and incubated with 70% ethanol and rinsed with sterile PBS prior to use. The instruments are dry-sterilized between animals.

1. Preprogram the pump for the appropriate delivery volume and rate.

[40] M. Spiegel, M. Bitzer, A. Schenk, H. Rossmann, W. J. Neubert, U. Seidler, M. Gregor, and U. Lauer, *J. Virol.* **72,** 5296 (1998).

[41] R. M. Taylor and J. H. Wolfe, *Exp. Cell Res.* **214,** 606 (1994).

[42] P. Moullier, V. Marechal, O. Danos, and J. M. Heard, *Transplantation* **56,** 427 (1993).

[43] A. Ghodsi, C. Stein, T. Derksen, G. Yang, R. D. Anderson, and B. L. Davidson, *Hum. Gene Ther.* **9,** 2331 (1998).

2. Anesthetize mice with ketamine/xylazine mix (0.1 ml per 10 g body weight), injected ip. Full anesthesia is achieved in approximately 10 to 15 min. Anesthesia is assessed by firmly pinching a toe; if the mouse exhibits a pedal reflex, anesthesia is not adequate and a second (1/3) dose of ketamine/xylazine is given.

3. Apply eye ointment to eyes. This helps prevent drying, as mice do not blink while under anesthesia.

4. Shave the dorsal aspect of the head and wipe with iodine tincture (be very careful to avoid eyes).

5. Using a scalpel, make a midline sagittal incision through the scalp over the skull to reveal the coronal, sagittal, and lambdoid sutures.

6. Lay the mouse on a small box (or something similar), to bring it up to an appropriate height for the mouse adaptor on the stereotaxic instrument. Firmly secure the head by means of the adaptor palate bar and the nose clamp. Adjust the angle of the palate bar to maintain the skull on a level plane.

7. With the (empty) syringe set into the injector unit, identify bregma as the zero coordinate. For a striatal injection, locate coordinates 0.4 mm rostral and 2.0 mm lateral to bregma, and mark the skull using an ultrafine-tip marker.

8. Remove the syringe, and drill small burr hole through the skull at the mark, being careful to maintain the integrity of the dura.

9. Draw vector into the syringe and set the syringe into the injector unit. Return the syringe to the set coordinates. Prior to lowering the syringe, "fast inject" just until a drop of vector is seen at the needle tip, to ensure proper syringe loading.

10. Lower the syringe through the burr hole until the tip of the needle touches the dura. Slowly insert the needle into the brain parenchyma to a depth of 3.0 mm. Start the microprocessor-controlled pump (preprogrammed to deliver 5 μl at a maximal rate of 500 nl per minute).

11. After vector delivery, leave the needle in place for 5 min, then slowly withdraw (over 5 min).

12. Remove the mouse from the stereotactic apparatus and close the incision with 4.0 silk suture.

13. Inject the mouse with 1 ml of lactated Ringer's subcutaneously to provide hydration during recovery.

14. Place the mouse in a cage with absorbent bedding, warmed by a lamp (75 watt bulb) 2.5 feet above the cage. Drape half the cage with a towel to allow the mouse voluntary escape from the warmth.

15. Monitor the mouse until it is ambulatory, then return it to animal housing.

Comments on Vector Injection into the Striatum. Consultation of a mouse brain atlas is necessary for identification of the skull sutures, bregma, and appropriate injection coordinates. Visualization of bregma may be facilitated by use of a dissecting microscope. Also, use of a dissecting microscope during injection allows visualization of the underlying dura while drilling, reducing the chance of puncture.

Vector Injection into Adult Mouse Cerebellum

The spinocerebellar ataxias (SCA) and other degenerative diseases of the cerebellum are potentially treatable by gene transfer if sufficient numbers of the affected neuronal types can be transduced. Cerebellar Purkinje cells are particularly affected, and as these neurons provide neuronal output for cerebellar control of movement, delivery of a therapeutic gene to these neurons may prevent Purkinje cell dysfunction/death and restore motor function. We have found that one injection of recombinant FIV encoding the reporter β-galactosidase into a cerebellar lobule transduces close to 100% of the Purkinje cells in that lobule.[14,44] Thus an FIV vector encoding a therapeutic molecule has potential clinical value.

For injection of a vector into a cerebellar lobule of an adult mouse, the procedure is essentially as outlined above for "injection into the striatum," with the following modifications. The scalp incision is more posterior with the lambdoid suture centrally revealed. The syringe is cemented with a pulled glass microcapillary tube, rather than a stainless steel needle. This type of needle is very fine, with exterior and interior diameters smaller than those of a 33-gauge stainless steel needle. The cerebellum is organized as a repeated lobular structure. However, the orientation of the lobules may differ between animals. Therefore, rather than using anterior–posterior and lateral coordinates, we choose a lobule in which to inject, then lower the needle into the brain parenchyma at a depth of 1 or 2 mm to dispense the vector (1 to 2 μl) into either the cerebellar cortex or the deep cerebellar nuclei, respectively.

Animal Perfusion/Fixation

The perfusion apparatus should be prepared prior to anesthetizing the first mouse. Set up the pump such that the rubber tubing draws PBS from a container and flushes it out the other end, which is attached to a 23 or 25 gauge butterfly i.v. catheter. Flush the line with PBS to remove all air bubbles from the circuit.

1. Anesthetize the mouse with 0.15 ml of ketamine/xylazine mix per 10 g of body weight, administered i.p. Once deep anesthesia is achieved (see above), place the mouse on its back on a Styrofoam board and tape each paw down.

2. Wet the fur over the abdomen and thorax with 70% ethanol. Using scissors, make a midline cut through the abdominal wall up from the intestinal area to just below the diaphragm, being careful to avoid cutting the underlying viscera. Continue the cut laterally, through the diaphragm (which causes an immediate tension pneumothorax), then up through the ribcage to expose the heart.

3. Using fine scissors, make a small snip in the right atrium.

---

[44] J. M. Alisky, S. M. Hughes, and S. L. Sauter, *NeuroReport* **11,** back cover (2001).

4. Quickly insert the i.v. catheter into the left ventricle, which can be done easily by poking through the apex of the heart.

5. Turn on the perfusion pump and flush PBS through the mouse, running the blood volume out through the cut right atrium. For a mouse, about 20–30 ml of PBS are sufficient to flush the blood volume out, which is evident by the rapid clearing of the liver from bright red to tan or brown.

6. At this point, turn off the pump and move the hose from the PBS into the paraformaldehyde. It is very important to turn off the pump while changing solutions or air emboli may be introduced that block penetration of the fixative.

7. Restart the pump and run 30–50 ml of paraformaldehyde through the mouse. Adequate fixation is indicated by the occurrence of rapid rigor mortis, due to protein cross-linking. If the mouse is limp, try carefully repositioning the tip of the needle within the left ventricle. Common errors include positioning of the needle bevel in the heart wall or entering the left atrium. Usually repositioning will save the perfusion.

Comments on Animal Perfusion/Fixation. For perfusion/fixation we use a peristaltic pump, which mimics the normal cardiac physiology, to provide thorough permeation of fixative without incurring tissue damage. Syringes or gravity (hanging solutions from the ceiling or from a high shelf) can also be used to deliver the fixative.

Removing the Brain

1. Following perfusion/fixation, decapitate the mouse with a razor blade or scalpel.

2. Use a blade to score the skull anterior to the olfactory bulbs and along the sagittal suture.

3. Using Ronguers, grasp and pull off pieces of skull from around the foramen magnum and surrounding the cerebellum. Continue pulling off pieces of skull until the brain is fully exposed laterally and as far anterior as the olfactory bulbs. Note that this procedure must be done carefully, to avoid damage to the underlying brain. A dissecting microscope may allow better visualization.

4. Use forceps and scalpel to carefully remove remaining dura from the brain surface. Scoop out the brain with a small curette (spoonlike instrument) or similar tool. The adhering olfactory nerves, optic nerves, and cranial nerves should give way with gentle pressure; if not, small scissors can be used.

5. Immerse the brain in paraformaldehyde and postfix overnight at 4°.

6. Transfer into 30% sucrose/PBS. The brain floats initially, but sinks as sucrose diffuses into the tissue. Place at 4° and allow the brain to sink. An intact mouse brain requires approximately 36 h to sink. Permeation of tissue with sucrose serves as a cryoprotectant.

7. Place the brain in a plastic mold. Cover the brain with OCT, and set the mold in a shallow dry ice/95% ethanol bath (do not allow ethanol to mix with the OCT), until the OCT is frozen (whitens upon freezing). Note that sections are cut from the bottom face of the block. Therefore before freezing, the brain can be cut first with a razor blade in the plane desired for sectioning (coronal or sagittal) and the pieces placed with cut sides down in the mold, providing a flat surface for cryosectioning. Alternatively, the whole brain can be immersed in OCT and held with forceps in a desired position until the OCT begins to harden around it.

8. Store the OCT-blocked brain at $-80°$ until it can be cryosectioned.

Tissue Sectioning

1. Peel the plastic mold away from the OCT-blocked brain.

2. Using a cryostat, cut sections (from 8 to 50 μm thick) off the block.

3. If sections are 20 μm or less in thickness, they can be captured directly from the cryostat stage onto glass slides. Two to three sections can be placed on one slide. Slides are kept in the cryostat until the brain has been completely sectioned. The slides are transferred to slide boxes and stored at $-20°$.

4. For thicker sections (20 to 50 μm), use forceps to grab an edge of the section, and quickly place into a well of a 24-well tissue culture dish with PBS. Store these sections in PBS (short-term) or in PBS/azide (long-term), at $4°$.

Dual Immunofluorescent Staining

We routinely use dual immunofluorescent staining and confocal microscopy for examination of transduced cells. Primary antibodies with specificity for neurons or glia can be used along with an antibody specific for the transgene product. We present methods for dual staining of β-galactosidase and NeuN (a marker common to most neuronal types), β-galactosidase and GFAP (a type II astrocyte marker), and β-galactosidase and calbindin (a common marker for cerebellar Purkinje neurons).

The following staining procedure is described for staining of 8 to 20 μm thick cryosections on glass slides. The procedure can be adapted to thicker sections (20 to 50 μm) (see "modifications for staining of free-floating sections").

All incubations are performed in a humidity chamber (which can be as simple as wet paper towels in a closed plastic container), and unless otherwise stated, incubations are at room temperature. Also, to minimize photobleaching of fluorochromes, all incubations subsequent to the addition of fluorochrome-labeled antibodies should be done in the dark.

1. Bring cryosections on glass slides to room temperature, about 10 min. This dries the sections and sticks them firmly to the slides.

2. Encircle sections with a PAP pen. Reagents are subsequently applied to slides such that the tissue is completely covered, but the reagent is held within the PAP-defined circle. Usually 100 μl volumes are sufficient to cover one section.

3. Pipette PBS onto slides and incubate for 5 min, to clear the tissue of OCT.

4. Aspirate off the PBS and add Block onto slides. Incubate for 1 h at room temperature.

5. Aspirate off Block and add primary antibodies (diluted in primary diluent). Incubate overnight at 4°.

6. Aspirate off primary antibodies and wash slides 3× with wash buffer, 10 min each.

7. Aspirate off wash and add fluorochrome-labeled secondary antibody(s). Incubate for 1–2 h.

8. Aspirate off secondary antibodies and wash slides 3× with wash buffer, 10 min each.

9. Aspirate off wash buffer and wash 1× with PBS, 10 min.

10. Aspirate off PBS. Add a small volume of mounting medium, and place a glass coverslip over the tissue. Excess mounting medium is removed by inverting the slide onto absorbent paper and applying gentle pressure. Analyze by confocal or standard fluorescence microscopy. Store in dark.

Modifications for Staining of Free-Floating Sections. The following modifications are applied for staining of thick (20 to 50 μm) free-floating sections.

1. Sections are stained (steps 3–9 above) free-floating in wells of a 24-well tissue culture dish. Sections are transferred to fresh wells for each step (use paint brush to transfer).

2. The Triton X-100 in the reagents is increased to 0.3%.

3. The primary antibody incubation time is extended to 48 h.

4. After staining, the sections are placed onto glass slides using a paintbrush. The sections are then coverslipped as described above.

Comments on Immunofluorescent Staining. The above staining procedure was optimized for FIV vector-transduced murine brain. It is important to emphasize that preliminary staining should be performed independently on each antibody to determine optimal concentrations (which can vary from lot to lot or with storage), and that appropriate control stainings be performed to confirm specificity. Negative controls should include concurrent staining of sections with (1) isotype control antibody in place of the antigen-specific primary antibody; and (2) secondary antibody alone (no previous addition of primary antibody). Fluorescence on these sections results from nonspecific binding and indicates that similar nonspecific binding is also occurring on the noncontrol (test) sections, making it difficult to distinguish true antigen-specific staining. Nonspecific binding can be reduced or eliminated by titrating down the antibody concentrations and/or modifying the block. When the primary antibody is of the same species as the tissue, the antibody may bind "nonspecifically" to Fc receptor-bearing cells. We have not observed this to be a problem with naive or FIV-injected murine brain, but we have observed high background staining in association with inflammation incurred by other vectors. In this case, it is wise to use mouse-derived primary antibodies directly conjugated to fluorochromes, and to include normal mouse serum in the block.

Analysis of Staining by Fluorescence Microscopy

To determine which cell types are transduced, the stained slides are analyzed first by standard upright fluorescence microscopy to evaluate the quality of the stains, and then by confocal microscopy. With confocal microcopy, fluorescence emission is collected from a subcellular plane (typically 0.3 to 0.5 μm) within a

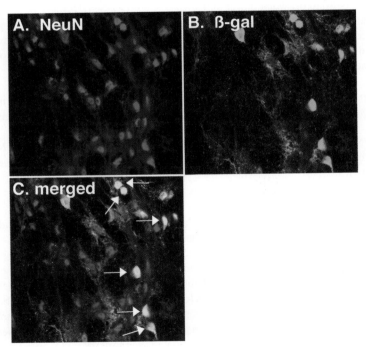

FIG. 2. Dual immunofluorescent staining for β-galactosidase and neurons. FIVβgal was injected into the striatum of a mouse brain. Eight weeks later, 50-μm coronal brain sections were dual stained with a neuronal-specific antibody (A, NeuN, red fluorescence) and a β-galactosidase-specific antibody (B, β-gal, green fluorescence). Overlapping red and green fluorescence appears yellow in the merged images (C) and identifies β-galactosidase-expressing transduced neurons (white arrows).

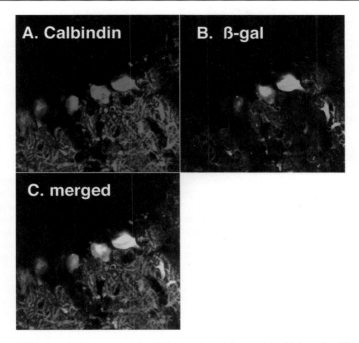

FIG. 3. Dual immunofluorescent staining for β-galactosidase and Purkinje cells. FIVβgal was injected into a cerebellar lobule of a mouse brain. Six weeks later, 50-μm coronal brain sections were dual stained with a calbindin-specific antibody (A, calbindin, red fluorescence), to detect the calbindin positive Purkinje neurons, and a β-galactosidase-specific antibody (B, β-gal, green fluorescence) to detect FIVβgal-transduced cells. Overlapping red and green fluorescence appears yellow in the merged images (C) and identifies β-galactosidase-expressing transduced Purkinje cells.

tissue section. This allows the user to unambiguously determine whether the signals from two different fluorochromes are arising from the same cell. Collection of a series of emission data from top to bottom of the tissue section allows for three-dimensional reconstruction and morphological assessment of the stained cells. Procedures for use of a confocal microscope and analysis of the data vary with the microscope and associated software.

When tissues are dual stained with one antibody that is FITC- or Alexa 488-conjugated (green fluorescence) and another that is rhodamine or CY-3-conjugated (red fluorescence), colocalization of signals appears yellow. Figure 2 shows colocalization of β-galactosidase and the neuronal cell marker NeuN in striatal neurons. Figure 3 shows colocalization of β-galactosidase and calbindin in cerebellar Purkinje cells.

Concluding Remarks

Gene transfer to the CNS for the treatment of neurodegenerative diseases is gradually progressing toward clinical application, as development of vectors advances. Successful therapy will ultimately depend on the use of a suitable vector and an effective therapeutic transgene. FIV-based vectors mediate stable gene transfer to cerebral neurons and cerebellar Purkinje cells and provide a sound basis for the further design of vectors applicable to human CNS disorders.

Acknowledgments

The authors thank our collaborators previously of Chiron Technologies, Center for Gene Therapy (Sybille Sauter, Julie Johnston, Phil Sheridan, Day Townsend, Tom Dubensky, and Doug Jolly) for development of the FIV constructs and vector production system, and for their scientific contribution. For technical support we thank Stephanie Hughes, Inês Martins, and The University of Iowa Gene Transfer Vector Core (Patrick Staber) and Cell Morphology Core. We thank Joe Alisky for critical comments and technical contributions, and Christine McLennan for manuscript preparation. These protocols were developed with support from the NIH (NS34568, HD33531, DK54759) and the Roy J. Carver Trust.

[26] Generation of HIV-1 Derived Lentiviral Vectors

By ANTONIA FOLLENZI and LUIGI NALDINI

Introduction

Gene transfer vectors derived from lentiviruses, such as human immunodeficiency virus 1 (HIV-1), are potentially important tools in clinical gene therapy for the treatment of a wide array of inherited and acquired diseases, because of their ability to transduce several types of target cells independently of their proliferation status both *ex vivo* and *in vivo*.

Lentiviral vectors (LV) are replication-defective, hybrid viral particles made by the core proteins and enzymes of a lentivirus, and the envelope of a different virus, most often the vesicular stomatitis virus (VSV) or the amphotropic envelope of MLV.[1] The use of VSV.G yields high vector titer and results in greater stability of the vector particles.[2]

[1] L. Naldini, U. Blomer, P. Gallary, D. Ory, R. Mulligan, F. H. Gage, I. M. Verma, and D. Trono, *Science* **272,** 263 (1996).

[2] J. C. Burns, T. Friedmann, W. Driever, M. Burrascano, and J. K. Yee, *Proc. Natl. Acad. Sci. U.S.A.* **90,** 8033 (1993).

Lentiviruses have a complex genome. In addition to the structural genes (*gag, pol,* and *env*) common to all retroviruses, lentiviruses also contain two essential regulatory (*tat* and *rev*) and several accessory genes involved in modulation of viral gene expression, assembly of viral particles, and structural and functional alterations in the infected cells. The lentiviruses replication is mediated, in part, by cis-acting viral sequences, which do not encode proteins. Most cis-acting sequences are essential for LV functioning and are usually included in the transfer construct (the part of LV which integrates in the host cell genome and encodes the gene of interest). The trans-acting sequences encode three groups of proteins: structural, regulatory, and accessory. Lentiviral vectors are defective for replication, so only the early steps of the lentivirus life cycle (attachment, entry, reverse transcription, nuclear transport, and integration) must be maintained in an LV. Since the early steps do not depend on viral protein synthesis, all trans-acting genes can be excluded from the transfer vector that encodes only the gene of interest. The required trans-acting proteins are provided by packaging plasmids (Fig. 1).

Early HIV-1 based vectors were generated by introducing deletions in the HIV-1 genome, rendering it replication-defective and separating helper from transfer functions in different constructs.[3-5] These vectors had very low titers, were restricted to CD4+ targets, and carried the risk of generating wild-type HIV-1 by recombination. In the first generation of hybrid LV described by Naldini *et al.,*[1] three constructs were used to produce pseudotyped HIV-derived vectors. The HIV genome was employed as a packaging construct under heterologous transcriptional control, after major deletions of the packaging signal and of the env gene. A separate construct expressed a heterologous envelope that was incorporated into the vector particles (pseudotyping) and allowed entry into a wide range of target cells. The third construct encoded the transfer vector RNA that did not contain any viral genes. This design broadened the possible uses of LV and made it impossible to generate wild-type HIV-1 by recombination. These vectors yielded titers of about 10^5 TU/ml when pseudotyped with the MLV envelope and about 5×10^5 when pseudotyped with VSV.G. The biosafety of these vectors was futher advanced by deleting unnecessary genes (*Vif, Vpr, Vpu,* and *Nef*) from the packaging construct, resulting in the production of so called multiply attenuated or "second-generation" packaging systems without reducing transduction efficiency.[6] In the most advanced vector design ("third-generation"), the essential regulatory genes *tat* and *rev* were eliminated from the core packaging construct without reducing transduction efficiency of the vector.[7-9] This construct expresses the HIV-1 *gag*

[3] M. Poznansky, A. Lever, L. Bergeron, W. Haseltine, and J. Sodroski, *J. Virol.* **65,** 532 (1991).
[4] T. Shimada, H. Fujii, H. Mitsuya, and A. W. Nienhuis, *J. Clin. Invest.* **88,** 1043 (1991).
[5] C. Parolin, T. Dorfman, G. Palu, H. Gottlinger, and J. Sodroski, *J. Virol.* **68,** 3888 (1994).
[6] R. Zufferey, D. Nagy, R. J. Mandel, L. Naldini, and D. Trono, *Nat. Biotechnol.* **15,** 871 (1997).
[7] V. N. Kim, K. Mitrophanous, S. M. Kingsman, and A. J. Kingsman, *J. Virol.* **72,** 811 (1998).

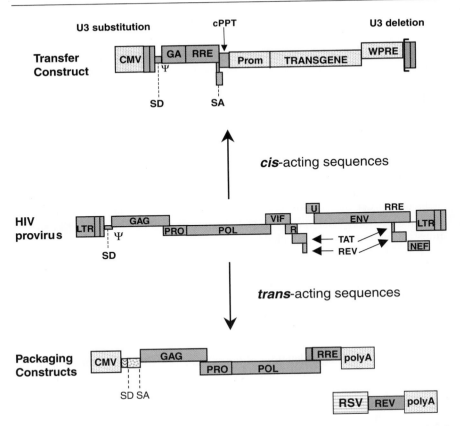

FIG. 1. Structure of the HIV-1 provirus and derived constructs. The typical design of lentiviral vectors is based on separating the sequences acting in *cis* from the viral genes that express proteins working in *trans* during transduction and that must be expressed in the producer cells. In the transfer construct are located all the *cis* functions allowing efficient encapsidation, reverse transcription, nuclear transport, and integration of the vector into the target cells. The packaging constructs express the indicated viral genes from the cytomegalovirus (CMV) and Rous sarcoma virus (RSV) promoters. Abbreviations: Long terminal repeats (LTRs); the 5′ LTR of transfer construct is chimeric, with the enhancer-promoter of the CMV replacing the U3 region and the 3′ LTR has an almost complete deletion of the U3 region (sin-18). Splice donor and acceptor sites (SD and SA), the packaging sequence (Ψ), 5′ portion of *gag* gene with truncated reading frame including extended packaging signal (GA), Rev Response Element (RRE), polyadenylation site (polyA), central polypurine tract (cPPT), internal promoter (Prom), posttranscriptional regulatory element from the genome of the woodchuck hepatitis virus (Wpre). The relevant parts of the constructs are shown.

and *pol* genes in the presence of the Rev protein, which is provided in trans by a separate nonoverlapping construct. Strong costitutive promoters upstream from the vector transcriptional start site could replace the function of the *tat* gene.[7,10] These results achieved an important biosafety gain because any replication-competent recombinant (RCR) virus that could be generated during vector manufacturing would lack all factors essential for HIV-1 replication and virulence *in vivo*. These vectors reached titers of about 10^7 TU/ml (end point titer on HeLa) demonstrating that the progress in the safety of vector design was accompanied by improvements in the production and transduction efficiency of the vectors.

Using the new minimal packaging constructs, it is now possible to generate stable producer cell lines that allow better characterization and scale-up of vector manufacturing. Various attempts at generating stable packaging cell lines for HIV-derived vectors have been reported, including the use of inducible regulatory elements to limit expression of toxic components from early generations of packaging constructs.[11–13]

Considerable progress in performance and biosafety was also achieved by the successful generation of self-inactivating LV.[14–16] These vectors are produced by transfer constructs that carry an almost complete deletion in the U3 region of the HIV 3′ LTR. The U3 region contains the viral enhancer and promoter. Transduction of vector deleted in the 3′ U3 results in duplication of the deletion in the upstream LTR and the transcriptional inactivation of both LTRs. Self-inactivating lentiviral vectors could be produced with infectious titers and levels of transgene expression driven by an internal promoter similar to vectors carrying wild-type LTRs, thus allowing the use of tissue specific or inducible promoter without interference from the activity of the LTR. Modifications have also been made to improve the efficiency of gene delivery and expression in target cells. One of these modifications involved inserting the posttranscriptional regulatory element from the genome of the woodchuck hepatitis virus (Wpre) into the 3′ end of the transfer vector. The Wpre acts at the posttranscriptional level, by promoting nuclear export of transcripts and/or by increasing the efficiency of polyadenylation of the nascent

---

[8] H. Mochizuki, J. P. Schwartz, K. Tanaka, R. O. Brady, and J. Reiser, *J. Virol.* **72**, 8873 (1998).

[9] M. Gasmi, J. Glynn, M. J. Jin, D. J. Jolly, J. K. Yee, and S. T. Chen, *J. Virol.* **73**, 1828 (1999).

[10] T. Dull, R. Zufferey, M. Kelly, R. J. Mandel, M. Nguyen, D. Trono, and L. Naldini, *J. Virol.* **72**, 8463 (1998).

[11] T. Kafri, H. van Praag, F. H. Gage, and I. M. Verma, *Mol. Ther.* **1**, 516 (2000).

[12] N. Klages, R. Zufferey, and D. Trono, *Mol. Ther.* **2**, 170 (2000).

[13] A. Consiglio, A. Quattrini, S. Martino, J. C. Bensadoun, D. Dolcetta, A. Trojani, G. Benaglia, S. Marchesini, V. Cestari, A. Oliverio, C. Bordignon, and L. Naldini, *Nat. Med.* **7**, 310 (2001).

[14] H. Miyoshi, U. Blomer, M. Takahashi, F. H. Gage, and I. M. Verma, *J. Virol.* **72**, 8150 (1998).

[15] R. Zufferey, T. Dull, R. J. Mandel, A. Bukovsky, D. Quiroz, L. Naldini, and D. Trono, *J. Virol.* **72**, 9873 (1998).

[16] T. Iwakuma, Y. Cui, and L. J. Chang, *Virology* **261**, 120 (1999).

transcript,[17] thus increasing the total amount of mRNA in cells. The addition of the Wpre to lentiviral vectors resulted in a substantial improvement in the level of transgene expression from several different promoters, both *in vitro* and *in vivo*.[17–19]

A recently described modification of the transfer vector backbone restored infectivity of vector particles to the level of wild-type virus. An additional copy of the polypurine tract (cPPT) is present in the middle of the HIV genome, included in the *pol* gene. The cPPT acts as a second origin of the *plus* strand DNA synthesis and introduces a partial strand overlap in the middle of the genome. The introduction of the cPPT sequence in the transfer vector backbone strongly increased the nuclear transport and the total amount of genome integrated into the DNA of target cells.[20,21]

The ability of LV to infect nondividing and terminally differentiated cells, including neurons, retinal photoreceptor cells, several types of epithelial tissues, and pluripotent hematopoietic stem cells (for a review see Ref. 22), may expand the range of disease targets and facilitate the development of gene therapy approaches.

Lentiviral vectors based on HIV-2 are also being developed.[23,24] It has been demonstrated that HIV-2-derived vectors can be cross-packaged with HIV-1 packaging functions.[25,26]

A different strategy to address the biohazards of LV was to develop vectors from nonprimate lentiviruses following similar approaches to those used for HIV-derived vectors. The rationale of this approach lies in the fact that parental viruses are not pathogenic to humans. However, these vectors must be pseudotyped by a different envelope and allow efficient entry and integration into human cells to be useful for gene therapy purposes. Thus, the possible biohazards of a RCR breaking out during their production remain to be investigated. Vectors derived from simian immunodeficiency virus (SIV),[27] feline immunodeficiency virus

[17] R. Zufferey, J. E. Donello, D. Trono, and T. J. Hope, *J. Virol.* **73**, 2886 (1999).

[18] N. Deglon, J. L. Tseng, J. C. Bensadoun, A. D. Zum, Y. Arsenijevic, D. A. Pereira, R. Zufferey, Trono D., and P. Aebischer, *Hum. Gene Ther.* **11**, 179 (2000).

[19] D. Farson, R. Witt, R. McGuinness, T. Dull, M. Kelly, J. P. Song, R. Radeke, A. Consiglio, A. Bukovsky, and I. Naldini, *Hum. Gen Ther.* **12**, 981 (2001).

[20] A. Follenzi, L. E. Ailles, S. Bakovic, M. Geuna, and L. Naldini, *Nat. Genet.* **25**, 217 (2000).

[21] V. Zennou, C. Petit, D. Guetard, U. Nerhbass, L. Montagnier, and P. Charneau, *Cell* **101**, 173 (2000).

[22] E. Vigna and L. Naldini, *J. Gene Med.* **2**, 308 (2000).

[23] E. Poeschla, J. Gilbert, X. Li, S. Huang, A. Ho, and F. Wong-Staal, *J. Virol.* **72**, 6527 (1998).

[24] S. K. Arya, M. Zamani, and P. Kundra, *Hum. Gene Ther.* **9**, 1371 (1998).

[25] E. Poeschla, P. Corbeau, and F. Wong-Staal, *Proc. Natl. Acad. Sci. U.S.A.* **93**, 11395 (1996).

[26] P. Corbeau, G. Kraus, and F. Wong-Staal, *Gene Ther.* **5**, 99 (1998).

[27] D. Negre, P. E. Mangeot, G. Duisit, S. Blanchard, P. O. Vidalain, P. Leissner, A. J. Winter, C. Rabourdin-Combe, M. Mehtali, P. Moullier, J. L. Darlix, and F. L. Cosset, *Gene Ther.* **7**, 1613 (2000).

(FIV),[28] equine infectious anemia virus (EIAV),[29,30] caprine arthritis/encephalitis virus (CAEV),[31] and the bovine jembrana disease virus[32] have been described. Minimal vector systems, similar to those described above for HIV vectors in which the accessory genes were found to be nonessential and were deleted from the packaging construct, have been reported to improve the predicted biosafety and gene transfer efficiency of FIV vectors[33,34] and SIV vectors.[35] However, because of the lower number and yet unclear role of accessory genes in nonhuman lentiviruses, the actual gain in biosafety of these advanced design remains to be established by *in vivo* testing. The availability of appropriate animal models for testing biosafety is a significant advantage of vectors derived from nonhuman lentiviruses.

In this chapter, we describe the methods of production and testing of pseudotyped HIV-derived vectors using a third-generation packaging system and an improved version of self-inactivating transfer vector. The methods described can be applied to the production of most versions and types of LV.

Methods

Recombinant Lentiviral Vector Plasmids

The expression constructs used for LV production are maintained in the form of bacterial plasmids and can be transfected into mammalian cells to produce replication-defective virus stocks. We found that to obtain high amounts of plasmids for transfection it is better to transform all vector plasmids in TOP 10 *Escherichia coli* cells and grow bacteria in Luria–Bertani (LB) medium in the presence of carbenicillin. It is important to use pure DNA for transfection. Plasmids should be prepared and purified by double CsCl gradient centrifugation or other commercially available column methods yielding endotoxin-free DNA.

[28] E. M. Poeschla, F. Wong-Staal, and D. J. Looney, *Nat. Med.* **4,** 354 (1998).

[29] J. C. Olsen, *Gene Ther.* **5,** 1481 (1998).

[30] K. Mitrophanous, S. Yoon, J. Rohll, D. Patil, F. Wilkes, V. Kim, S. Kingsman, A. Kingsman, and Mazarakis N., *Gene Ther.* **6,** 1808 (1999).

[31] L. Mselli-Lakhal, C. Favier, T. Da Silva, K. Chettab, C. Legras, C. Ronfort, G. Verdier, J. F. Mornex, and Y. Chebloune, *Arch. Virol.* **143,** 681 (1998).

[32] P. Metharom, S. Takyar, H. H. Xie, K. A. Ellem, J. Macmillan, R. W. Shepherd, G. E. Wilcox, and M. Q. Wei, *J. Gene Med.* **2,** 176 (2000).

[33] J. C. Johnston, M. Gasmi, L. E. Lim, J. H. Elder, J. K. Yee, D. J. Jolly, K. P. Campbell, B. L. Davidson, and S. L. Sauter, *J. Virol.* **73,** 4991 (1999).

[34] M. A. Curran, S. M. Kaiser, P. L. Achacoso, and G. P. Nolan, *Mol. Ther.* **1,** 31 (2000).

[35] P. E. Mangeot, D. Negre, B. Dubois, A. J. Winter, P. Leissner, M. Mehtali, D. Kaiserlian, F. L. Cosset, and J. L. Darlix, *J. Virol.* **74,** 8307 (2000).

Generation of Lentiviral Vectors

This section outlines the calcium phosphate method for cotransfecting 293T cells with four plasmids [third-generation core packaging plasmids: pMDLg/pRRE and pRSV.REV, self-inactivating (SIN) transfer vector plasmid pCCL-SIN-18PPT.Prom.EGFPWpre, and envelope plasmid pMD$_2$VSV.G (Fig. 2) [this last plasmid was constructed by cloning the VSV.G-encoding *Eco*RI fragment from the pMD.G to the pMD-2[1,10,15,20]], for packaging the vector genome carrying the gene of interest into hybrid replication-defective viral particles. LV are traditionally produced by transient cotransfection of human embryonic kidney 293T cells (originally called 293tsA1609ne[36] and derived from 293 cells, a continuous human embryonic kidney cell line transformed by sheared Type 5 adenovirus DNA, by transfection with the tsA 1609 mutant gene of SV40 large T antigen and the Neo[r] gene of *E. coli*), because these cells are good DNA recipients in transfection procedures and the backbones of the vector construct contain SV40 origin of replication.

We routinely use calcium phosphate precipitation to transfect plasmids; however, good transfection efficiency is also obtained by lipid-based methods: 24 hr before transfection seed 5×10^6 293T cells in a 10 cm dish, in IMDM, 10% fetal bovine serum (FBS), penicillin (25 U/ml), streptomycin (25 U/ml). Change medium 2 hr before transfection and prepare the plasmid DNA mix by adding 3 μg envelope plasmid (pMD2-VSV-G), 5 μg of core packaging plasmid (pMDLg/pRRE), 2.5 μg of pRSV-REV, and 10 μg of transfer vector plasmid (pCCLsin18.PPT.Prom.GFP.Wpre) together per dish. The plasmid solution is made up to a final volume of 450 μl with 0.1× TE/H$_2$O (2 : 1) in a polypropylene tube. Finally add 50 μl of 2.5 M CaCl$_2$ and mix the solution by vortexing. The precipitate is formed by dropwise addition of 500 μl of the 2× HBS (HEPES Buffered Saline) solution (280 mM NaCl, 100 mM HEPES, 1.5 mM Na$_2$HPO$_4$, pH 7.12; check pH of 2× HBS carefully) to the 500 μl DNA-TE-CaCl$_2$ mixture while vortexing at full speed. The precipitate should be added to the 293T cells immediately after 2× HBS addition. High magnification microscopy of the cells should reveal a very small granular precipitate of the calcium phosphate–DNA complexes, initially floating above the cell layer, and after incubation in the 37° incubator overnight, visible on the bottom of the plate in the spaces between the cells. The calcium phosphate–DNA precipitate should be allowed to stay on the cells for 14–16 hr, after which the medium should be replaced with fresh medium for vector collection to begin. Collect the cell supernatant at 24 and 48 hr after changing the medium. Centrifuge at 1000 rpm for 5 min at room temperature and filter supernatant through a 0.22-μm pore nitrocellulose filter. Sometimes a third

---

[36] R. B. DuBridge, P. Tang, H. C. Hsia, P. M. Leong, J. H. Miller, and M. P. Calos, *Mol. Cell Biol.* **7**, 379 (1987).

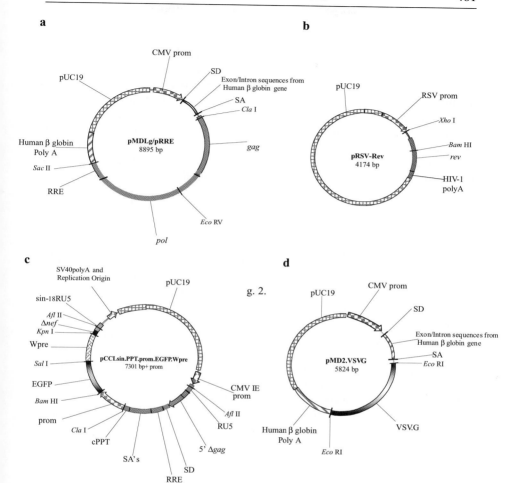

FIG. 2. Schematic drawing of the four constructs required to build VSV-pseudotyped third-generation HIV-1-derived lentiviral vectors. (a) The conditional packaging construct expressing the *gag* and *pol* genes driven the CMV promoter. (b) A nonoverlapping construct, RSV-Rev, expressing the *rev* cDNA under the RSV promoter. (c) The self-inactivating (SIN) transfer construct containing HIV-1 cis-acting sequences and an expression cassette for the transgene (enhanced green fluorescent protein or EGFP) driven by the internal promoter. (d) A fourth construct encoding a heterologous envelope to pseudotype the vector, here shown coding for the protein G of the vesicular stomatitis virus, Indiana serotype (VSV.G) under the control of the CMV promoter. Relevant restriction sites for cloning purposes are shown.

collection of vector after 72 hr can be carried out; in this case it is advisable to coat the dishes with poly-L-lysine before plating 293T cells for transfection.

Transient transfection can be reasonably reproducible by careful standardization of protocols and reagents. However, it may be difficult to scale up and adjust to the strict requirements of good manufacturing practices for producing vectors for human applications. Furthermore, cotransfection may increase the risk of recombination between plasmids, thus detracting from the biosafety devices built into the vector design. On the other hand, transient transfection allows use of the most updated and complex combinations of plasmids, and the short culture time of vector-producing cells may insure the culture against outbreak and propagation of recombinants. When stable packaging cell lines producing satisfactory amounts of late-generation vectors become available, the relative advantages and disadvantages of transient vs stable production will be addressed in detail.

Concentration of Lentiviral Vectors

Pseudotyped lentiviral particles containing the VSV.G glycoprotein have been shown to be quite stable and resistant to ultracentrifugation without significant loss in titer, allowing generation of highly concentrated vector stocks for *in vivo* applications. A variety of different envelope proteins, such as the amphotropic murine leukemia virus (MLV) 4070A or the rabies G glycoprotein, are currently being tested with regard to their capacity to pseudotype LV.[37] To obtain high-titer vector stock, it is best to concentrate the 293T conditioned medium by ultracentrifugation at 50,000g for 140 min at 20°. Discard the supernatant by decanting and resuspend the pellets in a small volume of PBS containing 0.5% BSA, pool in a small tube, and rotate on a wheel at room temperature for 1 hr. Store the concentrate vector from the first collection at 4° until the repeat of the procedure for the second collection the following day. Pool the vector suspension from the two collections, dilute in phosphate-buffered saline (PBS), and concentrate again by ultracentrifugation. Resuspend the final pellet in a very small volume (we usually do 1/500 of the starting volume of medium) of sterile PBS–0.5% bovine serum albumin (BSA). Resuspension of the second pellet requires prolonged incubation on a rotating wheel and pipetting at room temperature. Split into small aliquots (20 μl), store at −80°, and titer after freezing. For third-generation LV vector preparation, titers of more than 10[10] TU/ml can be reached after ultracentrifugation and concentration up to a thousandfold without significant loss of transducing activity.

Vector stock can be also concentrated using ultrafiltration devices such as Centricon Plus-80 following the instructions of the manufacturer and yielding titers of up to 10[9] TU/ml.[37] Other purification methods are based on anion-exchange chromatography. In an example of such application, LV could be concentrated in a continuous flow system from 1 liter of conditioned medium and eluted to free

[37] J. Reiser, *Gene Ther.* **7,** 910 (2000).

from most cellular contaminant yielding titers to 10^9 TU/ml (M. Scherr *et al.,* ASH, abstract 1848, page 430a, 2000).

Titration of Lentiviral Vectors

Relative concentrations of vectors are measured as a titer. The vector titer can be determined as the number of complete particles that are able to transduce cells. The transducing particles usually represent a small percentage of total particles and can vary between different preparations (see Chapter [29] by De Palma and Naldini). The determination of transducing particles is generally subject to great variation related to different methods used in different laboratories. The concentration of infectious particles is estimated by end-point titration of the vector preparation on a standard easily infectable cell line. For a standard titration of a vector containing GFP as a transgene and a ubiquitous promoter such as CMV or PGK we use HeLa cells (end-point titer on HeLa). Other easily infectable cell lines can be used for this purpose such as 293T or HOS cells. The day before the transduction we plate 5×10^4 HeLa cells per well in a 6-well plate. The following day prepare serial 10-fold dilutions of viral stocks in IMDM 10% FBS, approximately ranging from undiluted to 10^{-6} for 293T-conditioned medium and from 10^{-3} to 10^{-8} for concentrated vector. Of each dilution, 0.1 ml is added to the cells. Before titration the medium is replaced with 0.9 ml fresh medium containing 9 μg/ml Polybrene. Incubate the cells at 37° for 16 hr. The day after, change medium and grow the cells for 72 hr at 37°. Wash wells with PBS, add trypsin, and harvest cells with PBS in FACS tubes. Centrifuge, add 1 ml of fixing solution (1% formaldehyde, 2% FBS in PBS), and vortex the tubes (samples are stable at 4° for few days). The titer is performed by fluorescence-activated cytometric analysis for GFP using the FL1 channel on the fluorescence-activated cell sorting (FACS) measuring the percentage of cells expressing EGFP in the total population. The titer is expressed in TU/ml and calculated by the number of cells infected times the percentage of the EGFP(+) cells in the linear range of dilutions (between 1 and 20% of EGFP+ cells) times the corresponding dilution factor and divided by 100. Proof of linearity should be obtained showing that different dilutions in the optimal testing range yield linear increase in transduction frequency. The results of an end-point titration depend on several factors such as the type of promoter driving the expression of the transgene or the presence of regulatory elements. Acceptable titers should be in the range of 10^6–10^7 TU/ml in the 293T-conditioned medium.

HIV-1 p24 Assay

The Gag capsid protein p24 is a major structural component of the HIV-1 core and, consequently, of all HIV-1 derived vectors. The content of Gag proteins in vector particles can be assayed using antibodies recognizing the mature form of the protein and, with less affinity, the Gag precursor. This can be done by immunocapture and well-standardized kits are commercially available that allow specific and reproducible measurement of the Gag p24 concentration in the vector preparation. To perform the assay follow the instructions provided by the manufacturer, except for the sample preparation, as follows: Perform the test using serial dilutions from 10^{-3} to 10^{-6} if you have 293T-conditioned medium or dilutions from 10^{-5} to 10^{-8} if you have concentrated vector. The commercially available tests have a limit of sensitivity in the range of 3 pg/ml. A good vector preparation should have a p24 concentration in the range of 100–500 ng/ml in the 293T-conditioned medium.

The ratio between transduction units (TU) per ml and the nanograms of p24 per ml give the infectivity (or specific transducing activity), a reliable parameter to evaluate the quality of a vector preparation, where the transducing units (TU) per mililiter are obtained by end-point titration on HeLa and should be $>10^4$. A good vector preparation can reach up to 2×10^5 TU per ng of p24.

RNA Slot-Blot

The infectivity of vector particles is influenced by the efficiency of encapsidation of the transfer vector RNA. The size and specific sequence of the expression cassette may represent a limiting factor for the efficiency of RNA encapsidation. Methods to analyze RNA content can be used to verify the packaging of vector RNA into particles. This parameter can be monitored by RNA dot blot analysis. Match 0.22 μm-pore-size filtered vector suspensions for equal content of p24 and microcentrifuge at 14,000 rpm for 1 hr at room temperature in Eppendorf tubes to pellet the particles. Isolate total RNA from the particles, treat RNA with DNase for 30 min at 37°, denature in 50% formamide–17% formaldehyde, heat to 50° for 20 min, mix with 150 μl of 15× SSC, and bind to nitrocellulose by aspiration through a slot-blot apparatus and probe for sequences specific for the vector construct. Serial dilutions of p24 equivalent of sample and reference vectors are tested together with a standard curve of plasmid DNA.

Testing for the Absence of RCR

After preparation, vector batches should be checked that they do not contain RCR, although the occurrence of this event should be extremely rare in the hybrid, split design of late-generation LV. Advanced vector design minimized or abolished sequence overlaps between the different constructs, by deletion of unnecessary sequences[10,15] or by mismatching the viral source of packaging and transfer constructs.[38,39] In other versions of packaging constructs, recoding of viral

[38] S. M. White, M. Renda, N. Y. Nam, E. Klimatcheva, Y. Zhu, J. Fisk, M. Halterman, B. J. Rimel, H. Federoff, S. Pandya, J. D. Rosenblatt, and V. Planelles, *J. Virol.* **73**, 2832 (1999).

[39] J. F. Kaye and A. M. Lever, *J. Virol.* **72**, 5877 (1998).

sequences that have also a structural function maintained their coding potential for viral proteins while abolishing *cis*-acting activity, thus alleviating the risk for recombination. Moreover, optimized codon usage may improve gene expression in transfected cells for vector production.[40] One can predict that the only features of the parental virus shared by any of the eventual RCRs arising from late-generation systems would be those dependent on the *gag* and *pol* genes. Thus, RCR monitoring could be performed by assays based on HIV-1 *gag* detection, such as p24 immunocapture assay, and *gag* RNA-PCR assay.

The assays currently available for RCR detection need to be further characterized and their sensitivity needs to be determined. As a first approximation, one could verify the absence of expression of the p24 antigen in the culture medium of cells transduced by high input of vector. We use the easily infectable human T-cell line C8166. C8166 cells (2×10^5) are mixed with 100 ng of p24 equivalent of vector preparation in 1 ml medium (RPMI with 10% FBS) in the presence of 2 μg/ml Polybrene and incubated at 37° for 16 hr. The following day wash cells 3 times with medium. Resuspend the cells in fresh medium and incubate at 37°, 5% CO_2, for up to 3 weeks during which cells are split to avoid overconfluent culturing conditions. Measure p24 in the supernatant. The p24 Ag should become undetectable (<2 pg/ml) once the input antigen has been eliminated from the culture. Further development of this assay has been reported by Farson *et al.*,[19] who also developed a sensitive assay for the detection of recombinants between packaging and transfer vector constructs that are able to transfer the viral *gag pol* genes but lack an envelope gene (replication-defective recombinants). These types of recombinants may decrease the transduction efficiency of the vector by competing for encapsidation. Moreover, the contamination of a vector preparation by recombinants capable of transferring vector packaging function is of concern because it may lead to toxicity and immune reaction in the recipient. Sensitive detection of envelope-defective recombinants expressing the viral *gag pol* genes will thus be helpful to validate the safety of a vector preparation. These assays should address some of the strict requirements of possible future clinical applications of LVs.

Acknowledgments

The financial support of Telethon–Italy (Grant A.143), MURST (Grant 9905313431), the EU (QLK3-1999-00859), and the Italian Health Ministry (HIS, AIDS Research Program 40C.66) is gratefully acknowledged. A.F. is a recipient of an AIDS program fellowship from HIS, Italy.

---

[40] J. zur Megede, M. C. Chen, B. Doe, M. Schaefer, C. E. Greer, M. Selby, G. R. Otten, and S. W. Barnett, *J. Virol.* **74**, 2628 (2000).

[27] Design, Production, Safety, Evaluation, and Clinical Applications of Nonprimate Lentiviral Vectors

By Jonathan B. Rohll, Kyriacos A. Mitrophanous,
Enca Martin-Rendon, Fiona M. Ellard, Pippa A. Radcliffe,
Nicholas D. Mazarakis, and Susan M. Kingsman

Introduction

Lentivirus-based vectors are ideal tools for gene transfer because of their ability to stably integrate into the genome of dividing and nondividing cells and to mediate long-term gene expression.[1–4] The majority of research has been carried out using vectors derived from human immunodeficiency virus, type 1 (HIV-1), and has been greatly facilitated by the wealth of information and reagents that have been accumulated on this virus since its recognition as the causal agent of AIDS in 1984. However, vector systems based on other primate lentiviruses and the nonprimate lentiviruses, such as equine infectious anemia virus (EIAV)[4,5] and feline immunodeficiency virus (FIV),[6–8] are also highly developed. More recently a vector system based on Jembrana virus, which is similar to bovine immunodeficiency virus, has been reported; however, at present this system is not as well developed as those for EIAV and FIV.[9] Little progress has been made in establishing caprine arthritis–encephalitis virus- and Maedi-Visna virus-based systems.[10,11]

This review focuses on the EIAV vector system. First, the basic features of EIAV are discussed and then the features of a minimal EIAV system are described. Sections follow on production technology, methods for analysis of vector performance, and assays for replication-competent virus in vector preparations. Finally, potential applications are discussed.

[1] L. Naldini and I. M. Verma, *Adv. Virus Res.* **55**, 599 (2000).

[2] G. L. Buchschacher, Jr. and F. Wong-Staal, *Blood* **95**, 2499 (2000).

[3] E. Vigna and L. Naldini, *J. Gene Med.* **2**, 308 (2000).

[4] K. Mitrophanous, S. Yoon, J. Rohll, D. Patil, F. Wilkes, V. Kim, S. Kingsman, A. Kingsman, and N. Mazarakis, *Gene Ther.* **6**, 1808 (1999).

[5] J. C. Olsen, *Gene Ther.* **5**, 1481 (1998).

[6] M. A. Curran, S. M. Kaiser, P. L. Achacoso, and G. P. Nolan, *Mol. Ther.* **1**, 31 (2000).

[7] J. C. Johnston, M. Gasmi, L. E. Lim, J. H. Elder, J. K. Yee, D. J. Jolly, K. P. Campbell, B. L. Davidson, and S. L. Sauter, *J. Virol.* **73**, 4991 (1999).

[8] E. M. Poeschla, F. Wong-Staal, and D. J. Looney, *Nat. Med.* **4**, 354 (1998).

[9] P. Metharom, S. Takyar, H. H. Xia, K. A. Ellem, J. Macmillan, R. W. Shepherd, G. E. Wilcox, and M. Q. Wei, *J. Gene Med.* **2**, 176 (2000).

[10] L. Mselli-Lakhal, C. Favier, K. Leung, F. Guiguen, D. Grezel, P. Miossec, J. F. Mornex, O. Narayan, G. Querat, and Y. Chebloune, *J. Virol.* **74**, 8343 (2000).

[11] R. D. Berkowitz, H. Ilves, I. Plavec, and G. Veres, *Virology* **279**, 116 (2001).

EIAV: Host Range, Genome Structure, and Function

Host Range

EIAV causes a self-limiting, lifelong, but rarely fatal infection of all *Equidae*,[12] and is a worldwide disease of horses, most prevalent in warmer climates. In the United States most outbreaks occur in stables in the southeastern states, where transmission is as a result of insect bites, mostly by horseflies. Despite repeated exposure to the virus in such a setting there have been no reported cases of infection in humans, suggesting that it is an intrinsically safe virus on which to base vectors for use in the clinical setting. The principal productively infected cell in the horse is the monocyte/macrophage;[13] however, productive infection of equine endothelial cells *in vivo* has been desmonstrated.[14] *In vitro*, EIAV propagation in equine dermal cells and equine fetal kidney has been reported.[15] In addition, EIAV can infect canine cells.[16] EIAV cannot replicate in human cells, because of either the inability of the envelope to mediate entry into human cells or the low transcriptional activity of the long terminal repeat (LTR) in human cells, as discussed below.

Genome Structure and Function

The genomic structure of EIAV is quite distinct from the structures of other lentiviruses (Fig. 1A). It has the simplest genome, containing only three accessory genes: *tat, rev,* and *S2*. Unlike other lentiviruses, EIAV does not contain a *vif* open reading frame between the end of *pol* and the start of *env*.[17] In addition, EIAV has the shortest *pol–env* intergenic region (190 base pairs). This, together with the short long terminal repeat, accounts for the small size of the EIAV genome (8.0 kb) relative to other lentiviruses (HIV-1, 9.5 kb).

LTR. The EIAV LTR ($\sim$320 bp) is shorter than other retroviral LTRs and, in particular, is approximately 300 bp shorter than the LTRs of HIV ($\sim$630 bp). The sequence of the U3 region of different EIAV strains is highly variable compared to those of other lentiviruses.[18–20] The sequence variation results in the presence

---

[12] R. C. Montelaro, J. M. Ball, and K. E. Rushlow, *in* "The Retroviridae" (J. A. Levy, ed.), Vol. 2. Plenum Press, New York, 1993.

[13] W. Maury, *J. Virol.* **68**, 6270 (1994).

[14] W. Maury, J. L. Oaks, and S. Bradley, *J. Virol.* **72**, 9291 (1998).

[15] F. Li, B. A. Puffer, and R. C. Montelaro, *J. Virol.* **72**, 8344 (1998).

[16] T. R. Albrecht, L. H. Lund, and M. A. Garcia-Blanco, *Virology* **268**, 7 (2000).

[17] M. S. Oberste and M. A. Gonda, *Virus Genes* **6**, 95 (1992).

[18] S. Carpenter, S. Alexandersen, M. J. Long, S. Perryman, and B. Chesebro, *J. Virol.* **65**, 1605 (1991).

[19] D. L. Lichtenstein, J. K. Craigo, C. Leroux, K. E. Rushlow, R. F. Cook, S. J. Cook, C. J. Issel, and R. C. Montelaro, *Virology* **263**, 408 (1999).

[20] S. L. Payne, K. La Celle, X. F. Pei, X. M. Qi, H. Shao, W. K. Steagall, S. Perry, and F. Fuller, *J. Gen. Virol.* **80**, 755 (1999).

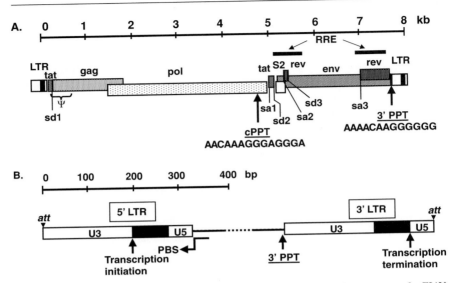

FIG. 1. Schematic drawing of the EIAV genome. (A) The location of coding sequences for EIAV proteins are indicated as are the sequence and location of the central polypurine tract (cPPT) and the 3′ polypurine tract (3′PPT). The positions of the splice donors (sd1, 2, and 3) and acceptors (sa1, 2, and 3) and the packaging signal (Ψ) are shown. The black bars show the location of the two sequences shown to have Rev response element (RRE) activity. A scale in kilobases (kb) is shown above. (B) A detailed schematic of the 5′- and 3′-LTRs. The position of the primer binding site (PBS) and sequences required for integration (*att*) are shown. The scale shown indicates base pairs (bp).

of different transcription factor binding sites in the LTRs of different strains and almost certainly is a contributing factor in determining the tropism and pathology of different strains.[19,20]

Tat. EIAV Tat protein is a 75 amino acid protein that acts in an analogous manner to the Tat protein of HIV-1. As for the HIV-1 Tat protein, it binds to a stem-loop at the 5′ end of the viral RNA and, in association with cyclin T1, causes a 100-fold increase in transcription. EIAV Tat is able to form a complex with both equine and canine cyclin T1; however, it interacts only poorly with human cyclin T1, accounting for its relatively weak transactivating activity in human cells.[21] A single amino acid change to the human cyclin T1 has been shown to render EIAV Tat fully active in human cells.[16,21,22] Tat is translated from exons 1 and 2 of a 4-exon mRNA, which also encodes Rev, an arrangement of genes unique to EIAV. Translation is initiated from a CUG triplet in a favorable Kozak

[21] P. D. Bieniasz, T. A. Grdina, H. P. Bogerd, and B. R. Cullen, *Mol. Cell Biol.* **19**, 4592 (1999).
[22] R. Taube, K. Fujinaga, D. Irwin, J. Wimmer, M. Geyer, and B. M. Peterlin, *J. Virol.* **74**, 892 (2000).

context within the leader region [upstream of the *gag* open reading frame (ORF)]. The functional activity of EIAV Tat is associated with the peptide encoded by exon 2.

Rev/RRE. EIAV Rev is a 165 amino acid protein that is translated from exons 3 and 4 of the same mRNA which encodes Tat, and is produced as a result of read-through of the Tat ORF.[23] EIAV Rev functions in an analogous way to HIV-1 Rev, by binding to a Rev-response element (RRE), thereby promoting the export of unspliced and partially spliced viral transcripts (reviewed in Refs. 24, 25). At a mechanistic level the EIAV Rev/RRE system is different to and more complex than that of HIV-1. First, the RRE function appears to reside in two elements located within the *env* gene. These elements can act independently; however, their activity is synergistic.[26,27] Second, EIAV Rev regulates its own concentration by alternative splicing.[26] This phenomenon is mediated by the presence of two purine-rich motifs (PU-A and PU-B) located in exon 3 and termed the exon splicing enhancer (ESE).[28–30] In the absence of Rev, SR proteins (pre-mRNA splicing factors[31]) bind to the ESE causing splicing reactions between splice donor 3 and splice acceptor 3; however, if Rev is present it binds to the ESE leading to displacement of the SR proteins. This leads to the preferential use of splice donor 2 in splicing reactions and concomitant down-regulation of Rev synthesis.[30] SR proteins have Ser/Arg-rich repeats.

S2. The *S2* gene is located between the second exon of *tat* and extends into the *env* ORF; it appears to be unique to EIAV. The S2 protein is a 65 amino acid protein which is thought to be translated by ribosomal read-through or leaky scanning of a tricistronic mRNA encoding Tat, S2, and Env. Although not required for the replication of EIAV *in vitro*[15] the *S2* sequence is highly conserved and is important for viral replication and pathogenicity *in vivo*,[32] and antibodies to S2 are readily detectable in sera from infected horses.[33] S2 is localized to the cytoplasm and can interact with the Gag precursor.[34] Using an EIAV-based vector system we

[23] R. Carroll and D. Derse, *J. Virol.* **67**, 1433 (1993).

[24] V. W. Pollard and M. H. Malim, *Annu. Rev. Microbiol.* **52**, 491 (1998).

[25] T. J. Hope, *Arch. Biochem. Biophys.* **365**, 186 (1999).

[26] L. Martarano, R. Stephens, N. Rice, and D. Derse, *J. Virol.* **68**, 3102 (1994).

[27] M. Belshan, M. E. Harris, A. E. Shoemaker, T. J. Hope, and S. Carpenter, *J. Virol.* **72**, 4421 (1998).

[28] R. R. Gontarek and D. Derse, *Mol. Cell Biol.* **16**, 2325 (1996).

[29] M. E. Harris, R. R. Gontarek, D. Derse, and T. J. Hope, *Mol. Cell Biol.* **18**, 3889 (1998).

[30] M. Belshan, G. S. Park, P. Bilodeau, C. M. Stoltzfus, and S. Carpenter, *Mol. Cell Biol.* **20**, 3550 (2000).

[31] B. R. Graveley, *RNA* **6**, 1197 (2000).

[32] F. Li, C. Leroux, J. K. Craigo, S. J. Cook, C. J. Issel, and R. C. Montelaro, *J. Virol.* **74**, 573 (2000).

[33] R. L. Schiltz, D. S. Shih, S. Rasty, R. C. Montelaro, and K. E. Rushlow, *J. Virol.* **66**, 3455 (1992).

[34] S. Yoon, S. M. Kingsman, A. J. Kingsman, S. A. Wilson, and K. A. Mitrophanous, *J. Gen. Virol.* **81**, 2189 (2000).

have demonstrated that the S2 gene is not required for efficient transduction of dividing or nondividing cells.[4]

dUTPase. The nonprimate lentiviruses are unique among the retroviruses in that they encode a dUTPase (*dut*) present in an additional domain within the *pol.*[35] The EIAV dUTPase has been shown to be important for the replication of EIAV in quiescent cells such as macrophages.[36–38] dUTPases lower the dUTP-to-dTTP ratio in cells in a cell cycle dependent manner, thereby reducing the incorporation of dUTP into DNA.[39,40] Cells contain both a nuclear and a mitochondrial dUTPase, but these are absent in cells that are quiescent (in G_0 of the cell cycle).[41] Cells in G_0, rather than actively cycling, have a relatively high dUTP to dTTP ratio; hence retroviral reverse transcription in such a cell will be relatively more prone to error. As a result there will an increased frequency of incorporation of dUTP into the genome, subsequently leading either to G to A mutations[42] or, on integration, to the genome being subject to uracil glycosylase-mediated excision repair. It has been proposed that the reason that primate lentiviruses do not contain a dUTPase is that this function is supplied from endogenous retroviruses.[41] The above considerations suggest that dUTPase is unlikely to be an important determinant of transduction of human cells by EIAV vectors. In line with these views we have previously demonstrated that our EIAV-based vectors do not require dUTPase for transduction of nondividing cells[4]; however, the current vector system does include an active dUTPase.

The EIAV Minimal Vector System

The aim when designing a vector system is threefold. The first is to minimize the amount of viral sequence in the transfer vector, thus increasing the capacity for transgenes and associated regulatory elements. The second is to eliminate the expression of viral sequences in the target cells. Viral genes may elicit immune responses that result in destruction of transduced cells and may have undesirable effects *per se.* The third is to express, in the production system, only those viral proteins required for efficient production of transduction competent vector.

[35] J. H. Elder, D. L. Lerner, C. S. Hasselkus Light, D. J. Fontenot, E. Hunter, P. A. Luciw, R. C. Montelaro, and T. R. Phillips, *J. Virol.* **66,** 1791 (1992).

[36] D. L. Lichtenstein, K. E. Rushlow, R. F. Cook, M. L. Raabe, C. J. Swardson, G. J. Kociba, C. J. Issel, and R. C. Montelaro, *J. Virol.* **69,** 2881 (1995).

[37] W. K. Steagall, M. D. Robek, S. T. Perry, F. J. Fuller, and S. L. Payne, *Virology* **210,** 302 (1995).

[38] D. S. Threadgill, W. K. Steagall, M. T. Flaherty, F. J. Fuller, S. T. Perry, K. E. Rushlow, S. F. Le Grice, and S. L. Payne, *J. Virol.* **67,** 2592 (1993).

[39] P. Turelli, G. Petursson, F. Guiguen, J. F. Mornex, R. Vigne, and G. Querat, *J. Virol.* **70,** 1213 (1996).

[40] P. Turelli, F. Guiguen, J. F. Mornex, R. Vigne, and G. Querat, *J. Virol.* **71,** 4522 (1997).

[41] E. M. McIntosh and R. H. Haynes, *Acta Biochim. Pol.* **43,** 583 (1996).

[42] D. L. Lerner, P. C. Wagaman, T. R. Phillips, O. Prospero-Garcia, S. J. Henriksen, H. S. Fox, F. E. Bloom, and J. H. Elder, *Proc. Natl. Acad. Sci. U.S.A.* **92,** 7408 (1995).

The underlying strategy used to establish the EIAV vector system, in line with the outlined criteria, has been the "deconstruction" strategy adopted for murine leukemia virus (MLV) vectors, that is, the separation of the *gag/pol,* vector, and envelope components onto different transcriptional units.[43] However, whereas for MLV vectors the homologous envelope (amphotropic) or other C-type retroviral envelopes have been commonly used, the envelope of choice for lentiviral vectors, the vesicular stomatitis virus G protein (VSV-G), is derived from the rhabdovirus group of enveloped viruses. Two EIAV-based vector systems have been developed using this approach to date.[4,5] All three components, vector genome, Gag/Pol, and envelope, are being optimized and the current status is summarized below.

Vector Genome

Genome Structure. In their most minimal forms the vector components of both published systems are very similar. In each, transcription of the vector genome is driven by the human CMV enhancer/promoter fused to the R region of EIAV so that the first base of the transcript is the same as that formed as a result of transcription from the EIAV U3 region. This configuration allows high titers to be obtained in the absence of Tat protein, which, as discussed above, does not operate effectively in human cells. The replacement of the retroviral U3 region has also proved a successful strategy for MLV, HIV, FIV, and other vector systems.[4-7,44-46] The EIAV sequences necessary for production of high-titer vector have been established by deletion analysis, and current vectors have a wild-type EIAV sequence extending from the beginning of the R region up to nucleotide (nt) 372 of the gag ORF[4] (Fig. 2A). Within this region of the pONY8 series of vectors we have mutated the first two ATG codons of *gag* to ATTG in order to prevent expression of (truncated) Gag proteins. At the 3' end of the vector genome both pONY8.0 and pONY8.1 vectors retain 73 nt upstream of the 3'LTR. This region includes the U-rich sequence immediately upstream of the 3'-polypurine tract which has been shown to be important for optimal replication of MLV[47,48] and may be desirable in EIAV vectors although this has yet to be proved. The pONY8.0 series of vectors also contain a 2.2 kb segment of noncoding EIAV sequence downstream of the reporter cassette (Fig. 2A).

[43] J. M. Coffin, S. H. Hughes, and H. E. Varmus (eds.), "Retroviruses." Cold Spring Harbor Laboratory Press, Cold Spring Harbor, NY, 1997.

[44] Y. Soneoka, P. M. Cannon, E. E. Ramsdale, J. C. Griffiths, G. Romano, S. M. Kingsman, and A. J. Kingsman, *Nucleic Acids Res.* **23,** 628 (1995).

[45] V. N. Kim, K. Mitrophanous, S. M. Kingsman, and A. J. Kingsman, *J. Virol.* **72,** 811 (1998).

[46] T. Dull, R. Zufferey, M. Kelly, R. J. Mandel, M. Nguyen, D. Trono, and L. Naldini, *J. Virol.* **72,** 8463 (1998).

[47] N. D. Robson and A. Telesnitsky, *J. Virol.* **74,** 10293 (2000).

[48] N. D. Robson and A. Telesnitsky, *J. Virol.* **73,** 948 (1999).

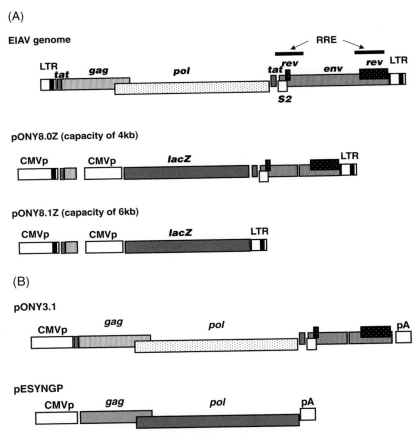

FIG. 2. (A) Minimal EIAV vector genomes, pONY8.0Z and pONY8.1Z. Both vectors have either deletions or mutations which abrogate the expression of any EIAV proteins. The alterations in pONY8.0Z: (1) prevented expression of Tat by an 83 nt deletion in the exon 2 of *tat;* (2) prevented S2 ORF expression by a 51 nt deletion; (3) prevented Rev expression by deletion of a single base within exon 1 of *rev;* and (4) prevented expression of the N-terminal portion of *gag* by insertion of T in ATG start codons, thereby changing the sequence to ATTG from ATG. With respect to the wild-type EIAV sequence Acc. No. U01866 these correspond to deletion of nt 5234–5316 inclusive, nt 5346–5396 inclusive, and nt 5538. The insertion of T residues was after nt 526 and 543. In pONY8.1Z the EIAV-derived sequence downstream of the reporter gene and up to nt 7895 with respect to wild-type EIAV, Acc. No. U01866 is deleted. The capacity of the vector systems for heterologous sequence is indicated and is calculated with respect to the length of the EIAV genomic RNA. Vectors with expression cassettes for β-galactosidase are shown. (B) EIAV *gag/pol* expression cassettes, pONY3.1 and pESYNGP. These are expression plasmids for wild-type *gag/pol* and codon-optimized EIAV *gag/pol,* respectively. Both cassettes are based on pCIneo (Promega) in which expression is driven from a human CMV enhancer/promoter (CMVp), and which has an intron upstream of the cloning site for heterologous sequences. The 5′-end of the EIAV sequence in pONY3.1 corresponds to the PBS, and the 3′-end, to the *Mlu*I site within the U3 region. *Env* expression is ablated by a 736 bp deletion in *env.*[4]

Influence of Rev/RRE on Vector Titer. The titers obtained from pONY8.0 vectors are 5- to 10-fold higher than from pONY8.1 vectors because of the inclusion of sequences that contain the RRE between the reporter expression cassette and the 3′ LTR. However, this reduces the capacity of the vector by approximately 2 kb (Fig. 2A). We have made attempts to reduce the extent of sequences required for complete RRE function by making vectors with deletions within this region; however, these have not been very successful, which likely reflects our poor understanding of the EIAV Rev/RRE system. In addition we have consistently seen lower titers from vectors in which the RRE-containing sequences are moved to a position upstream of the reporter expression cassette. Our current analysis of the types of transcripts made by vectors in the presence or absence of Rev will be used to design vector genomes with reduced Rev/RRE requirements.

Transgene Capacity. The capacity of vector genomes is potentially an important factor in determining their utility. The EIAV genomic RNA, at 7967 nucleotides (nt) [without the poly(A) tail] is somewhat smaller than the genomes of FIV (9.2 kb) and HIV (9.5 kb) suggesting that the load capacity of EIAV vectors may be somewhat less than for other lentiviral vectors. However, although we have not carried out experiments to determine the upper size limit, we have observed efficient packaging and transduction by a vector with an RNA which is 111% of wild-type genome size (data not shown). On the basis of recently published experiments with MLV-based vectors, which suggest that vector genomes much larger than 111% of the viral genome can be packaged, it is likely that EIAV vector genomes even larger than 111% of the viral genome will be packageable.[49]

gag/pol

Initially the Gag/Pol proteins were provided from an expression cassette termed pONY3.1[4] (Fig. 2B). This contains the natural EIAV *gag/pol* flanked by extensive EIAV sequences that include the leader region upstream of *gag* and the coding regions for all the EIAV accessory proteins, and also contains the sequences necessary for complete RRE function. Variants of pONY3.1 in which *tat* and *S2* were disrupted and the 3′LTR sequences deleted were also made, and functioned equivalently in transient production assays. However, these alterations do not address the two general problems with the Gag/Pol expression constructs in lentiviral vector systems. The first is the sequence overlap with the vector genome, and the second is that Gag/Pol expression is Rev-dependent. One method of overcoming both these problems is to "codon-optimize" the gene. This has been demonstrated with *gag/pol* genes of HIV-1 and SIV-1.[50,51] The optimization process makes the

[49] N. H. Shin, D. Hartigan-O'Connor, J. K. Pfeiffer, and A. Telesnitsky, *J. Virol.* **74,** 2694 (2000).

[50] E. Kotsopoulou, V. N. Kim, A. J. Kingsman, S. M. Kingsman, and K. A. Mitrophanous, *J. Virol.* **74,** 4839 (2000).

[51] R. Wagner, M. Graf, K. Bieler, H. Wolf, T. Grunwald, P. Foley, and K. Uberla, *Hum. Gene Ther.* **11,** 2403 (2000).

expression of *gag/pol* Rev-independent, possibly as a result of removing RNA instability sequences, and, for HIV-1, results in an increase in expression of approximately 10-fold over that achieved with the wild-type *gag/pol* in the presence of Rev.[50]

Changes in the primary structure of the RNA as a result of the codon-optimization procedure will inevitably lead to changes in secondary and tertiary structure almost certainly leading to the disruption of any structures required for efficient packaging.[51] Therefore the codon-optimization process increases the safety of the system in two ways. First, *gag/pol* mRNA is unlikely to be packaged as efficiently as the wild-type *gag/pol* RNA. Second, because of the lack of significant homology with the vector in the region of the packaging signal, it is unlikely to be involved in recombination reactions which might result in generation of replication competent retrovirus (RCR), even if incorporated into the vector particle. In our view all lentiviral vector systems should use this strategy of codon-optimization as a major safety feature (see later).

We have now constructed a codon-optimized EIAV *gag/pol* gene using the same strategy.[52] All but 309 nt of the *gag/pol* ORF were "optimized," the unchanged region corresponding to the frame-shift site region through which *pol* expression is achieved, and to the region of overlap between the sequences coding for the C terminus of Gag and the N terminus of Pol. This construct is termed pESYNGP (Fig. 2B). When compared to pONY3.1, slightly lower levels of expression were observed, however there were no significant differences in the pattern of proteolytic processing.[52] However, unlike the situation with the wild type *gag/pol,* expression was REV-independent. The codon-optimized *gag/pol* gene is now being utilized for construction of EIAV packaging and producer cell lines.

Envelope

To date the envelope most commonly used for lentiviral vectors is the G protein of the rhabdovirus, vesicular stomatitis virus, termed (VSV-G). Where envelopes are derived from a heterologous virus the process is generally referred to as pseudotyping.[53] The receptor for VSV-G is an abundant membrane phospholipid, hence the virus has a broad tropism, and the envelope permits transduction of an extremely wide range of cell types.[54] *In vivo,* VSV-G pseudotyped particles have been shown to transduce a wide range of tissue types.[4,55] High-titer viral stocks $[>10^9$ transducing units (tu)/ml] can be routinely achieved through concentration of transiently produced viral supernatants by ultra- or slow-speed centrifugation

[52] J. B. Rohll, F. J. Wilkes, F. M. Ellard, A. L. Olsen, M. Esapa, D. F. Baban, E. Martin-Rendon, R. D. Barber, N. Mazarakis, S. Kingsman, A. Kingsman, and K. A. Mitrophanous, in preparation (2001).

[53] D. N. Love and R. A. Weiss, *Virology* **57,** 271 (1974).

[54] R. Schlegel, T. S. Tralka, M. C. Willingham, and I. Pastan, *Cell* **32,** 639 (1983).

[55] L. Naldini, U. Blomer, F. H. Gage, D. Trono, and I. M. Verma, *Proc. Natl. Acad. Sci. U.S.A.* **93,** 11382 (1996).

TABLE I
TITERS OF UNCONCENTRATED VIRAL SUPERNATANTS

Pseudotype	Titer (tu/ml) on D17 cells with indicated core	
	EIAV	MLV
VSV-G	1×10^6	5×10^6
Rab-G	5×10^5	1×10^6
4070A	2×10^4	3×10^5
FeLV-B	10	5×10^4
FeLV-C	10	5×10^4
RD114	2×10^2	1×10^6
GALV	2×10^2	2×10^4

(see Addendum, Protocol 1). A major disadvantage of the envelope, however, is that the protein is cytotoxic to producer cells. Establishing stable packaging cell lines has been difficult, but inducible systems to overcome this problem have now been developed (see section on Producer Systems). Another problem is that VSV-G in the culture fluid harvested from producer cells, during either transient or stable production, is not exclusively associated with virus core particles, yet such "free" VSV-G, probably present in vesicles of cellular membrane, can be concentrated by centrifugation. The resulting concentrated vector preparation can consequently be toxic to target cells if particularly high multiplicity of infections is desired or if the cells are particularly sensitive. A further consequence of the ability of VSV-G to bud independently of viral cores is the phenomenon of pseudotransduction. This is where there is the illusion of effective transduction due to transfer of the marker protein directly to the target cell. It is a particular problem with membrane markers such as the low affinity nerve growth factor[56] but it also occurs with cytoplasmic protein markers, such as green fluorescent protein (GFP), at a high multiplicity of infection.[57] The envelope proteins of two other rhabdoviruses, rabies virus and the closely related Mokola virus, have also been used to pseudotype lentiviral vectors.[4,58] The titers obtained using the rabies virus G protein (Rab-G) are only slightly lower than observed for VSV-G with the EIAV vector system (Table I) and a wide host range is conferred in cell lines and *in vivo*.[4,59]

[56] L. Ruggieri, A. Aiuti, M. Salomoni, E. Zappone, G. Ferrari, and C. Bordignon, *Hum. Gene Ther.* **8**, 1611 (1997).

[57] D. L. Haas, S. S. Case, G. M. Crooks, and D. B. Kohn, *Mol. Ther.* **2**, 71 (2000).

[58] H. Mochizuki, J. P. Schwartz, K. Tanaka, R. O. Brady, and J. Reiser, *J. Virol.* **72**, 8873 (1998).

[59] N. Mazarakis, M. Azzous, J. B. Rohll, F. M. Ellard, F. J. Wilkes, A. L. Olsen, E. E. Carter, R. D. Barber, D. F. Baban, S. M. Kingsman, A. J. Kingsman, K. O'Malley, and K. A. Mitrophanous, *Hum. Mol. Genet.* **10**, 2109 (2001).

We have also tested the pseudotyping ability of a wide range of envelopes derived from type C retroviruses; however, this survey has not revealed any envelopes which pseudotype as well as VSV-G. The MLV 4070A, amphotropic envelope, is the most efficient of the C-type retroviral envelopes tested. Titers with the 4070A envelope are in excess of 10^4 tu/ml (Table I) and vector can be further concentrated 1000-fold by low-speed centrifugation. Very poor pseudotyping was observed using the envelopes of gibbon ape leukemia virus (GALV) and cat endogenous retrovirus RD114, both of which pseudotype MLV cores very effectively. This restriction is unfortunate in view of the ability of these envelopes to target hematopoietic stem cells and the ease with which packaging cells expressing these envelopes can be made. A potential solution to these problems is the use of hybrid envelopes, such as that between the GALV and MLV envelopes, reported recently to allow improved pseudotyping of HIV-1 cores.[60]

Other envelopes, which, like those of GALV and RD114, do not efficiently pseudotype lentiviral cores, include those of feline leukemia virus (FeLV)-B and FeLV-C. The FeLV-C envelope is acquired by both HIV-1 and EIAV cores; however, it appears to be nonfunctional in mediating cell entry. We also noted that high levels of FeLV-C envelope were shed into the media as has been observed for other retroviral envelopes.[61,62] High levels of shed envelope may result in decreased transduction of target cells,[63] although this is unlikely to account for the observed disparity in titers between FeLV-B/C pseudotypes of MLV and EIAV (Table I). In summary, VSV-G and Rab-G are currently the envelopes of choice for lentiviral vectors due to their efficient pseudotyping of lentiviral cores and ability to be easily concentrated.

Further Enhancements of Vector Components

cPPT/CTS: The "FLAP." Recent reports indicate that the central polypurine tract (cPPT) of HIV-1 is important for the efficient transduction of non-dividing cells by HIV-1 vectors.[64–66] In HIV-1 the cPPT is a direct copy of the 3′-polypurine tract (3′PPT) and is located approximately 200 bp to the 5′ side of the center of

[60] J. Stitz, C. J. Buchholz, M. Engelstadter, W. Uckert, U. Bloemer, I. Schmitt, and K. Cichutek, *Virology* **273**, 16 (2000).

[61] D. P. Bolognesi, A. J. Langlois, and W. Schafer, *Virology* **68**, 550 (1975).

[62] J. Schneider, O. Kaaden, T. D. Copeland, S. Oroszlan, and G. Hunsmann, *J. Gen. Virol.* **67**, 2533 (1986).

[63] J. H. Slingsby, D. Baban, J. Sutton, M. Esapa, T. Price, S. M. Kingsman, A. J. Kingsman, and A. Slade, *Hum. Gene Ther.* **11**, 1439 (2000).

[64] A. Follenzi, L. E. Ailles, S. Bakovic, M. Geuna, and L. Naldini, *Nat. Genet.* **25**, 217 (2000).

[65] A. Sirven, F. Pflumio, V. Zennou, M. Titux, W. Vainchenker, L. Coulombel, A. Dubart-Kupperschmitt, and P. Charneau, *Blood* **96**, 4103 (2000).

[66] V. Zennou, C. Petit, D. Guetard, U. Nerhbass, L. Montagnier, and P. Charneau, *Cell* **101**, 173 (2000).

the genome. It is speculated that the central positioning is important for efficient function of the element, and if this is the case this should be borne in mind in the construction of vector genomes. In EIAV, the cPPT is located approximately 500 bp to the 3′-side of the genome center, and is not identical to the 3′PPT (Fig. 1). The effect of the cPPT on transduction by EIAV vectors is being assessed and, while clearly not obligatory for transduction of many cell types, it may enhance titer in some specific cells.

Woodchuck Hepatitis Posttranscriptional Regulatory Element (WHPRE). The WHPRE is a 600 bp element that enhances the expression of proteins by increasing the half-life of mRNA through a mechanism involving enhanced polyadenylation. Its beneficial effect has been demonstrated in a number of vectors including HIV-1 based vectors[67–69] and has also been shown to increase expression from EIAV-based vectors (data not shown).

Production of EIAV Vectors

Transient Production

The production of vectors for evaluation purposes has been achieved using the HEK 293T transient system utilized extensively for MLV, HIV, and other retroviral vector systems.[43] In each case the ratio of plasmids for the vector genome, Gag/Pol and Envelope (and additional plasmids such as a Rev expression plasmid) is optimized for the particular vector system.[70] Transfection is most usually carried out by the calcium phosphate method; however, use of Fugene-6 (Boehringer) has been reported.[6,70] We have obtained very similar end-point titers with the EIAV vector system using both reagents, with Fugene-6 requiring lower quantities of DNA (1/10th). The procedure for transient vector production and downstream processing are described in Protocol 1 (see Addendum).

The use of transient production methods is unlikely to be feasible for production of vector for use in large/medium size animal experiments or for preclinical evaluation and is certainly very problematical for human clinical trials, due to the difficulty of successfully implementing the technique under Good Manufacturing Practice (GMP) conditions. However, it is feasible to produce enough vector for limited, small animal studies. Using a method based on 10 cm diameter dishes we regularly produce 3.5 ml lots with titers of $>10^9$ tu/ml when titrated on D17, canine osteosarcoma cells.

[67] R. Zufferey, J. E. Donello, D. Trono, and T. J. Hope, *J. Virol.* **73**, 2886 (1999).

[68] J. E. Loeb, W. S. Cordier, M. E. Harris, M. D. Weitzman, and T. J. Hope, *Hum. Gene Ther.* **10**, 2295 (1999).

[69] A. Ramezani, T. S. Hawley, and R. G. Hawley, *Mol. Ther.* **2**, 458 (2000).

[70] M. W. Yap, S. M. Kingsman, and A. J. Kingsman, *J. Gen. Virol.* **81** (Part 9), 2195 (2000).

Stable Producer Systems

Stable producer systems configured along the lines of those available for MLV vectors[43] are being avidly sought; however, there are several problems:

1. In some vector systems the high level expression of Gag/Pol is cytotoxic. Although not found to be a problem with EIAV Gag/Pol it is reported to be a problem for FIV and HIV Gag/Pol.

2. High-level expression of VSV-G, the most favored pseudotype, is generally cytotoxic.

3. There are concerns over producer cell stability due to self-transduction with the use of VSV-G.[71]

4. High titers are only likely to be achieved if transduction is used to introduce multiple transcriptionally active vector genomes into the packaging cells. This view is based on the report of producer cells for MLV vectors which make very high titers ($>10^8$ tu/ml in unconcentrated culture supernatant) and which were made by transduction of the vector genome into the packaging cell line. This strategy is not possible if, as in the case of EIAV, the LTR is inactive in the producer cell, or if a vector with a self-inactivating (SIN) configuration is used.

These difficulties may be resolved by using inducible systems to control expression of Gag/Pol and VSV-G as described initially by Kafri *et al.*[72] and more recently by Trono and co-workers,[73] who both utilized the tetracycline-inducible system to create HIV-1 producer systems. In addition, Kafri and co-workers have described a "conditional" SIN configuration for the HIV-1 vector system which will allow introduction of multiple copies of a transcriptionally active vector genome into a packaging cell line.[74] In this approach the "operator" region of the tetracycline (Tet)-inducible system is introduced into the 3'LTR of the vector genome. On transduction the tetracycline responsive element is relocated to the 5'LTR. This allows the transcriptional activity of the 5'LTR to be regulated in cells expressing the appropriate, components of the Tet-inducible system, such as a packaging cell line; however, its activity will be several orders of magnitude lower in target cells which do not express these factors.

The current producer system for EIAV-based vectors utilizes a vector genome which has a conditional SIN configuration to allow transduction of the vector into the packaging cell line. The packaging cell line expresses VSV-G under the control of a Tet-inducible system and EIAV Gag/Pol is expressed from the codon-optimized *gag/pol* expression construct described above (Fig. 3). In addition, as

---

[71] T. Arai, M. Takada, M. Ui, and H. Iba, *Virology* **260,** 109 (1999).
[72] T. Kafri, H. van Praag, L. Ouyang, F. H. Gage, and I. M. Verma, *J. Virol.* **73,** 576 (1999).
[73] N. Klages, R. Zufferey, and D. Trono, *Mol. Ther.* **2,** 170 (2000).
[74] K. Xu, H. Ma, T. J. McCown, I. M. Verma, and T. Kafri, *Mol. Ther.* **3,** 97 (2001).

pONY8-TRE

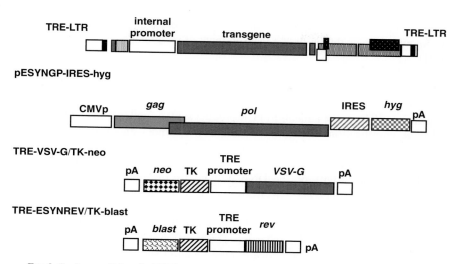

FIG. 3. Producer cell line for EIAV vectors. The system currently under development is based on HEK 293 cells. Expression of the codon-optimized EIAV *gag/pol* is achieved from pESYNGP-IRES-hyg which is based on the pIRES1hygro plasmid (Clontech), which allows for selection of transfected cells on the basis of hygromycin resistance. Envelope and Rev are expressed from TRE-VSV-G/TK-neo and TRE-ESYNREV/TK-blast cassettes, respectively. Transcription is driven by the Tet-inducible promoter, using the Tet[off] configuration of the system. Multiple copies of vector genome, pONY8-TRE, which has a conditional SIN configuration, are introduced by transduction, using VSV-G pseudotyped vector particles made by transient transfection of HEK 293T cells. The cell line also contains a CMV-driven expression cassette for tTA, the tetracycline-controlled transactivator (Clontech). IRES, internal ribosome entry site; *hyg,* hygromycin resistance gene; *blast,* blasticidin resistance gene; *neo,* G418 resistance gene; TK, thymidine kinase promoter; pA, polyadenylation signal; TRE-LTR, a hybrid LTR which contains 7 copies of the tetracycline-responsive element in place of the wild-type EIAV sequence.

noted above, the titers of vector are enhanced by the presence of Rev in the producer cell. Provision of Rev using a constitutive expression cassette is a solution to this problem. The producer system currently under development will utilize a Tet-inducible codon optimized *rev* expression cassette.

Titration of Unmarked Vector Genomes

The measurement of vector titer in the situation where the vector expresses antibiotic resistance genes, or markers such as β-galactosidase, alkaline phosphatase, or GFP, is readily achieved using well-known techniques. The underlying principles for detection of these markers are applicable to other transgenes, particularly those for which antibodies are available. When these are not available, or it is undesirable to include marker genes in the vector genome, alternative methods are required. We have utilized Taqman "real-time" PCR-based methods to measure the titer of "markerless" vector preparations, relative to a standard vector preparation of known biological titer, by the measurement of integrated vector DNA in target cells.

The key to quantitation by Taqman is the exploitation of the $5'$-nuclease activity of Taq DNA polymerase and the presence of a "probe" which is labeled at the $5'$ end with a fluorescent "reporter" dye, and at the $3'$ end with a fluorescence "quencher." The probe anneals to PCR product synthesized in a conventional type of PCR amplification reaction, between the two primers, and in its intact state is not fluorescent, because of the interaction of the quencher and the fluorescent reporter dye. During the extension phase of the PCR reaction the probe is digested by the nuclease activity of Taq DNA polymerase resulting in separation of the quencher and reporter dyes. This leads to an increase in fluorescence that directly correlates with the amount of PCR product and increases as more cycles of PCR are carried out. The PCR cycle number at which the fluorescence level increases above background is called the Ct value. The higher the amount of target molecules the lower the Ct value. Theoretically amplification should occur exponentially, resulting in a doubling of the amount of DNA per cycle. Thus, a difference of 1 in Ct value between two samples corresponds to a 2-fold difference in the concentration of the DNA. However, in practice this may not be seen because of competing amplification reactions and inhibitors present in the samples being tested. Further information on Taqman technology can be obtained from the Applied Biosystems Website (http://www.appliedbiosystems.com).

We developed the assay using D17 cells and initially measured the amount of EIAV vector sequences in transduced cells using primer/probes specific for the *gag* region of pONY8-type vectors. In preliminary experiments we determined that the level of EIAV vector DNA reached a steady state relative to β-actin by 7–8 days posttransduction, reflecting the decay of unintegrated vector sequences. At this time-point total cellular DNA is prepared and used as a template in Taqman assays (see Addendum, Protocol 2). The appropriate time point for the preparation of cellular DNA needs to be determined for each cell type used in this assay. The titer of the "markerless" vector genome is established relative to a standard EIAV vector genome of known titer that is transduced in parallel. The relative titers measured by this method and biological titration agreed well (Table II).

A further parameter that can be conveniently assessed using Taqman technology is the ratio of vector genome to particles in the vector preparation. This assessment of quality is made by measuring the concentrations of encapsidated vector RNA and reverse transcriptase. Respectively, these are measured relative to a standard preparation using a reverse transcription-PCR-based Taqman protocol (see Addendum, Protocol 3) and a Taqman-based product enhanced reverse transcriptase assay (PERT assay) (see Addendum, Protocol 4). The applications of

TABLE II
COMPARISON OF RELATIVE TITER ASSESSED BY MEASUREMENT OF BOTH
INTEGRATED VECTOR SEQUENCE AND GFP EXPRESSION

Vector[a]	ddCt Value[b]		Biological titer (tu/ml)
	10^{-3} Dilution	10^{-4} Dilution	
pONY8G, preparation 1	7.3 ± 0.4	4.5 ± 0.2	$2.0 \times 10^9 \pm 0.5$
pONY8G, preparation 2	7.6 ± 0.1	4.0 ± 0.3	$2.6 \times 10^9 \pm 1.2$

[a] Vector preparations were made independently.
[b] ddCt Value is the difference in Ct values obtained for the reaction to detect vector and β-actin. The primer/probe set used for this analysis was designed to specifically detect the EIAV packaging signal present in the vector RNA.

these techniques are numerous and include measurements of packaging efficiency of vector and gag/pol transcripts and measurement of packaging cell performance.

Measurement of Reverse Transcriptase by PERT Assay

The measurement of reverse transcriptase (RT) can be readily achieved using a Taqman-based PERT assay (Addendum, Protocol 4).[75,76] In this method, reverse transcriptase associated with vector particles is released by mild detergent treatment and used to synthesize cDNA using MS2 bacteriophage RNA as template (Fig. 4). MS2 RNA template and primer are present in excess; hence, the amount of cDNA is proportional to the amount of RT released from the particles. Therefore, the amount of cDNA synthesized is proportional to the number of particles. MS2 cDNA is then quantitated using Taqman technology.[77]

Preliminary PERT experiments showed that the amount of RT activity present when no particle disruption was carried out was approximately 1000-fold lower than when the particles were disrupted; hence, no adjustment for RT not associated with particles is required (Fig. 5). The assay also displayed a linear output over at least four orders of magnitude. The assay was also evaluated against supernatants from HEK 293T cells transfected with various combinations of vector system plasmids (Fig. 6). When the *gag/pol* expression plasmid was excluded from the transfection mix the Ct from PERT assays were reduced by at least 14 cycles, indicating that no RT activity is associated with any of the other components of the vector system or is derived from HEK 293T cells. The vector RNA content

[75] A. Lovatt, J. Black, D. Galbraith, I. Doherty, M. W. Moran, A. J. Shepherd, A. Griffen, A. Bailey, N. Wilson, and K. T. Smith, *J. Virol. Methods* **82**, 185 (1999).
[76] B. A. Arnold, R. W. Hepler, and P. M. Keller, *Biotechniques* **25**, 98 (1998).
[77] Y. S. Lie and C. J. Petropoulos, *Curr. Opin. Biotechnol.* **9**, 43 (1998).

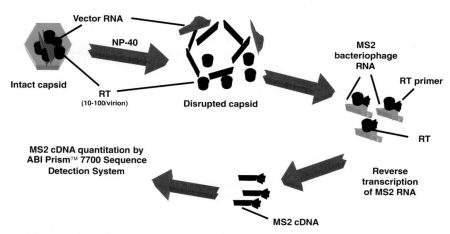

FIG. 4. Schematic representation of the product enhanced reverse transcription assay (PERT). Refer to text for explanation.

of the particles was also assessed, according to Protocol 3 (see Addendum). This analysis demonstrated that in the absence of Gag/Pol, but with VSV-G, no viral vector RNA was detected in the supernatant of transfected cells. This suggests that there is no significant transport of vector RNA by VSV-G vesicles. In additional preliminary studies it was also shown that the PERT assay was at least 50-fold more sensitive than RT assays based on the incorporation of radionucleotides and an ELISA-based assay in which incorporation of BrdUMP incorporation into DNA is measured by an ELISA (Fig. 7).

Replication-Competent Retrovirus (RCR) Assays

Safety Profile of EIAV Vector System

The safety of EIAV vector preparations for use in a clinical context is paramount to the acceptance of the vector system. Once the capability of the EIAV vector system was proved in principle, it was modified with safety in mind making use of the strategies used in the MLV vector field to avoid the production of RCR. The most important features for safety are

1. The partition of the components of the vector system in at least three independent expression cassettes.
2. The minimal sequence homology between the vector, *gag/pol,* and env components.
3. The use of human cell lines with low levels of endogenous retroviral sequences as the basis for producer cell lines.

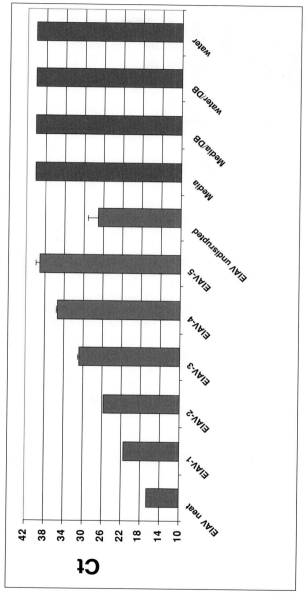

FIG. 5. PERT assay characteristics. EIAV vector pONY8.0Z was produced by transient transfection of 293T cells and had a titer of 2×10^6 tu/ml on D17 cells. At 48 hours the supernatant was harvested and after low speed centrifugation ($500 \times g$, 5 min) was filtered through a 0.45 μm filter. Tenfold dilutions of vector containing supernatant were made in complete medium (DMEM/glutamine/nonessential amino acids/10% fetal bovine serum) and then used in triplicate PERT assays. We have since modified the procedure and dilutions of vector are made in PBS. The bars corresponding to the EIAV dilutions are labeled EIAV neat or EIAV followed by a figure indicating the $\log_{10}$(1/dilution). PERT assays to control for the presence of reverse transcriptase or template in reagents were carried out as shown (bars labeled Media, Media/DB, water/DB, and water). DB indicates particle disruption buffer and mixtures were made at a ratio of 1 : 1. To test for the effect of the particle disruption buffer, a sample in which the disruption step was omitted was also tested (bar labeled EIAV undisrupted). The threshold cycle (Ct) value in this test was 10 cycles lower than observed for undisrupted vector sample and corresponded to the value observed for the 10^{-2} dilution (bar labeled EIAV-2), indicating that the particle disruption buffer efficiently releases RT from particles.

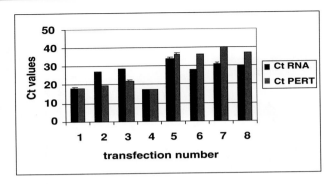

	transfection number							
	1	**2**	**3**	**4**	**5**	**6**	**7**	**8**
gag/pol	+	+	+	+	−	−	−	−
env	+	−	+	−	+	+	−	−
genome	+	−	−	+	−	+	+	−
Ct(RNA)	18.07	27.12	28.7	17.33	33.87	28	31.5	30.4
Ct(PERT)	18	19.2	21.85	17.03	36	36.09	40	37.1

FIG. 6. PERT and viral RNA assays on vector particles produced by transient transfection. 293T cells were transfected with various combinations of plasmids which encode the three components of infectious vector particles as indicated. Where one or more of these plasmids was omitted from the transfection, the same mass of pCIneo plasmid was included instead. Supernatants were harvested 48 hr posttransfection and samples used for RNA preparation and Taqman assay for vector RNA or in PERT assays. Efficient packaging of vector RNA was dependent on the presence of Gag/Pol. The results of vector RNA assays carried out in the absence of reverse transcriptase yielded Ct(RNA) values in the range 31–33.

Collectively, these types of measure have proved effective in minimizing the occurrence of RCR in MLV vector preparations.[78,79] Nevertheless, from examples in which RCR formation occurred in more basic MLV vector systems, it is clear that only small or partial regions of homology are sufficient for recombination.[80,81] Therefore the measures above cannot totally eliminate RCR formation.

Some additional features that will reduce the potential for recombinant formation and the potential pathological consequences arising from the presence of such recombinants in the EIAV vector system are

[78] P. L. Sheridan, M. Bodner, A. Lynn, T. K. Phuong, N. J. Depolo, D. J. de la Vega, J. O'Dea, K. Nguyen, J. E. McCormack, D. A. Driver, K. Townsend, C. E. Ibanez, N. C. Sajjadi, J. S. Greengard, M. D. Moore, J. Respess, S. M. Chang, T. W. Dubensky, D. J. Jolly, and S. L. Sauter, *Mol. Ther.* **2**, 262 (2000).

[79] D. Farson, R. McGuinness, T. Dull, K. Limoli, R. Lazar, S. Jalali, S. Reddy, R. Pennathur-Das, D. Broad, and M. Finer, *J. Gene Med.* **1**, 195 (1999).

[80] H. Chong and R. G. Vile, *Gene Ther.* **3**, 624 (1996).

[81] E. Otto, A. Jones-Trower, E. F. Vanin, K. Stambaugh, S. N. Mueller, W. F. Anderson, and G. J. McGarrity, *Hum. Gene Ther.* **5**, 567 (1994).

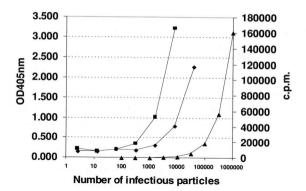

FIG. 7. Comparison of the sensitivity of conventional and ELISA-based reverse transcriptase assays. The same EIAV vector stock as used to characterize the PERT assay (Fig. 5) was used in a conventional reverse transcriptase assay (■) and the ELISA-based Cavidi Tech AB HS-kit Mg^{2+} reverse transcriptase assay. The conventional assay was based on a previously described assay[111] and was based on measurement of incorporation of ^{32}P-labeled thymidine triphosphate into DNA (c.p.m., counts per minute). Using the Calvidi Kit, reverse transcription reactions are performed using BrdUTP on template/primer complexes which are immobilized on multiwell plates. Incorporation of BrdUMP into DNA is then assessed using alkaline phosphatase-conjugated anti-BrdU binding antibody (◆) and absorbance measurement at 405nm (OD405nm). The sensitivity of the assay can be enhanced by prolonged RT reactions (■).

4. The use of a heterologous envelope component. Recombination events cannot generate wild-type EIAV.

5. The use of vectors which have SIN or conditional SIN configurations and therefore are almost transcriptionally silent in transduced cells.

6. The elimination of accessory proteins from the system. Tat, Rev, and S2 can be eliminated from the system without affecting transduction efficiency; however, Rev is maintained in the producer system since it improves titer by increasing cytoplasmic levels of vector RNA. Expression of Rev is from an independent expression cassette, in which *rev* is "codon-optimized" to minimize the chances of its involvement in recombination reactions.

These features provide a high safety margin. However, a scheme for the detection and validation of the sensitivity of detection is required. In developing a test for RCR derived from our EIAV vector system two main issues were borne in mind: first, the types of RCR that might be produced by the vector system and, second, the importance of detecting all classes of RCR using a single assay.

RCR Formation by Recombination

RCR formation takes place by recombination between different components of the vector system or by recombination of vector system components with nucleotide sequences present in the producer cells. Although recombination at the

DNA level during construction of producer cell lines is possible (perhaps leading to insertional activation of endogenous retroelements or retroviruses) it is thought that recombination to produce RCR occurs mainly between RNAs undergoing reverse transcription, hence occurs within the mature vector particles. In consequence, recombination will be more likely to occur between RNAs which contain packaging signals, such as the vector genome and the *gag/pol* mRNA. Usually, however, the *gag/pol* transcript is modified so that it is deleted with respect to some or all defined packaging elements, thereby reducing the chances of its involvement in recombination.

Another source of RNAs which may be involved in recombination are endogenous cellular RNA species. Despite lacking a packaging signal these are incorporated into virions at a low level.[82] In addition, packaging of endogenous retrovirus (ERV) sequences into MLV core particles has been noted. In particular, the RNA of the murine ERV VL30, present in both mouse and rat cells, and thus in packaging cells derived from mouse cells, is packaged and transmitted efficiently in MLV vector particles.[83] In addition, the presence of RCR, formed by recombinations involving murine ERVs, has been reported in rhesus macaques which were recipients of MLV vector preparations.[83] In contrast to the situation for producer cells based on murine cells, a lower level of ERV RNA and packaging of ERV RNA has been reported for packaging cell lines derived from human cells.[84] Furthermore, transfer of a human ERV (HERV-H) from a HEK 293–based MLV vector producer cell line has been reported to occur at extremely low frequency.[79] Nevertheless the packaging of these types of RNA does provide an opportunity for production of novel species through recombination events involving RNA of the vector system, in particular the vector genome RNA.[80]

Recombination between homologous sequences has been shown to require only short sequences with limited homology.[80,81] It is usually considered to involve the coding and LTR regions. However, in experiments to examine RCR formation by an HIV-1-based vector system, recombination has also been shown to occur between polyA sequences at relatively high frequencies.[85] Since poly(A) is a feature of most RNAs, only a single region of sequence overlap between the vector and *gag/pol* transcripts is effectively required for homologous recombination. Therefore, removal of all regions of overlap between the vector and *gag/pol* RNAs is preferable. The use of polyA sequence as a site for recombination also raises the chances of RCR formation as a result of nonhomologous recombination.

[82] M. L. Linial and A. D. Miller, *Curr. Top. Microbiol. Immunol.* **157,** 125 (1990).
[83] D. F. Purcell, C. M. Broscius, E. F. Vanin, C. E. Buckler, A. W. Nienhuis, and M. A. Martin, *J. Virol.* **70,** 887 (1996).
[84] C. Patience, Y. Takeuchi, F. L. Cosset, and R. A. Weiss, *J. Virol.* **72,** 2671 (1998).
[85] X. Wu, J. K. Wakefield, H. Liu, H. Xiao, R. Kralovics, J. T. Prchal, and J. C. Kappes, *Mol. Ther.* **2,** 47 (2000).

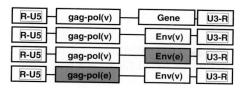

FIG. 8. Structures of all possible replication-competent retroviruses (RCRs). LTRs can be of vector or endogenous origin. All RCRs have either lentiviral gag/pol or the vector system envelope. (v) derived from vector system; (e) of endogenous origin. Assays for these components can be used to detect RCRs. The upper genome is termed a "pseudo" RCR and represents a mobilizable genome.

Taking the above into consideration a number of different types of RCR can be envisaged (Fig. 8). In summary they will have (1) LTRs of either vector or endogenous origin and (2) either *gag/pol* or *envelope* derived from the vector system. These latter characteristics underpin the methods we are developing for RCR detection. As discussed above, the frequency at which a particular RNA is involved in recombination is related to its concentration in the vector particle. Thus, recombinations involving vector, which is relatively enriched in the capsid because of its packaging signal, and other highly expressed RNAs, such as those of the vector system, are the most likely.

Relevance of Test System

In developing an RCR assay for the EIAV vector system we considered the properties of the currently used systems for RCR detection in MLV vector preparations. MLV producer cells were initially derived from mouse cells and more recently from human cells. The presence of RCR was assessed using a mouse cell line based assay, the S^+L^- assay. This assay has been employed in several forms, with the highest sensitivity being obtained from the PG4-S^+L^- assay, which can detect very low amounts of amphotropic MLV used to validate the assay.[86] However, there is the question of the relevance of using mouse cells to detect RCR present in vector made in mouse or human cells, and intended for transduction of human cells. For example, RCR formed by recombination between the vector components and endogenous human sequences may not be amplified in mouse cells due to restrictions at entry or later stages of the viral life cycle. Our aim has been to develop an RCR detection system that overcomes these limitations.

The EIAV-based vector system described above will use human cells for production and will utilize a heterologous envelope such as VSV-G or Rab-G. This system configuration presents two problems: first, the choice of cell line on which to detect RCR, and second, the validation of the assay sensitivity.

---

[86] J. V. Hughes, K. Messner, M. Burnham, D. Patel, and E. M. White, *Dev. Biol. Stand.* **88**, 297 (1996).

Choice of Cell Line for Amplification of RCR

Based on experience gathered from MLV vector systems, The Center for Biologics Evaluation and Research (CBER) recommends that vector preparation and producer cells "should be tested for RCR using a cell line permissive for the RCR most likely to be generated in a given producer cell line." In line with this we have selected the HEK 293 cell line for this purpose since it is competent for production and infection of VSV-G pseudotyped EIAV vector, and hence would be likely to support replication of an EIAV-based RCR carrying a VSV-G or Rab-G envelope. There are several other reasons why HEK 293 are particularly suitable:

1. They support efficient transduction by a wide range of vector core types (MLV, HIV, EIAV, FIV, SIV). Thus reverse transcription and integration of a wide range of RCR are likely to be supported.

2. The HEK 293T cell line, derived from HEK 293, can support transcription, expression and assembly of a wide variety of retroviruses (MLV, EIAV, HIV, FIV, SIV). This underpins its use as the cell of choice for transient production of retroviral vectors.

3. HEK 293 cells can be infected by core particles pseudotyped with a variety of envelopes (amphotropic MLV, GALV, 10A1, Rb-G, FeLV-B, VSV-G, LCMV). Therefore it is likely that the envelope carried by RCR will be able to mediate infection of HEK 293 cells.

The capacity of HEK 293 cells to support replication of different types of RCR is summarized in Table III. One concern over the detection of RCR carrying endogenous envelope is down-regulation of its receptor in target cells, resulting in lack of infection. This may be a particular problem if HEK 293 cells form the basis of the vector production and the RCR detection systems. The chance of HEK 293 cells expressing an envelope capable of pseudotyping EIAV cores is small; however, it could be assessed by transduction of a panel of human cell lines with supernatants from HEK 293 cells transfected with only the plasmids for expression of the vector and *gag/pol* components. Transduction of any of the cells would indicate acquisition of an envelope protein from HEK 293.

Regulatory Requirements for RCR Assay

The FDA has released Draft Guidance for RCR assays on retroviral vectors for clinical use (http://www.fda.gov/cber/guidelines.htm). Recommendations for the volume of supernatant and number of cells to be tested are made. For supernatant testing, 5% of the total supernatant is tested on a permissive cell line. However,

TABLE III
AMPLIFICATION OF DIFFERENT RCR TYPES BY HEK 293 CELLS

RCR Configuration		
Core derivation	Env derivation	Amplification by test cell?
EIAV	Vector	Yes, particles are phenotypically the same as vector particles. Production from and transduction of HEK293 is demonstrable.
EIAV	Endogenous	Yes, but down-regulation of receptor for an endogenous envelope is a possibility which would limit RCR spread in HEK293.
Endogenous	Vector	Yes, since the producer system is based on HEK 293, the LTR activity responsible for RNA encoding the endogenous core is functional. The ability of HEK 293 cells to support production of a wide variety of vector types suggests that formation of a core particle derived from an endogenous sequence will also be supported.
EIAV	None	No, transduction of HEK 293 cells is possible due to VSV-G pseudotyping, but no propagation will occur. This species is termed a "pseudo-RCR" and is not a genuine RCR.

for production volumes of greater than 6 liters an alternative assay is permitted. This assay must be sensitive enough to detect 1 RCR in 100 ml. To achieve this sensitivity with 95% confidence, 300 ml of culture supernatant must be tested. For cell line testing 1% of the producer cells or 10^8 cells, whichever is smaller, must be screened. An additional recommendation is that two independent end point assays be utilized. This advice is based on experience that shows that a single assay may not consistently detect an RCR.

Validation of RCR Assay Sensitivity Using Amphotropic MLV

Assays for detection of RCR in MLV vector preparations that use amphotropic envelope are well established and are validated using an amphotropic MLV stock of known titer available from the ATCC and created for this purpose (VR-1450). Since amphotropic MLV is capable of replication in HEK 293 cells, this stock will also be used to validate the HEK 293-based EIAV RCR assay. It is difficult to envisage a practical alternative. HIV is a human lentivirus but it seems unnecessary to use a pathogen as a positive control in an RCR assay. EIAV per se does not replicate in human cells and it would seem inappropriate to use equine cells to attempt to rescue a human tropic RCR.

Establishing the RCR Assay

An RCR assay for screening amphotropic MLV vector preparations which meets FDA regulations is illustrative. In this assay 300 ml of medium is screened using twelve 75 ml flasks of *Mus dunni* cells. Each flask is inoculated with 25 ml of medium. After five passages the supernatants are used in the PG4-S$^+$L$^-$ assay. The ability of a single RCR (amphotropic MLV) to be detected after inoculation into a single flask under identical conditions was established in prior experiments. It should be noted that the ability to detect a single RCR in 25 ml makes it practical to screen the 300 ml requirement using this approach.

A similar assay development procedure is being established for detection of single infectious particles of amphotropic MLV in HEK 293 cells.

Assays for RCR on Supernatants

A flow diagram for the assay on supernatants from EIAV vector producer cells is shown in Fig. 9. Cell-free supernatants are collected after five passages and used to infect fresh subconfluent HEK 293 cells. These cultures are then used in PERT or PCR-based assays.

Detection of RCR by PERT Assay. Supernatants are collected at daily intervals for use in PERT assays rather than in a cell-based indicator assay similar to the PG4-S$^+$L$^-$ assay. The application of the PERT assay to detection of reverse transcriptase of diverse origins has been demonstrated previously;[75,76] hence it is ideally suitable for use in an RCR assay where the properties of the RCR are unknown. Furthermore it is extremely sensitive. An increase in the signal observed in PERT assays carried out on supernatants collected daily is indicative of an increasing level of infection by RCR.

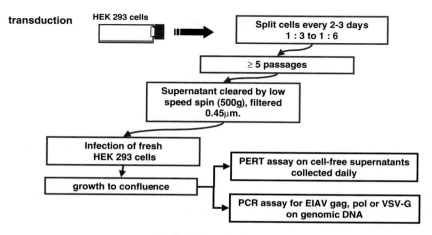

FIG. 9. RCR assay scheme.

Detection of RCR by PCR. All RCR will have sequences derived from either the vector system *gag/pol* or *env* (Fig. 8). Therefore detection of such sequences in total cellular DNA prepared once the fresh HEK 293 cell sheet has reached confluence is indicative of infection by RCR. Assays capable of detecting 1–10 copies of either EIAV synthetic *gag, pol,* or *VSV-G* in 5×10^5 cell equivalents of DNA have been developed by The Laboratory of the Government Chemist (Queens Road, Teddington, Middlesex, TW11 0LY, UK) using cell lines which have stably integrated copies of these sequences at known copy numbers.

Assays for RCR in Producer Cells

A similar assay will be used to screen producer cells for RCR. Producer cells, 1% or 10^8, will be cultured at a 1 : 1 ratio with HEK 293 cells. After five passages, cell-free supernatant will be passed to fresh HEK 293 cells and PERT and PCR assays employed to detect RCR as described above.

Applications

The ability of primate and nonprimate lentiviral vectors to deliver genes to nondividing cells and to provide long-term gene expression distinguishes them from other viral delivery systems, with perhaps the exception of adeno-associated virus vectors. As such there are many novel areas in which they will be used: as academic research tools, in gene discovery programs, and for gene therapy. In gene therapy there are three particularly promising areas of activity. These are for gene delivery to the brain, for transduction of hematopoietic stem cells, and for creation of "cellular factories" for the production of systemic factors.

In the brain, efficient delivery of reporter genes has been demonstrated for both HIV-1 and EIAV vectors and, for both, undiminished expression has been observed for at least 8 months (last time point checked).[4,55,59] The first demonstrations have been made of functionally relevant levels of transgene expression achieved using the HIV-1 vector system. These experiments include neuroprotection of substantia nigra dopaminergic neurons in mouse, rat and monkey Parkinsonian models using glial-derived neurotrophic factor,[87–90] rescue of facial motoneurons using the same

[87] J. C. Bensadoun, N. Deglon, J. L. Tseng, J. L. Ridet, A. D. Zurn, and P. Aebischer, *Exp. Neurol.* **164,** 15 (2000).

[88] N. Deglon, J. L. Tseng, J. C. Bensadoun, A. D. Zurn, Y. Arsenijevic, L. Pereira de Almeida, R. Zufferey, D. Trono, and P. Aebischer, *Hum. Gene Ther.* **11,** 179 (2000).

[89] C. Rosenblad, M. Gronborg, C. Hansen, N. Blom, M. Meyer, J. Johansen, L. Dago, D. Kirik, U. A. Patel, C. Lundberg, D. Trono, A. Bjorklund, and T. E. Johansen, *Mol. Cell Neurosci.* **15,** 199 (2000).

[90] J. H. Kordower, M. E. Emborg, J. Bloch, S. Y. Ma, Y. Chu, L. Leventhal, J. McBride, E. Y. Chen, S. Palfi, B. Z. Roitberg, W. D. Brown, J. E. Holden, R. Pyzalski, M. D. Taylor, P. Carvey, Z. Ling, D. Trono, P. Hantraye, N. Deglon, and P. Aebischer, *Science* **290,** 767 (2000).

growth factor,[91] rescue of photoreceptors in a retinitis pigmentosa mouse model by expressing the rod photoreceptor cGMP phosphodiesterase β subunit gene,[92] and reversal of pathology in the entire brain of a mucopolysaccharidosis type VII mouse model by expressing β-glucuronidase in multiple brain regions.[93] We are pursuing a Parkinson's disease therapy in which three important genes of the biosynthetic pathway for the neurotransmitter dopamine are expressed from a tricistronic EIAV vector. The strategy is to transduce cells in the striatum and to thereby create a source of dopamine in the region where the neurotransmitter is deficient because of the death of dopaminergic neurons in the substantia nigra.

Hematopoietic stem cells (HSC) have long been a target for MLV-based vector systems; however, despite achieving high levels of transduction following stimulation of cell division by cytokines, the levels of repopulation observed have not been high. This is not a particular problem if there is a strong selection for transduced cells once they are re-administered;[94] however, for some applications it presents a severe limitation.[95] HIV-1 vectors have been shown to efficiently transduce unstimulated HSC by many groups[65,96–103] and they have obvious potential for the treatment of diseases of the blood system. This view is confirmed by a recent report in which delivery of human β-globin to bone marrow of β-thalassemic mice resulted in restoration of globin tetramer to levels in the physiological range.[104]

Another application of lentiviruses is in the production of systemic factors. A good example of this principle is the demonstration of HIV-1 vector mediated delivery of human clotting factors VIII and IX to mouse liver, leading to therapeutic

[91] A. F. Hottinger, M. Azzouz, N. Deglon, P. Aebischer, and A. D. Zurn, *J. Neurosci.* **20,** 5587 (2000).

[92] M. Takahashi, H. Miyoshi, I. M. Verma, and F. H. Gage, *J. Virol.* **73,** 7812 (1999).

[93] A. Bosch, E. Perret, N. Desmaris, D. Trono, and J. M. Heard, *Hum. Gene Ther.* **11,** 1139 (2000).

[94] M. Cavazzana-Calvo, S. Hacein-Bey, G. de Saint Basile, F. Gross, E. Yvon, P. Naubaum, F. Selz, C. Hue, S. Certain, J. L. Casanova, P. Bousso, F. L. Deist, and A. Fischer, *Science* **288,** 669 (2000).

[95] M. Bhatia, D. Bonnet, U. Kapp, J. C. Wang, B. Murdoch, and J. E. Dick, *J. Exp. Med.* **186,** 619 (1997).

[96] R. K. Akkina, R. M. Walton, M. L. Chen, Q. X. Li, V. Planelles, and I. S. Chen, *J. Virol.* **70,** 2581 (1996).

[97] J. Reiser, G. Harmison, S. Kluepfel Stahl, R. O. Brady, S. Karlsson, and M. Schubert, *Proc. Natl. Acad. Sci. U.S.A.* **93,** 15266 (1996).

[98] H. Miyoshi, K. A. Smith, D. E. Mosier, I. M. Verma, and B. E. Torbett, *Science* **283,** 682 (1999).

[99] N. B. Woods, C. Fahlman, H. Mikkola, I. Hamaguchi, K. Olsson, R. Zufferry, S. E. Jacobsen, D. Trono, and S. Karlsson, *Blood* **96,** 3725 (2000).

[100] N. Uchida, R. E. Sutton, A. M. Friera, D. He, M. J. Reitsma, W. C. Chang, G. Veres, R. Scollay, and I. L. Weissman, *Proc. Natl. Acad. Sci. U.S.A.* **95,** 11939 (1998).

[101] R. E. Sutton, M. J. Reitsma, N. Uchida, and P. O. Brown, *J. Virol.* **73,** 3649 (1999).

[102] S. S. Case, M. A. Price, C. T. Jorden, X. J. Yu, L. Wang, G. Bauer, D. L. Haas, D. Xu, R. Stripecke, L. Naldini, D. B. Kohn, and G. M. Crooks, *Proc. Natl. Acad. Sci. U.S.A.* **96,** 2988 (1999).

[103] J. Douglas, P. Kelly, J. T. Evans, and J. V. Garcia, *Hum. Gene Ther.* **10,** 935 (1999).

[104] C. May, S. Rivella, J. Callegari, G. Heller, K. M. Gaensler, L. Luzzatto, and M. Sadelain, *Nature* **406,** 82 (2000).

levels of expression.[105] Another tissue that could be targeted as a good site for the formation of systemic factors is skeletal muscle for which transduction has been reported for HIV-1, FIV, and EIAV vectors[7,106] (our unpublished data). Another class of agents that might be amenable to delivery by lentiviral vectors are anti-angiogenic factors, which show promise for the treatment of cancer. Currently the use of such factors is limited by production difficulties. A way to overcome this is to generate the factors *in vivo*. This has been demonstrated using adenovirus-based vectors.[107,108] and HIV-1 based vectors.[109] The advantage of the lentivirus-based vectors is their ability to give long-term expression that appears to be the preferred method for administration of these factors.

Lentiviral vectors are also particularly suited for use in gene discovery and target validation programs by virtue of their ability to transduce many types of cells that are not amenable to transduction by other methods. Examples of such cells are highly differentiated cells, such as neuronal cells or granulocytes, and primitive, multipotent cells such as HSC or stem cells of other lineages. An additional advantage of lentiviral vectors, which is important for success in this application, is that they cause minimal disturbance to the recipient cell as a result of expressing only the selected transgene. In addition it would be possible to use lentiviral vectors to generate and screen cDNA or random peptide libraries as described previously for MLV vectors.[110] These are then used to transduce cells and assay for the phenotype of interest. The exploration of the range of applications of the nonprimate lentiviral vectors is still underway. It is anticipated that many, if not all, of the cells that can be transduced by HIV-based vectors will be amenable to vectors such as those based on EIAV. To date this is confirmed by studies with brain,[4,59] muscle, and various hematopoietic cells (unpublished data).

Conclusion

The development of minimal lentiviral vectors over the past half-decade places us now at the beginning of a period where the vectors will be used in the clinic.

[105] F. Park, K. Ohashi, and M. A. Kay, *Blood* **96,** 1173 (2000).

[106] T. Kafri, U. Blomer, D. A. Peterson, F. H. Gage, and I. M. Verma, *Nat. Genet.* **17,** 314 (1997).

[107] C. T. Chen, J. Lin, Q. Li, S. S. Phipps, J. L. Jakubczak, D. A. Stewart, Y. Skripchenko, S. Forry-Schaudies, J. Wood, C. Schnell, and P. L. Hallenbeck, *Hum. Gene Ther.* **11,** 1983 (2000).

[108] B. V. Sauter, O. Martinet, W. J. Zhang, J. Mandeli, and S. L. Woo, *Proc. Natl. Acad. Sci. U.S.A.* **97,** 4802 (2000).

[109] A. Pfeifer, T. Kessler, S. Silletti, D. A. Cheresh, and I. M. Verma, *Proc. Natl. Acad. Sci. U.S.A.* **97,** 12227 (2000).

[110] X. Xu, C. Leo, Y. Jang, E. Chan, D. Padilla, B. C. Huang, T. Lin, T. Gururaja, Y. Hitoshi, J. B. Lorens, D. C. Anderson, B. Sikic, Y. Luo, D. G. Payan, and G. P. Nolan, *Nat. Genet.* **27,** 23 (2001).

[111] S. T. Perry, M. T. Flaherty, M. J. Kelley, D. L. Clabough, S. R. Tronick, L. Coggins, L. Whetter, C. R. Lengel, and F. Fuller, *J. Virol.* **66,** 4085 (1992).

The minimal EIAV vector system described here is configured with clinical applications in mind. Clearly the most effective vector must be chosen for the specific therapeutic or analytical process. The EIAV vector system that we have described has a high safety profile and assays will soon be in place to monitor safety during production and in the clinic. Provided they achieve the required therapeutic index the EIAV vectors will be translated to the clinic. The immediate challenges are to solve the problems of vector production and to design vectors for gene delivery to specific tissues. This is most conveniently achieved using tissue-specific promoters. The use of regulatable vectors in which expression of the transgene can be completely ablated may also be required for some applications. Another way of achieving tissue specific expression following *in vivo* administration is by targeting the vector to the cells of interest. Despite much work in this area there are no standard methods for vector targeting. This still represents a major challenge.

Addendum

Protocol 1: Transient Vector Production

A detailed method for transfection of HEK 293T cells is not described here as most laboratories will have developed their own variations on the widely used calcium phosphate precipitation method for transfection. Perhaps less widely publicized are details of vector harvesting, concentration, and storage, and these are detailed below.

Harvesting and Concentration of Vector

Following transfection the cells are incubated overnight. The medium is then changed for complete medium (DMEM, 10% fetal calf serum, glutamine, nonessential amino acids) supplemented with sodium butyrate (10 mM). After 7 hr the medium is collected (first harvest) and medium without sodium butyrate placed on the cells. Following overnight incubation a second harvest is carried out.

Procedure

1. Carefully pipette off the supernatant and transfer to 500 ml Corning polypropylene conical centrifuge bottles (Fisher CFT-410-010A) (Support cushions for tubes Fisher CFT-410-030R).

2. Replenish the cells with another 6 ml of warmed medium and reincubate at 37° overnight.

3. Centrifuge the collected supernatant for 15 min at 1500 rpm and 4° in a benchtop centrifuge to remove cell debris.

4. Filter the supernatant through a prewetted filter [either Nalgene 158-0045 (Fisher, TKV-220-110C), 0.45 μm cellulose acetate, 1000 ml, or Corning PES 0.22 μm, 1000 ml (Fisher TKV-700-0611); note that the maximum capacity of these filters has been found to be 450 ml of medium containing 10% FCS].

5. Take two samples of 500 μl after filtration for titration, and store at $-80°$. These are used to assess recovery after centrifugation and are particularly important in the evaluation of novel envelopes.

6. Transfer the filtered supernatant to sterile Beckman centrifuge bottles (if using the JLA 10.500 rotor do not put more than 375 ml in any one bottle) and spin at the appropriate speed. (The centrifugation time can be calculated based on the total volume and adjusted k factor of the particular rotor used. Assuming a virus sedimentation coefficient of 600 Svedbergs and using a Beckman JLA 10.500 rotor the following values are obtained. For a volume of 375 ml and RPM of $6000g$, the adjusted k factor is 5516 and theoretically the virus should sediment in 9.2 hr. For convenience the spin can be carried out overnight.)

7. To make the position of the pellets easier to find, mark the bottles with an arrow and position them in the rotor buckets, with the arrows facing to the outside.

8. At the end of the spin, decant the supernatant into disinfectant and then carefully remove all the supernatant with a micropipette.

9. At this point in the procedure the virus may be directly resuspended and aliquotted for immediate use or for storage at $-80°$ or may be further concentrated by ultracentrifugation. The viral pellet is resuspended by gently pipetting in the appropriate volume of the desired buffer [e.g., phosphate buffered saline (PBS) or formulation buffer]. Vector stored in formulation buffer appears more stable than vector stored in PBS.

Formulation buffer (100 ml):

Tissue culture grade water	28.65 ml
19.75 mM Tris-HCl buffer pH 7.0	19.75 ml of a 0.1 M solution
40 mg/ml lactose	26.6 ml of a 150 mg/ml solution
37.5 mM sodium chloride	24.4 ml of a 154 mM solution
1 mg/ml human serum albumin [human serum albumin (20%): Albutein, Alpha therapeutics UK Ltd., Thetford, Norfolk]	5 ml of a 20% solution
5 μg/ml protamine sulfate (protamine sulfate 5 mg/ml: Prosulf, CP Pharmaceuticals, Wrexham, UK)	100 μl of a 5 mg/ml solution

Protocol 2: Titration by Measurement of Integrated Vector Sequences

Transductions are carried out in duplicate with the markerless vector(s) and a reference standard vector (for example pONY8.0Z) for which an accurate biological titer has been established. The appropriate time point for preparation of DNA from transduced cells needs to be determined for each target cell type. Measurement of DNA is carried out using a ABI Prism 7700 Sequence Detection System.

Procedure

1. Transductions of target cells are carried out in 12-well plates in duplicate with the test sample and a reference standard vector. For vector with an estimated titer of 10^6 tu/ml, dilutions of 10^{-1} and 10^{-2} are plated out and cells passaged as required up to the time of DNA isolation. For vector of higher titer corresponding dilutions are used.

2. Total cellular DNA is isolated from transduced cells using a QIAamp DNA Mini Kit. Using D17 cells we isolate DNA from approximately 5×10^5 cells and the concentration of DNA is determined by absorbance measurement at 260 nm.

3. Setting up Taqman reactions: Reactions are carried out separately to measure β-actin and the selected target sequence in the vector. We have used primer/probe sets to detect the internal CMV promoter, the transgene, and the "packaging signal." The last primer/probe set does not detect wild-type *gag/pol* sequence.

4. Preparation of reaction mix. Sufficient reaction mix for triplicate reactions for each sample should be prepared, preferably in a PCR clean room. The total number of samples to be analyzed: 4 reference standards, 2 "no template" DNA control samples, plus unknowns. Our experience is that triplicate reactions yield sufficiently accurate data.

5. Reaction mix components for a single 25 μl reaction: 100 ng of cellular DNA, 12.5 μl of 2× Master Mix (Applied BioSystems), x μl forward primer, y μl reverse primer, z μl probe and water to 25 μl. Optimal concentrations of primer/probe are determined according to methods recommended by Applied BioSystems. In practice a supermix containing primer/probe, master mix, and water is made up and an appropriate volume for 3.5 reactions added to 350 ng of cellular DNA in an appropriate volume of water. The mixture is then mixed well and 25 μl aliquots added to the optical plate. The plate is capped and spun at 1100 rpm for 3 min.

6. Cycling conditions on ABI Prism 7700 Sequence Detection System: hold, 50° for 2 min; hold, 95° for 10 min; 40 cycles, 95° for 15 sec; 60° for 1 min.

7. Data interpretation: As described in user bulletins available at http://docs.appliedbiosystems.com/ (Tech support/Documents on Demand/).

Protocol 3: Vector RNA Assay by Taqman

Vector RNA is isolated in a two-stage process. In the first, RNA is isolated from tissue culture fluid and in the second the isolated RNA is treated with DNAse to remove contaminating DNA, in particular plasmid DNA present as a result of the transfection. RNA is quantitated using a one-step reverse transcriptase-PCR Taqman assay perfomed on an ABI Prism 7700 Sequence Detection System. This method can be readily adapted for measurement of any RNA.

Procedure

1. Viral RNA isolations are carried using a Qiagen viral RNA extraction kit. For concentrated viral vector stocks (typically 5×10^8–3×10^9 tu/ml), a 10 μl aliquot is diluted to 140 μl in PBS prior to starting the extraction procedure. For vector stocks of lower titer 140 μl of harvested supernatant is used; 140 μl of the reference standard is required for the assay. A single preparation of RNA is made for each sample. Tests in which preparation of RNA from the same vector stock were carried out in parallel show that the reproducibility of the procedure is very good. The RNA is resuspended in a final volume of 120 μl and stored at $-80°$ in 20 μl aliquots.

2. DNase I treatment: 5 μl of viral RNA prep is added to an Eppendorf tube containing 5 μl of 10× DNase I reaction buffer (100 mM Tris-HCl pH 7.5, 25 mM MgCl$_2$, 1 mM CaCl$_2$), 37 μl of nuclease-free water, and 3 μl of (2 U/μl) DNase I from Ambion (DNA-free Kit), made up as a supermix. The reaction is incubated at 37° for 30 min.

3. Inactivation of the enzyme and removal of divalent cations is achieved by using the DNase I Inactivation Reagent supplied in the kit. Add 5 μl of this reagent to the reaction and flick the tube to mix the reagent. Incubate the tube for 2 min at room temperature. Flick the tube once more. Centrifuge the tube for ~1 min at 21,000g in a microfuge to pellet the DNase Inactivation Reagent. Remove the supernatant solution containing the viral RNA to a fresh tube.

4. Vector RNA dilutions prior to setting up Taqman reactions: Serial 10-fold dilutions of the template RNA in nuclease-free water are carried out. 10^{-3} and 10^{-4} dilutions of RNA from concentrated vector samples and 10^{-1} and 10^{-2} dilutions of RNA from unconcentrated vector stocks are made. For the reference standard vector 10^{-1}, 10^{-2}, 10^{-3}, and 10^{-4} dilutions are made. These and the undiluted vector RNA are used to construct the standard curve. This range of standards covers a broad Ct range.

5. Preparation of reaction mix: Sufficient reaction mix for triplicate reactions with each sample should be prepared, preferably in a PCR clean room. The total number of samples to be analyzed: 5 standards, 1 "no RNA" control sample, plus unknowns. Our experience is that triplicate reactions yield sufficiently accurate data. Two master mixes are prepared: plus and minus reverse transcriptase (RT). The reactions carried out in the absence of RT are included to allow an assessment of the contribution made to the signal by DNA.

6. Reaction mix components for triplicate reactions:

	Volume (μl)	Final concentration
10× TaqMan Buffer A	10	1×
25 mM Magnesium Chloride	22	5.5 mM
10 mM dATP	3	300 μM
10 mM dCTP	3	300 μM
10 mM dGTP	3	300 μM
20 mM dUTP	3	600 μM
Forward primer(*)	x	
Reverse primer(*)	y	
Probe(*)	z	
AmpliTaq Gold	0.5	0.025 U/μl
MultiScribe RT(**)	0.5	
RNase inhibitor	2	0.4 U/μl
Nuclease free water	Up to 90 μl	

*The concentration of a defined set of primers and probe has to be determined by optimizing the reaction prior to these experiments. As for Protocol 2 the primer/probe set is chosen according to the sequence which is to be detected

**In the minus RT reaction, this volume is replaced by nuclease-free water.

7. Setting up the Taqman reactions: 10 μl aliquots of diluted template RNA are added to tubes containing 90 μl aliquots of each reaction mix. The contents are mixed by vortexing and briefly collected by centrifugation, then 25 μl aliquots are transferred to the optical plate. The plate is capped and spun at 1100 rpm for 3 min.

8. Cycling conditions on the ABI Prism 7700 Sequence Detection System: hold, 48° for 30 min; hold, 95° for 10 min; 40 cycles, 95° for 15 sec, 60° for 1 min.

9. Data interpretation: as described in user bulletins available at http://docs. appliedbiosystems.com/ (Tech support/Documents on Demand/).

Protocol 4: PERT Assay Procedure

The rationale for the PERT assay is described in the main text.

Procedure

1. Template preparation: Dilutions of reference standard vector stock of known titer and the test vector preparations are made in PBS. The reference standard is diluted to 10^{-2}, 10^{-3}, 10^{-4}, and 10^{-5}. Dilutions of vector with titers of 10^{-6} tu/ml are used in the PERT assay at 10^{-1} and 10^{-2} dilutions and vector stocks with higher titer vector stocks are diluted to be in this range (for concentrated viral

stocks ($\approx 10^8$–10^9 tu/ml); prepare 10^{-4} and 10^{-5} dilutions). The particles are then disrupted by adding 25 μl of diluted vector to 25 μl of particle disruption buffer[76]: 40 mM Tris-HCl, pH 7.5 (20°), 50 mM KCl, 20 mM DTT, 0.2% NP-40. Made up using RNase-free reagents. The mixture is agitated briefly to mix and then centrifuged briefly.

2. Preparation of reaction mix: Sufficient reaction mix for triplicate reactions of each sample should be prepared, preferably in a PCR clean room. The total number of samples to be analyzed: 4 standards, 1 "no vector" RNA control, 1 no MS2 RNA control sample, plus unknowns. Duplicate reactions yield sufficiently accurate data.

3. Reaction mix components for duplicate PERT assay:

	Volume (μl)	Final [ ]
10× TaqMan Buffer A	7	1×
25 mM Magnesium chloride	15.4	5.5 mM
10 mM dATP	2.1	300 μM
10 mM dCTP	2.1	300 μM
10 mM dGTP	2.1	300 μM
20 mM dUTP	2.1	600 μM
PERT Forward primer	4.2	300 nM
PERT Reverse primer	4.2	300 nM
PERT Probe	2.1	150 nM
AmpliTaq Gold	0.35	0.025 U/μl
RNase inhibitor	1.4	0.4 U/μl
Nuclease free water	19.4	

All components of the reaction mix, except MS2 RNA, are dispensed in the PCR clean room. MS2 RNA is obtained from from Roche as a 0.8 mg/ml stock. It is aliquotted into 25 μl lots and stored at $-20°$. PERT forward primer: 5' TCCTGCT CAACTTCCTGTCGA; PERT reverse primer: CACAGGTCAAACCTCCTAG GAATG; PERT probe: CGAGACGCTACCATGGCTATCGCTGTAG-(TAMRA).

4. Setting up the reactions: A 54 μl aliquot of the reaction mix is removed to act as the "no MS2 RNA" control, then 0.525 μl of MS2 RNA per duplicate PERT assay is added. The mixture is vortexed briefly and collected by brief centrifugation, and then 54 μl of the final reaction supermix is aliquotted into the required number of microcentrifuge tubes. Six μl of the sample is then added and after gentle mixing, 25 μl aliquots of sample/reaction mix are added to each of 2 wells of the optical plate. Cap plate and spin at 1100 rpm for 3 min.

5. RT-PCR conditions on the ABI Prism 7700 Sequence Detection System are the same as in Protocol 3.

Acknowledgments

The authors thank Fraser J. Wilkes, Anna L. Olsen, and Lucy Walmsley and all those at Oxford BioMedica who have contributed to the work described herein. In addition, we thank Dr. John C. Olsen, University of North Carolina at Chapel Hill, for many useful discussions.

[28] Gene Transfer to Airway Epithelia Using Feline Immunodeficiency Virus-Based Lentivirus Vectors

By Guoshun Wang, Patrick L. Sinn, Joseph Zabner, and Paul B. McCray, Jr.

Introduction

Vector-mediated gene transfer to airway epithelia offers the potential to correct genetic lung diseases such as cystic fibrosis (CF). CF, an autosomal recessive disease, is the most common lethal genetic disease among Caucasians and as such serves as a prototype for the development of gene-based therapies for the lung.[1] Mutations in the cystic fibrosis transmembrane conductance regulator (CFTR) gene cause CF.[2] CFTR is an epithelial cAMP- and nucleoside-activated chloride (Cl) channel that plays an important role in regulating electrolyte and liquid transport across epithelia. For most CF patients, chronic bacterial infection resulting from defective pulmonary host defenses gradually destroys the lungs. The well-characterized genetic basis of CF and the perceived accessibility of the airway epithelium by aerosol delivery or direct instillation have generated significant interest in gene transfer-based approaches to correct CF pulmonary disease.

While the fundamental concepts of gene complementation for the treatment of CF were proved soon after the discovery of the CFTR gene,[3,4] the implementation of gene transfer-based therapies for patients with CF has proven to be a formidable task. As a critical interface between the outside environment and the host, the airway epithelium has evolved many defense barriers to hold microbes at bay. Not

[1] M. J. Welsh, T. F. Boat, L.-C. Tsui, and A. L. Beaudet, *in* "The Metabolic and Molecular Basis of Inherited Disease" (C. R. Scriver, A. L. Beaudet, W. S. Sly, and D. Valle, eds.), p. 3799. McGraw-Hill, Inc., New York, 1995.

[2] J. R. Riordan, J. M. Rommens, B. Kerem, N. Alon, R. Rozmahel, Z. Grzelczak, J. Zielenski, S. Lok, N. Plavsic, J. I. Chou, M. L. Drumm, M. C. Iannuzzi, F. S. Colin, and L.-C. Tsui, *Science* **245,** 1066 (1989).

[3] D. P. Rich, M. P. Anderson, R. J. Gregory, S. H. Cheng, S. Paul, D. M. Jefferson, J. D. McCann, K. W. Klinger, A. E. Smith, and M. J. Welsh, *Nature* **347,** 358 (1990).

[4] M. L. Drumm, H. A. Pope, W. H. Cliff, J. M. Rommens, S. A. Marvin, L. C. Tsui, F. S. Collins, R. A. Frizzell, and J. M. Wilson, *Cell* **62,** 1227 (1990).

surprisingly, the same barriers thwart the efficiency of gene transfer with viral and nonviral vectors.

Several vectors are under investigation in both preclinical and clinical studies. An ideal vector system for CF gene therapy would readily transduce airway cells with progenitor capacity from the luminal surface, correct enough cells to reverse the physiologic defect (estimated ~6–10%[5]), and exhibit low immunogenicity. Currently, no such vector exists. Our laboratory has focused on the development of lentivirus-based vectors for several reasons. Lentivirus-based systems have the ability to transduce nondividing cells,[6–9] an important consideration in the airways where most cells are mitotically quiescent.[10] Retroviral vectors integrate, and therefore a correct copy of the CFTR gene is passed on to daughter cells as well. Finally, retroviral vectors are less immunogenic than the present encapsidated vectors, allowing for readministration.[11] We and others have demonstrated the feasibility of correcting the CF Cl transport defect using retroviral vectors.[7,9,12,13] Thus lentiviral vectors are well suited for the job at hand.

Our approach focuses on the use of nonprimate retroviral vectors derived from feline immunodeficiency virus (FIV). A first-generation lentivirus vector derived from FIV was originally described by Poeschla et al.[14] Similar to HIV, FIV-based vectors transduce nondividing cells.[14,15] Since wild-type FIV is genetically and antigenically distinct from HIV and does not infect human cells or cause disease in humans,[16] FIV-based vectors may offer additional safety features over HIV-based systems.[15] Johnston and colleagues[15] developed a second-generation FIV vector in which unnecessary trans-acting elements (vif, orf 2) were deleted, further reducing the possibility of production of replication-competent virus. Pseudotyping the

[5] L. G. Johnson, J. C. Olsen, B. Sarkadi, K. L. Moore, R. Swanstrom, and R. C. Boucher, Nat. Genet. 2, 21 (1992).

[6] L. Naldini, U. Blomer, P. Gallay, D. Ory, R. Mulligan, F. H. Gage, I. M. Verma, and D. Trono, Science 272, 263 (1996).

[7] G. Wang, V. Slepushkin, J. Zabner, S. Keshavjee, J. C. Johnston, S. L. Sauter, D. J. Jolly, T. W. Dubensky, Jr., B. L. Davidson, and P. B. McCray, Jr., J. Clin. Invest. 104, R55 (1999).

[8] L. Johnson, J. Olsen, L. Naldini, and R. Boucher, Gene Ther. 7, 568 (2000).

[9] M. J. Goldman, P.-S. Lee, J.-S. Yang, and J. M. Wilson, Hum. Gene Ther. 8, 2261 (1997).

[10] S. G. Shami and M. J. Evans, in " Comparative Biology of the Normal Lung" (R. A. Parent, ed.), p. 145. CRC Press, Boca Raton, FL, 1991.

[11] J. E. McCormack, D. Martineau, N. DePolo, S. Maifert, L. Akbarian, K. Townsend, W. Lee, M. Irwin, N. Sajjadi, D. J. Jolly, and J. Warner, Hum. Gene Ther. 8, 1263 (1997).

[12] G. Wang, B. L. Davidson, P. Melchert, V. A. Slepushkin, H. H. van Es, M. Bodner, D. J. Jolly, and P. B. McCray, Jr., J. Virol. 72, 9818 (1998).

[13] J. C. Olsen, L. G. Johnson, M. J. Stutts, B. Sarkadi, J. R. Yankaskas, R. Swanstrom, and R. C. Boucher, Hum. Gene Ther. 3, 253 (1992).

[14] E. M. Poeschla, F. W-Staal, and D. L. Looney, Nat. Med. 4, 354 (1998).

[15] J. C. Johnston, M. Gasmi, L. E. Lim, J. H. Elder, J. K. Yee, D. J. Jolly, K. P. Campbell, B. L. Davidson, and S. L. Sauter, J. Virol. 73, 4991 (1999).

[16] K. Hartmann, Vet. J. 155, 123 (1998).

recombinant FIV with envelope glycoproteins such as VSV-G allows the vector to transduce a wide variety of mammalian cells, including human.[17] We will describe our methods for using the FIV-based vector to transduce airway epithelia *in vitro* and *in vivo*.

Methods

Model Systems for Investigating Gene Transfer to Airway Epithelia

A critical issue facing investigators examining gene transfer to the airway is the choice of an appropriate model system for conducting *in vitro* and *in vivo* experiments. The respiratory epithelium has evolved a series of barriers to prevent viral infections and an appropriate model system should manifest these obstacles. Such barriers include a mucous layer secreted by the epithelium designed to bind and clear foreign inhaled particles, a ciliated surface, airway surface liquid secreted by the cells, an extracellular matrix that may deter viral binding to cell surface receptors, and an apical cell membrane lacking receptors for many viral vectors (such as those for adenovirus[18,19] and AAV2.[20] In some cases, viral receptors are present only on the basolateral membrane of polarized airway epithelia. The list of potential barriers does not end with binding; viral entry may be further discouraged by the low basal rate of endocytosis at the apical surface. Calcium chelators have been successfully used to open epithelial junctions and provide access to the basolateral cell surface (discussed below). This approach offers an alternative strategy for achieving viral vector entry by increasing access to receptors.[7,12,19,21] Furthermore, a number of intracellular barriers may follow, including blocks to intracellular trafficking, translocation into the nucleus, and genomic integration steps.

In addition to these transduction barriers, the targeting of the appropriate cell types remains an important issue. In order to persistently correct a genetic pulmonary disease using integrating viral vectors it is necessary to target a population of cells with progenitor capacity. Therefore, the airway cells with progenitor capacity must be identified and viral vectors must be designed to target those cells. Such cell types may include basal cells, intermediate cells, and some ciliated surface cells.[22–25] The identification of envelopes with the capacity to target lentiviral

[17] J. C. Burns, T. Friedmann, W. Driever, M. Burrascano, and J.-K. Yee, *Proc. Natl. Acad. Sci. U.S.A.* **90,** 8033 (1993).

[18] R. J. Pickles, J. A. Fahrner, J. M. Petrella, R. C. Boucher, and J. M. Bergelson, *J. Virol.* **74,** 6050 (2000).

[19] R. W. Walters, T. Grunst, J. M. Bergelson, R. W. Finberg, M. J. Welsh, and J. Zabner, *J. Biol. Chem.* **274,** 10219 (1999).

[20] D. Duan, Y. Yue, P. B. McCray, Jr., and J. F. Engelhardt, *Hum. Gene Ther.* **9,** 2761 (1998).

[21] G. Wang, J. Zabner, C. Deering, J. Launspach, J. Shao, M. Bodner, D. J. Jolly, B. L. Davidson, and P. B. McCray, Jr., *Am. J. Respir. Cell. Mol. Biol.* **22,** 129 (2000).

[22] C. G. Plopper, S. J. Nishio, A. P. Kass, and D. M. Hyde, *Am. J. Respir. Cell Mol. Biol.* **7,** 606 (1992).

vector entry via the apical surface in these cell types will likely confer persistent gene expression.

The following is a brief list of commonly used model systems and some of their advantages and disadvantages.

In Vitro Models

In vitro models for studying gene transfer to airway epithelia simplify the number of variables that influence gene transfer efficiency *in vivo*. A key issue for conducting informative *in vitro* gene transfer experiments is the use of cell culture models appropriate to the questions being asked. Fundamentally, the cell culture system of choice should express the viral receptors of interest endogenously and respond appropriately to physiological stimuli known to accompany viral transduction (such as receptor down-regulation or relocalization). In the case of respiratory epithelial models, cells should be polarized and mirror transduction properties of the apical and basolateral membranes of the *in vivo* epithelium.

Immortalized Cell Lines. While many transformed cell lines derived from respiratory epithelia are available (A549, H441, IB-3, etc.), viral vector transduction is often investigated in polarized epithelial cell lines such as MDCK, CaCo-2, or Vero C1008. These cell lines may be readily cultured in a polarized fashion and have been used to demonstrate the asymmetric infection, maturation, and release of a number of viruses. These cell models have the advantages of ease of culture and availability; however, they may not possess the representative extracellular barriers and receptor distributions a viral vector encounters in the lung *in vivo*. In addition, they lack the differentiated cell diversity that is required to answer questions concerning the targeting of airway cells with progenitor capacity.

Primary Culture of Well-Differentiated Human Airway Epithelia at Air–Liquid Interface. A great deal of evidence has demonstrated that this culture model is representative of the *in vivo* airways.[26–28] The differentiated epithelia achieved with this model form a pseudostratified columnar epithelium consisting of an assortment of pulmonary cells representative of the conducting airways, including ciliated cells, intermediate cells, secretory cells (goblet cells), and basal cells. However, access to primary cultures is dependent on tissue availability and the cultures are more labor intensive to set up and maintain. Our laboratory uses this model extensively and a description of the methods is included here.

[23] N. F. Johnson and A. F. Hubbs, *Am. J. Respir. Cell Mol. Biol.* **3**, 579 (1990).

[24] Y. Inayama, G. E. Hook, A. R. Brody, G. S. Cameron, A. M. Jetten, L. B. Gilmore, T. Gray, and P. Nettesheim, *Lab. Invest.* **58**, 706 (1988).

[25] M. J. Evans, L. V. Johnson, R. J. Stephens, and G. Freeman, *Lab. Invest.* **35**, 246 (1976).

[26] M. Yamaya, W. E. Finkbeiner, S. Y. Chun, and J. H. Widdicombe, *Am. J. Physiol.* **262**, L713 (1992).

[27] J. Zabner, B. G. Zeiher, E. Friedman, and M. J. Welsh, *J. Virol.* **70**, 6994 (1996).

[28] H. Matsui, C. W. Davis, R. Tarran, and R. C. Boucher, *J. Clin. Invest.* **105**, 1419 (2000).

The methods and procedures were adapted and modified from those originally reported in 1992 by Yamaya, Finkbeiner, Chun, and Widdicombe.[26,27] Airway epithelial cells are obtained from human donor lungs, lungs resected at surgery, and nasal polyps. On arrival in the laboratory, the airways are dissected down to the third bronchial bifurcation under aseptic conditions in a laminar flow hood. Once the airways are dissected, they are transferred to 50 ml conical tubes containing dissociation enzymes (Pronase, Boehringer Mannheim; deoxyribonuclease 1, Sigma) in chilled dissociation solution [Ca^{2+}- and Mg^{2+}-free Minimal Essential Medium (MEM)]. The tissue is kept at 4° for 24–96 hr. Once the epithelial cells are dispersed, 1% (v/v) fetal bovine serum is added and the cell suspension is transferred to new polypropylene tubes.

The cell suspension is then plated on Primaria tissue culture dishes for 1 hr or longer to allow for removal of fibroblasts by differential adherence. Airway epithelial cells attach poorly to the plastic surface without collagen pretreatment and are collected in a 1 : 1 ratio of Dulbecco's MEM and Hams F-12 supplemented with 5% fetal bovine serum, and 1% nonessential amino acids solution (GIBCO-BRL) in MEM plus 100 U/ml penicillin and 100 μg/ml streptomycin. Then, the cell suspension containing 5×10^5 epithelial cells in 100 μl total volume is added to the top of a permeable membrane insert [Millipore PCF membrane, precoated with a 50 μg/ml solution of human placental collagen, Type VI (Sigma) for 24 hr], and 200 μl of medium is added to the bottom of the filters. The cells are allowed to attach for 24 hr at 37° in a 9% CO_2 incubator.

The day after seeding, the medium on the bottom is changed to a 1 : 1 ratio of Dulbecco's MEM and Ham's F-12 supplemented with 2% Ultroser G (Biosepra SA, Cedex, France), 100 U/ml penicillin, 100 μg/ml streptomycin, 1% nonessential amino acids, and 0.12 U/ml insulin, and the medium on the top is removed, leaving the epithelia at the air–liquid interface. Once the air interface is maintained by the cells (typically 4–6 days), the cultures are kept in a 5.5% CO_2 humidified atmosphere at 37° for the remainder of their time in culture, with the media replaced every third day. Ten days after seeding the cells are stimulated for 24 hr with 100 ng/ml keratinocyte growth factor (R & D Systems, Inc., Minneapolis, MN) to enhance differentiation.

Two weeks after seeding the epithelia are well differentiated, consisting of a pseudostratified columnar epithelium, with ciliated columnar, nonciliated columnar, goblet, and basal cells (see Fig. 1). Epithelia grown in this fashion can survive many months in culture at the air–liquid interface. Importantly, cells harvested from CF specimens maintain the characteristic defect in cAMP-stimulated Cl secretion and liquid transport, and thus can be readily used for CFTR gene transfer studies.[26,27]

Tracheal Explants. Isolated tracheal tissues for *ex vivo* experimentation provide the advantages of morphology and physiology that closely mimic the *in vivo* airways with the ease of applying a viral vector to a very specific region. Therefore, the question of which cell types are transduced can be readily addressed. Tracheal

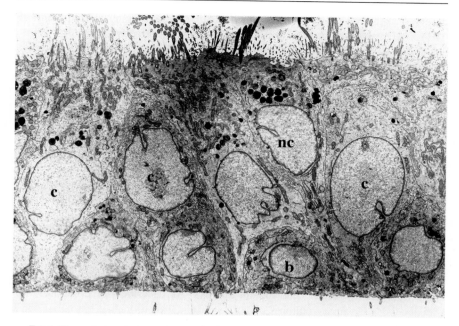

FIG. 1. Transmission electron micrograph of well-differentiated human airway epithelia grown in the air–liquid interface culture system. The epithelia form a pseudostratified columnar epithelium containing cells types representative of the intact airways including cilated surface cells (c), basal cell (b), and nonciliated surface cells (nc).

explants can be prepared from rabbit, ferret, sheep, nonhuman primates, and human tissues. We routinely place small full-thickness explants ($\sim$0.5 cm^2 in size) in airway culture medium (see above) in a 24-well culture dish for short-term culture. The mucosal surface of the explants is maintained above the medium. Tracheal and bronchial explants can be maintained in culture for a few days in this manner.

In Vivo Models

Animal Models for Gene Transfer to Airway Epithelia

In addition to the *in vitro* models described above, many *in vivo* models are available for preclinical studies. Through the combined use of *in vitro* and *in vivo* models, valuable and complementary information can be gained concerning gene transfer to the airway epithelia. Indeed, *in vivo* verification of *in vitro* findings is continually required.

Rodents. Rodents are the most widely used *in vivo* model for pulmonary gene transfer studies. Mice provide a convenient, low cost model for many experimental situations. The nasal epithelia of rodents are also representative of many cell types observed in the lower airways of humans.[29] Importantly, the nasal epithelium of CFTR null mice or mice with CFTR mutations exhibits defective Cl transport similar to that in humans with CF.[29,30] Additionally, catheters can be used to deliver viral vectors to the lower airways, and because of their small lung volumes, small quantities of vector are required. Rabbits and rats are also commonly used. A disadvantage of all rodent models is the absence of significant numbers of submucosal glands from the conducting airways, an important site of CFTR expression in humans.[31]

Other Animal Models. Ferrets, sheep, and nonhuman primates are also useful animal models for airway gene transfer studies. Ferrets have a larger trachea (relative to mice or rats) which may offer advantages for certain experiments. Unlike rodents, the physiology and morphology of the ferret respiratory system is more representative of that of humans, as ferrets have abundant submucosal glands in the conducting airways, an important site of CFTR expression in the human lung.[31,32] Sheep also have airway anatomy more similar to that of humans. Nonhuman primates such as rhesus monkeys and baboons play an important role in preclinical safety trials.

Bronchial Xenografts. Reconstituted bronchial xenografts are an alternative model for gene transfer to the human airways. As described by Engelhart and colleagues,[33–35] primary cultures of human tracheal or bronchial epithelial cells are seeded into denuded rat trachea and implanted into BALB/c (nu/nu) mice. The grafts develop a fully differentiated mucociliary epithelium that closely resembles *in vivo* airway morphology, including some degree of submucosal gland development.[36] Typically, the grafts can be maintained for approximately 6 weeks.

Gene Transfer

FIV Vector Production

Recombinant replication defective pseudotyped FIV vector particles are generated by transient transfection as previously reported.[15] The system consists of an FIV packaging construct, an FIV vector construct, and a plasmid encoding the envelope glycoprotein. The FIV packaging plasmid (pCFIVΔorfΔvif[15]) contains

[29] B. R. Grubb, A. M. Paradiso, and R. C. Boucher, *Am. J. Physiol.* **267**, C293 (1994).

[30] A. Fasbender, J. Zabner, M. Chillon, T. O. Moninger, A. P. Puga, B. L. Davidson, and M. J. Welsh, *J. Biol. Chem.* **272**, 6479 (1997).

[31] J. F. Engelhardt, J. R. Yankaskas, S. A. Ernst, Y. Yang, C. R. Marino, R. C. Boucher, J. A. Cohn, and J. M. Wilson, *Nat. Genet.* **2**, 240 (1992).

[32] A. Sehgal, A. Presente, and J. F. Engelhardt, *Am. J. Respir. Cell Mol. Biol.* **15**, 122 (1996).

[33] J. F. Engelhardt, E. D. Allen, and J. M. Wilson, *Proc. Natl. Acad. Sci. U.S.A.* **88**, 11192 (1991).

[34] J. F. Engelhardt, J. R. Yankaskas, and J. M. Wilson, *J. Clin. Invest.* **90**, 2598 (1992).

[35] D. Duan, J. F. Engelhardt, and Y. Zhang, *in* "Current Protocols in Human Genetics," p. 9.1. John Wiley & Sons, New York, 1998.

[36] J. F. Engelhardt, H. Schlossberg, J. R. Yankaskas, and L. Dudus, *Development* **121**, 2031 (1995).

the FIV packaging signal (Ψ), the *gag* and *pol* genes, and the *rev* sequence. FIV *rev* is analogous to HIV *rev* in enabling expression of late genes encoded by unspliced or singly spliced mRNAs containing the cis-acting Rev-responsive element (RRE). The proviral FIV 5′ LTR is replaced by the CMV promoter/enhancer and the 3′ LTR is replaced with the simian virus 40 polyadenylation signal. These changes are necessary to increase vector titers in nonfeline cells in which the wild-type FIV 5′ LTR is poorly active. A deletion in the *env* gene and mutations in FIV accessory genes *vif* and *orf 2* render these sequences inactive without negatively affecting vector titer. These changes in the proviral DNA further reduce the possibility of generating replication-competent FIV. The FIV vector plasmids (based on pVET$_L$[15]) consist of the FIV 5′ and 3′ LTR sequences flanking a portion of the *gag* sequence including the packaging signal, a transgene cassette, and the Rev-responsive element (RRE). The U3 region of the 5′ FIV LTR is replaced with the CMV promoter. Typical transgenes include the cytoplasmic *Escherichia coli* β-galactosidase or human CFTR genes. In our vector constructs, the CMV promoter directs β-galactosidase expression (FIV-βgal), while the MuLV LTR promoter directs CFTR expression (FIV-CFTR).[7] The envelope plasmid typically consists of a promoter directing transcription of an envelope cDNA, followed by the SV40 polyadenylation signal. The most common envelope protein utilized in lentiviral vector production is the VSV-G envelope glycoprotein.[17] Importantly, this vector production method is readily adapted for pseudotyping the viral particles with any envelope protein contained within a mammalian expression plasmid. Substitution of the envelope plasmid allows for experiments evaluating different envelope pseudotypes to be performed.

One day prior to transfection, 293T cells are plated at a density of 2.8×10^6 per 10 cm diameter tissue culture dish. Three plasmid cotransfections are performed at a 1 : 2 : 1 molar ratio of the packaging, vector, and envelope-expressing plasmids. The DNA complexes are prepared with calcium phosphate (Profectin, Promega, Madison, WI) and transfected into cells using the manufacturer's protocol. Cell culture supernatants are harvested at 24, 36, 48, and 60 hr after the transfection, filtered through a 0.45 μm filter, and concentrated by centrifugation for 16 hr at a speed of $7000g$ (SW-28 rotor, Beckman L-70 ultracentrifuge) at 4°. The vector pellet is resuspended in 19.5 mM Tris-HCl, pH 7.3 ± 0.2, 37.5 mM NaCl, and 40 mg/ml lactose. The osmolality of the vector buffer is $\sim$105 mmol/kg as measured using a vapor pressure osmometer. The βgal preparations are titered on HT1080 cells by limiting dilutions. To titer the FIV-CFTR vector, a real-time PCR-based assay system is used as previously described.[7] Titers of 10^8 to 10^9 TU/ml are routinely obtained using the VSV-G envelope.

Vector Formulation

We previously noted that the receptors for several retroviral envelope pseudotypes including VSV-G, amphotropic, xenotropic, 10A1, RD114, and GALV are functionally inaccessible from the apical surface of polarized human airway epithelia[7,12] (and unpublished observations). These findings stimulated the discovery that transiently disrupting the epithelial junction complex using vectors formulated in calcium chelators such as EGTA allows for successful retroviral gene transfer from the apical surface of airway epithelia *in vitro* and *in vivo*.[7,12,21]

To formulate FIV-based vectors with EGTA, the stock vector is diluted 1 : 1 (vol/vol) with 12 mM EGTA (Fisher Scientific) buffered in 20 mM HEPES/HCl (pH 7.4) to attain a final EGTA concentration of 6 mM. The final buffered EGTA solution is also hypotonic (osmolality, ∼40 mmol/kg). We have successfully used EGTA concentrations from 1.5 to 12 mM for *in vitro* and *in vivo* applications.[7,12,21] Other calcium chelators such as EDTA and BAPTA produce similar results.[21] The use of calcium chelators and other agents to transiently disrupt tight junctions has been reported for nonretroviral vectors as well.[19,20,37]

In Vitro Gene Transfer to Airway Epithelia

Well-differentiated human airway epithelia grown at the air–liquid interface (>2 weeks old, resistance >1000 ohm cm^2) are routinely used in gene transfer studies. A typical experiment with a β-galactosidase reporter construct is outlined below.

Procedure

1. Fifty μl of VSV-G pseudotyped FIV vector [multiplicity of infection (MOI) 10–50] is diluted with 50 μl of 12 mM EGTA in hypotonic buffer to attain a final EGTA concentration of 6 mM.

2. The EGTA formulated viral solution is applied to the apical surface of the epithelia for 4 hr at 37°. After transduction, the vector solution is aspirated off the apical surface.

3. The cells are maintained in culture for 3 to 5 days (or longer depending on the experimental protocol).

4. Cells are fixed in 2% paraformaldehyde for 20 min (see Appendix), and rinsed 2 times for 10 min with PBS.

5. The X-Gal staining solution (see Appendix) is applied to both apical and basal sides of the cells for 4 hr at 37°. After X-Gal staining, the cells are rinsed twice with PBS for 10 min.

6. The filter membrane is then excised from the plastic ring using a razor blade and mounted on a glass slide with Vectashield mounting media (Vector Laboratories, Inc., Burlingame, CA). Epithelia prepared this way can be viewed *en face* using a light microscope.

[37] C. B. Coyne, M. M. Kelly, R. C. Boucher, and L. G. Johnson, *Am. J. Respir. Cell Mol. Biol.* **23,** 602 (2000).

7. To prepare epithelia for sectioning, the membranes are embedded in paraffin (see Appendix). Following rehydration, the section is counterstained with nuclear fast red solution for 1 min (see Appendix). The sample is then dehydrated with 100% ethanol and mounted using Permount medium.

In Vivo Gene Transfer to Rabbit Airway Epithelia

Procedure for Gene Transfer to Rabbit Tracheal Epithelia

1. Adult animals of approximately 4 lb body weight are used. Rabbits are anesthetized using 32 mg/kg ketamine, 5.1 mg/kg xylazine, and 0.8 mg/kg acepromazine intramuscularly. This usually lasts for over 1 hr. If additional anesthesia is required, a half dose can be readministered.

2. The animal is placed on its back and gently restrained. Hair on the ventral neck is removed using a combination of electric hair clippers and Nair (Carter-Wallace Inc.).

3. The surgical area is disinfected with betadine followed by 70% alcohol.

4. Using standard sterile surgical techniques, a ventral midline incision is made and a segment of 1.5–2.0 cm length of trachea is exposed and isolated. A tracheotomy is performed using a High Temp cauterizing instrument (Accu Temp). Three incisions are made between the tracheal rings; the most caudal one is used for ventilation. The two cephalad incisions are each cannulated with PE 330 Intramedic polyethylene tubing (Clay Adams, Division of Becton Dickinson and Company, New Jersey), leaving a ~1.5 cm tracheal segment between the sites for the gene transfer experiment.

5. The isolated tracheal segment is rinsed with saline and then filled with a solution of 12 mM EGTA in 20 mM HEPES buffer (pH 7.2–7.4) for 30–60 min. The EGTA solution is then replaced with ~300 μl FIV-βgal vector. The vector is left in place for 45 min.

6. The cannulae are removed and the tracheal incisions are closed using 4–0 chromic gut cutting suture. Muscles and superficial fascia are sutured with a 3–0 black monofilament nylon cutting suture. The skin is closed using 0 Vicryl violet braided suture. Finally, a tissue adhesive and antibiotic ointment are applied.

7. Following the desired experimental interval, the animal is euthanized with pentobarbital sodium (300 mg/kg IV) and the tissues are prepared for fixation and staining as required (see Appendix). An example of results obtained using this approach is shown in Fig. 2.

Procedure for Gene Transfer to Rabbit Lower Airways

1. The animal is anesthetized as describe above and placed on the sternum, and the head and neck are extended.

2. A laryngoscope (Welch Allyn, New York) with a #0 blade is placed over the tongue and gentle pressure exerted. The epiglottis is visualized and a PE-50

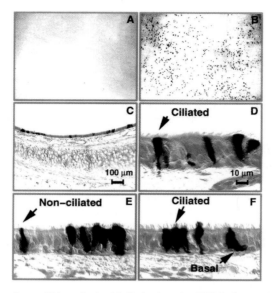

FIG. 2. Gene transfer to rabbit tracheal epithelia *in vivo* using FIV-βgal vector. Panels show results 5 days following gene transfer. (A) Low magnification *en face* view of X-Gal stained trachea from control (A) or FIV vector treated trachea (B). Blue cells were only seen in the trachea transduced with the FIV vector (B). (C) Low magnification view of X-Gal stained tracheal section. β-Galactosidase expressing cells are noted at both the surface and basal cell levels of the transduced epithelium. (D–F) Higher magnification views of tracheal epithelium showing cell types expressing β-Galactosidase. No inflammatory cells were noted in control or transduced specimens ($n = 4$ animals). Scale bar in panel D also applies to panels E and F. Reproduced with permission from G. Wang *et al., J. Clin. Invest.* **104,** R55 (1999).

Intramedic polyethylene tubing (Clay Adams, Division of Becton Dickinson and Company, New Jersey) is slowly advanced into the trachea and bronchi until resistance is felt. This procedure will wedge the tubing in a distal bronchus.

3. FIV vector (300 μl) formulated with 6 mM EGTA (final concentration) is introduced into the lung segment via the tubing using a syringe attached to a 25-gauge blunt needle. The PE tubing is then withdrawn and the animal allowed to recover. An example of results obtained using this approach is shown in Fig. 3.

End Points for Assessment of Gene Transfer Efficacy

Several end points are used to evaluate the success of FIV-mediated gene transfer in the model systems. For both *in vitro* and *in vivo* experiments using the β-galactosidase reporter construct, X-Gal staining and standard morphometric techniques will allow ready identification of the cell types transduced and

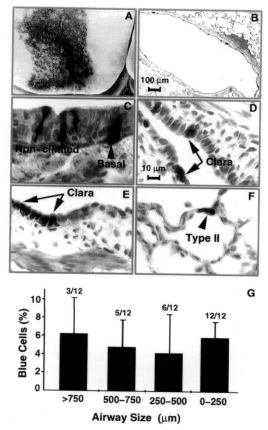

FIG. 3. FIV-βgal transduction of lower airway epithelia 5 days following gene transfer. (A) *En face* view of pleural surface of lung following fixation and X-Gal staining showing βgal expressing cells. All treated animals had similar segments of β-Galactosidase expressing cells extending to the pleural surface. (B–F) Higher magnification views of tissue sections showing lower airway and parenchymal cells transduced. (B) Low magnification view of a large bronchus (>750 μm diameter) demonstrating patches of β-Galactosidase expressing cells extending around the circumference of the epithelium. (C) High magnification view of expression in a large bronchus (>750 μm diameter) showing expression in ciliated cells and basal cells. (D) High magnification view of expression in a medium sized airway (500–750 μm diameter) demonstrating expression in nonciliated surface cells (Clara cells). (E) β-Galactosidase expression in a small bronchus (250–500 μm diameter) showing expression in nonciliated surface cells (Clara cells). (F) Distal lung sample (airways 0–250 μm diameter) showing expression in cuboidal cells consistent with alveolar type II cells. (G) Gene transfer expressed as a function of airway size. Numbers above each bar represent the number of animals with transduced cells in the corresponding region. Tissues from 12 animals were studied. Scale bar in panel D also applies to panels E and F. Reproduced with permission from G. Wang *et al.*, *J. Clin. Invest.* **104,** R55 (1999).

quantitation of the percentage of transgene expressing cells. We have also monitored the functional expression of the human CFTR in corrected CF epithelial cultures for approximately 1 year following gene transfer by measuring the bioelectric properties of the cells[7] (and unpublished observation). Duration of expression analysis of *in vivo* studies allows the identification of clonal expansion of targeted progenitor cells.

Concluding Remarks

A central issue for the successful correction of CF lung disease by integrating vectors is the identification of the cell types to target in order to attain long-term gene expression. At this time it is unknown whether the appropriate progenitor cell populations can be targeted by applying an FIV-based vector to the apical surface of epithelia *in vivo*. Long-term duration of expression studies comparing the efficacy of vectors formulated with agents that disrupt the junctional complex to the efficacy of vectors applied to the intact epithelium will be required to address this problem. In addition, ongoing work will likely identify new envelope pseudotypes that can transduce intact epithelia from the apical surface. Similar duration of expression studies will be required to evaluate the utility of such a vector.

Acknowledgments

We thank our colleagues Beverly Davidson, John Engelhardt, and Mike Welsh for support and critical comments related to this work. We acknowledge the technical assistance of Patrick Staber, Melissa Hickey, Greg Williams, Andrea Vivado, and Hong Shen. We thank collaborators at Chiron Technologies, Sybille Sauter, Julie Johnston, Phil Sheridan, Kay Townsend, Tom Dubensky, and Doug Jolly, for the development of the FIV vector system, scientific input, and support. We acknowledge the support of the Cell Culture Core, Vector Core, and Cell Morphology Core, partially supported by the Cystic Fibrosis Foundation, NHLBI (PPG HL51670), and the Center for Gene Therapy for Cystic Fibrosis (NIH P30 DK-97-010). This work was supported in part by Chiron Technologies, NIH RO1 HL61460 (P.B.M.), PPG HL-51670 (P.B.M. and J.Z.), and the Cystic Fibrosis Foundation (WangG99GO).

Appendix

1. 2% Paraformaldehyde for Fixation of Cells and Tissues

Place 2 grams of paraformaldehyde powder (Fisher Scientific) in a 200 ml flask and suspend the powder in 100 ml of PBS. Gently heat the flask on a hot plate with a stir bar. Once the solution clears, cool to room temperature before use.

2. X-Gal Staining Solution

To visualize the β-galactosidase reaction product, the X-Gal solution is prepared on the day that it will be used as follows. Cells or tissues are stained for 3–4 hours to overnight at 37° or at room temperature depending on the experimental design.

Final concentration	Stock Conc.	Vol./100 ml solution
PBS	1X	83 ml
35 mM $K_3Fe(CN)_6$	0.5 M	7 ml
35 mM $K_4Fe(CN)_6 \cdot 3H_2O$	0.5 M	7 ml
2 mM MgCl	1 M	0.2 ml
0.01% sodium deoxycholate	10%	0.1 ml
0.02% NP40	10%	0.2 ml
1 mg/ml X-Gal solution	40 mg/ml	2.5 ml

3. Paraffin Embedding of Epithelia and Lung Tissue

a. Airway Epithelia Cultured on Permeable Supports. Epithelia are dehydrated and embedded in paraffin using the following steps:

50% alcohol	30 minutes
70% alcohol	30 minutes
85% alcohol	30 minutes
95% alcohol	30 minutes
100% alcohol	30 minutes
100% alcohol	30 minutes
Clearing Solvent*	30 minutes
Clearing Solvent	30 minutes
Melted Paraffin	1 hour (62°)
Melted Paraffin	1 hour (62°)

*Clearing Solvent (Citrus Based, Stephens Scientific, Riverdale, NJ). Caution: Do not use xylenes because it will remove the β-galactosidase reaction product.

Paraffin blocks are cast and the epithelia cut to ~5–6 μm sections. To deparaffinize the sections, the clearing solvent is used.

b. Lung Tissues. Tracheal tissue is fixed overnight in 2% paraformaldehyde in PBS. The fixative is removed by rinsing with 3 changes of PBS over 4 hours. In studies where a β-galactosidase reporter is used, the tracheal segment is next placed in the X-Gal staining solution for 4–6 hours, then rinsed 3× with PBS over 4 hours. The tissue is then ready for en bloc photography or can be embedded in paraffin for sectioning.

Intact lung tissues are inflation fixed overnight with 2% paraformaldehyde introduced using a tracheal catheter. After fixation, the lung is placed on filter paper and gentle pressure applied with a finger to remove the fixative. The lung is then inflated with PBS ×4 over 8 hours to remove the fixative. Next, the lung is inflated with the X-Gal staining solution for 4–6 hours at 37°. The X-Gal solution is removed and the tissue again rinsed with PBS as above. The lung tissue can then be photographed en bloc or embedded in paraffin for sectioning.

4. Nuclear Fast Red Staining Solution

A nuclear fast red counterstain is commonly used in studies using the blue β-galactosidase reaction product. To prepare the solution, five grams of aluminum sulfate and 0.1 gram of nuclear fast red powder (Sigma) are suspended in 100 ml of double distilled water. The solution is heated while stirring until the powder dissolves. The solution is cooled to room temperature and filtered through a 0.45 μm filter for use.

[29] Transduction of a Gene Expression Cassette Using Advanced Generation Lentiviral Vectors

By MICHELE DE PALMA and LUIGI NALDINI

Introduction

Lentiviral vectors (LVs) provide a powerful tool for gene transfer into both dividing and non-dividing cells (for a review see Vigna and Naldini [1]). They are able to stably transduce primary and terminally differentiated cells such as lymphocytes, hematopoietic stem cells, macrophages, neurons, and hepatocytes from different species, including rodents and primates. Moreover, LVs pseudotyped with the G protein envelope of the vesicular stomatitis virus (VSV-G) can be concentrated to high titers and allow efficient transduction of a wide range of tissues in vivo. [2–4]

Pseudotyped LVs are currently produced by transient transfection into 293T cells of a combination of plasmids encoding the required lentiviral packaging functions, the envelope of an unrelated virus that pseudotypes the particle, and the transfer vector (see [26] by Follenzi and Naldini). Different versions of these constructs and their combinations have been described and characterized, from HIV, SIV, and nonprimate lentiviruses. Here we refer to a late-generation vector system that we have contributed to develop from HIV-1 and which has been extensively characterized for performance and biosafety. [5–7] However, the approaches

[1] E. Vigna and L. Naldini, J. Gene Med. 2, 308 (2000).
[2] L. Naldini, U. Blomer, P. Gallay, D. Ory, R. Mulligan, F. H. Gage, I. M. Verma, and D. Trono, Science 272, 263 (1996).
[3] L. Naldini, U. Blomer, F. H. Gage, D. Trono, and I. M. Verma, Proc. Natl. Acad. Sci. U.S.A. 93, 11382 (1996).
[4] T. Kafri, U. Blomer, D. A. Peterson, F. H. Gage, and I. M. Verma, Nat. Genet. 17, 314 (1997).

and methods described in this chapter can be applied to LVs produced by most types and sources of constructs, including by stable inducible packaging cell lines that are under advanced development in several laboratories.

Third-generation, replication-defective, pseudotyped HIV-1 derived vectors are produced by cotransfection of four types of plasmids into 293T cells[5]:

The core packaging construct, encoding the proteins and enzymes of the vector core, products of the HIV-1 *gag* and *pol* genes. Expression of this plasmid is conditional on expression of the Rev protein by a separate plasmid.

The Rev expression plasmid, containing the nonoverlapping HIV-1 *rev* cDNA.

The envelope construct, expressing the surface glycoprotein of an unrelated virus, most often VSV-G.

The transfer vector, containing the transgene expression cassette linked to the minimal HIV-1 sequences required for efficient encapsidation, reverse transcription, nuclear transport, and integration into the target cell chromatin.

At the time of transduction, only the transfer vector, which does not contain any viral genes, is integrated into the host cell genome.

This advanced vector design has alleviated most of the biosafety concerns associated with the use of vectors derived from HIV-1. In fact, the following crucial determinants of HIV-1 pathogenesis[8] have been removed from the constructs used to make vector:

The *env* gene, responsible for targeting T-helper lymphocytes.

All four accessory genes *vpr, nef, vif,* and *vpu,* whose crucial role in pathogenesis has been clearly demonstrated.

The *tat* gene responsible for the tremendous rate of expression of the HIV-1 genome in infected cells.

Consequently, the predicted biohazards of a replication-competent retrovirus originating during vector production by an unlikely series of illegitimate recombinations among the constructs would be substantially lower than, and different from,

[5] T. Dull, R. Zufferey, M. Kelly, R. J. Mandel, M. Nguyen, D. Trono, and L. Naldini, *J. Virol.* **72,** 8463 (1998).

[6] R. Zufferey, T. Dull, R. J. Mandel, A. Bukovsky, D. Quiroz, L. Naldini, and D. Trono, *J. Virol.* **72,** 9873 (1998).

[7] A. Follenzi, L. E. Ailles, S. Bakovic, M. Geuna, and L. Naldini, *Nat. Genet.* **25,** 217 (2000).

[8] R. C. Desrosiers, *Nat. Med.* **5,** 723 (1999).

those associated with HIV-1, or even attenuated versions of primate lentiviruses currently evaluated for vaccine purposes.[9]

A further gain in vector biosafety and performance is obtained using a self-inactivating (SIN) transfer vector.[6,10,11] In the SIN-18 transfer vector construct,[6] for example, the region spanning the transcriptional enhancers and promoter of HIV-1, including the TATA box, was deleted from the 3′ LTR. Since the 3′ LTR is used as a template to generate both copies of the LTR in the integrated proviral form of the vector, the deletion results in transcriptional inactivation of both LTRs and prevents its mobilization and recombination in transduced cells.[12]

In this chapter, we describe the methods required to transduce an expression cassette for a transgene of interest into a selected target. The following experimental steps are discussed:

> Construction of the transfer vector carrying the desired expression cassette.
> Generation of the vector stock (these methods are discussed in detail in [26] by Follenzi and Naldini).
> Quality control of the vector stock.
> Transduction of the selected target cells.

For each step, the most relevant experimental parameters and limitations, the potential pitfalls, and the troubleshooting approaches are highlighted.

Construction of Transfer Vector Carrying Desired Expression Cassette

The genome of lentiviruses encodes a protein regulator of its own expression (Tat), which is essential for high-level transcription from the LTR. In its absence, as occurs in cells transduced by a LV, the LTR has a low basal transcriptional activity. Thus, most LVs incorporate an exogenous promoter to drive expression of the transgene. The simplest type of expression cassette is made by a promoter and the cDNA for the gene of interest. The promoter is most often located downstream to the HIV-derived intron of the vector (internal position). Promoters successfully used in LV are strong viral promoters such as the immediate early enhancer/promoter of the human cytomegalovirus (CMV)[3,4,13] and those of

[9] M. S. Wyand, K. Manson, D. C. Montefiori, J. D. Lifson, R. P. Johnson, and R. C. Desrosiers, *J. Virol.* **73**, 8356 (1999).

[10] H. Miyoshi, U. Blomer, M. Takahashi, F. H. Gage, and I. M. Verma, *J. Virol.* **72**, 8150 (1998).

[11] T. Iwakuma, Y. Cui, and L. J. Chang, *Virology* **261**, 120 (1999).

[12] A. A. Bukovsky, J. P. Song, and L. Naldini, *J. Virol.* **73**, 7087 (1999).

[13] H. Miyoshi, K. A. Smith, D. E. Mosier, I. M. Verma, and B. E. Torbett, *Science* **283**, 682 (1999).

endogenous housekeeping genes, such as phosphoglycerate kinase 1 (PGK)[7,14] and elongation factor 1α (EF1α),[15] which are expressed more ubiquitously but less powerfully. Promoters derived from oncoretroviral LTR have also been used.[16,17] If any of the above promoters is selected, the cDNA of the gene of interest is cloned by conventional DNA technology downstream to the promoter in the available unique cloning sites. The expression cassette must rely on the HIV-1 polyadenylation site in the vector 3′ LTR. Interestingly, the splice-suppressor activity of the Rev-RRE axis in producer cells may be exploited to deliver a cassette containing at least one intron into target cells.[17] This feature could represent a unique advantage of lentiviral over oncoretroviral vectors given the role of intervening sequences in controlling the efficiency of processing, export, and translation of RNA transcripts. Considering that the vector backbone of a SIN-18 vector is around 2 kb and that the HIV-1 genomic RNA is 9.18 kb, the size of the expression cassette should not be more than 7.5 kb. However, the actual size limits of LV remain to be investigated. If novel promoters, such as inducible and tissue-specific promoters, are to be used, the required regulatory sequences will be cloned downstream to the HIV-1 intron. Additional control elements, such as enhancer and matrix binding regions, may be introduced. Again, Rev activity in producer cells may enable the faithful delivery of complex DNA sequences containing cryptic splice sites that preclude transfer by oncoretroviral vectors.[18] Posttranscriptional regulatory elements enhancing the expression of the transgene may also be incorporated, such as an element from the 3′ end of the genome of the woodchuck hepatitis virus reported to enhance nuclear export and/or polyadenylation of the transcript and, consequently to increase its steady state in transduced cells.[19] Bicistronic expression cassettes containing an internal ribosome entry site (IRES) or two different promoters in tandem can also be constructed. If the expression cassette requires a polyadenylation site different from that of the vector, insertion in reverse orientation may be attempted. However, the accumulation of antisense transcripts in vector-producer cells may interfere with vector production.

An alternative design of the expression cassette is to introduce the promoter in place of the deleted transcriptional sequences of HIV-1 in the vector LTR. The heterologous sequences are merged with the residual viral sequences or hybrid versions containing larger portions of the HIV-1 promoter can be tested. The

[14] G. Guenechea, O. I. Gan, T. Inamitsu, C. Dorrell, D. S. Pereira, M. Kelly, L. Naldini, and J. E. Dick, *Mol. Ther.* **1**, 566 (2000).

[15] P. Salmon, V. Kindler, O. Ducrey, B. Chapuis, R. H. Zubler, and D. Trono, *Blood* **96**, 3392 (2000).

[16] S. K. Kung, D. S. An, and I. S. Chen, *J. Virol.* **74**, 3668 (2000).

[17] A. Ramezani, T. S. Hawley, and R. G. Hawley, *Mol. Ther.* **2**, 458 (2000).

[18] C. May, S. Rivella, J. Callegari, G. Heller, K. M. Gaensler, L. Luzzatto, and M. Sadelain, *Nature* **406**, 82 (2000).

[19] R. Zufferey, J. E. Donello, D. Trono, and T. J. Hope, *J. Virol.* **73**, 2886 (1999).

potential advantages of this type of design remain to be investigated. While the presence of intronic sequences within the primary vector transcript may enhance its processing, export, and translation, it is also possible that the suboptimal consensus of the HIV-1 splice acceptor sites and the presence of upstream RNA structures such as the TAR loop and the polyadenylation site in the 5′ R region may inhibit optimal expression. Moreover, the self-inactivating feature of the original vector is compromised.

The integrity of the expression cassette may be initially validated by transient transfection of the transfer vector construct (without the other plasmids) into 293T cells and scoring for expression of the transgene. In this setting, however, expression of the transgene is influenced by the strong constitutive promoter inserted upstream of the transfer vector. The activity of the internal promoter can be properly assessed only after transduction of the vector.

Generation of Vector Stock

These methods are discussed in detail in Chapter [26].

Quality Control of Vector Stock

Once a batch of vector is produced, it must be assayed for transducing activity, for the content of vector particles, for the absence of RCR, and for sterility. Knowledge of these parameters is required to properly set up and optimize transduction of the desired target *ex vivo* or *in vivo*. The screening of a lentiviral vector stock for the absence of RCR is a challenging task and is discussed elsewhere in this book (see [26]). Sterility of a vector stock is evaluated by testing aliquots in standard microbiological assays used for tissue culture.

Assaying Transducing Activity

LVs integrate into the host cell genome, thus allowing stable maintenance of the transgene in the progeny of transduced cells and providing a basis for long-term expression of the transgene. The transducing activity of the vector stock can be measured more easily when ubiquitous promoters and reporter transgenes that can be detected within individual cells are transferred. Using such vectors, reproducible assays can be set up to measure the concentration of infectious particles and the maximal transduction efficiency of the stock. The concentration of infectious particles is estimated by end-point titration of the vector stock on a standard well-infectable cell line (see protocol below). The assay is designed for high sensitivity to best approximate the number of infectious particles and to calculate their ratio to physical particles in the vector stock. It is well acknowledged that end-point titer per se has a poor predictive value on vector performance in the transduction of primary targets. This parameter, however, can be combined with the measurement of physical particles to calculate the specific transducing activity, a useful indicator of vector performance (see below). End-point titration should always be accompanied by bulk vector assays that measure the maximal frequency of transduction obtained when high vector input is used (see protocol below). The gold standard of bulk vector assays is the measurement of the copy number of vector DNA integrated per genome of transduced cell (see protocol below). This assay provides conclusive evidence of transduction of the intact expression cassette and allows quantification independent of expression in the target cells. It permits proper comparisons to be made between the transducing activity of vectors differing for the promoter and the transgenes.

In most cases, the promoter of choice and/or the gene of interest do not allow proper evaluation of the vector stock in the titration assays mentioned above. In such case, it may be advisable to do one or more of the following:

Validate the production and purification methods using a reference vector construct and transducing a continuous cell line according to a standardized protocol.

Measure the particle content of the test vector stock (see below) and compare it to the reference vector.

When the gene product can be scored within the cells but the promoter of choice is expected to be selective, estimate transducing activity by titration in a permissive cell line in comparison with the reference vector stock.

In all other cases, verify bulk transduction activity by DNA analysis using a standard cell line as target.

Assaying Physical Components of Vector Particles

Several assays are available to measure the physical components of vector particles in a stock: the core viral proteins, the encapsidated vector RNA, and the reverse transcriptase (RT) activity. Because these assays are not dependent on transduction of cells, they provide no proof of activity of the stock. However, the measurement of the content of physical particles can be integrated with the information obtained by the transduction assays (end-point titer and DNA analysis) to calculate vector *infectivity* (or specific transducing activity), a reliable parameter to evaluate the quality of the vector stock.

The Gag capsid protein p24 is a major structural component of the HIV-1 core and, as a consequence, of all HIV-1 derived vectors. The total content of particles in the vector stock can be estimated by immunocapture of the mature core protein using highly sensitive and commercially available kits. Alternatively, the content of mature vector particles can be estimated by the RT activity of the suspension using an exogenous substrate. The advantages of the immunocapture assay are the specificity, reproducibility, and yield of an absolute figure for the p24 concentration. It is important to verify the extent of association of the measured protein with particles by checking its sedimentation in the pellet after ultracentrifugation. A tentative estimate of the actual number of physical particles may then be made by calculating that an average of 2000 Gag molecules assemble the virion core. However, only electron microscopy or fluorescence microscopy after microfiltration and immunolabeling of the core proteins allow direct counting of individual particles.

Immunocapture assays for p24 also score, although less efficiently, the p24-containing Gag precursor polyprotein in immature particles. The presence of excess immature particles in a vector stock indicates a poorly developed production system. When new packaging systems are introduced, they should be validated for the yield of mostly mature vector particles by Western blot analysis of pelleted particles using antibodies against the viral core proteins.

The Gag p24 and RT assays described above do not distinguish between complete vector particles and noninfectious particles lacking some components such as the envelope or the vector RNA. The type of envelope protein and its level of incorporation in the particle not only control the target range, but also have a major influence on the stability and transduction efficiency of the vector. The content of encapsidated vector RNA in a vector stock is a good predictor of its transduction efficiency. Poor expression of unspliced vector RNA in producer cells and large size of the expression cassette may represent limiting factors for the efficiency of encapsidation of vector RNA. Methods to analyze RNA content can then be used to verify the effective packaging of vector RNA into particles by comparison with reference standard stocks or by copy number calculation. RNA can be extracted from pelleted vector particles using any RNA isolation kits, treated with DNase, and spotted onto nitrocellulose through a slot-blot apparatus. The nitrocellulose filter is then hybridized with a vector-specific riboprobe. For instance, serial dilutions of p24 equivalent of test and reference vectors are tested together with standards prepared using plasmid and carrier DNA, as shown for DNA analysis (see below). If radiolabeled probes are used, signal acquisition by storage phosphor screens and analysis of digital images allow comparing the standards with the loaded samples, and determining the RNA content per nanogram of p24. A content of RNA much lower than expected can be indicative of poor infectivity of particles. Real-time reverse PCR approaches can also be developed to score transfer vector-specific sequences. In all these methods, particular care must be exercised to eliminate residual plasmid DNA in the vector stock if transient transfection has been used for production.

Protocol for End-Point Titration of Vector Transducing Activity

End-point titration is performed by transducing a target cell line with serial dilutions of the vector preparation. In the simplest case, a reporter gene is cloned downstream of a constitutive promoter, as in the standard vector SIN-18.PGK.EGFP.

The PGK promoter drives expression of the enhanced green fluorescent protein (EGFP). The transduced cells can be analyzed by fluorescence-activated flow cytometry measuring the percentage of cells expressing EGFP in the total population and the mean fluorescence intensity (MFI) of positive cells. Since the PGK promoter is constitutively active in most cell types, every well-infectable cell line (for example HeLa or 293T cells) can be used for this analysis.

Several experimental parameters affect transduction when titering a vector preparation. The following protocol can be reasonably modified keeping in mind that what mainly influences transduction of cells in a dish is the concentration and not the total number of vector particles in the transduction medium.

1. Seed 1×10^5 cells/well in a six-well cell culture plate for as many tenfold serial dilutions of vector preparation. The cells are allowed to adhere and incubated at 37° for 24 h in appropriate medium. Before titration, the medium is replaced with 0.9 ml fresh medium containing 9 μg/ml polybrene.

2. Prepare serial tenfold dilutions of the viral stock, approximately ranging from 10^{-2} to 10^{-7} for concentrated vector and from undiluted to 10^{-5} for conditioned medium. Add 0.1 ml of each dilution to the cells. Incubate the cells at 37° for 12 h to allow transduction.

3. Change the medium and incubate the cells at 37° for an additional 72 h.

4. Wash cells with phosphate-buffered saline (PBS), than detach and fix, if required, in 1 ml of fixing solution [1% formaldehyde, 2% fetal bovine serum (FBS) in PBS].

5. Analyze cells by flow cytometry (fixed cells can be stored at 4° for a few days). Unfixed cells can be scored also for viability by exclusion of propidium iodide.

Mock-transduced cells are used as a standard control to gate the population of negative events. The titer is defined as number of transducing units per milliliter (TU/ml) of vector preparation, based on the assumption that a single vector copy integrated in the host genome will give a positive cell. Assuming that all the cells are equally susceptible to transduction, following the Poisson distribution for random independent events, a single transduction event, and not more, has occurred in most positive cells when the the percentage of positive cells in the total population is below 25%. It follows that the titer must be calculated from a sample corresponding to a vector dilution where positivity of cells ranges between 1% (to ensure an acceptable signal over the instrument noise) to 25%, in order not to underestimate the titer when multiple transduction events per cell have occurred. Proof of linearity must be obtained showing that different dilutions in the optimal testing range yield linear increase in transduction frequency. The equation to calculate titer is:

$$\text{Titer (TU/ml)} = \text{(number of cells at the time of vector addition)}$$
$$\times \text{(\%EGFP-positive cells/100)} \times \text{(dilution factor)}$$

When the transgene is different from EGFP, end-point titer can be performed staining cells expressing the transgene at steady state for flow cytometry using specific fluorochrome-conjugated antibodies or microscopy analysis. In the latter case, higher vector dilutions must be tested to allow scoring of positive cells. Vectors expressing reporter genes encoding for enzymatic activity such as LacZ can be titered by histochemical staining of transduced cell using chromogenic substrates precipitating inside the cells. Vectors expressing selectable markers can be titered testing transduced cells for long-term resistance to the selector drug. Resistant cells form isolated colonies at the highest dilutions of vector stock; for titration it is assumed that each drug-resistant colony results from a single transduction event.

End point titration is strongly affected by (1) the type of cell transduced; and (2) the promoter in the expression cassette.

1. Some cell lines are transduced less efficiently than others, thus leading to underestimation of the titer of the vector preparation. In contrast, other cell lines are remarkably susceptible to transduction. Such discrepancy implies that when transducing relatively refractory cells (such as some types of primary cells) using vectors titered on easily transduced cells, one must empirically employ a high number of TU to get measurable expression of the transgene (see below, the section on transduction of target cells).

2. The transcriptional activity of the promoter is strictly dependent on the cell type. This means that vectors containing weak, regulatable, or tissue-specific promoters cannot be titered according to the above-reported protocol. In this case, titration should be performed on cell lines in which transcription driven by that particular promoter is favored. If the vector is to be used for transcriptional targeting with tissue-specific expression, end-point titer must be specifically performed on the target cell line. To overcome such problems, vectors containing a second independent expression cassette can be designed, as in the hypothetical construct SIN-18.Promoter.Transgene.PGK.EGFP. However, this type of vector should be carefully tested, since promoter interference is likely to occur between the two expression cassettes.

Bulk Assay of Vector Transducing Activity: Maximal Frequency of Transduction

End-point titration should always be accompanied by bulk vector assays, i.e., by measuring the maximal frequency of transduction that is obtained when high vector input is used. Dose-response analysis shows a linear trend, in which the percentage of positive cells increases with the number of TU added, and then reaches a plateau, corresponding to the maximal frequency of transduction. In the region of the curve approaching the plateau, the MFI of positive cells should increase with increasing vector doses, reflecting an increase in the average copy number of vector per transduced cell. Transduction of well-infectable cells with a vector stock of good infectivity should not plateau until it reaches a very high frequency of transduced cells, approaching 100%. If a vector stock of poor infectivity is used, the frequency of transduction may plateau at a lower percentage of cells and may not improve just by increasing the number of TU per cell, but only using a vector stock with higher infectivity. This limitation becomes particularly evident when target cells relatively resistant to transduction are used.

Bulk assays verify that excess noninfectious particles and other contaminants do not interfere with vector performance in experimental conditions more representative of the transduction of a gene of interest into most cells of a target population. Such interferences may be missed when highly diluted, low vector inputs are used and may explain the discrepancies mentioned above between a high end-point titer and a poor bulk transduction ability of a vector stock. If available, the cell types selected as target could be used directly to score bulk transduction activity. When using high input of vector, it is important to score cell viability (for instance by dye exclusion). Possible mechanisms of cytotoxicity at high vector dose are particle-mediated cell fusion (fusion-from-without), toxic levels of transgene expression, and excess integration events in the genome. If a fraction of bulk-transduced cells is lost, one could underestimate the transducing activity of a vector stock.

Bulk Assay of Vector Transducing Activity: Measurement of Copy Number of Vector Integrated per Genome

While end-point titer and maximal transduction frequency depend on both infectivity and transcriptional activity of the vector in the transduced cells, DNA analysis of transduced cells directly reveals the ability of the vector to integrate into the target cell genome, independent of transgene expression. DNA analysis is then crucial when vectors containing nonconstitutive promoters or transgenes that cannot be directly tested within the cells have to be validated. For instance, in some experimental conditions it is required to transduce the target cells with equal amounts of infectious particles of different vectors, as when testing transcription efficiency from a repertoire of expression cassettes in a given cell line. Real-Time PCR (TaqMan) and Southern blot analysis provide a method to normalize different vector stocks, independently of the type of construct, for the number of integration events per cell and thus can be used as basic titration assays more useful and reliable than end-point titer.

It has been shown that unintegrated lentiviral DNA can persist in transduced cells for few passages in culture and can serve as template for transgene expression during the first hours after transduction (our unpublished data and Ref. 20). Since

[20] D. L. Haas, S. S. Case, G. M. Crooks, and D. B. Kohn, *Mol. Ther.* **2,** 71 (2000).

unintegrated DNA is not responsible for long-term expression[2] and is lost with time, Southern blot analysis must be performed on DNA from cells that have been cultured for several passages. A stable cell line is transduced with two or more reasonably high doses of vector stock (for instance: 10 and 100 ng of p24 equivalent of vector per 10^5 HeLa cells). Transduction is performed as described for endpoint titer; 8 μg/ml of polybrene can be added to the medium to improve transduction efficiency. Cells are cultured for at least 5–7 passages and then lysed with TNE (10 mM Tris-HCl, pH 7.5; 100 mM NaCl; 1% SDS) in the presence of 200 μg/ml proteinase K overnight at 37°, followed by extraction and purification of genomic DNA with phenol-chloroform and ethanol; however, any standard or commercial kit-provided methods for genomic DNA extraction and purification can be adopted. For Southern analysis, genomic DNA is exhaustively digested with one or more restriction enzymes able to release a vector fragment spanning a sequence that, when probed, reveals its total content in transduced cells. Southern blot analysis can be performed using standard protocols. Ten to 20 μg of genomic DNA from each sample is separated on 1% agarose gel, transferred to a hybridization membrane, and probed for vector-specific sequences. A probe for an endogenous sequence is used to normalize DNA loading for each sample. Following hybridization, the membrane is washed with SSC/SDS solutions and exposed.

To calculate integrated proviral copy number one can use the DNA from a reference cell clone where a single or predetermined number of copies of vector is integrated in the genome. Alternatively, proviral copy number can be calculated reconstructing standard copy numbers with a vector plasmid DNA. To calculate the amount of plasmid DNA per microgram of genomic DNA to obtain a copy number equivalent of 1 per genome, determine the number of base pairs in the plasmid and perform the following calculation: μg of plasmid DNA equivalent to 1 copy per genome per μg of genomic DNA = base pairs in plasmid/base pairs in genome (6×10^9 bp/genome for euploid human cells and 12×10^9 bp/genome for euploid murine cells).

A standard curve is made by adding vector plasmid DNA equivalent to 10, 5, 3, 1, and 0.5 copies per genome to 10–20 μg of genomic DNA from untransduced cells, digesting the mixture as above, and comparing the standards with the loaded samples.

If radiolabeled probes are used, signal acquisition by storage phosphor screens and analysis of digital images allow quantification of vector DNA.

Calculation of Infectivity of Vector Stock

The infectivity of a vector preparation can be defined as the transducing activity per unit of physical particle, where the first parameter is expressed as transducing units (TU) per milliliter, as obtained by end-point titration, or copy number of integrated vector, as a result of real-time PCR or Southern blot analysis of bulk transduced cells.

Advanced versions of VSV-G pseudotyped LVs, carrying ubiquitous or strong viral promoters driving the expression of EGFP, can be routinely titered to $0.5–1.0 \times 10^8$ TU/ml (end-point titer on HeLa cells) in supernatants of transfected 293T cells. For such vectors, ultracentrifugation and concentration up to a thousandfold can be performed without significant loss of transducing activity, and titers of more than 10^{10} TU/ml can be obtained. Since 1 ng of p24 could theoretically contain 1.2×10^7 particles, if all the particles in the vector stock were infectious, a titer of 10^8 TU/ml would correspond to a p24 concentration of about 10 ng/ml. Indeed, a concentration in the range of 500–1000 ng p24/ml is more reasonably expected. This is due to two major reasons. The first reason is that end-point titer fails to estimate the real content of infectious particles in a vector preparation, because vector particles move by Brownian motion in the medium and only a fraction of them have the chance to get in contact with a cell in the monolayer and transduce it in the time window of the assay.[21] This operational limitation of the assay could be accounted by a correction factor calculated by mathematical means[22] and does not affect comparison of different batches of vectors. The second reason instead is crucially linked to the quality of the vector batch tested. In fact, the efficiency of packaging of infectious particles is lower than what is predicted by theoretical calculations, and a good fraction of the total p24 protein is not assembled into infectious virions. This is due to several factors. Some factors are intrinsic to the biological mechanism of viral assembly. Other factors are dependent on the limitations imposed by the vector design, which employs a fraction of the viral genome split among separate and independently regulated constructs. Optimization of the type and expression ratio of these constructs within vector producer cells is required to obtain a vector stock of acceptable infectivity. When a new transfer vector construct is introduced, further optimization of the production conditions may be needed.

All these factors add up in reducing the measurable infectivity of a vector stock, which can be calculated by the following equation:

$$\text{Infectivity} = (\text{TU/ml})/(\text{ng p24/ml}) = \text{TU/ng p24}$$

For instance, a titer of 1×10^8 TU/ml (end-point titer on HeLa cells) corresponding to a concentration of p24 of 500 ng/ml gives an infectivity of $(1 \times 10^8$ TU/ml$)/(500$ ng p24/ml$) = 2 \times 10^5$ TU/ng p24. Such a level of infectivity would

[21] S. P. Forestell, E. Bohnlein, and R. J. Rigg, *Gene Ther.* **2,** 723 (1995).

[22] S. Andreadis, T. Lavery, H. E. Davis, J. M. Le Doux, M. L. Yarmush, and J. R. Morgan, *J. Virol.* **74,** 3431 (2000).

be considered more than satisfactory, as values approaching 10^5 TU/ng p24 are acceptable for most applications. Infectivity can also be calculated referring to vector copy numbers. In this case, infectivity must be calculated multiplying the vector copy number/cell genome by the number of cell genomes present at the time of vector addition, per ng of p24.

Infectivity = (vector copy number/cell genome)
          × (number of cell genomes at the time of vector addition)/ng p24
        = copy number/ng p24

The ratio between the infectivity estimated from end-point titration and the infectivity estimated by DNA analysis indicates the fraction of integrated vectors that allow expression of the transgene, or the minimal number of vector copies allowing detectable expression of the transgene within that cell type. Transgene silencing may occur because of random integration of the vector into heterochromatin and it may be induced by poorly understood genome surveillance mechanisms, particularly following long-term follow-up of transduced cells or selection and expansion of clones, as shown for MLV-derived vectors.[23] In such cases, the use of tissue-targeted promoters, minimal viral sequences (such as in SIN vectors), and chromatin insulators has been shown to improve the expression performance of retroviral vectors.[24]

Transduction of Target Cells

Once the vector stock has been proved to display an infectivity comprised in the indicated range and to obey the above-mentioned quality controls, it can be utilized to transduce the desired target cell *in vitro* or *in vivo*.

Cells can be transduced *in vitro* directly in their culture medium by adding the required amount of vector preparation. Polybrene can help transduction, but some primary targets do not tolerate the standard concentrations; other compounds, such as dextran sulfate and the fibronectin fragment CH-296 (Takara Shuzo, Osaka, Japan) may be tested. It is likely that the choice of viral envelope will affect the efficacy, if any, of these cofactors. Some quiescent cell types, such as lymphocytes, can be more efficiently transduced with addition of a growth or activation stimulus. Indeed, cells can be treated with cytokines or growth factors without any interference with the transduction performance.

Information obtained by titration of the vector stock by end-pointx analysis or by calculation of integrated copy number is exploited to design the transduction

[23] S. Halene and D. B. Kohn, *Hum. Gene Ther.* **11,** 1259 (2000).
[24] D. W. Emery, E. Yannaki, J. Tubb, and G. Stamatoyannopoulos, *Proc. Natl. Acad. Sci. U.S.A.* **97,** 9150 (2000).

protocol. A typical experiment is performed by testing increasing doses of vector calculated as TU/ml or as p24 equivalent of vector. To achieve high-frequency transduction of most target cell types, dose ranges from 10^6 to 10^8 TU/ml (end-point titer on HeLa) can be tested. These doses correspond to 10 to 1000 ng p24/ml of a vector with an infectivity of 10^5 TU/ng p24.

Critical Factors Affecting Transduction *in Vitro*

Vector Concentration

When the vector titer is calculated on a permissive cell line, refractory targets may need to be transduced employing a high number of TU per ml. One should take into account the fact that the expression *TU/cell,* which is equivalent to MOI (multiplicity of infection), is an arbitrary definition, because it does not consider the volume in which the transduction is performed, and therefore the particle concentration. Within this context, it is well acknowledged that vector concentration in the transduction medium is more important than the absolute number of particles available for each cell, since only a fraction of them come into contact with the target, especially when cells are cultured in monolayers in a large volume of medium. To increase the transduction efficiency, using a given amount of vector, cells must be transduced at a reasonably high density in the least volume, so that the possibility that vectors and cells encounter each other is enhanced. Other maneuvers that increase the chances of vector particles to come into contact with target cells, such as prolonged centrifugation of vector together with the cells (spinoculation), have been shown to increase transduction.[25] Haas *et al.*[20] have shown that the level of gene transfer into CD34 + cells by VSV-G pseudotyped LVs is improved by increasing the concentration of vector particles, holding the MOI constant in the transduction medium, but not increasing the MOI when the vector concentration is held constant. This result supports the notion that the concept of MOI is misleading when defining the transduction conditions.

Transgene Expression and Effects of Pseudotransduction and Transient Expression from Unintegrated Vector

One crucial advantage of LVs is the independence of transduction from cell division. Thus, almost every cell in a population can be transduced by a single exposure to the vector. However, when saturation is observed below the required frequency of transduction, either a vector has a less than optimal infectivity, or some cells in a heterogenous population are refractory to transduction. These two

[25] U. O'Doherty, W. J. Swiggard, and M. H. Malim, *J. Virol.* **74,** 10074 (2000).

possibilities can be addressed performing more than a single round of transduction, which may result in a higher frequency of transduced cells.

Following transduction of a given target, the average expression level of the transgene is proportional to the vector input, provided that the transduction frequency has not reached the saturation threshold. This means that the average expression level depends on the copy number of integrated vectors. According to a random distribution of independent events, it is impossible to achieve the maximal transduction frequency with a predominance of single integration events. Therefore, a cell population transduced to the maximal frequency displays a relatively wide range of integration events per cell. As a consequence, the expression level of the transgene can be very variable and this may result in toxicity in a fraction of cells.

Evaluation of transgene expression should be obtained at steady-state level, a condition which is normally reached several days after transduction, based on the cell type, the promoter transcriptional activity, mRNA stability, and half-life of the transgene protein.

Pseudotransduction effects have been described. Pseudotransduction is due to direct transfer of the transgene protein by either its presence in vector supernatants or its incidental incorporation into the vector particles.[26] This phenomenon occurs with low efficiency and can be detected only at early stages, usually in the range of few hours, following transduction. Of major concern is the expression driven by unintegrated vectors. Nonintegrated circular vectors can persist in the cell nucleus and represent dead-end by-products of aborted vector transfer. These DNA molecules can express transgenes. The extent of expression from unintegrated vector can be directly demonstrated producing a vector stock by transfection of a packaging plasmid defective for the expression of the viral integrase.[2] Such a vector is unable to actively integrate into the host cell genome, except for the rare events due to non-integrase-mediated mechanisms. The absolute level of transgene expression from unintegrated vectors can be relatively high during the first days after transduction, but is shown to rapidly decrease with time when the cells are proliferating (transgene protein half-life plays a critical role in determining these kinetics).[20] In addition, integrated vs unintegrated vector DNA can be demonstrated by Southern analysis. To show unintegrated vector, an enzyme with a unique restriction site in the vector has to be used to digest the genome from transduced cells. Upon this treatment, while integrated DNA, because of random integration, appears in the form of a smear, unintegrated vector appears as a single band. DNA analysis can be therefore used to estimate the ratio between the two DNA forms and allow timing of clearance of the unintegrated vectors more reliably than by scoring expression.

[26] M. L. Liu, B. L. Winther, and M. A. Kay, *J. Virol.* **70,** 2497 (1996).

Concluding Remarks

Pseudotyped lentiviral vectors provide a powerful tool for several gene transfer applications. The above-discussed protocols serve to predict the potential performance of vector stocks, especially when they are intended for *in vivo* purposes, when high-titer and high-infectivity concentrated vector is required. Among the other parameters, the infectivity of a vector has proved to be a limiting-factor to efficient gene transfer in many applications, such as *ex vivo* transduction of lymphocytes and CD34 + hematopoietic progenitors and *in vivo* administration.[7] Late-generation of VSV-G pseudotyped LVs can be concentrated to titers of 10^{10} TU/ml, with an infectivity in the range of 10^5 TU/ng p24 (end-point titer on HeLa). When displaying such features and complying with the other quality controls discussed in this chapter, the vector preparations allow significant levels of gene transfer in challenging *in vivo* settings.

Acknowledgments

The financial support of Telethon–Italy (Grant A.143), MURST (Grant 9905313431), the EU (QLK3-1999-00859), and the Italian Health Ministry (AIDS Research Program 40C.66) is gratefully acknowledged.

[30] Construction, Purification, and Characterization of Adenovirus Vectors Expressing Apoptosis-Inducing Transgenes

By SEMYON RUBINCHIK, JAMES S. NORRIS, and JIAN-YUN DONG

Introduction

Construction and large-scale production of recombinant adenovirus (rAd) vectors expressing proapoptotic transgenes, or any cytotoxic product in general, presents a special challenge to researchers. On the one hand, high levels of expression are often desired in the target cells, especially if the rAd vector is to be used as a therapeutic agent and is therefore required to kill the cells it infects as efficiently as possible. On the other hand, high levels of cytotoxic gene expression may be strongly deleterious to the packaging cell line in which the rAd vector is developed and propagated. In the most severe cases, total failure to obtain a viable vector after transfecting packaging cells with vector DNA results. However, transgene-related cytotoxic activity that does not kill the packaging cells outright can be even more problematic, since it is likely to place a strong selective pressure

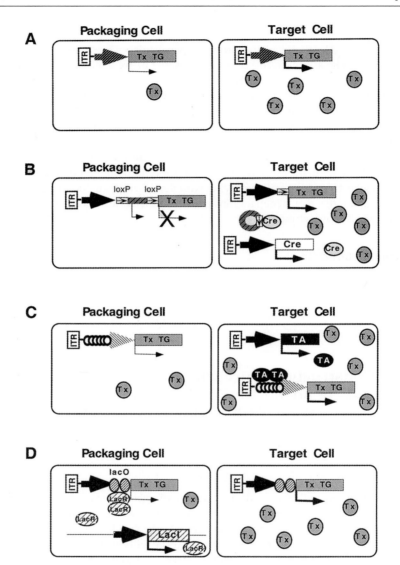

FIG. 1. Regulating Ad vector-delivered cytotoxic transgene expression in packaging and target cells. (A) Transgene regulation using tissue-specific promoters. Very low promoter activity in packaging cells results in minimal expression of the toxic transgene (TxTG). In target cells of specific origin, promoter activity is induced, and high levels of Tx protein are generated. (B) Cre/loxP switch. A stuffer sequence flanked by loxP sites in direct orientation is inserted between a strong ubiquitously active promoter and TxTG. The stuffer contains transcription and translation termination sites, so that Tx protein is not expressed in packaging cells. Target cells are coinfected with a second Ad vector, constitutively

on the replicating rAd vector to reduce or eliminate that activity. As a result, wild-type revertants as well as vector variants with mutations and small- or large-scale rearrangements in the transgene expression cassette can arise, all of which will have a replicative advantage over the desired vector.

Several approaches have been taken to solve this problem. Most of them involve regulation of transgene expression, so that in packaging cells no expression or very low levels of expression occur, but in the target cells expression is activated. A popular method is to use tissue-specific promoters that are not normally active in the packaging cell line but will drive toxic gene expression in target cells of specific origin (Fig. 1A).[1] However, this strategy is not always possible or desirable, since the activity of available promoters may not be sufficient to the task, or a ubiquitous expression pattern is preferable. In an alternative strategy, a stuffer sequence containing a transcription termination site is placed between a strong ubiquitous promoter and the cytotoxic transgene sequence.[2,3] The blocking element is flanked by the target sites of the site-specific recombinase Cre.[4] In the packaging cells, the blocking element completely prevents transcription of the transgene. Activation of expression in target cells requires coinfection with a second rAd vector, one that expresses Cre. This enzyme mediates excision of the blocking element and reattaches the transgene sequence next to the promoter, allowing unimpaired

[1] K. Aoki, L. M. Akyurek, H. San, K. Leung, M. S. Parmacek, E. G. Nabel, and G. J. Nabel, *Mol. Ther.* **1**, 555 (2000).

[2] T. Okuyama, M. Fujino, X. K. Li, N. Funeshima, M. Kosuga, I. Saito, S. Suzuki, and M. Yamada, *Gene Ther.* **8**, 1047 (1998).

[3] H. G. Zhang, G. Bilbao, T. Zhou, J. L. Contreras, J. Gomez-Navarro, M. Feng, I. Saito, J. D. Mountz, and D. T. Curiel, *J. Virol.* **72**, 2483 (1988).

[4] Y. Kanegae, G. Lee, Y. Sato, M. Tanaka, M. Nakai, T. Sakaki, S. Sugano, and I. Saito, *Nucleic Acids Res.* **23**, 3816 (1995).

FIG. 1. (*Continued*) expressing Cre protein. Cre-mediated site-specific recombination excises the stuffer element, allowing high rate of Tx expression. (C) Synthetic promoter/transcriptional activator systems. These rely on the exclusive interaction between unique binding sites in the UAS of the promoter and an engineered transcriptional activator (TA), which only binds to those sites. TA is not present in packaging cells, so that TxTG is expressed only with uninduced background promoter activity. However, amplification of Ad vector DNA to high copy numbers can still result in substantial levels of Tx. Target cells are coinfected with a second Ad vector which constitutively expresses TA. Promoter activity is induced and high levels of Tx are generated. An advantage of this system is that TA binding to the promoter (and therefore promoter activity) can usually be regulated by a small molecule such as a drug or a hormone that can be added exogenously. (D) Repressible promoter systems. These take advantage of high specificity and very low K_d of certain prokaryotic transcription repressor, such as LacR or TetR. The packaging cell line is modified to constitutively express the repressor protein. Meanwhile, a strong ubiquitous promoter is modified by having repressor-specific operator sequences inserted near the transcription initiation site. In these packaging cells, binding of the repressor to its operators interferes with RNA polymerase binding and/or initiation of transcription, so that very little Tx is expressed. In target cells repressor is absent, and expression levels are high.

high-level transcription (Fig. 1B). The major disadvantage of these methods is the requirement for coinfection with a second vector expressing Cre recombinase, which adds to the cost of vector production and reduces the efficiency of *in situ* target cell transduction.

Alternatively, regulatable expression systems can be used to reduce or shut off transgene expression in the packaging cell line. Regulation is achieved in two basic ways. The one most commonly encountered uses a protein with transcriptional activator activity to bind to specific DNA sites near the TATA region of the promoter in the presence (or absence) of a signal delivered by a chemical or hormonal regulator (Fig. 1C). The uninduced or background activity of such promoters is usually very low, but once transactivator binding occurs very high expression levels can be reached. Such systems use either mammalian transcriptional activators modified to respond to inducers that do not interact with the unmodified versions, or fusion proteins incorporating prokaryotic transcriptional regulators and eukaryotic transactivating domains.[5] Typically, a second vector is required to deliver transcriptional activator to the target cells. Another method involves the use of prokaryotic repressor proteins such as LacR to suppress the transcriptional activity of a strong constitutive promoter (Fig. 1D). In this case, operator-binding sites of the repressor are placed in a close proximity to the promoter, so that binding of the repressor to those sites will interfere with RNA polymerase II binding to the promoter or initiation of transcription.[6] This method has certain advantages, since packaging cell lines can be engineered to stably express the desired repressor,[6] thus ensuring low expression levels of toxic proteins during vector production. In the absence of repressor in the target cells, a high level of expression is achieved.

In addition to transcriptional regulation, packaging cell lines can be made more resistant to apoptosis-inducing signals, by developing variants that constitutively express protective molecules such as viral caspase inhibitors cytokine response modifier A (CrmA) from cowpox virus and p35 from baculovirus. However, it is important to realize that these protective molecules can be "flooded" by very high expression levels of apoptosis-promoting molecules in the course of vector replication. Therefore, a combination of regulated expression with a resistant packaging cell line is usually desirable.

Design and Assembly of Adenovirus Vectors Expressing FasL-GFP

Many strategies for construction of rAd vectors have been described.[7–14] We utilize an *in vitro* ligation reaction to assemble rAd vector genomes, which are then transfected into packaging cells for replication. One advantage of this method is

[5] M. Molin, M. C. Shoshan, K. Ohman-Forslund, S. Linder, and G. Akusjarvi, *J. Virol.* **72**, 8358 (1998).

[6] D. A. Mathews, D. Cummings, C. Evelegh, F. L. Graham, and F. L. Prevec, *J. Gen. Virol.* **80**, 345 (1999).

that since the rAd genome is preassembled prior to entry into the cell, no recombination is required. As a result, it is not necessary to screen plaques to obtain pure vector stocks. Another advantage of the method is the ability to insert sequences in both left and right ends of the Ad5 genome. In order to do so, we constructed two shuttle vectors, pL-Ad.CMV and pR-Ad.6MCS. The shuttle vector for the insertion of transgenes into the E1 region of adenoviral genome, pL-Ad.CMV, was constructed using PCR primers ITR-F (5′AATTC*ATTTAAAT*CATCATCAATAATATA CCTTAT3′) and ITR-R (5′GGATCCTCTAGAGTCGA*CTGACTATAAATAATAAA ACGCC*3′), and contains bp 1 to 450 of Ad5 sequence, which includes the left inverted terminal repeat (ITR) and the packaging signal (Ψ) of Ad5 (Fig. 2A). These sequences are followed by a human cytomegalovirus immediate early (hCMVie) enhancer/promoter element and a multiple cloning site (MCS) of pcDNA3 (Invitrogen). Immediately preceding the ITR sequences is the *Swa*1 cleavage site, introduced by the ITR-F primer (underlined). Following the MCS is the "adapter" element, which contains *Xba*1, *Avr*2 and *Spe*1 cleavage sites, any one of which can be used to attach pL-Ad.CMV sequences to the Ad5 genome. The shuttle vector for the insertion of transgenes between the E4 region and the right ITR of Ad5 was constructed to contain Ad5 sequences from bp 27331 to bp 35935, with a MCS inserted at bp 35825 (Fig. 2B). The E3 region (bp 28592–30470) and all E4 ORFs with the exception of ORF 6 were deleted from this construct, which was named pR-Ad.6MCS. The right ITR is followed by a *Swa*1 cleavage site, again introduced by PCR primer. The *Eco*R1 cleavage site is used to attach R-Ad sequence to the rest of the rAd vector genome. Transgene expression cassettes are cloned into MCS of either one or both of the shuttle vectors. The new constructs are then cleaved with *Swa*1 and either *Eco*R1 for pR-Ad vector, or one of the sites in the adapter for the L-Ad vector. The middle section of the rAd vector, containing most of the Ad5 sequence, is derived from the genome of an Ad5 variant, Ad5-sub360, which contains an *Xba*1 site at bp 1338 and an *Eco*R1 site at bp 27331 (Fig. 3). Purified Ad5-sub360 DNA is digested with these enzymes, the large 26 kb fragment purified and then ligated to L-Ad and R-Ad fragments (Fig. 3). Ligation products are transfected into 293 cells to generate replicating vectors.

Being able to clone a transgene expression cassette into a site near the right ITR of Ad5 offers additional advantages. One of them is increased separation from

[7] F. L. Graham and L. Prevec, *Mol. Biotechnol.* **3**, 207 (1995).

[8] Y. Kanegae, S. Miyake, Y. Sato, G. Lee, and I. Saito, *Acta Paediatr. Jpn.* **38**, 182 (1996).

[9] J. Bramson, M. Hitt, W. S. Gallichan, K. L. Rosenthal, J. Gauldie, and F. L. Graham, *Hum. Gene Ther.* **7**, 333 (1996).

[10] S. Fu and A. B. Diesseroth, *Hum. Gene Ther.* **8**, 1321 (1997).

[11] H. Kojima, N. Ohishi, and K. Yagi, *Biochem. Biophys. Res. Commun.* **246**, 868 (1998).

[12] H. Mizuguchi and M. A. Kay, *Hum. Gene Ther.* **9**, 2577 (1998).

[13] P. Ng, R. J. Parks, D. T. Cummings, C. M. Evelegh, U. Sankar, and F. L. Graham, *Hum. Gene Ther.* **10**, 2667 (1999).

[14] R. D. Anderson, R. E. Haskell, H. Xia, B. J. Roessler, and B. L. Davidson, *Gene Ther.* **7**, 1034 (2000).

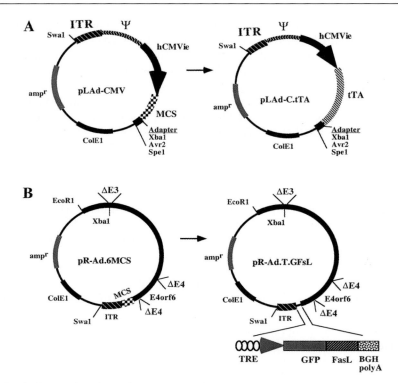

FIG. 2. Construction of pL-Ad and pR-Ad shuttle vectors. (A) pL-Ad.C.tTA. This plasmid is based on the pL-Ad.CMV shuttle vector, which contains the leftmost 450 bp of Ad5 genome followed by a strong hCMVie enhancer/promoter and an MCS with an adapter sequence containing restriction sites Xba1, Avr2 and Spe1. The tTA gene from pUHD15-1 was inserted into the MCS. After assembly into rAd vectors, E1A poly (A) is utilized for efficient tTA expression. A similar strategy is used to construct pL-Ad vectors containing other transgenes. (B) pR-Ad.T.GFsL. This plasmid is based on the pR-Ad.6MCS, which contains Ad5-sub360 sequences from the unique EcoR1 site (bp 27333) to the right ITR (bp 35935), with E3 (bp 28594 to bp 30484) and E4 deletions (the Orf6 of E4 is retained).[32] An MCS was inserted at bp 35825. The regulatable FasL-GFP expression cassette, consisting of the TRE promoter, FasL-GFP fusion protein, and bovine growth hormone (BGH) poly A site, was inserted into that MCS.

the packaging signal region near the left ITR. This region also contains the E1A enhancer, which in several cases was found to interfere with the activity of regulatable and tissue-specific promoters.[15,16] We find that placing such promoters near the right ITR results in minimal interference from the E1A enhancer.[17] Another

[15] C. J. Ring, J. D. Harris, H. C. Hurst, and N. R. Lemoine, *Gene Ther.* **3,** 1094 (1996).

[16] Q. Shi, Y. Wang, and R. Worton, *Hum. Gene Ther.* **8,** 403 (1997).

[17] S. Rubinchik, S. Lowe, Z. Jia, J. S. Norris, and J.-Y. Dong, *Gene Ther.* **8,** 247 (2001).

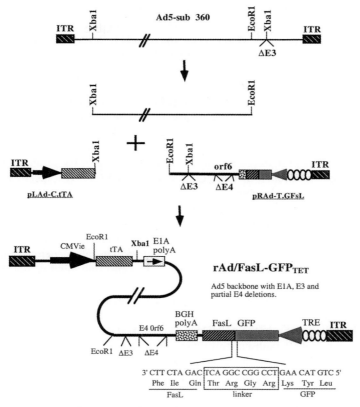

FIG. 3. *In vitro* assembly of the rAd/FasL-GFP$_{TET}$ vector genome. Purified Ad5-sub360 DNA is cut with *Eco*R1 and *Xba*1 restriction enzymes, and the large internal 26 kb fragment is purified. It is ligated to the fragments prepared from pL-Ad.C.tTA (*Swa*1 and *Xba*1 digest) and pR-Ad.T.GFsL (*Swa*1 and *Eco*R1 digest). After ligation, the products are transfected into 293 cells to generate the rAd/FasL-GFP$_{TET}$ vector. The region of the junction between the GFP and FasL reading frames is expanded. Other rAd vectors are generated using a similar strategy.

advantage is the ability to avoid negative selection pressure against cytotoxic genes in packaging cell lines. As mentioned in the introduction, deleterious effects of cytotoxic proteins such as FasL on cell viability can give a very strong selective advantage to those vectors that are able to get rid of them. If the transgene is cloned into the E1 region of Ad5, as is most commonly done, a recombination event can occur between vector sequences and wild-type Ad5 sequences integrated in 293 cell DNA. Such an event would replace transgene sequence with wild-type Ad5 sequence, creating a "revertant." These recombinations occur with all rAd vectors at low frequency, but in the absence of selective pressure revertant titers in stocks of most rAd vectors remain very low. If the transgene is cytotoxic, however, revertants have a strong replicative advantage and will quickly come to dominate the culture. If the transgene is cloned near the right ITR, revertants can still arise as the result of recombination with the left end, but since they are not able to get rid of the transgene, they have no advantage over the desired vector. Therefore, in our strategy transgene expression cassettes containing tissue-specific or regulatable promoters and/or cytotoxic genes are cloned into pR-Ad vectors.

We feel that it is very important to have a simple and rapid method of monitoring, both qualitatively and quantitatively, apoptotic transgene expression during each step of rAd vector development, culture expansion and functional analysis. Green fluorescent protein (GFP) was selected as a reporter based on the following features: fluorescence of the red-shifted variant EGFP (Clontech) is bright and resistant to photobleaching, and can be easily detected in living cells using a fluorescence microscope with commonly available fluorescein isothiocyanate (FITC) optics; GFP fluorescence can be analyzed quantitatively in whole cells or cell lysates.[18] GFP has no known activities that interfere with cellular functions, while at the same time no known process in mammalian cells generates a fluorescence emission profile similar to that of GFP, thus ensuring minimal interference and low background.

A FasL-GFP fusion protein was constructed.[19] In order to place the FasL-GFP gene under the control of the tet-regulated expression system,[20] it was excised from pC.GFsL and cloned into the pUHD10-3 vector that contains the TRE promoter to generate pT.GFsL. Subsequently, the expression cassette consisting of TRE promoter, FasL-GFP gene, and SV40 pA was removed from the vector and cloned into pR-Ad.6MCS to generate pR-Ad-T.GFsL. At the same time, the Tet-OFF fusion activation protein (tTA) expression cassette consisting of hCMVie promoter and the tTA gene was extracted from pUHD15-1 and inserted into pL-Ad-CMV to generate pL-Ad-C.TA. The two shuttle vectors were used to make fragments for ligation with the Xba1/EcoR1 Ad5-sub360 fragment (described in the methods sections) and the resulting rAd vector was named rAd/FasL-GFP$_{TET}$ (Fig. 3). In this vector, tTA protein is constitutively expressed from the hCMVie promoter, and, in the absence of tetracycline, binds to the TRE promoter to activate high levels of expression of FasL-GFP. If tetracycline (or its analogs such as doxycycline) is present, tTA fails to bind to the TRE and FasL-GFP expression remains uninduced. However, a low level of background expression inherent in the TRE promoter itself still exists.[21] We find that in 293 cells, where rAd vector DNA is amplified to

[18] L. M. Cashion, L. A. Bare, S. Harvey, Q. Trinh, Y. Zhu, and J. J. Devlin, *Biotechniques* **26,** 928 (1999).

[19] S. Rubinchik, R.-X. Ding, A. J. Qiu, F. Zhang, and J.-Y. Dong, *Gene Ther.* **7,** 875 (2000).

[20] M. Gossen and H. Bujard, *Proc. Natl. Acad. Sci. U.S.A.* **89,** 5547 (1992).

[21] M. Gossen, A. L. Bonin, and H. Bujard, *Trends Biochem. Sci.* **18,** 471 (1993).

very high copy numbers, uninduced TRE expression is also amplified, so that the efficiency of rAd/FasL-GFP$_{TET}$ replication in 293 cells is diminished compared to vectors with nontoxic genes.

Development of 293 Cell Lines Stably Expressing CrmA and Bcl2

The original packaging cell line developed for the production of rAd vectors is HEK293. It was derived by introducing sequences from the left end of the human adenovirus type 5 into human embryonic kidney cells and selecting for the immortalized phenotype.[22] These cells provide Ad5 E1A and E1B functions in trans to the replication-defective rAd vectors. Subsequently, other cell lines were derived from 293 that also provide E4, E2, or Ad fiber functions. Although 293 cells express Ad5 protein E1B 19K, which has Bcl-2-like anti-apoptosis activity,[23,24] the majority of variants of this cell line are sensitive to apoptosis induction through the Fas pathway,[25] although to varying extents. We found that although we obtained primary lysates of rAd/FasL-GFP$_{TET}$ after transfection of 293 cells, the titers were significantly lower than that of rAd/GFP, suggesting reduced replication activity. In order to facilitate development of vectors expressing apoptosis-inducing proteins, we established 293 cell line stably transformed with CrmA gene. CrmA inhibits the interleukin (IL)-1beta converting enzymes (ICE) known as caspases, although with varying efficiency.[26] It is a potent and specific inhibitor of certain apoptotic pathways, including Fas and TNFα-mediated apoptosis.[27]

The CrmA coding sequence (from pcDNA3-CrmA vector generously provided by Dr. Vishva M. Dixit at Genentech) is cloned into the MCS of pIRES1*neo* vector (Clontech) to generate pCrmA-I-Neo. In this vector, two genes are transcribed from a single hCMVie promoter, the second gene utilizing an internal ribosome entry site (IRES) from encephalomyocarditis virus (ECMV) to initiate translation from the middle of the mRNA transcript.[28] In pIRES1*neo,* the second gene codes for neomycin phosphotransferase, which is expressed at a reduced level compared to the gene immediately following the promoter.[29] Selection of transformed cells with high concentrations of G418 ensures that the surviving resistant cells also express high levels of the gene of interest. The cell lines were established and

[22] F. L. Graham, J. Smiley, W. C. Russell, and R. Nairn, *J. Gen. Virol.* **36**, 59 (1977).

[23] T. Subramanian, J. M. Boyd, and G. Chinnadurai, *Oncogene* **11**, 2403 (1995).

[24] C. A. Boulakia, G. Chen, F. W. Ng, J. G. Teodoro, P. E. Branton, D. W. Nicholson, G. G. Poirier, and G. C. Shore, *Oncogene* **12**, 529 (1995).

[25] A. T. Larregina, A. E. Morelli, R. A. Dewey, M. G. Castro, A. Fontana, and P. R. Lowenstein, *Gene Ther.* **85**, 563 (1998).

[26] P. G. Ekert, J. Silke, and D. L. Vaux, *EMBO J.* **18**, 330 (1999).

[27] M. Tewari and V. M. Dixit, *J. Biol. Chem.* **270**, 3255 (1995).

[28] V. Gurtu, G. Yan, and G. Zhang, *Biochem. Biophys. Res. Commun.* **229**, 295 (1996).

[29] B. Galy, *Trends Biochem. Sci.* **25**, 426 (2000).

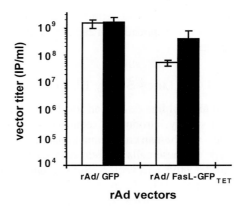

FIG. 4. Improved efficiency of rAd/FasL-GFP$_{TET}$ replication in 293CrmA cells. Twelve-well plates were seeded with 1×10^4 293 (open bars) or 293CrmA (filled bars) cells and 3 wells each were infected with rAd/FasL-GFP$_{TET}$ or rAd/GFP at an MOI of 5 one day later. Then, 48 hours posttransduction, cells from each well were collected and lysed in 1 ml of DMEM. Lysates were titrated for transduction of GFP fluorescence on 293CrmA cells in 96-well plates. Results are reported as IP/ml lysate and represent means and average errors of 2 sets of independent experiments.

characterized.[19] We find that 293CrmA cells transduced with rAd/FasL-GFP$_{TET}$ generate significantly higher titers than 293 cells (Fig. 4), thus, facilitating large-scale vector production. However, the titers are still not as high as those for vectors with nontoxic genes such as GFP. It would seem that CrmA expression is not sufficient to completely counteract high levels of FasL-GFP expression in 293 cells.

Methods for Propagation, Amplification, and Characterization of Adenovirus Vectors

Preparation of DNA Fragments for in Vitro Assembly of rAd Vector Genomes

Materials

1. Agarose and low-melt agarose (OmniPur from EM Science or other providers).

2. $50\times$ TAE buffer: Tris-HCL, sodium acetate, EDTA. pH with glacial acetic acid.

3. Restriction endonucleases, calf intestinal phosphatase (CIP), T4 ligase, and β-agarase (aka Gelase). New England Biolabs (NEB) or other providers.

4. Phenol (TE-saturated, pH 8.0), chloroform, 20% (w/v) sodium dodecyl sulfate (SDS), 3 M sodium acetate (pH 7.5), isopropanol, 70% (v/v) ethanol. Available from several providers.

5. DNase I (10 mg/ml in H_2O), proteinase K (10 mg/ml in H_2O), DNase-free ribonuclease A (1 mg/ml in H_2O). Available from Sigma or several other providers.

Relevant features of pL-Ad, pR-Ad, and Ad5.sub360 are shown in Fig. 2. The pL-Ad and pR-Ad plasmid DNA is purified using Qiagen Tip-100 or Tip-500 columns according to manufacturer's instructions. DNA concentrations are determined by obtaining the OD_{260} with a spectrophotometer (1 $OD_{260} = 50$ $\mu g/ml$ purified DNA). For digests, at least 2 μg (5 μg is preferable) of each plasmid should be used. Both pL-Ad and pR-Ad vectors are digested with *Swa*1 (5 μg DNA, 10 U *Swa*1, 1× NEB buffer 3, and BSA in 100 μl reaction) overnight at ambient temperature. Next, *Xba*1, *Avr*2, or *Spe*1 is added to the pL-Ad reaction, depending on which is unique (20 U enzyme, 10 μl 10× NEB buffer 2, 90 μl H_2O). *Eco*R1 is added to the pR-Ad reaction (20 U *Eco*R1, 15 μl 10× NEB *Eco*R1 buffer, 85 μl H_2O). Both reactions are put in a 37° water bath and incubated for 2 hr. Next, 6 U of CIP is added to each reaction, followed by an additional hour at 37°. A 0.5% (w/v) agarose gel is prepared (0.5 g of agarose in 100 ml of 1× TAE buffer) with large wells. Fifty μl of gel-loading buffer [80% (v/v) glycerol, 0.1% (w/v) bromphenol blue] is added to each reaction and mixed well, and the reactions are loaded on the gel which is run in 1× TAE buffer at 7 V per cm of distance between the electrodes. Once the bromphenol blue dye has migrated an appropriate distance (typically 2/3 of the length of the gel), the gel is soaked for 10 min in H_2O with 0.5 $\mu g/ml$ ethidium bromide. The DNA bands are visualized under long wavelength UV light (330 to 360 nm) and the appropriate fragments are excised by razor blade or scalpel and placed inside 2 ml snap-cap tubes. Several techniques can be used to purify the DNA fragments from the gel. We routinely use the "freeze–squeeze" method, which delivers good yields and quality of DNA.

1. Crush the gel slice into paste inside the 2 ml tube using a flame-sealed 1 ml pipette tip.
2. Add 1.5 gel volumes of TE-saturated phenol and vortex vigorously until the two are completely mixed (the mixture will become white).
3. Freeze the tube with gel/phenol mixture at −70° or in dry ice/ethanol bath for 5 min.
4. Take the tube out and centrifuge while still frozen at maximum speed for 5 min in a microcentrifuge.
5. Transfer the top aqueous phase to a new tube, being careful not to touch the interface.
6. Add an equal volume of chloroform, mix vigorously, and centrifuge at max speed for 2 min.
7. Transfer the top aqueous phase to a new tube; add 1/10 volume 3 M sodium acetate (pH 7.5) and 1 volume isopropanol. Centrifuge at max speed for 20 min in the microcentrifuge.

8. Aspirate the supernatant. Wash the pellet with 70% ethanol. Centrifuge at max speed for 5 min, aspirate supernatant, and place the open tube on the benchtop for 30 min to dry the pellet.

9. Add 20 ml of 10 mM Tris-HCl (pH 8.0) to dissolve the pellet.

To prepare the middle ligation fragment, Ad5-sub360 DNA must be purified. First, high-titer lysate is obtained (see following sections). The lysate is treated with DNase I (100 μg/ml final concentration) and RNase H (100 μg/ml final concentration) for 30 min at 37°. Next, 20% SDS, 0.5 M EDTA, and 10 mg/ml proteinase K is added to obtain 0.5% (v/v) SDS, 20 mM EDTA, and 125 mg/ml proteinase K final concentrations. The reaction is incubated at 37° overnight. Next, the reaction is extracted 2 times with equal volumes of phenol/chloroform (50 : 50 v/v) and once with 1 volume of chloroform. The DNA is precipitated with 1/10 volume of sodium acetate (pH 7.5) and 1 volume of isopropanol, and pelleted in a microcentrifuge (max speed for 30 min at 4°). The pellet is washed with 70% ethanol and dried in an open tube on the benchtop. Virus DNA is dissolved in TE (pH 8.0) at 25 μl per 1 ml of original lysate. DNA concentration is determined by OD$_{260}$ measurement. Typically, between 5 and 15 μg of viral DNA per ml of lysate is recovered. Five μg of virus DNA is digested with EcoR1 and Xba1 (20 U each enzyme, 10 μl 10× NEB EcoR1 buffer, BSA, 100 μl reaction volume) for 4 hr at 37°.

Meanwhile, 0.7% (w/v) low-melt agarose gel is prepared [0.7 g SeaKem (FMK) agarose in 100 ml of 1× TAE buffer] with a large well. The reaction is mixed with loading buffer and electrophoresis is performed in 1× TAE buffer overnight at 1.4 V/cm. The next day, the gel is soaked in 5 gel volumes of 10 mM Tris-HCl (pH 7.0) with 0.3 μg/ml ethidium bromide for 30 min. The DNA bands are visualized under long-wavelength UV light (330 to 360 nm) and the largest band (26 kb) is excised. The gel slice is placed in one or more 2 ml snap-cap tubes, which are placed at 65° for 5 to 10 min, until the agarose has completely melted. The tubes are then transferred to a 40° water bath, and β-agarase is added to 1 U per ml molten agarose final concentration. The tubes are mixed and incubated at 40° overnight. Next, 1/10 volume of 3 M sodium acetate (pH 7.5) is added, and the reactions are placed on ice for 20 min. The tubes are then spun at max speed in a microcentrifuge for 20 min at 4°. The supernatants are transferred to new 2 ml tubes, and DNA is precipitated with 1 volume of isopropanol as described above. The pellets are dissolved in 25 μl of 10 mM Tris-HCl (pH 8).

Once all of the ligation components are purified, they are checked for concentration by OD$_{260}$, and for fragment quality by running 1 μl of each fragment on a 0.5% agarose gel. The ligation reaction is set up by combining 1 μg of the Xba1/EcoR1 Ad5-sub360 fragment with 1 μg of the pL-Ad Swa1/Xba1 fragment and 2 μg of the pR-Ad Swa1/Xba1 fragment with 10 μl of 10× NEB T4 ligase buffer and 800 U T4 ligase from NEB in a 100 μl reaction volume. This generates approximate molar ratios between pL-Ad, sub360, and pR-Ad of 10 : 1 : 10. The reaction is placed at 4° overnight, and ligation products are precipitated the next day by the addition of 1/10 volume of 3 M sodium acetate (pH 7.5) and 1 volume of isopropanol. Pellets are washed and dried as described above, and dissolved in 20 μl of TE (pH 8). One ligation reaction is used to transfect two wells of a 6-well plate seeded with 293 cells (see following sections).

Cell Culture Techniques

Materials

1. 10× PBS: 1.37 M NaCl, 27 mM KCl, 14 mM KH$_2$PO$_4$, 43 mM Na$_2$HPO$_4$. Sterilized by autoclaving. To make 1× PBS, 50 ml of 10× stock are mixed with 450 ml of ddH$_2$O and filter-sterilized using a 500 ml Stericup 0.22 μm GP Express Membrane filter system (Millipore) in a laminar flow hood under sterile conditions. Store at ambient temperature.

2. Trypsin solution: 0.05% trypsin and 1 mM EDTA in 1× PBS. Ten ml of 2.5% trypsin in HBSS (Mediatech), 1 ml of 0.5 M EDTA (pH 8.0), 50 ml of 10× PBS, and 439 ml of ddH$_2$O are mixed and filter-sterilized using 500 ml Stericup filter unit in a laminar flow hood under sterile conditions. Ready-to-use solution is stored at 4°.

3. Complete medium: Dulbecco's modified of Eagle's medium (DMEM) with 4.5 g/liter glucose, L-glutamine, and sodium pyruvate (from Mediatech Inc., but also available from other suppliers) supplemented with 10% (v/v) of Cosmic Calf Serum (CCS) available from HyClone. DMEM is purchased ready to use in 1 liter bottles. We found it unnecessary to heat-inactivate CCS, which is aliquoted into 50 ml screw-cup tubes (USA Scientific) and stored at −20°. Thawed serum is added to DMEM under sterile conditions.

We use 293 cells that are generally high passage number, and are rapidly dividing, small size (approximately 2/3 the size of a HeLa cell) and loosely adherent to standard culturing dish surfaces. Cultures of these cells are maintained in 100-mm or 150-mm dishes (available from several sources) in complete medium and are typically subcultured when they reach 80% confluency. Culture medium is removed from dishes by aspiration, and trypsin solution (warmed to room temperature) is added at 5 ml per 100-mm dish or 12 ml per 150-mm dish. Dishes are gently rocked once or twice to ensure uniform cover with trypsin. Progress of cell detachment from dish surface is monitored by inverted microscope. When cells assume a rounded shape, but before they detach completely, trypsin solution is aspirated. Cells are then immediately dislodged by a sharp slap of the hand against the side of the dish. Complete culture medium is added at 5 ml per 100-mm dish or 10 ml per 150-mm dish. Cells are further dispersed by repeatedly aspirating the medium into and out of a 10 ml culture pipette with its tip firmly pressed to the bottom of the dish. New dishes are seeded with 1 ml of cell suspension per 100-mm dish or 2.5 ml per 150-mm dish (approximately 1.2×10^6 and 3×10^6 cells, respectively). Complete medium is added (9 and 20 ml per 100-mm and 150-mm dishes, respectively) and the cells are evenly dispersed by a gentle rocking motion. Dishes are placed in an incubator at 37° and 5% CO_2.

Transfections

Materials

1. DMEM and complete medium
2. SuperFect reagent (Qiagen)
3. Plasmid DNA, purified using Qiagen Tip-100 columns
4. Ethanol-precipitated ligation reactions

The 293 cell line and its derivatives can be transfected with high efficiency using a variety of methods. We find the SuperFect reagent to be efficient and convenient, but other reagents, such as LipofectAMINE (GIBCO BRL) and FuGENE 6 (Roche Diagnostics) have also been used successfully. Cells are seeded in 6-well plates at 1×10^5 per well with 2 ml of complete medium. The next day, transfection reactions are set up in the laminar flow hood by adding 2 μg of DNA in TE buffer (10 mM Tris-HCl and 1 mM EDTA, pH 8.0) to a 1.5 ml Eppendorf tube. Then, ambient temperature DMEM is added to the tube to bring the total volume to 90 μl. The contents of the tube are mixed by gentle vortexing. Then, 10 μl of SuperFect reagent is added to the tube, which is again gently vortexed to mix the contents. The reaction is incubated at ambient temperature for 15 to 30 min. Culture medium is removed from the 6-well plate, and immediately replaced with 1 ml of fresh complete culture medium, warmed to ambient temperature and added slowly so that the cells are not dislodged. Next, 1 ml of complete medium is added to the reaction and mixed well. The mixture is added drop-by-drop to the well and mixed by gentle rocking of the plate. After 12 to 24 hr, the transfection medium is replaced with complete medium. Transfection efficiency is determined at 48 hr by observing GFP fluorescence with an Axiovert 25 inverted microscope using a fluorescent lamp attachment (Zeiss Optical Systems Inc., Thornwood, NY) and an FITC excitation/emission filter set (Chroma Technology Corp, Brattleboro, VT).

Primary Virus Vector Lysates

Materials

1. DMEM and complete medium with 1 μg/ml doxycycline, purchased as the hydrochloride (Sigma)

2. SuperFect reagent and ethanol-precipitated rAd vector ligation products in 10 μl of TE
3. Dry ice/ethanol bath

In vitro ligated rAd vector genomes are transfected into 6-well plates containing 293 Crma$^+$ cells as described in the previous section, except that the complete medium contains 1 μg/ml doxycycline in order to prevent activation of the TRE promoter that induces higher levels of FasL-GFP expression and associated apoptosis. However, replication of rAd vector DNA in 293 cells results in fairly high levels of transgene expression even from uninduced TRE promoter. Transfection efficiency is monitored at 48 hr posttransfection with an inverted fluorescent microscope, with higher percent fluorescent cells corresponding to a more rapid development of replicating vectors in cell culture. After 72 hr, cells from each well are detached with a trypsin solution, dispersed, and transferred to a 150-mm dish with 25 ml of complete medium plus doxycycline. Dishes are monitored once a day for evidence of vector replication. Typically, replicating centers are first observed with fluorescent microscope as clusters of cells glowing faint green between 7 and 10 days posttransfection, with plaques formed of cells showing cytopathic effects (CPE) visible with brightfield a day or two later. If cells reach high confluency (85 to 95%) without vector replication being observed, they are detached with trypsin solution and dispersed, and 1/3 of the culture is passed to a new 150-mm dish for several additional days of development. If no replication is observed after 16 days, a new transfection should be attempted. Once vector replication is observed, the cell culture is split into four 150-mm dishes and monitored for 2 to 3 more days. Typically, cells are collected when approximately 50% or more of them show CPE. If some cells remain attached, they can be washed or scraped off the bottom of the dish using a pipette tip. Medium and the detached cells from each dish are collected into 50 ml screw-cap centrifuge tube (USA Scientific) and centrifuged at 800g for 10 min. Supernatants are aspirated and 1 ml of DMEM is added to each pellet. Cells are resuspended and transferred to a 1.5 ml Eppendorf tube. The tubes are placed in a dry ice/ethanol bath for 3 min, and then transferred to a 37° water bath for 3 min. This is repeated two more times. The tubes containing the lysates are centrifuged at 5000g (7300 rpm in Micromax centrifuge) for 1 min. Supernatant containing virus vector particles is then transferred to a labeled 2 ml Cryotube (US Scientific or other providers) and is stored at −70° or in a liquid nitrogen storage container.

Determination of Infectious Particle (IP) Titers of rAd Vector Stocks

Materials

1. DMEM and complete medium
2. Trypsin solution

3. 1–20 μl filter-tip pipette tips
4. 96-well tissue culture plates (Greiner or other providers)

We take advantage of vector-induced GFP fluorescence in successfully transduced cells to rapidly and accurately determine IP titers of primary lysates, large-scale lysate stocks, and purified vector stocks. On the day of the assay, 293 cells from a 100-mm or 150-mm stock plate are detached with trypsin solution, dispersed, and diluted with complete medium to approximately 5×10^4 cells/ml. Next, a multichannel pipetter (either 12- or 8-channel) is used to deliver 90 μl of cell suspension to each well of a 96-well culture plate. Ten μl of each vector solution is delivered to a well on row "A" of the plate. Primary lysates are measured directly, while large-scale lysates and purified vector stocks are typically prediluted 10- to 100-fold. Usually, each stock is tested in triplicate, so that four different vectors can be tested per plate. Pipette tips with filters are used in this case to prevent contamination of the pipetter and cross-contamination of vector stocks. A 12-channel pipetter is than used to mix the contents of row "A," followed by standard serial dilution. The plate is then placed in a tissue culture incubator at 37° and 5% CO_2 and examined by inverted fluorescent microscopy 48 hr postinfection. At higher dilutions, a 293 cell exhibiting GFP fluorescence is considered to be the result of an infection with a single infectious particle. The IP titers are determined according to the formula

$$\text{IP per ml of stock} = \text{predilution factor}$$
$$\times \text{\# of fluorescent cells per well} \times 10^{(\text{row \#}+1)}$$

For example, if the stock were prediluted 10-fold, and 12 fluorescent cells were seen in row "D," which is the fourth row, the titer is $10 \times 12 \times 10^{(4+1)}$ or 1.2×10^7 IP/ml. Typically, titers from all wells where accurate cell numbers can be determined are obtained, and an average value is calculated. Values from replicate dilution series are also used to determine average titer. The error associated with these measurements is usually within 20%.

Preparation of Large-Scale Lysates and Purified Vector Stocks

Materials

1. DMEM and complete medium.
2. Infection medium: DMEM plus 5% (v/v) CCS, 100 U/ml penicillin, 100 μg/ml streptomycin, 1 μg/ml doxycycline. Penicillin and streptomycin are purchased as a 100× working concentration stabilized solution (Sigma or other sources).
3. 1× PBS.
4. HEPES buffered saline (HBS): 21 mM HEPES, 140 mM NaCl, 5 mM KCl,

0.75 mM Na$_2$HPO$_4 \cdot$ 2H$_2$O, and 0.1% (w/v) dextrose. Adjust pH with NaOH to 7.5. Filter using 500 ml Stericup filter unit and store at 4°.
   5. Dry ice/ethanol bath.
   6. CsCl: high-purity powder available from several providers.

Primary lysate titers are determined as described in the previous section. To amplify vector stocks, 293 or 293CrmA cells are seeded at 3 × 10^6 cells per 150-mm dish with 20 ml of complete medium. When cells reach 70–80% confluency, medium is aspirated and replaced with 20 ml of warmed infection medium containing 1 × 10^7 IP/ml virus. This results in an MOI of between 20 and 30. Dishes are placed in the incubator. After 48 hr, cells exhibiting CPE and are either detached or can be easily detached by pipetting. They are collected into 50 ml tubes and centrifuged at 800g. The supernatant is aspirated and DMEM is added to the pellets at 0.5 ml per 150-mm dish culture. Cell suspensions are transferred to 2 ml snap-cap tubes and cells are lysed by three transfers between dry ice/ethanol bath and a 37° water bath as described in a previous section. Lysates are centrifuged at 5000g (7300 rpm in Micromax centrifuge) for 1 min. The supernatant, containing viral vector particles, is transferred to a 2 ml Cryotube and the pellet is discarded. The lysates are titrated as described in a previous section and are stored at −70°.

To prepare purified vector stocks, 150-mm dishes are seeded and infected as described above. For vectors expressing non-toxic genes, 20 to 40 dishes are used; for vectors expressing FasL-GFP, culture is scaled up to 30 to 80 dishes. Cells are collected and centrifuged as above, and supernatants are aspirated. Then, 45 ml of 1× PBS is added to each tube; cells are gently resuspended and then recentrifuged. PBS is removed, and HBS is added at 0.5 ml per 150-mm dish culture. Cell lysates are prepared as described above. These are collected in 15 ml screw-top centrifuge tubes (US Scientific or other providers) at 10 ml of lysate per tube. Next, 4.8 g of CsCl is added to each tube and completely dissolved. This results in a solution with density of 1.34 g/ml. CsCl-containing lysates are placed in clear ultracentrifuge tubes (Beckman) and the volumes are adjusted to recommended levels with HBS containing 0.48 g/ml CsCl. The tubes are loaded in a SW-1.1 swing-bucket rotor (Beckman) and CsCl gradients are developed overnight at 35,000 rpm at 4° in a Beckman L8-80M ultracentrifuge. The bands in the middle of the gradient are collected through the wall of the tube using a 5 ml syringe with 19 gauge needle. Banded material (1.5 to 2.5 ml) is placed in a new ultracentrifuge tube and the tube is filled with HBS plus 0.48 g/ml CsCl. Vectors are rebanded and reextracted. To remove CsCl, a PD-10 desalting column (Pharmacia Amersham) is equilibrated with 25 ml of HBS, and loaded with 2.5 ml of banded material. Once the flow-through containing vector particles is collected, the column is reequilibrated with 25 ml HBS, and the sample is reloaded. The flow-through from the second column is the purified desalted vector stock, and it is aliquotted into 2 ml Cryotubes at 250 μl/tube and stored in liquid nitrogen. An aliquot is used to determine IP/ml as described above, and also total particles are calculated by 1 : 20 dilution in HBS and measuring the OD_{260}. One OD_{260} unit equals approximately 10^{12} particles/ml.

Detection of Wild-Type Revertants in Vector Preparations

Materials

1. Purified DNA extracted from rAd vectors
2. dNTP mix [2.5 mM each of dATP, dCTP, dGTP and dTTP), 10× Pwo buffer with 15 mM MgSO$_4$, Pwo polymerase (2.5 U/μl) which is available from Roche Diagnostics]
3. LNCaP cells (Urocor, Oklahoma City, OK) which are cultured under the same conditions as the 293 cells
4. DMEM, complete medium, infection medium, and 1× PBS

We use PCR to determine if our primary lysates are contaminated with wild-type revertants. Vector DNA is prepared from crude or purified stocks as described in the "Fragment Purification" section above. The presence of wild-type sequences is detected using PCR primers E1A-1 (5′ATTACCGAAGAAATGGCCGC3′) and E1A-2 (5′CCCATTTAACACGCCATGCA3′), which amplify the region between Ad5 bp 590 and 1656. PCR primers E2B-1 (5′TCGTTTCTCAGCAGCTGTTG3′) and E2B-2 (5′CATCTGAACTCAAAGCGTGG3′) are used to amplify the region between Ad5 bp 3955 and 4815, which is present in both wild-type virus and our vectors. This serves as a control for total vector DNA. PCR reactions contain 5 μl of 10× Pwo buffer with MgSO$_4$, 2 μl of dNTP mix, 5 μl each of forward and reverse primers (20 pmol/μl), 100 ng vector DNA (equivalent to 2.6 × 10^9 particles), and 2.5 U Pwo polymerase in a 50 μl reaction volume. Reactions are placed in 0.2 μl thin-wall PCR tubes with domed caps (USA Scientific), which are loaded into a Perkin-Elmer 9600 thermocycler. Samples are heated at 97° for 1 min, then cycled 30 times (97° for 20 sec, 56° for 20 sec, 72° for 1 min). Ten μl volumes are taken from each sample and run on a 0.5% agarose gel in 1× TAE buffer. DNA from Ad5-sub360 is used as a positive control. Wild-type E1 sequences generate a band of 1066 bp, while PCR of the E2 region results in a band of 860 bp. If a PCR reaction has no detectable 1066 bp band, the sample DNA is serially diluted in 10-fold steps and PCR is repeated until the dilution at which the 860 bp band also disappears is determined. The maximum possible contamination level with wild-type revertants is then calculated based on this dilution factor. For example, if the lowest dilution at which 860 bp band is detected is 10^4-fold, then there is 1 or less wild-type revertants per 10^4 vector particles.

To determine levels of wild-type virus contamination in purified vector stocks, a more sensitive supernatant rescue assay is used.[30] In the case of vectors expressing

FasL-GFP, a cell line with high level of resistance to Fas-mediated apoptosis, such as prostate cell line LNCaP,[31] must be used, since the cells are infected at high MOI and, if FasL-sensitive, would die before wild-type virus replication can occur. Five 150-mm plates are seeded with 3×10^6 LNCaP cells and allowed to reach confluence. At this point, approximately 1×10^8 cells are present on each plate (determined by counting an extra plate). The medium is removed from each plate and replaced with 25 ml of infection medium containing 4×10^7 IP/ml rAd/FasL-GFP$_{TET}$.

Three days later, cells from each dish are individually collected into 50 ml tubes and centrifuged at 800g. Supernatants are aspirated and cells are washed in 45 ml $1 \times$ PBS. They are repelleted by centrifugation at 800g and 1 ml of DMEM is added to each pellet. The cells are lysed and clarified lysates are prepared as described in a previous section. LNCaP cells in 6-well plates (at approximately 4×10^5 cell/well) are infected with 100 μl of each lysate in triplicate. CPE indicating wild-type virus replication is scored with microscopic observation (both fluorescence and bright-field) 4 to 6 days later. This method is sensitive enough to detect 1 wild-type revertant in 5×10^9 vector particles.

[30] L. D. Dion, J. Fang, and R. I. Garver, *J. Virol. Methods* **56,** 99 (1996).

[31] T. Takeuchi, Y. Sasaki, T. Ueki, T. Kaziwara, N. Moriyama, K. Kawabe, and T. Kaizoe, *Int. J. Cancer* **67,** 709 (1996).

[32] E. Bridge and G. Ketner, *J. Virol.* **63,** 631 (1989).

Section V

Retrovirus

[31] Ligand-Inducible Transgene Regulation for Gene Therapy

By XIANGCANG YE, KURT SCHILLINGER, MARK M. BURCIN, SOPHIA Y. TSAI, and BERT W. O'MALLEY

Introduction

Steroid hormones, as ligands for nuclear receptors, regulate gene expression in target cells, and subsequently modulate physiological changes in animal growth, development, and homeostasis. The interaction of a specific ligand with its nuclear receptor induces both dissociation of heat shock proteins from the receptor and translocation of the receptor into the nucleus. This in turn allows receptors to bind their cognate DNA response elements and regulate the transcriptional activity of their target genes. Regulation can be influenced by the presence of either an agonist or an antagonist. For example, the progesterone antagonist mifepristone (RU486) induces an irregular conformational change in the progesterone receptor (PR) to suppress the transcriptional cofactor recruitment and the action of progesterone.[1] Knowledge of the mechanistic basis for the anti-progesterone activity of mifepristone has led to its clinical use.[2] Recently, however, a novel application for mifepristone has been described by our laboratory in the form of a mifepristone-dependent transcriptional regulatory system designed to control transgene expression for preclinical research and gene therapy.[3,4] Because of increasing concern over the safety and effectiveness of gene therapy in clinical trials, it is apparent that both targeted gene transfer and regulable gene expression will be essential for successful application of gene therapy in the future. In this chapter, we will discuss the principles underlying the mifepristone-dependent regulatory system, the methods for using this system, and the practical aspects involved in establishing an inducible expression system.

Generation of Ligand-Regulable Gene Switch GLVP

In studies of the structure of PR and its interaction with ligands, we found that the C terminus of PR is essential for its transcriptional activation activity in the presence of progesterone.[1] A carboxy-terminally truncated PR fails to bind to

---

[1] E. Vegeto, G. F. Allan, W. T. Schrader, M.-J. Tsai, D. P. McDonnell, and B. W. O'Malley, *Cell* **69**, 703 (1992).

[2] F. Cadepond, A. Ulmann, and E.-E. Baulieu, *Annu. Rev. Med.* **48**, 129 (1997).

[3] Y. Wang, S. Y. Tsai, and B. W. O'Malley, *Adv. Pharmacol.* **47**, 343 (2000).

[4] S. Y. Tsai, K. Schillinger, and X. Ye, *Curr. Opin. Mol. Ther.* **2**, 515 (2000).

progesterone and other endogenous steroids, but retains its affinity for progesterone antagonists such as mifepristone. Interestingly, this truncated PR becomes transcriptionally active when bound by mifepristone.[1] Further studies have revealed that 12 amino acids (residues 917–928) in the C terminus of PR contain a repressor domain, which is sufficient to prevent activation of full-length PR by mifepristone and other antiprogestins.[5]

Based on these early findings, the mifepristone-dependent chimeric regulator GLVP was created by fusing a truncated PR ligand-binding domain (residues 640–891) to the yeast GAL4 DNA-binding domain and the herpes simplex virus VP16 activation domain (Fig. 1). Because no homolog of the GAL4 exists in mammals, fusion proteins such as GLVP which contain the Gal4 DNA-binding domain do not interfere with endogenous gene expression and regulation in mammalian cells. However, in transient transfection assays, GLVP was able to specifically activate reporter genes placed in the context of both yeast GAL4 response elements (17-mer consensus sequence: CGGAGTACTGTCCTCCG) and either the minimal promoter of the thymidine kinase gene (*tk*) or the TATA box from the adenoviral major late gene *E1b*. Importantly, the transcriptional activation of these reporter genes by GLVP occurred only in the presence of a progesterone antagonist but not in the presence of a progesterone agonist.[6] The chimeric GLVP regulator was then further refined by replacement of the PR ligand-binding domain with a 19-amino acid-deletion variant of PR (residues 640–914) and rearrangement of the VP16 activation domain to the C terminus of the ligand-binding domain (Fig. 1). The result was a more potent mifepristone-dependent transactivator called $GL_{914}VPc'$.[7]

Properties of GLVP in Cell Culture and Transgenic Mice

The functional properties of the GLVP regulatory system have been extensively studied in various cell lines and transgenic mice. The results have shown that the mutated PR ligand-binding domain by itself has very low transcriptional activity, but its transactivation activity is greatly enhanced when it is fused with the VP16 transactivation domain. Importantly, GLVP transcriptional activity is activated only by the addition of mifepristone suggesting that the GLVP gene switch can be turned on by administration of mifepristone and is not interfered with by endogenous physiological conditions. The concentration of mifepristone required for induction of reporter gene expression in cell culture is usually at

[5] J. Xu, Z. Nawaz, S. Y. Tsai, M.-J. Tsai, and B. W. O'Malley, *Proc. Natl. Acad. Sci. U.S.A.* **93**, 12195 (1996).

[6] Y. Wang, B. W. O'Malley, Jr., S. Y. Tsai, and B. W. O'Malley, *Proc. Natl. Acad. Sci. U.S.A.* **91**, 8180 (1994).

[7] Y. Wang, J. Xu, T. Pierson, B. W. O'Malley, and S. Y. Tsai, *Gene Ther.* **4**, 432 (1997).

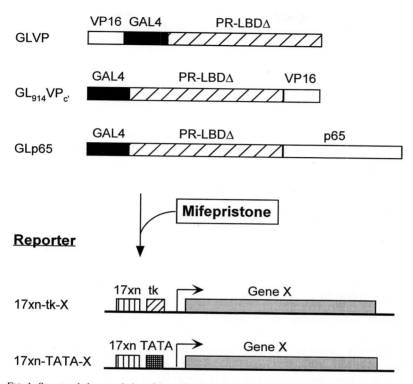

FIG. 1. Structural characteristics of the mifepristone-regulable gene switch. The chimeric regulator (transactivator) is composed of a GAL4 DNA-binding domain, a mutant PR ligand-binding domain, and a transcriptional activation domain of the herpes simplex virus VP16 or of the human p65 factor. Once activated by ligand mifepristone, it binds to the artificial promoter consisting of the GAL4 DNA response elements and the *tk* minimal promoter or TATA box to stimulate reporter/therapeutic gene expression.

10 nM, but concentrations as low as 0.1 nM, are also effective.[6] With the more potent GL$_{914}$VPc′ regulator, induction of reporter gene expression can occur with concentrations of mifepristone as low as 0.01 nM.[7] However, in animal studies, the dose of mifepristone used is slightly higher, possibly due to pharmacokinetics and altered bioavailability of the drug in target tissues. Based on our studies in mouse models, 50–250 μg/kg of mifepristone is the suitable dosage range. These doses are significantly lower than those required for use of mifepristone as a progesterone antagonist (4–10 mg/kg) in humans and are nontoxic according to the

established safety record.[8,9] Moreover, because mifepristone is a small hydrophobic molecule, it diffuses passively through cell membranes and distributes in a wide spectrum of tissues including the brain. Therefore, it can be used for rapid induction of transgene expression in almost all target tissues via oral, intraperitoneal, or subcutaneous administration. In cell culture, the expression signal is detectable within 2 hr after addition of mifepristone, and reaches a maximum at ~10 hr. Similarly, an induction peak is detected in the transgenic mouse serum at ~12 hr after a single dose of mifepristone is administered.[10] However, although the kinetics of induction are similar for different target genes, inducible levels of protein vary between different systems due to differences in the vector constructs, expression level of the regulator, cell types, and integration effects in transgenic animal lines. For example, mifepristone-induction of human growth hormone (hGH) by GLVP in transiently transfected HepG2 cells is ~37–fold, whereas in transgenic mice induction can reach ~1500-fold.[10] Most importantly, this mifepristone-inducible expression is not only effective but also highly specific, as the GAL4 DNA-binding domain of GLVP binds only to the tandem repeats of 17-mer GAL4 binding consensus sequence in the promoter of the target gene and does not interact with the promoter of any endogenous genes.

Development of Humanized Gene Switch GLp65 and Vector System

Although the mifepristone-regulable GLVP gene switch system has been successfully used in many animal studies, it has been necessary to humanize the regulator to facilitate future application of this system to human gene therapy. To this end, we replaced the viral VP16 activation domain with p65, a partner of NF-κB in the human RelA heterodimeric transcription factor complex. When compared with GLVP, the new regulator GLp65 has similar inducibility by mifepristone. In fact, the magnitude of the GLp65 induction is superior to that of GLVP because of the lower basal activity of GLp65.[11] However, because p65 is potentially less immunogenic as compared to the VP16 transactivation domain, the probability of an immune response in humans is markedly reduced. Additionally, GLp65 can be coupled to a tissue-specific promoter to limit its expression location, which can reduce the load of chimeric protein *in vivo*. For example, the liver-specific enhancer/promoter of the transthyretin gene (TTR) was chosen to control the regulator GLp65. Consequently, expression of the reporter gene

[8] I. M. Spitz and C. W. Bardin, *N. Engl. J. Med.* **329,** 404 (1993).

[9] S. Christin-Maitre, P. Bouchard, and I. M. Spitz, *N. Engl. J. Med.* **342,** 946 (2000).

[10] Y. Wang, F. J. DeMayo, S. Y. Tsai, and B. W. O'Malley, *Nat. Biotechnol.* **15,** 239 (1997).

[11] M. M. Burcin, G. Schiedner, S. Kochanek, S. Y. Tsai, and B. W. O'Malley, *Proc. Natl. Acad. Sci. U.S.A.* **96,** 355 (1999).

(i.e., 4×17-mer-*tk*-hGH) in response to mifepristone administration could only be detected in liver while the product of the reporter gene, hGH, was efficiently secreted from the liver and exhibited systemic effects.[11]

To enhance the efficiency of *in vivo* delivery of both the regulator and the reporter/therapeutic gene, a high-capacity adenoviral vector (HC-Ad) has been adapted to our system. Because this vector does not contain any viral protein coding sequences but retains the viral packaging signals, the toxicity and immunogenicity caused by viral proteins and resulting in short duration of target gene expression has been virtually eliminated.[12,13] Simultaneously, this "gutless" adenoviral vector possesses a large capacity for foreign DNA, allowing a total of up to 35 kb of target gene and regulatory elements to be inserted and packaged into virions in a helper-virus-dependent procedure. After incorporation of the hGH reporter gene and a TTR-driven GLp65 regulator in the same HC-Ad vector, simultaneous transfer of both target and regulator genes becomes more reliable. In a mouse model transduced with this vector, expression of hGH was very low in absence of mifepristone but increased dramatically after administration of mifepristone.[11] The induction profiles observed were comparable to those of the bigenic transgenic mice harboring both TTR-GLVP and 4×17-mer-*tk*-hGH.[10] Also, the ligand-dependent induction was repeated multiple times over a 9-month period by oral administration of mifepristone (Fig. 2A), although induction levels gradually decreased with time, presumably due to some combination of the loss of vector DNA or immune reaction to the foreign protein. Moreover, long-term sustained induction of transgene expression has also been achieved by transplantation of extended-release mifepristone pellets in mice (Fig. 2B). These results demonstrated that use of both the HC-Ad-mediated gene transfer strategy and the mifepristone-inducible gene regulatory system was capable of temporally and spatially controlling transgene expression in a mouse model and has great potential for future study in primates.

Methods of Using GLVP and GLp65 Regulatory Systems

Gene Switch Construction

Plasmids containing the regulator GLVP or GLp65 have been used broadly in combination with different reporter constructs such as chloramphenical acetyltransferase (CAT), luciferase (Luc), the human growth hormone gene (hGH), inhibin A, and transforming growth factor $\beta 1$ (TGF$\beta 1$). More recently, they have

[12] S. Kochanek, P. R. Clemens, K. Mitani, H.-H. Chen, S. Chan, and C. T. Caskey, *Proc. Natl. Acad. Sci. U.S.A.* **93,** 5731 (1996).
[13] G. Schiedner, N. Morral, R. J. Parks, Y. Wu, S. C. Koopmans, C. Langston, F. L. Graham, A. L. Beaudet, and S. Kochanek, *Nat. Genet.* **18,** 180 (1998).

A

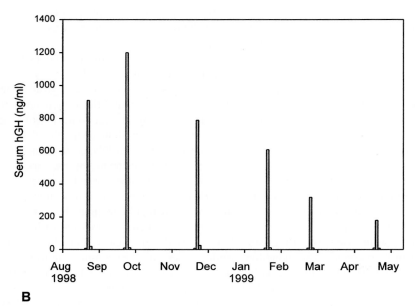

B

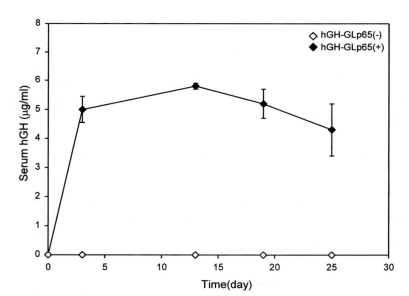

FIG. 2. Mifepristone-inducible expression of human growth hormone in mice transduced by the HC-adenovirus carrying 17 × 4-TATA-hGH and TTR-GLp65. The mice were subjected to (A) oral administration of mifepristone (250 μg/kg) for repetitive induction or (B) subcutaneous implantation of biodegradable pellets (360 μg mifepristone for 60-day release) for sustained expression.

also been used in conjunction with therapeutic genes such as vascular endothelial growth factor ($VEGF_{165}$) and erythropoietin (Epo). These reporter/therapeutic genes are placed under the control of the GAL4 binding element-linked miminal *tk* promoter or TATA promoter, which will respond to mifepristone-activated GLVP or GLp65. The regulators, by contrast, are usually linked to a general constitutive promotor such as CMV or RSV-LTR, but constructs linking the regulators to a promoter having tissue specificity such as the liver-specific TTR promoter or the muscle-specific SP promoter have been designed.

One of the most important properties of any gene-regulatory construct is the extent to which basal expression of the target gene, i.e., expression of the target gene in the absence of ligand, occurs. In case of the mifepristone-inducible gene regulatory system, basal activity is generally low. However, slight differences between the basal activities of the GAL4 binding element-linked TATA promoter and the corresponding *tk* promoter have been observed in cell transient transfection assays. In general, the TATA promoter has lower basal activity than the *tk* promoter.[6] As a result, while the TATA-driven promoter may not reach the same absolute expression levels as the *tk*-driven promoter, its relative induction is greater due to lower basal expression. Similarly, the basal expression mediated by the regulator GLp65 is lower than that of GLVP in transfected cells.[11] These differences provide a spectrum of potential combinations of transgene regulator and target promoter for use in different situations where certain levels of target gene expression are desired.

For long-term expression of a highly bioactive transgene *in vivo*, a more tightly regulated gene switch is desired for reason of safety. This requires both minimization of transgene expression prior to ligand-dependent induction and a reduction in initial levels of chimeric regulator accumulated in cells. An autoregulable switch has been created by placing a 4 × 17-mer-*tk* promoter in front of the GLp65 regulator while linking a reporter/therapeutic gene to the more stringent 6 × 17-mer-TATA.[14] Because of the low level of basal expression from the 4 × 17-mer-*tk* promoter in the absence of mifepristone, only a few molecules of the GLp65 protein are produced by the autoregulable switch system in transfected cells. Upon addition of mifepristone, the ligand simultaneously stimulates increased GLp65 expression via a positive feedback control circuit using the 4 × 17-mer-*tk* promoter and also induces reporter gene expression from the 6 × 17-mer-TATA promoter. Although the absolute target gene expression level derived from this switch system is generally lower than that of the conventional constructs, its relative fold of induction is often higher than others because of its extremely low basal activity.[14]

---

[14] R. V. Abruzzese, D. Godin, V. Mehta, J. L. Perrard, M. French, W. Nelson, G. Howell, M. Coleman, B. W. O'Malley, and J. L. Nordstrom, *Mol. Ther.* **2,** 276 (2000).

Cell Transfection

Regulable gene expression with GLVP or GLp65 has been investigated in several cell lines including Hela, CV1, HepG2, and primary chicken myoblast. The general procedure for transient transfection should be performed according to the manufacturer's instruction for transfection reagents such as the various liposomal derivatives. To achieve high induction levels in transfected cells, the structures of the regulator and reporter constructs, as described above, are crucial. However, adjustments in plasmid DNA ratios and mifepristone concentrations used during the transfection are also important for successful experiments. When two plasmids carrying the regulator and the reporter genes separately are used for cotransfection, the ratio of GLVP plasmid : reporter plasmid generally ranges from 1 : 10 to 1 : 2.5; whereas the ratio of GLp65 plasmid : reporter plasmid can be 1 : 1, due to the lower basal activity of GLp65. Nonetheless, the molar ratio of plasmids should always be adjusted empirically according to the properties of the regulator and the desired induction level. If an experiment is designed to use a single vector to carry both regulator and reporter genes, then the choice of an appropriate promoter for the regulator is even more important. Using a strong constitutive promoter to express the regulator, you may expect high basal expression of reporter genes and relatively low amounts of induction.

The mifepristone can be prepared as a 10^{-2} M, stock solution by dissolving the powder in DMSO and a working solution can be prepared by diluting the stock solution to 10^{-5} M, with 80% ethanol. Before addition to cells, the working solution is usually added to prewarmed culture medium with 5% stripped fetal calf serum (5% SFCS) to give a final mifepristone concentration of 10^{-8} M. Immediately after transfection, cells should be incubated in mifepristone-free culture medium for at least 3 hr. This medium can then be replaced with the prewarmed mifepristone-containing culture medium +5% SFCS. Depending on the properties of different cell lines and transfection efficiencies, incubation may then continue for 12 to 48 hr before the harvesting of cells for reporter activity assay or protein analysis.

Intramuscular Injection

Plasmid constructs encoding GLp65 and constructs encoding target genes have been successfully used for direct muscle injection and electroporation in mice.[14,15] Similar to what is seen in transient transfection with the two-plasmid system, use of the appropriate molar ratio of plasmids is critical to achieving high induction levels.

[15] R. V. Abruzzese, D. Godin, M. Burcin, V. Mehta, M. French, Y. Li, B. W. O'Malley, and J. L. Nordstrom, *Hum. Gene Ther.* **10**, 1499 (1999).

The plasmids should be purified and suspended in saline or polyvinyl pyrrolidine at concentration of 1 mg/ml. For injection of plasmid DNA, animals should first be anesthetized. Then, 25–50 μg of DNA is injected directly into either the tibials cranialis or gastrocnemius muscles. To enhance transfection efficiency, the injected sites can be subjected to electroporation using two parallel plate caliper electrodes connected to an Electro Square Porator. A typical voltage setting for the electroporation is 375 V/cm, and this is delivered in two square-wave pulses which each last for 25 μs. After gene transfer, the animals are allowed to recover from the minor inflammation at the injection site. It appears that more consistent inductions of transgene expression are obtained from the injected mice when they are allowed to rest for ∼14 days before the first administration of mifepristone. Mifepristone is usually dissolved in sesame oil at a concentration of 62.5 or 82.5 μg/ml. For a mouse with a body weight of 25 g, 0.1 ml of the mifepristone oil is then given orally or intraperitoneally. This corresponds to a daily dose of 250 or 330 μg/kg which is used routinely to obtain desired induction of target gene expression. Alternatively, the mifepristone can be incorporated into extended-release pellets for subcutanous implantation into animals. The pellets can be customized to release 180 μg or 360 μg over a 60-day period to obtain sustained induction of target gene expression. Based on analysis of SEAP in mouse serum samples, the transgene expression reaches its peak level at ∼2.5 days after administration of a single dose of mifepristone.[15]

Adenovirus-Mediated Gene Transfer

Unlike intramuscular injection and electroporation, gene transfer by recombinant adenovirus can provide high transduction efficiency both in cell culture and animals. Also unlike the intramuscular injection system, the simultaneous delivery of regulator and target gene using an HC-Ad vector means that the inducibility of the adenovirus-mediated gene switch system relies almost entirely on the intrinsic properties of the regulator. As a preclinical model for mifepristone-inducible expression of the target gene hGH in hepatocytes, we have successfully constructed and tested an HC-Ad vector that contains the GLp65 regulator driven by the TTR promoter and the hGH reporter gene in the context of the 4 × 17-mer-TATA promoter.[11] For transduction of HepG2 cells, 5×10^7 infectious units (IU) of the hGH-GLp65 virus were used with 6×10^5 cells. After removal of virus, the cells were incubated for 3 hr in fresh culture medium. Then, 10^{-8} M, mifepristone in culture medium with 5% SFCS was added to the culture medium. The induced expression of hGH was then analyzed in the samples of culture medium collected at 24 hr after addition of mifepristone. Because the basal level of hGH expression was undetectable by the assay, making an accurate calculation of the fold induction in response to mifepristone was difficult. However, it was estimated that fold-induction in HepG2 cells was greater than three orders of magnitude.[11] Similarly, the high inducibility of the mifepristone-regulable system was also observed in an independent study using the E1–deficient adenovirus vector to transduce HeLa cells.[16]

To use the same recombinant viral construct for regulable transgene expression *in vivo*, 2×10^9 IU of virus in saline was delivered to mice by tail vein injection.[11] Although a small amount of virus may be taken up by other tissues during tail vein infusion, a majority of the virus will be concentrated in the liver.[17] Also, because the GLp65 regulatory system is under control of the TTR promoter, induction of the transgenes will occur only in the liver. The mifepristone dose given to the adenovirally transduced mice was the same as that given to mice receiving intramuscular injection of plasmid DNA as described above. It is important to note that for 8 days immediately following the injection of adenovirus, expression of hGH was not induced by administration of mifepristone. However, after day 8, hGH expression in response to mifepristone gradually increased and reached a peak level at day 14. Moreover, after this 2-week period, hGH expression was repetitively inducible upon mifepristone administration. Additionally, some transduced mice expressed such low basal levels of hGH that their mifepristone-dependent induction levels reached four orders of magnitude.[11]

Transgenic Animal Studies

Our gene switch system has been shown to be very useful for transgenic animal studies such as conditional gene knockout, inducible rescue of mutation, and generation of disease models. To meet these purposes, the constructs encoding regulator and reporter/target gene should be integrated individually into separate transgenic lines. In our experience, addition of two copies of a chromosomal insulator sequence localized in the 5′ region of the chicken β-globin gene (HS4) to the flanking regions of constructs can potentially reduce integration position effects in tissues such as liver.[10] Before gene transfer, the plasmids need to be linearized and purified. Only the fragments of interest are delivered to pronuclei of donor mouse embryos by microinjection. Injected embryos are then transferred into pseudopregnant females and allowed to develop to term. Mice are screened at 3 to 4 weeks of age by Southern and PCR analysis of the genomic DNA isolated from the mouse tails. The positive lines are bred and the expression of the transgene in their offspring is analyzed by Northern or ribonuclease protection assay. Based on transgene expression profiles, founder lines are then selected

[16] M. Molin, M. C. Shoshan, K. Öhman-Forslund, S. Linder, and G. Akusjärvi, *J. Virol.* **72**, 8358 (1998).
[17] T. A. G. Smith, M. Mehaffey, D. B. Kayda, J. M. Saunders, S. Yei, B. C. Trapnell, A. McClelland, and M. Kaleko, *Nat. Genet.* **5**, 392 (1993).

and established. Using mice carrying the regulator or the reporter/target gene for breeding, bigenic mice harboring both transgenes are generated and selected by PCR analysis. For analysis of *in vivo* transgene expression, 5-week-old bigenic mice are given mifepristone in sesame oil at an oral dose of 250 μg/kg or are implanted with an extended release pellet having the corresponding dosage. Blood or tissue samples are then collected prior to and after administration of mifepristone to quantify target gene induction. Using the above strategy, the TTR-GLVP regulator line and the 4 × 17mer-tk-hGH reporter line have been established as the first successful transgenic models incorporating the mifepristone-regulable system. The bigenic mice derived from crossings between these two lines showed liver-specific and mifepristone-regulable expression of human growth hormone, which could be induced three to four orders of magnitude in a ligand-dependent manner.[10] More mifepristone-regulable transgenic mice have been or are being generated and studied to elucidate gene functions involved in growth, differentiation, and tumorigenesis *in vivo*.[18,19]

Summary

A synthetic ligand regulable system for gene transfer and expression has been developed in our laboratory based on mechanistic studies of steriod hormone receptor and transcriptional regulation. This gene switch system possesses most of the important features that are required for application of the system in biological research and clinical gene therapy in the future. As the primary ligand tested in this system, mifepristone can effectively turn on the regulatory circuit at doses much lower than those used in the clinic. By modification of the chimeric regulator and its feedback regulatory mode, this system has been optimized to produce very low basal activity with high inducibility in the presence of mifepristone. Also, improvements in regulator composition have been made to minimize immunogenicity and make the system more amenable to human gene therapy. Moreover, incorporation of this gene switch system into the HC-Ad vector system has further enhanced the efficiency of gene transfer and the long-term inducible expression of transgenes. However, for each application within a different biological system, the gene switch needs to be optimized to achieve appropriate inductions. In particular, the method used to deliver the transgenes and adjustment of ligand dosage are critical for *in vivo* gene expression.

[18] X. J. Wang, K. M. Liefer, S. Y. Tsai, B. W. O'Malley, and D. R. Roop, *Proc. Natl. Acad. Sci. U.S.A.* **96,** 8483 (1999).

[19] T. M. Pierson, Y. Wang, F. J. DeMayo, M. M. Matzuk, S. Y. Tsai, and B. W. O'Malley, *Mol. Endocrinol.* **14,** 1075 (2000).

[32] Large-Scale Production of Retroviral Vectors for Systemic Gene Delivery

By MATTHEW J. HUENTELMAN, PHYLLIS Y. REAVES,
MICHAEL J. KATOVICH, and MOHAN K. RAIZADA

Introduction

Since their development in the early 1980s oncoretroviral vectors such as the Moloney murine leukemia virus (MLV) have played a major role in the development of *in vitro* and *in vivo* gene transfer.[1,2] In fact, approximately 35% of all current gene therapy clinical trials make use of oncoretroviral-based vectors.[3] For this reason, the retroviral vector is perhaps the best clinically characterized gene transfer vehicle.

Oncoretroviruses have unique qualities that enable them to be highly amenable to gene transfer and therapy: (i) They have a large capacity for nucleic acid (~8 kb) to accommodate large therapeutic genes or entire regulateable expression systems in a single viral genome. (ii) They stably integrate into the host cell genome. This trait allows for long-term, inheritable transgene expression. (iii) Gene transfer is nonimmunogenic. No viral genes are expressed following target cell transduction and integration. (iv) They are easily produced, purified, and concentrated. Retroviral vectors are produced by "packaging cells" and secreted into the culture medium. This enables their simple collection and purification while their ability to be pseudotyped with very stable envelope glycoproteins (such as that from the vesicular stomatitis virus—VSV-G) broadens their tropism and allows them to be easily ultraconcentrated.[4,5]

However, oncoretroviral vectors have a major drawback that prevents many investigators from utilizing them. Ongoing cell division is required to enable the oncoretrovirus to integrate into the host cell genome and express transgene. Therefore, slowly dividing or quiescent cell types—such as neurons and stem cells—are not effectively transduced by MLV-based vectors. Such cell types represent the majority of targets important for the future of human gene therapy.

[1] T. Friedmann, *in* "The Development of Human Gene Therapy" (T. Friedmann, ed.), Chapter 1. Cold Spring Harbor Laboratory Press, New York, 1999.

[2] A. D. Miller, D. J. Jolly, T. Friedmann, and I. M. Verma, *Proc. Natl. Acad. Sci. U.S.A.* **80**, 4709 (1983).

[3] http://www.wiley.co.uk/genetherapy/clinical/

[4] N. Emi, T. Friedmann, and J. K. Yee, *J. Virol.* **65**, 1202 (1991).

[5] J. C. Burns, T. Friedmann, W. Driever, M. Burrascano, and J. K. Yee, *Proc. Natl. Acad. Sci. U.S.A.* **90**, 8033 (1993).

Vectors based on the lentivirus subfamily of retroviruses, such as the human immunodeficiency virus (HIV), sidestep the need for obligatory cell division following transduction. The lentivirus achieves this by exposing a nuclear localization signal (NLS) after gaining entry to the cell.[6] It is then actively transported into the nucleus by the endogenous cellular machinery. This enables very efficient transduction of postmitotic cell types and greatly improves the utility of these vectors.

By far the most common lentivirus backbone of these new systems is HIV-1; however, success has been achieved using simian (SIV), feline (FIV), and equine (EIAV) lentiviruses as well.[7-9] The newly developed lentiviral vectors have additional advantages over the classic oncoretroviruses in that they are somewhat larger (approaching a 9.5 kb capacity in the fully deleted versions) and inherently more resistant to inactivation.[10] These lentivectors are clearly the gene transfer vehicles of the future as they combine all the advantages of the retrovirus family plus the ability to infect both dividing and nondividing cell targets, all in a larger, more stable viral particle.

Many investigators produce both of these vectors in relatively small, concentrated amounts. However, in order for the vectors to be useful in a systemic gene transfer protocol they must be produced in large amounts. Here we will outline a novel protocol for the large-scale production and efficient concentration of the lentiviral vector. Additionally, we will describe a protocol for the systemic delivery of retroviral vectors to newborn rat pups. This protocol was developed in our laboratory and consists of direct intracardiac injection of vector. In neonatal animals this is the most effective way to introduce vector into the systemic circulation. We have used this method to deliver MLV encoding an antisense gene against a vasoconstrictor receptor (the angiotensin II type 1 receptor, AT_1R) and prevented the development of hypertension in the spontaneously hypertensive rat (SHR) model of the disease. Some results of these experiments will be introduced.

MLV Packaging Cell Production

A "packaging" cell line, such as the PA317 clone (ATCC CRL 9078), has the viral genes *env, gag,* and *pol* stably integrated into its genome. These genes are essential for proper production of virus. Normally the packaging cell line

[6] M. I. Bukrinsky and O. K. Haffar, *Front. Biosci.* **4,** D772 (1999).

[7] K. Mitrophanous, S. Yoon, J. Rohll, D. Patil, F. Wilkes, V. Kim, S. Kingsman, A. Kingsman, and N. Mazarakis, *Gene Ther.* **6,** 1808 (1999).

[8] P. E. Mangeot, D. Negre, B. Dubois, A. J. Winter, P. Leissner, M. Mehtali, D. Kaiserlian, F. L. Cosset, and J. L. Darlix, *J. Virol.* **74,** 8307 (2000).

[9] G. Wang, V. Slepushkin, J. Zabner, S. Keshavjee, J. C. Johnston, S. L. Sauter, D. J. Jolly, T. W. Dubensky, Jr., B. L. Davidson, and P. B. McCray, Jr., *J. Clin. Invest.* **104,** R55 (1999).

[10] D. Trono, *Gene Ther.* **7,** 20 (2000).

constitutively expresses these three genes; however, in some cases *env* expression is kept under the control of an inducible promoter. To produce recombinant MLV one needs to simply transfect these cells with a plasmid containing viral long terminal repeats (LTRs) and an intact packaging signal (Ψ) flanking an expression cassette encoding the gene of interest. Retroviral plasmids and packaging cell lines are readily available from several different commercial sources (Clontech, ATCC).

The following is an outline of the necessary steps to produce a stably transfected packaging cell line that can be used at any time to produce high-titer MLV vector:

1. Plate 200,000 packaging cells (PA317, ATCC CRL 9078, or equivalent) in a 60 mm diameter sterile culture dish in high glucose (4500 mg per liter) Dulbecco's modified Eagle's medium (DMEM) containing 10% fetal bovine serum (referred to as "complete" DMEH). Grow the cells overnight in a humidified incubator at 37° and 5% CO_2. The cells should reach approximately 50% confluence during this time.

2. Mix 3 μl of lipofectamine (GIBCO BRL) into 100 μl of room temperature, serum-free, antibiotic-free DMEM in a sterile, polystyrene tube. In a separate tube, dilute 3 μg of MLV plasmid DNA into 100 μl of room temperature serum-free, antibiotic-free DMEM. Mix both tubes well and incubate at room temperature for 45 min. Next, add the DNA containing solution to the LipofectAMINE mixture. Mix gently and incubate at RT for 15 minutes. During this incubation, rinse the PA317 cells once with serum-free DMEM and overlay with 1.8 ml of the same serum-free medium. After the 15-min incubation add the LipofectAMINE:DNA complexes dropwise onto the cells. Rock the culture dish to evenly distribute the complexes and return the cells to the 37°, 5% CO_2 incubator. After an overnight incubation in this transfection medium, rinse the cells once and overlay with 4 ml of complete DMEM. Return the cells to the incubator for another 24 hr. It is important to conduct a control reaction using a reporter gene construct (such as green fluorescent protein) in order to confirm high transfection efficiency was achieved. Generally, greater than 50% efficiency is acceptable.

3. Two days following transfection, detach the packaging cells with trypsin and subculture at a 1 : 20 ratio in 100 mm dishes containing complete DMEM + 1000 μg per ml gentamicin sulfate—G418 (referred to as "selection" medium). Note that the drug resistance gene encoded by the MLV packaging plasmid determines the drug used in the selection media. Continue this selection for 14 days taking care to completely change the medium every 3 days with medium containing *freshly added* G418.

4. After 14 days individual, drug-resistant clones will be seen in the dish. Of these colonies 18–24 should be separately harvested using sterile cloning cylinders. The stable clones should then be grown to confluency in selection medium in 6-well plates. Watch the colonies grow very carefully. The day before confluency is

achieved rinse the cells once and overlay with complete DMEM (*no* G418). On the following morning collect the supernatant and immediately freeze at $-80°$. Additionally, detach the packaging cells, centrifuge briefly, and resuspended in complete DMEM + 10% dimethyl sulfoxide (DMSO). Freeze the cells slowly using a commercially available isopropanol chamber (Nalgene) to $-20°$ and move them into liquid nitrogen storage the following morning.

Each individual clone will grow at a slightly different rate, but at the end of this procedure one should have between 18 and 24 stable cell lines in liquid nitrogen and their representative viral-containing supernatants frozen at $-80°$. It is important to realize that each of these cell lines will be producing virus at a different concentration thus necessitating a "titration" of the virus.

Titration of MLV Encoding a Drug Resistance Marker

1. Plate a total of 1×10^5 NIH3T3 (ATCC CRL 1658) cells in 60 mm dishes in complete medium. Allow to attach and grow overnight in a $37°$, 5% CO_2 incubator. The correct confluency on the following morning is between 80 and 90%. At least two separate dishes should be established for each virus to be titrated.

2. Remove the medium and overlay with 2 ml of complete medium containing 8 μg per ml of hexadimethrine bromide (Polybrene, Sigma). Polybrene is made as a $100 \times$ stock in water and is sterile filtered through a 0.2 μm syringe filter (Nalgene) before storage at $-20°$. Quick thaw the previously collected PA317 supernatants in a $37°$ water bath, mix thoroughly, and add 1 μl to each of the designated culture dishes. Rock the dish to evenly distribute the virus and return the cultures to the incubator.

3. Four hours later, add 2 ml of complete DMEM to each dish. Incubate overnight.

4. The following morning, detach the cells using trypsin and subculture into 60 mm dishes at 1 : 10 and 1 : 50 ratios in *selection* medium. As before, replace this every 3 days with medium containing freshly added G418. If titrating concentrated virus subculture at ratios of 1 : 50 and 1 : 100.

5. After 14 days in selection medium rinse the cells once with phosphate-buffered saline (PBS) and overlay with 2 ml of 0.1% crystal violet in 10% ethanol (prefiltered through Whatman paper). Allow the cells to stain for 20 min, rinse with distilled water, and count the total number of purple colonies per dish. Multiple dishes to which no virus was added must be included as controls. Subtract from the total number of virus positive colonies the number of colonies that can be seen in the control dish. Average the duplicates and calculate the retroviral titer using the equation detailed in Fig. 1. Clones whose titers range between 10^6 and 10^7 colony-forming units per ml of supernatant should be selected and saved for viral preparation.

$$\text{TITER} = \frac{\text{Colony Forming Units}}{\text{Milliliter of Virus}} = \frac{(\text{\# of Colonies})(\text{Dilution Factor})}{\text{Virus Volume (ml)}}$$

$$1 \times 10^6 \text{ colony forming units per ml} = \frac{(20 \text{ colonies})(1{:}50)}{0.001 \text{ ml of virus}}$$

FIG. 1. Equation to calculate the titer of a retroviral vector encoding a drug resistance gene. A sample calculation is also denoted. If 20 distinct colonies are evident following a 50-fold dilution of target cells previously infected with 1 μl of virus, an approximate titer of 1×10^6 colony-forming units per ml is assumed. See text for more details.

Large-Scale Production of MLV

For *in vivo* gene delivery it is important to concentrate virus to greater than 10^8 colony-forming units per ml. The following protocol is used to concentrate amphotropic-enveloped virus:

1. Thaw the desired packaging cell line and plate in two T-75 flasks in *selection* medium. After these flasks become confluent ($\sim$3–5 days), detach the cells using trypsin and evenly distribute into 30 T-75 flasks in *complete* medium. Observe the confluence of the cells closely over the next several days.

2. When the cells reach approximately 85–90% confluency ($\sim$3–5 additional days) remove the old old medium and overlay with 7.5 ml of fresh, complete medium.

3. Twenty-four hours later harvest the media with a sterile, 5 ml pipette into prechilled 50 ml polypropylene conical tubes (Falcon). Spin the tubes at 5000g for 5 min at 4°. This will pellet any cellular debris, but will not pellet the virus.

4. Immediately filter the virus through a 0.45 μm PES membrane (Nalgene) into a prechilled receiving flask. The virus may now be frozen at −20° and stored for future use, or one may continue with the concentration protocol. It is important to note that more than one freeze–thaw cycle will lower amphotropic retroviral titer approximately 2-fold. This decrease is still present, but to a much lower degree, in VSV-G enveloped virus.

5. Centrifuge the filtered retrovirus at 50,000g for 2 hr at 4°. Be sure to monitor this ultracentrifugation and immediately proceed with the next step as soon as the run is complete.

6. Decant the supernatant from each centrifuge tube and remove any remaining traces of medium with gentle vacuum suction. Be aware that the virus is now pelleted at the bottom of the tube, but will not be visible.

7. Resuspend the virus in 1/300th to 1/500th of the centrifuge tube volume in Hanks' balanced salt solution (HBSS) by pipetting gently. The suspension procedure should be performed as gently as possible and the introduction of

bubbles during this process should be avoided as they can denature viral envelope proteins and decrease the titer. After pipetting for several minutes, shake the virus for 5–18 hr on an orbital shaker at 200 rpm at 4°. Periodically stop the shaker and repipette the virus to aid in resuspension. Take care to keep the virus at 4° as much as possible.

8. Pipette the virus once more and dispense aliquots into prechilled microfuge tubes on the following morning. Immediately set aside a 5–10 μl aliquot for titration and freeze the rest of the concentrated virus at −20°.

A typical large-scale preparation of virus involves the concentration of 180 ml of PA317 supernatant to a volume of 500 μl. This results in an approximate 100-fold increase in titer with final values approaching 1×10^9 colony-forming units per ml.

Systemic Delivery of MLV to Neonatal Rat Pups

1. Quick thaw a viral aliquot in a 37° water bath. Immediately place on ice and keep there until injected into the animal.

2. Anesthetize a 5-day-old pup by short incubation in a sealed chamber containing methoxyflurane. Remove the pup and place on a preheated thermal pad. Using a tuberculin syringe, remove ∼35 μl of the chilled virus. Clear the syringe of all visible bubbles.

3. Lay the pup on its back and gently pull its chest skin taut. Locate the heart visually or by applying pressure to the chest. In animals of this age the location of the heart is clear because their skin is still somewhat translucent. Insert the needle into the left ventricle and draw a small portion of blood back into the syringe. If the needle has been inserted correctly, blood will flow easily back into the syringe and will pulse in time with the animal's heartbeat. Deliver the virus slowly taking approximately 15 sec to deliver the entire 35 μl volume.

4. Identify the animal by ear clipping or notching.

5. After the pup recovers from the anesthesia, lightly coat it with peanut oil and return it immediately to its mother. This coating procedure is important as it increases the mother's reacceptance of her offspring. This systemic delivery protocol yields a greater than 96% survival rate.

Antihypertensive Actions of Systemic AT$_1$R Antisense Gene Delivery

In this section we will summarize our observations with the use of systemically delivered MLV containing an antisense gene against the angiotensin II type 1 receptor (AT$_1$R). The AT$_1$R antisense gene was cloned according to the protocols

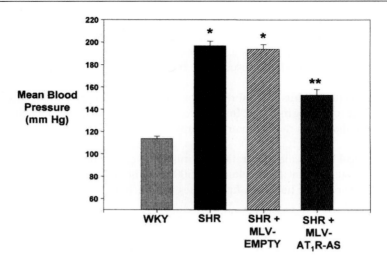

FIG. 2. Systemic delivery of MLV encoding an angiotensin II type 1 receptor antisense gene (MLV-AT_1R-AS) prevents the development of hypertension in the spontaneously hypertensive rat (SHR). MLV was produced, concentrated, and delivered systemically to 5-day-old rat pups as described in this review. Indirect blood pressures were measured at 60 days of age using the fail cuff method. Six animals were utilized per group. $*P < 0.05$ compared to WKY control rats. $**P < 0.03$ compared to MLV-EMPTY treated SHR.

detailed in Reaves *et al.*[11] The antisense transcript was designed to be approximately 1.3 kb in length and to begin hybridization approximately 130 nucleotides upstream of the target's start codon. The rat has two subtypes of AT_1R (A and B) and this antisense gene was designed to inhibit both forms of the receptor. Concentrated viral particles were made according to the procedures outlined above. Treatment of 5-day-old rats with an AT_1R antisense encoding vector was able to prevent the development of high blood pressure exclusively in the spontaneously hypertensive rat (SHR, Fig. 2). Additionally, antisense gene transfer prevented the development of cardiac hypertrophy (Fig. 3) and endothelial dysfunction–two additional cardiovascular pathophysiologies associated with hypertension.[12,13] Transgene was able to be detected in several tissues including the heart, liver,

[11] P. Y. Reaves, H-W. Wang, M. J. Katovich, C. H. Gelband, and M. K. Raizada, *Methods* **22**, 211 (2000).
[12] C. H. Gelband, P. Y. Reaves, J. Evans, H. Wang, M. J. Katovich, and M. K. Raizada, *Hypertension* **33**, 360 (1999).
[13] J. R. Martens, P. Y. Reaves, D. Lu, M. J. Katovich, K. H. Berecek, S. P. Bishop, M. K. Raizada, and C. H. Gelband, *Proc. Natl. Acad. Sci. U.S.A.* **95**, 2664 (1998).

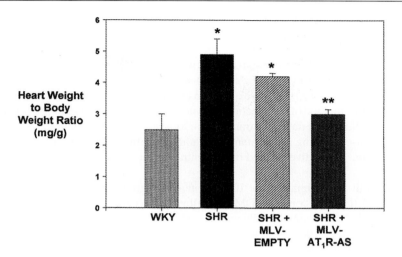

FIG. 3. Neonatal delivery of MLV-AT_1R-AS prevents the development of cardiac hypertrophy in the SHR. MLV was produced, concentrated, and delivered systemically to 5-day-old rat pups as described in this review. At 200 days of age, the animals were sacrificed and whole hearts were excised, rinsed in physiological saline, and weighed. The results are shown as a ratio of heart weight (mg) to body weight (g). Six to 8 animals were utilized per group. *$P < 0.05$ compared to WKY control rats. **$P < 0.02$ compared to MLV-EMPTY treated SHR.

kidney, lung, and adrenal gland.[12–14] This expression was shown as early as 3 days postinjection and as late as 200 days.

Large-Scale HIV-1 Lentivector Production and Concentration

As discussed earlier, lentiviral-based vectors possess important advantages over traditional oncoretroviral vectors. These advantages include the ability to infect quiescent cell types, a larger capacity for nucleic acid, and increased resistance to inactivation. As with the other retroviral-based vectors, lentivirus is generally made in small, concentrated quantities. No efficient, high-titer packaging cell line exists for the production of HIV-1 based lentivirus. Therefore, HIV vector is produced by the transient transfection of at least three different plasmid constructs into an easily transfectable cell line (293T cells are the line of choice). This transfection is usually mediated by calcium phosphate coprecipitation of the DNA. The use of calcium phosphate for transfection has several drawbacks. Calcium phosphate coprecipitation requires a large quantity of pure plasmid DNA, is harsh on the transfected cells, is somewhat difficult to scale up, and usually

[14] S. N. Iyer, D. Lu, M. J. Katovich, and M. K. Raizada, *Proc. Natl. Acad. Sci. U.S.A.* **93**, 9960 (1996).

results in slower gene expression when compared to alternative chemical-based transfections. For these reasons we designed a method for producing recombinant HIV-1 using the activated dendrimer chemistry of Superfect transfection reagent (Qiagen). Superfect transfection requires five times less DNA and is much gentler on the cells when compared to calcium phosphate coprecipitation. Additionally it is much easier to scale up, gene expression can occur in 12 hr, and we have shown that using this method of transfection does not lower viral titer. Our laboratory routinely uses Superfect mediated transfection to produce several hundred milliliters of lentiviral-containing supernatant. This unique method of lentiviral production including our innovative method of concentrating the virus is described below:

1. Split approximately 1×10^7 exponentially growing 293T cells into each of 25 T-75 flasks in complete DMEM. Allow these cells to attach and grow for at least 20 hr. The optimal confluency at transfection is between 90 and 95%. An extremely high cell density at the time of transfection does not seem to markedly decrease the efficiency.

2. Remove the media and overlay the cells with 5 mls of fresh, prewarmed complete DMEM. Return the cells to the incubator.

3. In a 50 ml polystyrene conical tube (Fisherbrand) mix 400 μl of room temperature, serum-free DMEM with 21 μg of *gag/pol/rev/tat* encoding plasmid plus 8.5 μg of LTR/Ψ/transgene encoding plasmid and 8μg of envelope plasmid per flask to be transfected. Mix and incubate at room temperature for 3 min.

4. Add 60 μl of Superfect (Qiagen) per flask to be transfected. Mix, spin briefly to gather the contents of the tube into the bottom, and incubate at room temperature for 5 min.

5. Add 500 μl of the Superfect:DNA complexes to each flask. Rock the flask to evenly distribute the reagent and return to the incubator.

6. Three and one-half hr later remove the transfection mixture with gentle vacuum suction, carefully rinse the cells with 3 ml of complete DMEM, and overlay with 7.5 ml of fresh, complete DMEM. Return to the incubator. Transfection using a GFP reporter gene should be set up in parallel in order to assess the efficiency. Transfection efficiency must approach 80% or better.

7. The medium is removed the following morning from all the T-75 flasks into prechilled 50 ml conical tubes. Overlay each flask with an additional, fresh 7.5 ml of complete DMEM and return to the incubator. Spin the chilled conical tubes at 5000g for 5 min at 4° to pellet any cellular debris. Immediately filter the supernatant through a 0.45 μm PES membrane (Nalgene) into a prechilled collecting flask.

8. Add 80 ml of the filtrate into one Centricon Plus-80 centrifuge filter (NMWL membrane, 100,000 kDa cutoff, Amicon Bioseperations). Add the remaining 80 ml into a second unit. Spin the Centricon units at 1000g for 2.5 hr in a swinging bucket rotor at 4°. Invert the unit into the concentrate-collecting cup

(according to the manufacturer's instructions) and spin at 900g for 2 min at 4°. The collected volume should be less than 1 ml per filter unit. Immediately freeze concentrated aliquots at −20°. It is advisable to collect aliquots during each step of the concentration process in order to assay the titer of the virus after each step. The titer should increase markedly following each concentration step.

9. Repeat this procedure for a total of four viral media collections. Virus should be collected 12, 36, 60, and 72 hr after the initial transfection. After 72 hr the titer of the virus being produced drops rapidly. The highest titer collections are between 12 and 36 hr posttransfection. After 5 days approximately 5 ml of filter-concentrated virus is obtained. This can be stored at −20° until the second concentration step is ready to be performed.

10. Quick-thaw the filter-concentrated aliquots in a 37° water bath. Once thawed, immediately place on ice. Add the total 5 ml into a prechilled ultracentrifuge tube. The best results are obtained when the smallest possible tube capable of holding the *entire* 5 ml is used. The "Ultra-Clear" Beckman tubes (13 × 51 mm) are ideal for this procedure.

11. Pellet the virus in a swinging bucket rotor at 50,000g for 2.5 hr at 4°.

12. Immediately remove the supernatant using vacuum suction. Take care not to disturb the viral pellet in the bottom of the tube. Add between 100 and 500 μl of HBSS to the tube. Resuspend the virus by gently pipetting up and down until no precipitate is visibly attached to the bottom of the tube. Cover the tube opening with tape and transfer to an orbital shaker. Shake the virus at 200 rpm for 5 hr at 4°. Periodically stop the shaker and pipette the solution up and down to aid in resuspension. As before, avoid the introduction of bubbles during all pipetting steps. Immediately freeze the virus at −20° following at least 5 hr of shaking. It is essential to set aside a small aliquot (5–10 μl) of the virus to be used in the future for titration. This virus is now concentrated for *in vivo* usage. Titers greater than 1 × 10^8 are generally obtained by this protocol. Figure 4 shows neuronal cells successfully transduced *in vitro* with concentrated lentivirus vector encoding the gene for enhanced green fluorescent protein.

Discussion

Gene therapy advances in leaps and bounds every year. As researchers we are responsible for discerning not only what genes are the best suited targets or therapeutic agents, but also what is the best method to deliver these molecules. This review has introduced the concept of systemic gene delivery through the use of retroviral vectors. Key emphasis was placed on the large-scale production and concentration of the virus. Such developments help to move the field of gene transfer forward as they open new avenues of vector preparation and delivery. There are still many questions to be answered regarding systemic viral delivery. Where *can* the virus go? Can it enter the brain or germ cells if delivered

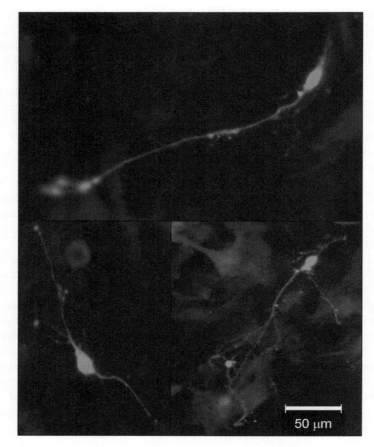

FIG. 4. Recombinant HIV-1 vector efficiently transduces neuronal cells *in vitro*. Neurons were cultured from the brainstem and hypothalamus regions of 1-day-old rat pup brains. 5×10^6 particles of lentivirus encoding enhanced green fluorescent protein (EGFP) were added directly into the culture medium 10 days following the initial plating of the cells and 24 hr later the cells were fed with fresh medium. EGFP expression was assayed 48 hr following transduction, but was still evident 14 days posttransduction at the termination of the experiment.

at an age before the barriers that protect these areas develop? How far can the virus penetrate out from the circulatory system? In other words, does the endothelial cell layer prevent virus from reaching the underlying smooth muscle cells? The answers to these questions dictate the future use of such retroviral vectors and their utility in systemic gene transfer. However, other questions surrounding gene therapy still abound. Is it safe? This single question will haunt users of the

HIV-1 based lentiviral vectors for years to come. We believe the future holds great promise for the lentivectors. They are clearly as safe or even safer than all of their oncoretroviral predecessors. We look forward to the future advances gene therapy will make and hope to play a part in both answering some of the questions posed here and asking some new ones that others must seek to answer.

Acknowledgments

The authors thank Jason Coleman of the Department of Neuroscience at the University of Florida McKnight Brain Institute for invaluable assistance during the development of the large-scale lentiviral production and concentration protocols.

[33] Oncoretroviral and Lentiviral Vector-Mediated Gene Therapy

By Thierry VandenDriessche, Luigi Naldini, Desire Collen, and Marinee K. L. Chuah

Introduction

The historical aspects of oncoretroviral or lentiviral vector development and the biological properties of the viruses from which they are derived have been extensively reviewed previously and are not the subject of this chapter. Instead, only the essential features of vector biology relevant for understanding current vector production protocols are summarized. One of the hallmarks of current oncoretroviral and lentiviral vectors is the lack of viral sequences that encode viral proteins, rendering them incapable of viral replication. Consequently, *de novo* expression of potentially immunogenic viral proteins in the transduced target cells can be excluded. Oncoretroviral and lentiviral vectors typically contain only the minimal cis-acting elements required for vector genome production, encapsidation, reverse transcription, genomic integration, and transgene expression. The production of oncoretroviral and lentiviral vector particles exploits the principle of trans-complementation. The missing viral proteins are essential for generating functional viral vector particles are hereby supplied in trans using packaging cells.

Production of Oncoretroviral Vectors

Oncoretroviral vector particles can be generated by transfecting the recombinant vector DNA into a packaging cell line derived from a variety of different

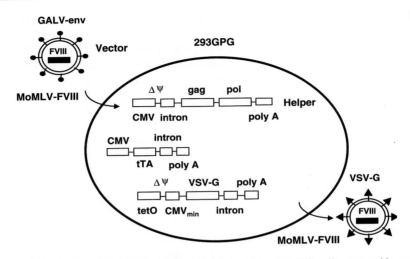

FIG. 1. Production of MoMLV-based oncoretroviral vectors. 293GPG cells were stably transduced with a GALV-env pseudotyped FVIII oncoretroviral vector (MFG-FVIIIΔB). The 293GPG cells constitutively express MoMLV gag and pol whereas VSV-G is conditionally expressed in a tetracycline-regulated fashion. The human cytomegalovirus (CMV) promoter is used to drive the tetracycline-regulated transcriptional activator (tTA). In the presence of tetracycline, tTA represses the minimal CMV promoter, which drives the VSV-G gene, by binding on the upstream tet operator (tetO) sequences. In the absence of tetracycline, this repression is relieved and VSV-G expression is consequently induced giving rise to functional VSV-G pseudotyped oncoretroviral particles. The packaging sequence is deleted in the helper construct (ΔΨ) but is the present in the vector (Ψ+). See text for details.

cell lines, including murine fibroblastic cell lines (e.g., NIH-3T3)[1] or human embryonic kidney cell lines (293).[2] The oncoretroviral vectors contain the cis-acting regulatory sequences required for encapsidation (Ψ⁺), reverse transcription, and integration (long terminal repeat, LTR) (Fig. 1). The transgenes of interest can be driven from the MoMLV long terminal repeats (LTR) or from any other internal promoter and can be combined with several other genes using additional promoters or preferably internal ribosome entry sites (IRES) to circumvent promoter suppression. The packaging cell line harbors helper proviral sequences that express the oncoretroviral structural proteins (Gag, Pol) in trans as well as an envelope protein (Env) essential for cellular entry via specific membrane receptors. The vector titer depends at least partly on the choice of promoters used to drive these viral genes,

[1] Y. Yang, E. F. Vanin, M. A. Whitt, M. Fornerod, R. Zwart, R. D. Schneiderman, G. Grosveld, and A. W. Nienhuis, *Hum. Gene Ther.* **9,** 1203 (1995).

[2] D. S. Ory, B. A. Neugeboren, and R. C. Mulligan, *Proc. Natl. Acad. Sci. U.S.A.* **93,** 11400 (1996).

and strong promoters such as the cytomegalovirus (CMV) promoter are obviously preferred. However, these *helper* sequences cannot be encapsidated into virions because they do not contain the cis-acting elements required for packaging ($\Delta \Psi$). Instead, as a consequence of complementation in trans, only recombinant viral vectors containing the Ψ^+ packaging signal are encapsidated into viral vector particles. These viral vector particles are released into the medium by the producer cells and can achieve gene transfer by transducing and integrating into the target cell's genome, but because they lack the viral structural genes, they are unable to replicate. The biosafety of the oncoretroviral vectors has been improved significantly by minimizing the sequence homology between vector and helper sequences, by expressing the helper sequences on separate constructs (so-called split configuration), and by using heterologous promoters and/or cis-acting elements, such as heterologous polyadenylation signals. This makes it most unlikely that replication-competent oncoretroviruses would arise through homologous recombination.

To generate oncoretroviral vectors several variations exist based on this principle. For instance, the vector and helper sequences can be transiently cotransfected into the desired packaging cell or the vector can be transiently transfected into a packaging cell line that already harbors stably integrated helper sequences. This method allows for the rapid but transient production of oncoretroviral vectors. However, the preferred condition for generating oncoretroviral vectors relies on the use of stable producer cell lines. The availability of such a line facilitates large-scale vector production and characterization, including biosafety assessment. To achieve this, both helper and vector sequences have to be stably transfected into the desired packaging cell and individual vector producer clones can be selected. Alternatively, if vector particles are already available, they can be used to stably transduce a packaging cell line that contains stably integrated helper sequences. This strategy is particularly useful to generate vector particles with distinct viral envelopes (pseudotyping). Regardless of whether the vector was introduced by stable transfection or stable transduction, production of functional vector particles typically occurs in a constitutive fashion, unless the vesicular stomatitis virus G glycoprotein (VSV-G) envelope is employed which requires conditionally regulated expression (see below).

The envelope protein used to pseudotype the oncoretroviral vectors can be derived from different types of retroviruses and include the murine oncoretroviral ecotropic envelope (which restricts vector transduction to murine cells) or the amphotropic murine oncoretroviral 4070A or 10A1 envelope, the feline oncoretroviral RD114, or gibbon ape leukemia virus (GALV) envelope. These different pseudotypes exhibit different tropisms which are at least partly determined by the levels of the cognate receptors on the target cell membrane. In particular, cells are relatively refractory to oncoretroviral transduction if the cognate cellular receptor is expressed at only very low levels.

Alternatively, oncoretroviral vector particles can be pseudotyped with the envelope glycoprotein G of vesicular stomatitis virus (VSV-G) (Fig. 1). One particular advantage of using VSV-G pseudotyped oncoretroviral vectors is that it permits efficient transduction of an expanded range of different cell types from distinct species. It has been proposed that the relative abundance of this elusive receptor may explain the broad tropism of VSV-G pseudotyped vectors. The VSV-G receptor appears to involve phosphatidylserine moieties on the lipid bilayer. However, the expanded host range of VSV-G pseudotyped vectors is a double-edged sword: though it permits gene transfer in cell types that would otherwise be refractory to oncoretroviral (or lentiviral) transduction, this may sometimes be undesirable in the context of gene therapy applications which require a more refined and selective targeting of gene transduction into specific cell types. Another advantage associated with the use of VSV-G pseudotyped oncoretroviral vectors includes the ability to concentrate the vectors to high titer by ultracentrifugation while retaining viral vector infectivity, which is compromised when other envelope pseudotypes are employed. Though stable VSV-G expression is toxic to most mammalian cells, this limitation can be circumvented either by transiently transfecting a VSV-G expression plasmid in the packaging cells or by generating a stable cell line that expresses VSV-G conditionally. Human packaging cell lines derived from NIH-3T3 or 293 cells have been developed that express VSV-G conditionally in a tetracycline-dependent manner.[1,2] These cell lines express a tetracycline-regulated chimeric transcription factor (designated as tTA), composed of the tetracycline repressor (tetR) and the VP16 trans-activating sequences of the herpes simplex virus VP16 gene product. VSV-G expression in these cell lines is driven from a synthetic promoter (TetO-CMV$_{min}$) consisting of the CMV minimal promoter coupled to seven tandem repeats of the tetracycline responsive operator (TetO) (Fig. 1). tTA only binds to TetO and transactivates the CMV$_{min}$ promoter in the absence of tetracycline. Conversely, tetracycline relieves the binding of tTA on TetO and hereby prevents expression of VSV-G. VSV-G pseudotyped viral particles can thus be generated conditionally upon removal of tetracycline from the culture medium.

This chapter focuses on the generation of MoMLV-based oncoretroviral vectors using large-scale production cell factories based on a stable VSV-G producer cell lines derived from 293 human embryonic carcinoma cells (293GPG). This cell line was designed to express VSV-G in a conditional, tetracycline-regulated manner and was used to generate high-titer oncoretroviral vectors that express a potentially therapeutic protein, in casu the human coagulation factor VIII (FVIII) gene.

Production of Lentiviral Vectors

Lentiviral vector particles are typically obtained by transiently cotransfecting the recombinant vector DNA into a human embryonic kidney cell line (293T cells)

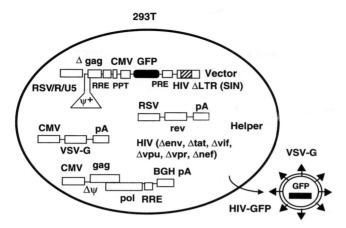

FIG. 2. Production of HIV-based lentiviral vectors based on quadruple transient transfection of 293T cells with the self-inactivating (SIN) HIV transfer vector along with Rev-, VSV-G-, and gag/pol-expression plasmids. The transfer vector expresses GFP from the internal *CMV* promoter and contains the Woodchuck posttranscriptional regulatory element (*PRE*) and central polypurine tract (*PPT*) that enhance gene expression and facilitate nuclear import, respectively. Gag and pol expression conditionally depends on the Rev-*RRE* interaction, and VSV-G is constitutively expressed. Only the transfer vector contains the packaging sequence (Ψ+) which is deleted from the helper construct (ΔΨ). The full-length vector RNA is driven from the chimeric *RSV/HIV-LTR (RU5)*.

along with helper constructs (Fig. 2). The lentiviral vectors contain the cis-acting regulatory sequences required for encapsidation (Ψ+), reverse transcription, and integration.[3,4] The 5′ end of the *gag* gene and *Rev-responsive element (RRE)* are essential to obtain high-titer viral vector particles. In addition, splice donor (*SD*) and splice acceptor (*SA*) are included in the vector backbone. The transgene is driven from an internal promotor since the mutated HIV long terminal repeat (LTR) self-inactivates following transduction as an additional safety feature. The HIV-derived helper sequences express the lentiviral structural proteins (Gag, Pol) in a conditional Rev-dependent fashion from the CMV promoter, by virtue of the presence of RRE downstream of *gag* and *pol* genes. The Rev-dependency provides an additional safety feature since *gag* and *pol* expression cannot occur in the absence of Rev. These helper sequences cannot be encapsidated into virions because they do not contain the cis-acting elements required for packaging (ΔΨ). The HIV accessory *vif, vpr, vpu, tat,* and *nef* genes are not necessary for transduction of most, if not all, cell types and are deleted from the helper sequences to improve

[3] L. Naldini, U. Blomer, P. Gallay, D. Ory, R. Mulligan, F. H. Gage, I. M. Verma, and D. Trono, *Science* **2,** 263 (1996).

[4] E. Vigna and L. Naldini, *J. Gene Med.* **5,** 308 (2000).

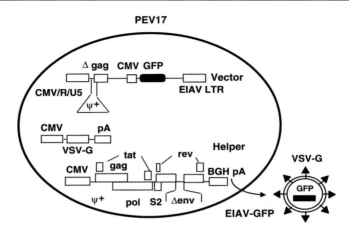

FIG. 3. Production of EIAV-based lentiviral vectors based on dual transient transfection of PEV17 cells with the EIAV transfer vector along with a VSV-G expression plasmid. The EIAV helper construct contains the full EIAV complement except for the EIAV envelope and has been stably transfected into the packaging cells. The transfer vector expresses GFP from the internal CMV promoter. Both the transfer vector and the helper construct contain the packaging sequence ($\Psi+$). However, in future helper constructs, the packaging sequence should preferably be deleted. The full-length vector RNA is driven from the chimeric *CMV/EIAV-LTR (RU5)*.

their safety profile. The HIV envelope is replaced by a heterologous viral envelope (see below) that is supplied in trans. As a consequence of complementation, only recombinant lentiviral vectors containing the Ψ^+ packaging signal are encapsidated into viral vector particles. These viral vector particles are released into the medium by the producer cells. The resultant replication-defective lentiviral vectors can achieve gene transfer by transducing and integrating into the target cell's genome. The generation of the nonprimate lentiviral vectors essentially mirrors that of the HIV-based vectors, with few differences inherent in the characteristic molecular biological features that distinguish these distinct lentiviruses. For instance, lentiviral vectors have been derived from the equine infectious anemia virus which are trans-complemented by helper construct that express the missing EIAV proteins (Gag, Pol, Tat, Rev, S2) except for Env (Fig. 3).

The development of stable lentiviral packaging cell lines has been hampered by the inherent cytotoxicity of some of the HIV proteins necessary to generate functional vector particles. However, conditional expression systems have been developed to circumvent this potential limitation. Nevertheless, titers from lentiviral vectors obtained using the currently available stable packaging cell lines are typically 10-fold lower compared to the titer levels that can be achieved by most transient transfection production systems. It is likely that this caveat will be overcome in the near future. In analogy with the onco-retroviral vectors, the biosafety profile of the lentiviral vectors has been improved significantly by minimizing the sequence homology between vector and helper sequences, by expressing the helper sequences on separate constructs (so-called split configuration), and by using heterologous promoters and/or heterologous cis-acting elements, such as polyadenylation signals.

In analogy with oncoretroviral vectors, lentiviral vectors can be pseudotyped with the murine oncoretroviral ecotropic envelope, VSV-G, or the amphotropic murine oncoretroviral 4070A or 10A1 envelopes.[5] Pseudotyping lentiviral vectors with the GALV envelopes appears to compromise viral vector infectivity and hybrid envelopes are required instead that include the external domain of the GALV envelope combined with the cytoplasmic domain of the MoMLV amphotropic envelope.[6] In this chapter, production of EIAV and HIV-based vectors will be compared and a large-scale HIV-based vector production system based on transient transfection of helper and vector sequences will be described. In addition, viral vector concentration based on centrifugation and ultrafiltration methods will be outlined.

Materials and Methods

Production of Oncoretroviral Vectors

Cell Lines and Culture Conditions. PG13 packaging cells are used to generate GALV-env pseudotyped oncoretroviral vectors whereas the 293GPG packaging cell line is employed to produce VSV-G pseudotyped oncoretroviral vectors. PG13 packaging cells (kindly provided by Dr. A.D. Miller, Fred Hutchinson Cancer Research Center, Seattle, WA) constitutively express MoMLV gag and pol along with the GALV-env.[7] The PG13 packaging cell line and its derivatives including the PG13-F8 producer cell lines are grown in Dulbecco's modified Eagle's medium (DMEM) with 2 mM L-Gln, 100 IU/ml penicillin, 100 μg/ml streptomycin, and 10% heat-inactivated fetal bovine serum (designated as D10 medium) (Life Technologies, Merelbeke, Belgium). The 293 human embryonic carcinomaderived conditional VSV-G packaging cell line 293GPG (kindly provided by Dr. D. Ory, Washington University, St. Louis) constitutively expresses MoMLV gag and pol, whereas VSV-G is conditionally expressed in a tetracycline-dependent manner (Fig. 1).[2] The 293GPG packaging cell line and its derivatives, the LXSN-transduced producer clone 293GPG-LXSN and the MFG-FVIIIΔB-transduced producer clone 293GPG-F8, are cultured in D10 medium. VSV-G expression is

---

[5] T. VandenDriessche, D. Collen, and M. Chuah, unpublished observations.

[6] J. Stitz, C. J. Buchholz, M. Engelstadter, W. Uckert, U. Bloemer, I. Schmitt, and K. Cichutek, *Virology* **273,** 16 (2000).

[7] A. D. Miller, J. V. Garcia, N. von Suhr, C. M. Lynch, C. Wilson, and M. V. Eiden, *J. Virol.* **65,** 2220 (1991).

repressed when the medium is supplemented with tetracycline (1 μg/ml), whereas its expression is induced when tetracycline is omitted. The murine fibroblast-like cell line NIH-3T3 is also grown in D10. All cells are grown at 37°, except where noted, in an incubator with 95% humidity and 5% CO_2.

Generation and Characterization of GALV-Pseudotyped Oncoretroviral Vectors. The *LXSN* vector was described previously[8] and was kindly provided by Dr. A. D. Miller. The corresponding amphotropic LXSN vector particles are generated in PA317 packaging cells as described previously.[9] The *FVIII B-domain deleted cDNA* is obtained from the *pMT2LA* plasmid (provided by Genetics Institute, Cambridge, MA).[10] The splicing vector, *MFG-FVIIIΔB,*[11] is generated by cloning a *B-domain deleted FVIII cDNA* downstream of the *MoMLV-LTR* into the *MFG* oncoretroviral backbone[12] which contains the native *MoMLV env* intron. The *FVIII cDNA* in this *MFG-FVIIIΔB* vector is obtained from plasmid *pAGF8* (Chiron Corporation, Emeryville, CA) and was provided by Somatix (Alameda, CA). The corresponding amphotropic MFG-FVIIIΔB vector particles are generated in Ψ-CRIP-derived packaging cells as described previously,[11] (provided by Somatix, Inc., Alameda, CA).

To generate GALV-env pseudotyped vectors, PG13 packaging cells are transduced with MFG-FVIIIΔB vectors. The resulting PG13-F8 producer cells are cloned by limiting dilution. FVIII production by each of the PG13-F8 clones is quantified using a functional chromogenic assay as described below. Clones that express the highest levels of FVIII are further screened for viral production by RNA dot-blot analysis and subsequently subjected to Southern blot analysis to exclude the presence of rearranged proviral sequences (see below).

Generation and Characterization of VSV-G Pseudotyped Oncoretroviral Vectors. To generate VSV-G pseudotyped oncoretroviral vectors, subconfluent 293GPG packaging cells are transduced with PA317-LXSN and PG13-F8 vectors (Fig. 1).[13] To achieve this, viral-vector containing supernatant is first collected at 32° over 24 hr from a confluent plate of PA317-LXSN or PG13-F8 producer cells and filtered through a 0.45 μm filter to exclude the presence of residual producer cells. 293GPG cells are transduced successively with PA317-LXSN or PG13-F8 vectors (3 and 7 rounds, respectively) in the presence of Polybrene (8 μg/ml) and tetracycline (1 μg/ml), followed by a centrifugation step at 32° for 1 hr at

[8] A. D. Miller and G. J. Rossman, *Biotechniques* **7,** 980 (1989).

[9] M. K. Chuah, T. VandenDriessche, and R. A. Morgan, *Hum. Gene Ther.* **6,** 1363 (1995).

[10] J. J. Toole, D. D. Pittman, E. C. Orr, P. Murtha, L. C. Wasley, and R. J. Kaufman, *Proc. Natl. Acad. Sci. U.S.A.* **83,** 5939 (1986).

[11] V. J. Dwarki, P. Belloni, T. Nijjar, J. Smith, L. Couto, M. Rabier, S. Clift, A. Berns, and L. K. Cohen, *Proc. Natl. Acad. Sci. U.S.A.* **92,** 1023 (1995).

[12] O. Danos and R. C. Mulligan, *Proc. Natl. Acad. Sci. U.S.A.* **85,** 6460 (1988).

[13] T. VandenDriessche, V. Vanslembrouck, I. Goovaerts, H. Zwinnen, M. L. Vanderhaeghen, D. Collen, and M. K. Chuah, *Proc. Natl. Acad. Sci. U.S.A.* **96,** 10379 (1999).

1400g (2600 rpm) directly on six-well plates, and an overnight incubation at 32°. The resulting producer cells are designated as 293GPG-LXSN and 293GPG-F8, respectively, and individual clones are obtained by limiting dilution. The 293GPG-LXSN producer clones are selected in G418 (0.8 mg/ml). FVIII production by each of the individual 293GPG-F8 clones is quantified using a functional chromogenic assay as described below. Clones that express the highest levels of FVIII are further screened for viral production by RNA dot-blot analysis and subsequently subjected to Southern blot analysis to exclude the presence of rearranged proviral sequences (see below).

Southern Blot Analysis. For Southern blot analysis, genomic DNA is extracted with the high-pure PCR template preparation kit (Boehringer, Mannheim, Germany) and 23 μg of DNA was restricted with *Sma*I or *Nhe*I. Hybridizations are performed by probing the Southern blot membrane with a FVIII-specific probe corresponding to a random primed 1095 bp *Bgl*II–*Spe*I restriction fragment of plasmid *pMT2LA*,[10] which was provided to us by Genetics Institute, Inc. The membranes are washed stringently at 65° in 2× SSC and 0.1% SDS for 30 min followed by 0.5× SSC and 0.2% SDS and for an additional 30 min.[9,14] After autoradiography, the presence of rearranged or nonrearranged vector sequences can be visualized.

Vector Production and Concentration. For small-scale production, it suffices to seed the cells in 20 ml D10 without tetracycline at a concentration of 8×10^7 cells per 75 cm^2 tissue culture flask. For large-scale production, 293GPG-LXSN and 293GPG-F8 producer cells are first grown to 50% confluency in a single tray (Nalge Nunc International, Roskilde, Denmark) containing 100 ml D10 per tray supplemented with tetracycline. Subsequently, semiconfluent cells are trypsinized, pooled, and seeded in a cell factory containing 1000 ml D10 with tetracycline until they reach 90–95% confluency (typically after 4 days). The conditioned medium is then removed and the cells are washed with phosphate-buffered saline (PBS) to remove residual traces of tetracycline. Subsequently 1000 ml D10 without tetracycline is added per cell factory to induce VSV-G expression. The supernatant is harvested every 24 or 48 hr over 2 weeks, frozen immediately on dry ice prior to storing at −80°, and filtered using a 0.45 μm filter before use. Viral vector particles from filtered supernatant are sedimented by centrifugation at 4500 rpm overnight at 4° with a SLA1500 rotor. The pelleted viral vector particles are resuspended thoroughly in PBS and frozen in dry ice before storing at −80°. Titer and yield of concentrated vector preparations are determined by RNA dot-blot analysis (see below).

Oncoretroviral Vector Titration. The functional titer of the 293GPG-LXSN can be determined by transducing NIH-3T3 cells as described previously.[9,15] Briefly,

[14] M. K. Chuah, H. Brems, V. Vanslembrouck, D. Collen, and T. VandenDriessche, *Hum. Gene Ther.* **9**, 353 (1998).

[15] M. K. Chuah, T. VandenDriessche, H. K. Chang, B. Ensoli, and R. A. Morgan, *Hum. Gene Ther.* **5**, 1467 (1994).

2.5×10^4 NIH-3T3 cells are seeded in 2 ml D10 per well in a 6-well plate. The next day, the medium is replaced with serially diluted vector-containing 293GPG-LXSN supernatant supplemented with 8 μg/ml Polybrene and 24 hr later the vector-containing medium is replaced with D10 supplemented with 0.8 mg/ml G418. Ten to 12 days postselection, G418^R colonies are first stained with methylene blue in methanol and then counted. Multiplication with the dilution factor yields the end-point functional titer in cfu (colony-forming units) per ml. Since the 293GPG-F8 vector does not contain a selectable marker, vector titer is determined by RNA dot-blot analysis instead. To establish a standard curve, serially diluted viral vector supernatants based on 293GPG-LXSN with known functional titer (G418^R cfu/ml) are used. The viral particles in 1 ml of culture medium or serially diluted viral supernatant are precipitated by adding 0.5 ml of PEG solution (30% PEG8000, 1.5 M NaCl). After 30 min on ice, the samples are centrifuged for 5 min at 4°. The pellets are resuspended in 0.2 ml of VTR buffer [10 mM Tris-HCl pH 8.0, 1 mM EDTA, 20 mM vanadyl ribonucleoside complex (VRC), 100 μg/ml yeast tRNA] and then lysed by adding 0.2 ml of 2× lysis buffer (1% SDS, 0.6 M NaCl, 20 mM EDTA, 20 mM Tris, pH 7.4). The viral RNA is extracted once with phenol and chloroform and precipitated with ethanol. The RNA pellets are dissolved in 0.25 ml of 15% formaldehyde solution followed by addition of 0.25 ml of 20× SSC. The RNA samples are then loaded onto a Hybon N nylon membrane (Amersham, Buckinghamshire, UK) using a vacuum-operated dot-blot apparatus (Minifold System Dot/Slot blot Unit, Schleicher & Schuell). After loading, the membrane is rinsed in 10 × SSC, air-dried, UV-cross-linked, and prehybridized in Quickhyb (Stratagene/Westburg, Leusden, The Netherlands) to which 100 μg/ml denatured salmon sperm DNA is added.

Hybridizations are performed by probing the membrane with a random primed vector-specific probe which is derived by PCR using the MFG-FVIIIΔB plasmid as target and primers spanning a 418-bp region within the packaging sequence (5'-GGGCCAGACTGTTACCACTCCC-3' and 5'-GGCGCCTAGAGAAGGAGT GAGGG-3') which recognized a sequence common to the LXSN and MFG-FVIIIΔB vectors. This probe is generated by PCR using Taq polymerase and a Progene thermocycler by denaturation for 8 min at 95°, followed by 28 cycles of 1 min at 95°, 1 min at 59°, 2 min at 72°, and a final extension for 5 min at 72°. Amplified samples are separated by gel electrophoresis on 1.5 % agarose gels and the PCR fragment is extracted from the agarose using Geneclean. After hybridization, the membranes are washed stringently at 65° in 2× SSC for 30 min followed by 0.5× SSC for an additional 30 min. After background subtraction, signal intensities are quantified using a Phosphorimager (Molecular Dynamics, Sunnyvale, CA). By comparing signal intensities of the 293GPG-F8 vector-containing supernatant with those signals obtained with the 293GPG-LXSN vectors, the functional titer of 293GPG-F8 can be estimated.

FVIII Quantification. FVIII activity in transduced NIH-3T3 cells and viral producer cell clones or mouse plasma is quantified by measuring the FVIII-dependent generation of factor Xa from factor X using a chromogenic assay (Coatest FVIII, Chromogenix, Molndal, Sweden) following the manufacturer's instructions.[9] Briefly, 24 hr-conditioned culture medium is harvested in phenol-red free media to avoid colorimetric interference in the FVIII chromogenic assay. All samples are sufficiently diluted to yield values that fell within the linear range of the assay. Human plasma purified FVIII (Octapharma, Langenfeld, Germany) of known activity is used as a FVIII standard and 1 U is defined as 200 ng FVIII/ml. The lowest level of detectable FVIII is 0.01–0.03 ng/ml, and media containing heat-treated FBS did not yield any detectable FVIII activity.

Production of Lentiviral Vectors

293T or PEV17 cells are used for production of HIV-based and EIAV-based lentiviral vectors, respectively (Figs. 2 and 3). Both cell lines are derived from the 293 human embryonic kidney cell line. The 293T cells also contain stably transfected SV40 large T antigen. PEV17 cells are obtained by stably transfecting a 293-derivative with an EIAV helper construct that encodes all EIAV genes (*gag, pol, S2, rev, tat*) (PEV53[16]) except EIAV *env* as described previously (the PEV17 cell line was kindly provided by Dr. Olson, University of North Carolina, USA[16]) (Fig. 3). 293T or PEV17 cells are seeded in a cell factory and/or single tray unit (Nalge Nunc International, Roskilde, Denmark), expanded in D10 medium (described previously[14]), and transfected when 90–100% confluent using either GIBCO (Life Technologies, Merelbeke, Belgium) or 5 Prime/3 Prime (Sanvertech, Boechout, Belgium) calcium phosphate transfection kits according to manufacturer's instructions. Plasmid DNA to be transfected is extracted using Qiagen Maxi or endo-free plasmid Mega kits (Westburg, Leusden, NL) according to manufacturer's instructions. Plasmid DNA is sterilized by ethanol precipitation at a concentration of 1 μg/μl prior to transfection. Per cell factory, a total volume of 655 ml D10 medium (or 65.5 per cell tray unit) is used during transfection. For making the HIV vectors, the following amount of DNA is used per cell factory: 3 mg of lentiviral-GFP vector, 1.5 mg of pMDL gag/pol RRE helper plasmid, 1.5 mg of Rev expressing plasmid, and 1.5 mg pCI-VSV-G envelope-encoding plasmid, which were previously described[17,18] (Fig. 2). The lentiviral-GFP vector is similar to the published construct,[17,18] except that an additional polypurine tract

[16] J. C. Olsen, *Gene Ther.* **5,** 1481 (1998).
[17] T. Dull, R. Zufferey, M. Kelly, R. J. Mandel, M. Nguyen, D. Trono, and L. Naldini, *J. Virol.* **72,** 8463 (1998).
[18] R. Zufferey, T. Dull, R. J. Mandel, A. Bukovsky, D. Quiroz, L. Naldini, and D. Trono, *J. Virol.* **72,** 9873 (1998).

(PPT) and posttranscriptional regulatory element from woodchuck hepatitis virus (WPRE) had been introduced in the vector backbone to augment transgene expression and/or viral vector titer.[19] For making the EIAV vectors, the following amount of DNA is used per cell factory: 3 mg of EIAV-GFP vector and 1.5 mg of VSV-G envelope-encoding plasmid (pCI-VSV-G), which were previously described[16] (Fig. 3). Since the helper construct is already stably integrated into the PEV17 cells, it does not have to be transfected again.

The next day, medium is removed and replaced with 100 ml D10 containing 1.1 μg/ml soduim-butyrate (Sigma, Bornem, Belgium) per cell tray. Conditioned medium is collected every 24 or 48 hr during the subsequent 3–4 days until significant cell death and/or cell detachment occur. Viral vector concentration is performed after filtration with a 0.45 μm filter (Corning, Elscolab, Kruibeke, Belgium) at 4500 rpm overnight at 4° with a SLA1500 rotor.[13] The pelleted viral vector particles are resuspended thoroughly in PBS and frozen in dry ice before storing at −80°. Typically, vectors are concentrated 1000-fold using this concentration procedure. Alternatively, vector concentration is performed by ultrafiltration. To achieve this, conditioned medium containing viral vectors is collected in DMEM with L-gln and penciciline/streptromycin supplemented with 10% Nu-Serum IV (Becton Dickinson, Erembodegem, Belgium) instead of FCS. After prefiltration using a 0.45 μm filter, the conditioned medium is applied on a Centricon Plus-80 centrifugal filter device (Millipore, 100 kDa nominal molecular mass cutoff) and centrifuged for 100 min at 1500 rpm, (10°). The retentate contained in the upper container of the Centricon device can be retrieved by centrifuging for 2 min at 2000 rpm (10°). A 200-fold concentration can be achieved using this method. Titer and yield of concentrated vector preparations are subsequently determined (see below).

Lentiviral Vector Titration

NIH-3T3 fibroblasts are seeded in a 6-well plate (10^5 cells per well) and transduced the next day with serially diluted unconcentrated or concentrated vector-containing supernatant supplemented with 8 μg/ml Polybrene. Cells are washed twice the next day, with PBS and fresh D10 without Polybrene was added.[14] Since GFP is used as a marker, the number of fluorescent cells per microscopic field (at least 3–10 fields) is counted 48 hr posttransfection, on which basis the total number of fluorescent cells per ml viral vector-containing supernatant can be calculated (= titer in TU/ml, transducing units per ml). To further corroborate the functionality of the HIV-GFP vectors, human umbilical vein endothelial cells (HUVEC) and primary human smooth muscle cells (10^5 cells/ml) are transduced overnight with concentrated HIV-GFP vector (multiplicity of infection or MOI = 40) supplemented with Polybrene (8 μg/ml). GFP expression is monitored 48 hr later.

[19] A. Follenzi, L. E. Ailles, S. Bakovic, M. Geuna, and L. Naldini, *Nat. Genet.* **25,** 217 (2000).

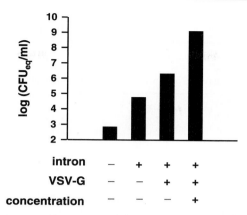

FIG. 4. Titer of FVIII-oncoretroviral vectors. The presence or absence of the MoMLV env intron in the vector was indicated ($+/-$). The intron-containing vector corresponded to the MFG-FVIIIΔB vector. The vector which lacked a functional intron corresponded to the LF8SN vector, which was described previously.[9] The MFG-FVIIIΔB vector was produced either by Ψ–CRIP amphotropic packaging cells (VSV-G: -) or by 293GPG cells (VSV-G: +). The titer of nonconcentrated ($-$) and concentrated ($+$) VSV-G pseudotyped MFG-FVIIIΔB vector was shown. Viral vector concentration was carried out by low-speed centrifugation. Titer was expressed as log (CFUeq/ml)] and was determined by RNA dot-blot analysis (MFG-FVIIIΔB) or by functional titration in NIH-3T3 cells (LF8SN). See Materials and Methods for details.

Results and Discussion

Production of Oncoretroviral Vectors

To generate the VSV-G pseudotyped MFG-FVIIIΔB vector, 293GPG packaging cells were first transduced with the MFG-FVIIIΔB vector (Fig. 1) and subsequently cloned by limiting dilution. The titer of the VSV-G pseudotyped MFG-FVIIIΔB vector (designated as 293GPG-F8) produced by the highest producer clone was equivalent to $1.8 \pm 0.7 \times 10^6$ cfu/ml (mean $\pm$ SEM, $n = 7$) and was significantly increased (30-fold, $p < 0.001$) compared to the original amphotropic MFG-FVIIIΔB splicing vector produced by Ψ-CRIP cells from which it was originally derived ($6 \pm 3 \times 10^4$ G418^R cfu/ml, mean $\pm$ SEM, $n = 23$) and to a standard nonsplicing FVIII-vector such as LF8SN (7×10^2 G418^R cfu/ml) which was previously described[9] (Fig. 4). In addition, the 293GPG derived FVIII vector could be concentrated 1000-fold further by centrifugation, yielding final average titers equivalent to $1.1 \pm 0.4 \times 10^9$ cfu/ml with yields of 63 ± 9 % (mean $\pm$ SEM, $n = 7$). Viral concentration by prolonged low-speed centrifugation yielded concentrated viral preparations with yields up to 70% (Table I).

These relatively high titers could possibly be attributed to the use of an intron-based vector, the ability to concentrate the VSV-G pseudotyped FVIII vector by

TABLE I
TITRATION OF VSV-G PSEUDOTYPED MFG-FVIIIΔB VECTOR

Titer (cfu$_{eq}$/ml)			
Preconcentration	Postconcentration	Fold concentration	Yield(%)
3.2×10^6	2.1×10^9	$1000 \times$	65
1.5×10^6	9.2×10^8	$1000 \times$	61
2.0×10^6	1.2×10^9	$1000 \times$	69
1.7×10^6	1.2×10^9	$1000 \times$	67

centrifugation, the high vector copy number, and the inherent improved characteristics of the 293GPG packaging cell line, which drives the MoMLV *gag* and *pol* genes from a relatively potent CMV promoter, as opposed to the more commonly used, but weaker MoMLV LTR or SV40 promoter.[1] When MFG-FVIIIΔB vectors were stably produced from NIH-3T3 –derived packaging cells that conditionally express VSV-G in addition to *gag, pol* from the SV40 promoter,[1] titers were typically 15-fold lower than when the 293GPG cells were used (titers equivalent to $1–1.7 \times 10^5$ cfu/ml).

The use of a conditional VSV-G packaging cell line obviates the need for repeated transient transfections permitting the use of cell factories which further facilitates upscaling of viral production. Production of oncoretroviral vectors in human packaging cells such as 293T cells has several advantages compared to that in murine packaging cell lines. First, human cells lack endogenous murine retroviral sequences and retroviral-like RNAs (VL30 retrotranposons), thereby reducing the likelihood for the formation of replication-competent retroviruses (RCR) and eliminating the risk of copackaging these sequences into oncoretroviral vector particles, thereby increasing their safety profile. In addition, human cells do not express α1-4 galactosyltransferase, resulting in increased resistance of viral vector particles to human complement though other mechanisms may also be involved, possibly increasing their efficacy for *in vivo* gene therapy. Another potential advantage of using 293 cells to produce FVIII oncoretroviral vectors is that oncoretroviral vector particles produced from stable 293 cell-derived amphotropic packaging cell lines transduce target cells at higher efficiencies than when standard murine 3T3-derived packaging cells, such as Ψ-CRIP, are used. This difference could be attributable, at least partially, to the presence of a transferable inhibitor in the Ψ-CRIP cell line, possibly shedded amphotropic envelope.

This high-titer vector was injected into young FVIII-deficient mice with targeted inactivation of the FVIII gene, which mimic the clinical hemophilia A phenotype.[12] Hepatic expression of FVIII caused long-term supraphysiologic and physiologic functional FVIII production and correction of the bleeding disorder in about 50% of the recipient mice. The remaining animals expressed either no or transient FVIII expression levels which correlated with the induction of neutralizing antibodies against the human FVIII xenoprotein. Human FVIII-expression in the FVIII-expresser mice ranged between 20 and 1250% of normal human FVIII levels, and expression was sustained for at least 14 months in recipients that did not develop inhibitory antibodies, which represent the highest stable levels of FVIII reported after gene therapy. Efficient gene transfer occurred in the active reticuloendothelial organs, including liver, spleen, and lungs, with predominant FVIII mRNA expression in the liver. To our knowledge, this is probably the first demonstration that a disease, in particular hemophilia, could be corrected by *in vivo* gene therapy using oncoretroviral vectors. These findings underscore the importance of developing high-titer vector preparations to achieve therapeutic effects in preclinical animal models.

Production of Lentiviral Vectors

To generate the VSV-G pseudotyped lentiviral vectors, 293T or PEV17 cells were transiently transfected with the HIV and EIAV helper and vector constructs (Figs. 2 and 3). The functional vector titer was determined by transducing NIH-3T3 cells. The titer of the nonconcentrated HIV-GFP vector preparations corresponded to approximately 3 to 4×10^7 TU/ml (Table II), whereas titers of the EIAV-GFP vector were typically about 100-fold lower ($3.2 \pm 2.6 \times 10^5$ TU/ml). After concentration using ultrafiltration, HIV-GFP vector titers were significantly and reproducibly increased (about 100-fold) to about 3 to 4×10^9 TU/ml, typically corresponding to a 50–60% yield. These HIV-GFP vectors could efficiently transduce human umbilical vein endothelial cells and smooth muscle cells *in vitro* (Fig. 5).

The reason why HIV-GFP yields higher vector titers compared to EIAV-GFP is not fully understood. However, it is possible that this is at least partly due to the inherent biological differences that distinguish these different lentiviruses. In

TABLE II
TITRATION OF VSV-G PSEUDOTYPED HIV-GFP LENTIVIRAL VECTOR

Titer (TU/ml)			
Preconcentration	Postconcentration	Fold concentration	Yield(%)
3.8×10^7	4.3×10^9	$185 \times$	62
3.8×10^7	4.1×10^9	$174 \times$	63
3.8×10^7	3.6×10^9	$185 \times$	52
3.0×10^7	3.0×10^9	$180 \times$	56

HIV LENTIVIRAL TRANSDUCTION

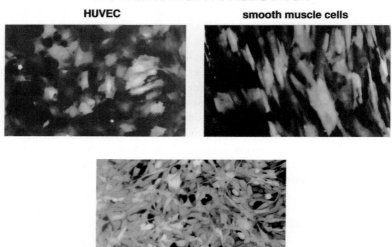

NIH-3T3 fibroblasts

Fig. 5. HIV-lentiviral transduction of HUVEC, primary human smooth muscle cells, and NIH-3T3 fibroblasts.

addition, the HIV-vector backbone contains several elements that improves vector performance such as the cPPT and WPRE elements, which are absent in the EIAV vector backbone. However, even HIV-based vectors which do not contain these elements have elevated titers compared to EIAV-based vectors, suggesting that other factors may account for the titer difference. It is possible that the implementation of a self-inactivation configuration in the HIV vector does contribute to improved vector performance compared to the EIAV vector by alleviating potential promoter suppression effects on the internal CMV promoter. Finally, the presence of the encapsidation sequence in this early generation EIAV helper construct may lead to possible competition of the helper RNA with the vector RNA for encapsidation into the vector particles, hereby reducing vector titers.

We have tested whether 293G cells[2] can be used for producing EIAV vectors by transient transfection of EIAV vector and helper plasmids.[20] A minimal helper construct was designed from which the packaging sequence was deleted and which only contained and expressed the EIAV gag and pol genes as confirmed

[20] T. VandenDriessche, D. Collen, and M. Chuah, unpublished observations.

by RT-PCR (data not shown). EIAV vector titers produced with this system corresponded to about 10^4 TU/ml. These vector titers are somewhat lower than what has been achieved using the PEV17 cells for reasons which may reflect differences in transfection efficiency. It is likely that some of the present limitations of using EIAV vectors may eventually be resolved by further improving the properties of the packaging cells.

Summary

Oncoretroviral vectors and lentiviral vectors offer the potential for long-term gene expression by virtue of their stable chromosomal integration and lack of viral gene expression. Consequently, their integration allows passage of the transgene to all progeny cells, which makes them particularly suitable for stem cell transduction. However, a disadvantage of oncoretroviral vectors based on Moloney murine leukemia virus (MoMLV) is that cell division is required for transduction and integration, thereby limiting oncoretroviral-mediated gene therapy to actively dividing target cells. In contrast, lentiviral vectors can transduce both dividing and nondividing cells. Lentiviral vectors have been derived from either human or primate lentiviruses, with the human immunodeficiency virus (HIV) as prototype, or from nonprimate lentiviruses, such as the equine infectious anemia virus (EIAV). The ability to pseudotype oncoretroviral and lentiviral vectors with the vesicular stomatitis virus G glycoprotein (VSV-G) allowed for the production of high-titer vectors (10^9–10^{10} transducing units/ml). These high-titer vector preparations were shown to effectively cure genetic diseases in experimental animal models and constitute an essential step toward direct *in vivo* gene therapy applications. This chapter focuses on different methods that permit large-scale production of high-titer VSV-G pseudotyped oncoretroviral and primate or nonprimate lentiviral vectors and highlights their importance for achieving therapeutic effects in preclinical animal models.

Acknowledgments

The authors are particularly grateful to Dr. Daniel Ory (Washington University) and Dr. Richard Mulligan (Harvard University), Dr. Haig Kazazian (University of Pennsylvania), Dr. Dusty Miller (Fred Hutchinson Cancer Research Center), Dr. John Olson (University of North Carolina), Dr. E. Vanin (University of Tennessee), Genetics Institute, Inc., and Somatix, Inc., for providing us with some of the important reagents used in this study. We are also very grateful to Dr. Antonia Follenzi for her contributions. Finally, we thank Hans Zwinnen, Els Vangoidsenhoven, and Veerle Vanslembrouck for their technical assistance. This work is supported by a grant of the Flemish government (VIB dotatie) and the Flanders Research Foundation (FWO).

Section VI

Other Strategies

[34] HSV-1 Amplicon Vectors

By Sam Wang, Cornel Fraefel, and Xandra Breakefield

Introduction

Over the past several years, many different viral vectors including adenovirus, adeno-associated virus (AAV), retrovirus (including lentivirus), and herpes simplex virus (HSV) have been used to deliver genes into a variety of tissue types. These vectors differ in tropism, transgene carrying capacity, titer, toxicity, and transgene fate in host cells. Thus, the vector can be tailored for the intended use. In this chapter, the preparation of HSV-1 amplicon vectors will be discussed. These vectors have broad tropism (with high infectivity of neural cells), large transgene capacity, moderate titers, and low toxicity. Plasmid DNA packaged inside the virion is efficiently delivered to the cell nucleus and transgene fate can be controlled through incorporation of DNA informational elements.[1]

HSV-1 amplicons were first described by Spaete and Frenkel.[2] In essence, the amplicon contains two minimal elements from the HSV-1 virus genome acting in cis. These include an origin of DNA replication (*ori-S* or *ori-L*) and a DNA cleavage/packaging signal ("a" or *pac*), which allow the amplicon DNA to be replicated and to be packaged into virions when HSV-1 helper virus functions are provided, e.g., by a coreplicating helper virus. The amplicon also contains a bacterial origin of DNA replication and an antibiotic resistance gene that make the amplicon easily engineered using recombinant technology. The lack of any other viral coding sequences provides the opportunity to generate amplicon-containing HSV-1 virions (amplicon vectors), which do not express any viral genes and have little to no immunogenicity and toxicity.

Amplicon vectors should be able to accommodate large sequences of foreign DNA based on the size of HSV-1 genomic DNA, 152 kb. The range of packagable DNA is assumed to be from about 140 to 160 kb with the amplicon DNA consisting of a concatenate within the virion. Thus far, amplicon vectors containing transgene cassettes up to 40 kb have been described,[3] and amplicon elements in large capacity cloning plasmids (F-plasmids) are available.[4] These vectors have effected nontoxic and efficient gene delivery into a variety of tissues including brain, liver, and muscle using the green fluorescent protein (GFP) and lacZ as markers.[5–8] The techniques currently available for amplicon vector production give titers of

---

[1] M. Seña-Esteves, Y. Saeki, C. Fraefel, and X. O. Breakefield, *Mol. Ther.* **2,** 9 (2000).

[2] R. R. Spaete and N. Frenkel, *Cell* **30,** 295 (1982).

[3] X. Wang, G. R. Zhang, T. Yang, W. Zhang, and A. I. Geller, *Biotechniques* **28,** 102 (2000).

[4] Richard Wade-Martins, E. Antonio Chiocca, and Yoshinaga Saeki, unpublished results (2000).

[5] A. I. Geller and X. O. Breakefield, *Science* **241,** 1667 (1988).

10^6–10^8 amplicon transducing units/ml (tu/ml), and are suitable for most experiments in tissue culture and in rodent models designed to investigate the effects of focal gene expression. Current packaging methodology described in this chapter requires the rate-limiting step of transfection of DNA components (amplicon and helper-virus genome) into producer cells for vector production, which limits yields but is easily carried out by most laboratories under biosafety level 2(BL-2) conditions.

Helper-Virus Free Packaging System

Because amplicons contain no viral genes, they require helper-virus functions to produce the necessary components required for DNA replication, virion assembly, and packaging. These functions can be provided using replication-deficient HSV-1 mutants in producer cells stably transduced with the defective viral gene. However, this preparation typically results in the production of helper virus at comparable levels to the amplicon vector, which makes this system suboptimal for gene transfer studies, as gene expression even from the replication-defective helper virus can cause cytotoxic effects and immune responses.[9,10]

Amplicon vector preparation has been improved by the development of helper-virus free packaging systems. By using replication-competent but packaging-defective HSV-1 mutants, amplicon vectors can be prepared in the absence of or with minimal helper virus contamination. Two distinct systems have been developed using this paradigm: one using an overlapping five-cosmid set spanning the entire HSV-1 genome (C6Δa48Δa)[11,12] and another using an F-plasmid bacterial artificial chromosome (BAC) encoding the entire genome.[13,14] Both can be used to produce high titer amplicon vectors (Fig. 1). In each case, almost the entire HSV-1 genome is present, allowing the production of all viral

[6] L. C. Costantini, D. R. Jacoby, S. Wang, C. Fraefel, X. O. Breakefield, and O. Isacson, *Hum. Gene Ther.* **10**, 2481 (2000).

[7] C. Fraefel, D. R. Jacoby, C. Lage, H. Hilderbrand, J. Y. Chou, F. W. Alt, X. O. Breakefield, and J. A. Majzoub, *Mol. Med.* **3**, 813 (1997).

[8] Y. Wang, C. Fraefel, F. Protasi, R. A. Moore, J. D. Fesseneden, I. N. Pessah, A. DiFrancesco, X. O. Breakefield, I. A. Alion, and P. D. Allen, *Am. J. Physiol. Cell Physiol.* **278**, C619 (2000).

[9] A. I. Geller, K. Keyomarski, J. Bryan, and A. B. Pardee, *Proc. Natl. Acad. Sci. U.S.A.* **87**, 8950 (1990).

[10] F. Lim, D. Hartley, P. Starr, P. Lang, S. Song, L. Yu, Y. Wang, and A. I. Geller, *Biotechniques* **20**, 469 (1996).

[11] C. Cunningham and A. J. Davison, *Virology* **197**, 116 (1993).

[12] C. Fraefel, S. Song, F. Lim, P. Lang, L. Yu, Y. Wang, P. Wild, and A. I. Geller, *J. Virol.* **70**, 7190 (1996).

[13] Y. Saeki, T. Ichigawa, A. Saeki, E. A. Chiocca, K. Tobler, M. Ackermann, X. O. Breakefield, and C. Fraefel, *Hum. Gene Ther.* **9**, 2787 (1998).

[14] T. A. Stravopoulus and C. A. Strathdee, *J. Virol.* **72**, 7137 (1998).

A. BAC Helper   B. Cosmid Set C6Δa48Δa

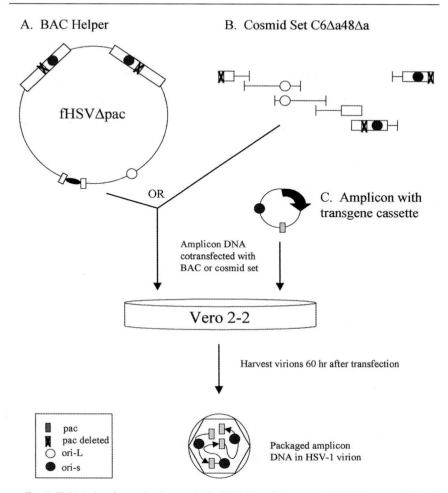

FIG. 1. Helper-virus free packaging system for HSV-1 amplicon vector. The HSV-1 genome deleted for pac is contained in either the five-cosmid (C6Δa48Δa) or BAC helper (fHSVΔpac). BAC (A) or cosmid (B) helper virus DNA is cotransfected with amplicon plasmid DNA (C) into Vero 2-2 cells. Amplicon vector is harvested 60 hr later. Both packaging methods provide all the trans-acting functions necessary for replication and packaging of the amplicon DNA, but do not provide the cis-acting functions necessary for packaging of the helper virus genome.

proteins necessary for DNA amplification and virion packaging. However, the packaging signals have been deleted from the helper virus genome to prevent it from being packaged; this results in virtually helper-virus free preparations of amplicon vectors. The ablation of helper virus in these preparations results in a vector stock that is safe for use in a wide variety of experimental protocols.

Materials and Methods

Cosmid Amplification

Helper virus functions to prepare helper-free amplicon vectors can be provided by either the five-cosmid set (C6Δa48Δa)[11,12] or an F-plasmid (e.g., pBAC-V2 or fHSVΔpac).[13,14] Cosmids are inherently unstable and great care must be exercised when preparing DNA from them. Single-colony amplification from plates for each cosmid is not recommended. Furthermore, to facilitate fidelity of cosmid replication each cosmid is cultured at 30°. Typically, for each of the five cosmids, a 10-ml starter culture of SOB[15] supplemented with 50 μg/ml ampicillin (SOB^{amp}) is amplified throughout the day beginning with a scraping from a frozen bacterial stock. The starter cultures are then transferred at the end of the day into 1-liter SOB^{amp} for overnight amplification. In contrast, although F-plasmids are much larger than cosmids, they are much more stable during the amplification stage. The F-plasmid DNA can be amplified similarly to the five-cosmid set with two major differences. Because the F-plasmid exists as a single copy in bacteria, large amounts of culture, e.g., 5 liters, must be prepared to generate useful quantities of DNA. Also, because of its large size, great care must be used to isolate the F-plasmid without shearing from bacterial genomic DNA. In this chapter, the procedure for amplicon preparation using the five-cosmid set is detailed.

DNA Purification

All cosmid and amplicon DNA samples are prepared using highly purified DNA, which has been banded twice in cesium chloride (CsCl) gradients using standard molecular biology protocols.[15] To ensure the highest titers of amplicon vector production, only DNA of the highest purity should be used in these protocols. In lieu of preparing twice-banded purified DNA, a smaller quantity of each cosmid DNA can be prepared quickly by growing a 300 ml culture overnight and purifying the DNA in two stages. The first stage involves the purification of DNA through an anion-exchange resin, provided by the Qiagen-tip 500 column (Qiagen, Valencia, CA), using a modified protocol. Transfection quality DNA for vector production is then obtained by a single CsCl centrifugation.

For the small quantity DNA purification, three 1-ml aliquots of the overnight culture are collected into three separate 1.5 ml cryogenic storage vials and 70 μl dimethyl sulfoxide (DMSO) is added to each vial. These stocks can be stored for many years at $-80°$. The remaining culture is centrifuged for 10 min at 4000g at 4° in a 250-ml polypropylene centrifuge bottle and then DNA is prepared using a modified Qiagen protocol as follows:

[15] J. Sambrook, E. F. Fritsch, and T. Maniatis, "Molecular Cloning," 2nd Ed. Cold Spring Harbor Press, Cold Spring Harbor, New York, 1989.

1. Decant the supernatant and invert each bottle on a paper towel. Allow the bottle to drain for 1–2 min.

2. Carefully resuspend the pellet in 15 ml cold P1 (the composition of all solutions is provided within the Qiagen kit; P1 contains 50 mM Tris, pH 8.0/10 mM EDTA/100 μg/ml RNaseA). Next, add 15 ml of P2 [0.2 N NaOH/1% sodium dodecyl sulfate (SDS)] and invert the bottle several times. Allow the cells to lyse completely, which should take no more than 5 min at room temperature. Add cold P3 (3 M potassium acetate, pH 5.5) to the mixture, gently invert several times, and incubate 30 min on wet ice, mixing occasionally.

3. Invert the tube once just before centrifugation at 25,000g at 4° for 30 min. While the sample is being centrifuged, equilibrate the column with 10-ml QBT (750 mM NaCl, 50 mM MOPS pH 7, 15% isopropanol) and allow the column to drain by gravity flow.

4. Once centrifugation is completed, carefully transfer the supernatant into the column by filtering through two layers of cheesecloth and allow the column to drain by gravity flow.

5. Wash the column twice with 30-ml QC (1 M NaCl, 50 mM MOPS pH 7, 15% isopropanol) and elute DNA into a 30-ml centrifuge tube by passing 15 ml QF (1.6 M NaCl, 50 mM Tris pH 8.5, 15% isopropanol), preheated to 65°, through the column.

6. Precipitate the DNA with 0.7 volume of isopropanol (10.5 ml) and immediately centrifuge 30 min at 25,000g at 4°. Carefully wash the pellet with 70% ethanol and centrifuge as before. Aspirate off the supernatant completely, but do not allow the pellets to dry.

7. Resuspend the pellet in 3 ml TE, pH 8.0.

The second phase of this DNA preparation involves a short centrifugation in a CsCl gradient as follows:

8. Add 3 g CsCl to the DNA solution followed by 300 μl 10 mg/ml ethidium bromide (EtBr) and carefully transfer into a Beckman UltraClear tube (144075). Seal the tube using the Beckman heat-sealing apparatus.

9. Centrifuge the mixture for 4 hr in a Vti65.2 Beckman rotor at 354,413g (60,000 rpm), 16°.

10. Following centrifugation, ventilate the sealed tube by puncturing the top with an 18-gauge needle. Then, carefully extract the lower band, which contains supercoiled DNA, using an 18-gauge needle attached to a 3 ml syringe inserted on the side of the tube just under the band.

11. Transfer the solution to a clear polystyrene tube and clean the solution by performing a minimum of three TE-saturated n-butanol extractions. Add one volume of TE-saturated n-butanol to the sample, thoroughly mix by inversion several times, and remove the top layer. Repeat until the bottom aqueous layer is clear.

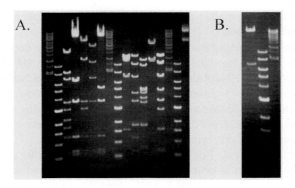

FIG. 2. Restriction analysis of cosmid set. (A) Diagnostic 0.4% analytical gel to confirm cosmid integrity. The samples are as follows: lanes 1, 8, 15, high molecular weight DNA marker (GIBCO); lanes 2, 9, 16, 1 kb ladder (NEB); lanes 3–7, *Dra*I digests of cosmids 6Δa, 14, 28, 48Δa, and 56, respectively; lanes 10–14, *Kpn*I digests of cosmids 6Δa, 14, 28, 48Δa, and 56, respectively; and lane 17, representative uncut cosmid. (B) Diagnostic 0.4% analytical gel representing the mixed cosmid set digested with *Pac*I restriction enzyme. Lane 1 represents the *Pac*I digested cosmid set. Lanes 2 and 3 are DNA size markers, 1 kb ladder (NEB) and high molecular weight DNA marker (GIBCO), respectively.

12. Once the solution has cleared, add 2.5 volumes deionized, double-distilled water to dilute the CsCl. Next, precipitate the DNA by adding 1/10 volume 3 M sodium acetate (NaAc), pH 5.2 and 2.5 combined volumes (sample volume + NaAc) 100% ethanol, and incubate at $-20°$ for 30 min. Centrifuge the mixture for 10 min at 12,000g, 4°. Then, carefully remove the supernatant and add 2 ml 70% ethanol to the pellet. Wash the pellet by swirling the tube and centrifuge again. Remove the supernatant and allow the pellet to air dry for a few minutes. At this point the DNA sample is considered clean. Remember to use aseptic technique from this point forward since the samples will be used in cell culture.

13. Do not allow the pellet to dry completely and resuspend the DNA pellet in 200 μl sterile TE overnight at 4°.

14. Take a small aliquot of the resuspended DNA and determine concentration using a spectrophotometer (μg/ml $= A_{260} \times 50 \times$ dilution factor). Dilute the DNA to 1 μg/μl using sterile TE. The A_{260}/A_{280} ratio should be greater than 1.8.

15. Verify each cosmid by digestion using either *Kpn*I or *Dra*I restriction enzymes and compare the size of fragments to those in Fig. 2A.

Vector Production

Cosmid Preparation. To prepare the DNA for transfection, add 10 μg of each cosmid into a single tube (50 μg, total) and digest with *Pac*I restriction enzyme overnight, according to the manufacturer's instructions. On the following day, take an aliquot and characterize the digest on a 0.4% agarose gel to confirm liberation of the 35–40 kb HSV-1 fragment from the cosmid backbone, ~7 kb (Fig. 2B). This *Pac*I restriction digest is crucial as it frees the HSV-1 inserts from the cosmid backbone and facilitates homologous recombination of the five different fragments to form replication-competent, packaging-defective HSV-1 genomes. The remaining DNA digest is cleaned with two phenol:chloroform:isoamyl alcohol (PCI, 25 : 24 : 1) extractions followed by a single chloroform:isoamyl alcohol (CI, 24 : 1) extraction in the following manner. One volume of either PCI or CI is added to the digest, mixed well, and centrifuged for 1 min at 12,000*g*; the upper, aqueous layer is collected for further manipulations. The samples should not be vortexed during the extractions, but rather gently tapped or inverted. Also, the use of gel-lock phase tubes (Eppendorf) to minimize DNA manipulation is helpful, but not necessary if extreme care is used during the extractions. Once the digests have been cleaned, the DNA is precipitated using 3 *M* sodium acetate and ethanol as described previously, centrifuged, washed with 70% ethanol, centrifuged again, and the pellet resuspended in 100 μl sterile TE. The DNA concentration of the digested cosmids is now ready to be quantitated by spectrophotometry, aliquotted, and stored at $-20°$ until ready for use. Typically, 12 μg of *Pac*I-digested cosmid DNA is aliquotted in 12–20 μl TE.

Tissue Culture. Vero2-2[16] and 293T/17 cells[17] are passed twice a week on 100 mm plates with 10% fetal bovine serum (FBS)/Dulbecco's modified Eagle's medium (DMEM) and 100 U/100 μg penicillin/streptomycin (with or without the presence of 250 μg/ml G418, respectively). The use of Vero2-2 cells is described here for packaging, although any cell type that is known to replicate HSV can be used to package amplicon vectors (e.g., BHK-21 cells). Cells should be used within a 15-passage interim to assure consistency.

Transfection. One day prior to transfection, Vero2-2 cells are passed 1 : 4 in the absence of G418 (approximately 3×10^6 cells/100-mm plate). A typical amplicon vector preparation utilizes two 100-mm plates. On the day of transfection, the plates should be confluent. Described below is a protocol for Lipofectamine (GIBCO)-mediated transfection. Other transfection reagents can be used, but they must be optimized for maximum transfection efficiency. For each 100-mm plate, carefully add 6 μg of the *Pac*I-digested cosmid DNA and 1.2 μg amplicon DNA into 300 μl OptiMEM in a polypropylene tube. In a separate tube, add 36 μl of Lipofectamine to 300 μl OptiMEM, also in a polypropylene tube. Mix both individual tubes well by gentle tapping, and then slowly drop the Lipofectamine solution into the tube containing the DNA mixture. Tap the tubes again to mix and allow the DNA–lipid complex to incubate at room temperature for 45 min. Just before the addition of the DNA–lipid complex to the cells, the cultures are

[16] I. L. Smith, M. A. Hardwicke, and R. M. Sandri-Goldin, *Virology* **186**, 74 (1992).

[17] W. S. Pear, G. P. Nolan, M. L. Scott, and D. Baltimore, *Proc. Natl. Acad. Sci. U.S.A.* **90**, 8392 (1993).

FIG. 3. Status of amplicon vector production at critical time points. (A) Representative transfection efficiency (typically 20–40%) approximately 16 hr after cotransfection of the cosmid set and a GFP-expressing amplicon using fluorescence microscopy. (B) Light photomicrographic representation of the desired cytopathic effect 60 hr after transfection. (C) Same field as in (B) visualized using fluorescence microscopy.

washed twice with OptiMEM and 3 ml fresh OptiMEM is added to each plate. Carefully transfer the 600 μl DNA–lipid complex into the plate (~3.6 ml final incubation volume) by slowly dropping the solution into the plate and incubate 6 hr at 37°. Rotate the plates 1/3 turn every 2 hr to ensure an even distribution over the cells. After the incubation period, the plates are washed three times with OptiMEM, recovered in 6 ml 6% FBS/DMEM/25mM HEPES, and maintained at 34° for 60 hr. One day posttransfection, most cells should show expression of the marker gene, comparable to Fig. 3A.

Virus Harvesting and Concentration. The virus is typically ready to be harvested approximately 60 hr posttransfection when the cells exhibit cytopathic effect, i.e., the transfected cells should be rounded in shape, clustered, and slightly detached from the plate (Fig. 3B and 3C). At this point the cells can be collected into polypropylene centrifuge tubes using a cell scraper and then sonicated (Fisher 550 Sonic Dismembrator, Pittsburgh, PA) for 30 s at 4°, using 5% power (Setting 3.5) to liberate the virus. Alternatively, the lysate can be freeze–thawed (dry-ice/ethanol bath and a 37° water bath) over three cycles. The lysate is then cleared by centrifugation at 3000g for 15 min at room temperature, and the supernatant is collected, aliquotted, and titered for experiments in cell culture. However, if higher titers or more purified vector is desired, the supernatant should be concentrated immediately. This is performed by adding 18 ml of DMEM to the cleared lysate (12 ml) and layering this solution onto a 25% sucrose cushion (5 ml). The virus is centrifuged for 3 hr at 87,275g (22,000 rpm) at 16° using the Beckman SW-28 rotor and the supernatant carefully aspirated off. PBS (300 μl) is added to the pellet and allowed to sit overnight at 4°. On the following day, the pellet is carefully resuspended in the PBS and stored in 25 μl aliquots. In both cases, the virus stock is quickly frozen in a dry ice–ethanol bath and transferred into −80° freezer for long-term storage.

Titering. Amplicon vector concentration is typically quantitated by determining the transducing units/ml (tu/ml), or the number of marker-expressing cells per unit vector volume one day after infection. Vectors that use GFP as a marker allow for easy determination of viral titer. However, any other histochemical or immuno-cytochemical marker (e.g., LacZ, gene of interest) can also be used. One day prior to titering, approximately 1×10^5 293T/17 cells in 0.5 ml/well are plated into a 24-well plate. On the following day, the cells are washed once with phosphate-buffered saline and then incubated with virus stock at various dilutions (e.g., 0.1 λ, 1 λ) in triplicate in 250–300 μl 2% FBS/DMEM overnight. GFP$^+$ cells are typically counted 18 hr later using a fluorescent microscope. Alternatively, the cultures can be fixed and stained using histochemical assays or immunocytochem-istry. The number of gene-expressing cells per well is multiplied by its dilution factor and expressed as (tu/ml) vector. Typical yields from two 100-mm plates are 10^{5-6} tu/ml for lysate supernatant (total volume, 12 ml) and from concentrated virus, 10^{7-8} tu/ml (total volume, 300 μl).

Gene Transfer Using Amplicon Vectors

Most cultured cells can be infected using unconcentrated viral lysates, where cellular debris, which can be toxic, is diluted out in the medium. The amount of vector required to result in efficient gene transfer must be determined experimen-tally for each cell type. Typically, cells are initially infected at several different multiplicities of infection (MOI) to determine the optimal dosage. In this case, one MOI represents one transducing unit (determined on the titering cells) per test cell. For example, Gli36, a glioblastoma-derived cell line (kindly provided by Dr. David Louis, Massachusetts General Hospital), can be infected at an MOI of 0.5 at a transduction efficiency of up to 80%,[18] whereas immunature hNT2/N neurons[19] at an MOI of 10 transduce significantly fewer cells, <5%.[20] To increase virus–cell interactions, cell cultures are typically infected in a minimal volume under low-serum (2%) or serum-free conditions for 1–2 hr at 37° followed by virus medium removal and cell recovery in normal growth media. Depending on cell type and MOI, the virus may be kept in the culture vessel without removal with, in many cases, no obvious signs of toxicity.

Current State of Amplicon Vectors

One of the rate-limiting factors for amplicon vector usage is the ability to generate large volumes of high titered stock. With the initial step dependent on a transfection, which can be inefficient (less than 50% of the cells transfected), large-scale virus preparation becomes expensive, time-consuming, and variable from stock to stock. However, for most experimental paradigms in culture or

[18] Paula Lam, unpublished results (2000).
[19] S. J. Pleasure, C. Page, and V. M. Lee, *J. Neurosci.* **12**, 1802 (1992).
[20] Sam Wang, unpublished results (2000).

rodent models, the virus generated using these protocols is sufficient. New BACs are currently being developed to increase vector production.[21] Alternatively, the development of packaging cell lines and helper virus variants, being investigated, should also boost amplicon vector production.

Several HSV-1 amplicon-based hybrid vectors are currently being developed to extend and control transgene expression. On infection, amplicon vectors typically deliver the transgene cassette as a free, extrachromosomal element into the cell nucleus. This "episome" is lost over time in dividing cells, as it has no means of replication without helper virus functions and no means of distribution to daughter cells. In nondividing cells, expression of the transgene decreases over time and eventually disappears because of degradation of the vector and/or inactivation of the promoter. HSV/AAV hybrids have been designed for long-term integrative gene delivery.[22] By incorporating the transgene between the AAV inverted terminal repeat elements, it is expected that the transgene will be able to integrate into the host chromosome, and inclusion of the AAV *rep* open reading frame within the amplicon should result in favored integration at the AAVS1 site in human chromosome 19,[23] thereby stabilizing the transgene cassette as part of the host genome.

Alternatively, amplicons incorporating Epstein–Barr virus (EBV) elements have also been developed to extend transgene expression.[24] In this case, the ori-P and EBNA-1 elements from EBV promote amplicon replication in synchrony with the host cell genome as an episomal element, which allows distribution to the daughter cells and extends gene expression in the cells. Additionally, HSV/EBV/retrovirus tribrids have been developed to extend transgene expression further through the production of retrovirus vectors by amplicon-infected cells over an extended period, and stable integration of retrovirus sequences into dividing cells.[25]

The HSV-1 amplicon vector is a highly efficient tool for delivering genes to cells both in culture and *in vivo*. These vectors have a large transgene capacity, are easy to manipulate, do not appear to stimulate an immune response or be toxic, and can infect a wide range of different types of cells. Of all the virus vectors currently in use, they are the simplest to generate and package in a research laboratory and provide an exceptionally versatile tool for elucidating structure–function relationships and metabolic/physiologic functions. One current limitation of this system is the difficulty in scaling up amplicon vector production and to

[21] Y. Saeki, C. Fraefel, T. Ichikawa, X. O. Breakefield, and E. A. Chiocca, *Mol. Ther.* **3**, 591 (2001).

[22] K. M. Johnston, D. Jacoby, P. Pechan, C. Fraefel, P. Borghesani, D. Schuback, R. J. Dunn, F. I. Smith, and X. O. Breakefield, *Hum. Gene Ther.* **8**, 359 (1997).

[23] R. M. Kotin, M. Siniscalo, R. J. Samulski, X. D. Xhu, L. Hunter, C. A. Laughlin, N. Muzyczka, M. Rocchi, and K. I. Berns, *J. Virol.* **87**, 2211 (1990).

[24] S. Wang and J. Vos, *J. Virol.* **70**, 8422 (1996).

[25] M. Sena-Esteves, Y. Saeki, S. M. Camp, E. A. Chiocca, and X. O. Breakefield, *J. Virol.* **73**, 10426 (2000).

generate highly purified vector for large-scale studies in large animals or clinical trials. In time this drawback will be rectified and amplicon vectors will move into the mainstream of the gene therapy repertoire.

Acknowledgment

This work was supported by NIH CA69246 and NIH CA99004.

[35] Microencapsulation of Genetically Engineered Cells for Cancer Therapy

By J.-Matthias Löhr, Robert Saller, Brian Salmons, and Walter H. Günzburg

Introduction

Gene therapy is usually regarded as the transfer of new genetic information into target cells with resulting therapeutic potential. However, in practice, this conventional definition is often widened to include forms of therapeutic vaccination and cell therapy in which nonautologous genetically modified cells are transplanted into a patient, rather than attempting to modify the patient's own cells. On a conceptual level, such an approach could offer major advantages over conventional gene therapy since it would allow the controlled placement of the production site of therapeutic molecules (viral vectors, antibodies, cytokines, other proteins) and prohibit spillover of the transgene into the general system (systemic circulation) of the host.[1] However gene therapy in general and cancer gene therapy in particular are hampered by several obstacles including (1) the controlled expression of the therapeutic gene at the right time and place and (2), in the case of transplantation of genetically modified cells, the maintainance of the integrity of the transplanted cells and their protection from the host immune system. The first problem is generally tackled by searching for tissue or cell type-specific promoters and engineering them to control the therapeutic gene, usually in the context of a vector.[2-6] The second problem has not yet been satisfyingly tackled and, depending

[1] A. M. Sun, *Ann. N.Y. Acad. Sci.* **831,** 271–279 (1997).
[2] W. H. Günzburg and B. Salmons, *Mol. Med. Today* **1,** 410 (1995).
[3] K. W. Peng and S. J. Russell, *Curr. Opin. Biotechnol.* **10,** 454 (1999).
[4] R. J. Yanez and A. C. Porter, *Gene Ther.* **5,** 149 (1998).
[5] D. M. Nettelbeck, V. Jerome, and R. Müller, *Trends Genet.* **16,** 174 (2000).
[6] W. H. Günzburg and B. Salmons, *J. Mol. Med.* **74,** 171 (1996).

on the origin of the transplanted cells, they have recently been found to harbor additional risks such as the production of potentially pathogenic species-jumping viruses.[7]

One solution to the problem of maintaining transplanted cells could be to gather the transfected cells in a distinct containment and, by the same token, shield the genetically engineered cells from the host immune system. Microencapsulation was envisioned more than 40 years ago,[8,9] however, because of the lack of biocompatible procedures for preparing microcapsules capable of containing viable cells, it took another 30 years before living cells (islets of Langerhans) could be encapsulated.[10]

We have established a system to encapsulate genetically engineered cells expressing the 2B1 isomer of the rat cytochrome p450 gene (CYP2B1), which is capable of converting ifosfamide into its active compounds, phosporamide mustard and acrolein.[11] In contrast to the widely used alginate/L-polylysine system,[27] we utilized sodium cellulose sulfate (NaCS) as polyanion and poly(diallyldimethyl-ammonium chloride) (PDADMAC) as polycation.[12] We have exploited this system for cancer gene therapy,[13,14] but it represents a basic technology for a whole variety of other clinical aplications.

Pancreatic cancer served as our model system since it is a devastating disease[15] in which death is usually caused by the primary tumor, and even emerging therapeutic concepts[16,17] have not yet contributed to a significant improvement of patient survival or quality of life. Chemotherapy is limited in its effectivity by (1) access to the tumor, (2) the degree of drug sensitivity of the tumor, and (3) local and systemic toxicity of the chemotherapeutic agent. In pancreatic carcinoma, chemotherapy is mainly delivered systemically,[18] although recent developments include the placement of intraarterial catheters delivering the cytotoxic drugs into the celiac

---

[7] W. H. Günzburg and B. Salmons, *Mol. Med. Today* **6,** 199 (2000).

[8] T. M. S. Chang, "Hemoglobin Corpuscles." Research report for Honours Physiology, Medical Library, McGill University, McGill University Press, Montreal (1957).

[9] T. M. S. Chang, *Ann. N.Y. Acad. Sci.* **831,** 249 (1997).

[10] F. Lim and A. M. Sun, *Science* **210,** 908 (1980).

[11] H. A. A. M. Dirven, B. van Ommen, and P. J. van Blaederen, *Chem. Res. Toxicol.* **9,** 351 (1996).

[12] H. Dautzenberg, U. Schuldt, G. Grasnik, P. M. P. Karle, M. Löhr, M. Pelegrin, M. Piechacyk, K. v. Rombs, W. H. Günzburg, B. Salmons, and R. M. Saller, *Ann. N.Y. Acad. Sci.* **875,** 46 (1999).

[13] M. Löhr, P. Müller, P. Karle, J. Stange, S. Mitzner, H. Nizze, S. Liebe, B. Salmons, and W. H. Günzburg, *Gene Ther.* **5,** 1070 (1998).

[14] T. Kammertöns, W. Gelbmann, P. Karle, B. Salmons, W. H. Günzburg, and W. Uckert, *Cancer Gene Ther.* **7,** 629 (2000).

[15] S. Rosewicz and B. Wiedenmann, *Lancet* **349,** 485 (1997).

[16] W. H. Günzburg and B. Salmons, *Trends Mol. Med.* **7,** 30 (2000).

[17] L. Rosenberg, *Drugs* **59,** 1071 (2000).

[18] J. D. Ahlgren, *Cancer* **78,** 654 (1996).

trunk.[19–22] On the other hand, instillation of solid microspheres for chemoembolization is a routine procedure for hepatic masses not suitable for surgery.[23–25]

With this background, we reasoned that local application of microencapsulated cells capable of locally converting ifosfamide to its active, antitumor metabolites might result in good antitumor efficacy. Since much of the systemically applied ifosfamide should be locally converted at the tumorsite, rather than relying, as in conventional chemotherapy, on conversion at the natural site of expression of the CYP2B1 enzyme, the liver, followed by systemic distribution of the relatively short-lived antitumor metabolites, we expected to be able to reduce the dose of ifosfamide used, resulting in reduced systemic side effects, without losing antitumor efficacy.[29] For application in humans, a supraselective intraarterial instillation in tumor-vascularizing arteries seemed to provide a minimally invasive approach for the exact placement of encapsulated cells.

Methodology and Results

Vector Construct

The cDNA coding for the CYP2B1 enzyme[26] was cloned into the plasmid pcDNA3 and the resulting expression plasmid pC3/2B1, in which the 2B1 cytochrome isoform is expressed from a linked cytomegalovirus promoter,[31] is amplified in *Escherichia coli*. Preparation was performed using the Maxi Preparation Kit and the Pyrogen Extraction Kit (both Qiagen, Germany). Sequencing confirmed the integrity of the insertion and of the plasmid vector DNA. Subsequent preparations were quantified by photometric and agarose gel analysis, and identity was confirmed by restriction enzyme digestion.

Plasmid pC3/2B1, which carries a neomycin (G418) resistance gene, was transferred into qualified (tested free of a panel of viruses and other adventitious agents,

[19] R. Kawasaki, S. Morita, Y. Noda, and A. Tsuji, *Cardiovasc. Interventional Radiol.* **21,** 152 (1998).

[20] O. Ishikawa, H. Ohigashi, Y. Sasaki, K. Masao, T. Kabuto, H. Furukawa, and S. Imaoka, *Hepatogastroenterology* **45,** 644 (1998).

[21] C. A. Maurer, M. M. Borner, J. Lauffer, H. Friess, K. Z'graggen, J. Triller, and M. W. Buchler, *Int. J. Pancreatol.* **23,** 181 (1998).

[22] F. K. Wacker, J. Boese-Landgraf, A. Wagner, D. Albrecht, K. J. Wolf, and F. Fobbe, *Cardiovasc. Intervent. Radiol.* **20,** 128 (1997).

[23] P. Berghammer, F. Pfeffel, F. Winkelbauer, C. Wiltschke, T. Schenk, J. Lammer, C. Muller, and C. Zielinski, *Cardiovasc. Intervent. Radiol.* **21,** 214 (1998).

[24] F. Florio, M. Nardella, S. Balzano, E. Caturelli, D. Siena, and M. Cammisa, *Cardiovasc. Intervent. Radiol.* **20,** 23 (1997).

[25] Groupe d'Etude et de Traitement du Carcinome Hepatocellulaire, *N. Engl. J. Med.* **332,** 1256 (1995).

[26] K. M. Kedzie, C. A. Balfour, G. Y. Escobar, S. W. Grimm, Y. A. He, D. J. Pepperl, J. W. Regan, J. C. Stevens, and J. R. Halpert, *J. Biol. Chem.* **266,** 22515 (1991).

Q-One, Glasgow, UK) 293 cells (human embryonic kidney cell line; ATCC) by transfection using a Pharmacia Transfection Kit according to the manufacturer's instructions. Subsequently the cells were grown for 2 days and then selected for G418 resistance. Resistant clones were isolated and tested for the presence and activity of the vector.

The expression of biologically active CYP2B1 in the transfectants was determined using a biochemical assay, which is specific for the cytochrome P450 isoforms 1A1 and 2B1.[27] In this assay 7-pentoxy resorufin is converted by CYP2B1 into resorufin. The amount of produced resorufin was measured with a fluorometer at 530 nm excitation and 590 nm emission.[13] A standard curve was produced using different amounts of purified resorufin (Sigma). The cell clone 22P1G,[28] which showed the highest enzymatic activity, was chosen for further experimentation.

Encapsulation

For encapsulation, CYP2B1 transfected 22P1G cells are suspended in sodium cellulose sulfate (NaCS) or phosphate-buffered saline (PBS, pH 7) containing 2–5% NaCS (Fig. 1A) depending on the degree of sulfation (polyanionic solution). The suspension is passed through an adjustable droplet generation system (Inotech AG, Switzerland). The cell-containing droplets eventually pass into a precipitation bath containing 3–4% polydiallyldimethylammonium (PDADMAC; Fig. 1B) in NaCl depending on the concentration of the NaCS (polycationic solution). On contact of the polyanion with the polycationic solution, a polyelectrolyte complex starts forming producing a capsule membrane, which forms from the outside toward the center (Fig. 2A). Histological sectioning of the capsules followed by microscopic analysis reveal that the cells are distributed throughout the capsule, however they are more concentrated in the center, while the polymerized cellulose sulfate is concentrated toward the surface of the capsule (Fig 2B). The capsules are then washed twice with saline before they are stored at 4°.[13]

Survival of cells is measured according to the manufacturer's recommendations with the two-color LIVE/DEAD Viability/Cytotoxicity Kit (Molecular Probes), which shows a green fluorescence for living cells and a red fluorescence for dead cells in confocal laser microscopy (Fig. 2C).

The capsules had an average diameter of 0.7–0.85. They can withstand considerable physical pressure and are able to withstand delivery through an angiography catheter. *In vitro* and *in vivo* monitoring revealed that the capsules are flexible enough to take different shapes while remaining intact. Preparations of capsules were tested in several ways. One experiment consisted of forcing capsules through different diameters of needles attached to a syringe. The capsules

[27] V. Kurowski and T. Wagner, *Cancer Chemother. Pharmacol.* **33**, 36 (1993).
[28] W. H. Günzburg, P. Karle, R. Renz, B. Salmons, and M. Renner, *Ann. N.Y. Acad. Sci.* **880**, 326 (1999).

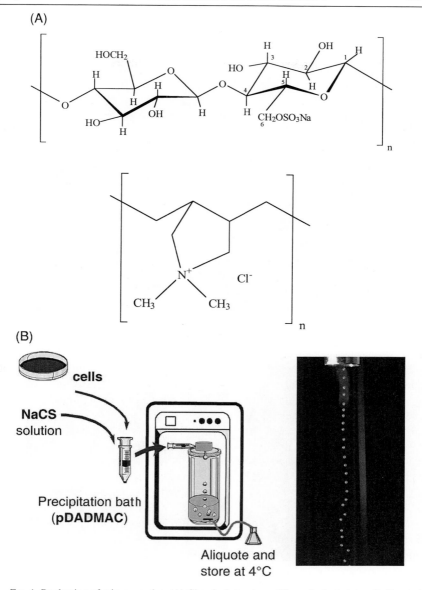

FIG. 1. Production of microcapsules. (A) Chemical structure of the polyelectrolytes. Sodium cellulose sulfate (NaCS; upper structure) and poly(diallyldimethylammonium chloride) (PDADMAC; lower structure). (B) Schematic of the encapsulation process. A suspension of a minimum of 3×10^8 cells is mixed with NaCs in a loading vessel and injected into the vibrating nozzle of an encapsulation machine. The resultant droplets pass into a polymerization solution of PDADMAC where they solidify and then are washed and aliquotted.

(A)

(B)

(C)

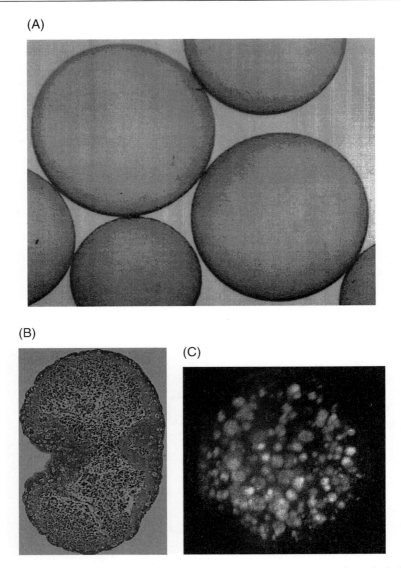

Fig. 2. Capsules and encapsulated cells *in vitro*. (A) Phase contrast microscopy of capsules lacking cells (empty) with a diameter of ~0.7 mm. (B) H&E staining of a section of encapsulated cells. (C) LIVE/DEAD assay of encapsulated cells in confocal laser microscopy 24 hr after encapsulation demonstrating that most of the cells retain viability (white cells).

could pass through needle diameters down to 15-gauge without demonstrating leakage of cells in subsequent tissue culture. Furthermore, capsules were instilled into angiography catheters in isotonic contrast medium (Visipaque) to test the pressure that resulted after loading up to 1000 capsules into a standard 2.3F/140 cm angiography catheter (Cordis). Capsules could be easily pushed by hand through the catheter. The capsules were returned to tissue culture after this experiment. They retained integrity, i.e., no outgrowth of cells could be observed. The cells remained viable (LIVE/DEAD assay) at the same level as prior to the experiment. The same was true for the enzyme activity (resorufin assay).

Animal Models

Initial studies were aimed at detection of immediate or delayed toxic effects in rodents. In order to assess acute toxicity, ground capsule material was injected at a concentration of 2000 mg/kg body weight into 10 immunocompetent mice. Animals were followed for 14 days; however, no toxicity was observed in any of the animals. Chronic toxicity studies were carried out in rabbits, which were injected intraperitoneally with up to 10 times the planned clinical dose. The animals were monitored for weight loss, fever, or any abnormal behavior, but no toxicity could be detected, even at the histological level, after a period of 3 months, indicating that the capsule material is well tolerated over the long term. To ascertain the tolerance to the capsule material at the planned site of clinical application, empty capsules were injected orthotopically into the pancreas of both nude and immunocompetent mice.[29] A mild foreign body reaction was seen surrounding the capsules; however, no systemic reaction or granuloma formation could be observed[30] (Fig. 3A).

The antitumor effect of the administration of microencapsulated CYP2B1-producing cells followed by systemic ifosfamide application was investigated using encapsulated CYP2B1-expressing feline kidney cells and has been reported previously.[31] Briefly, a suspension of 1×10^6 PANC-1 cells was injected subcutaneously into athymic nude mice.[31] Once tumors reached a volume of 1 cm^3, capsule implantation took place. The capsules were carefully aspirated from a container with normal medium without supplements or fetal calf serum into a 1 ml syringe (insulin syringe; Braun, Melsungen, Germany) without a needle attached to it. Twenty to 40 capsules were injected into the preformed tumor through a 21-gauge needle (Microlance 3, inner diameter 0.6 mm) that was shortened to a 1 cm needle length, resharpened, and sterilized. Animals were then treated with low-dose ifosfamide and MESNA (both at 100 mg/kg) i.p. every 3 days for 2 weeks.[13] This

[29] X. Fu, F. Guadagni, and R. M. Hoffman, *Proc. Natl. Acad. Sci. U.S.A.* **89**, 5645 (1992).

[30] P. Müller, R. Jesnowski, P. Karle, R. Renz, R. Saller, H. Stein, K. Püschel, K. von Rombs, H. Nizze, S. Liebe, T. Wagner, W. Günzburg, B. Salmons, and M. Löhr, *Ann. N.Y. Acad. Sci.* **880**, 337 (1999).

[31] M. Löhr, B. Trautmann, M. Göttler, S. Peters, I. Zauner, B. Maillet, and G. Klöppel, *Br. J. Cancer* **69**, 144 (1994).

(A)

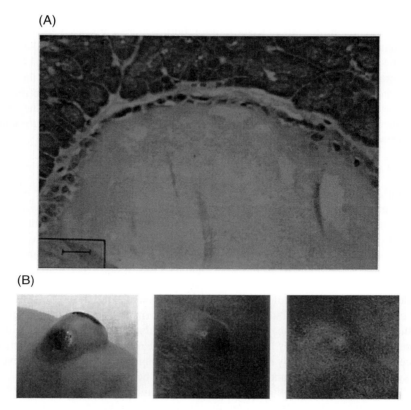

(B)

FIG. 3. Capsules and encapsulated cells *in vivo*. (A) Mouse pancreas 1 week after implantation of an empty capsule (H&E stain). (B) Gross appearance of xenotransplanted human pancreatic carcinoma on nude mouse without treatment (left), partial remission (middle), and complete remission (right) after instillation of CYP2B1-expressing, microencapsulated cells and ifosfamide treatment for 3 weeks.

resulted in a complete remission of the established tumors in almost 20% of the animals and partial remission in another 50%[13] (Fig. 3B).

In order to quantitate the added value of the local conversion of ifosfamide, the former experiment was repeated, but this time 40 capsules were injected on one side of the tumor only, which was marked. Animals were then treated with a single dose of ifosfamide (100 mg/kg). After 30 min, i.e., reaching the plateau phase in blood plasma (Fig. 4A), animals were anesthetized and tumor tissue was removed in three parts: side of the capsule implantation, middle portion, and the side opposite to the capsule implantation. Ifosfamide and its active metabolites were measured in the snap-frozen tissue as well as in blood plasma. Although the levels of ifosfamide were the same at all three sites (Fig. 4B), 4-hydroxyifosfamide was at least two times higher where the capsules were implanted (Fig. 4C).

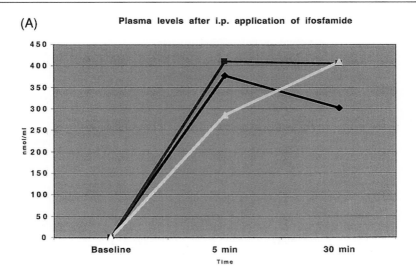

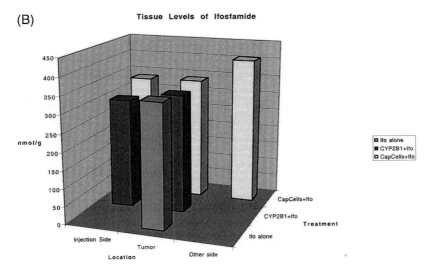

FIG. 4. Pharmacokinetics of ifosfamide in xenografted human pancreatic carcinomas on the nude mouse. (A) Plasma levels of ifosfamide in three groups of animals carrying tumors and having received ifosfamide only (ifo alone ■), naked CYP2B1-expressing cells (CYP2B1+ifo △), or microencapsulated CYP2B1-expressing cells (CapCells +ifo ◆). (B) Tissue levels of ifosfamide for the three groups described in panel A. In all groups, the levels are about the same. (C) Tissue levels of the metabolized, activated ifosfamide, i.e., 4-hydroxyifosfamide, in the three groups. The levels are almost three times higher in the group treated with the encapsulated CYP2B1 cells and 50% higher at the site of the capsule implantation.

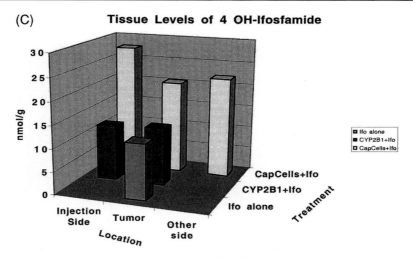

FIG. 4. (*continued*)

Angiographic Route of Delivery

Adolescent pigs (mean age 90 days, mean weight 46 kg) were kept fasting overnight. Capsules were produced as described above.[13,32,33] After encapsulation, the capsules were washed, aliquotted at 100 capsules per vial, and kept at 4° until further use. A parallel sample was assessed for viability and enzyme activity as described.[32] For angiography, animals were sedated and placed in a supine position on the angiography table. Animals were intubated and ventilated. The femoral artery was punctured after open preparation with a 20-gauge angiography needle. A 4F introducer system (Terumo) was placed in by the Seldinger technique.[34] Under fluoroscopy, the celiac trunc was catheterized with a 4F Cobra 2 guiding catheter (Cordis). Supraselective cannulation was achieved by further advancement of the guidewire and a coaxial 2.3F microcatheter system (Cordis). In general, the splenic lobe was cannulated (Fig. 5A). Initial angiography was performed with Visipaque 270 (Nycomed). After positive placement of the catheter in the main vessel leading into the splenic lobe of the pancreas, 100 capsules were instilled slowly. At the end, control angiography was performed. The introductory set was removed, the wound closed, and the animals were allowed to

[32] P. Karle, P. Müller, R. Saller, K. von Rombs, R. Renz, H. Nizze, S. Liebe, W. H. Günzburg, B. Salmons, and M. Löhr, *Adv. Exp. Med. Biol.* **451,** 97 (1998).

[33] J. Stange, S. Mitzner, H. Dautzenberg, W. Ramlow, M. Knippel, M. Steiner, B. Ernst, R. Schmidt, and H. Klinkmann, *Biomat. Art. Cells. Immob. Biotech.* **21,** 343 (1993).

[34] J. C. Kröger, H. Bergmeister, A. Hoffmeyer, M. Ceijna, P. Karle, R. Saller, I. Schwendenwein, K. von Rombs, S. Liebe, W. H. Günzburg, B. Salmons, K. Hauenstein, U. Losert, and M. Löhr, *Ann. N.Y. Acad. Sci.* **880,** 374 (1999).

(A)

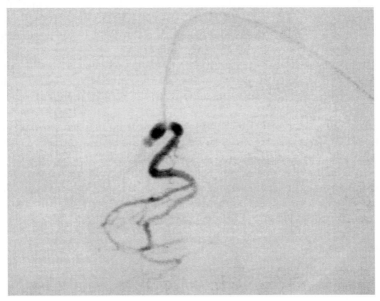

(B)

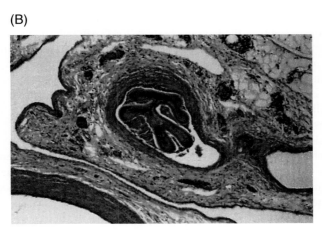

Fig. 5. Preclinical feasibility studies for angiographic access in the pig. (A) Supraselective angiography of the splenic lobe of the pig pancreas. (B) Histological appearance of the intra-arterially instilled microcapsules (H&E stain).

wake. Animals were monitored clinically and by lab tests for a week. During the first 24 hr, the pigs were kept NPO (nothing *per os*) but received saline infusions (500 ml/12 hr).

A total of 18 animals were investigated. In all animals, it was possible to cannulate both the splenic and the duodenal lobe arteries selectively. In all animals, instillation of capsules was successful. Manipulation with the guidewire, injection of contrast dye, and instillation of capsules resulted in a significant and prolonged vascular spasm that resolved spontaneously within 6 to 15 minutes. None of the animals developed pancreatic symptoms, but the first animal experienced slight tenderness of the abdomen. The amylase levels in all animals remained within normal limits. At the end of the observation period, another control angiography was performed. In addition, ifosfamide was administered systemically at low dosage (1 g/m^2). Finally, the animals were sacrificed and the pancreas and other inner organs were removed. There was no visible macroscopic or microscopic damage to the pancreas. The capsules were found in capillaries, partly in conjunction with thrombotic material but never occluding the entire vessel (Fig. 5B).

Studies in Humans

A nonrandomized, phase I/II study was designed for patients with advanced-stage pancreatic carcinoma not suitable for curative surgery.[29] The study was approved by the state Ethics Committee, the Somatic Gene Therapy Commission of the German Federal Medical Association (KSG-BAK), the Working Party for Oncology (AGO) of the German Society for Gastroenterology (DGVS), and the German Working Party for Gene Therapy (DAG-GT). For the clinical study, the entire process of culturing the cells carrying the therapeutic gene and the encapsulation process was transferred to a contract research organization to ensure compliance with GCP-GCH guidelines. The cells were cultured under constant GMP conditions. After passing the requested quality assurance tests, the capsules were released for clinical use (Fig. 6A).

The selected patients underwent angiography after giving informed consent. Angiography for celiac trunc visualization was performed in a standard manner via the femoral route. A 4F introducer system (Terumo) was placed by the Seldinger technique. Under fluoroscopy, the celiac trunk and splenic and common hepatic arteries, as well as the superior mesenteric artery (SMA), were visualized with Visipaque 270 (Nycomed). Relating the localization of the tumor in the CAT scan to the vessel anatomy, the most appropriate access to the tumor was determined. Supraselective catheterization of the transversal artery was performed with a coaxial 2.3F microcatheter system (Cordis).[34] After documentation of the correct placement, 300 microcapsules were slowly instilled through the catheter one by one. For this procedure, a translucent Luer-lock connector on the catheter is mandatory (Fig. 6B). To have best control over the capsule instillation, 1-ml insulin syringes were utilized. The capsule instillation was followed by another control angiography and documentation of the correct catheter placement (Fig. 6C). Finally, the catheters and the introducer set were withdrawn.

Patients rested for 48 hr. During this period, the patient was monitored clinically by follow-up lab tests and abdominal ultrasound in order to detect any abnormality in the upper abdomen, e.g., pancreatitis or ischemia.

Patients received ifosfamide at 1 g/m^2 body surface for three consecutive days (day 2–4). This regimen was repeated starting at day 22. The application

(A)

FIG. 6. Clinical study in humans. (A) Aspiration of capsules with an insulin syringe from a container from a clinical batch. Capsules freely float in the syringe. (B) Instillation of the capsules into a 2.3F microcatheter. Note the translucent adaptor piece of the catheter allowing the visual control of the capsule instillation (arrow). (C) Supraselective angiography of the arteria transversalis (branch of the A. pancreatoduodenalis).

(B)

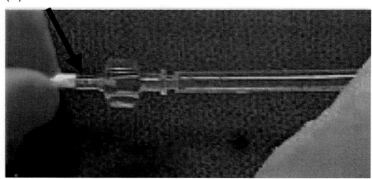

(C)

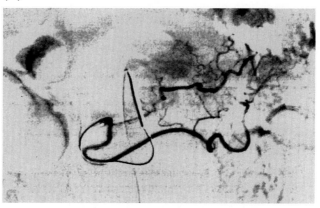

FIG. 6. (*continued*)

of ifosfamide was accompanied by monitoring of the drug and its metabolites. Routine laboratory monitoring according to oncological standards was performed during the entire period. Patients were followed up for a total of 5 months at least. A total of 14 patients were treated according to protocol. The procedure was well tolerated. None of the patientsdeveloped any signs of allergic reaction or pancreatitis. The treatment resulted in two partial remissions and stabilization in the remainder of the patients. The survival time was extended from 25 months (recent control group at study center) to 44 months.[35]

[35] M. Löhr, A. Hoffmeyer, J. C. Kröger, M. Freund, J. Hain, A. Holle, P. Karle, W. T. Knöfel, S. Liebe, P. Müller, H. Nizze, M. Renner, R. M. Saller, B. Salmons, T. Wagner, K. Hauenstein, and W. H. Günzburg, *Lancet* **357**, 1591 (2001).

Discussion

With the series of experiments described, we could demonstrate positive proof for the use of genetically engineered, microencapsulated cells in cancer therapy with pancreatic carcinoma as a model system. Although several gene therapy approaches have been described in experimental pancreatic cancer,[36–40] this is one of the very few gene/cell therapy protocols that reached clinical trial.[16]

In animal models, encapsulation using biodegradable polylactic acid microspheres releasing recombinant interleukin-12 (IL-12) has been used for *in situ* tumor immunotherapy leading to tumor reduction.[41] However, this was a cell-free system relying on a relatively short-term release of IL-12. The microcapsules described in this paper have also been shown to be capable of secreting antibodies produced by encapsulated hybridoma cells[42] even resulting in protection from a virally induced disease[43] and could be used to deliver other biomolecules. Another new and promising delivery technique encompasses the application of gene activated matrix (GAM) directly to target tissue.[44] Although this is feasable for superficial sites such as wounds, it does not allow delivery to inner parts of the body.

The general route of *in vivo* delivery of therapeutic agents, including gene therapy reagents, is direct injection into a body compartment (e.g., abdominal cavity) or into a peripheral vein. In addition, catheter-directed gene delivery has been investigated with application of adenoviral vectors to swine lung arteries demonstrating reporter gene for several weeks.[45] The vascular delivery of genetic vectors for prevention of restenosis and coronary angiogenesis has used a wide range of devices and methods.[46–48] Although gene transfer has been demonstrated

[36] K. Makinen, S. Loimas, J. Wahlfors, E. Alhava, and J. Janne, *J. Gene. Med.* **2,** 361 (2000).

[37] K. Makinen, S. Loimas, V. M. Kosma, J. Wahlfors, S. Yla-Herttuala, E. Alhava, and J. Janne, *Ann. Chir. Gynaecol.* **89,** 99 (2000).

[38] H. Kijima and K. J. Scanlon, *Mol. Biotechnol.* **14,** 59 (2000).

[39] D. Evoy, E. A. Hirschowitz, H. A. Naama, X. K. Li, R. G. Crystal, J. M. Daly, and M. D. Lieberman, *J. Surg. Res.* **69,** 226 (1997).

[40] J. M. DiMaio, B. M. Clary, D. F. Via, E. Coveney, T. N. Pappas, and H. K. Lyerly, *Surgery* **116,** 205 (1994).

[41] N. K. Egilmez, Y. S. Jong, M. S. Sabel, J. S. Jacob, E. Mathiowitz, and R. B. Bankert, *Cancer Res.* **60,** 3832 (2000).

[42] M. Pelegrin, M. Marin, D. Noel, M. Del Rio, R. Saller, S. Stange, S. Mitzner, W. H. Günzburg, and M. Piechaczyk, *Gene Ther.* **5,** 828 (1998).

[43] M. Pelegrin, M. Marin, A. Oates, D. Noel, R. Saller, B. Salmons, and M. Piechaczyk, *Hum. Gene Ther.* **11,** 1407 (2000).

[44] J. Bonadio, *J. Mol. Med.* **78,** 303 (2000).

[45] S. Badran, S. K. Schachtner, H. S. Baldwin, and J. J. Rome, *Hum. Gene Ther.* **11,** 1113 (2000).

[46] M. Simons, R. O. Bonow, J. Chrono, H. K. Hammond, R. J. Laham, W. Li, M. Pike, F. W. Sellke, T. J. Stegmann, J. E. Udelson, and T. K. Rosengart, *Circulation* **102,** e73 (2000).

[47] S. Nikol, T. Huehns, E. Krausz, S. Esin, M. G. Engelmann, D. Winder, B. Salmons, and B. Höfling, *Gene Ther.* **6,** 737 (1999).

for each device,[49] most studies of catheter-based gene transfer have revealed low efficiency except in the context of prolonged vessel occlusion with ligated branches or a double-balloon catheter.[50] However, this concept is not applicable to anything else but a blood vessel, certainly not for cancer gene therapy of tumors in solid organs. Here, the placement of cells contained in a microcapsule represents a definite advantage.

Outlook

The capsule technology employed here, though already mature, has still considerable room for optimization. Furthermore, application of capsules containing CYP2B1 expressing cells in conjunction with ifosfamide, as used in our experiments, may prove effective for other cancers, e.g., peritoneal metastasis/malignant ascites, soft tissue sarcoma, and hepatic masses (primary and metastatic).

Acknowledgments

This work was supported by the Danish Government (Væstfond), a project Grant by Bavarian Nordic A/S, Copenhagen, Denmark, the Minister of Science and Education, Berlin, Germany (bmbf BOE 21–03113673), and EC Grant BIO4-CT-0100. This work is dedicated to our patients with pancreatic carcinoma who joined our inaugural study, and to Professor Horst Dauzenberg, the entrepreneur of capsule technology, who died much too early. The work with the capsules owes much to many investigators who contributed substantially to the many aspects of this project. Their input is gratefully acknowledged: Zoltan Bago, Stephan Benz, Mathias Freund, Johannes Hain, Karlheinz Hauenstein, Anne Hoffmeyer, Ralf Jesnowski, Jens Kröger, Karle, Stefan Liebe, Udo Losert, Petra Müller, Steffen Mitzner, Horst Nizze, Alexander Probst, Matthias Renner, Regina Renz, Jan Stange, Hartmut Stein, Kerstin von Rombs, Thomas Wagner, and Inge Walter. We also thank Asger Åmund and Peter Wulff for their encouragement and continuous support.

[48] T. Huehns, E. Krausz, S. Mrochen, M. Schmid, S. Esin, M. G. Engelmann, P. K. Schrittenloher, B. Höfling, W. H. Günzburg, and S. Nikol, *Atherosclerosis* **144,** 135 (1999).

[49] S. Baek and K. L. March, *Circ. Res.* **82,** 295 (1998).

[50] J. J. Rome, V. Shayani, K. D. Newman, S. Farrell, S. W. Lee, R. Virmani, and D. A. Dichek, *Hum. Gene Ther.* **5,** 1249 (1994).

[36] HVJ (Hemagglutinating Virus of Japan; Sendai Virus)-Liposome Method

By RYUICHI MORISHITA and YASUFUMI KANEDA

Introduction

Toward the success of human gene therapy, numerous viral and nonviral (synthetic) methods of gene transfer have been developed,[1,2] but each has its own limitations as well as advantages. Therefore, to develop *in vivo* gene transfer vectors with high efficiency and low toxicity, several groups have attempted to overcome the limitations of one vector by combining them with the strengths of another. Especially for the treatment of cardiovascular disease, it is necessary to develop safe and efficient gene delivery methods. The HVJ-liposome method appears to possess many ideal properties for *in vivo* gene transfer such as (1) efficiency, (2) safety, (3) simplicity of handling, (4) brevity of incubation time, and (5) no limitation of inserted DNA size. In this method, foreign DNA is complexed with liposomes, a nuclear protein, and the viral protein coat of HVJ (Fig. 1). The HVJ method has been successfully employed for gene transfer *in vivo* to many tissues including liver, kidney, and vascular wall.[3–18]

[1] R. C. Mulligan, *Science* **260**, 926 (1993).

[2] F. D. Ledley, *Hum. Gene Ther.* **6**, 1129 (1995).

[3] Y. Kaneda, *Biogenic Amines* **14**, 553 (1998).

[4] Y. Kaneda, Y. Saeki, and R. Morishita, *Mol. Med. Today* **5**, 298 (1999).

[5] V. J. Dzau, M. Mann, R. Morishita, and Y. Kaneda, *Proc. Natl. Acad. Sci. U.S.A.* **93**, 11421 (1996).

[6] T. Hirano, J. Fujimoto, T. Ueki, H. Yamamoto, T. Iwasaki, R. Morishita, Y. Sawa, Y. Kaneda, H. Takahashi, and E. Okamoto, *Gene Ther.* **5**, 459 (1998).

[7] Y. Saeki, N. Matsumoto, Y. Nakano, M. Mori, K. Awai, and Y. Kaneda, *Hum. Gene Ther.* **8**, 1965 (1997).

[8] Y. Sawa, Y. Kaneda, H-Z. Bai, K. Suzuki, J. Fujimoto, R. Morishita, and H. Matsuda, *Gene Ther.* **5**, 1472 (1998).

[9] G. Yamada, S. Nakamura, R. Haraguchi, M. Sakai, T. Terashi, S. Sakisaka, T. Toyoda, Y. Ogino, H. Hatanaka, and Y. Kaneda, *Cell. Mol. Biol.* **43**, 1165 (1997).

[10] Y. Yonemitsu, Y. Kaneda, A. Muraishi, T. Yoshizumi, K. Sugimachi, and K. Sueishi, *Gene Ther.* **4**, 631 (1997).

[11] E. Mabuchi, K. Shimizu, Y. Miyao, Y. Kaneda, H. Kishima, M. Tamura, K. Ikenaka, and T. Hayakawa, *Gene Ther.* **4**, 768 (1997).

[12] Y. Kaneda, K. Iwai, and T. Uchida, *Science* **243**, 375 (1989).

[13] Y. Kaneda, K. Iwai, and T. Uchida, *J. Biol. Chem.* **264**, 12126 (1989).

[14] K. Kato, M. Nakanishi, Y. Kaneda, T. Uchida, and Y. Okada, *J. Biol. Chem.* **266**, 3361 (1991).

[15] N. Tomita, J. Higaki, R. Morishita, S. Tomita, M. Aoki, T. Ogihara, and Y. Kaneda, *Gene Ther.* **3**, 477 (1996).

[16] N. Tomita, J. Higaki, R. Morishita, K. Kato, Y. Kaneda, and T. Ogihara, *Biochem. Biophys. Res. Comm.* **186**, 129 (1992).

HVJ-Liposome Gene Transfer Method

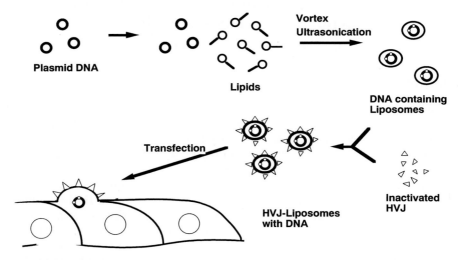

FIG. 1. Gene transfer by HVJ-liposomes. HVJ-liposomes bind to cell surface sialic acid receptors and associate with lipids in the lipid bilayer to induce cell fusion. By the fusion of the envelope of HVJ-liposomes with cell membrane, DNA in the HVJ-liposomes can be directly introduced into the cytoplasm.

Development of HVJ-Liposomes

Our basic concept is the construction of novel, hybrid-type liposomes with functional molecules inserted into them.[3,4] Based on this concept, DNA-loaded liposomes were fused with UV-inactivated HVJ (hemagglutinating virus of Japan; Sendai virus) to form HVJ-liposomes (approximately 400 to 500 nm in diameter). These viral liposomes bind to cell surface sialic acid receptors and fuse with cell membranes to directly introduce DNA into the cytoplasm without degradation (Fig. 1). The HVJ-liposomes can encapsulate DNA smaller than 100 kb. RNA, oligodeoxynucleotides including antisense, decoy, or ribozyme, proteins, and drugs can also be enclosed and delivered to cells. HVJ-liposomes are useful for *in vivo* gene transfer.[5] When HVJ-liposomes containing the LacZ gene were injected directly into rat blood vessels, approximately 80–90% of cells expressed LacZ gene activity, and no pathological hepatic changes were observed.[6] One advantage of HVJ-liposomes is allowance for repeated injections. Gene transfer to rat liver cells was not inhibited by repeated injections up to 8 times.[6,17] Of importance,

---

[17] R. Morishita, G. H. Gibbons, Y. Kaneda, T. Ogihara, and V. J. Dzau, *Biochem. Biophys. Res. Commun.* **273,** 666 (2000).

[18] R. Morishita, G. Gibbons, Y. Kaneda, T. Ogihara, and V. Dzau, *J. Clin. Invest.* **91,** 2580 (1993).

after repeated injections, anti-HVJ antibody generated in the rat was not sufficient to neutralize HVJ-liposomes. In addition, cytotoxic T cells recognizing HVJ were not detected in the rat transfected repeatedly with HVJ-liposomes.[6]

The HVJ-liposome method also enhances the efficiency and prolongs the half-life of antisense ODN *in vitro* and *in vivo*.[19-21] The HVJ-liposome method is suitable for an intraluminal molecular delivery system that has several advantages over the peri-adventitial polymer delivery approach. ODN transfected by the HVJ-liposome method substantially increases the efficiency of uptake of ODN and appears to increase the stability of the ODN within intracellular compartments.[19-21] HVJ is well known to cause cell fusion, and, thus, the complex in the viral envelope might enter the cells directly via cell membrane-liposome membrane fusion. However direct transfer of ODN has several unresolved problems such as (1) significant intracellular degradation via the lysosomal pathway and (2) efflux from endosomes after trapping. Modification of ODN pharmacokinetics with the use of HVJ-liposome complex will facilitate the potential clinical utility of the agents by (1) allowing a shorter intraluminal incubation time to preserve organ perfusion, (2) prolonging the duration of biological action, and (3) enhancing efficacy such that the nonspecific effects of high doses of ODN can be avoided. Since the virus is inactivated by ultraviolet light, there is little potential for biological hazard with this method as compared to the retroviral *in vivo* gene transfer approach.

Improvements of HVJ-Liposomes

The HVJ-liposome gene delivery system has several advantages, although further modification is necessary for it to be used in humans. To increase the efficiency of gene delivery, we investigated the lipid components of liposomes.[7] First, the most efficient gene expression was achieved with a phosphatidylcholine, phosphatidylethanolamine, and sphingomyelin molar ratio of 1 : 1 : 1. Second, anionic HVJ-liposomes should be prepared using phospatidylserine (PS) as the anionic lipid. Finally, the ratio of phospholipids to cholesterol should be 1 : 1. Accordingly, we also developed new anionic liposomes called HVJ-AVE liposomes, i.e., HVJ-artificial viral envelope liposomes. The lipid components of AVE liposomes are very similar to the HIV envelope and mimic the red blood cell membrane.[22] HVJ-AVE liposomes have yielded gene expression in liver and muscle 5 to 10 times higher than that observed with conventional HVJ-liposomes.[7] Interestingly, HVJ-AVE liposomes were most effective for gene transfer to mouse skeletal

[19] R. Morishita, G. H. Gibbons, M. Horiuchi, M. Nakajima, K. E. Ellison, W. Lee, Y. Kaneda, T. Ogihara, and V. J. Dzau, *J. Cardiovasc. Pharmacol. Ther.* **2,** 213 (1996).

[20] R. Morishita, G. H. Gibbons, K. E. Ellison, M. Nakajima, H. V. L. Leyen, L. Zhang, Y. Kaneda, T. Ogihara, and V. J. Dzau, *J. Clin. Invest.* **93,** 1458 (1994).

[21] R. Morishita, G. H. Gibbons, Y. Kaneda, T. Ogihara, and V. J. Dzau, *Gene* **149,** 13 (1994).

[22] R. Chander and H. Schreier, *Life Sci.* **50,** 481 (1992).

muscle in various nonviral gene transfer methods. HVJ-AVE liposomes were also very effective for gene delivery to isolated rat heart via the coronary artery. LacZ gene expression was observed in the entire heart, whereas expression was not observed with empty HVJ-AVE liposomes.[8] Safety of HVJ-liposomes has been tested and evaluated in monkeys. There were no significant pathological signs after injection of HVJ-liposomes into skeletal muscle or saphenous vein of cynomolgus monkeys. Messenger RNAs for fusion proteins of HVJ were not detected in monkey tissues after the injection.[23]

Another improvement was construction of cationic-type HVJ-liposomes using cationic lipids. Of the cationic lipids, positively charged 3β-[N-(N',N'-dimethylaminoethane)-carbamoyl] cholesterol hydroxide (DC)[24] has been the most efficient for gene transfer. For luciferase expression, HVJ-cationic DC liposomes were 100 times more efficient than were conventional HVJ-anionic liposomes.[7] Although it has been very difficult to transfer genes to bone marrow and spleen cells using conventional HVJ-liposomes, HVJ-cationic liposomes have been shown to be effective for gene transfer to both types of cells. However, when introduced into mouse muscle or liver, total luciferase expression after transfection with HVJ-cationic liposomes was shown to be 10 to 150 times lower than that with conventional anionic HVJ-liposomes,[7] which were less efficient for *in vitro* transfection. Furthermore, AVE liposomes were modified further to create AVE+DC10 (contains 10% PS and 10% DC), AVE+DC20 (containing 10% PS and 20% DC), and AVE-PS (containing neither PS nor DC) liposomes. We examined *in vivo* gene transfection efficiency with these liposomes after conjugation with the HVJ envelope. AVE yielded the highest luciferase expression in liver. AVE-PS and AVE+DC10 liposomes, which have a net neutral charge, showed intermediate luciferase activities. AVE+DC20 liposomes, which have an excessive amount of cationic lipid, yielded luciferase activities similar to those of HVJ-DC liposomes. Nevertheless, we have found HVJ-cationic liposomes to be more effective in some cases for *in vivo* gene transfer. High expression of the LacZ gene was obtained in restricted regions of chick embryos after injection of HVJ-cationic liposomes,[9] whereas HVJ-anionic liposomes were ineffective. In addition, when HVJ-cationic liposomes containing the LacZ gene were administered to rat lung with a jet nebulizer, more efficient gene expression in the epithelium of the trachea and bronchus was observed compared to that found with HVJ-anionic liposomes.[10] HVJ-cationic liposomes were also much more effective for gene transfer to tumor masses or disseminated cancers[3,11] in animal models compared to HVJ-AVE liposomes. Therefore, HVJ anionic and cationic liposomes can complement each other, and each liposome should be used for proper targeting.

[23] M. Kawauchi, J. Suzuki, R. Morishita, Y. Wada, A. Izawa, N. Tomita, J. Amano, Y. Kaneda, T. Ogihara, S. Takamoto, and M. Isobe, *Circ. Res.* **87,** 1063 (2000).
[24] K. Goyal and L. Huang, *J. Liposome Res.* **5,** 49 (1995).

Materials

Preparation of HVJ

1. Seed of HVJ: One hundred microliter aliquots of the chorioallantoic fluid containing HVJ (Z strain) in 10 % dimethy sulfoxide were stored in liquid nitrogen.

2. Polypeptone solution (1% polypeptone, 0.2% NaCl, pH 7.2) and BSS (137 mM NaCl, 5.4 mM KCl, 10 mM Tris-HCl pH7.6) were sterilized by autoclaving and stored at 4°.

3. Embryonated chick eggs (10 to 14 days after fertlization).

4. Incubator: The temperature and the moisture were set at 36.5° and at 30 to 40%, respectively.

5. Centrifuge tubes including 50 ml conical tube (Becton-Dickinson, Lincoln Park, NJ), 35 ml centrifuge tube (Beckman), and 10 ml ultracentrifuge tube (Hitachi, Tokyo, Japan) were sterilized.

6. A photometer (Spectrophotometer DU-68, Beckman Instruments, Tokyo, Japan), low-speed centrifuge (05PR-22, Hitachi, Tokyo, Japan), and centrifuge with JA-20 rotor (J2–HS, Beckman Instruments, Tokyo, Japan) were used.

Preparation of Lipid Mixtures

1. Chromatographically pure bovine brain phosphatidylserine-sodium salt (PS) (Avanti Polar Lipids Inc., Birmingham, AL), dioleoyl-L-α-phosphatidylethanolamine (DOPE) (Sigma, St. Louis, MO), sphingomyelin (Sph) (Sigma), egg yolk phosphatidylcholine (PC) (Sigma), 3β-[N-(N',N'-dimethylaminoethane) carbamoyl] cholesterol hydroxide (DC) (Sigma), and cholesterol (Chol) (Sigma) were stored at −20°.

2. Glass tubes (24 mm caliber and 12 cm long) were custom-made (Fujiston 24/40, Iwaki Glass Co. Ltd., Tokyo, Japan). The fresh tubes were immersed in saturated KOH–ethanol (180 g KOH in 500 ml ethanol) solution for 24 hr, rinsed with distilled water, and heated at 180° for 2 hr before use.

3. A rotary evaporator with a water bath (Type SR-650, Tokyo Rikakikai Inc., Tokyo, Japan) and vacuum pump with a pressure gauge (Type Asp-13, Iwaki Glass Co. Ltd., Tokyo, Japan) were used.

Preparation of HVJ-Liposomes

1. Plasmid DNAs were purified by a column procedure (Qiagen, Germany). The preparations were dissolved in BSS. The final concentration of DNA should be more than 1 mg/ml, and stored at −20°.

2. Cellulose acetate membrane filters (0.45 μm, 0.20 μm; Iwaki glass) were used for sizing liposomes.

3. BSS and 30% (w/v) sucrose in BSS were sterilized by autoclaving and stored at 4°.

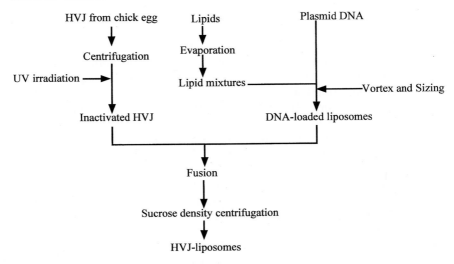

FIG. 2. Flow chart for the preparation of HVJ-liposomes.

4. A water bath (Thermominder Jr 80, TAITEC, Saitama, Japan) for preparing liposomes, a water bath shaker (Thermominder, TAITEC, Saitama, Japan) for fusing liposomes with HVJ, and an ultracentrifuge with RPS-40T rotor (55P-72, Hitachi, Tokyo, Japan) for purifying HVJ-liposomes and an ultraviolet cross-linker (Spectrolinker XL-1000, Spectronics Co.) for inactivating HVJ were used.

Methods

A flow chart for the preparation of HVJ-liposomes is shown in Fig. 2.

Preparation of HVJ in Eggs

1. The seed was rapidly thawed, and diluted to 1000 times with polypeptone solution. The seed diluted should be kept at 4° before proceeding to the next step.

2. Embryonated eggs were observed under illumination in a dark room, and an injected point was marked at about 0.5 mm above the chorioallantoic membrane. The eggs were disinfected with tincture of iodine and punctured at the point marked.

3. The diluted seed of 0.1 ml was injected into each egg using 1 ml disposable syringe with a 26-gauge needle. The needle should be inserted vertically so as to stab the chorioallantoic membrane.

4. After inoculation of the seed, the hole punctured on the egg was covered with melted paraffin. Then the eggs were incubated for 3 days at 36.5° in 30 to 40% moisture.

5. The eggs were chilled at 4° for more than 6 hr before harvesting the virus.

6. The eggshell was partially removed, and the chorioallantoic fluid was removed to an autoclaved bottle using 10 ml syringe with an 18-gauge needle. The virus in the fluid stored at 4° is stable for at least 3 months.

Steps 2, 3, and 6 can be carried out at room temperature.

Purification of HVJ from Chorioallantoic Fluid

1. Two hundred ml of the chorioallantoic fluid was transferred into four 50 ml disposable conical tubes, and span at 3000 rpm (1000*g*) for 10 min at 4° in a low-speed centrifuge.

2. Then, the supernatant was aliquotted into 6 tubes (Beckman JS-20), and centrifuged at 15,000 rpm (27,000*g*) for 30 min at 4°.

3. About 5 ml of BSS was added to the pellet in one of the tubes, and the material was kept at 4° overnight.

4. The pellets were gently suspended, collected in two tubes, and centrifuged as described in step 2. The resultant pellet in each tube was kept at 4° in 5 ml of BSS for more than 8 hr.

5. The pellets were gently suspended and subjected to rotation at 3000 rpm in a low-speed centrifuge.

6. The supernatant was removed to an aseptic tube and stored at 4°.

7. Virus titer was indicated by measuring the absorbbance at 540 nm of the 10 times-diluted supernatant using a photometer. An optical density at 540 nm corresponded to 15,000 hemagglutinating units (HAU), which was well correlated with fusion activity. The supernatant prepared as above usually showed 20,000 to 30,000 HAU/ml. A virus solution aseptically prepared maintains the fusion activity for 3 weeks.

Preparation of Lipid Mixture

1. Dry reagents of DOPE (12.2 mg), Sph (11.8 mg), and Chol (24.0 mg) were dissolved in 3870 μl of chloroform. Then, 130 μl of PC (13.0 mg) chloroform solution was added to the 3870 μl lipids solution. This 4000 μl lipids solution was added is called a basal mixture for liposomes. The basal mixture is ready to prepare anionic or cationic liposomes described below or can be stored at −20° after infusing nitrogen gas.

2. To prepare an anionic lipid mixture, 10 mg of PS was added to the basal mixture. To obtain a cationic lipid mixture, 6 mg DC was added to the mixture.

3. The lipid solution of 0.5 ml was aliquotted into 8 glass tubes. The tubes were kept on ice or $-20°$ in nitrogen gas before evaporation. The lipid solution should be evaporated as soon as possible.

4. The tubes were connected to a rotary evaporator. The tubes should be immersed in a 40° water bath at the tip.

5. The organic solvent was evaporated in a rotary evaporator under vacuum. The lipids were dried usually for about 5 to 10 min. Lipids appropriate for liposome preparation were those in the sides of the tubes in a thin layer. Those accumulated at the bottom of the tubes were inappropriate.

Preparation of HVJ-Liposomes Containing DNA

1. Plasmid DNA (200 μg) in 200 μl was added to a lipid mixture in the glass tube prepared as above, and agitated intensely by vortexing for 30 sec followed by an incubation at 37° for 30 sec. This cycle was repeated 8 times. By this method, plasmid DNA was enclosed at the ratio of 10–30% in anionic liposomes or 50–60% in cationic liposomes.

2. For preparing sized unilamellar liposomes, the liposome suspension is filtered with 0.45 μm pore size cellulose acetate filter and then with 0.2 μm filter. Sizing by an extruder with polycarbanate filters is better for preparing sized liposomes.

3. In the meantime, inactivate the HVJ virus and keep on ice. Add 15,000 HAU of the HVJ virus to the liposome suspension and leave the tube on ice for 5 to 10 min. Then, incubate the sample at 37° for 1 hr with shaking (120/min) in a water bath.

4. Add 7 ml 30% sucrose solution to a centrifuge tube and overlay the HVJ-liposome mixture on it. Separate the HVJ-liposome complexes from the free HVJ by sucrose gradient density centrifugation at 62,000g for 90 min at 4°.

5. Stop the centrifuge and gently remove only the conjugated liposomes. The final volume of the HVJ-liposome suspension should be approximately 1 ml.

Only in the case of HVJ-anionic liposomes, after centrifugation, collect the complexes, add 4 volumes of chilled BSS, spin at 27,000g for 30 min, and suspend the pellet in 0.5 to 1.0 ml of appropriate buffer.

Applications

Transfer of DNA into Cultured Cells

1. HVJ-cationic liposomes should be used for *in vitro* gene transfer because HVJ-cationic liposomes are approximately 100 times more efficient in gene transfer

to cultured cells than HVJ-anionic liposomes.[7] Ten μl of 1ml HVJ-cationic liposome suspension was added to 10^5 cells in serum-containing culture medium.[25]

2. Incubate the cells with the liposomes at 37° for 2 hr. Then change the medium and continue the culture.

Gene Transfer in Vivo by HVJ-Liposomes

1. For gene transfer to tissues, HVJ-anionic liposomes are recommended. The liposomes are useful for gene transfer to liver, skeletal muscle, heart, lung, artery, brain, spleen, eye, and joint space of rodent, rabbit, dog, lamb, and monkey. For example, for gene transfer to rat carotid artery, a lumen of a segment of the artery was filled with 0.5 ml anionic HVJ-liposome complex for 20 min at room temperature using a cannula.[18] For gene transfer into rat kidney, 1 ml of anionic HVJ-liposome suspension was injected into the renal artery.[16,17] To introduce DNA into rat liver, 2 to 3 ml of HVJ-anionic liposomes was injected into portal vein using a 5 ml syringe with a butterfly needle[12,13] or directly into liver under the perisplanchnic membrane using a 5 ml syringe with a 27-gauge needle.[14,15]

2. For gene transfer to tumor masses or disseminated tumors, direct injection of cationic HVJ-liposomes (0.1 ml to 0.5 ml) is recommended.

Notes

1. UV-inactivated HVJ can be stored for more than 6 months in 10% dimethyl sulfoxide at −80°. Do not store it at 4° for more than 1 day.

2. Lipid vials should be left at room temperature for about 30 min before opening the lids. Many lipids are highly hygroscopic.

3. The lipid mixture in the glass tube can be stored at −20° in nitrogen gas for 1 month after evaporation.

4. HVJ-liposomes can be stored for 3 weeks at 4°, and for more than 3 months with 10% dimethyl sulfoxide at final concentration in a freezer (below −20°).

5. Gene transfer efficiency of HVJ-liposomes is greatly affected by the fusion activity of HVJ envelope. Hemagglutinating ability should be frequently checked by hemagglutination of chick red blood cells.[26]

6. Reconstituted fusion liposomes can be prepared using isolated fusion proteins derived from HVJ instead of from inactivated whole viral particles.[27]

[25] T. Nishikawa, D. Du. X. L. Edelstein, S. Yamagishi, T. Matsumura, Y. Kaneda, M. A. Yorek, D. Beebe, P. J. Oates, H. P. Hammes, I. Giardino, and M. Brownlee, *Nature* **404,** 787 (2000).

[26] Y. Okada and J. Tadokoro, *Exp. Cell Res.* **26,** 98 (1962).

[27] K. Suzuki, H. Nakashima, Y. Sawa, R. Morishita, H. Matsuda, and Y. Kaneda, *Gene Ther. Reg.* **1,** 65 (2000).

[37] Gene Transfer with Foamy Virus Vectors

By Grant Trobridge, George Vassilopoulos, Neil Josephson, and David W. Russell

I. Introduction

Retroviral vectors have been used for many gene transfer applications because of their ability to efficiently deliver genes via a precise integration mechanism into the genomes of target cells. Retroviral vector systems based on the well-studied oncoviruses and lentiviruses are established, but the less-studied spumaviruses [or foamy viruses (FVs)] also hold promise as potential gene transfer vectors.

A. Effects of Foamy Virus Infection

FVs were named for their impressive cytopathic effects on cultured cells, which include a characteristic vacuolation or "foaming" during infection. Despite their cytotoxicity *in vitro,* FVs have not yet been shown to cause disease in any host.[1,2] The prototype human foamy virus (HFV, also called human spumaretrovirus or HSRV) was originally isolated from a nasopharyngeal carcinoma of a Kenyan African after culture *in vitro.*[3] This nomenclature is unfortunate as HFV is now considered to be a chimpanzee virus that contaminated the human cell culture,[4-7] and a comprehensive study failed to detect the presence of FVs in human populations.[8] Humans can be infected with FVs through animal bites, but no disease has been associated with these exposures and human-to-human transfer has not been documented.[2,9] Although many FVs have been identified in a variety of mammalian hosts, unless otherwise stated we will focus on vector development with the HFV isolate.

The safety of retroviral vectors used in clinical protocols and/or laboratory research is an important consideration, and the FVs have an advantage over their

[1] J. J. Hooks and C. J. Gibbs, Jr., *Bacteriol. Rev.* **39,** 169 (1975).

[2] M. Schweizer, V. Falcone, J. Gange, R. Turek, and D. Neumann-Haefelin, *J. Virol.* **71,** 4821 (1997).

[3] B. G. Achong, P. W. Mansell, M. A. Epstein, and P. Clifford, *J. Natl. Cancer Inst.* **46,** 299 (1971).

[4] P. Brown, G. Nemo, and D. C. Gajdusek, *J. Infect. Dis.* **137,** 421 (1978).

[5] O. Herchenroder, R. Renne, D. Loncar, E. K. Cobb, K. K. Murthy, J. Schneider, A. Mergia, and P. A. Luciw, *Virology* **201,** 187 (1994).

[6] G. J. Nemo, P. W. Brown, C. J. Gibbs, Jr., and D. C. Gajdusek, *Infect. Immun.* **20,** 69 (1978).

[7] M. Schweizer and D. Neumann-Haefelin, *Virology* **207,** 577 (1995).

[8] M. Schweizer, R. Turek, H. Hahn, A. Schliephake, K. O. Netzer, G. Eder, M. Reinhardt, A. Rethwilm, and D. Neumann-Haefelin, *AIDS Res. Hum. Retroviruses* **11,** 161 (1995).

[9] W. Heneine, W. M. Switzer, P. Sandstrom, J. Brown, S. Vedapuri, C. A. Schable, A. S. Khan, N. W. Lerche, M. Schweizer, D. Neumann-Haefelin, L. E. Chapman, and T. M. Folks, *Nat. Med.* **4,** 403 (1998).

onco- and lentiviral cousins in this respect. Despite their lack of proven pathogenicity, care should still be taken when working with FVs and vectors derived from them. As described below, current vector stock production methods have greatly reduced the risk of replication-competent retrovirus (RCR) formation. However, there is always a remote possibility that a rare, complex recombination event could generate RCR, so vector stocks should be handled and disposed of as if they contained infectious virus. Another safety consideration relates to the biological effects of the vector transgene, which can be efficiently delivered to human cells.

B. Properties of FVs Relevant for Vector Development

1. FV Genome and Genes. FV genomes are the largest of all the retroviruses, ranging from 12 to 13 kb. For comparison, the HIV-1 genome is just under 10 kb and the MLV genome is 8.3 kb. A complex cis-acting sequence mapped to a 1-2 kb region of the FV *pol* gene needs to be retained for efficient vector production,[10–12] as do essential regions of the LTRs (approximately 0.5 kb each in our current deleted vectors) and the packaging signal that extends into the 5′ region of *gag* (0.6 kb).[10–12] Assuming a cDNA genome size of 13 kb (the functional genome of FVs is DNA; see below), there should still be room for packaging at least 9 to 11 kb of foreign DNA. Future deletion analysis may further reduce the size of essential cis-acting sequences, and the maximum genome size that FV vectors can efficiently package has not been determined.

FVs are complex retroviruses that contain three overlapping open reading frames (ORFs) in addition to the *gag, pol,* and *env* genes present in all retroviruses (see Fig. 1A). These ORFs were originally designated Bel1–3 for between envelope and LTR; however, Bel1 has been renamed Tas for transactivator of spumavirus. The Tas protein is essential for virus replication[13] and acts as a transactivator of transcription.[14–16] Following integration into the host genome the internal promoter in *env* has basal transcriptional activity and produces an mRNA encoding Tas, which subsequently activates both the internal and LTR promoters, allowing a bimodal mechanism of regulation.[13,17–19] The additional accessory proteins Bel2,

---

[10] O. Erlwein, P. D. Bieniasz, and M. O. McClure, *J. Virol.* **72,** 5510 (1998).

[11] M. Heinkelein, M. Schmidt, N. Fischer, A. Moebes, D. Lindemann, J. Enssle, and A. Rethwilm, *J. Virol.* **72,** 6307 (1998).

[12] M. Wu, S. Chari, T. Yanchis, and A. Mergia, *J. Virol.* **72,** 3451 (1998).

[13] M. Lochelt, H. Zentgraf, and R. M. Flugel, *Virology* **184,** 43 (1991).

[14] A. Rethwilm, O. Erlwein, G. Baunach, B. Maurer, and V. ter Meulen, *Proc. Natl. Acad. Sci. U.S.A.* **88,** 941 (1991).

[15] A. Keller, K. M. Partin, M. Lochelt, H. Bannert, R. M. Flugel, and B. R. Cullen, *J. Virol.* **65,** 2589 (1991).

[16] L. K. Venkatesh, P. A. Theodorakis, and G. Chinnadurai, *Nucleic Acids Res.* **19,** 3661 (1991).

[17] M. Lochelt, W. Muranyi, and R. M. Flugel, *Proc. Natl. Acad. Sci. U.S.A.* **90,** 7317 (1993).

[18] M. Lochelt, R. M. Flugel, and M. Aboud, *J. Virol.* **68,** 638 (1994).

[19] Y. Kang, W. S. Blair, and B. R. Cullen, *J. Virol.* **72,** 504 (1998).

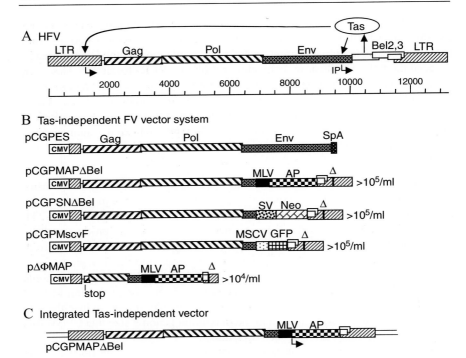

FIG. 1. Tas-independent FV vectors. (A). The wild-type FV genome depicted is the 13,242 bp HFV provirus, which contains the complete U3 region of HFV [M. Schmidt, O. Herchenroder, J. Heeney, and A. Rethwilm, *Virology* **230**, 167 (1997)] with the positions of the LTRs and open reading frames indicated. Arrows indicate transcription start sites. The size in bp is shown below. The transcriptional transactivator Tas (Bel1) is expressed from the internal promoter (IP) in *env* at basal levels and activates transcription both at the IP and the otherwise silent HFV LTR. LTR transcription produces the full-length genomic RNA and mRNAs for expression of the different viral genes. (B). The relevant portions of Tas-independent constructs are shown in their plasmid forms (indicated by a "p" prefix). The helper plasmid pCGPES expresses Gag, Pol, and Env from a constitutive CMV-LTR fusion promoter, followed by an SV40 polyadenylation site (SpA). Vector plasmids contain transgene cassettes in place of *env* and *bel* sequences, and the vector genome is transcribed during vector production by a CMV-LTR fusion promoter. The transgenes consist of internal MLV, murine stem cell virus [MSCV; R. G. Hawley, F. H. Lieu, A. Z. Fong, and T. S. Hawley, *Gene Ther.* **1**, 136 (1994)] or SV40 (SV) promoters expressing human placental alkaline phosphatase (AP), neomycin phosphotransferase (Neo), or green fluorescent protein (GFP) as indicated. The 3′ LTRs of pCGPMAPΔBel, pCGPSNΔBel, and pCGPMscvF were derived from the pHSRV13 clone [M. Lochelt, H. Zentgraf and R. M. Flugel, *Virology* **184**, 43 (1991)] which contains a deletion (Δ) in the U3 region [M. Schmidt, O. Herchenroder, J. Heeney, and A. Rethwilm, *Virology* **230**, 167 (1997)]. The pΔΦMAP vector has a larger U3 LTR deletion, a deletion in *gag* and *pol,* and a stop codon in the 5′ portion of the *gag* reading frame (unpublished results). Typical titers of unconcentrated stocks are indicated for each vector in transducing units/ml. (C). Following integration of Tas-independent vectors, the U3 and R regions of the 5′ vector LTR are derived from the 3′ LTR in the vector plasmid, forming a Tas-dependent promoter that is not expressed in transduced cells. Transgenes are expressed from the internal vector promoter.

Bel3, and Bet (a fusion between the amino-terminal 88 amino acids of Bel1 and the Bel2 ORF) are dispensable for viral replication *in vitro*.[20–22] The functions of Bel2 and 3 are currently unknown, but the Bet protein has been shown to confer resistance to infection by an as yet unknown mechanism.[23] For further information on FV biology and replication the reader is referred to Refs.1, 24–26.

2. Host Range. An attractive property of foamy vectors is their broad host cell tropism. FVs are able to infect cells from all vertebrate species tested (ranging from reptiles to humans) and many different tissues, including cells of fibroblastic, epithelial, neural, and hematopoietic origin.[1,27–31] This suggests that the cellular receptor(s) for FVs is evolutionarily conserved and ubiquitously expressed. FV vectors do not efficiently pseudotype with oncoviral envelope proteins or the vesicular stomatitis virus glycoprotein (VSV-G).[32] Receptor interference assays suggest that several primate FV isolates including HFV share a common receptor on the BHK-21 cell line.[31] While the broad host range of FVs increases the utility of FV vectors, the lack of FV-resistant cell lines has prevented expression cloning of the receptor, which remains unidentified.

3. Cell Cycle Requirements. Given that many gene transfer applications require the transduction of quiescent, non-dividing cells, the cell cycle dependency of viral vectors is a crucial issue. Oncoviruses such as murine leukemia virus (MLV) require breakdown of the nuclear membrane during mitosis for integration,[33] while lentiviruses are able to actively enter intact, nonmitotic nuclei via redundant, cooperative mechanisms (reviewed in Naldini[34]). There are indications that FVs may also have mechanisms for entering the nucleus independently of mitosis. FVs contain a nuclear localization signal in the nucleocapsid protein that is able to translocate heterologous proteins to the nucleus,[35] and two additional nuclear

[20] G. Baunach, B. Maurer, H. Hahn, M. Kranz, and A. Rethwilm, *J. Virol.* **67**, 5411 (1993).

[21] S. F. Yu and M. L. Linial, *J. Virol.* **67**, 6618 (1993).

[22] A. H. Lee, H. Y. Lee, and Y. C. Sung, *Virology* **204**, 409 (1994).

[23] M. Bock, M. Heinkelein, D. Lindemann, and A. Rethwilm, *Virology* **250**, 194 (1998).

[24] R. M. Flugel, *J. Acquir. Immune Defic. Syndr.* **4**, 739 (1991).

[25] A. Rethwilm, *J. Acquir. Immune Defic. Syndr. Hum. Retrovirol.* **13**, S248 (1996).

[26] M. L. Linial, *J. Virol.* **73**, 1747 (1999).

[27] D. Neumann-Haefelin, A. Rethwilm, G. Bauer, F. Gudat, and H. zur Hausen, *Med. Microbiol. Immunol.* **172**, 75 (1983).

[28] A. Mergia, N. J. Leung, and J. Blackwell, *J. Med. Primatol.* **25**, 2 (1996).

[29] D. W. Russell and A. D. Miller, *J. Virol.* **70**, 217 (1996).

[30] S. F. Yu, J. Stone, and M. L. Linial, *J. Virol.* **70**, 1250 (1996).

[31] C. L. Hill, P. D. Bieniasz, and M. O. McClure, *J. Gen. Virol.* **80**, 2003 (1999).

[32] T. Pietschmann, M. Heinkelein, M. Heldmann, H. Zentgraf, A. Rethwilm, and D. Lindemann, *J. Virol.* **73**, 2613 (1999).

[33] T. Roe, T. C. Reynolds, G. Yu, and P. O. Brown, *Embo. J.* **12**, 2099 (1993).

[34] L. Naldini, *Curr. Opin. Biotechnol.* **9**, 457 (1998).

[35] A. W. Schliephake and A. Rethwilm, *J. Virol.* **68**, 4946 (1994).

localization domains have been described in Pol.[36] Whether these domains mediate entry of the FV preintegration complex in a manner analogous to the HIV matrix, Vpr and integrase proteins is unknown. Saib *et al.*[37] demonstrated that incoming FV Gag antigens and the FV genome were localized near the centrosome in infected cells and that the FV genome could translocate into the nucleus of G_1/S but not G_0 arrested cells.

A unique property of the FV genus of retroviruses is that the viral genome is composed of double-stranded DNA.[38] Although replication occurs through an RNA intermediate, the genome is converted to a cDNA copy in the cell producing the FV virion rather than the cell being infected.[39,40] The presence of a DNA genome may be an advantage when transducing quiescent cell types if reverse transcription within the target cell is a rate-limiting step, as has been suggested for lentiviral vectors.[41]

The central polypurine tract (cPPT) of lentiviral vectors has been shown to be required for efficient entry of vector genomes into the nucleus.[42,43] During reverse transcription of HIV a DNA flap is created when second-strand DNA synthesis terminates at a sequence 99 bp downstream of an internal cPPT initiation site for DNA synthesis. HIV vectors that contain the cPPT and termination sequence are efficiently transported into the nucleus, have a higher titer, and produce more transgene expression.[42,43] A cPPT was identified in HFV,[44] which mapped to a central S_1 nuclease sensitive region, and three additional polypurine-rich cPPT sites nearby have since been identified.[11] A DNA flap may thus form at one or more of these cPPTs in FV genomes, but their contribution to nuclear entry has not yet been determined. The cPPT identified by Kupiec *et al.*[44] has been removed from the *pol* cis-acting region in some deleted vectors with only a modest (approximately 50%) drop in titer.[11] It is unclear if the other putative cPPT sequences in FV genomes are important for efficient gene transfer.

The cell cycle requirements of human foamy virus have been compared to both onco- and lentiviruses. In one study, infection with wild-type HFV was reduced in G_1/S-arrested and G_2-arrested cells, and the HFV dependency on cell proliferation was significantly greater than that observed with HIV, but slightly less than that of

[36] H. Imrich, M. Heinkelein, O. Herchenroder, and A. Rethwilm, *J. Gen. Virol.* **81,** 2941 (2000).

[37] A. Saib, F. Puvion-Dutilleul, M. Schmid, J. Peries, and H. de The, *J. Virol.* **71,** 1155 (1997).

[38] S. F. Yu, D. N. Baldwin, S. R. Gwynn, S. Yendapalli, and M. L. Linial, *Science* **271,** 1579 (1996).

[39] A. Moebes, J. Enssle, P. D. Bieniasz, M. Heinkelein, D. Lindemann, M. Bock, M. O. McClure, and A. Rethwilm, *J. Virol.* **71,** 7305 (1997).

[40] S. F. Yu, M. D. Sullivan, and M. L. Linial, *J. Virol.* **73,** 1565 (1999).

[41] R. E. Sutton, M. J. Reitsma, N. Uchida, and P. O. Brown, *J. Virol.* **73,** 3649 (1999).

[42] A. Follenzi, L. E. Ailles, S. Bakovic, M. Geuna, and L. Naldini, *Nat. Genet.* **25,** 217 (2000).

[43] V. Zennou, C. Petit, D. Guetard, U. Nerhbass, L. Montagnier, and P. Charneau, *Cell* **101,** 173 (2000).

[44] J. J. Kupiec, J. Tobaly-Tapiero, M. Canivet, M. Santillana-Hayat, R. M. Flugel, J. Peries, and R. Emanoil-Ravier, *Nucleic Acids Res.* **16,** 9557 (1988).

MLV.[45] These experiments required both provirus integration and the synthesis of viral proteins in infected cells, so it is not clear at what stage of the viral life cycle cell proliferation played a role. In a study of vectors expressing reporter genes from heterologous promoters, HFV vectors transduced stationary phase cultures at reduced rates but much more efficiently than MLV vectors.[29] When quiescent cells were stimulated to proliferate after exposure to FV vectors, transduction rates were similar to those on dividing cells. The relative importance of nuclear transport, cDNA synthesis, and genome persistence in nondividing cells has not yet been dissected in a careful comparison of the three classes of retroviral vectors.

II. Vector Production Strategies

A. Early Attempts

Replication-competent FV-based vectors were first generated by Schmidt and Rethwilm[46] at titers in excess of 10^5 focus-forming units (FFU)/ml. These vectors were capable of delivering a small transgene cassette, but also expressed Gag, Pol, Env, and Tas in transduced cells and were highly cytopathic. Replication-incompetent vectors were generated by Russell and Miller which lacked the FV *env* gene and internal promoter, and prevented expression of the remaining viral gene products in transduced cells.[29] Env and Tas supplied in trans by cotransfection of a *gag-pol*-deleted helper plasmid allowed for Gag and Pol expression from the FV LTR promoter in vector constructs during stock production. These replication-incompetent vectors were able to transduce a wide variety of vertebrate cells via integration of the vector genome. Although primary hematopoietic cells were relatively refractory to transduction, this could be overcome by cocultivation with vector-producing cells.[47] This vector production method was also limited by the cytopathic effects of replication-competent retrovirus, which was produced at significant levels by recombination between vector and helper constructs.

B. FV Packaging Cells

Packaging cell lines that express essential viral proteins such as Gag, Pol, and Env in trans offer the potential for convenient and large-scale retroviral vector production.[48] Vector stocks can be produced from these cell lines either by transient transfection with vector plasmids or by stable transfection and selection of vector-expressing producer cell lines. Cell lines which continuously express oncoviral envelope glycoproteins are stable; however, expression of VSV-G is cytotoxic so

---

[45] P. D. Bieniasz, R. A. Weiss, and M. O. McClure, *J. Virol.* **69,** 7295 (1995).

[46] M. Schmidt and A. Rethwilm, *Virology* **210,** 167 (1995).

[47] R. K. Hirata, A. D. Miller, R. G. Andrews, and D. W. Russell, *Blood* **88,** 3654 (1996).

[48] A. D. Miller, *Hum. Gene Ther.* **1,** 5 (1990).

this gene must be under the control of an inducible promoter.[49,50] The FV Env protein causes syncytium formation via its interaction with cellular receptors.[51,52] Packaging cells that constitutively or inducibly express the simian foamy virus (SFV) structural proteins Gag, Pol, Env, and Tas have been described.[53] When transfected with vector plasmids the constitutive cell line produced vector titers of 3.5×10^3/ml, while the tetracycline-inducible cell line produced titers of up to 1.1×10^4/ml. HFV Tas-dependent producer cell lines have also been described that exhibit some syncytia and produce Tas-deleted vectors at titers over 10^5/ml.[54] Some of these cell lines generated RCR, likely through recombination between the integrated Tas construct and the transfected vector. Based on these reports, it is not clear if FV Env cytotoxicity will be an obstacle to the production of a high titer ($>10^6$/ml), stable packaging cell line. Although such a cell line would be highly desirable, in our opinion the Tas-independent transient transfection system described below represents the best currently available method for producing replication-incompetent FV vector stocks that express a variety of different transgenes and are free of RCR.

C. Tas-Independent Vector Production

The unique replication strategy of FVs allowed us to design a Tas-independent vector production method that eliminates the generation of RCR by recombination during transient transfections.[55] Since Tas is required for FV replication, wild-type replication-competent FV formation cannot occur in its absence. In this system, the *bel* genes (including the transactivator *bel1* or *tas*) have been removed from both the helper and vector plasmid constructs (Fig. 1B), and the FV *gag, pol,* and *env* genes were placed under control of a constitutive CMV promoter on the helper plasmid. The vector constructs also have a large deletion in *env*, where a variety of transgene expression cassettes can be incorporated. Transcription of the vector genome is driven by a Tas-independent CMV-LTR fusion promoter. During reverse transcription, the 5′ HFV LTR is regenerated and integrated vector proviruses have LTRs that are silent in the absence of Tas, allowing strict control of transgene expression by the internal vector promoter (Fig. 1C). Although the *gag* and *pol* genes were included in the initial vector constructs used with this

[49] Y. Yang, E. F. Vanin, M. A. Whitt, M. Fornerod, R. Zwart, R. D. Schneiderman, G. Grosveld, and A. W. Neinhuis, *Hum. Gene Ther.* **6,** 1203 (1995).

[50] T. Kafri, H. van Praag, L. Ouyang, F. H. Gage, and I. M. Verma, *J. Virol.* **73,** 576 (1999).

[51] P. A. Goepfert, K. L. Shaw, G. D. Ritter, Jr., and M. J. Mulligan, *J. Virol.* **71,** 778 (1997).

[52] O. Herchenroder, D. Moosmayer, M. Bock, T. Pietschmann, A. Rethwilm, P. D. Bieniasz, M. O. McClure, R. Weiss, and J. Schneider, *Virology* **255,** 228 (1999).

[53] M. Wu and A. Mergia, *J. Virol.* **73,** 4498 (1999).

[54] P. D. Bieniasz, O. Erlwein, A. Aguzzi, A. Rethwilm, and M. O. McClure, *Virology* **235,** 65 (1997).

[55] G. D. Trobridge and D. W. Russell, *Hum. Gene Ther.* **9,** 2517 (1998).

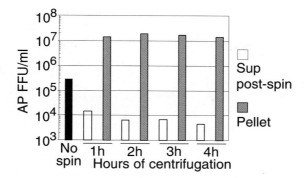

FIG. 2. Concentration of FV vector. Four 11.5 ml aliquots of unconcentrated CGPMAPΔBel vector stock were centrifuged at 50,000g at 20° in a Beckman SW-41 rotor for the specified amount of time. Following centrifugation, each supernatant (Sup post spin) was carefully collected with a pipette, the pellet was immediately resuspended in 0.5 ml of fresh medium, and both were titered on BHK-21 cells. In this particular experiment, the total yield of transducing particles actually increased slightly after centrifugation. AP FFU/ml: focus-forming units/ml. No spin: vector stock before concentration.

system, the genes are not expressed from the silent FV LTR in transduced cells, and newer, deleted vectors lack most of these sequences (see below). This method can generate high titer ($>10^5$ FFU/ml) vector stocks by transient cotransfection of 293T cells with helper and vector plasmids, followed by collection of cell-free, filtered medium 3 days later. If necessary, the stocks can be further concentrated by ultracentrifugation (Fig. 2). This production method has allowed us to generate helper-free vector stocks of sufficient titer to evaluate the ability of FV vectors to transduce hematopoietic stem cells in animal studies.

D. Helper Virus Testing

Tas-independent vector stocks do not contain detectable RCR as measured by two different assays.[55] The FAB cell line has an integrated HFV LTR upstream of a nuclear-localizing β-Gal reporter gene, so after infection by wild-type HFV (or any Tas-expressing construct) the integrated HFV LTR is activated by Tas and nuclear β-Gal is produced.[21] Not surprisingly, stocks made by the Tas-independent system are free of RCR by this assay. A more stringent method is required to identify Tas-independent RCR. These RCRs could conceivably form by complex recombination events between the FV LTR, viral genes, and other promoters such as CMV or cellular sequences that combine all the functions required to complete the viral life cycle in the absence of transactivation by Tas. Such a Tas-independent RCR has been constructed artificially,[56] suggesting that this type of RCR is possible, although

[56] T. Schenk, J. Enssle, N. Fischer, and A. Rethwilm, *J. Gen. Virol.* **80,** 1591 (1999).

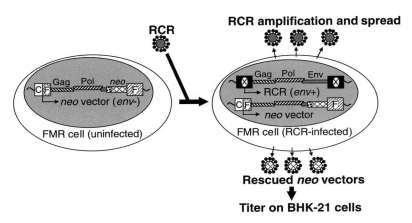

FIG. 3. Foamy marker rescue assay. An uninfected FMR cell is depicted on the left with its inte-grated, Tas-independent, *env*-deficient *neo* vector provirus. On the right, an FMR cell is shown after infection by a hypothetical, Tas-independent, *env*+ RCR virion, containing both RCR and vector proviruses. This cell produces additional RCR particles that can spread through the culture, and *neo* vector particles that can transduce BHK-21 cells. The positions of *gag, pol, env,* and *neo* genes in the proviruses are indicated. The CMV-HFV LTR fusion promoter (C|F), the HFV LTR (F) and a novel, Tas-independent LTR (X) are indicated. This assay can be used to detect contaminating Env-expressing virions in FV vector stocks as long as the vector does not express *neo* itself. Reproduced from G. D. Trobridge and D. W. Russell, *Hum. Gene Ther.* **9**, 2517 (1998) with permission.

this mutant virus replicates very poorly. In order to assay FV vector stocks for this type of RCR, the foamy marker rescue (FMR) cell line[55] can be used (Fig. 3). FMR cells contain an integrated CMV-driven foamy vector genome that expresses Gag and Pol and can be rescued by any viral particle expressing Env. Although, we have never detected RCR by the FMR assay in Tas-independent stocks, this method should be used to definitively rule out the presence of RCR when necessary.

E. Deleted Vectors

We have generated deleted vectors that lack most of the viral gene sequences of FV, but retain essential cis-acting sites in the LTRs, 5′ *gag* region, and *pol*. The 3′ LTR in these constructs has a deletion in the U3 region, to increase the packaging capacity and further disable the HFV LTR promoter in vector proviruses. We call vectors made from this deleted backbone Δ Φ vectors (deleted foamy), such as that produced by the pΔΦMAP construct (Fig. 1B). Using the Tas-independent system, Δ Φ vectors can be produced at approximately 20% of the titer of their parental vectors, which contain the entire *gag* and *pol* genes (G. Trobridge *et al.,* unpublished results, 2001). Future experiments will be required to determine why this drop in titer occurs, and whether it can be overcome by the inclusion of additional sequences. The deletions present in $\Delta\Phi$ vectors increase their theoretical packaging capacity to approximately 9 kb of foreign DNA without exceeding the genome size of wild-type HFV.

III. Protocols

A. Tas-Independent Vector Stock Production by CaPO₄ Transfection

Reagents. 293 T cells[57] are cultured in D10 medium, which is Dulbecco's modified Eagle's medium (DMEM) supplemented with 100 IU/ml penicillin G, 100 μg/ml streptomycin, 1.25 μg/ml amphotericin B, and 10% heat-inactivated fetal bovine serum (FBS), in a 5% CO_2 atmosphere at 37°. 2× HEPES-saline is 280 mM NaCl, 50 mM HEPES, adjusted to pH 7.1 with 0.5 N NaOH. 0.1× TE is 1 mM Tris pH 8.0, 0.1 mM EDTA. Stock chloroquine is 100 mM in H_2O, stored at −20° and diluted to 10 mM in D10 before use. Stock sodium butyrate is 500 mM in DMEM, stored at −80° (sodium butyrate). Plasmid preparations are heated to 65° for 10 min prior to use.

Procedure

Day 1. In the afternoon, plate three 15 cm dishes, each containing 30 ml of 293 T cells at 4×10^5 cells/ml (1.2×10^7 cells/dish). Treat well with trypsin so that when counting there are single cells, not clumps. After plating, rock dishes to evenly distribute cells. Use 293 T cells that were plated 2 days previously at 5×10^6 cells/15 cm dish and evenly distributed to ensure the cells are actively growing. In general, do not allow 293 T cell cultures to remain confluent.

Day 2. In the afternoon, add 75 μl of 10 mM chloroquine to each dish of 293 T cells and mix by gently swirling the supernatant in the dish and then prepare $CaPO_4$ transfection Solutions A and B in 50 ml tubes. Solution A is 0.1 ml 0.15 M Na_2HPO_4 pH 7.1 and 10 ml 2× HEPES-saline. Solution B is 1.25 ml of 2.0 M $CaCl_2$, 8.75 ml of 0.1× TE, 125 μg of helper plasmid (pCGPES), and 125 μg of vector plasmid (i.e., pCGPMAPΔBel). These volumes are enough for transfecting three 15 cm dishes. Vortex Solutions A and B separately. To produce the $CaPO_4$ precipitate, drip Solution B into Solution A with a small-tip pipette (we use sterile transfer pipettes, Samco, San Fernando, CA) while holding Solution A on the vortex at medium setting. Add so there are individual droplets entering as opposed to a stream. Approximately 10–20 min after generating the precipitate, vortex the mix at high speed for a few seconds and add 6.5 ml of this $CaPO_4$ precipitate to each chloroquine-treated plate of 293 T cells by dripping with a pipette. Distribute the precipitate evenly throughout the dish.

[57] R. B. DuBridge, P. Tang, H. C. Hsia, P. M. Leong, J. H. Miller, and M. P. Calos, *Mol. Cell Biol.* **7,** 379 (1987).

Day 3. In the morning [approximately 17 hours (h) posttransfection], carefully remove most of the culture medium containing the transfection reagents from each dish, leaving approximately 4 ml/dish. Add 25 ml D10 with 10 mM sodium butyrate (made by diluting sodium butyrate stock 50-fold in D10) to each dish. Twelve h later, change medium to 29 ml D10 without sodium butyrate.

Day 5. Dislodge the cells, mix in the medium with a 25 ml pipette, and transfer the contents to two 50 ml tubes. Centrifuge at 200 g for 5 min to pellet cells and debris. Filter the supernatant through a 0.45 μm low protein-binding filter. Do not substitute a 0.2 μm filter for a 0.45 μm filter as this will severely reduce the titer. The titer should be approximately 2–8 × 10^5/ml.

Notes. It is difficult to change the medium of 293 T cells because they do not attach well. It is especially important with large (15 cm) dishes not to disturb the cells while moving the plates, so move the dishes slowly and minimize rocking of the plates while handling them. Alternatively, the protocol can be scaled down by decreasing the surface area of the plate, media volume, and CaPO$_4$ volume in equal proportions. The transfection efficiency of the CaPO$_4$ precipitates generated will vary and some investigators prefer mixing Solutions A and B in smaller volumes (3.3 ml total) and then pooling the precipitates before adding them to the cells. We have also generated high-titer stocks by increasing the final medium volume on day 3 to 35 ml. The addition of both chloroquine and sodium butyrate increases titers approximately 4-fold using the above protocol. Chloroquine and sodium butyrate can be omitted if the cell density plated on Day 1 is reduced to 1 × 10^6 cells per 10 cm dish. The lower cell density will result in increased toxicity to the transfected cells but unconcentrated vector titers should still be in excess of 1 × 10^5 AP FFU/ml.

B. *Vector Concentration by Ultracentrifugation*

Procedure. Centrifuge vector stock for 2 h at 50,000g, 20°. An SW-28 rotor can be used with 6 Beckman polyallomer tubes for a stock from eight 15 cm dishes. Carefully remove the supernatant down to a small translucent pellet by aspirating with a Pasteur pipette. Immediately add 0.1 ml of phosphate-buffered saline (PBS) and resuspend the pellet vigorously by pipetting up and down. Let sit 5 min, repeat the vigorous resuspension, and transfer to a microfuge tube. Rotate the microfuge tube in a rotating rack for 30 min at room temperature to allow the pellet to fully resuspend. To remove unwanted particulates, spin the microfuge tube at 2000g for 2 min. Remove the vector-containing supernatant to a new tube, being careful not to disturb the pellet of particulate debris. At this point the stock can be used or frozen.

Notes. Recovery is generally greater than 50% for a 100-fold concentration, so final titers of 1–8 × 10^7/ml can be achieved. Recovery decreases for concentrations over 100-fold, presumably due to difficulties in fully resuspending the pelleted virions. Cell culture medium can be used to resuspend the vector stock instead of PBS. The removal of particulate debris by low-speed centrifugation may result in a reduction of vector titers, and for high concentrations this loss may be up to 75%. However, it is advisable to perform this low-speed centrifugation to reduce the toxicity of the vector stock preparation to transduced cells (likely due to transfection debris). We have not defined optimal procedures for resuspending the vector-containing pellet but vigorous pipetting alone can result in recovery of over 50% of the original vector titer.

C. Freezing and Thawing of Stocks

Procedure. Measure the vector stock volume. Add a volume of dimethyl sulfoxide (DMSO) equal to 5% of the stock volume to a new microfuge tube. Add the stock to this tube so that it mixes with the DMSO immediately, then freeze at −80°. The rapid addition of the 95% volume vector stock to the 5% DMSO volume results in improved mixing and minimizes the toxic effects of uneven DMSO distribution on stock viability. To thaw, place in a 37° water bath until there is no frozen material left and then immediately transfer to room temperature until use.

Notes. With this method, there should be little or no loss in titer as tested on BHK-21 cells.[58] For unclear reasons, transductions of primary hematopoietic cells are sometimes superior when freshly prepared stocks are used (see below). It has been reported that freeze–thawing can release intracellular FV in infected BHK-21 cells and increase viral titer.[21] However, FV vectors are also sensitive to freeze–thaw damage and we have found that production of FV vectors in 293 T cells is not enhanced by freeze–thawing.[55] DMSO (5%) protects FV vector stocks from titer loss due to freeze–thawing. We have not tested the stability of vector stocks after repeated freeze–thawing in the presence of DMSO but recommend only freeze–thawing stocks once. The presence of 5% DMSO in the vector stock may be an issue if large volumes are added to cells. Usually stocks can be diluted to reduce the DMSO concentration in the cell culture medium to an acceptable level. Alternatively, DMSO can be removed by ultracentrifugation and resuspension of the vector stock in medium without DMSO (Section III,B), but this may lead to poor recovery (sometimes less than 10%), and there is increased toxicity associated with stocks that have been concentrated twice.

D. Helper Assay with Foamy Marker Rescue (FMR) Cells

Reagents. Coomassie blue fixative/stain is 300 ml methanol, 100 ml glacial acetic acid, 600 ml H$_2$O, and 1.5 g Coomassie Brilliant Blue G. Cell culture medium (D10) and conditions are as described in Section III,A above. BHK-21[58] and FMR[55] cells have been described.

---

[58] I. Macpherson and M. Stoker, *Virology* **16,** 147 (1962).

Procedure

Day 1. Plate 1×10^5 FMR cells/well in 1 ml of D10 in 12-well tissue culture plates. Plate one well for each sample dilution to be tested, one well for each wild-type HFV dilution to be tested as positive controls, and one well for a negative control.

Day 2. In the morning add vector samples to wells. If testing unconcentrated stocks replace medium in well with 1 ml of vector stock (in D10). If testing concentrated vector stocks, replace medium with 1 ml of fresh D10 and add a small volume of vector stock (typically in PBS). As a positive control add 10-fold serial dilutions of wild-type HFV, covering a predicted dose of 10^{-2} to 10^3 infectious units.

Day 4. Plate BHK-21 cells at 1×10^5 cells/well in 1 ml of D10 in a 12-well plate. Plate one well for each well of FMR cells.

Day 5. Remove medium from the infected FMR cell wells (and the negative control well) and filter through 0.45 μm filters. Replace the medium in wells containing BHK-21 cells with these FMR-conditioned samples. Discard FMR cells.

Day 6. Twenty-four h after replacing the medium on BHK-21 cells, treat the wells with trypsin and resuspend the cells from each well in 1 ml of D10. Plate both 0.1 ml and 0.9 ml of each cell suspension in a 10 cm dish in a final volume of 10 ml D10.

Day 7. Twenty-four h after plating add 10 ml G418 to a final concentration of 0.7 mg/ml active compound to each dish.

Day 10. Change the medium to 10 ml of fresh D10 with G418.

Day 15. Remove medium, then rinse each dish with 5–10 ml of PBS (taking care not to dislodge any colonies). Add 5–10 ml Coomassie blue fixative/stain for 10 min. Carefully rinse plates in H_2O and count colonies.

Notes. Any G418-resistant colonies formed are due to marker rescue of the *neo*-containing provirus in the FMR cell line and indicate that the sample contained replication-competent FV. To date, we have never obtained a positive FMR assay with stocks made by the Tas-independent transfection method. The sensitivity of the assay is between 3 and 92 blue cell focus-forming units as measured with wild-type HFV on FAB cells.[55]

E. Production and Titration of Wild-Type FV as a Control for RCR Detection

Reagents. The pHSRV13 plasmid is a full-length infectious clone of HFV[13] that can be used to generate wild-type FV following $CaPO_4$-mediated transfection of BHK-21 cells. FAB cells contain a Tas-dependent HFV LTR promoter driving nuclear β-Gal[21] and can be used to titer wild-type HFV. Ten ml of β-Gal substrate is 8.8 ml PBS with 500 μl 100 mM $K_3Fe(CN)_6$, 500 μl 100 mM $K_4Fe(CN)_6$, 20 μl 1.0 M $MgCl_2$, and 200 μl 50× X-Gal. $K_3Fe(CN)_6$ (100 mM) and 100 mM $K_4Fe(CN)_6$ are made separately in H_2O and stored at room temperature protected from light. X-Gal ($50\times$) is 50 mg/ml 5-bromo-4-chloro-3-indolyl β-D-galactopyranoside in N,N-dimethylformamide, stored in a glass vial at $-20°$.

Day 1. Plate 4 ml of BHK-21 cells at 1×10^5 cells/ml in a 6 cm tissue culture dish in D10.

Day 2. In the morning prepare CaPO$_4$ transfection reagents as outlined in Protocol A (Section III,A) except use 0.5 ml of Solution A (0.5 ml of $2\times$ HEPES-saline and 5 μl of 0.15 M NaHPO$_4$) and 0.5 ml of Solution B with the pHSRV13 plasmid [62.5 μl of 2.0 M CaCl$_2$, pHSRV13 DNA (10 μg in 20 μl), and 420 μl of $0.1\times$ TE]. Drip Solution B into Solution A while agitating Solution A, let sit 10 min, vortex, and add the mix to the dish of BHK-21 cells. Eight h later remove the medium from the transfected cells and add 4 ml of D10.

Day 6. Plate 2.5 ml of FAB cells per well at 1×10^5 cells/ml in a 6-well tissue culture plate to titer the viral stock. Six wells are adequate to titer serial dilutions of the virus stock.

Day 7. Remove the medium from the pHSRV13-transfected cells, pass through a 0.45 μm filter, and store in 200 μl aliquots at $-80°$. After the virus has been frozen for 2 h, remove an aliquot by thawing briefly in a $37°$ water bath and placing at room. Add five 10-fold serial dilutions of virus to the FAB cells corresponding to 100 μl down to 0.01 μl of the original viral stock. Add medium used to dilute the viral stock to one well as a negative control.

Day 9. Prepare FAB cells for β-gal assay as outlined in Protocol F (Section III,F, below), but do not heat inactivate. Remove PBS after second wash following fixation and add 1.5 ml of β-Gal substrate per well. Incubate 5 min to 24 h in the dark at $37°$ and wash cells with H$_2$O when the nuclei of positive cells are dark blue. Calculate the titer in blue cell-forming units (BCFU).

Notes. We have prepared HFV stocks with titers over 10^5 BCFU/ml using this method. If titers are low, the viral stock can be amplified by an additional passage on BHK-21 cells plated the day before adding virus, followed by growth for another 5 days and harvest as above. We have not used DMSO when freezing wild-type HFV, so there may be some reduction in titer as compared to fresh stocks. The titers obtained are more than adequate for use as a positive control in the FMR assay.

F. Transduction of Adherent Cells with Alkaline Phosphatase (AP) Vectors

Reagents. Fixative is 0.5% glutaraldehyde or 3.7% formaldehyde in PBS. AP substrate is made from a $50\times$ stock of Nitro Blue Tetrazolium (NBT, diluted to $1\times$) and a $100\times$ stock of 5-bromo-4-chloro-3-indolyl phosphate (BCIP, diluted to $1\times$) in 100 mM Tris pH 8.5, 100 mM NaCl, 50 mM MgCl$_2$. NBT ($50\times$) is 50 mg/ml NBT in 70% dimethylformamide and 30% H$_2$O prepared in a glass vial and stored at $-20°$. BCIP ($100\times$) is 10 mg/ml BCIP in H$_2$O stored at $-20°$. Cell culture medium (D10) and conditions are as in Protocol A (Section III,A).

Procedure

Day 1. Plate cells in a 6-well tissue culture plate at a density that will result in confluence approximately 2 days later. For BHK-21 cells, plate at a concentration of 2.5×10^5 cells/well in 2.5 ml D10. Use one well for each vector dilution to be tested, plus one well for a negative control.

Day 2. Change to fresh medium, add vector stock dilutions, and mix thoroughly by rocking the 6-well plate. Typically a small volume ($<100 \mu l$) of vector stock is added to each well.

Day 4. Rinse each well carefully with 2 ml PBS for 5 min. Add fixative for 10 min at room temperature. Wash 3 times with PBS. Let sit in PBS for 5–10 min each time. Add fresh PBS prewarmed to 68° and move plate to a 68° oven. Heat-inactivate endogenous AP for 1 h at 68°. Remove PBS and add AP substrate. Use only 1–1.5 ml AP substrate per well. Stain 10 min to overnight at room temperature. Stop staining by rinsing with H_2O and air-drying wells when AP cells are clearly stained purple and before background staining appears.

Notes. Many, if not all, cell types can be transduced with FV vectors, and transduction does not require the addition of polycations such as Polybrene. This assay detects AP expression from human placental alkaline phosphatase transgenes (such as those in the vectors shown in Fig. 1), since this protein is heat-resistant, and can be used to titer AP-expressing vector stocks. It is necessary to confirm that the cell line to be transduced does not contain endogenous heat-resistant AP by including a negative control in the assay. Sometimes staining due to endogenous AP can be further reduced by increasing the heat-inactivation time.

G. Transduction of Mouse Hematopoietic Stem Cells

Reagents. The stock solution of recombinant fibronectin fragment CH296 (Panvera, Madison, WI) is 1 mg/ml in H_2O. Hypotonic red blood cell (RBC) lysis solution is 155 mM NH_4Cl, 7.3 mM $NaHCO_3$, 126 μM EDTA. D20 with cytokines is DMEM with 20% heat-inactivated FBS, 20 ng/ml IL3, 50 ng/ml IL6, and 50 ng/ml stem cell factor (SCF). Pen/Strep (100×) is 10,000 IU penicillin G sodium and 10,000 μg/ml streptomycin sulfate in 0.85% saline (GIBCO-BRL, Rockville, MD). Agar solution is 1.2% Noble agar (Sigma, St. Louis, MO) in H_2O. Stock cytokine solutions are 20 μg/ml IL3, 50 μg/ml IL6, and 50 μg/ml SCF. All cytokines were murine, except for IL6 which was of human origin. 5-Fluorouracil solution is provided as a sterile 50 mg/ml solution (Pharmacia & Upjohn, Kalamazoo, MI) and is diluted to 5 mg/ml in room temperature PBS just before use.

Procedure

Day 1. Inject 5 mg/ml 5-fluorouracil intraperitoneally (150 μg/g body weight) into bone marrow donors. Use five 8- 10-week-old mice for a typical experiment with 3–4 transplant recipients. Begin preparation of vector stock (Day 1 of

Protocol A, Section III,A), and continue to completion. Each experiment will require approximately 230 ml of crude vector stock concentrated to a final volume of 3 ml in DMEM. A total of eight 15-cm dishes (scaled up from the three dishes used in protocol A) will generate one vector stock that fills a SW28 rotor with six polyallomer tubes to capacity during concentration.

Day 3. Euthanize animals and harvest bone marrow cells by flushing femurs with Hanks' balanced salt solution (HBSS) using a 23-gauge needle. Make a single cell suspension by passing the cells through the needle several times. Spin the cells down (200 g for 5 min), wash once with HBSS, and lyse red blood cells by resuspension in 10 ml hypotonic RBC lysis solution for 5 min at room temperature. Spin the cells down again and resuspend in D20 with cytokines at a concentration of 1–2 × 10^6 cells/ml. Plate the cells in 6-well dishes with 2 ml/well. This procedure should yield approximately 5 × 10^7 cells from 5 donors, plated in 15–18 wells.

Day 4. Plate indicator cells for titering the vector. When using AP-expressing vectors follow Protocol F above (Section III,F) with BHK-21 cells.

Day 5. Coat non-tissue culture-treated 6-well plates with recombinant fibronectin fragment CH296 by diluting stock to 50 μg/ml in H_2O, adding 1 ml to each well, incubating for 15 min at room temperature, then aspirating and air-drying. Harvest vector stock and concentrate by ultracentrifugation as described in Protocol A above (Section III,A), except resuspend the final stock in 3 ml of DMEM. Titer an aliquot of stock using cells plated on Day 4. Collect bone marrow cells by aspirating several times with a pipette, wash in HBSS, resuspend in DMEM at approximately 1 × 10^7 cells/ml, and set on ice. To 1 ml of cells, add 1.6 ml FBS, 3 ml of vector stock, 2.4 ml DMEM, 80 μl of 100× Pen/Strep solution, and 8 μl each of 20 μg/ml IL3, 50 μg/ml IL6, and 50 μg/ml SCF. Add the mixture to CH296-coated wells (2 ml/well) and culture for 2 days. Set up an extra well with 2.5 × 10^6 cells in D20 with cytokines as a nontransduced control for colony assays.

Day 7. Use a rubber policeman to collect all bone marrow cells, including those that are loosely attached to the CH296-coated wells. Wash cells in HBSS, resuspend at approximately 1 × 10^6 cells/ml in HBSS, and then set on ice. Save about 50,000 cells for colony assays to assess transduction efficiency at the progenitor cell level, and use the rest of the cells for transplantation. Irradiate recipient mice with a single dose of 1100 rads, then administer the transduced bone marrow cells by tail vein injection within 2 h of irradiation. Typically, there are enough cells to inject 3–4 recipient animals with 0.5–1 × 10^6 cells each. For colony assays, make the agar by briefly boiling the 1.2% agar solution, then cooling to 56° in a water bath. Transfer enough agar solution for the plating to a 40° heating block in the tissue culture hood. Make a culture mix by combining the following reagents (volumes for one dish): 300 μl DMEM, 250 μl 2× DMEM, 200 μl heat-inactivated FBS, 10 μl of 100× Pen/Strep, and 1 μl each of 20 μg/ml IL3, 50 μg/ml IL6, and 50 μg/ml SCF. To the above, add 250 μl warm agar and vortex briefly. Add transduced cells (0.5, 1, or 2×10^4 cells), vortex gently, and plate in one 3.5 cm tissue culture dish. Perform a similar plating with nontransduced control cells. Work swiftly to avoid solidification of the agar before plating. When AP-expressing vectors are used, cells should be plated in 24 mm Transwells (Costar #3412, Corning, NY), so they can be stained *in situ* for AP expression. For other reporter genes such as GFP, conventional tissue culture dishes are adequate. After the cells are plated, refrigerate the dish for one min to solidify the agar. If Transwells are used, add 3 ml of D20 with cytokines to the immersion well after the agar has solidified. Culture soft agar dishes at $37°$, 5% CO_2, for 7 days.

Day 14. The progenitor transduction rates are calculated by determining the number of hematopoietic colonies grown in soft agar that express the vector-encoded transgene. For GFP vectors, determine the percentage of GFP+ colonies by visualization with a fluorescent dissecting microscope. For AP vectors, stain the Transwell cultures as follows, using sequential 3 ml volumes added to each immersion well and taking care not to disturb the agar layer in the Transwell. Aspirate medium, add PBS to the immersion well, and incubate at room temperature for 15 min. Repeat this step three times. Aspirate PBS, add 0.5% gluteraldehyde in PBS and incubate at room temperature for 20 min. Aspirate, add fresh 0.5% gluteraldehyde in PBS, and incubate for an additional 70 min. Wash with PBS four times (30 min each). Aspirate and add PBS prewarmed to $68°$ and incubate at $68°$ for 1 h to inactivate endogenous phosphatases. Aspirate, add PBS at room temperature, and let the plate cool. Aspirate, add AP substrate (Protocol F, Section III,F) and stain overnight.

Day 15. Aspirate AP substrate and rinse wells with PBS. Count colonies under a dissecting microscope. AP+ CFU have a deep-purple color compared to the light brown color of the negative colonies.

Days 28 on. Bone marrow and peripheral blood samples of transplant recipients can be analyzed at the desired time points for transduced cells by DNA analysis for FV vector genomes, or by appropriate assays for reporter gene expression.

Notes. Using this protocol and FV vector CGPMAPΔBel, a high percentage of murine hematopoietic stem cells can be transduced, with a majority of peripheral blood and bone marrow cells expressing the vector-encoded transgene in many transplant recipients.[58a] This protocol has been adapted from earlier experiments with MLV vectors,[59] and it is not known if treatment with 5-fluorouracil or prolonged *ex vivo* culture in the presence of cytokines is important for transduction with FV vectors. We routinely use mouse strain C57BL6/J as donors and the congenic strain B6.SJL-Ptprc^{a}Pep3^b/BoyJ (Jackson Labs) as recipients to facilitate engraftment analysis based on distinct CD45 alleles. The protocol can easily be

[58a] G. Vassilopoulos, G. Trobridge, N. C. Josephson, and D. W. Russell, *Blood* **98**, 604 (2001).

[59] D. M. Bodine, K. T. McDonagh, N. E. Seidel, and A. W. Nienhuis, *Exp. Hematol.* **19**, 206 (1991).

applied to other mouse strains, which may have different hematopoietic cell numbers and/or radiation sensitivity. All cell counts refer to viable cells as determined by trypan blue exclusion.

H. Transduction of Human CD34+ Cells

Reagents. Transduction medium is DMEM supplemented with 20% heat inactivated FBS, 100 IU/ml penicillin G, 100 μg/ml streptomycin, 1.25 μg/ml amphotericin B, 100 ng/ml recombinant human Flt-3 ligand (rhFlt-3lig), and 100 ng/ml recombinant human stem cell factor (rhSCF). Posttransduction medium is DMEM supplemented with 20% FBS, 100 IU/ml penicillin G, 100 μg/ml streptomycin, 1.25 μg/ml amphotericin B, 100 ng/ml rhFlt-3lig, 100 ng/ml rhSCF, 20 ng/ml recombinant human IL-3 (rhIL-3), and 20 ng/ml recombinant human IL-6 (rhIL-6). Prior to vector stock production, CD34+ positive cells can be isolated from umbilical cord blood using the Milteny Biotec CD34+ Progenitor Isolation Kit following the manufacturer's instructions, and frozen in 100% FBS with 10% DMSO at a concentration of $5-10 \times 10^6$/ml.

Procedure

Day 1. Begin preparation of a vector stock (Protocol A, Section III,A). Each experiment will require approximately 230 ml of crude vector stock from eight transfected 15 cm dishes (scaled up from the three dishes used in Protocol A).

Day 4. Plate indicator cells for titering the vector. When using AP-expressing vectors follow Protocol F (Section III,F) with BHK-21 cells.

Day 5. Transductions are performed in CH296-coated non-tissue culture-treated 6-well dishes (prepared as per the manufacturer's instructions except at a concentration of 5 μg/cm^2). A single well can be used to transduce $0.75-1.5 \times 10^6$ cells. Concentrate the vector stock by ultracentrifugation (Protocol B, Section IV,B) to a final volume of 1–6 ml in transduction medium (1 ml/transduction well). Frozen CD34+ cord blood cell samples are thawed in a 37° water bath until the ice pellet melts and then placed on ice. Ten min later the thawed cells should be diluted in 10 ml DMEM supplemented with 20% heat inactivated FBS, pelleted at 200g for 5 min, and resuspended in transduction medium at a concentration of $1.5-3.0 \times 10^6$ cells/ml. A typical thaw will yield 50–80% recovery of frozen cells. One ml of vector stock and 0.5 ml of cells are added to each CH296-coated well and gently rocked to ensure even settling of cells. A negative control with an equivalent number of cells in non-vector containing medium should also be plated. The wells are incubated overnight (10–16 h) at 37°, in a 5% CO$_2$ atmosphere.

Day 6. The weakly adherent cells are harvested by repeated pipetting of medium onto the plate surface. After removal of medium and cells, 3.0 ml PBS is added to the plate and additional pipetting onto the plate surface is performed.

Both cell harvests are combined, and collection of nearly all cells should be confirmed by examining the wells under the microscope. Spin the cells down and resuspend in posttransduction medium (the volume depends on subsequent steps). When the GFP reporter gene is used, transduction rates can be determined both in cells grown in liquid culture and in methylcellulose progenitor assays. For liquid culture, 10^5 cells are plated in 1.0 ml of posttransduction medium. After 3–5 days, transduction rates can be determined as the percentage of GFP+ cells by using flow cytometry. For methylcellulose progenitor assays, 1500 cells diluted in 350 μl of Iscove's modified Dulbecco's medium (IMDM) are added to 3.5 ml of MethoCult H4230 medium (Stem Cell Technologies) as prepared per the manufacturer's instructions and additionally supplemented with 100 IU/ml penicillin G, 100 μg/ml streptomycin, 1.25 μg/ml amphotericin B, 50 ng/ml rhSCF, 10 ng/ml rhIL-3, 10 ng/ml recombinant human granulocyte macrophage colony stimulating factor, and 2 U/ml human recombinant erythropoietin. After vortexing, 1 ml of the above mixture (which should contain 390 cells) is plated in 3.5 cm tissue culture dishes in triplicate. At this concentration approximately 70–125 clonogenic progenitors should grow in each plate. To ensure that the methylcellulose does not dry out, multiple 3.5 cm dishes should be cultured inside a covered 15 cm dish containing a small reservoir of distilled water. The methylcellulose cultures are maintained for 2 weeks at 37°, in a 5% CO_2 atmosphere. Progenitor cells transduced with AP vectors should be assayed in agar as previously described (Day 7, Protocol G, Section III,G) except that the medium used (DMEM) should contain the same cytokines as present in the methylcellulose assays, the cell numbers plated should be 1500–3000 cells per well (there is decreased plating efficiency in agar as compared to methylcellulose), and the Transwells are cultured for 2 weeks before staining.

Day 20. The progenitor colonies grown in methylcellulose can be counted and scored for GFP expression on a fluorescent dissecting microscope. The Transwell cultures can be stained for AP expression as described [Days 14–15, Protocol G (Section III,G)].

Notes. This protocol was developed for CD34+ cells isolated from umbilical cord blood. It should also work well with progenitors from bone marrow or peripheral blood. A single vector stock of CGPMAPΔBel or CGPMscvF (Fig. 1) can be used to infect up to 5 × 10^6 CD34+ cells at an MOI > 4, and transduce 60–90% of human progenitors. For larger cell numbers, multiple stocks can be pooled. We typically use fresh vector stocks because they lead to more reproducibly efficient transduction of CD34+ cells than thawed vector stocks. Our protocol calls for an overnight (10–16 h) transduction. However, we have found efficient transduction of CD34+ cord blood progenitors with as little as 4 h of exposure to vector stocks. We have not experimented with different cytokine cocktails to any significant degree, and those listed here may not be optimal. All cell counts refer to viable cells as determined by trypan blue exclusion.

IV. Concluding Remarks

A. Applications

FV vectors are promising gene delivery vehicles with potential applications in basic science and gene therapy. The current generation of vector constructs and production methods allows the generation of high-titer ($>10^7$/ml), helper-free vector stocks by a simple cotransfection procedure followed by vector concentration. No mammalian cell line has been identified that is resistant to transduction by FV vectors, and they can be used for gene delivery into primary cell lines that are difficult to transfect. Unlike most retroviral vectors, FV vectors do not require the addition of polycations such as Polybrene for efficient transduction and are resistant to inactivation by human serum,[29] so they may be useful for *in vivo* gene delivery, which is an area that remains to be explored. The improved performance of FV vectors in nondividing cells (as compared to oncoviral vectors) may be a major advantage when attempting to transduce quiescent stem cells or terminally differentiated epithelial cells. The large packaging capacity and silent proviral LTRs of Tas-independent deleted vectors make them well suited for delivering large transgene cassettes that are regulated independently, without the influence of viral enhancers found in many LTRs. Finally, one of the most important advantages of FV vectors is their safety. This stems from the lack of proven pathogenicity of wild-type foamy viruses in multiple animal hosts, as well as features incorporated into the Tas-independent vector system. These features include a lack of RCR production, silent LTRs to prevent the activation of downstream genes (such as oncogenes) after integration, and the removal of the viral Tas protein itself, which has been shown to influence the expression of multiple cellular genes.[60]

One application that we have focused on is the use of FV vectors for the transduction of primary hematopoietic cells, including hematopoietic stem cells (HSCs). HSCs can repopulate the entire hematopoietic system after transplantation into myeloablated recipients, so *ex vivo* transduction and reinfusion of HSCs represents an ideal treatment strategy for many genetic and hematological diseases. Hirata *et al.*[47] originally showed that FV vectors compared favorably to oncoviral vectors in their ability to transduce human, baboon, and mouse hematopoietic progenitor cells. More recently, murine bone marrow transplant experiments have shown that FV vectors can transduce HSCs, with transgene expression persisting long-term in all the hematopoietic lineages of transplant recipients (over 50% of circulating peripheral blood cells were transduced in some animals).[61] Similarly, FV vectors can efficiently transduce human CD34$^+$ hematopoietic cells after a

---

[60] A. Wagner, A. Doerks, M. Aboud, A. Alonso, T. Tokino, R. M. Flugel, and M. Lochelt, *J. Virol.* **74,** 4441 (2000).

[61] G. Vassilopoulos, G. Trobridge, and D. W. Russell, *Blood* **94–10,** S611a (1999).

single 10 h exposure to vector stocks, including NOD/SCID repopulating cells thought to represent a very primitive hematopoietic cell type.[62] Given the recent demonstrations of impressive developmental plasticity in different stem cell populations, including HSCs,[63] we are optimistic that *ex vivo* stem cell transduction by FV vectors will be useful for treating nonhematologic diseases as well.

B. Next-Generation Vectors

Further development is needed to increase the carrying capacity, titer, and purity of FV vectors. Potential for improvements exists both in the design of vector genomes and in stock production methods. It is likely that essential *cis*-acting sites will be more finely mapped so that additional viral sequences can be removed from vector genomes, and non-FV sequences may ultimately be incorporated to increase titer and performance. The development of high-titer, helper-free packaging and producer cells should facilitate the large-scale production of FV vector stocks with greater purity and reliability than those generated from transfections. This is especially important when stocks are concentrated by ultracentrifugation, as cellular debris and calcium phosphate precipitate present during cotransfection tend to pellet along with virions and limit the potential for concentration. Improved purification methods more specific for FV virion characteristics could also eliminate unwanted contaminants in stocks, and thereby increase the capacity for further concentration. Another potential problem is the highly cell-associated nature of FV. During *in vitro* FV infection most virus remains cell-associated, budding from intracellular membranes.[64] The foamy Env has an ER retrieval signal that is thought to be in part responsible for this phenomenon.[51] One challenge for improving FV stock titers will be to enhance the egress of vectors from vector-producing cells. Finally, given the relative lack of basic research to date on FV biology (as compared with other types of retroviruses), there will no doubt be significant advances in our understanding of the molecular mechanisms of FV infection, replication, packaging, and assembly, all of which may lead to improvements in vector design.

Acknowledgments

The authors thank Jaclynn Mac for expert technical assistance and Roli Hirata for invaluable discussions on methodologies. This work was supported by grants from the U.S. National Institutes of Health (D.R., G.V., and N.J.) and the Medical Research Council of Canada (G.T.).

[62] N. Josephson, G. Vassilopoulos, G. D. Trobridge, G. V. Priestly, T. Papayannopoulou, and D. W. Russell, *Blood* **96–11,** S525a (2000).

[63] I. L. Weissman, *Science* **287,** 1442 (2000).

[64] N. Fischer, M. Heinkelein, D. Lindemann, J. Enssle, C. Baum, E. Werder, H. Zentgraf, J. G. Muller, and A. Rethwilm, *J. Virol.* **72,** 1610 (1998).

[38] Infectious Epstein–Barr Virus Vectors for Episomal Gene Therapy

By Jianlong Wang and Jean-Michel H. Vos*

Introduction

The Epstein–Barr virus (EBV) is a B-lymphotropic γ-herpesvirus with a 172 kb double-stranded DNA genome. The virus primarily infects human B-lymphocytes by specific binding of the viral protein gp350/220 to the receptor (CD21) present on the B cell surface followed by receptor-mediated endocytosis. A lytic infection begins with viral DNA, RNA, and protein synthesis, followed by the assembly of viral proteins and lysis of the host cells. Alternatively, the more common latent, nonlytic infection can occur, in which the virus establishes itself in the nucleus as a large circular extrachromosomal replicating plasmid known as an episome that is maintained through subsequent cell division.

Ever since it was shown that a vector carrying the EBV latent replication origin (OriP) and the viral nuclear antigen gene (EBNA1) was sufficient for plasmid replication in lymphoblastoid cell lines (LCLs), investigators have tried to develop EBV-based vectors for gene transfer.[1–4] In our laboratory, we have previously developed a human artificial episomal chromosome (HAEC) system based on the EBV OriP for propagation and stable maintenance of large DNA inserts ranging from 60 to 330 kb as circular minichromosomes in human cells.[5,6] Such a large cloning capacity of an EBV-based vector is ideal for gene therapy and led us to develop infectious viral vectors for delivery of therapeutic genes as HAECs into human cells.

Gene therapy based on HAECs has the following advantages: first, the autonomous replication of HAECs reduces the potential pathological risks associated with "uncontrolled" chromosomal integration occurring with retroviral and adeno-associated viral systems. Second, an episomal vector may also avoid the variability

---

* Deceased.

[1] J. Yates, N. Warren, D. Reisman, and B. Sugden, *Proc. Natl. Acad. Sci. U.S.A.* **81**, 3806 (1984).

[2] J. M. Vos, *in* "Viruses in Human Gene Therapy" (J. M. H. Vos, ed.), pp. 109–140. Carolina Academic Press and Chapman & Hall, Durham, NC, and London, 1995.

[3] J. M. Vos, E. M. Westphal, and S. Banerjee, *in* "Gene Therapy" (N. R. Lemoine and D. N. Cooper, eds.), pp. 127–153. BIOS Scientific Publishers, Ltd., Oxford, UK, 1996.

[4] J. M. Vos, K. B. Quattrocchi, and B. J. Wendelburg, *in* "Gene Therapy for Neurological Disorders and Brain Tumors" (E. A. Chiocca and X. O. Breakefield, eds.), p. 93. Humana Press Inc., Totowa, NJ, 1997.

[5] T. Q. Sun and J. M. H. Vos, *in* "Methods in Molecular Genetics," Vol. 8, p. 167. Academic Press, San Diego, 1996.

[6] T. Q. Sun, D. A. Fernstermacher, and J. M. Vos, *Nat. Genet.* **8**, 33 (1994).

in expression frequently observed with integrated sequences. Third, autonomous replication is expected to increase the long-term persistence of a transferred gene in cells in comparison to nonreplicating epichromosomal systems based on adenoviruses and herpes simplex virus-1 (HSV-1). Fourth, gene therapy requiring the stable maintenance of large DNA molecules in target cells may be more effective following an episomal strategy than the rather inefficient process of chromosomal integration. Finally, the large cloning capacity of HAECs allows for delivery of multiple therapeutic genes in tandem or large human genes with their native regulatory sequences. This is particularly beneficial for designing vectors with a multigenic strategy to attack polygenic disorders such as cancers, or vectors with large therapeutic genes that require their own regulatory elements to be functional.

Infectious EBV vectors developed in our laboratory deliver such therapeutic HAECs into target cells by transduction. Such a transduction strategy is based on a recombinant amplicon carrying a transgene expression cassette and the minimal viral *cis* elements required for replication (i.e., the origins for latent and lytic replication, OriP and OriLyt) and packaging (i.e., terminal repeat sequences, TR) as infectious particles (referred to as miniEBV) (Fig. 1A). The miniEBV vector is dependent on the nontransforming packaging cell line HH514 (P3HR1 EBV strain)[7] for providing *in trans* the lytic replication and packaging functions to process and package the recombinant amplicon DNA. When miniEBV virions are produced and purified from this helper cell line, they are mixed with the endogenous P3HR1 EBV (Fig. 1B), which is replication-competent and transforming-incompetent.[7,8]

Because of the natural B lymphotropism of EBV in humans, such miniEBV vectors naturally target inherited B-cell diseases and B-cell lymphomas. Infectious miniEBV virions were shown not only to express reporter genes such as *lacZ*[9] and GFP[10] in the infected human disease B cells, but also to correct phenotypically the defects in LCLs established from Fanconi anemia and Lesch–Nyhan patients.[11,12] It was demonstrated that this infectious miniEBV was able to carry inserts in the range of 140–160 kb and was also episomal in the infected B cells.[9,12]

A rapidly emerging technology for the treatment of various cancers involves the delivery of suicide genes to the affected cells, followed by prodrug treatment that induces cells expressing suicide gene products to undergo cell death.[13,14] We have

[7] M. Rabson, L. Gradoville, L. Heston, and G. Miller, *J. Virol.* **44,** 834 (1982).

[8] E. S. Robertson, T. Ooka, and E. D. Kieff, *Proc. Natl. Acad. Sci. U.S.A.* **93,** 11334 (1996).

[9] T. Q. Sun and J. M. Vos, *Int. J. Genome. Res.* **1,** 45 (1992).

[10] J. Wang and J. M. Vos, *Mol. Ther.* **3,** 976 (2001).

[11] T. Q. Sun, E. Livanos, and J. M. Vos, *Gene Ther.* **3,** 1081 (1996).

[12] S. Banerjee, E. Livanos, and J. M. Vos, *Nat. Med.* **1,** 1303 (1995).

[13] S. M. Freeman, R. Ramesh, A. Munshi, K. A. Whartenby, J. L. Freeman, and A. Marrogi, *in* "Gene Therapy of Cancer: Methods and Protocols" (W. Walther and U. Stein, eds.), p. 155. Humana Press, Totowa, NJ, 2000.

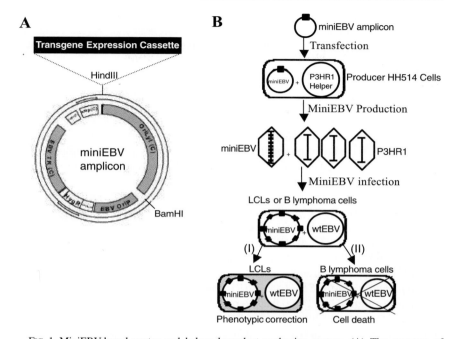

FIG. 1. MiniEBV-based vector and helper-dependent packaging system. (A) The structure of miniEBV amplicon. The shaded regions are minimal *cis* elements required for episomal replication (OriP), amplification (OriLyt), and packaging of the vector (TR). OriP, plasmid origin of replication; OriLyt, lytic replication origin; TR, terminal repeat sequences. "C" indicates the complementary strand in this vector relative to its presence in the native EBV genome. Hygromycin B resistance gene (Hyg^R) and ampicillin resistance gene (amp) were engineered for selection in mammalian and prokaryotic cells, respectively. Two restriction sites (*Hind*III and *Bam*HI) used for cloning of the transgene expression cassette are indicated. The transgenes could be a reporter gene such as *lacZ* or GFP, a therapeutic gene for an inherited disease, and a suicide gene for cancer therapy. (B) Helper-dependent miniEBV packaging and delivery of transgenes by miniEBV infection. An octameric miniEBV genome that is packaged into virions is shown. The black bar stands for a transgene cassette. LCLs, B lymphoblastoid cell lines; wtEBV, the helper P3HR1 or the endogenous EBV associated with LCLs or lymphoma cells, or both. Two possible applications of infectious miniEBV vectors are shown: (I) treatment of inherited human diseases that required prolonged expression of the transgene to correct the defective phenotype of disease cells (indicated by shaded LCLs); (II) treatment of cancers using suicide strategy to kill the malignant cells (indicated by a cross on lymphoma cells).

developed such infectious miniEBV-based suicide vectors for possible use in gene therapy of B lymphomas: an infectious, recombinant minEBV carrying the HSV-1 thymidine kinase (TK) gene expression cassette has been shown to successfully transduce both EBV positive and negative lymphoma cells. Moreover, on treatment

[14] F. L. Moolten and P. J. Mroz, *in* "Gene Therapy of Cancer: Methods and Protocols" (W. Walther and U. Stein, eds.), p. 209. Humana Press, Totowa, NJ, 2000.

with the prodrug ganciclovir (GCV), the growth inhibition of the miniEBV/TK-transduced lymphoma cells cultured *in vitro* or implanted in severe combined immunodeficiency (SCID) mice has been observed.[15]

Methods

Generation and Titration of MiniEBV Vector Stocks

Transfection of Helper Cell Line HH514 with MiniEBV Amplicon DNA. HH514 cells were grown in RPMI-1640 with 10% fetal bovine serum (FBS) and supplemented with 1% L-glutamine and 0.1% penicillin/streptomycin. Five million exponentially grown HH514 cells and 20 μg of miniEBV amplicon DNA were mixed in 0.3 ml of complete growth medium in a 0.4 cm electroporation cuvette (Bio-Rad) and incubated on ice for 10 min. Electroporation was carried out at 200 V and 960 μF with a Bio-Rad Gene Pulser. After electroporation, cells were incubated on ice for another 10 min, then seeded into 10 ml of complete growth medium, and incubated at 37°/5% CO_2. Three days after electroporation, hygromycin B (Boehringer-Mannheim Products) was added to a final concentration of 200 μg/ml to select for stable cell transformants.

Preparation of MiniEBV (and EBV) Virions. The stable HH514 cell transformants were cultured in a T-300 flask to near confluence ($2 - 4 \times 10^6$ /ml) in a total volume of 500 ml. The lytic phase of EBV was induced by adding to the medium 20 ng/ml 12-O-tetradecanoyphorbol 13-acetate (TPA, Sigma) and 1 mM n-butyric acid (sodium butyrate, Sigma). Five days later, supernatants were collected by centrifugation at 6000g for 10 min at 4°. The cell pellet was resuspended with 10 ml complete growth medium and subjected to three cycles of freeze/thaw process in a dry ice/ethanol bath and a 37° water bath. Cell debris was spun down at 6000g for 10 min at 4° and discarded. The supernatants were recovered, combined with previous collection, and treated with DNase I (20 U/ml, Sigma) for 30 min at room temperature to destroy nonpackaged DNA. The DNase I reaction was stopped with addition of 20 mM EDTA (pH 8.0) and 0.1% of sodium azide to the medium. The supernatants were then filtered through a 0.45-μm filter (Corning) to eliminate all cell debris. Virions were pelleted by centrifugation at 12,500g for 2 hr at 4° and resuspended in 2.5 ml RPMI-1640 without serum. Virions can be used for titration or infection assay directly.

The virus suspension can be aliquotted in glass vials, then quickly frozen in a dry ice/ethanol bath and stored at −80°. To minimize the loss of viral titer, we emphasize the use of glass vials other than plastic tubes and low temperature (−80°) for storage. When removing the virus stock from the freezer for use, we recommend rapid thawing in a 37° water bath with swirling of the tube to avoid localized warming. Repeated freezing and thawing of the virus stock should also be avoided.

---

[15] J. Wang, S. Banerjee, and J. M. Vos, unpublished data (2001).

Determination of Physical Titer of MiniEBV Vectors. In the absence of a lysis plaque- or colony-transforming assay for the infectious nontransforming miniEBV virions, we used the number of physical particles (i.e., physical titer) as a measure of the MOI (multiplicity of infection). Packaged DNA was isolated from virions by proteinase K digestion at 42° for 1 hr in a solution containing 100 μg/ml proteinase K, 0.2% SDS, 8 mM EDTA (pH 8.0), and 50 mM NaCl, followed by phenol extraction and ethanol precipitation. The purified DNA was subjected to pulse-field gel electrophoresis (PFGE). PFGE was carried out on a CHEF apparatus in 1% agarose and 0.5× TBE at 200 V for 20 hr at a switching time of 15 sec. After electrophoresis, DNA was transferred from agarose gels to nitrocellulose membranes and hybridized with radioactively labeled probes specific for miniEBV or helper EBV. After hybridization, radioactive DNA signals were detected by autoradiography. The yield of the viral vector was established by using a copy number standard of plasmid DNA carrying miniEBV- and EBV-specific sequences. Because of the concatemeric nature of packaged miniEBV DNA (Fig. 1B), the miniEBV signal was normalized to the monomeric level. An average of 4–6 × 10^5 recombinant miniEBV particles were produced per milliliter of virus suspension.[12]

Gene Transduction in Human Disease B Cells with Infectious MiniEBV

We tested the potential of infectious miniEBV vectors for stable transfer of a therapeutic gene in LCLs from a Fanconi anemia group C patient (HSC536) and an HRPT-deficient Lesch–Nyhan patient (RJK853), and in an EBV-positive lymphoma-derived cell line (Raji).[11,12] Five million exponentially grown cells were harvested by centrifugation at 500g for 5 min and resuspended in 1 ml of fresh RPMI-1640 without serum. Virus particles were added at an MOI of 10 into cells, and the mixture was incubated at 37°/5% CO$_2$ for 1 to 2 hr. After incubation, the cells were seeded into 10 ml complete growth medium and incubated at 37°/5% CO$_2$. Three days after infection, hygromycin B was added to the cells (final concentration 200 μg/ml) to select stable transformants for analysis of the intranuclear fate of the miniEBV DNA described below. To determine the transduction efficiency of miniEBV on target cells, a miniEBV carrying the reporter gene (*lacZ* or GFP) was produced from HH514 cells and used for infection of target cells. Cells were subsequently subjected to X-Gal staining (for *lacZ* expression)[11,12] or flow cytometry (for GFP expression, Fig. 2).

Intranuclear Fate of MiniEBV DNA in Transduced Cells

Episomal Maintenance of MiniEBV DNA in Stably Transduced Cells. Approximately 2 months after hygromycin B selection, the potential for episomal maintenance of the transduced miniEBV DNAs was examined in the stable transformants by agarose gel electrophoresis. Episomal DNA was extracted from stably

A   **B**

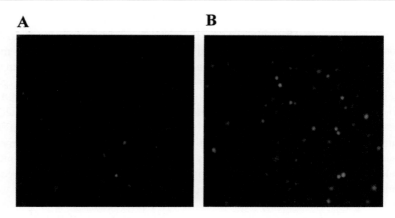

FIG. 2. Infection of human disease cells with infectious miniEBV vectors carrying the GFP expression cassette. (A) EBV-transformed B lymphoblastoid cells (HSC536). (B) EBV-positive B lymphoma cells (Raji). Five million exponentially grown cells were infected by miniEBV with an MOI of 10, i.e., 10 miniEBV particles per cell. Three days after infection, cells were subjected to fluorescence microscopy and photography. The percentage of GFP positive cells of HSC536 or Raji was determined by flow cytometry, which revealed 15% and 40% transduction efficiency for HSC536 and Raji cells, respectively (J. Wang and J. M. Vos, 2001).

transduced cells as previously described with modifications.[11,12] Cells (1×10^8) were collected by centrifugation at $500g$ for 5 min, washed once with 2 ml of phosphate-buffered saline (PBS), and then mixed thoroughly with 10 ml of lysis solution (50 mM NaCl, 8 mM EDTA, 1% SDS, pH 12.45) by vigorous vortex for 2 min. The mixture was incubated at 30° for 30 min. Two milliliters of 1 M Tris-Cl (pH 7.0) was added and mixed by gently swirling the tube, followed by addition of 1.32 ml of 5 M NaCl and 0.12 ml of 10 mg/ml proteinase K. The mixture was incubated at 37° overnight. The lysate was then extracted three times with 13 ml of phenol saturated with 0.2 M NaCl and 0.2 M Tris-Cl (pH 8.0) and one time with equal volume of chloroform by gently swirling the tube (vigorous vortex will break the large episomes and thus should be avoided). After centrifugation, the aqueous phase was collected, mixed with 30 ml of ethanol, and incubated at −20° overnight. The episomal DNA was spun down and resuspended in 50 μl of 10 mM Tris-Cl (pH 8.5). The episomal DNA was then separated on a 0.85% agarose gel that was run in 1× TBE at 5.4 V/cm for 10 to 11 hr at 4°. The samples were transferred to nylon membranes for hybridization with a probe specific for miniEBV or EBV. As shown in Fig. 3A, the supercoiled episomal DNA that is close to natural EBV genome can be found between linear DNA (running ahead) and nicked DNA (running behind or retained in wells). MiniEBV DNAs were maintained as 160–180 kb circular DNA molecules in stably transduced cells.[11,12]

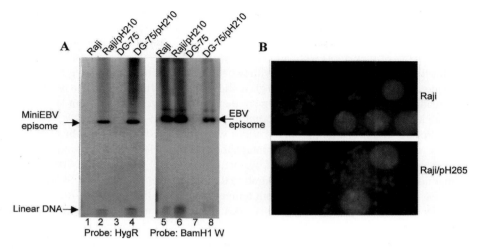

FIG. 3. Intranuclear fate of the transduced miniEBV DNA. (A) Episomal maintenance of miniEBV DNA in transduced cells. Infectious miniEBV (pH210) (J. Wang, S. Banerjee, and J. M. Vos, unpublished data) was produced from the nontransforming helper cell line HH514 and used for infection of EBV-positive lymphoma line Raji and EBV-negative Burkitt's lymphoma line DG-75. Episomes were extracted and hybridized with a probe specific for miniEBV (hygromycin resistance gene, HygR), or a probe specific for EBV (BamHI W). The signals in lanes 5 and 6 reflect the EBV from both the helper HH514/P3HR1 and the endogenous Raji/EBV, whereas the signal in lane 8 indicates the helper HH514/P3HR1 EBV only. (B) Chromosomal association, but not integration of the miniEBV episomal DNA. The B lymphoma Raji cells were stably transduced with miniEBV (pH265). Interphase and metaphase spreads from uninfected (top) or infected (bottom) cells were prepared and hybridized with a Fluor-12-dUTP labeled probe using FISH technique to detect miniEBV (pH265) only. [This figure is adapted with permission from T.-Q. Sun, E. Livanos, and J.-M. H. Vos, *Gene Ther.* **3**, 1081 (1996).]

Chromosomal Association, but Not Integration of MiniEBV Episomal DNA. To distinguish integrated from episomal miniEBV DNA, fluorescence *in situ* hybridization (FISH) with vector-specific probes was performed on metaphase spreads of the miniEBV-transduced cells. Integrated miniEBV DNA is expected to result in double, closely spaced, symmetrical signals localized on both sister chromatids while episomal miniEBV DNA will generate single FISH signals.

Preparation of interphase and metaphase spreads was described previously.[16] Ten milliliters of miniEBV-transduced cells (1×10^6/ml) was cultured in complete RPMI-1640 medium containing 200 μg/ml hygromycin B for 2 to 3 days. Two hours prior to cell harvest, 0.1 ml of cocemid solution was added (final concentration 0.1 μg/ml). Cells were then harvested by centrifugation at 500*g* for

[16] D. E. Rooney and B. H. Czepulkowski, "Human Cytogenetics: A Practical Approach." IRL Press, Oxford, 1986.

5 min and resuspended thoroughly in 10 ml of prewarmed 0.075 M KCl, followed by incubation at 37° for 15 min. Cells were spun down at 500g for 5 min and supernatant removed. Ten milliliters of fresh fixative (1 part of acetic acid and 3 parts of methanol) were added dropwise and mixed thoroughly with cells. This fixation process was repeated with two more changes of fresh fixative. Cells in suspension were dropped onto clean glass slides in a humid environment to promote spreading of chromosomes. Slides were then air dried overnight at room temperature and stored at −80° with desiccant until further use.

Prior to hybridization, prepared slides were incubated at 60° for 3 hr to promote further drying and harden metaphase spreads. Metaphase chromosomes on prepared slides were denatured by incubation at 70° for 2 min in 70% formamide/2× SSC, followed by dehydration through a series of cold 70%, 90%, and 100% ethanol (10 min each) at room temperature. The slides were air-dried and subjected to hybridization as previously described.[17] The miniEBV-specific probe was labeled with Fluor-12-dUTP by Prime-It-Fluor Kit (Stratagene) to detect the miniEBV DNA only. A total of 50 ng of labeled miniEBV probe was prehybridized for 30 to 60 min at 37° with 10–1000 ng of Cot-1 DNA (GIBCO BRL) and 500 ng of human placental DNA (Sigma, sonicated to less than 500 bp) in a solution containing 2× SSC (0.3 M sodium citrate buffer), 1% BSA, and 10% dextran sulfate (Sigma). Hybridization was performed at 37° in a humidified chamber for at least 18 hr by covering the samples on slides with the solution containing the probe described above. Posthybridization washes were performed by rinsing the sample in 50% formamide/2× SSC at 37° and 50° for 30 min each.

Samples were directly visualized at 1000 × magnification on an Olympus IMT-2 microscope (NY, US) equipped with epifluorescence for fluorescein detection. Color pictures were photographed using Kodak Ultra Gold film (400 ASA) (Kodak, NY, US). As shown in Fig. 3B, some of the miniEBV-specific FISH signals appeared to be associated with human chromosomes while others were located some distance from them.[11] The quantitative analysis by scanning of 100 metaphase signals revealed that the vast majority (approximately 99%) of the FISH signals was detected as single nonintegrated episomes.[11] In addition, an average number of miniEBV episomes was estimated to be 3 to 6 copies per nucleus in the corrected HSC536 cells from Fanconi anemia patients and one to two copies per nucleus in the corrected RJK851 cells from Lesch–Nyhan patients.[11,12]

Long-Term Transgene Expression and Phenotypic Correction
in Human Disease Cells

To demonstrate that stable expression of a transgene is possible from the miniEBV episomal DNA and that therapeutic effects of the transgene could also

[17] J. B. Lawrence, C. A. Villnave, and R. H. Singer, *Cell* **52**, 51 (1988).

be achieved, we transduced miniEBV vectors carrying FACC cDNA and HPRT cDNA into B-lymphoblastoid cells from an FA-C patient and an HPRT-deficient Lesch–Nyhan patient, respectively.[11,12] The expression of the transgene from stably transduced cells was examined by Northern blot analysis and functional complementation analysis.

Northern Blot Analysis of Transgene Expression. After miniEBV infection, cells were selected and maintained in the presence of 200 μg/ml hygromycin B for more than 2 months. Messenger RNA was isolated from cells by the PolyATtract System 1000 (Promega, WI, US) and the concentration determined by spectrophotometry (OD_{260}). The mRNA (7 μg) was separated on a 1.5% agarose formaldehyde gel at 5 V/cm for 4 hr and transferred to a nylon membrane. The Northern blot was hybridized with the transgene cDNA and β-actin (as internal loading control) probes. An example is given showing that human HPRT mRNA was detected in the miniEBV-transformed Lesch–Nyhan cells, but absent from nontransformed cells (Fig. 4A).[11]

Functional Complementation Assay in Human Disease Cells. The therapeutic effects of the expressed transgenes detected by Northern blot analysis described above were further confirmed by phenotypic correction of a FA-C cell defect by measuring hypersensitivity to diepoxybutane exposure, and of an HPRT-deficient

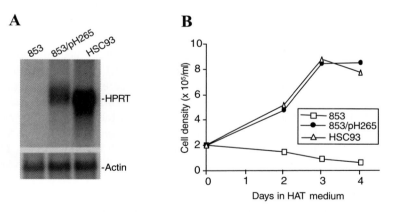

FIG. 4. HPRT expression and phenotypic correction of HRPT deficiency in Lesch–Nyhan cells by miniEBV(pH265) infection. (A) Northern blot analysis of HPRT gene expression. Messenger RNAs were extracted from uninfected Lesch–Nyhan lymphoblastoid cell line RJK853 (853) cells, miniEBV(pH265) infected cells (853/pH265), and an HPRT(+) control human lymphoblastoid cell line (HSC93). RNA samples (7 μg each) were separated on a formaldehyde agarose gel and hybridized to a human HPRT cDNA probe (top) or a human β-actin probe to verify equal loading (bottom), respectively. (B) Analysis of HPRT function by growing cells in HAT medium. Cell growth was determined at days 3, 4, and 5 by trypan blue exclusion. This figure is reproduced with permission from T.-Q. Sun, E. Livanos, and J.-M. H. Vos, *Gene Ther.* **3,** 1081 (1996).

cell type by growing cells in hypoxanthine–aminopterin–thymidine (HAT) medium.[11,12] We take the latter as an example. Lymphoblastoid cells growing actively in complete RPMI-1640 medium were counted, pelleted, and resuspended at 2×10^5 cells/ml in fresh medium supplemented with $1\times$ HAT (GIBCO BRL). Cells were then seeded into 24-well plates (2×10^5 cells per ml per well). Living cells were counted at day 3, 4, and 5 after seeding by trypan blue exclusion. As shown in Fig. 4B, the uninfected cells died quickly under HAT selection, whereas the miniEBV-transduced cells grew like the normal HPRT(+) HSC93 cells. This indicates that normal cellular HPRT function had been restored in the majority of the Lesch–Nyhan cell clones stably transduced by the episomal miniEBV. It is therefore concluded that functional human HPRT mRNA and protein were produced from the large HAECs transduced into the human disease B cells by miniEBV infection.[11]

Suicide Strategy for Gene Therapy of B Lymphomas with Infectious MiniEBV

We have designed an infectious, recombinant miniEBV/TK for delivering and expressing HSV-1 TK in both EBV positive and negative lymphoma cells.[15] The effects of this infectious miniEBV-based suicide vector on the prodrug-mediated destruction of B lymphoma cells have been manifested by growth inhibition of the transduced cells cultured *in vitro* or preimplanted *in vivo* in a SCID mouse model.

In Vitro Cytotoxicity Assay. A total of 1×10^6 miniEBV/TK-transduced or nontransduced lymphoma cells were plated in a well of 12-well culture dish in 1 ml of complete RPMI-1640 medium with varying concentrations ($0–400~\mu M$) of the prodrug GCV (InvivoGen Products). On day 3, 1 ml of fresh medium containing corresponding concentration of the prodrug in each well was added. Viable and nonviable cells were counted on day 5 by trypan blue exclusion. As shown in Fig. 5A, both EBV positive and negative lymphoma cell lines transduced with miniEBV/TK were remarkably sensitive to low concentrations of GCV, whereas the nontransduced cells were resistant to high concentrations of GCV. These results demonstrate the feasibility of using infectious miniEBV carrying HSV-1 TK to effectively eliminate EBV positive as well as negative B lymphomas *in vitro* at low concentrations of the prodrug GCV (i.e., $20–40~\mu M$).

In Vivo Tumor Growth Inhibition Assay. As a first step in testing the applicability of this system for treating human B cell derived tumors *in vivo,* we analyzed the ability of this infectious suicide gene delivery vector to inhibit the growth of the stably transduced B lymphoma cells in a SCID mouse model. Lymphoma cells stably transduced with miniEBV/TK were injected subcutaneously (1×10^7 cells) into the inguinal region of SCID mice. GCV was injected into the same site every other day at 20 mg/kg in half of the mice, while PBS was injected every other day in the other half mice until tumor burden required euthanasia. Tumor growth was monitored and tumor volume recorded over a 3-week period. As shown in Fig. 5B,

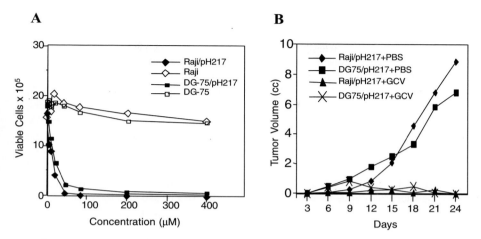

FIG. 5. Targeted destruction of human B cell lymphomas by infectious miniEBV carrying the HSV-1 thymidine kinase gene (pH217, J. Wang, S. Banerjee, and J. M. Vos, unpublished data). (A) Inhibition of growth *in vitro* of human B lymphoma cells stably transduced by the miniEBV. One million EBV-positive (Raji) and EBV-negative (DG-75) lymphoma cells were cultured in the presence of varying concentrations of GCV (0, 0.1, 10, 20, 40, 80, 200, and 400 μM), and viable cells were counted by trypan blue exclusion 5 days after incubation. (B) Inhibition of human B cell tumor growth *in vivo* by infectious miniEBV(pH217). EBV-positive Raji or EBV-negative DG-75 cells (1×10^7) stably transduced with miniEBV/pH217 were injected subcutaneously into the inguinal region of SCID mice. GCV (20 mg/kg) was injected at the initial site of tumor cell delivery every other day in half of the mice, and PBS was injected every other day in the other half of the mice until tumor burden required euthanasia. Tumor growth was monitored and volume recorded over a 3-week period.

mice receiving GCV had no detectable tumor even 3 weeks after inoculation, whereas mice receiving PBS had substantial tumor growth by visual inspection and tumor necropsy assay.[15]

Summary and Prospects

The development of infectious EBV vectors for therapeutic purposes is still at an early stage. The B lymphotropism of EBV suggests that it may be particularly well adapted for the treatment of diseases involving circulating and/or diffusible gene products. Thus, inherited recessive monogenic disorders of serum proteins such as blood clotting factors (e.g., hemophilia), hormones such as insulin (diabetes), or enzymes such as glucocerebrosidase (Gaucher disease), α_1-antitrypsin (inherited emphysema), and β-glucuronidase (Sly syndrome) may be suitable candidates for EBV-based gene therapy. In addition, EBV may also be useful for treatment of acquired diseases such as cancer (especially B-cell lymphomas) and infectious diseases.

However, several potential difficulties will have to be overcome before EBV can be safely and effectively used in human clinical trials. The current helper-dependent packaging cell line produces miniEBV vectors mixed with the helper virus that has the potential for transcriptional activation and production of onco-genic LMP1 and/or EBNA3C. In addition, a recombination event could potentially occur between the miniEBV amplicon DNA and the helper P3HR1 genome. These considerations may not be critical in suicide strategy for gene therapy of cancers, because any cancer cell coinfected with the helper P3HR1 EBV and miniEBV carrying suicide gene(s) will eventually be ablated with the administration of the prodrug. However, gene therapy of inherited diseases requires prolonged expression of a therapeutic gene product, and thus it is important that a packaging cell line free of infectious helper virus be created. Finally, of paramount importance for the success of gene therapy is the availability of disease-specific delivery systems that direct the activity of therapeutic/suicide genes specifically to the sites of disease/malignancy. The transductional targeting by the infectious miniEBV vectors restricts the transgene expression to B lymphocytes and probably some other cell types, including those of epithelial origin.[18] More restricted transgene expression could be achieved by transcriptional targeting using tumor- or tissue-specific promoters to drive the transgene expression.[19] Therefore, development of a helper-free packaging system and enhancement of the vector targeting specificity would deserve our future endeavor in perfecting the infectious miniEBV vectors for use in gene therapy.

Acknowledgments

We thank Eliane Wauthier in the laboratory for critical reading of the manuscript and Drs. T.-Q. Sun and S. Banerjee for their partial contribution to the data cited in this chapter. Research in the principal investigator's laboratory was supported by grants from the Fanconi Anemia Research Foundation, the Leukemia Society of America, the Department of Energy, and the National Institute of Health. J. Wang is a postdoctoral fellow supported by the Lymphoma Research Foundation of America.

This chapter is dedicated to the memory of Dr. Jean-Michel H. Vos, who succumbed to cancer on November 29, 2000, at the early age of 44. For more than 10 years, Dr. Vos had been a pioneer in the fields of chromosome engineering and gene therapy, and he had made important contributions to the development of mammalian artificial chromosomes and mini-herpesvirus based episomal vectors for animal transgenesis and gene therapy.

[18] G. Miller, in "Virology" (B. N. Fields and D. K. Knipe, eds.), 2nd Ed., p. 1921. Raven Press, New York, 1990.
[19] D. M. Nettelbeck, V. Jerome, and R. Muller, *Trends Genet.* **16,** 174 (2000).

Author Index

Numbers in parentheses are footnote reference numbers and indicate that an author's work is referred to although the name is not cited in the text.

A

Abdallah, B., 24
Aberle, A. M., 75
Abordo-Adesida, E., 297, 301(56), 302(56), 303(56)
Aboud, M., 629, 647
Abruzzese, R. V., 557, 558, 558(14), 559(15)
Achacoso, P., 151, 435, 466, 471(6)
Achong, B. G., 628, 633(3)
Ackermann, M., 591, 593(13)
Acsadi, G., 72, 73, 92, 104(39), 105, 125, 134, 158, 202
Adam, M. A., 200
Adams, R., 336
Addison, C. L., 187, 230, 267, 277(15), 278
Adesanya, M. R., 211
Adey, N. B., 170
Adib, A., 115
Adler, R., 361
Aebischer, P., 457, 491
Afione, S. A., 336, 355, 394
Agata, J., 4(42), 12, 258, 259(22), 260(22), 261, 261(22), 262, 262(30)
Aggoun-Zouaoui, D., 385
Agrawal, R. S., 312, 313(9), 316(11), 317(9), 319(11)
Aguilar-Cordova, E., 194, 266, 268(12), 293
Aguzzi, A., 633
Ahlgren, J. D., 604
Ahn, M., 219
Aiello, L. P., 332
Aihara, H., 129, 202
Ailles, L. E., 458, 459(20), 476, 514(7), 515, 517(7), 529(7), 584, 632
Aird, W., 147
Airenne, K. J., 320
Aiuti, A., 474

Akazawa, C., 434
Akbarian, L., 501
Akimoto, H., 222
Akita, G. Y., 146
Akkina, R. K., 492
Akli, S., 292
Akusjarvi, G., 532, 560
Akyurek, L. M., 208, 531
Alam, J., 220, 399
Albelda, S. M., 213
Albrecht, D., 604
Albrecht, T. R., 467, 468(16)
Alejandro, R., 73
Alestrom, P., 109, 110(16)
Alexander, I. E., 125, 230
Alexander, R. B., 80
Alexandersen, S., 468
Alexeev, V., 15
Alhava, E., 312, 313, 313(9), 317(9), 318(13), 616
Alino, S. F., 65, 67
Alion, I. A., 590(8), 591, 598(8)
Alisky, J. M., 434, 435(14), 448, 448(14)
Alitalo, K., 315
Allan, G. F., 551, 552(1)
Allawi, H. T., 111
Allen, E. D., 292(3), 293, 506
Allen, H., 179, 187(25), 188(25), 295, 305(46)
Allen, J. M., 394, 412
Allen, M., 222
Allen, P. D., 590(8), 591, 598(8)
Alon, N., 500
Alonso, A., 647
Alston, J., 355
Alt, F. W., 591, 598(7)
Altman, S., 358
Altschuler, M., 377
Alvira, M., 158, 207
Amado, R. G., 359

O

P

Subject Index

A

T

V

ISBN 0-12-182247-8

90051

9 780121 822477

DATE DUE

DEMCO 13829810